ND | Springer Series in **Nonlinear Dynamics**

Solitons – Introduction and Applications
Editor: M. Lakshmanan

What Is Integrability?
Editor: V.E. Zakharov

Rossby Vortices and Spiral Structures
By M.V. Nezlin and E.N. Snezhkin

Algebro-Geometrical Approach to Nonlinear Evolution Equations
By E.D. Belokolos, A.I. Bobenko, V.Z. Enol'skii, A.R. Its, and V.B. Matveev

Darboux Transformations and Solitons
By V.B. Matveev and M.A. Salle

Optical Solitons
By F. Abdullaev, S. Darmanyan, P. Khabibullaev

Wave Turbulence Under Parametric Excitation
Applications to Magnets
By V.S. L'vov

Kolmogorov Spectra of Turbulence I Wave Turbulence
By V.E. Zakharov, V.S. L'vov and G. Falkovich

Nonlinear Processes in Physics
Editors: A.S. Fokas, D.J. Kaup, A.C. Newell and V.E. Zakharov

Important Developments in Soliton Theory
Editors: A.S. Fokas and V.E. Zakharov

ND | Springer Series in **Nonlinear Dynamics**

Solitons – Introduction and Applications
Editor: M. Lakshmanan

What Is Integrability?
Editor: V. E. Zakharov

Rossby Vortices and Spiral Structures
By M. V. Nezlin and E. N. Snezhkin

Algebro-Geometrical Approach to Nonlinear Evolution Equations
By E. D. Belokolos, A. I. Bobenko, V. Z. Enolsky, A. R. Its and V. B. Matveev

Darboux Transformations and Solitons
By V. B. Matveev and M. A. Salle

Optical Solitons
By F. Abdullaev, S. Darmanyan and P. Khabibullaev

Wave Turbulence Under Parametric Excitation
Applications to Magnetics
By V. S. L'vov

Kolmogorov Spectra of Turbulence I Wave Turbulence
By V. E. Zakharov, V. S. L'vov and G. Falkovich

Nonlinear Processes in Physics
Editors: A. S. Fokas, D. J. Kaup, A. C. Newell and V. E. Zakharov

Important Developments in Soliton Theory
Editors: A. S. Fokas and V. E. Zakharov

A. S. Fokas V. E. Zakharov (Eds.)

Important Developments in Soliton Theory

With 59 Figures

Springer-Verlag Berlin Heidelberg GmbH

Professor A. S. Fokas

Clarkson University, Potsdam, NY 13699-5815, USA

Professor V. E. Zakharov

Landau Institute for Theoretical Physics, ul. Kosygina 2
117334 Moscow, Russia and
University of Arizona, Tucson, AZ 85721, USA

ISBN 978-3-642-63450-5 ISBN 978-3-642-58045-1 (eBook)
DOI 10.1007/978-3-642-58045-1

Originally published by Springer-Verlag Berlin Heidelberg New York in 1993
Softcover reprint of the hardcover 1st edition 1993

Typesetting: Camera ready by authors/editors
54/3140 - 5 4 3 2 1 0 - Printed on acid-free paper

Preface

The mathematical modeling of a great variety of nonlinear phenomena arising in physics and biology leads to certain nonlinear equations. It is quite remarkable that many of these universal equations are "integrable". Roughly speaking, this means that one can extract from these equations almost as much information as one can extract from the corresponding linear equations. Furthermore, precisely because these equations are nonlinear, they exhibit richer phenomenology than the linear equations. In particular many of them support solitons, i.e. localized solutions with particle-like properties.

Following the discovery of the concepts of solitons and of integrable behavior by Kruskal and Zabusky in the late 1960s, there was an explosion of results associated with integrable equations. Most of the results obtained in the decade 1970–1980 can be found in several books.

Important developments in the theory of integrable equations continued in the decade of 1980–1990. In particular this decade has been marked by the impact of soliton theory in many diverse areas of mathematics and physics such as algebraic geometry (the solution of the Schottky problem), group theory (the discovery of quantum groups), topology (the connection of Jones polynomials with integrable models), quantum gravity (the connection of the KdV with matrix models) etc. There exists no book covering the important developments of the last decade. We have collected in this volume most of these important developments.

This book should be of interest to a wide audience. This includes: Experts in the field of integrable equations, since it provides a compendium of the up to date developments; experts in the field of nonlinear phenomena in general, since integrable equations can be used as a first approximation to more complex equations including those exhibiting chaotic behavior; applied mathematicians, mathematical physists, engineers, and biologists, since it gives a comprehensive picture of a new area in mathematical physics with a tremendous range of applications.

Part of this project has been supported by the Air Force Office of Scientific Research – Arje Nachman (Program Manager), whose support is gratefully acknowledged. The authors are thankful to Cindy Smith for her help in completing this volume.

Spring 1993

A.S. Fokas
V.E. Zakharov

Contents

Part V **Near-Integrable Models and Computational Aspects**

Introduction

A.S. Fokas and V.E. Zakharov

Many apparently disparate nonlinear systems, have the commonality that they exhibit a similar behavior. "Chaotic" and "integrable" behaviors are among the best studied such behaviors arising in a variety of physically important equations. This book contains some of the most important developments in the theory of integrable equations that occurred in the last decade.

The fascinating new world of solitons and of integrable behavior was discovered by Kruskal and Zabusky. These investigators, in their effort to explain some curious numerical results of Fermi-Pasta-Ulam, were lead to study the Korteweg-deVries (KdV) equation. This equation is a nonlinear evolution equation in one spatial and one temporal dimension. It was first derived in 1885 by Korteweg and deVries (who were trying to explain the famous experiments of J. Scott Russell of 1834) and had already reappeared in the 1960s in a number of other physical contexts (plasma physics, stratified internal waves, etc.). The KdV possesses an exponentially localized traveling-wave solution. Kruskal and Zabusky studied numerically the interaction properties of these localized waves and found a unexpected behavior: After interaction, these waves regained their initial amplitude and velocity, and the only residual effect of interaction was a "phase shift" (i.e. a change in the position they would have reached without interaction). This particle-like property led them to call these waves Solitons. The next challenge was the analytical understanding of these numerical results. The search for additional conservation laws (which perhaps were responsible for the stability properties of solitons) led Miura to the discovery of the modified KdV as well as to a Riccati-type relationship between KdV and modified KdV. The linearization of this Riccati equation, led Gardner, Green, Kruskal, and Miura to discover the connection between the KdV and the time-independent Schrödinger scattering problem. In this way, the solution of the Cauchy problem for the KdV had been reduced to the reconstruction of a potential from knowledge of appropriate scattering data: Let KdV describe the propagation of a water wave and suppose that this wave is frozen at a given instant of time. By bombarding this water wave with quantum particles, one can reconstruct its shape from knowledge of how these particles scatter. In other words, the scattering data provide an alternative description of the wave at a fixed time. The mathematical expression of this description takes the form of a linear integral equation and was found by Faddeev, who generalized the earlier results of Gel'fand and Levitan and of Marchenko on the radial Schrödinger equation. The time evolution of the water wave satisfies the KdV which is a nonlinear equation. The above alternative description of the shape of the wave would be useful if the evolution of the scattering data is linear. This is indeed the case, and hence this highly nontrivial change of variables from the physical to scattering space, provides a linearization of the KdV equation. The essence of this ingenious discovery was extracted by Lax in 1968, who introduced the so called Lax pair formulation of the KdV. Following Lax's formulation, Zakharov and Shabat in 1972 solved the nonlinear Schrödinger equation. Soon thereafter the sine Gordon equation and the modified KdV equation were solved by Ablowitz, Kaup, Newell, and Segur and by Wadati in 1972. The mathematical technique of Gardner, Greene, Kruskal, and Miura, although it was developed for solving a special physical problem, was applicable to a wide range of other physical problems. In this way a powerful new method in mathematical physics, the so-called inverse scattering or inverse spectral method was born. Following these developments there was an explosion of results associated with integrable equations. Most of the results obtained in the 1970-1980 decade can be found in several books.

Many important developments in the theory of integrable equations have also occurred in the decade of 1980-1990. In particular this decade has been marked by the impact of soliton theory in many diverse areas of mathematics and physics such as algebraic geometry (the solution of the Schottky problem), group theory (the discovery of quantum groups), topology (the connection of Jones polynomials with integrable models), quantum gravity (the connection of the KdV with matrix models) etc. There exists no book covering the important developments of the last decade. We have collected in this volume most of these important developments. For pedagogical reasons we present these results in five sections.

I. Methods of Solution

The Cauchy problem on the line for evolution equations in one spatial variable was studied extensively in the 1970s. However, the scattering theory associated with systems larger than 2×2 or with rational dependence on the spectral parameter, presents several technical difficulties and was not worked out in full until the 1980s. This work is reviewed by Beals, Deift and Zhou. A method for studying the Cauchy problem on the plane for evolution equations in two spatial variables was discovered in the 1980s. This method, which is usually referred to as the $\bar{\partial}$ method, is reviewed by Coifman and Fokas. Although a general method for finding periodic solutions for evolution equations in two spatial variables was discovered in the 1970s, a method for studying the relevant Cauchy problem was found by Krichever only recently. The study of periodic solutions of KP has led Dubrovin and others to the solution of the Schottky's classical problem of characterizing Riemann matrices in the theory of Riemann surfaces. The extension of the IST method for solving ODE's of the Painlevé type has led to the discovery of the so-called isomonodromy method. This method, its use for obtaining connection formulae for the Painlevé equations, and the recent appearance of Painlevé equations in 2D quantum gravity, is reviewed by Fokas and Its. The investigation of integrable lattice systems by the so-called direct linearizing method is reviewed by Capel and Nijhoff, while the study of singular integro-differential equations is reviewed by Santini. Calogero reviews C-integrable equations, i.e. nonlinear equations which can be solved by a change of variables, and he also discussed their universal character.

II. Asymptotic Results

Several important asymptotic results associated with integrable equations are included in this volume. Witham's method from the soliton point of view is reviewed by S.P. Novikov. The zero dispersion limit of several dispersive systems is reviewed by Lax, Levermore and Venakides. The long time asymptotics of the Cauchy problem on the line is reviewed by Deift, Its and Zhou.

III. Algebraic Aspects

It was established in the 1970s that the KdV and other integrable evolution equations in one spatial variable are bi-Hamiltonian systems. The formulation of integrable ODE's, and of evolution equations in two spatial variables as bi-Hamiltonian systems was achieved in the 1980s and is reviewed by McKean and by Fokas and Gel'fand respectively. Integrable equations possess infinitely many symmetries; actually this property can be used as a test to identify integrable equations, as reviewed by Mikhailov and Shabat. Beautiful connections between τ-function and representation theory is discussed by Kac and J.W. van de Leur. The relationship between integrability and the form of the associated dispersion laws and connections with the theory of webs is discussed by Zakharov and Schulman.

IV. Quantum and Statistical Mechanical Models

One of the most important developments in the field of integrable equations has been the understanding of certain quantum field theoretic analogues of the integrable classical models. It turns out that these quantum models are closely related to certain two dimensional statistical mechanical models. For example, the classical nonlinear Schrödinger equation is gauge equivalent to the classical Heisenberg model; the quantum version of this equation is related to the XXX Heisenberg spin chain studied by Bethe in the 1930s. The so-called Yang-Baxter equation and the R matrix play a prominent role in this theory, which is usually referred to as quantum inverse scattering. After the discovery of Jones of the connection between statistical mechanics and knot theory, it became clear that techniques from soliton theory could be used to generate higher Jones polynomials. Indeed solutions of the Yang-Baxter equation lead to certain representation of braid groups whose invariants are the Jones polynomials. Truly new mathematical structures have emerged from quantum inverse scattering. In particular, Drin'feld introduced the quantum groups by writing the relevant formalism in the framework of Hopf algebras. These results are summarized in the articles of Faddeev, of McCoy, of Takhtajan and of Wadati. Also Its, Izergin, Korepin, and Slavnov show that the quantum correlation function can be described as the τ-function of classical integral equations.

V. Near Integrable Models and Computation Aspects

Ablowitz and Herbst show that the standard discretizations of the nonlinear Schrödinger equation for large enough value of the mesh size, behave chaotically. This numerically induced chaos is associated with the study of homoclinic orbits. The geometry of such structures is described by McLaughlin who constructs "whiskered tori" using the periodic spectral theory of soliton equations. This construction is useful for the understanding of near-integrable PDE's. Finally, Deift, Li, and Tomei study the symplectic aspects of several important numerical algorithms by using the Hamiltonian structure associated with certain integrable equations.

Part I

Methods of Solution

The Inverse Scattering Transform on the Line

R. Beals[1], *P. Deift*[2], *and X. Zhou*[1]

[1]Mathematics Department, Yale University,
Box 2155 Yale Station, New Haven, CT 06520, USA
[2]Mathematics Department, Courant Institute,
New York University, New York, NY 10012, USA

1. Introduction

The 1967 discovery of the inverse scattering method for integrating the KdV equation [GGKM] was followed in the early 1970s by similar results for many other nonlinear wave equations and systems. We recall very briefly the principles, in the case of one space and one time variable.

The nonlinear equations and systems integrable by this method occur as the *compatibility conditions* for the coefficients of an overdetermined *linear* system depending on a *spectral parameter*. Part of the overdetermined system is a *spectral problem* in the spatial variable, and the compatibility conditions are time-evolution equations for its coefficients. Given an appropriate scattering-inverse scattering theory for this spectral problem, the *scattering data* evolve in a simple explicit way. Thus the *initial value problem* for the coefficients can be solved if the *direct* map (coefficients to scattering data) and the *inverse* map (scattering data to coefficients) are understood.

In the case of KdV the linear spectral problem (ignoring time dependence) is the Schrödinger equation on the line

$$-\psi_{xx}(x,\xi) + q(x)\psi(x,\xi) = \xi^2\psi(x,\xi), \qquad x,\ \xi \in \mathbf{R}. \tag{1.1}$$

The scattering theory had been developed by Faddeev [Fa2], completing work in the 1950s by Gelfand-Levitan, Marchenko, Krein, and others; see [Fa1].

Some features of the theory for real q with smoothness and decay are

(i) the standard bounded eigenfunctions suffice for the direct problem, i.e. they determine adequate scattering data;

(ii) the standard *scattering matrix*, relating asymptotics of the bounded eigenfunctions at $x = -\infty$ and $x = +\infty$ exists and contains all the necessary information (apart from finite *discrete data*);

(iii) the direct problem gives rise immediately to Volterra equations;

(iv) the inverse problem can also be solved via an integral equation of standard type, known as the Gelfand-Levitan-Marchenko equation, which is automatically solvable.

Acknowledgment. This work was supported by National Science Foundation grants DMS-8916968, DMS-9001857, and DMS 9196033.

The next such nonlinear wave equations–the modified KdV equation [Wa], the cubic nonlinear Schrödinger equation [ZS1], and the sine-Gordon equation [AKNS1]–were associated to the 2×2 *ZS-AKNS spectral problem*

$$
\begin{aligned}
(1.2) \qquad & \frac{d\psi_1}{dx}(x,\xi) = -i\xi\psi_1(x,\xi) + q_1(x)\psi_2(x,\xi) \\
& \frac{d\psi_2}{dx}(x,\xi) = i\xi\psi_2(x,\xi) + q_2(x)\psi_1(x,\xi), \quad x,\ \xi \in \mathbf{R}.
\end{aligned}
$$

A scattering theory analogous to that for (1.1) was developed by Zaharov and Shabat [ZS2] and Ablowitz, Kaup, Newell, and Segur [AKNS2]. In full generality, however, one loses features (ii), (iv): there may be infinitely many discrete singularities, and the inverse problem for formal scattering data is not necessarily solvable.

It was found that other nonlinear equations of interest arise in connection with linear equations of higher order, and with systems larger than 2×2 or with other (rational) z-dependence. The corresponding scattering theory presented new difficulties and was not worked out in full until the 1980s. In particular, all of the features (i)-(iv) fail in general.

The key to the general problem is the observation of Shabat [Sa1] that the inverse problem is a *Riemann-Hilbert problem*. For full development of this idea one must

(v) identify appropriate *normalized eigenfunctions* for the spectral problem and determine their existence and properties;

(vi) determine the form and properties of the associated *scattering data*;

(vii) recover the potential q from its scattering data–the *inverse problem*;

(viii) determine criteria for the global solvability of the inverse problem, for applications to the initial-value problem for nonlinear wave equations.

Each of these problems presents conceptual and/or technical difficulties. In the direct problem, the most natural approach leads to a Fredholm equation and to eigenfunctions which may have infinitely many singularities. The inverse problem can involve a self-intersecting contour, and always involves seeking estimates uniform with respect to a parameter.

In the next four sections we give a concise outline of the scattering theory for $n \times n$ ZS-AKNS systems and scalar equations of order n, and some indications of related developments. Basic conceptual proofs are sketched, but limits of space rule out technical arguments. Many of the results stated below have been found (in varying degrees of generality and rigor) by several authors. References in the statements are given in chronological order, so that the result *as stated* is usually to be found in the last reference, or the final few.

Sects. 2 and 3, respectively, are devoted to the direct problem and to the inverse problem for the $n \times n$ system which generalizes (1.2). Sects. 4 and 5 outline the direct and inverse problems for the n-th order equation which generalizes (1.1). A few historical indications are given in Sect. 6, as well as some references to other developments for problems on the line.

2. The Direct problem for n × n ZS-AKNS Systems

The linear spectral problem here has the form

$$(2.1)\quad \frac{d}{dx}\psi(x,z) = [zJ + q(x)]\psi(x,z), \quad x \in \mathbf{R},\ z \in \mathbf{C},\ \psi(x,z) \in GL(n,\mathbf{C}).$$

It is assumed that J is a given diagonal matrix with distinct eigenvalues:

$$(2.2)\quad J = \mathrm{diag}(\lambda_1, \lambda_2, \ldots, \lambda_n), \quad \lambda_j \neq \lambda_k \text{ if } j \neq k.$$

The function q takes values in $M_n(\mathbf{C})$, the space of $n \times n$ complex matrices, and is off-diagonal. For ease of exposition assume that the entries belong to the Schwartz class $\mathcal{S}(\mathbf{R})$:

$$(2.3)\quad q \in \mathcal{Q} \equiv \{q \in \mathcal{S}(\mathbf{R}; M_n(\mathbf{C})) :\ q_{jj}(x) \equiv 0\}.$$

For application to nonlinear wave equations one seeks solutions of (2.1) which are *uniquely specified by their asymptotic behavior as* $x \to \pm\infty$. The *free equation* with $q \equiv 0$ has the exponential solution e^{xzJ}, and we seek ψ in the form $\psi(x,z) = m(x,z)e^{xzJ}$. The equation for m and the normalization we choose are

$$(2.4)\quad \frac{d}{dx}m(x,z) = [zJ,\ m(x,z)] + q(x)m(x,z);$$

$$(2.5)\quad \lim_{x \to -\infty} m(x,z) = I, \qquad \sup_x \|m(x,z)\| < \infty.$$

Here I is the $n \times n$ identity matrix and $\|\ \|$ is the matrix norm inherited from the action on the unitary space $\mathbf{C}^n$ with the standard inner product.

The *singular set* is the union of lines through the origin

$$(2.6)\quad \Sigma = \Sigma(J) = \{z \in \mathbf{C} : \mathrm{Re}(z\lambda_j) = \mathrm{Re}(z\lambda_k),\ \text{some } j \neq k\}.$$

The basic result for the *small norm* problem is the following.

Theorem 2.1 [Sh2, BC1] *Suppose* $\int_{\mathbf{R}} \|q(x)\| dx < 1$. *Then for each complex* z *the problem (2.4), (2.5) has a unique solution. For fixed* $x \in \mathbf{R}$ *the function* $m(x,\cdot)$ *is holomorphic in* $\mathbf{C} \setminus \Sigma$ *and has limits*

$$(2.7)\quad m_\pm(x,z) = \lim_{\varepsilon \downarrow 0} m(x, z \pm i\varepsilon z), \quad z \in \Sigma \setminus 0.$$

These limits are related to m *on* $\Sigma \setminus 0$ *by functions* $v_\pm(z)$:

$$(2.8)\quad m_\pm(x,z) = m(x,z)e^{xzJ}v_\pm(z)e^{-xzJ}, \quad z \in \Sigma \setminus 0, \quad \det v_\pm = 1.$$

Moreover $m(x,\cdot)$ *is bounded and* $m(x,z) = I + O(z^{-1})$ *as* $z \to \infty$.

If Σ consists of a single line, we modify the definitions so that $m_\pm$ refer to limits from opposite sides of the line, not its two rays separately.

Sketch of Proof: Uniqueness. If $m(\cdot, z)$ exists, x−differentiation shows that the determinant is constant, hence 1, so the inverse is bounded. If m_1, m_2 are solutions,

$$\frac{d}{dx}[m_1^{-1}m_2] = [zJ, m_1^{-1}m_2] \tag{2.9}$$

so $m_2 = m_1 e^{xzJ} a e^{-xzJ}$. Boundedness and (2.5) imply $a = 1$.

Existence. Consider the integral equation

$$m(x,z) = I + \int^x e^{(x-y)zJ} q(y) m(y,z) e^{-(x-y)zJ} dy. \tag{2.10}$$

More explicitly, this is a *system* of integral equations

$$m(x,z)_{jk} = \delta_{jk} + \int_{\epsilon_{jk}\infty}^{x} e^{(x-y)z(\lambda_j-\lambda_k)}[q(y)m(y,z)]_{jk}dy, \quad \epsilon_{jk} = \pm 1. \tag{2.11}$$

The conditions of boundedness and asymptotics in (2.5) clearly require

$$\epsilon_{jk} \operatorname{Re} z(\lambda_j - \lambda_k) \geq 0; \quad \epsilon_{jk} = -1 \text{ if } \operatorname{Re} z(\lambda_j - \lambda_k) = 0. \tag{2.12}$$

Writing (2.10) as $m(\cdot,z) = I + K_{z,q}m(\cdot,z)$ one sees that $K_{z,q}$ maps the space of bounded matrix functions $L^\infty(\mathbf{R}; M_n(\mathbf{C}))$ to itself, with norm

$$\|K_{z,q}\| \leq \int_{\mathbf{R}} \|q(x)\| dx. \tag{2.13}$$

Existence follows. Holomorphy and existence of the limits (2.7) are easily proved. We leave the discussion of large z to Theorem 2.4. The uniqueness argument proves (2.8). □

We take as *scattering data* for q in Theorem 2.1 the matrix function

$$v : \Sigma \setminus 0 \to SL(n, \mathbf{C}); \quad v(z) = v_-(z)^{-1}v_+(z). \tag{2.14}$$

Theorem 2.2 [Sh2, BC1] *The potential matrix q of Theorem 2.1 is uniquely determined by its scattering data v.*

Proof. It is enough to show that $m(x,\cdot)$ is determined by being holomorphic, bounded, $= I$ at ∞, and having limits on $\Sigma \setminus 0$ with

$$m_+(x,z) = m_-(x,z)e^{xzJ}v(z)e^{-xzJ}. \tag{2.15}$$

These conditions imply that the determinant is bounded, holomorphic on $\mathbf{C}\setminus\Sigma$, continuous at $\Sigma \setminus 0$, and 1 at ∞, hence $\equiv 1$. Thus the inverse is bounded. If m_1, m_2 are two such functions, $m_2m_1^{-1}$ is bounded; the singularities on Σ are removable and the limit at $z = \infty$ is I, so $m_2 = m_1$. □

This shows that the *inverse problem*, determining q from v, can be solved by solving a *Riemann-Hilbert factorization problem* with data on Σ.

We turn to the large-norm problem. Note that the uniqueness argument in the proof of Theorem 2.1 applied without restriction on the norm.

Theorem 2.3 [BC1, Zh1] *For $q \in \mathcal{Q}$ there are solutions m^+, m^- of (2.4) with the properties: $m^\pm(x,\cdot)$ is bounded and holomorphic on $\mathbf{C} \setminus \Sigma$, has limits on Σ, and*

$$\lim_{x \to \pm\infty} m^\pm(x,z) = I \tag{2.16}$$

Proof. Choose $q_0 \in \mathcal{Q}$ and $x_0 \in \mathbf{R}$ such that $\int ||q_0|| < 1$ and $q_0(x) = q(x)$ when $x \leq x_0$. Let m^- coincide with the normalized solution for q_0 on the interval $(-\infty, x_0]$, and continue m^- as the unique solution of (2.4) on $[x_0, +\infty)$ with the given value at x_0. Construction of m^+ is similar. □

Theorem 2.4 [Sh2, BC1] *For $q \in \mathcal{Q}$ the solution of (2.4), (2.5) is meromorphic with respect to z on $\mathbf{C} \setminus \Sigma$, and holomorphic for large z. It has an asymptotic expansion*

$$m(x,z) \sim \sum_{k=0}^{\infty} z^{-k} m_k(x) \quad \text{as } z \to \infty; \qquad m_0 = I. \tag{2.17}$$

Sketch of Proof. Choose $m^\pm$ as in Theorem 2.3. These are related to each other and to the normalized m, whenever it exists, by

$$m^-(x,z) \equiv m^+(x,z) e^{xzJ} a(z) e^{-xzJ}, \tag{2.18}$$

$$m(x,z) \equiv m^\pm(x,z) e^{xzJ} a^\pm(z) e^{-xzJ}. \tag{2.19}$$

These relations and the asymptotic conditions are equivalent to

$$a(z) = a^+(z) a^-(z)^{-1}; \tag{2.20}$$

$$\lim_{x \to -\infty} e^{xzJ} a^-(z) e^{-xzJ} = I, \quad \sup_x ||e^{xzJ} a^+(z) e^{-xzJ}|| < +\infty. \tag{2.21}$$

Problem (2.20), (2.21) is a standard algebraic factorization problem. In fact change basis in $\mathbf{C}^n$, for a given $z \in \mathbf{C} \setminus \Sigma$, so that

$$\mathrm{Re}(z\lambda_1) > \mathrm{Re}(z\lambda_2) > \ldots > \mathrm{Re}(z\lambda_n). \tag{2.22}$$

Then (2.21) becomes: a^- is upper triangular with diagonal part $= I$; a^+ is lower triangular. The obstructions to the factorization (2.20) are the zeros of the upper principal minors of $a(z)$, and meromorphy follows.

For large z, consider a formal solution of (2.4), (2.5) given by an asymptotic expansion (2.17). The corresponding recursion for the m_k determines the diagonal and off-diagonal parts as

$$[J, m_{k+1}^{off}] = (\frac{d}{dx} - q) m_k; \quad m_{k+1}^{diag}(x) = \int_{-\infty}^{x} [q m_{k+1}^{off}]^{diag} dy. \tag{2.23}$$

The actual solution m can be obtained as $m_{(k)} \phi_k$, where

$$m_{(k)} = \sum_{j=0}^{k} z^{-j} m_j; \qquad q_k(x,z) \equiv m_{(k)}^{-1}[\frac{d}{dx} - z\,\mathrm{ad}\,J - q]m_{(k)};$$

$$[\frac{d}{dx} - z\,\mathrm{ad}\,J - q_k]\phi_k = 0, \qquad \lim_{x\to-\infty} \phi_k(x,z) = I, \quad \sup_x \|\phi_k(x,z)\| < \infty.$$

Now $q_k = O(z^{-k})$, so this becomes a small-norm problem for ϕ_k. □

Note that in (2.23), $q = -[J\ , m_1]$, showing that *the expansion at $z = \infty$ is valid if and only if q is off-diagonal.*

Theorem 2.5 [Zu] *If $z_0 \in \mathbf{C} \setminus \Sigma$ is a pole of $m(x,\cdot)$, there is a unique rational matrix function $v(\cdot, z_0)$ which is I at ∞, regular except at $z = z_0$, and has the properties*

$$(2.24) \qquad m(x,z)[e^{xzJ} v(z,z_0) e^{-xzJ}]^{-1} \text{ is regular at } z = z_0;$$

$$(2.25) \qquad \lim_{x\to-\infty} e^{xzJ} v(z,z_0) e^{-xzJ} = I.$$

Proof. Let m^- be as in Theorem 2.3, and let $a^-(z)$ be as in (2.19). For simplicity, assume (2.22), so a is upper triangular with diagonal I and has a pole at z_0. Choose $b(z)$ rational, upper triangular, regular except at z_0 and $= 1$ at ∞, such that $(a^-)^{-1} = b + O(z - z_0)$. Then $m_1 = me^{xzJ}be^{-xzJ}$ has a pole of lower order at z_0 than does m. Iterating this construction, we eventually obtain $v(\cdot, z_0)^{-1}$ as a product. If v_1, v_2 are two such functions, then $v_1^{-1}v_2$ is entire and $= I$ at ∞, so $v_1^{-1}v_2 \equiv 1$. □

For a dense open subset of $\mathcal{Q}$, $m(x,\cdot)$ has only finitely many poles and m has limits on $\Sigma \setminus 0$ as in Theorem 2.1 [BC1]. Moreover the *soliton, multi-soliton*, and other special solutions of the nonlinear wave equations are of this type; indeed $m(x,\cdot)$ is rational. One strategy is to confine attention to these "generic" potentials q, and take v on $(\Sigma \setminus 0) \cup \{\text{poles of } m\}$ as the scattering data; the inverse problem is a Riemann-Hilbert problem with singularities [BC1]. However, even for $q \in \mathcal{Q}$, m can have countably many singularities in $\mathbf{C} \setminus \Sigma$ and/or extremely bad behavior on Σ [Zh2]. To avoid these problems, one can utilize the original m, which has good behavior for large z, together with $m^\pm$ of Theorem 2.3.

It is convenient to introduce the following subsets of $M_n(\mathbf{C})$, depending on a *diagonal* matrix A:

$$[[A]]^\pm = \{a : \lim_{x\to\pm\infty} \|e^{xA} a e^{-xA}\| < \infty\};$$

$$(2.26) \qquad [[A]]_0^\pm = \{a \in [[A]]^\pm : \mathrm{diag}(a) = I\};$$

$$[[A]] = [[A]]^+ \cap [[A]]^-; \quad [[A]]_0 = [[A]]_0^+ \cap [[A]]^-.$$

Theorem 2.6 [Zh2] *Given $q \in \mathcal{Q}$, let m be the normalized eigenfunction, and let $m^\pm$ be as in Theorem 2.3. Let $\Gamma = \{|z| = r\}$, where r is chosen so that m is regular on $\{z \in \mathbf{C} \setminus \Sigma : |z| > r\}$ and has limits on the boundary from each component. Let $m_\pm$ denote the limits on $\Sigma \cap \{|z| > r\}$, let $m_\pm^-$ denote* *the limits of m^- on $(\Sigma\setminus 0)\cap\{|z|<r\}$. Let S be the set of self-intersections of $\Sigma\cup\Gamma$: $S=\Sigma\cap\Gamma$ if Σ is a single line, $S=(\Sigma\cap\Gamma)\cup\{0\}$ otherwise. There is a function $v^-:(\Sigma\cup\Gamma)\setminus S\to SL(n,\mathbb{C})$ such that*

$$m_+(x,z)=m_-(x,z)e^{xzJ}v^-(z)e^{-xzJ},\ z\in\Sigma\cap\{|z|>r\};$$
$$m_+^-(x,z)=m_-^-(x,z)e^{xzJ}v^-(z)e^{-xzJ},\ z\in\Sigma\cap\{|z|<r\};$$
$$m(x,z)=m^-(x,z)e^{xzJ}v^-(z)e^{-xzJ},\ z\in\Gamma\setminus\Sigma.$$

Moreover $v^-(z)\in[[zJ]]_0^-$, $z\in\Gamma$ and v^- can be factored as

$$v^-(z)=[v_-^-(z)]^{-1}v_+^-(z),\ z\in\Sigma; \tag{2.27}$$
$$v_\pm^-\in[[(z\pm i\epsilon z)J]]_0\cap[[zJ]]_0,\qquad |z|>r,\ z\in\Sigma;$$
$$v_\pm^-\in[[(z\pm i\epsilon z)J]]_0^-,\qquad |z|<r,\ z\in\Sigma.$$

There is a corresponding function v^+ relating m, m^+.

Theorem 2.6 follows from earlier arguments and the form of the integral equation. Note that on $\Gamma\setminus\Sigma$, v^- coincides with the piecewise meromorphic function a^- of (2.19); thus the poles of m are the poles of this meromorphic extension of v^- from $\Gamma\setminus\Sigma$.

Theorem 2.7 [BC1, Zh2] *The functions $v^- - I$, $v_\pm^- - I$ belong to the Schwartz class on each component of their domains, in the sense that the entries are smooth, each derivative has limits at the endpoints, and the derivatives decay rapidly as $z\to\infty$.*

The function v^- of Theorem 2.6 can be taken as the ***revised scattering data*** for $q\in\mathcal{Q}$. Again it determines q via a Riemann-Hilbert factorization which gives the ***revised eigenfunction***, defined to be m^- inside Γ and m outside. However v^- is not unique, since neither Γ nor m^- is. In fact m^- is unique only up to right multiplication by $e^{xzJ}b(z)e^{-xzJ}$, where b is holomorphic on $(\mathbb{C}\setminus\Sigma)\cap\{|z|<r\}$, smooth up to the boundary on each component, and $b(z)$ is in $[[zJ]]_0^-$. Changing m^- in this way induces obvious changes in v^-.

Our final technical result here is a ***matching condition*** at points of self-intersection. This condition reflects the fact that the Riemann- Hilbert factorization problem for $e^{xzJ}v^-(z)e^{-xzJ}$ is "smooth", even though the contour $\Sigma\cup\Gamma$ has self-intersections at which the limits of v^- may differ.

At any finite endpoint of a component of $(\Sigma\cup\Gamma)\setminus S$, the function v^- has a Taylor series, which can be taken as a formal power series. On a given component v^- represents the multiplicative jump of the revised eigenfunction across this component; numbering components in order of increasing argument around a given point of S, and taking v^- and $(v^-)^{-1}$ for each component, as the case may be, one obtains the multiplicative jumps in order around a point of S: at the origin take v^- itself on each ray; at points of $\Sigma\cap\Gamma$ start with the bounded part of the ray and take $(v^-)^{-1}$, then v^- twice, then $(v^-)^{-1}$ again. We denote the corresponding formal power series by V_1, V_2, ... , V_N, where $N\le n(n-1)$ at the origin and $N=4$ at points of $\Sigma\cap\Gamma$.

Theorem 2.8 [BC1, Zh2] *At each self-intersection point of* $\Sigma \cup \Gamma$, *the product of the corresponding formal power series is* $V_1 \cdot \ldots \cdot V_N \equiv I$.

To conclude, we consider *reductions*, symmetries which reduce the number of independent entries of q.

Theorem 2.9 *Suppose* $a \to a^\sigma$ *is an algebra automorphism (resp. antiautomorphism) of* $M_n(\mathbf{C})$ *and suppose* $q^\sigma = q$ *and* $J^\sigma = \alpha J$ *for some* $\alpha \in \mathbf{C} \setminus 0$ *(resp.* $q^\sigma = -q$ *and* $J^\sigma = -\alpha J$*). Then there is a choice of* v^- *such that* $v^-(\alpha z) = v^-(z)^\sigma$ *(resp.* $v^-(\alpha z) v^-(z)^\sigma = I$*).*

In fact $m(x,z)^\sigma = m(x, \alpha z)$ *(resp.* $m(x,z)^\sigma m(x, \alpha z) = I$*), etc.*

Theorem 2.10 *Suppose that* $a \to a^\sigma$ *is an antilinear antiautomorphism and* $q^\sigma = -q$, $J^\sigma = -\alpha J$. *Then* v^- *can be chosen so that* $v^-(z)^\sigma = v^-(\alpha \bar{z})$.

The simplest examples are the modified KdV equation for Theorem 2.9, and the two forms of the cubic nonlinear Schrödinger equation and the 3-wave interaction equation for Theorem 2.10.

3. The Inverse Problem for $n \times n$ ZS-AKNS Systems

In the broad sense the inverse problem is to *characterize* functions $v^- : \Sigma \cup \Gamma \to SL(n, \mathbf{C})$ which are scattering data, and to investigate properties of the inverse map. Some of the necessary properties are listed in Theorems 2.6, 2.7, 2.8 (and in 2.9 or 2.10 for reductions). In this section, given a choice of a circle $\Gamma = \{|z| = r\}$, we set $\Sigma^* = \Sigma \cup \Gamma$ and form from the components of $\mathbf{C} \setminus (\Sigma^*)$ two disjoint sets $\Omega_\pm$ so that each has Σ^* as its boundary; there are two such choices. We give Σ^* the positive orientation with respect to Ω_+. We also *change notation from Sect. 2*, by taking $m_\pm$ and $m_\pm^-$ to refer to the limits from $\Omega_\pm$, etc.

Let $\mathcal{F}^-(\Sigma^*_\pm)$ be the space of matrix functions f with the properties: (i) $f - I$ is in the Schwartz class in the sense of Theorem 2.6;

(ii) on the boundary of any component of $\Omega_\pm$, the two Taylor series of f in $z - z_0$ at a corner point z_0 coincide;

(iii) the algebraic condition $f(z) \in [[zJ]]$, $|z| > 2r$;

(iv) the algebraic condition $f(z) \in [[z'J]]_0^-$, $z' \in G$, for z in the boundary of a component G of $\Omega_\pm$.

Define $\mathcal{F}^+(\Sigma^*_\pm)$ similarly, with $[[z'J]]_0^+$ in the last condition.

The following factorization of v^- follows (not immediately) from Theorems 2.6–2.8. The result is proved in [Zh1] with a finite amount of smoothness; a classical result of E. Borel allows one to replace formal power series factors with smooth functions to obtain the result here.

Theorem 3.1 [BC1, Zh1] *For* $q \in \mathcal{Q}$, *scattering data* v^- *has a factorization* $v^- = (I - w_-)^{-1}(I + w_+)$, *where* $I \pm w_\pm$ *belong to* $F^-(\Sigma^*_\pm)$.

Passing to the normalization at $x = +\infty$, recall the relationship (2.18) between m^- and m^+. There is a similar relationship between m and $\tilde{m}$ normalized at $x = +\infty$, given in this case by a diagonal, bounded matrix function δ holomorphic for $z \notin \Sigma$, $|z| > r$. Combining this with the analogue of Theorem 3.1 gives the following.

Theorem 3.2 *For $q \in \mathcal{Q}$, there is a matrix function δ, bounded and holomorphic on $\Omega = \Omega_+ \cup \Omega_-$, smooth up to the boundary on each component, and diagonal for $|z| > r$, such that $v^- = (I - w_-)^{-1}(I + w_+)$ has a factorization*

$$\delta_- v^- \delta_+^{-1} = (I - w_-^{\pm})^{-1}(I + w_+^{\pm}), \tag{3.1}$$

where $w_+^{\pm}$ and $w_-^{\pm}$ are boundary values of functions in $\mathcal{F}^+(\bar{\Omega}_\pm)$.

One converts the Riemann-Hilbert problem for the revised eigenfunction to a singular integral equation on Σ^*. The *Cauchy projections* for Σ^*, with the orientation as above, are the maps $\pm C_\pm : L^2(\Sigma^*) \to L^2(\Sigma^*)$,

$$C_\pm f(z) = \lim_{w \in \Omega_\pm, w \to z} Cf(w); \quad Cf(z) = \frac{1}{2\pi i}\int_{\Sigma \cup \Gamma} \frac{f(\zeta)}{\zeta - z} d\zeta$$

Then $C_+ - C_- = Id$, $C_\pm^2 = Id$, and the range of $C_\pm$ consists of boundary values of functions holomorphic on $\Omega_\pm$.

Let $\mathcal{H}^k(\Sigma^*)$ denote the space of functions $f \in L^2(\Sigma^*; M_n(\mathbf{C}))$ such that f and its derivatives to order $k-1$ are absolutely continuous on the boundary of each component of $\mathbf{C} \setminus \Sigma^*$ and the k-th derivative is also square integrable. Let $\hat{\mathcal{H}}^k(\Sigma^*)$ be the direct sum of $\mathcal{H}^k(\Sigma^*)$ and the constant functions.

Theorem 3.3 [Zh1] *Given $w = (w_+, w_-)$ from Theorem 3.1 and $f \in \hat{\mathcal{H}}^k(\Sigma^*)$, $k \geq 1$, set*

$$C_{w,x} f = C_-[f w_+^x] + C_+[f w_-^x], \quad w_\pm^x(z) = e^{xzJ} w_\pm(z) e^{-xzJ}, \ x \in \mathbf{R}. \tag{3.2}$$

Then for each x the equation for $m_w(x, \cdot) \in \hat{\mathcal{H}}^k(\Sigma^*)$,

$$m_w(x, \cdot) = I + C_{w,x} m_w(x, \cdot), \tag{3.3}$$

has a unique solution and the revised eigenfunction is

$$I + \frac{1}{2\pi i}\int_{\Sigma \cup \Gamma} \frac{m_w(x,\zeta)[w_+^x(\zeta) + w_-^x(\zeta)]}{\zeta - z} d\zeta. \tag{3.4}$$

In fact equations $m_\pm = m_w(I \pm w_\pm^x)$, $|z| > r$ and $m_\pm^- = m_w(I \pm w_\pm^x)$, $|z| < r$ define m_w on $\Sigma \cup \Gamma$. Now the revised eigenfunction is the Cauchy integral of its additive jump across $\Sigma \cup \Gamma$, so

$$m_w(I \pm w_\pm^x) = I + C_\pm(m_w w_+^x + m_w w_-^x)$$

which is equivalent to (3.3).

There is a partial converse.

Theorem 3.4 [BC1, Zh1, Zh2] *Suppose the functions $I \pm w_{\pm}$ belong to $\mathcal{F}^{-}(\Sigma^{*}_{\pm})$, and let $C_{w,x}$ be defined by (3.3). Then $Id - C_{w,x}$ is a Fredholm operator with index 0 in $\hat{\mathcal{H}}^{k}(\Sigma^{*})$ and is invertible for $x < x_0$, some $x_0 \in (-\infty, +\infty]$. The solution m_w of (3.3) and the function defined by (3.4) are solutions of an equation $dm/dx = [zJ,\ m] + qm$, where $q = q(x)$ is smooth, off-diagonal valued. Moreover, q and its derivatives converge rapidly to zero as $x \to -\infty$.*

Sketch of Proof. The Fredholm property follows from the fact that the operators

$$(Id - C_{\hat{w},x})(Id - C_{w,x}), \quad (Id - C_{w,x})(Id - C_{\hat{w},x}), \quad \hat{w}_{\pm} = w_{\pm}(w_{\pm} - I)^{-1}$$

have the form $Id +$ (compact); e.g. $(Id - C_{\hat{w},x})(Id - C_{w,x})f$ is

$$f + C_{-}[(C_{+}g)\hat{w}^{x}_{+}] - C_{+}[(C_{-}g)\hat{w}^{x}_{-}] \equiv f + T_{w,x}f,$$

where $g = f(w^{x}_{+} - w^{x}_{-})$. Compactness follows from the fact that commutators $[C_{\pm},\ w_{\pm}]$ are compact and $C_{+}C_{-} = 0 = C_{-}C_{+}$. The form of $w_{\pm}$ implies that the entries are either exponentially decaying or rapidly oscillating as $x \to -\infty$. From this and the algebraic form one deduces that the operator norm of $T_{w,x}$ tends to zero as $x \to -\infty$ so (3.3) is solvable and the Fredholm index is zero. Technical arguments imply smoothness of m_w. The function m defined by (3.4) has the property that $(d/dx - z\,\mathrm{ad}\,J)m$ satisfies the same multiplicative jump relationships as m, so $[(d/dx - z\,\mathrm{ad}\,J)m]m^{-1}$ is bounded and continuous on $\Sigma \setminus 0$, hence $= q(x)$, independent of z. □

If there is a transformation (3.2) from $w = (w_{+}, w_{-})$ in Theorem 3.4 to data w^{+} with δ as in Theorem (3.2), and if Eq. (3.3) is solvable for all x, then the relationship between solutions is given by an analogue of (2.18) and one concludes that both equations give the same q and that q belongs to $\mathcal{Q}$. Thus the following gives a full characterization.

Characterization Theorem 3.5 [Zh2] *If q belongs to $\mathcal{Q}$, then there are a contour $\Sigma \cup \Gamma$ and matrix functions $v^{\pm}$, δ, $w_{\pm}$, $w^{+}_{\pm}$ as in Theorems 3.1 and 3.2 such that (3.3) and its counterpart for w^{+} are solvable for all real x.*

Conversely, suppose that $I \pm w_{\pm}$ belong to $\mathcal{F}^{-}(\Sigma^{*}_{\pm})$, $I \pm w^{+}_{\pm}$ belong to $\mathcal{F}^{+}(\Sigma^{*}_{\pm})$, δ is bounded and holomorphic on Ω, smooth up to the boundary on each component, and diagonal for $|z| > r$, and (3.1) holds. Then (3.3) and its counterpart for w^{+} are solvable for $|x|$ large. If (3.3) is solvable for all real x then it gives normalized eigenfunctions of a function $q \in \mathcal{Q}$.

There are no completely general effective criteria for solvability of (3.3). Nevertheless the following is useful.

Theorem 3.6 [BC3, Zh1] *Suppose $J + J^{*} = 0$ and $v^{-}(z)^{*} = v^{-}(\bar{z})$, all $z \in [\Sigma^{*}] \setminus (\mathbb{R} \cup S)$ and suppose $\mathrm{Re}\, v(z) > 0$, $z \in \mathbb{R}$. Then (3.3) is solvable for all real x.*

The proof uses the Fredholm property: it is enough to show that the homogeneous equation $f(x,\cdot) = C_{w,x}f(x,\cdot)$ has only the trivial solution. The assumption on v^- off $\mathbf{R}$ implies that $g(x,z) = f(x,z)f(x,z)^*$ has limits $g_+ = g_-$ on $\Sigma^* \setminus \mathbf{R}$ and therefore is holomorphic on $\mathbf{C} \setminus \mathbf{R}$. By Cauchy's theorem the integral over $z \in \mathbf{R}$ of the trace of g_+ vanishes, and positivity of $\mathrm{Re}\, v^-$ on $\mathbf{R}$ implies that g_+ vanishes. □

Redundancy; scattering data in special cases. Since $q \in \mathcal{Q}$ is uniquely determined by v^-, the data δ and the associated v^+ are redundant. Nevertheless, there is no effective way to determine δ (which is not unique) from v^-. This is not an obstacle for solving the associated nonlinear wave equations: see the discussion below. In principle one can replace data (v, δ) by the corresponding *equivalence class* to overcome redundancy [Zh1].

A related problem is that, starting with v^- which has the analytical and algebraic properties discussed above, there is no effective way to determine the asymptotics of (3.3) as $x \to +\infty$.

Generically these difficulties may be overcome, as follows. As noted in Sect. 2, for a dense open set in $\mathcal{Q}$ the original $m(x, \cdot)$ have finitely many singularities and have limits on Σ from each component. If Z is the set of singularities, there is a corresponding matrix function $v : (\Sigma \setminus 0) \cup Z \to M_n(\mathbf{C})$ which plays the role of scattering data and which necessarily satisfies certain algebraic and analytic constraints. Here the transformation to data associated to the normalization at $x = +\infty$ is effective, given v; it is made by solving a Riemann-Hilbert problem for a diagonal matrix-valued function. The obstruction to solving this problem consists of winding-number relationships between principal minors of v on $\Sigma \setminus 0$ and the singularities. For details in the case of singularities of special form, see [BC1]; for general singularities, see [Zh2].

Partial smoothness and partial decay. We have assumed for smoothness of exposition that q is of Schwartz class, although Theorem 2.1 required only $q \in L^1$. The scattering transform $q \mapsto w = (w^+, w^-)$ is a nonlinear Fourier transform; finite decay of q implies finite smoothness of w, finite smoothness of q implies finite decay of w, and conversely. For details, see [Sh2, BY, BC1].

Multiple eigenvalues of J. All the previous results have exact counterparts when J has repeated eigenvalues. If a basis is chosen so that repeated diagonal elements of J occur in blocks, then there is a corresponding block structure for any matrix in $M_n(\mathbf{C})$. The potential matrix q should have diagonal blocks zero, equivalent to $q(x)$ lying in the range of ad J, etc.; see [Zh1,Zh2].

Rational $J(z)$ and $q(x,z)$. In some applications one considers a diagonal matrix $J(z)$ and/or a matrix function $q(x,z)$ which depend rationally on z. For the case of polynomial dependence, with q of smaller degree than J, see [Le1,Le2]. Results of this type can be obtained when each pole of q has order at most the order of the pole of J at the same point; see [Zh1,Zh2,Zh3].

Application to nonlinear wave equations. The nonlinear wave equations associated to the spectral problem (2.1) are those which come about as the compatibility conditions for systems

$$\frac{\partial}{\partial x}\psi(x,t,z) = [zJ + q(x,t)]\psi(x,t,z), \tag{3.5}$$

$$\frac{\partial}{\partial t}\psi(x,t,z) = R(x,t,z)\psi(x,t,z), \tag{3.6}$$

where $R(x,t,\cdot)$ is rational with some conditions on behavior at poles. We mention particularly the polynomial case

$$R(x,t,z) = z^k J_1 + \sum_{j=0}^{k-1} z^j r_j(x,t), \qquad [J,\ J_1] = 0 \tag{3.7}$$

A formal argument, which in appropriate cases can be made rigorous, shows that the compatibility condition for (3.5)-(3.7) is

$$\frac{\partial q}{\partial t} = [J,\ F_{k+1}], \tag{3.8}$$

where $F_j = F_j(q)$ are the coefficients of the asymptotic expansion as $z \to \infty$ of $m(x,t,z)J_1 m(x,t,z)^{-1}$; $m(\cdot,t,\cdot)$ is the normalized eigenfunction for $q(\cdot,t)$. The $F_j(q)$ are polynomials in q and its derivatives [Sa].

The evolution of scattering data is given, formally, by

$$\frac{\partial v^-(z,t)}{\partial t} = [z^k J_1,\ v^-(z,t)] \tag{3.9}$$

and similarly for δ, w, w^+, v^+. The necessary and sufficient condition that the solution to (3.9) have the appropriate properties for large z, for any initial scattering data $v^-(\cdot,0)$, is

$$a \in [[zJ]] \ \text{ implies } \ \sup_t \|e^{tz^k J_1} a^{-tz^k J_1}\| < \infty. \tag{3.10}$$

Under this assumption, both existence and uniqueness of the solution to the nonlinear wave equation can be deduced from the inverse scattering method:

Theorem 3.7 [BC2] *Suppose that J_1 is diagonal and satisfies (3.10). Then for each q in Q the solution of (3.8) with initial value $q(\cdot,0) = q$ exists on an interval $|t| < T$ with*

$$\limsup_{|t|\to T} \int_{\mathbf{R}} \|q(x,t)\|^2 dx = +\infty \ \ \textit{if} \ \ T \neq \infty.$$

Scattering data evolves according to (3.9), up to equivalence.

Condition (3.10) can be eliminated on rays of Σ where v^- has compact support, and can be weakened in some cases for the one-sided problem $t \geq 0$ or $t \leq 0$. Note that redundancy of scattering data is not a problem, here; one computes what is necessary at time $t = 0$, and lets it evolve by (3.9).

Again, there are generalizations to other problems (3.5), (3.6), and to other forms for $J(z)$, $q(x,z,t)$; [ZS2, Ca2], and the following.

Applications to geometric imbedding problems. One of the first applications of the 2×2 theory was to the sine-Gordon equation, which comes about geometrically in certain imbeddings of a curvature -1 surface in R^3. Generalizations to imbeddings of hyperbolic m-manifolds in R^{2m+1} and flat m-manifolds in S^{2m-1} lead to the generalized sine-Gordon equations (GSGE) and generalized wave equations (GWE) respectively. Solutions can be obtained by the inverse scattering method, by solving a $2m \times 2m$ Riemann-Hilbert problem as above for

$$e^{x\cdot J(z)}v(z)e^{-x\cdot J(z)}, \quad x\cdot J(z) = \sum_{j=1}^{m} x_j J_j(z),$$

where J is rational; [ABT, BT].

4. The direct problem for n-th order operators

Here we consider the scattering and inverse scattering theory for the n-th order equations

$$Lu \equiv D^n u + p_{n-2}(x)D^{n-2}u + \cdots + p_0(x)u = z^n u, \tag{4.1}$$

where $p_i(x) \in \mathcal{S}(\mathrm{R})$ and $D = \frac{1}{i}\frac{d}{dx}$. The free equation $L_0\psi \equiv D^n\psi = z^n\psi$ has a fundamental set of solutions $e^{i\alpha_j zx}, 0 \le j \le n-1$, where $\alpha_1, \ldots, \alpha_n$ are the n-th roots of unity. For $p_i \not\equiv 0$, we seek a fundamental set of solutions $\{u_i(\cdot, z)\}_{i=1}^n$ of (4.1) satisfying

$$\lim_{x\to-\infty} u_j(x,z)e^{-ix\alpha_j z} = 1, \quad \sup_x |u_j(x,z)e^{-ix\alpha_j z}| < \infty, \tag{4.2}$$

(cf. (2.4), (2.5)). The singular set is now given by the union of $2n$ rays

$$\Sigma = \{z \in \mathrm{C} : \mathrm{Re}\, i\alpha_j z = \mathrm{Re}\, i\alpha_k z, \text{ for some } j \ne k\}.$$

(When $n = 2$ there are only two rays–the real axis. We shall assume that $n > 2$.) Equation (4.1) can be written in the system form

$$D\psi = J_z\psi + q\psi \tag{4.3}$$

where

$$J_z = \begin{pmatrix} 0 & 1 & & & \\ & 0 & 1 & & \\ & & \ddots & \ddots & \\ & & & \ddots & 1 \\ z^n & & & & 0 \end{pmatrix} \quad \text{and} \quad q = \begin{pmatrix} 0 & & & \\ & 0 & & \\ & & \ddots & \\ -p_0 & \ldots & -p_{n-2} & 0 \end{pmatrix}. \tag{4.4}$$

Diagonalizing $J_z \to zJ \equiv z\,\mathrm{diag}(\alpha_1, \ldots, \alpha_n)$, (4.3) can be changed into a first order system of type (2.1), but now $q(x) = q(x,z)$ is a rational function of z with a pole at $z = 0$, and the techniques of Sects. 2 and 3 (as well as the extended techniques described in the remark on rational $J(z)$ and $q(x,z)$ in Sect. 3) do not apply without substantial modification.

In terms of $m \equiv \psi e^{-izxJ}$, we seek solutions of

$$Dm = J_z m - zmJ + qm \tag{4.5}$$

with

$$\lim_{x\to-\infty} m(x,z) = \Lambda_z, \quad \sup_{x\in\mathbf{R}} \|m(x,z)\| < \infty \tag{4.6}$$

where

$$\Lambda_z = \operatorname{diag}(1,z,\ldots,z^{n-1})(\alpha_j^{i-1})_{1\le i,j\le n} \equiv d_z\Lambda. \tag{4.7}$$

Note that the first row of me^{ixzJ} gives the desired solutions $(u_1,\ldots,u_n)$.

Fix $z \in \mathbf{C}\setminus\Sigma$ and *z-order* the roots of unity according to

$$\operatorname{Re} i\alpha_1 z > \operatorname{Re} i\alpha_2 z > \cdots > \operatorname{Re} i\alpha_n z. \tag{4.8}$$

To emphasize the z-order, we will sometimes write $\alpha_j = \alpha_j(z)$.

In studying $m_k \equiv me_k$, it is convenient (see [BDT]) to work with the tensors

$$f_k = m_1 \wedge\cdots\wedge m_k, \quad 1\le k\le n,$$

which satisfy

$$Df_k = (J_z^{(k)} - (\alpha_1+\cdots+\alpha_k)zI + q^{(k)})f_k \tag{4.9}$$

with

$$\lim_{x\to-\infty} f_k(x,z) = \Lambda_z e_1\wedge\cdots\wedge\Lambda_z e_k. \tag{4.10}$$

Here $A^{(k)}(u_1\wedge\cdots\wedge u_k) = \sum_{j=1}^{k} u_1\wedge\cdots\wedge u_{j-1}\wedge Au_j\wedge u_{j+1}\wedge\cdots\wedge u_k$. The point is that for each k, f_k satisfies a Volterra equation for each k, and this enables us to analyze the singularity at $z=0$ in an effective way. By contrast, the m_k's satisfy a general integral equation of type (2.10), and the behavior of the solution at $z=0$ is extremely difficult to extract, because of degeneracy as $z\to 0$; e.g. $\det\Lambda_z = O(z^N)$, $N = n(n-1)/2$.

Remark. The tensor method can of course also be used to construct the normalized solution for the first order systems of Sect. 2 and 3.

Having obtained $f_1,\ldots,f_n$ and analogous tensor solutions $\{g_k\}$ normalized at $x=+\infty$,

$$\lim_{x\to+\infty} g_k = \Lambda_z e_k\wedge\cdots\wedge\Lambda_z e_n, \tag{4.11}$$

one obtains the m_k's by solving the linear algebraic system

$$f_{k-1}\wedge m_k = f_k, \quad m_k\wedge g_k = 0. \tag{4.12}$$

This system is solvable if $f_{k-1}\wedge g_k \ne 0$: if $f_{k-1}\wedge g_k = 0$ for some $z=z_0$, then z_0 is a singularity for m_k as in Theorem 2.4.

For a dense open set $Q_{\text{gen}} \subset Q = (\mathcal{S}(\mathbf{R}))^{n-1}$ of *generic potentials* q, the solution $m(x,\cdot)$ has only finitely many simple poles $Z=\{z_j\}\subset\mathbf{C}\setminus\Sigma$, of type (2.24), (2.25) and $m(x,\cdot)$ has limits on Σ from each component of $\mathbf{C}\setminus\Sigma$.

Introduce the $n \times n$ block diagonal matrics (for odd n)

$$(4.13) \qquad \begin{cases} \pi_0 &= \mathrm{diag}(1, \sigma_1, \dots, \sigma_1), \\ \pi_1 &= \mathrm{diag}(\sigma_1, \dots, \sigma_1, 1), \\ \pi_j &= \pi_{j-2}, \qquad 2 \le j \le 2n-1, \end{cases}$$

where $\sigma_1 = \left(\begin{smallmatrix} 0 & 1 \\ 1 & 0 \end{smallmatrix}\right)$. Number the $2n$ rays $\Sigma_0, \Sigma_1, \dots, \Sigma_{2n-1}$ of Σ counterclockwise, starting from $\Sigma_0 = i\mathbf{R}_-$. The jumps across Σ are described by matrics $v(z)$,

$$(4.14) \qquad m_+(x,z)\pi_j = m_-(x,z)e^{ixzJ_-}v(z)e^{-ixzJ_-}, z \in \Sigma_j \setminus 0,$$

where again $m_\pm$ are defined as in (2.7) and $J_- = J_-(z)$ corresponds to $(z - i\epsilon z)$-order for the roots of unity, where $\epsilon > 0$. Furthermore at points $z_j \in Z$,

$$(4.15) \qquad v(z, z_j) = I + (z - z_j)^{-1}c(z_j)e_{k,k+1} \equiv I + \frac{w_j}{z - z_j},$$

(cf. 2.25) for some constant $c(z_j) \neq 0$ and some integer $1 \le k = k(z_j) \le n - 1$. (Here $e_{k,k+1}$ is the matrix with 1 in the $(k, k+1)$ position, and zero elsewhere). Let $\alpha = e^{2\pi i/n}$. Eqn. (4.1) is invariant under the map $z \to \alpha z$, and this leads to the *α-symmetry* of m

$$(4.16) \qquad m(x, \alpha z) = m(x, z)$$

which leads in turn to a corresponding α-symmetry for $v(z)$ and $v(\cdot, z_j)$ (see below).

We now characterize *scattering data* $(v, \{v(\cdot, z_j)\})$ for generic potentials when $n = 2\ell + 1$ is odd: there is a similar and slightly more complicated characterization when $n = 2\ell$ is even (see [BDT]). The main difference between the odd order case and the even order case $n = 2l \ge 4$, is that generically in the even order case one can have negative eigenvalues, which implies that $\Sigma \cap Z \neq \emptyset$ and that the diagonal elements of v can have zeros (cf. (4.21) below).

(4.17) $v(z) - I$ belongs to Schwartz class on each component of $\Sigma \setminus 0$, in the sense of Theorem 2.7.

(4.18) $v(\alpha z) = v(z), \quad z \in \Sigma \setminus 0$.

(4.19) $\alpha Z = Z$ and $v(\alpha z, \alpha z_0) = v(z, z_0)$.

(4.20) On $\Sigma_j \setminus 0, v(z)$ has the same block structure as π_j, and each block has determinant 1.

(4.21) If one of the 2×2 diagonal blocks is

$$\begin{pmatrix} v_{j-1,j-1}(z) & v_{j-1,j}(z) \\ v_{j,j-1}(z) & v_{jj}(z) \end{pmatrix},$$

then $v_{j-1,j-1}(z) \neq 0$ and $v_{jj}(z) = 1$.

(4.22) As $z \to 0$ in Σ_k, the 2×2 block in (4.20) above, has leading asymptotics

$$\begin{pmatrix} \gamma_{kj}z^2 & \rho_j \\ -\rho_j^{-1} & 1 \end{pmatrix} + \begin{pmatrix} O(|z|^3) & O(|z|) \\ O(|z|) & 0 \end{pmatrix} \qquad \text{as} \quad z \to 0 \quad \text{in} \quad \Sigma_k,$$

where $\gamma_{kj} \neq 0$ and $\rho_j = (-1)^j \alpha^{-(j-1)(j-2)/2}$.

(4.23) Let V_j be the formal power series of $v\pi_j \restriction \Sigma_j$ at $z = 0$. Then $V_1 V_2 \dots V_{2n} = I$.

If $L = L^*$ is self-adjoint, then in addition we have:

(4.24) If $z_j \in Z$, the inequalities $\operatorname{Re} \alpha_{k(z_j)} z_j \geq 0 \geq \operatorname{Re}(\alpha_{k(z_j)+1} z_j)$ cannot hold.

(4.25) For $z \in \Sigma_j \backslash 0$, $v(\bar{z}) = R\pi_j J_-(z)^* v(z)^* J_-(z)\pi_j R$, where R is the matrix with 1's on the antidiagonal, and zero elsewhere.

(4.26) For $z_j \in Z$, $v(z, \bar{z}_j) = RJ(z_j)^*(v(\bar{z}, z_j)^{-1})^* J(z_j)R$.

Theorem 4.1 [BDT] *For an open dense subset of potentials in the odd order case, the scattering data v satisfy conditions (4.17)–(4.26). Conversely, given a function v satisfying conditions (4.17)–(4.26), there exists a unique self-adjoint operator with Schwartz class coefficients and with scattering data given by v.*

Whereas the potential $q = 0$ is generic for first order system (2.1), the operator with $p_j = 0$ in (4.1) is *not* generic.

We turn now to a discussion of the *general* (i.e. possibly non-generic) case. We need the following non-trivial result.

Lemma 4.2 *There exists a generic self-adjoint operator $L^\#$ with $Z = \emptyset$.*

Proof. See Remark 5.5 below. □

Corollary 4.3 *For every $q \in Q$ there exists a solution $m(x, z)$ of (4.5) satisfying,*

$$\lim_{x \to -\infty} m(x, z) = \Lambda_z, \tag{4.27}$$

holomorphic in $\mathbb{C} \setminus \Sigma$, and smooth up to the boundary in each sector.

Proof. Given $L = L_q$, let $L^\#$ be a generic (self-adjoint) operator for which $Z = \emptyset$, as in Lemma 4.1. Since the generic operators form an open set in the Schwartz topology, we may modify $L^\#$ to a generic operator L^g (self-adjoint if L is self-adjoint) such that $L \equiv L^g$ for $x \leq x_0$ ($x_0 << 0$), and for which $Z(L^g)$ is again empty. Let m^g be the bounded, normalized eigenfunction for L^g satisfying (4.5), (4.6): as $Z(L^g) = \emptyset, m^g(x, \cdot)$ is analytic in $\mathbb{C} \setminus (\Sigma \cup Z)$ and smooth up to the boundary in each sector. Let $n = n(x, z)$ be the solution of (4.5) normalized at x_0 by $n(x_0, z) = I$; then

$$m^-(x, z) \equiv n(x, z)(e^{i(x - x_0)zJ} m^g(x_0, z) e^{-i(x - x_0)zJ})$$

is the desired solution for q. (Note that $m^-(x, z) = m^g(x, z)$ for $x \leq x_0$). □

Given $q \in Q$, solutions $m(x, z)$ of (4.5) satisfying (4.6) exist and are analytic in $(\mathbb{C} \setminus \Sigma) \cap \{|z| > r/2\}$, for some sufficiently large $r = r(q) > 0$. Let $\Gamma = \{|z| = r\}$. Let $m(x, z)$ denote the above normalized eigenfunction for $|z| > r$, and let $m(x, z) = m^-(x, z)$ for $|z| < r$. Orient $\Sigma \cup \Gamma$ so that all the rays in Σ point outwards and so that the circle is transversed clockwise. (The $+(-)$ side of an oriented contour lies to the left (right)). We find

$$m_+(x, z)\pi_j = m_-(x, z) e^{ixzJ_-} v(z) e^{-ixzJ_-} \qquad z \in \Sigma_j, \tag{4.28}$$

$$(4.29)\qquad m_+(x,z)=m_-(x,z)e^{ixzJ_-}v(z)e^{-ixzJ_-}\qquad z\in\Gamma.$$

The matrix function $v:\Sigma\cup\Gamma\to M_n(\mathbf{C})$ defines the *pre-scattering data* for general L.

Now as noted in the proof of Corollary 4.2, $m^-=m^g$ for $x<x_0$. Thus

(4.30) v has all the properties of generic scattering data on Σ, apart from the discontinuities at $\Sigma\cap\Gamma$.
(4.31) $v-I$ belongs to Schwartz class on each component of $(\Sigma\cup\Gamma)/\{$self-intersections$\}$ in the sense of Theorem 2.7.
(4.32) $v(\alpha z)=v(z)\qquad z\in\Sigma\cup\Gamma$.
(4.33) $v(z)-I$ is strictly upper triangular for $z\in\Gamma\setminus\Sigma$.

In addition to (4.23), we have conditions at self-intersections of $\Gamma\cap\Sigma$.

(4.34) For every j, let $\Gamma_1,\Gamma_2,\Gamma_3,\Gamma_4$ be the four curves intersecting at the point $\Sigma_j\cap\Gamma$, numbered counterclockwise with $\Gamma_1=\Sigma_j\cap\{|z|<r\}$. Let V_1,V_2,V_3,V_4 be the formal power series at $\Sigma_j\cap\Gamma$ of $(v\pi_j)^{-1}\restriction\Gamma_1, v\restriction\Gamma_2, v\pi_j\restriction\Gamma_3$ and $v^{-1}\restriction\Gamma_4$ respectively. Then

$$V_1V_2V_3V_4=I$$

If $L=L^*$ is self-adjoint, then we retain (4.25) and replace (4.26) with

$$(4.35)\qquad v(\bar z)=RJ(z)^*(v(z)^{-1})^*J(z)R,\quad z\in\Gamma\setminus\Sigma.$$

As indicated in Sects. 2 and 3, the inverse problem for first order systems reduces to analyzing a Fredholm equation of index zero. The problem is then solvable if the kernel of the equation is empty. Results which guarantee that the kernel is empty go under the general name of Vanishing Lemmas. For self-adjoint generic odd order operators, conditions (4.17)-(4.26) (there are additional, explicit conditions when n is even) are sufficient for proving a Vanishing Lemma [BDT], and hence the inverse problem for generic, self-adjoint (odd order) case can be solved completely. However, for the general self-adjoint case, conditions (4.30)-(4.35) are not sufficient for a Vanishing Lemma, and the list must be supplemented in the following way.

Note first that if $h(z)=(h_1(z),\dots,h_n(z))$ is α-symmetric, i.e. $h(\alpha z)=h(z)$, then

$$F(h)(\lambda)\equiv z^{1-n}\sum_{Re(i\alpha_j(z)z)<0}\alpha_j(z)h_j(z)\bar h_{n-j+1}(\bar z),\quad \lambda=z^n\notin\mathbf{R},$$

is unambiguously defined for $\lambda\in\mathbf{C}\setminus\mathbf{R}$.

In addition to (4.30)–(4.35), the pre–scattering data has the following property:

(4.36) If $h(z)$, defined for $|z|<r$, is an analytic vector in each sector of $\mathbf{C}\setminus\Sigma$, which is α-symmetric, continuous up to the boundary (including $|z|=r$), and has a Taylor expansion at 0 in each sector of order $n-2$, and if h satisfies the jump relation

$$h_+\pi_j=h_-v(z),\quad |z|<r,\ z\in\Sigma_j,$$

then

$$\int_{C_r} F(hv)d\lambda \geq 0,$$

where C_r denotes the clockwise oriented circle, $|\lambda| = |z|^n = r^n$, in the λ-plane.

To complete the list of properties of pre-scattering data we need a definition. An *auxiliary scattering matrix* δ is a matrix function $\delta : \mathbf{C} \setminus (\Sigma \cup \Gamma) \to M_n(\mathbf{C})$ such that

(i) $Rd_z R\delta d_z^{-1}$ is piecewise holomorphic and smooth up to the boundary.
(ii) $\det \delta = 1$ and $\delta(z)$ is diagonal for $|z| > r$.
(iii) $\delta(\alpha z) = \delta(z)$.
(iv) $\delta(z) - I$ and its derivatives decay as $|z|^{-1}$ as $z \to \infty$.
(v) The upper principal minors of δ do not vanish near Γ.

The final condition satisfied by pre-scattering data is the following.

(4.42) There exists an auxiliary scattering matrix δ as above for which $\tilde{v} \equiv \delta_- v \delta_+^{-1}$ has the following properties: $\tilde{v} - I$ belongs to Schwartz class on each component of $(\Sigma \cup \Gamma)/\{\text{self-intersections}\}$; $\tilde{v}$ has the same block structure as π_j on Σ_j and is lower triangular on Γ; all the upper principal minors of $\tilde{v}$ are equal to 1.

In terms of the direct problem, the matrix δ relates the eigensolution normalized at $x = -\infty$ to the analogous eigensolution $\tilde{m}$ normalized at $x = +\infty$: $m = \tilde{m} e^{ixz \operatorname{ad} J} \delta$. In terms of the inverse problem, the existence of the auxiliary matrix δ follows, in the generic case, directly from properties (4.17)–(4.23) for v. In the non-generic case, however, the existence of the auxiliary matrix δ does not follow from (4.30)–(4.34), and we must add in δ as part of the characterization of the pre-scattering data v (see also the remark on Redundancy in Sect. 3).

The pre-scattering data v is not uniquely defined and depends on the choice of $m(x, z)$ for $|z| < r$. We define the *scattering data* as the equivalence class $[v]$ of the pre-scattering data, under the following equivalence relation: $v \sim v'$ if there exists a matrix function c such that

(4.37) c is holomorphic in $\mathbf{C} \setminus (\Sigma \cup \Gamma)$ and smooth up to the boundary in each component of $\mathbf{C} \setminus (\Sigma \cup \Gamma)$.
(4.38) $c = I$ for $|z| > r$.
(4.39) $c - I$ is strictly upper triangular.
(4.40) $c(\alpha z) = c(z)$.
(4.41) $v' = c_-^{-1} v c_+$.

Theorem 4.4 [DZ] *The scattering transform $L \mapsto m \mapsto v$ induces a well defined map $L \mapsto [v]$ from the space of self-adjoint operators with Schwartz class coefficients to the space of equivalence classes of pre-scattering data.*

5. The inverse problem for n-th order operators

As in Sect. 3, the inverse problem for (v, δ) consists in solving a Riemann-Hilbert factorization problem for a piecewise holomorphic matrix function with jumps (4.28), (4.29) across $\Sigma' = \Sigma \cup \Gamma$, but in place of $m(x, z) \to I$ as $z \to \infty$, we must have $m(x,z) \to \Lambda_x$. Now the first row $\mu(x,z) = e_1^T m(x,z)$ determines the operator L: the remaining rows of $m \to \Lambda_x$ can be obtained by differentiation with respect to x. Thus it is sufficient to consider an α-symmetric row vector factorization, $\mu(\alpha z) = \mu(z)$,

$$\mu_+(x,z)\pi_j = \mu_-(x,z)v(z) \qquad z \in \Sigma_j, \tag{5.1}$$

$$\mu_+(x,z) = \mu_-(x,z)v(z) \qquad z \in \Gamma, \tag{5.2}$$

with

$$\mu(x,z) \to \mathbf{1} \equiv (1,1,\ldots,1). \tag{5.3}$$

Remark. One may try to solve the above row factorization problem by obtaining a bounded matrix factorization $m(x,z) \to I$, as in Sect. 3: μ is then obtained by adding up the rows of m. However the bounded matrix factorization may not exist, whereas the Vanishing Lemma guarantees the existence of the row factorization μ, in the self-adjoint case (see [BDT] §38).

In the direct problem of Sect. 4, it was convenient to work in the *local* z-order. For the inverse problem, however, it is helpful to work in a fixed *global* basis which we choose, for convenience, as follows. Let $z_0 = -i - \epsilon$, and order the roots according to

$$\operatorname{Re} i\alpha_1(z_0)z_0 > \operatorname{Re} i\alpha_2(z_0)z_0 > \cdots > \operatorname{Re} i\alpha_n(z_0)z_0. \tag{5.4}$$

Number the sectors $\Omega_0, \Omega_1, \ldots, \Omega_{2n-1}$ of $\mathbb{C}\setminus\Sigma$, counterclockwise, starting with the sector containing z_0. For $z \in \mathbb{C}\setminus\Sigma$, let $P(z)$ denote the permutation matrix

$$(\alpha_1(z),\ldots,\alpha_n(z))P(z) = (\alpha_1(z_0),\ldots,\alpha_n(z_0)). \tag{5.5}$$

In global order, for $z \in \mathbb{C}\setminus\Sigma$,

$$\mathbf{m}(z) = m(z)P(z), \quad \boldsymbol{\mu}(z) = \mu(z)P(z). \tag{5.6}$$

Orient Σ' as in Sect. 3. By $P_\pm$ we denote the boundary values of $P(z)$ relative to this orientation. The factorization problem (5.1)-(5.3) takes the form

$$\boldsymbol{\mu}_+(x,z) = \boldsymbol{\mu}_-(x,z)e^{izx \operatorname{ad} J(z_0)}\mathbf{v}, \quad z \in \Sigma', \tag{5.7}$$

$$\boldsymbol{\mu}(x,\alpha z) = \boldsymbol{\mu}(x,z)\pi_1\pi_0, \quad \boldsymbol{\mu}(x,z) \to \mathbf{1} \quad \text{as} \quad z \to \infty, \tag{5.8}$$

where

(5.9) on Σ, $\mathbf{v} \equiv P_-^{-1}vP_-$ (resp. $P_-^{-1}v^{-1}P_-$), if the orientation is outwards (resp. inwards), and

(5.10) on Γ, $\mathbf{v} \equiv P^{-1}vP$ (resp. $P^{-1}v^{-1}P$), if the arcs are in the even (resp. odd) sectors.

Also

$$\boldsymbol{\delta} = P^{-1}\delta P. \tag{5.11}$$

Proposition 5.1 *The data* $\mathbf{v}$ *can be factored as*

$$\mathbf{v} = (I - w_-)^{-1}(I + w_+) \tag{5.12}$$

with $I \pm w_\pm \in \mathcal{F}^-(\Sigma'_\pm)$ *and* α*-symmetry*

$$w_\pm(\alpha z) = \pi_0\pi_1 w_\pm(z)\pi_1\pi_0. \tag{5.13}$$

Also

$$\tilde{\mathbf{v}} = \delta_- \mathbf{v} \delta_+^{-1} = (I - \tilde{w}_-)^{-1}(I + \tilde{w}_+) \tag{5.14}$$

with $I \pm \tilde{w}_\pm \in \mathcal{F}^+(\Sigma'_\pm)$ *and* α*-symmetry*

$$\tilde{w}_\pm(\alpha z) = \pi_0\pi_1 \tilde{w}_\pm(z)\pi_1\pi_0. \tag{5.15}$$

Remark. In the definition of $\mathcal{F}^\pm$, the matrix J in Sect. 3 must now be replaced with $-iJ(z_0)$. Let $\mathcal{H}^k(\Sigma')$ denote the Sobolev space of Sect. 3 with matrix–valued functions replaced now by vector–valued functions. Let $\mathcal{H}^k_\alpha(\Sigma') \equiv \{f \in \mathcal{H}^k(\Sigma') \mid f(\alpha z) = f(z)\pi_1\pi_0\}$ and $\widehat{\mathcal{H}}^k_\alpha(\Sigma') = \mathbf{C}1 + \mathcal{H}^k_\alpha(\Sigma')$.

For $w = (w_-, w_+)$, define $C_{w,x} = C_-(f w_+^x) + C_+(f w_-^x)$ as before, with $w_\pm^x \equiv e^{ixz\,\mathrm{ad}\,J(z_0)}\tilde{w}_\pm$.

Theorem 5.2 *The operator* $C_{w,x}$ *maps* $\widehat{\mathcal{H}}^k_\alpha$ *into* $\mathcal{H}^k_\alpha$ *and* $Id - C_{w,x}$ *is Fredholm of index zero on* $\widehat{\mathcal{H}}_\alpha$. *In the self-adjoint case, for* k *sufficiently large, the operator* $Id - C_{w,x}$ *has 0 kernel, and hence is invertible. Whenever* $Id - C_{w,x}$ *is invertible, the unique solution* $\mu_w(x,\cdot) \in \widehat{\mathcal{H}}^k_\alpha$ *of*

$$\mu_w(x,\cdot) = 1 + C_{w,x}\mu_w(x,\cdot) \tag{5.16}$$

gives rise to the solution $\mu(x,\cdot)$ *of the factorization problem (5.7), (5.8) through*

$$\mu(x,z) = 1 + \frac{1}{2\pi i}\int_{\Sigma'} \frac{\mu_w(x,\zeta)}{\zeta - z}(w_+^x(\zeta) + w_-^x(\zeta))d\zeta, \quad z \in \mathbf{C}\setminus\Sigma. \tag{5.17}$$

Proof. The proof is similar to that of Theorem 3.3, apart from the claim about the kernel of $Id - C_{w,x}$ in the self-adjoint case. If $h = C_{w,x}h$ for some $h \in \widehat{\mathcal{H}}^k_\alpha$, then $h \in \mathcal{H}^k_\alpha$ and

$$\mathrm{h}(x,z) = \frac{1}{2\pi i}\int_{\Sigma'} \frac{h(x,\zeta)}{\zeta - z}[w_+^x(\zeta) + w_-^x(\zeta)]d\zeta \qquad z \in \mathbf{C}\setminus\Sigma,$$

solves the Riemann-Hilbert factorization problem (5.7), (5.8) with $\mathbf{h}(x,z) \to 0$ as $z \to \infty$ instead of $\mathbf{h} \to \mathbf{1}$. Then the Vanishing Lemma of [DZ], using (4.36), implies $\mathbf{h} = 0$. □

The appropriate modifications of the methods of Sect. 3, (see [BDT]) show that $\mu(x,z)$ solves an n-th order equation of type (4.5), with a smooth potential

$q(x)$ which decays, together with its derivatives, rapidly as $x \to -\infty$. To infer that $q(x)$ and its derivatives also decay rapidly as $x \to +\infty$, it suffices to show that $\mu(x,z)\delta^x(z)^{-1}$ solves the analogous Riemann-Hilbert problem for $\tilde{v}$. The key technical point is to show that $\mu(x,z)(\delta^x(z))^{-1}$ is smooth at $z = 0$. But this follows from (4.37) and the following result, whose proof depends on the detailed structure of $v(z)$ near the origin (see (4.22)).

Lemma 5.3 *$\mu P^{-1} d_z^{-1}$ is continuous as $z \to 0$ in each sector.*

Proof. See [BDT]. □

The function $\mu(x,z)$of (5.17) determines a unique self-adjointness operator L with Schwartz class coefficients, $L\mu = z^n\mu$. The main result is the following.

Theorem 5.4 *The map $\Phi : L \mapsto [v]$ of Theorem 4.4 is a bijection. The map $v \mapsto \mu \mapsto L$ described above, lifts to a map Ψ from scattering data to self-adjoint operators, taking $[v] \mapsto L$. We have $\Psi = \Phi^{-1}$.*

As in Sect. 3, the above results can be used to prove local existence and uniqueness results for the Cauchy problem for the (formally well-posed) Gel'fand-Dikii flows with general initial data in Q. If, in addition, the initial data are self-adjoint, then the solutions exist globally in time (see [BDT], [DZ]). As is well known, the Gel'fand-Dikii flows include the celebrated Korteweg–de Vries and Boussinesq equations of shallow water wave theory.

Remark. In the second order, self-adjoint case, Z is always finite. One may ask whether Z is in fact always finite in the higher order, self-adjoint case. In [DZ], the authors construct a self-adjoint third order operator with Schwartz space coefficients, for which Z is indeed infinite. In [DZ], the authors also construct a second order non-self-adjoint operator with a Schwartz space potential, for which Z is again infinite.

Remark 5.5 Finally we indicate how to prove Lemma 4.1. By the results of [BDT], there exists a generic self-adjoint operator L_n^0, with scattering data $(v, \{v(\cdot, z_i)\})$. In the odd order case simply discard the discrete data $\{v(\cdot, z_j)\}$ and then $v \restriction \Sigma$ clearly satisfies all the conditions for odd order scattering data and produces a generic self-adjoint operator without poles by the inverse map. In the even order case, there is some interaction between the poles and the continuous data, and the above procedure must be slightly modified (see [DZ]; see also the discussion preceding (4.17)). □

6. Historical Remarks; Related Developments

It is impossible in a short space to recount in any detail the history of soliton equations and the inverse scattering transform, even omitting the major related topics. We concentrate here on developments most closely related to the analytic theory outlined in Sects. 2–5 and refer to the books [BuC, Ko, Ne2, NM] for general accounts through the mid 80s.

As noted in the introduction, the subject began with the linear Schrödinger equation, (related to KdV) where the theory was already developed by Faddeev [Fa2]. This case was further completed by Marchenko (cf. [Mc]), Deift and Trubowitz [DT], and Melin [Me].

The approach of Lax [Lx] in writing KdV as an evolution equation for the Schrödinger operator itself led to a search for "Lax pairs" associated to other equations and systems. A theory was developed (somewhat formally) for 2×2 systems, in close analogy to the Schrödinger case, by Zakharov and Shabat [ZS2], Ablowitz, Kaup, Newell, and Segur [AK2], Flaschka and Newell [FS].

It soon became clear that a number of interesting model equations in two dimensions (space or space-time) were associated to equations of order 3 or higher and to systems of size 3×3 or larger: see below. Formal scattering theories were developed during the late 70s and early 80s.

In the following discussion we use the notation of Sects. 2 and 3, using J to denote the $N \times N$ diagonal matrix which characterizes the spectral problem, and using $J(z)$, $q(x, z)$ to denote the matrices which play the roles of zJ and $q(x)$ in more general spectral problems.

3×3 and $N \times N$ systems with iJ real. Zakharov and Manakov [ZM1] related the three-wave interaction equation to a 3×3 spectral problem; unlike equations associated to 2×2 and second-order operators, this equation allows soliton interactions which are non-trivial asymptotically [Ma]. Higher order systems were also associated to two-dimensional classical field models by Zakharov and Mikhailov [ZMi1, ZMi2]; again, non-trivial soliton interactions can occur.

In both cases, iJ is real; this means that the singular set Σ of Sect.2 is the real line, a full classical scattering matrix relating asymptotics at $\pm\infty$ exists, and one can try to model the theory on the 2×2 case.

Formal scattering transforms were developed for such higher order systems by Zakharov and Manakov [ZM2], Kaup [Ka1], Ablowitz and Haberman [AH], Miodek [Mio], Newell [Ne1]. The relation to Riemann-Hilbert factorization was emphasized by Zakharov and Shabat [ZS2].

A rigorous theory for $N \times N$ systems with iJ real was given by Shabat [Sh2], who found the relationship between the multiplicative jump matrix of Sect. 2 and the classical scattering matrix $S(\xi)$ which relates the limits

$$\lim_{x \to -\infty} e^{-x\xi J} \Psi(x, \xi), \qquad \lim_{x \to +\infty} e^{-x\xi J} \Psi(x, \xi), \quad \xi \in \mathbf{R}.$$

Shabat also identified the linearization of the scattering transform, i.e. the expression of a variation δS in terms of δq, and showed that both the scattering transform and its linearization could be viewed as eigenfunction expansions, thus clarifying the role of the "squared eigenfunctions" familiar from the Schrödinger equation–cf [DT]. Further results for this case were obtained by Bar Yaacov [BY].

Gerdzhikov and Kulish [GK], and Gerdzhikov [Gd] used the Shabat results to discuss the general nonlinear evolutions and Hamiltonian structures associated to these systems.

Third-order equations. Zakharov [Za] noted that the Boussinesq equation is related to a third-order spectral problem. Formal scattering theories were developed by Caudrey [Ca1] and Kaup [Ka2]. A rigorous treatment (but under some restrictions on scattering data which rule out the "generic case") was given by Deift, Tomei, and Trubowitz [DTT].

$N \times N$ systems with general J. It was found by Mikhailov [Mi] and Fordy and Gibbons [FG1] that some field theory models and new nonlinear evolution equations are associated to systems with iJ complex. Formal results were obtained by these authors, and a more complete theory was developed by Caudrey [Ca2]. Rigorous results, particularly in connection with the inverse problem, were first obtained by Beals and Coifman [BC1] for the "generic" case. This work was refined and extended to the general case by Zhou [Zh1,Zh2].

Equations of order n. Gelfand and Dikii [GD] showed that there is a hierarchy of nonlinear evolution equations of Lax form associated to each scalar differential operator of order $n \geq 2$. A formal scattering theory was developed by Caudrey [Ca2]. A rigorous theory for the "generic case" was begun by Beals [Be] and completed by Beals, Deift, and Tomei [BDT]. The self-adjoint case was also studied by Sukhanov [Su], who assumes no discrete data and assumes also, in order to get properties of the solution of the inverse problem, that scattering data trivializes for $z \sim 0$. The general (not necessarily generic) self-adjoint case was solved by Deift and Zhou [DZ] with heavy use of results from [BDT, Zh1, Zh2].

We conclude with a few references to generalizations and to closely related questions.

More general z–dependence. Various nonlinear evolution equations and field equations of interest are associated to linear systems with rational z–dependence of $J(z)$ and $q(x, z)$. We cite [AK1] and [FT] for the sine-Gordon equation, [KM] for the Thirring model and [KN] for the derivative nonlinear Schrödinger equation, and also [BLP]. Formal discussions are due to Zakharov and Shabat [ZS2], and in more detail to Gerdzhikov, Ivanov, and Kulish [GIK], Konopelchenko and Formusatic [KF], Gerdzhikov and Ivanov [GI], and particularly Caudrey [Ca2]. The analysis was worked out in detail by Lee [Le] for polynomial dependence. For the general case, see Zhou [Zh3]. Martinez-Alonso [MA] considered the Schrödinger equation with z-dependent potential.

Bäcklund-Darboux transformations, eigenvalues, and scattering data. A procedure going back at least to Moutard [Mo] and Darboux [Da] relates eigenfunctions of a given equation to those of a related equation. A special case, for the Schrödinger equation, inserts or removes an eigenvalue by a Bäcklund transformation of the potential; see [DT]; for higher order equations see [FG2]. From the point of view of scattering data, such transformations alter the discrete data (and possibly the continous data, but in a very regular way). This procedure is discussed in detail for n-th order operators in [BDZ]. For analogous results for systems, see [SZ], [BT].

References

ABT M. J. Ablowitz, R. Beals, and K. Tenenblat, *On the solution of the generalized wave and generalized sine-Gordon equations*, Stud. Appl. Math. **74** (1977), 177-203.

AH M. J. Ablowitz and R. Haberman, *Resonantly coupled nonlinear evolution equations*, J, Math. Phys. **16** (1975), 19-31.

AK1 M. J. Ablowitz, D. J. Kaup, A. C. Newell, and H. Segur, *Method for solving the sine-Gordon equation*, Phys. Rev. Lett. **30** (1973), 1262-1264.

AK2 M. J. Ablowitz, D. J. Kaup, A. C. Newell, and H. Segur, *The inverse scattering transform–Fourier analysis for nonlinear problems*, Stud. Appl. Math. **53** (1974), 249-315.

BY D. Bar Yaacov, *Analytic properties of scattering and inverse scattering for first order systems*, Dissertation, Yale 1985.

Be R. Beals, *Problemes inverses pour les equations differenticlles sur la droite*, Seminaire Goulaouic-Meyer-Schwartz 1982-1983, exposé 1, École Polytechnique, Palaiseau; *The inverse problem for ordinary differential operators on the line*, Amer. J. Math. **107** (1985), 281-366.

BC1 R. Beals and R. R. Coifman, *Scattering, transformations spectrales, et equations d'evolution nonlineaires*, Seminaire Goulaouic-Meyer-Schwartz 1980-1981, exposé 22, École Polytechnique, Palaiseau; *Scattering and inverse scattering for first order systems*, Comm. Pure Appl. Math. **37** (1984), 39-90.

BC2 R. Beals and R. R. Coifman, *Inverse scattering and evolution equations*, Comm. Pure Appl. Math. **38** (1985), 29-42.

BC3 R. Beals and R. R. Coifman, *Scattering and inverse scattering for first order systems, II*, Inverse Problems **3** (1987), 577-593.

BDT R. Beals, P. Deift, and C. Tomei, *Direct and Inverse Scattering on the Line*, Math. Surveys and Monographs no. 28, Amer. Math. Soc., Providence 1988.

BT R. Beals and K. Tenenblat, *Inverse scattering and the Bäcklund transformation for the generalized wave and generalized sine-Gordon equations*, Stud. Appl. Math **78** (1988), 227-256.

BLP M. Boiti, J. J-P. Leon, and F. Pempinelli, *A recursive generalization of local higher-order sine-Gordon equations and their Bäcklund transformations*, J. Math. Phys. **25** (1984), 1725- 1734.

BuC R. K. Bullough and P. J. Caudrey, eds., *Solitons, Topics in Current Physics no. 117*, Springer, Berlin 1980.

Ca1 P. J. Caudrey, *The inverse problem for the third order equation* $u''' + q(x)u' + r(x)u = -i\zeta^3 u$, Phys. Lett. **79A** (1980), 264-268.

Ca2 P. J. Caudrey, *The inverse problem for a general* $n \times n$ *spectral equation*, Physica D6 (1982), 51-66.

Da G. Darboux, *La Theorie Generale des Surfaces*, Gauthier Villars, Paris 1889, livre IV, Ch. IX.

DTT P. Deift, C. Tomei, and E. Trubowitz, *Inverse scattering and the Boussinesq equation*, Comm. Pure Appl. Math. **35** (1982), 567-628.

DT P. Deift and E. Trubowitz, *Inverse scattering on the line*, Comm. Pure Appl. Math. **32** (1979), 121-251.

DZ P. Deift and X. Zhou, *Direct and inverse scattering on the line with arbitrary singularities*, Comm. Pure Appl. Math. **44** (1991), 485-533.

Fa1 L. D. Faddeev, *The inverse problem in the quantum theory of scattering*, Uspehi Matem. Nauk **14** (1959) 57-119; J. Math. Physics **4** (1963), 72-104.

Fa2 L. D. Faddeev, *Properties of the S-matrix of the one-dimensional Schrödinger equation*, Trudy Matem.Inst. Steklov **73** (1964), 314-333; Amer. Math. Soc. Translations, series 2, vol. 65, 139-166.

FT L. D. Faddeev and L. A. Takhtajan, *Essentially nonlinear one-dimensional model of classical field theory*, Teor. Mat. Fiz. **21** (1974), 160-174; Theor. Math. Phys. **21** (1974), 1046-1057.

FN H. Flaschka and A. C. Newell, "Integrable systems of nonlinear evolution equations," in Dynamical Systems, Theory and Applications, J. Moser, ed., Lecture Notes in Physics, no. 38, Springer, Berlin, 1975, 355-440.

FG1 A. P. Fordy and J. Gibbons, *Integrable nonlinear Klein-Gordon equations and Toda lattices*, Comm. Math. Phys. **77** (1980), 21-30.

FG2 A. P. Fordy and J. Gibbons, *Factorizations of operators, I. Miura transformations*, J. Math. Phys. **21** (1980), 2508-2510; *II*, ibid. **22** (1981), 1170-1175.

Ga C. S. Gardner, *Korteweg-de Vries equation and generalizations, IV. The Korteweg-de Vries equation as a Hamiltonian system*, J. Math. Physics 12 (1971), 1548-1551.

GG C. S. Gardner, J. M. Greene, M. D. Kruskal, and R. M. Miura, *Method for solving the Korteweg-de Vries equation*, Phys. Rev. Lett. 19 (1967), 1095-1097.

GD I. M. Gelfand and L. A. Dikii, *Asymptotic behavior of the resolvent of Sturm-Liouville equations and the algebra of the Korteweg-De Vries equations*, Uspehi Mat, Nauk 30 (1975), 67-100; Russian Math. Surveys 30 (1975), 77-113; *Fractional powers of operators and Hamiltonian systems*, Funct. Anal. Appl. 10 (1976) 259-273.

Gd V. S. Gerdzhikov, *On the spectral theory of the integro-differential operator Λ generating nonlinear evolution equations*, Lett. Math. Physics 6 (1982), 315-323.

GK V. S. Gerdzhikov and P. P. Kulish, *Expansion in "squares" of eigenfunctions of a matrix linear system*, Zap. Naucn. Sem Leningrad. Otdel. Mat. Inst. Steklov 101 (1981), 46-63, 206; *The generating operator for the $n \times n$ linear system*, Physica D3 (1981), 549-564.

GI V. S. Gerdzhikov and M. I. Ivanov, *A quadratic pencil of general type and nonlinear evolution equations I, Expansion in "squares" of solutions–generalized Fourier transforms*, Bulg. J. Phys. 10 (1983), 13-26; *II, Hierarchies of Hamiltonian structures*, ibid., 130-143.

GIK V. S. Gerdzhikov, I. M. Ivanov, and P. P. Kulish, *Quadratic bundles and nonlinear equations*, Teor. Mat. Fiz. 44 (1980), 342-358; Theor. Math. Phys. 44 (1980), 784-795.

Ka1 D. J. Kaup, *The three-wave interaction–a non-dispersive phenomenon*, Studies Appl. Math. 55 (1976), 9-44.

Ka2 D. J. Kaup, *On the inverse scattering problem for cubic eigenvalue problems of the class $\psi_{xxx} + 6Q\psi_x + 6R\psi = k^3\psi$*, Stud. Appl. Math. 62 (1980), 189-216.

KN D. J. Kaup and A. C. Newell, *An exact solution for a derivative nonlinear Schrödinger equation*, J, Math. Phys. 19 (1978), 798-801.

Ko B. G. Konopelchenko, *Nonlinear Integrable Equations*, Lecture Notes in Physics, no. 270, Springer, Berlin 1987.

KF B. G. Konopelchenko and I. B. Formusatic, *On the structure of nonlinear evolution equations integrable by the Z_2 graded bundle*, J. Physics 15A (1982), 2017-2040.

KM E. A. Kuznetsov and A. V. Mikhailov, *On the complete integrability of the two-dimensional classical Thirring model*, Teor. Mat. Fiz. 30 (1977), 303-315; Theor. Math. Phys. 30 (1977), 193-200.

Lx P. D. Lax, *Integrals of nonlinear equations of evolution and solitary waves*, Comm. Pure App. Math. 31 (1968), 467-490.

Le J-h. Lee, *Analytic properties of a Zakharov-Shabat inverse scattering problem, I, II*, Chinese J. Math. 14 (1986), 225-248; ibid. 16 (1988), 81-110.

Ma S. V. Manakov, *An example of a completely integrable nonlinear wave field with non-trivial dynamics (Lee Model)*, Teor. Mat. Fiz. 28 (1976), 172-179; Theor. Math. Phys. 28 (1976), 709-714.

Mc V. A. Marchenko, *Sturm-Liouville Operators and Applications*, Birkhäuser, Basel 1986.

MA L. Martinez-Alonso, *Schrödinger spectral problems with energy dependent potentials as sources of nonlinear Hamiltonian evolution equations*, J. Math. Phys. 19 (1978), 2342-2349.

Me A. Melin, *Operator methods for inverse scattering on the real line*, Comm. Partial Diff. Equ. 10 (1985), 677-766.

Mi A. V. Mikhailov, *Reduction in integrable systems. The reduction group*, Pism. Zh. Eksp. Teor. Fiz. 32 (1980), 187-192; JETP Lett. 32 (1980), 174-178; *The reduction problem and the inverse scattering method*, Physica 3D (1981), 73-117.

Mio I. Miodek, *"IST–solvable" nonlinear evolution equations and existence – an extension of Lax's method*, J. Math. Phys. 19 (1978), 19-31.

Mt T. F. Moutard, *Note sur les equations differentielles lineaires du second ordre*, C. R. Acad. Sci. Paris 80 (1876), p. 729.

Ne1 A. C. Newell, *The general structure of integrable evolution equations*, Proc. Royal Soc. A365 (1979), 283-311.

Ne2 A. C. Newell, *Solitons in Mathematics and Physics*, Society for Ind. and Appl. Math., Philadelphia 1985.

NM S. Novikov, S. V. Manakov, L. P. Pitaevskii, and V. E. Zakharov, *Theory of Solitons, the Inverse Scattering Method*, Consultants Bureau, New York 1984.

Sa D. H. Sattinger, *Hamiltonian hierarchies on semisimple Lie algebras*, Stud. Appl. Math. 72 (1984), 65-86.

SZ D. H. Sattinger and V. D. Zurkowski, "Gauge theory of Bäcklund transformations I," in *Dynamics of Infinite-Dimensional Systems*, J. K. Hale and S. M. Chow eds., NATO ASI series F, Springer-Verlag, Berlin 1986. *II*, Physica D26 (1987), 225-250.

Sh1 A. B. Shabat, "One-dimensional perturbations of a differential operator and the inverse scattering problem", in *Problems in Mechanics and Mathematical Physics*, Nauka, Moscow, 1976.

Sh2 A. B. Shabat, *An inverse scattering problem*, Diff. Uravn. **15** (1979), 1824-1834; Diff. Equ. **15** (1980), 1299-1307.

Su V. V. Sukhanov, *An inverse problem for a self-adjoint operator on the line*, Math. Sbornik **137** (1988), 242-259; Math. USSR Sbornik **65** (1990), 249-266.

Wa M. Wadati, *The modified Korteweg-de Vries equation*, J. Physical Soc. Japan **34** (1973), 380-384.

Za V. E. Zakharov, On stochastization of one-dimensional chains of nonlinear oscillators, Zh. Eksp. Teor. Fiz. **65** (1973), 219-225; Soviet Physics JETP **38** (1974), 108-110.

ZM1 V. E. Zakharov and S. V. Manakov, *On resonant interaction of wave packets in non-linear media*, Pisma Zh. Eksp. Teor. Fiz. Lett. **18** 1973), 413-417; JETP Letters **18** (1973), 243-247.

ZM2 V. E. Zakharov and S. V. Manakov, *The theory of resonant interaction of wave packets in nonlinear media*, Zh. Eksp. Teor. Fiz. **69** (1975), 1654-1673; Soviet Physics JETP **42** (1976), 842-850.

ZMi1 V. E. Zakharov and A. V. Mikhailov, *Example of nontrivial interaction of solitons in two-dimensional classical field theory*, Pisma Zh. Eksp. Teor. Fiz. **27** (1978), 47-51; JETP Letters **27** (1978), 42-46.

ZMi2 V. E. Zakharov and A. V. Mikhailov, *Relativistically invariant 2-dimensional models of field theory which are integrable by means of the inverse scattering problem method*, Zh. Eksp. Teor. Fiz. **74** (1978), 1953-1973; Soviet Physics JETP **47** (1978), 1017-1027.

ZS1 V. E. Zakharov and A. B. Shabat, *Exact theory of two-dimensional self-focussing and one-dimensional self-modulation of waves in nonlinear media*, Zh. Eksp. Teor. Fiz. **61** (1971) 118-134; Soviet Physics JETP **34** (1972), 62-69.

ZS2 V. E. Zakharov and A. B. Shabat, *A scheme for integrating nonlinear equations of mathematical physics by the method of the inverse scattering transform, I*, Funk. Anal. Prilozh. **8** (1974), 54-56; Funct. Anal. Appl. **8** (1974), 226-235.

ZS3 V. E. Zakharov and A. B. Shabat, Integration of nonlinear equations of mathematical physics by the method of inverse scattering, II, Funk. Anal. Prilozh. **13** (1979), 13-23. Funct. Anal. Appl. **13** (1980), 166-174.

Zh1 X. Zhou, *Riemann Hilbert problem and inverse scattering*, SIAM J. Math. Anal. **20** (1989), 966-986.

Zh2 X. Zhou, *Direct and inverse scattering transforms with arbitrary spectral singularities*, Comm. Pure Appl. Math **42** (1989), 895-938.

Zh3 X. Zhou, it Inverse scattering transform for systems with rational spectral dependence, preprint.

Zu V. D. Zurkowski, *Scattering for first order linear systems on the line and Bäcklund transformations*, Dissertation, Univ. of Minnesota 1987.

C-Integrable Nonlinear Partial Differential Equations

F. Calogero

*Dipartimento di Fisica, Università di Roma "La Sapienza",
I-00185 Roma, Italy
Istituto Nazionale di Fisica Nucleare, Sezione di Roma,
I-00185 Roma, Italy
*On leave while serving as Secretary General, Pugwash Conferences on Science and World Affairs, Geneva London Rome.

A nonlinear partial differential equation is called *C-integrable*, if it can be solved by a *Change of variables*. Some of these equations have a *universal* character and are therefore of interest both from the theoretical and the applicative points of view. Techniques to manufacture nonlinear evolution equations of this kind (in N+1 dimensions) are reviewed, and examples reported.

1. Introduction

Recently the notions of C-integrable and S-integrable equations have been introduced [1]. *S-integrable* equations are those solvable via the *Spectral transform* (or, equivalently, via the *inverse Scattering method*); most of this book deals with such equations. *C-integrable* equations are those solvable via an appropriate *Change of variables*; they are therefore generally easier to investigate, indeed the study of their solutions might be considered a rather trivial exercise.

The discovery of S-integrability [2] has constituted a major advance in mathematics and mathematical physics, as demonstrated by the many ramifications of this finding (see, for instance, the contents of this book). But the importance of S-integrable equations is also grounded in their applicative relevance, which is remarkably wide. Indeed, the fact that certain nonlinear partial differential equations are both widely applicable *and* integrable is highly significant. As it has been recently emphasized [3], this phenomenon can be traced to the *universal* character of certain nonlinear PDEs,which are associated, via a limiting process, with a *large* class of nonlinear PDEs.

Indeed, the fact that a specific equation is associated, via some limiting process, with a *large* class of equations, attributes to it a *universal* character. Moreover, the fact that the limiting process that underpins such a connection may correspond to a situation relevant in many applicative contexts, implies the *wide applicability* of these universal equations; since the large classes of equations to which they are associated contain - as it were, by definition: since they are *large* - *many* equations, among which there presumably are several having applicative relevance. Finally, the fact that the limiting process that distils the *universal* equation from the *large* class of equations is, in some asymptotic sense, *exact*, and therefore preserves integrability, implies the expectation that these universal equations be *integrable*; since it is justified to expect that a *large* class of equations contain *at least one* integrable equation, and this is sufficient to guarantee the integrability of the universal equation associated with the large class (since this equation is obtained, by a limiting process that preserves the property of integrability, from *all* the equations of the *large* class, and therefore, in particular, from an integrable specimen likely to be contained in the large class) [3].

This notion applies to S-integrable and C-integrable equations. S-integrability is generally a less stringent requirement than C-integrability: indeed generally a C-integrable equation can also be considered to be S-integrable, but not vice versa. This corresponds to a natural classification of equations in terms of the difficulty of their solution: see, for instance, the definitions of S-integrability and C-integrability proposed in the Addendum in Ref.[3]. From this point of view, the fact that certain *universal* equations (such as, for instance, the nonlinear Schrödinger equation in 1+1 dimensions) are S-integrable but not C-integrable, is remarkable and indeed somewhat surprising. (This fact, incidentally, may be conveniently exploited to evince necessary conditions for C-integrability, which may be useful to exclude that some specific nonlinear evolution equation of applicative relevance be C-integrable, thereby implying that any search for a change of variables to solve it is doomed to failure) [4].

It stands to reason, in this frame of thinking, that universal equations should be a principal focus in the investigation of nonlinear partial differential equations, and that some of them are likely to be C-integrable. The validity of these notions motivates our interest in the investigation of C-integrable equations. Moreover, in view of the ease to study in explicit detail the phenomenology characterizing the solutions of this type of equations, C-integrable equations provide a convenient theoretical environment to elucidate various features of nonlinear evolution equations, including certain remarkable properties possessed by S-integrable equations but less transparently evidenced in that more difficult context (see, for instance, [5-8]).

In the following Section we review some techniques to manifacture C-integrable equations. Our presentation below is more in the guise of a guide to the literature, than a display of results, since there would be little point in our reporting here findings already published elsewhere [3,9,10]. We do, however, exhibit a few instances of C-integrable nonlinear evolution equations in $N+1$ dimensions, since these results [11] had not yet appeared in print when this contribution was completed (December 1991).

2. C-integrable equations

A prototypical C-integrable evolution equation is Burgers equation:

$$u_t = u_{xx} + 2u_x u = (u_x + u^2)_x\,, \qquad u \equiv u(x,t). \tag{2.1}$$

Its solvability by an appropriate Change of dependent variable was discovered over four decades ago [12,13]:

$$v(x,t) = u(x,t) \quad exp[\int_{-\infty}^{x} dx'\, u(x',t)]\,, \tag{2.2a}$$

$$u(x,t) = v(x,t) \quad /[1 + \int_{-\infty}^{x} dx'\, v(x',t)]\,, \tag{2.2b}$$

$$v_t = v_{xx}, \qquad v \equiv v(x,t). \tag{2.3}$$

In writing these equations, we have assumed for simplicity that $u(x,t)$ and $v(x,t)$, as well as their x-derivatives, vanish (sufficiently fast) as $x \to -\infty$.

The linearizing transformation(2.2) implies that the initial value problem (on the whole line) for the nonlinear C-integrable equation (2.1) can be solved via the following 3 steps. (i) Given the initial datum $u(x,0)$, compute $v(x,0)$ from (2.2a) (at $t=0$). (ii) From $v(x,0)$ evaluate $v(x,t)$, by solving the *linear* evolution equation (2.3); this can be done via the Fourier transform (which was indeed invented just for this purpose!). (iii) From $v(x,t)$ obtain $u(x,t)$ via (2.2b), thereby achieving the solution of the Cauchy problem.

This technique to solve the Burgers equation (2.1) is typical of those nonlinear C-integrable equations which can be linearized by a change of dependent variables. Other examples of such equations are [5]

$$u_t = u_{xxx} + 3(u_{xx}u^2 + 3u_x^2 u) + 3u_x u^4 , \quad u \equiv u(x,t) , \tag{2.4}$$

and the Eckhaus equation [1,6]

$$i\psi_t + \psi_{xx} + [2(|\psi|^2)_x + |\psi|^4]\psi = 0 , \quad \psi \equiv \psi(x,t) . \tag{2.5}$$

Other examples of such equations are reported in [3], and many more in [9].

All these equations are in 1+1 dimensions (one "space " and one "time" variable). Recently a class of C-integrable equations in $N+1$ dimensions (N "space" and one "time" variables) have been reported. Three instances of such equations read as follows [11]:

$$u_t + \Delta u + 2\mathbf{f}\cdot\nabla u + [Bu^p + \mathbf{f}^2 + \nabla\cdot\mathbf{f}]\,u = 0 , \tag{2.6a}$$

$$\mathbf{f}_t = p\,B\,u^{p-1}\nabla u ; \tag{2.6b}$$

$$u_{tt} - \Delta u + m^2 u + (2+p)Bu^p u_t - 2\mathbf{f}\cdot\nabla u + [B^2u^{2p} - \mathbf{f}^2 - \nabla\cdot\mathbf{f}]\,u = 0 , \tag{2.7a}$$

$$\mathbf{f}_t = p\,B\,u^{p-1}\nabla u ; \tag{2.7b}$$

$$i\psi_t + \Delta\psi + V(\mathbf{r})\psi + 2\mathbf{f}\cdot\nabla\psi + [iB|\psi|^p + \mathbf{f}^2 + \nabla\cdot\mathbf{f}]\,\psi = 0 , \tag{2.8a}$$

$$\mathbf{f}_t = p\,B\,|\psi|^{p-1}\nabla|\psi| . \tag{2.8b}$$

In these equations, B is a real constant,$u \equiv u(\mathbf{r},t)$ is a real scalar, $\mathbf{f} \equiv \mathbf{f}(\mathbf{r},t)$ is a real N-vector, $\psi \equiv \psi(\mathbf{r},t)$ is a complex scalar, $\mathbf{r}$ an N-vector (the "space" coordinate), and ∇ resp. $\Delta \equiv \nabla^2$ the standard (first-order) N-vector resp. (second order) Laplacian differential operators in N-dimensional space. The real quantity p in (2.6,7) is arbitrary ($p \neq 0$), and $V(\mathbf{r})$ in (2.8a) is an arbitrarily assigned "external potential". These three nonlinear evolution equations are C-integrable, but only provided the initial datum for the auxiliary field $\mathbf{f}$ satisfies the constraint

$$\nabla \times \mathbf{f}(\mathbf{r},0) = 0 , \tag{2.9}$$

namely only provided $\mathbf{f}(\mathbf{r},0)$ is irrotational. Note that, since the right-hand-sides of (2.6b), (2.7b) and (2.8b) are gradients, the *initial* condition (2.9) is sufficient to guarantee that $\mathbf{f}(\mathbf{r},t)$ remain irrotational for *all* time.

To illustrate the technique of solution for these C-integrable equations, let us report the Change of dependent variable that linearizes (2.6):

$$v(\mathbf{r},t) = C(\mathbf{r})\, u(\mathbf{r},t) exp\{B \int_0^t dt'[u(\mathbf{r},t')]^p\} \ , \qquad (2.10a)$$

$$u(\mathbf{r},t) = [v(\mathbf{r},t)/C(\mathbf{r})]\{1 + pB \int_0^t dt'[v(\mathbf{r},t')/C(\mathbf{r})]^p\}^{-1/p} \ , \qquad (2.10b)$$

$$v_t(\mathbf{r},t) + \Delta v(\mathbf{r},t) = 0 \ . \qquad (2.11)$$

For a detailed discussion of the solution of the Cauchy problem for these equations, as well as the display of some explicit solutions, see [11]. Other C-integrable equations are also reported in [11], together with a discussion of the technique to manufacture equations of this type. In this connection it is illuminating to note the analogies and differences between (2.2) and (2.10) (for a detailed discussion, see Appendix A of [11]).

The C-integrability of the PDEs discussed above is associated with a Change of *dependent* variables. Another route to manufacture nonlinear C-integrable PDEs is instead based on a Change of *independent* variables. Some such techniques are discussed in [3] and used in [10] to obtain and display many instances of C-integrable equations (in 1+1 dimensions). Analogous techniques to manufacture C-integrable equations in N+1 dimensions are discussed in [11]. An example of such equations (in 2+1 dimensions) reads as follows [11]:

$$\begin{aligned} u_t + &\{a_1[F^{(11)}]^2 + 2a_2F^{(11)}F^{(12)} + a_3[F^{(12)}]^2\}u_{xx} \\ &+ \{a_1[F^{(21)}]^2 + 2a_2F^{(22)}F^{(21)} + a_3[F^{(22)}]^2\}u_{yy} \\ &+ 2\{a_1F^{(11)}F^{(21)} + a_2[F^{(11)}F^{(22)} + F^{(12)}F^{(21)}] + a_3F^{(22)}F^{(12)}\}u_{xy} \\ &+ \{a_1[F_x^{(11)}F^{(11)} + F_y^{(11)}F^{(21)}] + 2a_2[F_x^{(11)}F^{(12)} + F_y^{(11)}F^{(22)}] \\ &\quad + a_3[F_x^{(12)}F^{(12)} + F_y^{(12)}F^{(22)}] + a_4F^{(11)} + a_5F^{(12)} + h^{(1)}[u]\}u_x \\ &+ \{a_1[F_y^{(21)}F^{(21)} + F_x^{(21)}F^{(11)}] + 2a_2[F_y^{(21)}F^{(22)} + F_x^{(21)}F^{(12)}] \\ &\quad + a_3[F_y^{(22)}F^{(22)} + F_x^{(22)}F^{(12)}] + a_4F^{(21)} + a_5F^{(22)} + h^{(2)}[u]\}u_y \\ &\quad + a_6u = 0 \ , \end{aligned} \qquad (2.12a)$$

$$F_t^{(11)} = F^{(11)}h^{(1)'}[u]\,u_x + F^{(21)}h^{(1)'}[u]\,u_y - F_x^{(11)}h^{(1)}[u] - F_y^{(11)}h^{(2)}[u] \ , \qquad (2.12b)$$

$$F_t^{(22)} = F^{(22)}h^{(2)'}[u]\,u_y + F^{(12)}h^{(2)'}[u]\,u_x - F_y^{(22)}h^{(2)}[u] - F_x^{(22)}h^{(1)}[u] \ , \qquad (2.12c)$$

$$F_t^{(12)} = F^{(12)}h^{(1)'}[u]\,u_x + F^{(22)}h^{(2)'}[u]\,u_y - F_x^{(12)}h^{(1)}[u] - F_y^{(12)}h^{(2)}[u] \ , \qquad (2.12d)$$

$$F_t^{(21)} = F^{(21)}h^{(2)'}[u]\,u_y + F^{(11)}h^{(1)'}[u]\,u_x - F_y^{(21)}h^{(2)}[u] - F_x^{(21)}h^{(1)}[u] \ . \qquad (2.12e)$$

In these equations the scalar $u \equiv u(x,y,t)$ and the (2×2)-matrix $\mathbf{F} \equiv \mathbf{F}(x,y,t)$, of elements $F^{(jk)}$, are the dependent variables, the 6 quantities a_j are arbitrary constants (they could actually depend on t without spoiling the C-integrability), and the two functions $h^{(1)}(u)$, $h^{(2)}(u)$ are also arbitrary. The C-integrability of these nonlinear evolution PDEs is subject to the validity of a constraint on the matrix $\mathbf{F}$, which however again need only be imposed at the initial time $t = 0$ (it then holds automatically for any time t). It reads:

$$F_x^{(11)}F^{(12)} + F_y^{(11)}F^{(22)} = F_x^{(12)}F^{(11)} + F_y^{(12)}F^{(21)}\,, \tag{2.13a}$$

$$F_y^{(22)}F^{(21)} + F_x^{(22)}F^{(11)} = F_y^{(21)}F^{(22)} + F_x^{(21)}F^{(12)}\,. \tag{2.13b}$$

A detailed analysis of the technique to solve the initial-value ("Cauchy") problem for this evolution equation is given in Appendix F of [11].

References

1. F.Calogero and W.Eckhaus: *"Nonlinear Evolution Equations, Rescalings, Model PDEs and their Integrability. I & II"*. Inverse Problems **3**, 229-262 (1987) & **4**, 11-33 (1988).
2. C.S.Gardner, J.M.Greene, M.D.Kruskal and R.M.Miura: *"Method for Solving the Korteweg-de Vries equation"*. Phys.Rev.Lett. **19**, 1095-1097 (1967).
3. F.Calogero: *"Why Are Certain Nonlinear PDEs Both Widely Applicable and Integrable?"*. In: *What Is Integrability?*, edited by V.E. Zakharov. Springer-Verlag, Berlin Heidelberg, 1991, pp.1-62.
4. F.Calogero and W.Eckhaus: *"Necessary Conditions for Integrability of Nonlinear PDEs"*. Inverse Problems **3**, L27-L32 (1987).
5. F.Calogero: *"The Evolution PDE $u_t = u_{xxx} + 3(u_{xx}u^2 + 3u_x^2 u) + 3u_x u^4$"*. J.Math.Phys. **28**, 538-555 (1987).
6. F.Calogero and S.De Lillo: *"The Eckhaus PDE $i\psi_t + \psi_{xx} + 2(|\psi|^2)_x\psi + |\psi|^4\psi = 0$"*. Inverse Problems **3**, 633-681 (1987).
7. F.Calogero and S.De Lillo: *"Cauchy Problems on the Semiline and on a Finite Interval for the Eckhaus Equation"*. Inverse Problems **4**, L33-L37 (1988).
8. F.Calogero and S.De Lillo: *"The Burgers Equation on the Semiline with General Boundary Conditions at the Origin"*. J.Math.Phys. **32**, 99-105 (1991).
9. F.Calogero and Ji Xiaoda: *"C-Integrable Nonlinear Partial Differential Equations. I"*. J.Math.Phys. **32**, 875-887 (1991).
10. F.Calogero and Ji Xiaoda: *"C-Integrable Nonlinear Partial Differential Equations. II"*. J.Math.Phys. **32**, 2703-2717 (1991).
11. F.Calogero: *"C-Integrable Nonlinear Partial Differential Equations in N+1 Dimensions"*. J.Math.Phys. **33**, 1257-1271 (1992).
12. J.D.Cole: *"On a Quasilinear Parabolic Equation Occurring in Aerodynamics"*. Quart.Appl.Math. **9**, 225-236 (1950).
13. E. Hopf: *"The Partial Differential Equation $u_t + u\,u_x = \mu u_{xx}$"*. Comm. Pure Appl. Math. **3**, 201-230 (1950).

Integrable Lattice Equations

H.W. Capel[1] *and F. Nijhoff*[2]

[1]Institute of Theoretical Physics, University of Amsterdam, Valckenierstraat 65, NL-1018 XE Amsterdam, The Netherlands
[2]Department of Mathematics and Computer Science and The Institute for Nonlinear Studies, Clarkson University, Potsdam, NY 13699-5815, USA

1 Introduction

Integrable Lattice systems are space- and time-discretizations of integrable partial differential equations (PDE's). The most convenient way of introducing them is based on the direct linearization method (DLM), which was introduced by Fokas and Ablowitz (1981),[1]. This method employs singular linear integral equations with general integration measure and contour. The integral equation introduced in [1] is of the form

$$u_k + \rho_k \int_C d\lambda(\ell) \frac{u_\ell}{k+\ell} = \rho_k. \tag{1.1}$$

Here u_k is a wave function to be solved from the integral equation depending on a complex spectral parameter k and on the coordinates of the system. As we shall note later, these coordinates can be chosen to be discrete as well as continuous, and it is the freedom in this choice that makes integral equations of the type (1.1) a convenient tool to develop discrete integrable systems. Furthermore, in eq. (1.1) C is a contour in the complex k-plane and $d\lambda(k)$ is a suitably chosen integration measure, whereas ρ_k is a free-wave function depending in a given way on k and on the coordinates of the system. The contour C and measure $d\lambda(k)$ need to be chosen to be such that the solution u_k of the integral equation for gives ρ_k is unique.

Consider now, for example, the free wave function

$$\rho_k = e^{kx+w(k)t} \tag{1.2}$$

with $w(k) = k^3$ and let u_k be the solution of (1.1) for some choice of C and $d\lambda(k)$, then the potential

$$u(x,t) = \int_C d\lambda(k) u_k(x,t) \tag{1.3}$$

with the same C and $d\lambda(k)$ satisfies

$$(\partial_t - \partial_x^3)u - 3(\partial_x u)^2 = 0, \tag{1.4}$$

i.e. $v = \partial_x u$ satisfies the Korteweg-deVries (KdV) equation.

The linear integral equation is closely related to the Gel'fand-Levitan-Marchenko (GLM) equation in the inverse scattering method [2]-[7].

Bäcklund transformations (BT's) can be generated by a singular transformations of the measure [8, 9], cf. also [2], or equivalently by a transformation of the free-wave function

$$\rho_k \to \tilde{\rho}_k = \frac{p+k}{p-k}\rho_k \,. \tag{1.5}$$

In fact, it can be shown that $\tilde{u} = \int_C d\lambda(k)\tilde{u}_k$ with $\tilde{u}_k$ being the solution of (1.1) with ρ_k replaced by $\tilde{\rho}_k$, is again a solution of (1.4).

Consider now two different BT's, one given by $\tilde{}$ and $p = p_1$ as in equation (1.5), the other one being given in (1.5) with the $\tilde{}$ and p_1 replaced by $\hat{}$ and p_2. From the combination of the two BT's one obtains

$$(p_1 + p_2 - \hat{\tilde{u}} + u)(p_2 - p_1 - \hat{u} + \tilde{u}) = p_2^2 - p_1^2 \,, \tag{1.6}$$

which is a so-called Bianchi identity expressing the commutativity of the two BT's. The main point is now that as the two BT's can be independently iterated retaining the commutativity, leading to a *lattice* of Bäcklund transformed fields u, we can interprete the Bianchi identity (1.6) as a consistency condition on a lattice, i.e. a partial *difference* equation. Thus, associating the two BT's $\tilde{}$ and $\hat{}$ with the basic translations T_1 and T_2 respectively of a two-dimensional lattice one finds [10]

$$(p_1 + p_2 - (T_1T_2 - 1)u)\,(p_2 - p_1 - T_2u + T_1u) = p_2^2 - p_1^2 \,, \tag{1.7}$$

which is an integrable equation in the sense that solutions can be obtained solving a linear integral equation. Furthermore, this interpretation makes also 'physical' sense, because one can show that by two appropriate consecutive continuous limits eq. (1.7) goes over into the potential KdV equation (1.4), implying that (1.7) indeed may be regarded as a lattice version of the (potential) KdV [10], cf. also [11]. Thus, integrable lattice equations can be obtained by applying BT's of the form (1.5) to the free-wave function in a linear integral equation of the type (1.1), and identifying each BT with an elementary translation on a lattice.

The above scheme to introduce lattice equations starting from the DLM can be generalized along various directions. We will mention some of the important ones here.

Infinite-matrix structures. Equation (1.1) can be generalized to integral equations for basic functions $u_k^{(i)}$ with the source term on the right-hand side being given by $k^i\rho_k$, $i \in \mathbb{Z}$. Then an infinite matrix of potentials can be introduced by $u^{(i,j)} = \int_C d\lambda(k)u_k^{(i)}k^j$. The infinite-matrix structures have been introduced in [12] and have been applied many times afterwards. Usually one can derive closed evolution equations for a few of the matrix elements like $u^{(0,0)}$, $u^{(1,0)}$, $u^{(1,1)}$ together with the Miura-transformations connecting the solutions of these equations. The integral equation with $k+\ell$ replaced by $k - \omega^N\ell$, $(\omega^N = 1)$ gives the direct linearization of the Gel'fand-Dikii hierarchy, with as special cases for $N = 2$ the KdV and for $N = 3$ the Boussinesq equation [13].

Coupled integral equations. Equation (1.1) is appropriate to investigate other the KdV-like equations. Similar results are obtained by considering other types of singular integral equations. For instance the direct linearization of the non-linear Schrödinger equation (NLS) and the isotropic Heisenberg Ferromagnet (IHSC) is obtained by considering two coupled integral equations for wave functions [14]-[16]. More generally one can study integral equations for general $N \times N$ matrices, leading to multicomponent integrable evolution equations, [17, 18].

Two-dimensional lattice equations. By introducing Bäcklund transformations [8, 9] to the various types of integral equations, one can derive a variety of integrable equations on the two-dimensional lattice, [10, 11],[19]-[22]. Applying continuum limits to the free-wave function in these integral equation as well as to the lattice equation one obtains integrable differential-difference equations with one continuous variable at the sites of a one-dimensional chain as well as the partial differential equations (PDE's) that one starts out with. A treatment of associated Hamiltonian structures for these differential-difference equations have been given in [23, 24], cf. also [25, 26]. We mention at this point other early treatments on two-dimensional lattice equations, which are given in [27]-[30].

Three-dimensional lattice equations. Integrable equations in $2 + 1$ dimensions are obtained replacing the contour C and the measure $d\lambda(k)$ by a surface D in the space of two complex variables ℓ, ℓ' and a measure $d\zeta(\ell, \ell')$. With this replacement equation (1.1) provides the direct linearization of the Kadomtsev-Petviashvili (KP) equation [31]-[33], cf. also [34, 35] for the continuous case. Introducing again in this case BT's one obtains integrable equations on a three-dimensional lattice. A systematic approach to continuum limits was given in [36]-[39], by using vertex operators $e^{\partial} = T$ associated with a translation T on the lattice, in which $\partial = \sum_j \varepsilon^j \partial_{\tau_j}$, i.e. each power of the lattice parameter ε is associated with a derivative with respect to a continuous variable τ_j. In this way one obtains the hierarchies of integrable equations with $m = 1, 2$ continuous variables at the sites of $3 - m$ dimensional lattice. Conserved quantities and Hamiltonian structures for these systems were found in [37]-[39]. Other approaches to integrable 3-dimensional lattice equations can be found in [40]-[42].

General-reference integral equations. Eq. (1.1) describes a transformation from a free-wave basis function ρ_k to a solution u_k of the Lax representation associated with the evolution equation for the potential u. More generally one can investigate 'general-reference' integral equations describing the transformation from an arbitrary solution u_k^0 of the Lax representation to a new basis function u_k of the same Lax representation. Such investigations were performed for the 2-dimensional lattices in [20, 43], and in [33, 44, 45] for the 3-dimensional ones, cf. also [46]. A connection between the DLM and the τ-function approach was given in [47] for this case. Similar general-reference treatments for the continuous case were done in [34, 35].

In this chapter we shall focuss on the lattice equations such as eq. (1.7). As, historically, the DLM has proven to be very useful in constructing many of the integrable lattice equations, we shall treat these equations mainly from that point of view, and refer to the literature for the instances of other treatments.

Let us mention here first some typical other examples of integrable lattice equations, that have been constructed along similar lines as (1.7). For the sake of notational convenience, let us adopt the same notation as in eq. (1.6). An equation that is Miura-related to the lattice KdV eqution is the lattice MKdV (or Toda) equation, [10, 11],

$$v\,[(q-r)\,\tilde{v} \;-\; (p-r)\,\hat{v}] \;=\; \hat{\tilde{v}}\,[(q+r)\,\hat{v} \;-\; (p+r)\,\tilde{v}]\ , \tag{1.8}$$

in which $p_1 = p, p_2 = q$ and r is a fixed parameter. The lattice sine-Gordon equation, [29],

$$\sin\left(\theta + \hat{\theta} + \tilde{\theta} + \hat{\tilde{\theta}}\right) \;=\; pq \sin\left(\theta - \hat{\theta} - \tilde{\theta} + \hat{\tilde{\theta}}\right)\ , \tag{1.9}$$

is closely related to the lattice MKdV, cf. also [10]. Other examples are provided by the lattice equations of the socalled lattice Gel'fand-Dikii hierarchy, cf. [48], the simplest case $N = 3$ being the lattice Boussinesq (BSQ) equation

$$\begin{aligned}&\frac{p^3 \;-\; q^3}{p-q+\hat{\tilde{u}}-\tilde{\tilde{u}}} \;-\; \frac{p^3 \;-\; q^3}{p-q+\hat{\hat{u}}-\hat{\tilde{u}}} \;-\; (2p+q)\,(\tilde{u}+\hat{\hat{\tilde{u}}}) \;+\; (p+2q)\,(\hat{u}+\hat{\tilde{\tilde{u}}})\\ &+\; \hat{\hat{\tilde{\tilde{u}}}}\,(p-q+\hat{\hat{\tilde{u}}}-\hat{\tilde{\tilde{u}}}) \;+\; u\,(p-q+\hat{u}-\tilde{u}) \;-\; \hat{u}\hat{\tilde{\tilde{u}}} \;+\; \tilde{u}\hat{\hat{\tilde{u}}} \;=\; 0\ .\end{aligned} \tag{1.10}$$

which is Miura-related to the lattice modified BSQ equation, [48],

$$\begin{aligned}&\frac{(p^2+pr+r^2)\,\hat{\tilde{v}} \;-\; (q^2+qr+r^2)\,\hat{\hat{v}}}{(p-r)\,\hat{\hat{v}} \;-\; (q-r)\,\hat{\tilde{v}}}\,\frac{\hat{\hat{\tilde{v}}}}{\hat{v}} - \frac{(p^2+pr+r^2)\,\tilde{\tilde{v}} \;-\; (q^2+qr+r^2)\,\hat{\tilde{v}}}{(p-r)\,\hat{\tilde{v}} \;-\; (q-r)\,\tilde{\tilde{v}}}\,\frac{\hat{\tilde{\tilde{v}}}}{\tilde{v}}\\ &\quad= (p-r)\left(\frac{v}{\tilde{v}} \;-\; \frac{\hat{\hat{\tilde{v}}}}{\hat{\hat{\tilde{\tilde{v}}}}}\right) \;-\; (q-r)\left(\frac{v}{\hat{v}} \;-\; \frac{\hat{\tilde{\tilde{v}}}}{\hat{\hat{\tilde{\tilde{v}}}}}\right)\ .\end{aligned} \tag{1.11}$$

Note that eqs. (1.10) and (1.11) involve not only nearest neighbour, but also farther neighbour points on the lattice.

Lattice systems related to the AKNS scheme, ([49]), of PDE's are, for example, the lattice nonlinear Schrödinger equation (NLS), [11],

$$|\alpha|^2+|\beta|^2+2Re\,\left(\theta\alpha\beta\,\phi\hat{\phi}^*\right) = |\alpha|^2(1+|\hat{\phi}|^2)\frac{\beta^*\hat{\tilde{\phi}}-\theta\beta\phi}{\theta\alpha\tilde{\phi}-\alpha^*\hat{\phi}}+|\beta|^2(1+|\phi|^2)\frac{\theta\alpha\tilde{\phi}-\alpha^*\hat{\phi}}{\beta^*\hat{\tilde{\phi}}-\theta\beta\phi} \tag{1.12}$$

in which α and β are parameters with $Re(\alpha) = Re(\beta)$, depending on the lattice parameters, θ is a parameter with modulus 1, $|\theta| = 1$, and the $*$ denotes complex conjugation, and the lattice isotropic heisenberg spin chain (IHSC) equation [11, 20], cf. also [30],

$$\begin{aligned} \hat{\tilde{\mathbf{S}}} - \tilde{\mathbf{S}} + \hat{\mathbf{S}} - \mathbf{S} &= \Delta\left(\frac{\gamma(\mathbf{S})(\mathbf{S}+\hat{\mathbf{S}}) + \alpha\hat{\mathbf{S}}\times\mathbf{S}}{1+\hat{\mathbf{S}}\cdot\mathbf{S}}\right) , \\ \gamma(\mathbf{S})^2 &= (\hat{\mathbf{S}}\cdot\mathbf{S})^2 + \beta\hat{\mathbf{S}}\cdot\mathbf{S} + \beta - \alpha^2 - 1 \ , \ \ \mathbf{S}\cdot\mathbf{S} = 1 , \end{aligned} \tag{1.13}$$

where $\Delta f \equiv \tilde{f} - f$, and α, β are arbitrary constants, is gauge-related to the lattice NLS analogous to the continuous situation. An anisotropic version of (1.13), i.e. the lattice Landau-Lifschitz equation, was presented in [50].

Next, there are lattice equations in 2+1 dimensions, which involve three Bäcklund transformations (or equivalently lattice translations) with parameters $p_1 = p, p_2 = q, p_3 = r$. Specific examples are the lattice KP equation, [31],

$$\left(p - r + \hat{u}' - \hat{\tilde{u}}\right)(q - r + u' - \hat{u}) = \left(q - r + \tilde{u}' - \hat{\tilde{u}}\right)(p - r + u' - \tilde{u}) , \tag{1.14}$$

the lattice sine-Gordon equation in 2+1 dimensions, [32], which is a coupled system, namely

$$\begin{aligned} p\left(\frac{\tilde{v}'}{v'} - \frac{\hat{w}}{\hat{\tilde{w}}}\right) + q\left(\frac{\tilde{w}}{\hat{\tilde{w}}} - \frac{\hat{v}'}{v'}\right) + r\left(\hat{v}'\hat{w} - \tilde{w}\tilde{v}'\right) &= 0 \\ p\left(\frac{\tilde{v}}{v} - \frac{\hat{w}}{\hat{\tilde{w}}}\right) + q\left(\frac{\tilde{w}}{\hat{\tilde{w}}} - \frac{\hat{v}}{v}\right) &= 0 , \end{aligned} \tag{1.15}$$

and the lattice modified KP (MKP) equation, [31, 32],

$$(p+s)\left(\frac{\tilde{v}'}{v'} - \frac{\hat{\tilde{v}}}{\hat{v}}\right) + (q+s)\left(\frac{\hat{\tilde{v}}}{\tilde{v}} - \frac{\hat{v}'}{v'}\right) + (r+s)\left(\frac{\hat{v}'}{\hat{v}} - \frac{\tilde{v}'}{\tilde{v}}\right) = 0 , \tag{1.16}$$

in which s is an additional fixed parameter. We can of course add other equations to this small list, such as the IHSC in 2+1 dimensions, the lattice Davey-Stewartson equation and other equations that have appeared in the literature, [40]- [42], notably the so-called DAGTE (discrete analogue of generalized Toda equation), which is a special case of the lattice MKP for $s = -p$. We will come back below more in detail to some of these equations, where we shall develop a general formalism for such equations from the DLM point of view. All of the equations mentioned above (1.6- 1.16) possess familiar integrability characteristics, in particular the existence of an associated linear problem, and an inverse (scattering) scheme. However, many of the integrability aspects have still to be investigated, and some interesting new features, due to the lattice aspect, come into play in these systems, that we do not encounter in the PDE's and the traditional differential-difference systems.

In section 2 we shall present the DLM from a general point of view i.e. in the context of a general-reference integral equation in the infinite-matrix structure. A *universal* three-dimensional lattice equation in the infinite-matrix structure is presented in section 3. The remaining sections are concerned with various reductions of the universal lattice equation. Section 4 contains various finite-matrix reductions and associated Miura transformations. The dimensional reduction to (matrix) lattice equations of AKNS-type ([64]) is treated in section 5. Section 6 is concerned with the reduction of two-dimensional lattice equations to finite-dimensional integrable *mappings*.

2 Integral Transform

In this section we describe the general structure of the DLM approach for 2+1-dimensional $N\times N$ systems, cf. also [45]. This approach consists of the following ingredients.

- A block vector Φ_k consisting of $N\times N$ matrices $\phi_k^{(i)}$, $i\in\mathbb{Z}$ depending on a (complex) spectral parameter k and on the sites $\mathbf{r}=n_1\mathbf{e}_1+n_2\mathbf{e}_2+n_3\mathbf{e}_3$ ($n_1,n_2,n_3\in\mathbb{Z}$, $\mathbf{e}_1,\mathbf{e}_2,\mathbf{e}_3$ three independent vectors) of a 3-dimensional lattice.
- A block matrix $\mathbf{H}$ consisting of $N\times N$ matrices $H^{(i,j)}$, $i,j\in\mathbb{Z}$ depending on the sites of the 3-dimensional lattice.
- An index raising operator Λ defined by
$$(\Lambda\cdot\Phi_k)^{(i)}=\phi_k^{(i+1)},\quad (\Lambda\cdot\mathbf{H})^{(i,j)}=H^{(i+1,j)}. \tag{2.1}$$
- A projection operator $\mathbf{O}$ defined by
$$(\mathbf{O}\cdot\Phi)_k^{(i)}=\phi_k^{(0)}\delta_{i,0},\ (\mathbf{H}\cdot\mathbf{O})^{(i,j)}=\delta_{j,0}H^{(i,0)},\ (\mathbf{O}\cdot\mathbf{H})^{(i,j)}=\delta_{i,0}H^{(0,j)}. \tag{2.2}$$

Now the construction of an integrable lattice system proceeds along the following line.

* Consider the 3 basic primitive translations on the three-dimensional lattice

$$T_\mu\mathbf{r}=\mathbf{r}-\mathbf{e}_\mu\ ,\ T_\mu\Phi_k(\mathbf{r})=\Phi_\mathbf{k}(\mathbf{r}+\mathbf{e}_\mu)\ ,\ T_\mu\mathbf{H}(\mathbf{r})=\mathbf{H}(\mathbf{r}+\mathbf{e}_\mu)\ ,\ \mu=1,2,3. \tag{2.3}$$

** Impose transformation properties of Φ_k under T_μ given by

$$T_\mu\Phi_k=[I_\mu+(\Lambda+T_\mu\mathbf{H}\cdot\mathbf{O})\cdot J_\mu]\cdot\Phi_k\ ,\ \ \mu=1,2,3, \tag{2.4}$$

I_μ and J_μ, $\mu=1,2,3$, are 6 commuting $N\times N$ matrices; multiplication by which does not affect the superscripts i and (i,j) of Φ_k and $\mathbf{H}$.

*** Consider the linear integral transform

$$\Phi_k=\Phi_k^0+\int\int_D\Phi_\ell\cdot d\zeta(\ell,\ell')\cdot G^0_{k\ell'} \tag{2.5}$$

transforming a reference-state for the $N\times N$ matrices $\phi_k^{(i)}$, i.e. $\phi_k^{(i)0}$, into a new state $\phi_k^{(i)}$.

The integrations in (2.5) are performed over a region D over the space of two complex variables ℓ and ℓ' with an $N \times N$ matrix measure $d\zeta(\ell,\ell')$. The kernel $G^0_{k\ell'}$ is an $N \times N$ matrix depending on the Φ^0_k, ${}^t\Phi^0_{\ell'}$ in the reference state in a way to be specified below. We restrict ourselves to regions D and measures $d\zeta(\ell,\ell')$ such that the solution Φ_k of (2.5) for given reference state Φ^0_k is unique.

Let us now assume that we have a reference state consisting of Φ^0_k and $\mathbf{H}^0$ in such a way that the solution Φ_k of (2.5) together with $\mathbf{H}$ satisfies the linear relations (2.4) as well, i.e. the linear relations (2.4) are invariant under the integral transformation (2.5).

¿From the integral equation (2.5) and the linear relation (2.4) for the reference system it is found that

$$\begin{aligned} \Psi_k &- \int\!\!\int_D \Psi_\ell \cdot d\zeta(\ell,\ell') \cdot G^0_{k\ell'} \\ &= \Psi^0_k - \left(T_\mu(\mathbf{H}-\mathbf{H}^0)\right) \cdot \mathbf{O}J_\mu \cdot \Phi^0_k \\ &+ \int\!\!\int_D (T_\mu \Phi_\ell) \cdot d\zeta(\ell,\ell') \cdot (T_\mu - 1) G^0_{k\ell'} \quad , \quad \mu = 1,2,3. \end{aligned} \tag{2.6}$$

in which

$$\Psi_k \equiv T_\mu \Phi_k - [I_\mu + (\Lambda + T_\mu \mathbf{H} \cdot \mathbf{O}J_\mu)] \cdot \Phi_k \ . \tag{2.7}$$

The first term in the right-hand side vanishes because of the linear relation (2.4) for the reference state. The second and third term can be made to cancel choosing $G^0_{k\ell'}$ and $\mathbf{H}$ such that

$$(T_\mu - 1) G^0_{k\ell'} = \left(T_\mu {}^t\Phi^0_{\ell'}\right) \cdot \mathbf{O}J_\mu \cdot \Phi^0_k \ , \tag{2.8}$$

$$\mathbf{H} - \mathbf{H}^0 = \int\!\!\int_D \Phi_\ell \cdot d\zeta(\ell,\ell') \cdot {}^t\Phi^0_{\ell'}. \tag{2.9}$$

Here ${}^t\Phi^0_{k'}$ is an 'adjoint' block vector consisting of $N \times N$ matrices ${}^t\Phi^{(i)}_{k'}$, $i \in \mathbb{Z}$, associated with the reference state in a way to be specified and $({}^t\Phi^0_{\ell'} \cdot \mathbf{O})^{(0)} = {}^t\phi^{(0)^0}_{\ell'}$. With (2.8) and (2.9) it is clear that the right-hand side of (2.6) vanishes. Because of the uniqueness condition the solution Φ_k of (2.5) together with $\mathbf{H}$ defined by (2.9) satisfy the linear relations (2.4).

Imposing the commutativity of the 3 basis translations $T_\mu, \mu = 1,2,3$ with respect to the reference system, i.e.

$$\begin{aligned} T_\mu T_\nu \Phi^0_k = T_\nu T_\mu \Phi^0_k \quad , \quad T_\mu T_\nu {}^t\Phi^0_{k'} = T_\nu T_\mu {}^t\Phi^0_{k'}, \\ T_\mu T_\nu G^0_{k\ell'} = T_\nu T_\mu G^0_{k\ell'}, \quad \mu,\nu = 1,2,3 \end{aligned} \tag{2.10}$$

one obtains the condition

$$\left\{T_\mu \left[{}^t\Phi^0_{\ell'} - (T_\nu {}^t\Phi^0_{\ell'}) \cdot (I_\nu - \mathbf{O}J_\nu \cdot \mathbf{H})\right]\right\} \cdot \mathbf{O}J_\nu \cdot \Phi^0_k \ - \ (\nu \longleftrightarrow \mu) = 0 \ , \tag{2.11}$$

where $(\nu \leftrightarrow \mu)$ denotes the term with μ and ν interchanged. The condition (2.11) is automatically valid provided that the adjoint block vector ${}^t\Phi^0_{k'}$ together with $\mathbf{H}^0$ in the reference state satisfy the (adjoint) linear relation

$$ {}^t\Phi_{k'} = (T_\mu {}^t\Phi_{k'}) \cdot \left[I_\mu + J_\mu({}^t\Lambda - \mathbf{O} \cdot \mathbf{H}) \right] . \tag{2.12} $$

It is straightforward to show that equation (2.12) is invariant under the integral transform

$$ {}^t\Phi_{k'} = {}^t\Phi^0_{k'} + \int\!\!\int_D G^0_{\ell k'} \cdot d\zeta(\ell, \ell') \cdot {}^t\Phi_{\ell'} \tag{2.13} $$

noting that $\mathbf{H} - \mathbf{H}^0$ can also be expressed as

$$ \mathbf{H} - \mathbf{H}^0 = \int\!\!\int_D \Phi^0_\ell \cdot d\zeta(\ell, \ell') \cdot {}^t\Phi_{\ell'} . \tag{2.14} $$

With the uniqueness condition for (2.5) and (2.13) it can be shown that equations (2.4) and (2.12) are invariant under the integral transforms (2.5) and (2.13) meaning that the commutativity of the translations holds also for the new state with Φ_k, ${}^t\Phi_{k'}$, $G_{k\ell'}$. Hence we have shown that the linear relations (2.4) and (2.12) are invariant under the integral transform (2.5, 2.13), i.e. starting from

$$ \Phi^0_k \ , \ {}^t\Phi^0_{k'} \ , \ G^0_{k\ell'} $$

with $\mathbf{H}^0$ satisfying (2.4) and (2.12), the solutions Φ_k, ${}^t\Phi_k$, of (2.5) and (2.13) with $\mathbf{H}$ defined by (2.9) satisfy (2.4) and (2.12).

We refer to the integral equations (2.5) and (2.13) as *general reference* integral equations, because $\mathbf{H}^0$ may be a general matrix compatible with (2.4) and (2.12). In a *free reference* system we have $\mathbf{H}^0 = 0$. (The case that $H^{(i,0)}$ is a constant matrix independent of i and invariant under the T_μ can be reduced to the case that $\mathbf{H}^0 = 0$ by introducing new matrices I_μ). The free-reference integral equations, cf. e.g. [12], are obtained by choosing

$$ \phi^{(i)0}_k = (-k)^i \rho_k \ , \quad {}^t\phi^{(i)0}_{k'} = {}^t\rho_{k'} k'^{-i} \ , \tag{2.15} $$

implying that

$$ T_\mu \rho_k = (I_\mu - k J_\mu) \cdot \rho_k \ , \quad {}^t\rho_{k'} = (T_\mu {}^t\rho_{k'}) \cdot (I_\mu + k' J_\mu) \ , \tag{2.16} $$

$$ G^0_{k\ell'} = -\frac{{}^t\rho_{\ell'} \cdot \rho_k}{k - \ell'} \ . \tag{2.17} $$

In relation to the integral transforms one can also formulate a *superposition principle*, which under suitable conditions enables one to carry out the successive steps between repeated transformations of the state of the system in one single step, cf. [43, 46].

Related treatments on general reference integral equations for 3 dimensional lattice problems are given in [32, 33],[44],[22]. For general reference integral equations in the case of 3 continuous variables see e.g. refs. [34, 35].

3 The Universal Lattice Equation

From equations (2.4,2.12,2.9) and (2.14) it can be shown that

$$(I_\mu + \Lambda J_\mu)\cdot \mathbf{H} - T_\mu \mathbf{H}\cdot(I_\mu + J_\mu {}^t\Lambda) + (T_\mu \mathbf{H}) J_\mu \cdot \mathbf{O}\cdot \mathbf{H} = \mathbf{C} \ , \quad \mu = 1,2,3, \tag{3.1}$$

in which $\mathbf{C}$ denotes an invariant under the integral transforms (2.9) and (2.13), meaning that if eq. (3.1) is satisfied for $\mathbf{H}^0$, then eq. (3.1) is satisfied for $\mathbf{H}$ as well with the same $\mathbf{C}$. More specifically we have $\mathbf{C} = 0$ when the potential $\mathbf{H}$ can be obtained via a series of integral transform from a free-reference state with trivial potential.

Considering two translations T_μ, T_ν and eliminating ${}^t\Lambda$ we have

$$[\Lambda J_\nu (T_\mu \mathbf{H}) J_\mu - I_\mu (T_\nu \mathbf{H}) J_\nu \ - \ (T_\mu T_\nu \mathbf{H})\cdot(I_\nu J_\mu + J_\mu \cdot \mathbf{O}\cdot(T_\nu \mathbf{H}) J_\nu)] - [\nu \leftrightarrow \mu] = 0, \tag{3.2}$$

$\mu, \nu = 1,2,3$. Equation (3.2) implies that

$$\sum_{\mu,\nu,\rho} \varepsilon_{\mu\nu\rho} J_\rho \left[(T_\mu T_\nu \mathbf{H})\cdot(I_\nu J_\mu + J_\mu \mathbf{O}\cdot(T_\nu \mathbf{H}) J_\nu) + I_\mu (T_\nu \mathbf{H}) J_\nu\right] = 0 \tag{3.3}$$

in which $\varepsilon_{\mu\nu\rho}$ is the Levi-Civita tensor ($\varepsilon_{\mu\nu\rho} = 1$, if μ,ν,ρ is an even permutation of $1,2,3$; $\varepsilon_{\mu\nu\rho} = -1$, if μ,ν,ρ is an odd permutation of $1,2,3$; $\varepsilon_{\mu\nu\rho} = 0$ otherwise). Equation (3.3) is an integrable equation on the 3 dimensional lattice generated by T_1, T_2, T_3 in the sense that solutions of (3.3) can be obtained starting from a trivial solution (e.g. in the free-reference state) by solving linear integral equations like (2.5) and (2.13). Equation (3.3) for the matrices $H^{(i,0)}$ and $H^{(0,j)}$ can be derived as the compatibility condition of the 3 linear relations (2.4) or of the 3 adjoint linear relations (2.12) respectively. Therefore (2.4) and (2.12) can be referred to as the Lax representation, or adjoint Lax representation of (3.3).

Equation (3.3) is an integrable equation on the 3-dimensional lattice generated by T_1, T_2, T_3, meaning that solutions of eq. (3.3) can be obtained starting from a trivial solution (e.g. in the free-reference state) by solving the linear integral equations (2.9) and (2.13). For the matrices $H^{(i,0)}$ and $H^{(0,j)}$, we can obtain eq. (3.3) as the compatibility conditions of the 3 linear relations (2.4) and (2.12) respectively. Therefore, eqs. (2.4) and (2.12) can be referred to as the Lax representation of eq. (3.3). Eq. (3.3) is universal in the sense that it holds for all matrices $H^{(i,j)}$ and therefore can give rise to various finite-matrix integrable equations on the 3-dimensional lattice. Such reductions to finite-matrix systems will be treated in section 4.

Remarks:

i) Equation 3.3 is a closed equation not containing Λ and ${}^t\Lambda$ but with the help of (3.1) and (3.2) one may investigate closed equations in terms of $\Lambda^{-1}\cdot\mathbf{H}$ and $\mathbf{H}\cdot{}^t\Lambda^{-1}$ in which $(\Lambda^{-1}\cdot H)^{(i,j)} = H^{(i-1,j)}$. The relation (3.2) after applying Λ^{-1}

to both sides can be regarded as the *Miura-transformation* relating a solution of the lattice equation (3.3) to a solution of the lattice equation for $\Lambda^{-1} \cdot \mathbf{H}$.

ii) Starting from the universal lattice equation (3.3) with $I_\mu = \mathbf{1}$, for $\mu = 1, 2, 3$, one may derive various hierarchies with 1 (or 2) continuous variables at the site of a 2-(respectively 1-) dimensional lattice, together with their associated general-reference integral equations and Lax representations [32]. The hierarchies of PDE's with 3 continuous variables depend on the interdependence of the matrices J_1, J_2, J_3. If these matrices are linearly independent one obtains the equation

$$\sum_{\alpha,\beta,\gamma} \varepsilon_{\alpha\beta\gamma} \left[\mathcal{J}_\alpha(\partial_{\xi_\beta}\mathbf{H})\mathcal{J}_\gamma + \mathcal{J}_\alpha\mathbf{H} \cdot \mathbf{O} \cdot \mathcal{J}_\beta\mathbf{H} \cdot \mathbf{O}\mathcal{J}_\gamma\right] = 0 \tag{3.4}$$

which is the N-wave interaction system studied e.g. in ref. [34]. In the case that 2 of the matrices are independent, we find after 3 consecutive continuum limits equations of Davey-Stewartson type, namely

$$\begin{aligned} -\mathcal{J}_1(\partial_{x_1}\partial_{\eta_1}\mathbf{H})\mathcal{J}_1 + \frac{1}{2}\mathcal{J}_1\left((\partial_{x_2} + \partial_{x_1}^2)\mathbf{H}\right)\mathcal{J}_2 - \frac{1}{2}\mathcal{J}_2\left((\partial_{x_2} - \partial_{x_1}^2)\mathbf{H}\right)\mathcal{J}_1 & \\ -\mathcal{J}_1(\partial_{x_1}\mathbf{H}) \cdot \mathbf{O} \cdot (\mathcal{J}_2 \cdot \mathbf{H} \cdot \mathcal{J}_1 - \mathcal{J}_1 \cdot \mathbf{H} \cdot \mathcal{J}_2) & \\ +(\mathcal{J}_2 \cdot \mathbf{H} \cdot \mathcal{J}_1 - \mathcal{J}_1 \cdot \mathbf{H} \cdot \mathcal{J}_2) \cdot \mathbf{O} \cdot \partial_{x_1}\mathbf{H} \cdot \mathcal{J}_1 &= 0 . \end{aligned} \tag{3.5}$$

Finally, in the case that all matrices J_1, J_2, J_3 are dependent we find matrix versions of the KP equation, in the form

$$\begin{aligned} -\frac{1}{3}\partial_{x_1}\partial_{x_3}\mathbf{H} + \frac{1}{4}\partial_{x_2}^2\mathbf{H} + \frac{1}{12}\partial_{x_1}^4\mathbf{H} + \frac{1}{2}\partial_{x_1}(\partial_{x_1}\mathbf{H} \cdot \mathbf{O} \cdot \partial_{x_1}\mathbf{H}) & \\ +\frac{1}{2}\left((\partial_{x_1}\mathbf{H} \cdot \mathbf{O}) \cdot \partial_{x_2}\mathbf{H} - (\partial_{x_2}\mathbf{H} \cdot \mathbf{O}) \cdot \partial_{x_1}\mathbf{H}\right) = 0 & \end{aligned} \tag{3.6}$$

Equation (3.6) is a matrix generalization of the Kadomtsev-Petviashvili (KP) equation.

iii) The conserved quantities and associated hamiltonian structures of the equations that are obtained from the lattice equations after one or more continuum limits have been studied systematically in the scalar case (i.e. $N = 1$) in [23, 37] for the hierarchies arising from the lattice KdV or MKdV equation, cf. also [39, 24], as well as for the hierarchies coming from the lattice KP, [43]-[38]. In the non-scalar case ($N \neq 1$) the monodromy factors give also rise to a comprehensive description of higher-order flows, [45]. However, due to the problem of matrix ordering, the derivation of hamiltonian structures in this case is more difficult, and this problem is still under investigation.

4 Finite-Matrix Lattice Equations

Taking the $(0,0)$ block of equation (3.3) one obtains a closed equation for the $N \times N$ matrices $H = H^{(0,0)}$. This equation is the (matrix) lattice analogue of the Kadomtsev-Petviashvili (KP) equation, the latter eqution being obtained after 3 consecutive continuum limits. The Lax representation can be found from (2.4) eliminating Λ and reads

$$J_\mu^{-1} \cdot T_\mu \phi_k - J_\nu^{-1} \cdot T_\nu \phi_k = W_\rho \cdot \phi_k \tag{4.1}$$

with

$$W_\rho = J_\mu^{-1} + J_\mu^{-1} \cdot T_\mu H \cdot J_\mu \;-\; (\mu \to \nu) \tag{4.2}$$

Here μ, ν, ρ is a cyclic permutation of $1,2,3$ and we have chosen $I_\mu = 1$ for convenience. Defining the $N \times N$ matrices

$$g_k = \left((\alpha + \Lambda \cdot J)^{-1} \cdot \Phi_k\right)^{(0)}, \tag{4.3}$$

$$g^{-1} = J^{-1} + \left((\alpha + \Lambda \cdot J)^{-1} \cdot \mathbf{H}\right)^{(0,0)} \tag{4.4}$$

with α an arbitrary complex parameter and J an $N \times N$ matrix commuting with J_μ for $\mu = 1,2,3$, $(I_\mu = 1)$, it can be shown that

$$(T_\mu g_k) - (1 - \alpha J_\mu \cdot J^{-1}) \cdot g_k = (T_\mu g)^{-1} \cdot J_\mu \cdot \phi_k. \tag{4.5}$$

For g we obtain the equation

$$\sum_{\mu,\nu,\rho} \varepsilon_{\mu\nu\rho} J_\mu^{-1} \cdot J_\nu^{-1} \cdot (T_\mu T_\nu g) \cdot (1 - \alpha J_\nu \cdot J^{-1}) \cdot (T_\mu g)^{-1} \cdot J_\mu = 0. \tag{4.6}$$

Equation (4.6) can be derived as compatibility condition of a Lax representation in terms of the g_k which can be obtained from (4.5) eliminating the ϕ_k with the use of two different translation operators T_μ, T_ν. The Lax representation can also be expressed in the form (4.1) with ϕ_k and W_ρ replaced by

$$\phi_k = g \cdot g_k \tag{4.7}$$

$$W_\rho = \left\{J_\mu^{-1} \cdot T_\mu g \cdot (1 - \alpha J_\mu \cdot J^{-1}) - J_\nu^{-1} \cdot T_\nu g \cdot (1 - \alpha J_\nu \cdot J^{-1})\right\} \cdot g^{-1} \tag{4.8}$$

and cyclic. Equation (4.6) is the matrix lattice analogue of the modified Kadomtsev-Petviashvili (MKP) equation, whereas eq. (4.7) is the gauge transformation connecting the eigenfunctions ϕ_k and g_k of the Lax representations of lattice KP and lattice MKP. Eq. (4.8) is the Miura transformation relating solutions of the lattice KP and lattice MKP [31, 32]. The lattice MKP with $\alpha = 0$ is also satisfied by ϕ_k and ${}^t\phi_k^{-1}$, in which $\phi_k = \phi_k^{(0)}$ and ${}^t\phi_{k'} = {}^t\phi_{k'}^{(0)}$ are the $i = 0$ blocks of the vectors Φ_k and ${}^t\Phi_{k'}$ of the Lax representations (2.4) and (2.12).

For general J eq. (4.6) is a sum of six terms determined by the nonzero values of $\varepsilon_{\mu\nu\rho}$. In the special case $J = \alpha J_\mu$ for one of the values $\mu = 1,2,3$ eq. (4.6) reduces to a sum of only 4 terms. This equation we refer to as the (matrix) Toda-lattice equation, since after two continuum limits it changes into the well-known two-dimensional Toda field equation.

A more general equation can be derived for the $N \times N$ matrices K defined by

$$K = \left[(\alpha + \Lambda J)^{-1} \cdot \mathbf{H} \cdot (\alpha' + J^t\Lambda)^{-1}\right]^{(0,0)} \tag{4.9}$$

with α and α' being complex parameters and J an $N \times N$ matrix commuting with J_μ, $\mu = 1,2,3$. In the special case $\alpha = \alpha' = 0$ this equation reads

$$\prod_{\mu=1}^{3} \left(J_\rho \cdot J^{-1} + (1 - T_\rho) \cdot T_\nu K\right) \cdot \left(J_\nu \cdot J^{-1} + (1 - T_\nu) \cdot T_\rho K\right) = 1 , \tag{4.10}$$

with (μ, ν, ρ) being cyclic permutations of $(1,2,3)$. For general α, α' this equation is obtained as the compatibility of the Lax representation

$$P_\mu^{-1} \cdot T_\mu g_k - (1 - \alpha J_\mu \cdot J^{-1}) \cdot g_k = (\mu \leftrightarrow \nu) \tag{4.11}$$

with

$$\begin{aligned} P_\mu &= J_\mu \cdot J^{-2} + (1 - T_\mu)K - \alpha J_\mu \cdot J^{-1} \cdot K + \alpha'(T_\mu K) \cdot J_\mu \cdot J^{-1}) \\ &= (T_\mu g)^{-1} \cdot J_\mu \cdot {}^t g, \end{aligned} \tag{4.12}$$

in which

$${}^t g \equiv J^{-1} - \left(\mathbf{O} \cdot \mathbf{H} \cdot (\alpha' + J^t\Lambda)^{-1}\right)^{(0,0)} \tag{4.13}$$

satisfies equation (4.6) with α replaced by α'. The two expressions (4.12) are the Miura transformation connecting a solution of eq. (4.10) with two solutions of the lattice MKP (4.6) with α and α' respectively.

Eq. (4.10) is the (matrix) lattice version of the isotropic Heisenberg ferromagnet in 2 + 1 dimensions [51], but contains also a (matrix) lattice version, [31, 32], of the 2 + 1-dimensional generalization of the Krichever-Novikov equation, [52]. In general the left-hand side of (4.10) is a product of six matrices determined by the three cases with $\varepsilon_{\mu\nu\rho} = 1$. In the special case that $J = \alpha J_\mu$, for one of the values $\mu = 1,2,3$, the left-hand side of (4.10) reduces to a product of 4 matrices and may be regarded as a product version of the (matrix) lattice Toda equation. The kernel $G^0_{k\ell'}$ in the integral equation (2.5) satisfies equation (4.10) with $\alpha = \alpha' = 0$. Finally in the limit $\alpha' \to \infty$, it can be shown that the matrix K^{-1} after some simple transformations satisfies equation (4.6).

5 Two-Dimensional Lattice Equations

Characteristic in the description of three dimensional systems is the measure $d\zeta(\ell, \ell')$ in integral equations depending on 2 complex variables ℓ and ℓ', and the property that the kernel $G^0_{k\ell'}$ at a point P may be evaluated starting from an arbitrary reference point 0 and summing up the various contributions of the type (2.8) connected with nearest neighbor links on a path going from 0 to P.

The reduction to 2 dimensions is obtained choosing a surface D which is a product of 2 contours C and C' (C' surrounding C) and a measure $d\zeta(\ell, \ell')$ containing a simple pole at $\ell = \ell'$, i.e.

$$D = C \times C' \quad , \quad d\zeta(\ell,\ell') = \frac{1}{2\pi i}\frac{d\lambda(\ell)d\ell'}{\ell'-\ell} \quad , \quad (\ell \in C, \ell' \in C'). \tag{5.1}$$

so that all integrations over D with measure $d\zeta(\ell,\ell')$ are reduced to integrations over C with measure $d\lambda(\ell)$. Apart from (5.1) both integral equations (2.5) and (2.13) should essentially contain the same information and this will put constraints on the possible choices of the kernel $G^0_{k\ell'}$. Rather than pursuing this from a systematic point of view, we shall present a specific choice leading to well known integrable two-dimensional systems of AKNS type

Choosing

$$G_{k\ell} = \frac{{}^t\Phi_\ell \cdot \mathbf{O} \cdot \Phi_k}{k-\ell} \tag{5.2}$$

in equation (2.5) together with (5.1) it can be shown that

$$k\,\Phi_k - \int_C \ell\,\Phi_\ell \cdot d\lambda(\ell) \cdot \frac{{}^t\Phi^0_\ell \cdot \mathbf{O} \cdot \Phi^0_k}{k-\ell} = k\,\Phi^0_k + (\mathbf{H}-\mathbf{H}^0)\cdot\mathbf{O}\cdot\Phi^0_k \tag{5.3}$$

in which $\mathbf{H}-\mathbf{H}^0$ is given by (2.9) and (2.14) with (5.1). Assuming that $k\,\Phi^0_k - \mathbf{H}^0\cdot\mathbf{O}\cdot\Phi^0_k = \Lambda\cdot\Phi^0_k$ in the reference system, the uniqueness condition of the integral equation yields

$$k\,\Phi_k = (\Lambda + \mathbf{H}\cdot\mathbf{O})\cdot\Phi_k. \tag{5.4}$$

This relation is valid for any state that can be obtained from a free-reference state with $k\,\Phi_k = \Lambda\cdot\Phi_k$ by a series of successive integral transforms. From the adjoint integral equation one obtains in a similar way

$$k\,{}^t\Phi_k = {}^t\Phi_k \cdot ({}^t\Lambda - \mathbf{O}\cdot\mathbf{H}). \tag{5.5}$$

Both algebraic relations (5.4),(5.5) are special cases of more general relations connecting $k^p\Phi_k$ and $\Lambda^p\cdot\Phi_k$ as well as $k^p\,{}^t\Phi_k$ and ${}^t\Phi_k\cdot{}^t\Lambda^p$, see ref. [43], for $p = 1,2,3,\cdots$. Inserting (5.4) and (5.5) in the linear relations (2.4) and (2.12) it is found that

$$\begin{aligned} T_\mu\Phi_k &= (I_\mu + kJ_\mu + [(T_\mu\mathbf{H})\cdot\mathbf{O}\cdot J_\mu - J_\mu\cdot\mathbf{H}\cdot\mathbf{O}])\cdot\Phi_k \\ {}^t\Phi_k &= (T_\mu\,{}^t\Phi_k)\cdot[I_\mu + kJ_\mu + (T_\mu\mathbf{H})J_\mu\cdot\mathbf{O}\cdot - J_\mu\mathbf{H}\cdot\mathbf{O}], \end{aligned} \tag{5.6}$$

implying that ${}^t\phi^{(0)}_k$ can be identified with $\phi^{(0)^{-1}}_k$. The relation between ${}^t\phi^{(i)}_k$ and $\Phi^{(i)}_k$ can be inferred taking the ith block of (5.6).

With the use of the algebraic relations (5.4),(5.5), and equations (2.9),(2.14) and (5.2) it is straightforward to show that

$$\Lambda\cdot\mathbf{H} - \mathbf{H}\cdot{}^t\Lambda + \mathbf{H}\cdot\mathbf{O}\cdot\mathbf{H} = 0. \tag{5.7}$$

Considering (5.7) in combination with (3.1) it is possible to eliminate the Λ and ${}^t\Lambda$ by only 2 translations T_1, T_2. The result is

$$\begin{aligned}(J_2I_1 - I_2J_1)\cdot \mathbf{H} \;&+\; (T_1T_2\mathbf{H})\cdot(J_2I_1 - I_2J_1) \\ +[(I_2\cdot(T_1\mathbf{H})\cdot J_1 \;&-\; J_2\cdot(T_1\mathbf{H})\cdot I_1) \;-\; (1\leftrightarrow 2)] \\ +[((T_1T_2\mathbf{H})\cdot J_1 \;&-\; J_1(T_2\mathbf{H}))\cdot \mathbf{O}\cdot((T_2\mathbf{H})\cdot J_2 - J_2\mathbf{H} \;-\; (1\leftrightarrow 2)] \;= 0\end{aligned} \tag{5.8}$$

Eq. (5.8) is the universal equation for the general matrix $\mathbf{H}$ consisting of the blocks $H^{(i,j)}$, $i,j \in \mathbb{Z}$, on the two-dimensional lattice generated by T_1, T_2. Associated with (5.8) one can investigate the finite matrix equations in terms of the $N\times N$ matrices $H = H^{(0,0)}$, g, K, cf. (4.4) and (4.9). The Lax representation of (5.8) is given by (5.6). The compatibility of (5.6) for $\mu = 1,2$ gives equation (??) for the matrices $H^{(i,0)}$ and $H^{(0,j)}$. Equation (5.6) is invariant under the integral transform (2.5) (or (2.13)) with (5.1) and (5.2) for the kernel $G^0_{k\ell}$. The case of the $N\times N$ matrix $H^{(0,0)}$ has been treated before in ref. [20]. Starting from (5.8) one can apply continuum limits yielding hierarchies of partial difference equations with one continuous variable on the sites of a one-dimensional chain or hierarchies of PDE's with two continuous variables. An extensive treatment of the hierarchies of PDE's including bilinear identities, conserved quantities and Hamiltonian structure has been given in [43].

Equation (5.8) has been obtained for a general measure $d\lambda(\ell)$. By specializing the measure to be anti-hermitean, [20], i.e. $d\lambda^\dagger(\ell) = -d\lambda(\ell^*)$ one obtain the (matrix) lattice versions of the nonlinear Schrödinger equations (NLS), the isotropic Heisenberg spin chain (IHSC) and the complex sine-Gordon equation (CSG). The (matrix) lattice versions of the MKdV and the sine-Gordon equation (SG) are obtained using a measure with $d\lambda(\ell) = d\tilde{\lambda}(-\ell)$, in which the tilde denotes the transposed matrix. The scalar versions have been discussed in [21, 11] in the framework of free-reference integral equation. The integrable versions of the discrete NLS [53] and the isotropic Heisenberg spin chain [54, 7] are obtained after one continuum limit.

Remark: Other reductions can also be obtained from the DLM, for example by imposing

$$\Phi_k \cdot {}^t\rho_k = {}^t\tilde{\Phi}_k \cdot \rho_k = \Psi_k \ ,$$

where ${}^t\rho_k$ and ρ_k are the free-reference basis functions (2.15,2.16) and where the tilde denotes the transposed matrix. The scalar case ($N = 1$) of this reduction has been treated in [33], and contains the general-reference and free-reference integral equations of the KdV and its lattice analogues. Furthermore, instead of (5.1) we could have chosen a more general measure satisfying

$$d\zeta(\ell,\ell') = \frac{1}{2\pi i}\frac{d\lambda(\ell)}{\ell + \omega\ell'}d\ell' \ , \quad \omega^N = 1 \ .$$

The free-reference integral equations associated with this case can be shown to describe the lattice Gel'fand-Dikii hierarchy [48] with for $N = 2,3,4$, the lattice version of KdV, Boussinesq and the Hirota-Satsuma equations. Finally not all 2 dimensional integrable lattice equations have been shown to be obtainable by reduction from a 3 dimensional lattice equation. An example is the lattice version of the Landau-Lifshitz equation [50].

6 Integrable Mappings

In the theory of chaotic phenomena there is much interest in dynamical mappings finite number of degrees of freedom, cf. e.g. [55, 56], that serve as time-discrete analogues of systems of ordinary differential equations, describing the dynamics of physical systems. Recently, also integrable mappings have become a focusspoint of attention, cf. e.g. [57]-[67], and especially mappings that are related to soliton systems . These mappings turn out to be *symplectic* with respect to a properly chosen Poisson bracket, i.e. they are canonical transformations, and furthermore they exhibit enough invariants in involution to enable one to exactly integrate the discrete-time flow. For integrable *discrete-time* systems, there are still many interesting open problems. Therefore it is useful to have at ones disposal explicit examples of mappings that are simple enough to study aspects of the discrete-time dynamics. The integrable lattice systems that we have discussed so far, form a natural starting point for the construction of such examples, together with their Lax representation and inverse scheme provided by the DLM.

We discuss here a simple example of how integrable mappings with $2P$ degrees of freedom $P = 1, 2, \cdots$ can be obtained by systematic reduction from integrable 2-dimensional lattice equations of the type discussed in this chapter, following refs. [63]-[65]. We take the lattice KdV (1.7) as a starting point. Noting that this equation involves the fields u, T_1u, T_2u, T_1T_2u at the sites of an elementary square. (If u is the field at a site ρ, then T_1u, T_2u are the fields at the sites obtained by a horizontal and vertical shift respectively). A 'local' initial value problem for the lattice KdV can be given by specifying values $a_0, a_1, a_2, a_3, \ldots$ of the field u on a staircase consisting of alternating horizontal and vertical steps, see Figure 1.

The complete solution of (1.7) and at the left of the staircase can be found by completing square. We now consider periodic solutions satisfying $(T_1T_2)^P u = u$

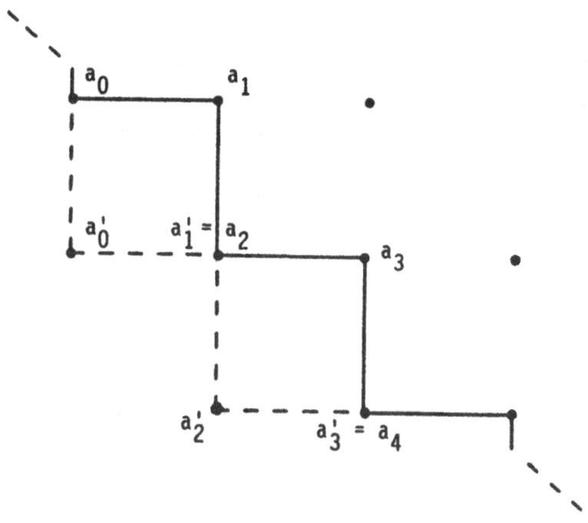

Fig.1: *Configuration of periodic initial data on the lattice.*

for the field on the staircase, or equivalently $a_j = a_{j+2P}$, $P = 1, 2, 3, \cdots$. The solution of (1.7) then satisfies the same periodicity property $(T_1T_2)^P u = u$. To solve equation (1.7) in the periodic case we can consider the $2P$ dimensional mapping corresponding to the vertical shift $T_2 \equiv V$ in downward direction. From (1.7) and Figure 1 we have, [63],

$$\begin{cases} Va_{2j+1} = a_{2j+2}, & j = 0, 1, \cdots, P-1 \\ Va_{2j} = a_{2j+1} - \delta + \dfrac{\varepsilon\delta}{\varepsilon + a_{2j} - a_{2j+2}}, & \end{cases} \tag{6.1}$$

where $\delta = p - q, \varepsilon = p + q$. This mapping can be reduced to a $2P-2$ dimensional mapping in terms of the fields $v_j = \varepsilon + a_{j-1} - a_{j+1}$, namely

$$\begin{cases} Vv_{2j} = v_{2j+1}, & j = 0, 1, \cdots, P-2 \\ Vv_{2j+1} = v_{2j+2} + \dfrac{\varepsilon\delta}{v_{2j+3}} - \dfrac{\varepsilon\delta}{v_{2j-1}}, & \end{cases} \tag{6.2}$$

and v_{2P-1} and v_{2P-2} are expressed in terms of the v_j with $0 \le j \le 2P-3$. Here, for simplicity, we have considered periodic boundary conditions but more general boundary conditions can be treated as well [64].

The lattice KdV is found from the linear relation

$$(p_1 - k)T_1 \begin{pmatrix} u_k \\ v_k \end{pmatrix} = \begin{pmatrix} p_1 - T_1 u & 1 \\ k^2 - p_1^2 + * & p_1 + u \end{pmatrix} \begin{pmatrix} u_k \\ v_k \end{pmatrix} \tag{6.3}$$

corresponding to a horizontal shift and a similar relation with $T_1 \to T_2$, $p_1 \to -p_2$ for the vertical shift. In (6.3) the $*$ indicates the product of the diagonal entries, and the v_k is the solution of the integral equation (1.1) with ρ_k replaced by $k\rho_k$. The compatibility of both relation yields (1.7).

The integrals of the mapping are directly obtained [64] with the use of the Lax representation (6.3). We compare the basis functions u_k, v_k at the points $(0,0)$ and (P,P) at the beginning and at the end of the staircase. We have

$$(p-k)^P(q-k)^P \begin{pmatrix} u_k(P,P) \\ v_k(P,P) \end{pmatrix} = \mathcal{T}_k \begin{pmatrix} u_k(0,0) \\ v_k(0,0) \end{pmatrix} \tag{6.4}$$

in which the monodromy matrix $\mathcal{T}_k$ is the product of the Lax matrices along the staircase, i.e.

$$\mathcal{T}_k(a_0, a_1, \cdots, a_{2P-1}) = \overset{\curvearrowleft}{\prod_{j=0}^{2P-1}} L_k(a_j, a_{j+1}; p_{j+1}) \tag{6.5}$$

where $L_k(a_j, a_{j+1}; p)$ denotes the matrix in the left-hand side of eq. (6.4) and the $\curvearrowleft$ indicates that the matrices are ordered from the left to the right. By noting that we can equally well connect the points $(0,0)$ and (P,P) by first going down from a_0 to Va_0, going through the shifted staircase from Va_0 to Va_{2p} and finally going up from Va_{2P} to a_{2P}, we obtain the following consistency relation for the monodromy matrix $\mathcal{T}_k$

$$\mathcal{T}_k = L_k^{-1}(a_{2P}, Va_{2P}, p_2)\left(\prod_{j=0}^{\overset{\curvearrowleft}{2P-1}} L_k(Va_j, Va_{j+1}, p_j)\right) L_k(a_0, Va_0, p_2). \qquad (6.6)$$

Because of the periodicity, the Lax matrices L_k involving a_0, V_0 and a_{2P}, Va_{2P} are the same. Hence,

$$tr(\mathcal{T}_k) = \sum_{j=0}^{P} k^{2j} I_j, \qquad (6.7)$$

in which the quantities I_j are the invariants under the mapping. In the case under considerations the I_P and I_{P-1} lead to trivial constants. The remaining coefficients $I_0, I_1, \cdots, I_{P-2}$ are $(P-1)$ nontrivial integrals of the mapping (6.2).

To show that the mapping is symplectic we introduce the action [64]

$$S = \sum_n V^n L(\{a_{2j}\}, \{Va_{2j}\}) \,, \qquad (6.8)$$

in which

$$L(\{a_{2j}\}, \{Va_{2j}\}) = \sum_{j=0}^{P-1} \{Va_{2j}(a_{2j+2} - a_{2j}) +$$
$$+\varepsilon\delta \log(\varepsilon + a_{2j} - a_{2j+2}) + \frac{1}{2}(Va_{2j})^2 - \frac{1}{2}a_{2j}^2\} . \qquad (6.9)$$

The condition that S is invariant under infinitesimal variations δa_{2j} leads to the Euler-Lagrange equations

$$\frac{\partial L}{\partial a_{2j}} + V^{-1}\frac{\partial L}{\partial Va_{2j}} = 0 \qquad (6.10)$$

which with (6.9) gives the mapping (6.1). Having established the Lagrangian structure we can apply a Legendre transformation to obtain the 'discrete-time Hamiltonian', i.e. the generating function of the canonical transformation which is the mapping. The canonically conjugate momenta and this Hamiltonian are defined by

$$Vp_{2j} = \frac{\partial L}{\partial Va_{2j}} \qquad (6.11)$$

$$\mathcal{H}(\{Vp_{2j}\}, \{a_{2j}\}) = \sum_{j=0}^{p-1}(Vp_{2j})(Va_{2j} - a_{2j}) - L. \qquad (6.12)$$

The mapping (6.1) is then obtained from the Hamiltonian equations

$$Va_{2j} - a_{2j} = \frac{\partial \mathcal{H}}{\partial Vp_{2j}} \quad , \quad Vp_{2j} - p_{2j} = -\frac{\partial \mathcal{H}}{\partial a_{2j}} \qquad (6.13)$$

which follows from 6.10, but can be checked also from the explicit form for $\mathcal{H}$ that can be derived from (6.12) and (6.9). We note that the Hamiltonian in (6.13) is not invariant, and in this respect the discrete-time situation differs from the continuous-time case. As a consequence of (6.13) the standard Poisson brackets

$$\{a_{2j}, p_{2j'}\} = \delta_{j',j} \quad , \quad \{a_{2j}, a_{2j'}\} = \{p_{2j}, p_{2j'}\} = 0 \tag{6.14}$$

are invariant under the mapping. From (6.10) and (6.14) we have the Poisson brackets

$$\{v_j, v_{j+1}\} = \delta_{j',j-1} - \delta_{j',j+1}. \tag{6.15}$$

The involution property

$$\{tr(\mathcal{T}_k), tr(\mathcal{T}_{k'})\} = 0 \tag{6.16}$$

can be proved by relating $tr\mathcal{T}_k$ to the determinant of a $2P \times 2P$ matrix related to a Lax representation in terms of the u_k along the staircase, but a more fundamental justification is obtained via a novel r-matrix structure, details of which were presented in [65].

Conclusive Remark Very recently, the considerations of refs. [63]-[65] have been extended to the quantum case, [68], as well as to the mappings arising from the lattice Gel'fand-Dikii hierarchy [48, 69]. It is the first time that integrable quantum structures with discrete-time evolution have been investigated, and some new interesting insights on integrable lattice equations seem to emerge from this. As a conclusion we express our feeling that integrable lattice equations form a rich subject in the theory of integrable systems, in which we expect that some novel developments are still to be expected. Unfortunately, we have had in this small introduction no room for going into the many applications of the systems that we have studied here, applications that range from numerical integration of PDE's and cellular automata all the way to string theory and knots. We hope to come back to these issues at some other point.

References

[1] A.S. Fokas and M.J. Ablowitz, Phys. Rev. Lett. **47** (1981) 1096.

[2] F. Calogero and A. Degasperis, Spectral Transform and Solitons, Vol. 1, (North Holland, Amsterdam, 1982).

[3] R.K. Bullough and P.J. Caudrey, Selected Topics in Current Physics **17**, (Springer, Berlin, 1980).

[4] W. Eckhaus and A. van Harten, The Inverse Scattering Transformation and the Theory of Solitons, (North Holland, Amsterdam, 1981).

[5] M.J. Ablowitz and H. Segur, Solitons and the Inverse Scattering Transform, (SIAM, Philadelphia, 1981).

[6] S. Novikov, S.V. Manakov, L.P. Pitaevskii and V.E. Zakharov, Theory of Solitons, The Inverse Scattering Method (Plenum, New York, 1984).

[7] L.D. Faddeev and L.A. Takhtajan, Hamiltonian Methods in the Theory of Solitons, (Berlin, Springer, 1987).

[8] F.W. Nijhoff and H.W. Capel, Phys. Lett. **91A** (1982) 431.

[9] G.R.W. Quispel, F.W. Nijhoff, H.W. Capel and J. van der Linden, Physica **123A** (1984) 319.

[10] F.W. Nijhoff, G.R.W. Quispel and H.W. Capel, Phys. Lett. **97A** (1983) 125.
[11] G.R.W. Quispel, F.W. Nijhoff, H.W. Capel and J. van der Linden, Physica **125A** (1984) 344.
[12] F.W. Nijhoff, G.R.W. Quispel, J. van der Linden and H.W. Capel, Physica **119A** (1983) 101.
[13] G.R.W. Quispel, F.W. Nijhoff and H.W. Capel, Phys. Lett. **91A** (1982) 143.
[14] F.W. Nijhoff, J. van der Linden, G.R.W. Quispel and H.W. Capel, Phys. Lett. **89A** (1982) 106.
[15] F.W. Nijhoff, J. van der Linden, G.R.W. Quispel, H.W. Capel, and J. Velthuizen, Physica **116A** (1982) 1.
[16] F.W. Nijhoff, H.W. Capel, G.R.W. Quispel and J. van der Linden, Phys. Lett. **93A** (1983) 455.
[17] J. van der Linden, F.W. Nijhoff, H.W. Capel and J. van der Linden, Physica **137A** (1986) 44.
[18] J. van der Linden, H.W. Capel and F.W. Nijhoff, Physica **A160** (1989) 235.
[19] F.W. Nijhoff, H.W. Capel and G.R.W. Quispel, Phys. Lett. **98A** (1983) 83.
[20] F.W. Nijhoff, H.W. Capel, G.L. Wiersma and G.R.W. Quispel, Phys. Lett. **103A** (1984) 293.
[21] H.W. Capel, in Magnetic Excitations and Fluctuations, eds. S.W. Lovesey, U. Balucani, F. Borsa, V. Tognetti, Springer Series in Solid State Sciences **54**, (Springer, Berlin, 1984) 86.
[22] F.W. Nijhoff, in Topics in Soliton Theory and Exactly Solvable Nonlinear Equations, eds. M.J. Ablowitz, B. Fuchssteiner and M. Kruskal (World Scientific, Singapore, 1987) p. 150.
[23] G.L. Wiersma and H.W. Capel, Physica **142A** (1987) 199.
[24] H.W. Capel, in Topics in Soliton Theory and Exactly Solvable Nonlinear Equations, eds. M.J. Ablowitz, B. Fuchssteiner and M. Kruskal (World Scientific, Singapore, 1987) p. 125.
[25] S.V. Manakov, Sov. Phys. JETP **40** (1975) 269.
[26] F. Kako and M. Mugibayashi, Progr. Theor. Phys. **60** (1978) 975; **61** (1979) 778.
[27] M.J. Ablowitz and J.F. Ladik, Stud. Appl. Math. **55** (1976); **57** (1977) 1.
[28] R. Hirota, J. Phys. Soc. Japan **43** (1977) 1424, 2074, 2079; **45** (1978) 321, **46** (1977) 312.
[29] S. Orfanidis, Phys. Rev. **D21** (1980) 1507.
[30] E. Date, M. Jimbo and T. Miwa, J. Phys. Soc. Japan **51** (1983) 4125.
[31] F.W. Nijhoff, H.W. Capel, G.L. Wiersma and G.R.W. Quispel, Phys. Lett. **105A** (1984) 267.
[32] F.W. Nijhoff, H.W. Capel and G.L. Wiersma, in Geometric Aspects of the Einstein Equations and Integrable Systems, R. Martini ed., (Springer, Berlin, 1985), p. 263.
[33] H.W. Capel, G.L. Wiersma and F.W. Nijhoff, Physica **138A** (1986) 276.
[34] A.S. Fokas and M.J. Ablowitz, in Nonlinear Phenomena, ed. B. Wolf, Lecture Notes in Physics **189** (Berlin, Springer, 1984) p. 167.
[35] P.M. Santini, A.S. Fokas and M.J. Ablowitz, J. Math. Phys. **25** (1984) 261.
[36] G.L. Wiersma and H.W. Capel, Phys. Lett. **124A** (1987) 124.
[37] H.W. Capel, in Magnetic Excitations and Fluctuations II, eds. S.W. Lovesey, U. Balucani and V. Tognetti, Springer Proceedings in Physics **23** (Springer, Berlin, 1988) p. 6.
[38] G.L. Wiersma and H.W. Capel, Physica **149A** (1988) 49,75.

[39] H.W. Capel, in Proceedings International Conference on Singular Behavior and Nonlinear Dynamics, eds. St. Pnevmatikos, T. Bountis, Sp. Pnevmatikos, (World Scientific, Singapore, 1989) p. 20.
[40] J. Hirota, J. Phys. Soc. Japan **50** (1981) 3785.
[41] E. Date, M. Jimbo and T. Miwa, J. Phys. Soc. Japan **52** (1983) 388, 766.
[42] D. Levi, L. Pilloni and P.M. Santini, J. Phys. **14A** (1981) 1567.
[43] F.W. Nijhoff, Physica **31D** (1988) 339.
[44] F.W. Nijhoff, Lett. Math. Phys. **9** (1985) 235; Physica **18D** (1986) 303.
[45] F.W. Nijhoff and H.W. Capel, Inv. Problems **6** (1990) 567.
[46] F.W. Nijhoff and J.M. Maillet, in Proc. of VIth Int. Workshop on Nonlinear Evolution Equations and Dynamical Systems, ed. J. Leon, (World Scientific, Singapore, 1987) p. 281.
[47] F.W. Nijhoff, Phys. Lett. **110A** (1985) 10.
[48] F.W. Nijhoff, V.G. Papageorgiou, H.W. Capel and G.R.W. Quispel, *The Lattice Gel'fand-Dikii Hierarchy*, Preprint INS # 154/91.
[49] M. Ablowitz, D.J. Kaup, A.C. Newell and H. Segur, Stud. Appl. Math. **53** (1974) 249.
[50] F.W. Nijhoff and V.G. Papageorgiou, Phys. Lett. **141A** (1989) 269.
[51] Y. Ishimori, Progr. Theor. Phys. **72** (1984) 33.
[52] J.M. Krichever and S.P. Novikov, Russ. Math. Surv. **35** (1980).
[53] M.J. Ablowitz and F. Ladik, J. Math. Phys. **17** (1976) 1011.
[54] Y. Ishimori, J. Phys. Soc. Japan **51** (1982) 3417.
[55] P. Cvitanovic, ed., *Universality in Chaos, A Reprint Selection* , (Hilger, Bristol, 1987).
[56] R.S. MacKay and J.D. Meiss, eds., *Hamiltonian Dynamical Systems*, (Hilger,Bristol, 1987).
[57] E.M. McMillan, in Topics in Physics, eds. W.E. Brittin and H. Odabasi, (Colorado Associated Univ. Press, Boulder, 1971) p. 219.
[58] A.P. Veselov, Funct. Anal. Appl. **22** (1988) 83, Theor. Math. Phys. **71** (1987) 446.
[59] J. Moser and A.P. Veselov, Preprint ETH (Zurich), 1989.
[60] P.A. Deift and L.C. Li, Commun. Pure Appl. Math. **42** (1989) 963.
[61] G.R.W. Quispel, J.A.G. Roberts and C.J. Thompson, Phys. Lett **128A** (1988) 419; Physica **D34** (1989) 183.
[62] Yu.B. Suris, Phys. Lett. **145A** (1990) 113.
[63] V.G. Papageorgiou, F.W. Nijhoff and H.W. Capel, Phys. Lett. **A147** (1990) 106.
[64] H.W. Capel, F.W. Nijhoff and V.G. Papageorgiou, Phys. Lett **A155** (1991) 377.
[65] F.W. Nijhoff, V.G. Papageorgiou and H.W. Capel, Integrable Time-Discrete Systems: Lattices and Mappings, preprint Clarkson University INS #166/90, Proceedings Second International Workshop on Quantum Groups, Leningrad, November 1990.
[66] G.R.W. Quispel, H.W. Capel, V.G. Papageorgiou and F.W. Nijhoff, Physica **173A** (1991) 243.
[67] M. Bruschi, O. Ragnisco, P.M. Santini and G.-Z. Tu, Physica **49D** (1991) 273.
[68] F.W. Nijhoff, H.W. Capel and V.G. Papageorgiou, *Integrable Quantum Mappings*, Preprint Clarkson University, INS# 168/91, Phys. Rev. **A** in print.
[69] F.W. Nijhof and H.W. Capel, *Integrable Quantum Mappings and Non-ultralocal Yang-Baxter Structures*, Preprint Clarkson University, INS # 171/91.

The Inverse Spectral Method on the Plane

R. Coifman[1] *and A.S. Fokas*[2]

[1]Department of Mathematics, Yale University,
New Haven, CT 06520, USA

[2]Department of Mathematics and Computer Science and
The Institute for Nonlinear Studies, Clarkson University,
Potsdam, NY 13699-5815, USA

1 Introduction

A method for solving the Cauchy problem for decaying initial data for integrable equations in $1+1$ (i.e. evolution equations in one spatial variable) was discovered in 1967 [1] in connection with the Korteweg-deVries (KdV) equation, and is usually referred to as the inverse spectral method or the inverse scattering transform method. This method reduces the solution of the Cauchy problem to the solution of an inverse scattering problem for an associated linear eigenvalue equation. This inverse problem can be formulated as a Riemann-Hilbert (RH) problem. This formulation expresses the important fact that the underlying eigenfunction is sectionally meromorphic.

There exist two spatial dimensional versions of most of the one spatial dimensional equations. $2+1$ analogues of the KdV, of the nonlinear Schrödinger , and of the N-wave interactions are called Kadomtsev-Petviashvili (KP), Davey-Stewartson (DS) and N-wave interactions in $2+1$ respectively. Another $2+1$ version of the KdV equation was derived in [2]. Depending on a certain sign (which in the context of water waves defines whether surface tension is dominant) there exist KPI and KPII, as well as DSI and DSII. It was discovered in the early 70's that there exist direct methods, such as the dressing method [3], which are capable of generating large classes of solutions of some of these multidimensional equations. However, a method for solving the Cauchy problem for decaying initial data for equations in $2+1$, emerged only in the early 80's. This method is sometimes referred to as the $\bar{\partial}$ (DBAR) method. It was mentioned earlier that analyticity (or sectional meromorphicity to be precise) plays a central role in the $1+1$ theory, since it is the basis of the RH formulation. Analyticity still "survives" in some multidimensional problems: It was shown formally by Manakov [4] and by Fokas and Ablowitz [5] that KPI leads to a **nonlocal RH problem**. However, for other multidimensional problems, such as KPII, the underlying eigenfunctions are analytic nowhere, and the entire framework upon which the inverse spectral method of equations in $1+1$ was based breaks down. Fortunately there exist a proper generalization of the RH problem called the $\bar{\partial}$ **problem** capable of dealing with lack of analyticity.

Actually a $\bar{\partial}$ problem had already appeared in the study of integrable equations in the work of Beals and Coifman [6], who noted that the RH problem appearing in the inverse scattering of one dimensional systems could be viewed as a special case of a $\bar{\partial}$ problem. Soon thereafter Ablowitz, BarYaacov and Fokas [7] showed that KPII required the essential use of the $\bar{\partial}$ problem. The situation with DS is analogous to the situation of KP [8].

The first results of the application of the above method to equations in 2+1 were formal. A review of these formal results can be found in [9]. A rigorous methodology for solving the Cauchy problems in 2+1 has recently emerged after the works of Beals and Coifman [10]-[13], of Wickerhauser [14] and of Fokas and Sung [15]-[16] (see also [17]-[18]). Because of these new developments the original formalism, in addition to being now rigorous it is also considerably sharper. In this paper we shall first review this rigorous methodology; the N-waves in 2+1, DSI, DSII, are used as illustrative examples.

The rigorous methodology involves the following main steps:

1. Investigate the solvability of the direct problem, i.e. find conditions on the potential $q(x,y)$ such that the linear integral equation characterizing an eigenfunction $\mu(x,y,k)$ of the associated linear eigenvalue problem has a unique solution. Investigate the regularity of μ and its large k behavior.

2. Investigate the solvability of the inverse problem, i.e. find conditions on the inverse data (which are $T(k,l)$, $k, l \epsilon R$ for the case of a nonlocal RH problem and $T(k_R, k_I)$ in the case of a $\bar{\partial}$ problem) such that the linear integral equation characterizing the solution $\mu(x,y,k)$ of the associated nonlocal RH or the $\bar{\partial}$ problem has a unique solution. Investigate the regularity of μ and its large k behavior.

3. Prove that the inverse problem solves the direct.

4. Estimate the norms of T in terms of the norms of q.

5. Prove that a given time-evolution of the inverse data T preserves the relevant norms, and derive the associated time-evolutions of μ and of q.

We note that the step 5 involves ideas similar to those of the dressing method and of the direct linearizing method [19]. The nonlinear equations obtained in this procedure are nonlocal. We emphasize that in the process of implementing the dressing method the precise meaning of the nonlocal operator is found. For the DSI this was first done in [20] and for the KP in [21].

Some of the rigorous results are "small norm" results, i.e. are derived under the assumption that certain norms are sufficiently small. However, there exist some rigorous results which are valid without the small norm assumption; such results are derived either when the integral equation under study is of the Volterra type, or when a "vanishing lemma" can be proved. A vanishing lemma means that the homogeneous version of the integral equation under study admits only the zero solution. Two types of vanishing lemmas are reviewed here:

one uses a certain symmetry of a nonlocal RH problem, and the other uses a certain symmetry of a $\bar{\partial}$ problem to reduce the $\bar{\partial}$ problem to a generalized analytic function [22].

Throughout this paper we assume that $q(x,y)$ belongs to the Schwartz class. In this way we avoid the complication of specifying the class of functions that the derivatives of $q(x,y)$ belong. It is possible to extend the results presented here to other classes of functions (see for example [23]).

The following concrete results are reviewed in this paper: In §2 we review the direct and inverse problem for $N \times N$ elliptic systems on the plane,

$$\Psi_x + \sigma J \Psi_y = Q\Psi, \quad \sigma = \sigma_R + i\sigma_I, \quad \sigma_I \neq 0, \tag{1.1}$$

where J is a constant, real, diagonal $N \times N$ matrix with entries J_j, $J_j \neq J_i$ for $j \neq i$, $J_j \neq 0$, and Q is off-diagonal. This problem is a special case of a problem in n-dimensions studies in [15]. If the L^1 and the L^∞ norms of Q are sufficiently small then the direct problem is solvable. Similarly, if the L^1 and the L^∞ norms of T are sufficiently small the inverse problem is solvable. The L^1 and the L^∞ norm of T in terms of the L^1 and L^∞ norms of Q. The particular case of $N = 2$, $\sigma = i$, $J = diag(1,-1)$ contain DSII. There exist two subcases of DSII, the so called defocusing and focusing DSII. It turns out that for the defocusing DSII both the direct and the inverse problems can be reduced to a generalized analytic function and hence solvability follows without a small norm assumption [12]. For the focusing case, if the L^1 and L^∞ norms of the initial data are sufficiently small, then DSII has a global in t solution which belongs to the Schwartz class with respect to the spatial variables.

In §3 we review the direct and inverse problems for $N \times N$ hyperbolic systems on the plane, i.e. equation (1.1) with $\sigma = -1$. This problem was studied in [15] (see also [24]-[26]). The direct problem can be formulated as a Volterra integral equation and hence it is always solvable. Furthermore, if the L^1 norm of the fourier transform with respect to k of $T(l+k,k)$ is sufficiently small then the inverse problem is solvable. This norm of T can be expressed in terms of the analogous norm of Q. In the particular case that Q satisfies the symmetry condition $Q^* = -Q$ (Q^* is the complex conjugate of the transpose of Q) then there exists a vanishing lemma and solvability is guaranteed without the small norm assumption. Using this result it follows that if $\|\!|Q(x,y,0)_{ab}|\!\| < \frac{1}{2N}$, $1 \leq a,b \leq N$, where $\|\!|f|\!\|$ means the L^1 norm of the Fourier transform of $f(x,y)$ w.r.t y, or if $Q^*(x,y,0) = -Q(x,y,0)$, then the N-waves interactions in $2+1$ have a global solution in t which belongs to the Schwartz class w.r.t. the spatial variables. Similarly, since DSI satisfies $Q^* = -Q$ the DSI is always solvable without a small norm assumption.

The direct problem associated with the KPI equation is solvable only under a small norm assumption [27], while for $q \epsilon R$ the inverse problem is always solvable [28]. Similar considerations apply to KPII [14], [16].

The discussion regarding the nonlinear equations in $2+1$ mentioned above can be summarized as follows:

	Direct	Inverse
N waves in $2+1$	always solvable	small norm
N waves in $2+1$ with $Q^* = -Q$	always solvable	always solvable
DSI	always solvable	always solvable
DSII defocusing	always solvable	always solvable
DSII focusing	small norm	small norm
KPI, q real	small norm	always solvable
KPII, q real	small norm	always solvable

Unfortunately there exist several important integrable evolution equations which can not be covered by this scheme. The most well known among them are the Veselov-Novikov and the Benjamin-Ono equation (this equation was treated formally in [29] and [30]). The analysis of the direct and inverse problems associated with these equations leads to singular $\bar{\partial}$ problems [31], [32]. These problems can be mapped to regular ones but this requires several tedious change of variables. Rather than presenting these transformations in detail we only derive in §6 the relevant singular $\bar{\partial}$ problems. We hope that a more direct approach of dealing with such singular problems will be developed in the future.

Remarks

1. Even if the direct and inverse problems for an equation in $2+1$ are always solvable, (i.e. the underlying discrete spectrum is empty), it is still possible for some equations in $2+1$ to support coherent structures [20], [33]. These structures, named dromions in [20], can be generated if there is a mechanism (for example via appropriate boundary conditions) of adding energy to the system. This is the case for DSI with nontrivial boundary conditions (see equation (4.2.1) with $u_1 \neq 0$, $u_2 \neq 0$). The time evolution of the inverse data T satisfies now a nontrivial evolution equation (see equation (4.2.3)). The discrete spectrum of this equation is responsible for the generation of the dromions.

2. Equations in $2+1$ support line-solitons. These solutions, which are straightforward generalizations of the solitons of the corresponding equations in $1+1$, do not decay on certain lines of the (x, y) plane. Hence, these solutions are not covered by the theory reviewed here. An appropriate extension of the method described here is in the process of being developed (see for example [34] and [35]).

3. For the cases that the direct and the inverse problems are solvable only under a small norm assumption there exist a possibility of a discrete spectrum which would give rise to solitons in $2+1$. It was shown formally in [5] and [7] that this is indeed the case for KPI and the focusing DSII, but the solutions obtained in this way (called lumps) decay like $\frac{1}{r}$ as $r = \sqrt{x^2+y^2}$ tends to ∞. The result about DSII was improved in [36], where lumps were obtained decaying like $\frac{1}{r^2}$ for large r.

We hope that some of the results reviewed here will be of interest to the PDE community. It should be noted that, in contrast to the $1+1$ case where existence and uniqueness can be established using PDE theory independently of the inverse spectral method, in the case of $2+1$ most PDE results are of small norm type [37].

2 The Direct and Inverse Problem for $N \times N$ Elliptic Systems on the Plane

We review the solvability of the direct and inverse problems associated with

$$\Psi_x + \sigma J \Psi_y = Q\Psi, \quad \sigma = \sigma_R + i\sigma_I, \quad \sigma_I \neq 0, \tag{2.1}$$

where $(x,y)\epsilon \boldsymbol{R}^2$, Ψ is an $N \times N$ matrix-valued function in $\boldsymbol{R}^2 (N \geq 2)$, J is a constant real, diagonal $N \times N$ matrix with entries J_j, $J_i \neq J_j$ for $i \neq j$, $J_i \neq 0$, and Q is off-diagonal with rapidly decreasing (Schwartz) component function. We denote by $S^0_{N\times N}(\boldsymbol{R}^2)$ the space of such potentials.

Letting $\Psi = \mu \exp[ik(y - \sigma x J)]$, equation (2.1) becomes

$$\mu_x + \sigma J \mu_y + i\sigma k[J,\mu] = Q\mu. \tag{2.2}$$

For the moment we concentrate on the abth component of this equation. It can be simplified by the change of variables

$$x = 2\xi, \quad y = 2J_a(\sigma_R \xi + \sigma_I \eta). \tag{2.3}$$

In the (ξ,η)-coordinates the abth component of equation (2.2) becomes

$$\partial_{\bar{z}} \tilde{\mu}_{ab} + i\sigma k_{ab}\tilde{\mu}_{ab} = (\widetilde{Q\mu})_{ab}, k_{ab} = k(J_a - J_b), z = \xi + i\eta, \tag{2.4}$$

where $\tilde{\mu}$ and $\tilde{Q}$ denotes μ and Q in the (ξ,η) coordinates and $\bar{z} = \xi - i\eta$ (throughout this paper bar denotes complex conjugate). The Green's function of the LHS of this equation is

$$G(z,k_{ab}) \doteq \frac{1}{\pi z} e^{-i(\sigma k_{ab}\bar{z} + \overline{\sigma k_{ab}} z)}, \quad z = \xi + i\eta, \tag{2.5}$$

where we have included the term $exp[-i\overline{\sigma \kappa_{ab}} z]$ in order to make the Green's function bounded. We look for a solution of (2.2) such that

$$\lim_{|x|+|y|\to\infty} \mu = I.$$

Then,

$$\tilde{\mu}_{ab}(\xi,\eta,k) = \delta_{ab} + \int_{\boldsymbol{R}^2} G(\xi',\eta',k_{ab})(\widetilde{Q\mu})_{ab}(\xi-\xi',\eta-\eta',k)d\xi' d\eta',$$

or in the original coordinates,

$$\mu_{ab}(x,y,k)=\delta_{ab}+\int_{\mathbf{R}^2} G(\xi,\eta,k_{ab})(Q\mu)_{ab}(x-2\xi,y-2J_a(\sigma_R\xi+\sigma_I\eta),k)d\xi d\eta. \tag{2.6}$$

Equation (2.6), where G is given by (2.5) is the basic equation of the direct problem. We denote this equation by

$$\mu = I + N_Q\mu. \tag{2.7}$$

In order to formulate the inverse problem we relate $\partial\mu/\partial\bar{k}$ with μ. Taking the $\bar{\partial}$ of equation (2.7) we find

$$\frac{\partial\mu}{\partial\bar{k}} = H + N_Q\frac{\partial\mu}{\partial\bar{k}}, \tag{2.8}$$

where

$$(H)_{ab}=\frac{\bar{\sigma}(J_a-J_b)}{i\pi}\int_{\mathbf{R}^2} e^{-i(\sigma k_{ab}\bar{z}+\overline{\sigma k_{ab}z})}(Q\mu)_{ab}(x-2\xi,y-2J_a(\sigma_R\xi+\sigma_I\eta),k)d\xi d\eta.$$

Defining β_a and γ_a by

$$\beta_a(x,y,k)=\frac{1}{\sigma_I}\left[|\sigma|^2k_Ix-\frac{(\sigma k)_I}{J_a}y\right], \quad \gamma_a=\frac{\bar{\sigma}}{4\pi i|J_a||\sigma_I|}, \tag{2.9}$$

the equation for $(H)_{ab}$ becomes

$$(H)_{ab}=\gamma_a(J_a-J_b)\int_{\mathbf{R}^2} e^{i\beta_a(x-\xi,y-\eta,k_{ab})}(Q\mu)_{ab}(\xi,\eta,k)d\xi d\eta. \tag{2.10}$$

Equation (2.10) suggests the following definition of the inverse data T_{ab}

$$T_{ab}(k)\doteqdot\int_{\mathbf{R}^2} e^{-i\beta_a(\xi,\eta,k_{ab})}(Q\mu)_{ab}(\xi,\eta,k)d\xi d\eta, \quad a\neq b. \tag{2.11}$$

Then equation (2.8) yields

$$\frac{\partial\mu}{\partial\bar{k}}(x,y,k)=\sum_{a,b=1}^{N}\gamma_a(J_a-J_b)e^{i\beta_a(x,y,k_{ab})}T_{ab}(k)\mu(x,y,\lambda_{ab}(k))E_{ab}, \tag{2.12}$$

where E_{ab} is the $N\times N$ matrix with 1 at the abth position and zeros everywhere else and

$$\lambda_{ab}(k)=k-\frac{(\sigma k_{ab})_I}{\sigma_I J_a}. \tag{2.13}$$

Conversely, given inverse data T, equation (2.12) and the inversion formula of the $\bar{\partial}$ operator imply the basic equation of the inverse problem,

$$\mu(x,y,k)=I+\frac{1}{\pi}\int_{\mathbf{R}^2}\frac{\frac{\partial\mu}{\partial\bar{k}}(x,y,k')}{k-k'}dk'_Rdk'_I. \tag{2.14}$$

Comparing the large k asymptotics of equations (2.2) and (2.14) it follows that Q can be reconstructed from

$$Q(x,y) = \frac{i\sigma}{\pi}\left[J, \int_{\boldsymbol{R}^2} \frac{\partial \mu}{\partial \bar{k}}(x,y,k)dk_R dk_I\right]. \tag{2.15}$$

With respect to the *direct problem* it can be shown that:

(a) If Q satisfies the small norm condition

$$\| Q \|_\infty \| Q \|_1 < \frac{\pi|\sigma_I|}{2(N-1)^2} \min_{1\leq a\leq N} |J_a|, \tag{2.16}$$

where

$$\| Q \|_\infty \doteqdot \max_{1\leq a,b\leq N} \sup_{(x,y)\epsilon \boldsymbol{R}^2} |Q_{ab}(x,y)|,$$

and

$$\| Q \|_1 \doteqdot \max_{1\leq a,b\leq N} \int_{\boldsymbol{R}^2} |Q_{ab}(x,y)|\, dxdy,$$

then equation (2.7) has a unique bounded solution for all $k\epsilon \boldsymbol{C}$.

(b) If T is defined by (2.11) then $T\epsilon S^0_{N\times N}(\boldsymbol{R}^2)$, and the nonlinear map $\mathbf{S}$ that takes Q to T is continuous with respect to the topology of $S^0_{N\times N}(\boldsymbol{R}^2)$.

With respect to the *inverse problem*: Now we use equations (2.14) and (2.15) (where $\frac{\partial \mu}{\partial \bar{k}}$ is given by (2.12)) to define μ and Q in terms of T. It can be shown that: (a) If $T\epsilon S^0_{N\times N}$ and if it satisfies the small norm condition

$$\| T \|_\infty \|T\|_1 < \frac{\pi}{8w^2(N-1)^2}, w \doteqdot \max_{1\leq a,b\leq N} |\gamma_a(J_a - J_b)| \tag{2.17}$$

then the inverse equation (2.14) is uniquely solvable.

(b) The nonlinear map $\hat{\mathbf{S}}$ that takes T to Q is continuous with respect to the topology of $S^0_{N\times N}(\boldsymbol{R}^2)$.

(c) The functions μ and Q, as defined by equations (2.14) and (2.15), solve the direct problem (equation (2.7)).

It can also be shown that $\hat{\mathbf{S}}$ is the inverse of $\mathbf{S}$ and the norms of T can be evaluated in terms of the norms of Q,

$$\| T \|_\infty \leq \frac{\| \hat{Q} \|_\infty}{1-\tau}, \quad \| T \|_1 \leq \frac{\rho\gamma \| \hat{Q} \|_1}{1-\tau}, \tag{2.18}$$

where

$$\rho \doteqdot \frac{\sigma_I}{|\sigma|^2} \max_{\substack{1\leq a,b\leq N \\ a\neq b}} \frac{|J_a|}{|J_a - J_b|^2}, \quad \gamma \doteqdot \max_{\substack{1\leq a,b\leq N \\ a\neq b}} \left|\frac{J_a}{J_b}\right|^{\frac{1}{2}}, \tag{2.19}$$

and we assume that

$$\tau \doteqdot \left(\frac{2}{\pi|\sigma_I|}\right)^{\frac{1}{2}} \frac{(N-1)}{\min_{1\leq a\leq N} |J_a|^{\frac{1}{2}}} \left[\frac{1}{(2\pi)^2} \| \hat{Q} \|_1 \| \hat{Q} \|_\infty\right]^{\frac{1}{2}} < 1. \tag{2.20}$$

Note that condition (2.16) follows from condition (2.20).

2.1 A 2 × 2 Particular Case

We let $N = 2$ and consider the particular case of

$$\sigma = i, \;\; J = \begin{pmatrix} 1 & 0 \\ 0 & -1 \end{pmatrix}, \;\; Q = \begin{pmatrix} 0 & q \\ \lambda\bar{q} & 0 \end{pmatrix}, \;\; \lambda = \pm 1. \tag{2.2.1}$$

Then equation (2.5) implies

$$G_{11} = G_{22} = \frac{1}{z}, \;\; G_{12} = \frac{e^{2(k\bar{z}-\bar{k}z)}}{z}, \;\; G_{21} = \frac{e^{2(\bar{k}z-k\bar{z})}}{z}, \;\; z = x + iy, \tag{2.2.2}$$

and equation (2.7) becomes

$$\mu_{11} = 1 + \frac{1}{\pi}\int_{R^2} \frac{1}{\zeta}(q\mu_{21})(x-2\xi, y-2\eta, k)d\xi d\eta,$$

$$\mu_{12} = \frac{1}{\pi}\int_{R^2} \frac{e^{2(k\bar{\zeta}-\bar{k}\zeta)}}{\zeta}(q\mu_{22})(x-2\xi, y-2\eta, k)d\xi d\eta$$

$$\mu_{21} = \frac{1}{\pi}\int_{R^2} \frac{e^{2(\bar{k}\zeta-k\bar{\zeta})}}{\zeta}\lambda(\bar{q}\mu_{11})(x-2\xi, y+2\eta, k)d\xi d\eta,$$

$$\mu_{22} = 1 + \frac{1}{\pi}\int_{R^2} \frac{\lambda}{\zeta}(\bar{q}\mu_{12})(x-2\xi, y+2\eta, k)d\xi d\eta, \tag{2.2.3}$$

where $\zeta \doteqdot \xi + i\eta$. These equations imply that there exists a symmetry relationship between the eigenfunctions,

$$(\mu_{12}(k), \mu_{22}(k)) = \left(\lambda\bar{\mu}_{21}(-\bar{k}), \bar{\mu}_{11}(-\bar{k})\right). \tag{2.2.4}$$

Also the $\partial_{\bar{z}}$ and ∂_z derivatives of equation (2.2.3) yield

$$\frac{\partial\mu_{11}}{\partial\bar{z}} = q\mu_{21}, \;\; \frac{\partial\mu_{21}}{\partial z} = \bar{k}\mu_{21} + \lambda\bar{q}\mu_{11},$$

or

$$\frac{\partial\psi_1}{\partial\bar{z}} = q\psi_2, \;\; \frac{\partial\psi_2}{\partial z} = \lambda\bar{q}\psi_1, \tag{2.2.5}$$

where $\psi_1 = \mu_{11}e^{-\bar{k}z}$, $\psi_2 = \mu_{21}e^{-\bar{k}z}$. These equations are the basic equations of the direct problem.

Equation (2.12) also simplifies in this case: Using

$$\lambda_{21} = \lambda_{12} = -\bar{k}, \;\; \gamma_1 = -\gamma_2 = -\frac{1}{2\pi}, \;\; i\beta_1 = k\bar{z} - \bar{k}z, \;\; i\beta_2 = \bar{k}\bar{z} - kz,$$

we find

$$\frac{\partial\mu}{\partial\bar{k}} = -\frac{1}{\pi}\mu(-\bar{k})T(k,\bar{k}), \;\; T(k,\bar{k}) = \begin{pmatrix} 0 & e^{k\bar{z}-\bar{k}z}T_{12} \\ e^{\bar{k}\bar{z}-kz}T_{21} & 0 \end{pmatrix}. \tag{2.2.6}$$

Because of the symmetry reduction (2.2.4) we only count the first column of this equation:

$$\frac{\partial}{\partial \bar{k}} \begin{pmatrix} \mu_{11} \\ \mu_{21} \end{pmatrix} = -\frac{1}{\pi} e^{\bar{k}\bar{z}-kz} T_{21}(k) \begin{pmatrix} \mu_{12}(-\bar{k}) \\ \mu_{22}(-\bar{k}) \end{pmatrix} = -\frac{1}{\pi} e^{\bar{k}\bar{z}-kz} T_{21}(k) \begin{pmatrix} \lambda\overline{\mu_{21}}(k) \\ \overline{\mu_{11}}(k) \end{pmatrix}. \tag{2.2.7}$$

Equation (2.2.6) is the basic equation of the inverse problem.

Equations (2.2.5) and (2.2.6) imply that if $\lambda = 1$ then both the direct and the inverse problem are solved without the small norm assumption. Indeed,

$$\frac{\partial}{\partial \bar{z}}(\psi_1 \pm \bar{\psi}_2) = \pm\overline{(\psi_1 \pm \bar{\psi}_2)}, \quad \frac{\partial}{\partial \bar{k}}(\mu_{11} \pm \mu_{21}) = \mp\frac{1}{\pi} e^{\bar{k}\bar{z}-kz} T_{21}(k)\overline{(\mu_{11} \pm \mu_{21})},$$

which shows that $\psi_1 \pm \bar{\psi}_2$ and $\mu_{11} \pm \mu_{21}$ are generalized analytic functions.

3 Inverse Problem for $N \times N$ Hyperbolic Systems on the Plane

We review the solvability of the direct and inverse problems associated with

$$\Psi_x - J\Psi_y = Q\Psi, \tag{3.1}$$

where $(x, y)\epsilon \boldsymbol{R}^2, \Psi$ is an $N \times N$ matrix-valued function in $\boldsymbol{R}^2 (N \geq 2)$, J is a constant, real, diagonal $N \times N$ matrix with entries $J_1 > J_2 > \cdots > J_N$, and Q is off-diagonal with rapidly decreasing (Schwartz) component functions. We denote the space of such potentials by $S^0_{N\times N}(\boldsymbol{R}^2)$.

Letting $\Psi = \mu \exp[ik(xJ + yI)]$, equation (3.1) becomes

$$\frac{\partial \mu}{\partial x} - J\frac{\partial \mu}{\partial y} = ik\hat{J}\mu + Q\mu, \quad \hat{J}A = [J, A]. \tag{3.2}$$

Writing equation (3.2) in component form we find

$$(\mu_{ab})_x - J_a(\mu_{ab})_y - ik(J_a - J_b)\mu_{ab} = (Q\mu)_{ab}. \tag{3.3}$$

Let $\tilde{f}_{ab}(\xi, \eta_a) := f_{ab}(\xi, \eta_a - J_a\xi)$, then equation (3.3) becomes

$$(\tilde{\mu}_{ab})_\xi - ik(J_a - J_b)\tilde{\mu}_{ab} = (\widetilde{Q\mu})_{ab},$$

which can be converted into the integral equation

$$\tilde{\mu}_{ab}(\xi, \eta_a, k) = \zeta_{ab}(\eta_a, k) + \int_{\alpha_{ab}}^{\xi} d\xi' (\widetilde{Q\mu})_{ab}(\xi', \eta_a, k) e^{ik(J_a - J_b)(\xi - \xi')},$$

where α_{ab} are arbitrary constants and ζ_{ab} are arbitrary functions of (η_a, k). Writing this equation in the original coordinates we obtain

$$\mu_{ab}(x,y,k) = \zeta_{ab}(y+xJ_a,k)e^{ik(J_a-J_b)x}$$
$$+\int_{\alpha_{ab}}^{x} dx'(Q\mu)_{ab}(x',y+(x-x')J_a,k)\, e^{ik(J_a-J_b)(x-x')}. \tag{3.4}$$

By a proper choice of ζ_{ab} and α_{ab} we can define eigenfunctions $\mu^{\pm}$ which are holomorphic in $\boldsymbol{C}^{\pm}$ (i.e. in the upper and lower complex plane):

$$\mu^{\pm}(x,y,k) = I+$$
$$\left[(\Pi_0+\Pi_{\pm})\int_{-\infty}^{x} dx' + \Pi_{\mp}\int_{+\infty}^{x} dx'\right] e^{ik(x-x')\hat{J}}(Q\mu^{\pm})(x',y+(x-x')J,k) \tag{3.5$^\pm$}$$

for $k \in \boldsymbol{C}^{\pm}$, where $f_{ab}(x',y+(x-x')J) := f_{ab}(x',y+(x-x')J_a)$. Hereafter we shall use Π_+A, Π_-A and $\Pi_0 A$ to denote the strictly upper triangular, the strictly lower triangular and the diagonal part of the matrix A, respectively. We will denote equation $(3.5)^{\pm}$ by

$$\mu^{\pm} = I + N^{\pm}_{Q,k}\mu^{\pm}.$$

Note that $(3.5)^{\pm}$ are equivalent to (3.2) with the asymptotic conditions

$$\lim_{x\to-\infty} \Pi_0\mu^{\pm} = I, \quad \lim_{x\to-\infty} \Pi_{\pm}\mu^{\pm} = 0, \quad \lim_{x\to+\infty} \Pi_{\mp}\mu^{\pm} = 0. \tag{3.6$^\pm$}$$

Let the scattering data $T_{\pm}$ be defined by

$$T_{\pm}(l,k) = \frac{1}{2\pi}\Pi_{\pm}\int_{\boldsymbol{R}^2} d\xi d\zeta e^{-il(\xi J+\zeta I)}Q(\xi,\zeta)\mu^{\mp}(\xi,\zeta,k)e^{ik(\xi J+\zeta I)} \tag{3.7$^\pm$}$$

We note that

$$\int_{\boldsymbol{R}} dx' f(x',y+(x-x')J) = \frac{1}{2\pi}\int_{\boldsymbol{R}^3} dx'dy'dl' e^{il'(x-x')J+il'(y-y')I} f(x',y'),$$

thus equation $(3.5)^{\pm}$ becomes

$$\mu^{\pm}(x,y,k) = I + \frac{1}{2\pi}\left(\int_{-\infty}^{x} dx' - \Pi_{\mp}\int_{\boldsymbol{R}} dx'\right)\int_{\boldsymbol{R}^2} dy'dl' e^{i(l'+k)(x-x')J+il'(y-y')I}$$
$$\cdot(Q\mu^{\pm})(x',y',k)e^{-ik(x-x')J}.$$

Letting $l'+k=l$ in these equations and taking $x\to-\infty$ we obtain

$$\lim_{x\to-\infty}\left(\mu^{\pm}(x,y,k) - I\right) e^{ik(xJ+yI)} = -\int_{\boldsymbol{R}} dl e^{il\eta} T_{\mp}(l,k). \tag{3.8$^\pm$}$$

The eigenfunction μ^0 is defined by

$$\mu^0(x,y,k) = I + \int_{-\infty}^{x} dx' e^{ik(x-x')\hat{J}}(Q\mu^0)(x',y+(x-x')J,k),\, k\in\boldsymbol{R}. \tag{3.9}$$

We will denote equation (3.9) by

$$\mu^0 = I + N^0_{Q,k}\mu^0.$$

Note that (3.9) is equivalent to (3.2) with the asymptotic condition

$$\lim_{x\to-\infty} \mu^0 = I. \tag{3.10}$$

Using the asymptotics of $\mu^\pm$ and μ^0 as $x \to -\infty$ it is straightforward to relate these eigenfunctions:

$$\mu^\pm(x,y,k) = \mu^0(x,y,k) - \int_{\boldsymbol{R}} dl \mu^0(x,y,l)e^{il(xJ+yI)}T_\mp(l,k)e^{-ik(xJ+yI)}. \qquad (3.11)^\pm$$

Adding $\boldsymbol{P}_\mp$ $(3.11)^\pm$, we find

$$\mu^0(x,y,\cdot) = I + \boldsymbol{T}_{(x,y)}\mu^0(x,y,\cdot); \boldsymbol{T}_{(x,y)} := \boldsymbol{P}_+\boldsymbol{T}^+_{(x,y)} + \boldsymbol{P}_-\boldsymbol{T}^-_{(x,y)}, \tag{3.12}$$

where $\boldsymbol{P}_\pm$ are the usual projection operators in k and

$$(\boldsymbol{T}^\pm_{(x,y)}f)(k) := \int_{\boldsymbol{R}} dl f(l)e^{il(xJ+yI)}T_\pm(l,k)e^{-ik(xJ+yI)}. \qquad (3.13)^\pm$$

Conversely, given inverse data $T_\pm$, equation (3.12) yields μ^0. Then the potential Q can be reconstructed from

$$Q(x,y) = \frac{\hat{J}}{2\pi}\int_{\boldsymbol{R}^2} dkdl\mu^0(x,y,l)e^{ik(xJ+yI)}\left[T_+(l,k) - T_-(l,k)\right]e^{-ik(xJ+yI)}. \tag{3.14}$$

This equation follows by comparing the large k asymptotics of equations (3.9) and (3.12).

With respect to the *direct problem* it can be shown that: (a) The integral equations for $\mu^\pm$, μ^0 can be solved without a small norm assumption for Q. (b) If $T_\pm$ are defined by $(3.7)^\pm$, then $T\epsilon S^0_{N\times N}(\boldsymbol{R}^2)$ and the nonlinear map $\mathbf{S}$ that takes Q to T is continuous with respect to the topology of $S^0_{N\times N}(\boldsymbol{R}^2)$. (c) If $\hat{L}_{N\times N}(\boldsymbol{R})$ denotes the space of functions which are the Fourier transform of functions in $L^1 \cap L^\infty$ with

$$\| f \|_{\hat{L}_{N\times N}(\boldsymbol{R})} := \frac{1}{2\pi}\max_{1\le a,b\le N} \| \hat{f}_{ab} \|_{L^1} + \max_{1\le a,b\le N} \| \hat{f}_{ab} \|_{L^\infty}, \tag{3.15}$$

where $\hat{f}$ is the Fourier transform of f, then $\mu^0(x,y,\cdot) - I \in \hat{L}_{N\times N}(\boldsymbol{R})$ and $\mu^\pm(x,y,\cdot) - I \in \hat{L}^\pm_{N\times N}(\boldsymbol{R}) := \mathbf{P}_\pm\hat{L}_{N\times N}(\boldsymbol{R})$. Using this fact the derivation of (3.12) and (3.14) are rigorously justified.

With respect to the *inverse problem*: Now we use equations (3.12) and (3.14) to define μ^0 and Q in terms of $T_\pm$. It can be shown that: (a) If

$$\mathbf{I} - \mathbf{T}_{(x,y)} \text{ is invertible } \forall (x,y) \in \boldsymbol{R}^2,$$

and

$$\sup_{(x,y)\in \boldsymbol{R}^2} \left\| \left(\mathbf{I} - \mathbf{T}_{(x,y)}\right)^{-1} \right\|_{\hat{L}_{N\times N}(\boldsymbol{R}) \to \hat{L}_{N\times N}(\boldsymbol{R})} < \infty, \tag{3.16}$$

then (3.12) has a unique solution in $\hat{L}_{N\times N}(\boldsymbol{R})$. Let

$$\|g\| := \frac{1}{2\pi} \int_{\boldsymbol{R}^2} dl d\zeta \left| \int_{\boldsymbol{R}} dk e^{-i\zeta k} g(l,k) \right|, \quad \check{T}(l,k) = T(l+k,k), \tag{3.17}$$

then

$$\max_{1\le b\le N} \sum_{a=1}^{N} \|\check{T}_{ab}\| < 1 \tag{3.18}$$

provides a sufficient condition for (3.16) to be valid.
(b) $Q \in S^0_{N\times N}(\boldsymbol{R}^2)$ and the nonlinear map $\hat{\mathbf{S}}$ that takes T to Q is continuous with respect to the topology of $S^0_{N\times N}(\boldsymbol{R}^2)$. (c) μ^0 and Q satisfy the integral equation (3.9) of the direct problem.

It can also be shown that $\hat{\mathbf{S}}$ is the inverse of $\mathbf{S}$, i.e., $\hat{\mathbf{S}} \circ \mathbf{S} = \mathbf{I}$ on the domain of $\hat{\mathbf{S}} \circ \mathbf{S}$, and $\mathbf{S} \circ \hat{\mathbf{S}} = \mathbf{I}$ on the domain of $\mathbf{S} \circ \hat{\mathbf{S}}$. Furthermore a norm of T can be estimated from a norm of Q,

$$\max_{1\le a,b\le N} \|\check{T}_{ab}\| \le \frac{\max_{1\le a,b\le N} \|Q_{ab}\|}{1 - \max_{1\le a\le N} \left(\sum_{b=1}^{N} \|Q_{ab}\|\right)}. \tag{3.19}$$

From the above it follows that in general the inverse problem is solvable only under a small norm assumption. However if Q is skew Hermitian then the small norm assumption is not necessary. Indeed if Q is skew Hermitian, i.e., $Q^* = -Q$ (Q^* is the complex conjugate transpose of Q), then T enjoys the symmetry property

$$T_-(l,k) + T_+^*(k,l) = \int_{\boldsymbol{R}} dp T_+^*(p,l) T_-(p,k). \tag{3.20}$$

Using this fact it can be shown that $\mathbf{I} - \mathbf{T}_{(x,y)}$ satisfies (3.16) without a small norm assumption. Furthermore it can be shown that Q reconstructed from T satisfying (3.20) is skew Hermitian.

3.1 A 2 × 2 Particular Case

We let $N = 2$ and consider equation (3.1) in the particular form

$$\Psi_x + J\Psi_y + Q\Psi = 0; \quad J = \begin{pmatrix} 1 & 0 \\ 0 & -1 \end{pmatrix}, \quad Q = \begin{pmatrix} 0 & q \\ -\bar{q} & 0 \end{pmatrix}. \tag{3.28}$$

We introduce the characteristic coordinates

$$\xi = x + y, \qquad \eta = x - y. \tag{3.29}$$

The analytic eigenfunction μ^+ satisfies the Volterra integral equation

$$\begin{aligned} \mu_{11}^+ &= 1 - \tfrac{1}{2}\textstyle\int_{-\infty}^{\xi} d\xi' q \mu_{21}^+, \qquad & \mu_{12}^+ &= -\tfrac{1}{2}\textstyle\int_{-\infty}^{\xi} d\xi' q \mu_{22}^+ e^{ik(\xi-\xi')}, \\ \mu_{21}^+ &= -\tfrac{1}{2}\textstyle\int_{\eta}^{\infty} d\eta' \bar{q} \mu_{11}^+ e^{ik(\eta'-\eta)}, \qquad & \mu_{22}^+ &= 1 + \tfrac{1}{2}\textstyle\int_{-\infty}^{\eta} d\eta' \bar{q} \mu_{12}^+. \end{aligned} \tag{3.30}$$

The eigenfunction μ^- satisfies equations similar to (3.30), with the integral in μ_{21}^+ and μ_{12}^+ replaced by $-\int_{-\infty}^{\eta}$ and $-\int_{\xi}^{\infty}$, respectively. In this particular case it is not necessary to introduce μ^0, since the eigenfunction μ^+ and μ^- are simply related by

$$\hat{\mu}(k) - \mu(k) = \int_{\boldsymbol{R}} dl \mu(l) T(k,l), \tag{3.31}$$

where

$$\mu = \begin{pmatrix} \mu_{11}^- & \mu_{12}^+ \\ \mu_{21}^- & \mu_{22}^+ \end{pmatrix}, \quad \hat{\mu} = \begin{pmatrix} \mu_{11}^+ & \mu_{12}^- \\ \mu_{21}^+ & \mu_{22}^- \end{pmatrix},$$

$$T(k,l) = \begin{pmatrix} 0 & s(k,l) e^{il\eta + ik\xi} \\ -\bar{s}(l,k) e^{-ik\eta - il\xi} & 0 \end{pmatrix}. \tag{3.32}$$

$$\mathrm{s}(k,l) = \frac{1}{4\pi} \int_{\boldsymbol{R}^2} d\xi d\eta q(\xi,\eta) e^{-il\eta - ik\xi} \mu_{22}^-(\xi,\eta,k). \tag{3.33}$$

Given $s(k,l)$, the nonlocal RH problem (3.31) yields μ^-, then q can be reconstructed from

$$q(\xi,\eta) = \frac{1}{\pi} \int_{\boldsymbol{R}^2} dk dl s(k,l) e^{il\eta + ik\xi} \mu_{11}^-(\xi,\eta,l). \tag{3.34}$$

Since Q satisfies the symmetry relationship $Q^* = Q$, the inverse problem can be solved without a small norm assumption. Indeed, it is easily shown that equations (3.31) cannot have a homogeneous solution: Consider the first row of equation (3.31).

$$\mu_{11}^+(k) - \mu_{11}^-(k) + - \int_{\boldsymbol{R}} dl \bar{\mathrm{s}}(l,k) e^{-il\xi - ik\eta} \mu_{12}^+(l) = 0 \tag{3.35a}$$

$$\mu_{12}^+(k) - \mu_{12}^-(k) + - \int_{\boldsymbol{R}} dl \mathrm{s}(k,l) e^{il\eta + ik\xi} \mu_{11}^-(l) = 0 \tag{3.35b}$$

and the scattering data $\mathrm{s}(k,l)$ is defined by equation (3.33).
Multiplying equation (3.35a) by $\overline{\mu_{11}^-(k)}$, equation (3.35b) by $\overline{\mu_{12}^+(k)}$, integrating over k, and adding these equations we find

$$\int_{\boldsymbol{R}} dk (|\mu_{11}^-(k)|^2 + |\mu_{12}^+(k)|^2) =$$

$$= \int_{\boldsymbol{R}^2} dk dl \left[\bar{s}(l,k) e^{-il\xi - ik\eta} \mu_{12}^+(l) \overline{\mu_{11}^-(k)} - s(l,k) e^{il\xi + ik\eta} \overline{\mu_{12}^+(l)} \mu_{11}^-(k) \right],$$

where we have used that $\overline{\mu_{11}^-(k)}$, $\overline{\mu_{12}^+(k)}$ are analytic in the upper, lower half k-complex plane, tending to zero for large k. The RHS of the above equation is imaginary which implies $\mu_{11}^- = \mu_{12}^+ = 0$.

4 Nonlinear Integrable Equations on the Plane

In this section we use the results of §2-4 to solve the Cauchy problem for decaying initial data for the N-waves in $2+1$, the DSI, and the DSII.

We recall that the dressing method provides a powerful method for finding the nonlinear evolution equations associated with a given RH or a $\bar{\partial}$ problem. Alternatively, instead of starting with a RH problem, it is possible to start with a linear integral equation (compare with the direct linearizing method). Here we use similar ideas to associate nonlinear equations, to the nonlocal RH and $\bar{\partial}$ problems derived in §2-4.

4.1 The N-Wave Equations in $2+1$

We consider the Cauchy problem for the N-wave equation

$$\frac{\partial Q}{\partial t} = C\frac{\partial Q}{\partial y} + \hat{C}\hat{J}^{-1}\left(\frac{\partial Q}{\partial x} - J\frac{\partial Q}{\partial y}\right) + [\hat{C}\hat{J}^{-1}Q, Q], \tag{4.1.1}$$

$$Q(x,y,0) = Q_0(x,y)\epsilon S^0_{N\times N}(\boldsymbol{R}^2),$$

where C is a constant real diagonal matrix and $\hat{C}f = [C, f]$. We shall show that if $\|\!|(Q_0)_{ab}\|\!| < \frac{1}{2N}$, $1 \leq a, b \leq N$, or if $Q_0^* = -Q_0$, then (4.1.1) has a global solution in $C^\infty(\boldsymbol{R}, S^0_{N\times N(\boldsymbol{R}^2)})$. Furthermore, this global solution is the only solution of (4.1.1) within the class $C^1([0,t^*), S^0_{N\times N(\boldsymbol{R}^2)})$ for any $t^*\epsilon\boldsymbol{R}$ (the norm $\|\!|\cdot\|\!|$ is defined in equation (3.17)). This solution can be obtained from equation (3.14) (μ^0 solves (3.12)), where T solves the linear Cauchy problem

$$\frac{\partial T}{\partial t} = ilCT - ikTC, \quad T(l,k,0) = \mathbf{S}(Q_0)(l,k). \tag{4.1.2}$$

To derive this result we first find the evolutions of μ, and of Q associated with the evolution of T specified by equation (4.1.2).

Proposition 4.1 Let μ^0 and Q be defined by equations (3.12) and (3.14), and let T evolve according to equation (4.1.2), then the evolutions of μ^0 of Q are given by

$$\frac{\partial \mu^0}{\partial t} - C\frac{\partial \mu^0}{\partial y} - ik\hat{C}\mu^0 = (\hat{C}\hat{J}^{-1}Q)\mu^0, \tag{4.1.3}$$

and by equation (4.1.1) respectively.

Proof Let the operators D_t, D_y be defined by

$$D_t f = \frac{\partial f}{\partial t} + ikfC, \quad D_y f = \frac{\partial f}{\partial y} + ikf.$$

Then, equation (3.12) implies

$$D_t\mu^0 = ikC - \hat{J}^{-1}QC + \boldsymbol{T}_{(x,y)}(D_t\mu^0),$$

$$D_y\mu^0 = ik - \hat{J}^{-1}Q + \boldsymbol{T}_{(x,y)}(D_y\mu^0).$$

Also

$$(\hat{C}\hat{J}^{-1}Q)\mu^0 = \hat{C}\hat{J}^{-1}Q\mu^0 + \boldsymbol{T}_{(x,y)}(\hat{C}\hat{J}^{-1}Q\mu^0).$$

Thus $(D_t - CD_y - (\hat{C}\hat{J}^{-1}Q))\mu^0$ satisfies the homogeneous version of equation (3.12), and since it belongs to $\hat{L}_{N\times N(\boldsymbol{R})}$, it must be zero. This is equation (4.1.3).

To derive equation (4.1.1), we note that it can be written in the form

$$\left(\frac{\partial}{\partial t} - C\frac{\partial}{\partial y} - (\hat{C}\hat{J}^{-1}Q)\right)Q = \left(\frac{\partial}{\partial x} - J\frac{\partial}{\partial y} - Q\right)(\hat{C}\hat{J}^{-1}Q). \tag{4.1.5}$$

But, the compatibility of equations (4.1.3) and of (3.2) yields

$$\left\{\left(\frac{\partial}{\partial t} - C\frac{\partial}{\partial y} - (\hat{C}\hat{J}^{-1}Q)\right)Q\right\}\mu^0 = \left\{\left(\frac{\partial}{\partial x} - J\frac{\partial}{\partial y} - Q\right)(\hat{C}\hat{J}^{-1}Q)\right\}\mu^0,$$

which implies (4.1.1), since $\mu^0 \to I$ as $k \to \infty$.

4.2 The DSI

We consider an initial value problem for the DSI equation

$$iq_t + q_{\xi\xi} + q_{\eta\eta} + \frac{1}{2}\left(\int_{-\infty}^{\xi} d\xi'|q|^2_\eta + \int_{-\infty}^{\eta} d\eta'|q|^2_\xi\right)q + (u_1(\eta,t) + u_2(\xi,t))\,q = 0, \tag{4.2.1}$$

where $q(x,y,0) = q_0(x,y)$, $u_1(\eta,t)$, and $u_2(\xi,t)$ are given. This equation is occasionally referred to as the focusing DSI. The defocusing DSI corresponds to replacing $\frac{1}{2}$ by $-\frac{1}{2}$ in (4.2.1). Considering only the focusing DSI is without loss of generality since one can pass from the focusing to the defocusing DSI via the simple transformation $\xi \to \xi$, $\eta \to -\eta$. Indeed, recall that equation (4.2.1) is associated with

$$\Psi_x + J\Psi_y + \begin{pmatrix} 0 & q \\ -\bar{q} & 0 \end{pmatrix}\Psi = 0. \tag{4.2.2}$$

Multiplying this equation by J from the left we find

$$\Psi_y + J\Psi_x - \begin{pmatrix} 0 & q \\ \bar{q} & 0 \end{pmatrix} \Psi = 0,$$

which corresponds to the defocusing case with $x \leftrightarrow y$ (thus $\xi \to \xi$, $\eta \to -\eta$).

We shall show that if $q_0(x,y)$ belong to the class of Schwartz functions in $\boldsymbol{R}^2$, and if u_1, u_2 are in the Schwartz class with respect to the spatial variables and continuous in t, then equation (4.2.1) has a unique, global in t solution which, for each fixed t, belongs to the Schwartz class in the spatial variables. This solution can be obtained by solving the RH problem (3.35) and using equation (3.34). The evolution of s is given by

$$i\hat{s}_t + \hat{s}_{\xi\xi} + \hat{s}_{\eta\eta} + (u_1(\eta,t) + u_2(\xi,t))\hat{s} = 0, \tag{4.2.3}$$

where $\hat{s}(\xi,\eta,t)$ is the Fourier transform of $s(k,l,t)$, $s(k,l,o) = \mathbf{S}(q_0)(k,l)$.

Proposition 4.2 Let μ and q be defined by equations (3.35) and (3.34) respectively, and let T evolve according to equation (4.2.3). Then the evolution of μ is given by

$$\mu_t + \int_{\boldsymbol{R}} dl \mu(k-l)\Gamma(l) = iJ(\mu_{\xi\xi} + \mu_{\eta\eta} - 2\mu_{\eta\xi}) + iQ(\mu_\xi - \mu_\eta) - 2k(E_2\mu_\xi + E_1\mu_\eta) + A\mu, \tag{4.2.4}$$

where,

$$Q = \begin{pmatrix} 0 & q \\ -\bar{q} & 0 \end{pmatrix}, E_1 = \begin{pmatrix} 1 & 0 \\ 0 & 0 \end{pmatrix}, E_2 = \begin{pmatrix} 0 & 0 \\ 0 & 1 \end{pmatrix},$$

$$A = \begin{pmatrix} -\frac{1}{2i}\int_{-\infty}^{\xi} d\xi' |q|_\eta^2 + iu_1 & -iq_\eta \\ -i\bar{q}_\xi & \frac{1}{2i}\int_{-\infty}^{\eta} d\eta' |q|_\xi^2 - iu_2 \end{pmatrix}, \tag{4.2.5}$$

$$\Gamma(k) = diag(e^{-ik\eta}\gamma_1(k,t), e^{ik\xi}\gamma_2(k,t)), \gamma_1(k,t)$$

$$= \frac{i}{2\pi}\int_{\boldsymbol{R}} d\eta e^{ik\eta} u_1(\eta,t), \gamma_2(k,t) = -\frac{i}{2\pi}\int_{\boldsymbol{R}} d\eta^{-ik\xi} u_2(\xi,t). \tag{4.2.6}$$

The evolution of q is given by (4.2.1).

Proof We note that

$$T_\xi = ikTE_2 - ilE_1T, \quad T_\eta = -ikTE_1 + ilE_1T. \tag{4.2.7}$$

These equations, and the time evolution of the scattering data, motivate the introduction of the operators

$$D_\xi\mu = \mu_\xi - ik\mu E_2, D_\eta\mu = \mu_\eta + ik\mu E_1, D_t\mu = \mu_t - ik^2\mu J + \int_{\boldsymbol{R}} dl\mu(k-l)T(l). \tag{4.2.8}$$

Commutativity among D_ξ, D_η, and D_t implies

$$\Gamma_\eta + ik\Gamma E_1 = 0, \quad \Gamma_\xi - ik\Gamma E_2 = 0,$$

thus

$$\Gamma_1 = e^{-ik\eta}\gamma_1(k,t), \quad \Gamma_2 = e^{ik\xi}\gamma_2(k,t). \tag{4.2.9}$$

The operator D_ξ and D_η commute with the RH problem (3.35) (see equations (4.2.7)). The commutativity of D_t with the RH problem (3.35) follows from equation (4.2.3). Indeed, commutativity occurs iff

$$T_t = -il^2 JT + ik^2 FJ + \int_{\boldsymbol{R}} d\nu[\Gamma(\nu - l)T(k,\nu) - T(k-\nu,l)\Gamma(\nu)].$$

The Fourier transform of these equations implies

$$i\hat{s}_t + \hat{s}_{\xi\xi} + \hat{s}_{\eta\eta} + \left(\int_{\boldsymbol{R}} dk e^{-ik\eta}\gamma_1(k,t) + \int_{\boldsymbol{R}} e^{ik\xi}\gamma_2(k,t)\right)\hat{s} = 0,$$

which is equation (4.2.3).

The expression $E_1 D_\xi \mu - E_2 D_\eta \mu + \frac{1}{2}Q\mu$ solves the RH problem (3.35), and tends to $O(\frac{1}{k})$ as $k \to \infty$, thus it must be zero, hence

$$2E_1\mu_\xi - 2E_2\mu_\eta - ik[J,\mu] + Q\mu = 0. \tag{4.2.10}$$

We consider the expression

$$-D_t\mu + iJ(D_\xi^2\mu + D_\eta^2\mu - 2D_\xi D_\eta\mu) + B(D_\xi - D_\eta)\mu + A\mu,$$

and choose A, B in such a way that this expression tends to $O(\frac{1}{k})$ as $k \to \infty$. The $O(k)$ term implies $B = iQ$. Using equation (4.2.10) to eliminate the $ik^2[J,\mu]$ term we find

$$\mu_t + \int_{\boldsymbol{R}} dl\mu(k-l)\Gamma(l) = iJ(\mu_{\xi\xi} + \mu_{\eta\eta} - 2\mu_{\eta\xi}) + A\mu + iQ(\mu_\xi - \mu_\eta) - 2k(E_2\mu_\xi + E_1\mu_\eta). \tag{4.2.11}$$

The $O(1)$ term of this equation implies

$$A = \begin{pmatrix} iu_1 & 0 \\ 0 & -iu_2 \end{pmatrix} + 2\begin{pmatrix} \mu^{(1)}_{11_\eta} & \mu^{(1)}_{12_\eta} \\ \mu^{(1)}_{21_\xi} & \mu^{(1)}_{22_\xi} \end{pmatrix}. \tag{4.2.12}$$

But the $O(\frac{1}{k})$ term of equation (4.2.10) yields

$$\begin{pmatrix} 0 & 2i\mu^{(2)}_{12} \\ -2i\mu^{(2)}_{21} & 0 \end{pmatrix} = 2\begin{pmatrix} \mu^{(1)}_{11_\xi} & \mu^{(1)}_{12_\xi} \\ \mu^{(1)}_{21_\eta} & \mu^{(1)}_{22_\eta} \end{pmatrix} + \begin{pmatrix} q\mu^{(1)}_{21} & q\mu^{(1)}_{22} \\ -\bar{q}\mu^{(1)}_{11} & -\bar{q}\mu^{(1)}_{12} \end{pmatrix},$$

or

$$2\mu^{(1)}_{11_\xi} + q\mu^{(1)}_{21} = 0,\quad 2\mu^{(1)}_{22_\eta} = \bar{q}\mu^{(1)}_{12},\quad 2i\mu^{(2)}_{12} = 2\mu^{(1)}_{12_\xi} + q\mu^{(1)}_{22}. \tag{4.2.13}$$

Equations (4.2.13a) and (4.2.13b) together with $q = 2i\mu^{(1)}_{12}$, $\bar{q} = 2i\mu^{(1)}_{21}$ imply that A as defined by equation (4.2.12), equals with the A given by equation (4.2.5). It should be emphasized again that the specific form of the integrals follows from the large k behavior of equations (3.30): $\mu^+_{11} \sim 1 - \frac{1}{2ik}\int_{-\infty}^{\xi} d\xi' |q|^2$, $\mu^+_{22} \sim 1 + \frac{1}{2ik}\int_{-\infty}^{\eta} d\eta' |q|^2$.

Using the expression of A in equation (4.2.11) and analyzing the 12 term of the order $(\frac{1}{k})$ we find

$$\mu^{(1)}_{12_t} - i\mu^{(1)}_{12}u_2 = i\left(\mu^{(1)}_{12_{\xi\xi}} + \mu^{(1)}_{12_{\eta\eta}}\right) - \frac{1}{2i}\left(\int_{-\infty}^{\xi} d\xi' |q|^2_\eta\right)\mu^{(1)}_{12} + iu_1\mu^{(1)}_{12} + \frac{q}{4}\left(\int_{-\infty}^{\eta} d\eta' |q|^2_\xi\right),$$

which is equation (4.2.1), since $\mu^{(1)}_{12} = \frac{q}{2i}$.

4.3 The DSII

We consider the Cauchy problem for the DSII equation

$$-iq_t + \frac{1}{2}q_{yy} - \frac{1}{2}q_{xx} + \frac{\lambda}{2}\left(\partial_{\bar{z}}^{-1}(|q|^2_z + \partial_z^{-1}|q|^2_{\bar{z}}\right)q = 0, \partial_{\bar{z}}^{-1}f = \frac{1}{\pi}\int_{\boldsymbol{R}^2} dz'_R dz'_I \frac{f(z',\bar{z}')}{z - z'}, \tag{4.3.1}$$

$$z = x + iy,\quad \lambda^2 = 1,\quad q(x,y,0) = q_0(x,y).$$

We shall show that if q_0 belongs to the class of Schwartz functions in $\boldsymbol{R}^2$, and if for $\lambda = -1$ in addition the L^1 and L^∞ norms of q_0 are sufficiently small, then equation (4.3.1) has a unique global in t solution which belongs to the Schwartz class in the spatial variables. This solution can be obtained by solving the $\bar{\partial}$ problem (2.27) and then using equation

$$q(x,y) = \frac{2}{\pi^2}\int_{\boldsymbol{R}^2} dz_R dz_I \mu_{11}(-\bar{k})T_{12}e^{k\bar{z}-\bar{k}z}. \tag{4.3.2}$$

The time evolution of T_{21} is given by

$$\frac{\partial T_{21}}{\partial t}(k,\bar{k},t) + i(k^2 + \bar{k}^2)T_{21} = 0,\quad T_{21}(k,\bar{k},0) = \mathbf{S}(q_0)(k,\bar{k}). \tag{4.3.3}$$

Proposition 4.3 Let μ and q be defined by equations (2.26) and (2.15) respectively and let T_{21} evolve according to equation (4.3.3). Then the evolution of μ is given by

$$-i\mu_t + J\mu_{yy} + ikJ\mu_y - k\mu_x + iQ\mu_y + A\mu, \tag{4.3.4}$$

where

$$A = \lambda\begin{pmatrix} \frac{1}{2}\partial_{\bar{z}}^{-1}(|q|^2)_z & \lambda q_z \\ -\bar{q}_{\bar{z}} & -\frac{1}{2}\partial_z^{-1}(|q|^2)_{\bar{z}} \end{pmatrix}, z = x + iy \tag{4.3.5}$$

and $\partial_{\bar{z}}^{-1}$, ∂_z^{-1} are defined in equation (4.3.1). The evolution of q is given by equation (4.3.1).

We note that the inverse data, appearing in the $\bar{\partial}$ problem (2.26), satisfy

$$T_x + \bar{k}JT + kTJ = 0, \quad T_y + i(\bar{k}+k)T = 0. \tag{4.3.6}$$

These equations motivate the introduction of the operators

$$D_x\mu = \mu_x + k\mu J, \quad D_y\mu = \mu_y + ik\mu. \tag{4.3.7}$$

The operators D_x and D_y commute with the $\bar{\partial}$ problem (2.26). We also introduce the operator D_t,

$$D_t\mu = \mu_t + ik^2\mu J, \tag{4.3.8}$$

which commutes with the $\bar{\partial}$ problem (2.26) iff

$$T_t = i\bar{k}^2JT - ik^2TJ \tag{4.3.9}$$

which implies equation (4.3.3).

The expression $D_x\mu + iJD_y\mu - Q\mu$ solves the $\bar{\partial}$ problem (2.26), and tends to $O(\frac{1}{k})$ as $k \to \infty$, thus it must be zero,

$$\mu_x + iJ\mu_y - k[J,\mu] = Q\mu. \tag{4.3.10}$$

We consider the expression $-iD_t\mu + JD_y^2\mu + BD_y\mu + A\mu$ and choose A, B in such a way that this expression tends to $O(\frac{1}{k})$ as $k \to \infty$. The $O(k)$ term implies $B = iQ$. Using equation (4.3.10) to eliminate the $-k^2[J,\mu]$ term we find

$$-i\mu_t + J\mu_{yy} + ikJ\mu_y + iQ\mu_y + -k\mu_x + A\mu = 0. \tag{4.3.11}$$

The $O(1)$ term of this equation implies

$$A = -\mu_x^{(1)} + iJ\mu_y^{(1)} = -2\begin{pmatrix} \mu_{11_z}^{(1)} & \mu_{12_z}^{(1)} \\ \mu_{21_z}^{(1)} & \mu_{12_z}^{(1)} \end{pmatrix}. \tag{4.3.12}$$

But the $O(\frac{1}{k})$ term of equation (4.3.10) yields

$$\begin{pmatrix} 0 & 2\mu_{12}^{(2)} \\ -2\mu_{21}^{(2)} & 0 \end{pmatrix} = \begin{pmatrix} q\mu_{21}^{(1)} & q\mu_{22}^{(1)} \\ \lambda\bar{q}\mu_{11}^{(1)} & \lambda\bar{q}\mu_{12}^{(1)} \end{pmatrix} + \mu_x^{(1)} + J\mu_y^{(1)}, \tag{4.3.13}$$

or

$$2\frac{\partial\mu_{11}^{(1)}}{\partial\bar{z}} + q\mu_{21}^{(1)} = 0, \quad 2\frac{\partial\mu_{22}^{(1)}}{\partial z} + \lambda\bar{q}\mu_{12}^{(1)} = 0, \quad 2\mu_{12}^{(2)} = q\mu_{22}^{(1)} + 2\frac{\partial\mu_{12}^{(1)}}{\partial\bar{z}}. \tag{4.3.14}$$

Equation (4.3.14a) and (4.3.14b) together with $q = -2\mu_{12}^{(1)}$ and $\lambda\bar{q} = 2\mu_{21}^{(1)}$ imply that A as defined by equation (4.3.12) equals with the A given by equation (4.3.5). It is again noted that the unique form of the integrals ∂_z^{-1} and $\partial_{\bar{z}}^{-1}$ follows from the large k behavior of equations (2.2.3).

Using the expression of A in equation (4.3.11) and analyzing the 12 term of the order $(\frac{1}{k})$ we find equation (4.3.1).

5 Dromion Solutions

In this section we summarize the results of [20] regarding the dromion solutions of DSI. Given $q(\xi,\eta,0)$, one first computes $\hat{s}(\xi,\eta,0)$; then, since $u_1(\eta,t)$ and $u_2(\xi,t)$ are given, equation (4.2.3) yields $\hat{s}(\xi,\eta,t)$. Then the RH problem (3.35) yields μ, and equation (3.35) yields $q(\xi,\eta,t)$. In order to investigate the structure of the solution $q(\xi,\eta,t)$, and in particular its behavior as $t\to\infty$, one needs to analyze equation (4.2.3). We distinguish two cases, depending on whether u_1 and u_2 are time-dependent.

5.1 Time-Independent Boundaries

We first consider the case when u_1, u_2 are time-independent. Further we assume that $u_1(\eta), u_2(\xi)\epsilon L_2^1$, where $u\epsilon L_2^1$ iff $\int_{-\infty}^{\infty}(1+|x|)|u|^2dx$ exists. Looking for separable solutions $\hat{s} = T(t)X(\xi)Y(\eta)$ of (4.2.3), it follows that

$$T' + i(k^2+k'^2)T,\quad X'' + (u_2+k^2)X = 0,\quad Y''+(u_1+k'^2)Y = 0,$$

i.e., the analysis of (4.2.3) is intimately related to the spectral theory of the stationary Schrödinger equation

$$\psi_{xx} + \left(u(x)+k^2\right)\psi = 0. \tag{5.1}$$

We recall that the above equation plays an important role in the integration of the Korteweg-deVries equation. Here we will make an other use of equation (5.1); we shall utilize it to define a generalized Fourier transform which then can be used for the integration of (4.2.3). Let $k_j = ip_j, p_j\epsilon R^+$ and $\varphi_j, 1\le j\le N$ be the discrete eigenvalues and eigenfunctions associated with $u(x)$; let $\varphi(x,k)$ be the associated continuous eigenfunctions. Then any function $f(x)\epsilon L_2$ can be expanded in terms of this orthonormal set,

$$f(x) = \sum_{n=1}^{N}\rho_n\varphi_n(x) + \int_R dk\rho(k)\varphi(x,k);\ \rho_n$$

$$\div \int_R dx\varphi_n^*(x)f(x), \rho(k) \div \int_R dx\varphi^*(x,k)f(x). \tag{5.2}$$

If the reflection coefficient of the potential u is zero then the discrete eigenfunctions can be found in closed form,

$$\varphi_n + \sum_{j=1}^{N} \frac{c_n c_j}{p_n + p_j} e^{-(p_n+p_j)x} \varphi_j = c_n e^{-p_n x}; \quad u = -2 \sum_{n=1}^{N} c_n \left(e^{-p_n x} \varphi_n(x)\right)_x. \quad (5.3)$$

Using the above results about the time-independent Schrödinger equation we can solve equation (4.2.3). Let

$$\begin{aligned} &u_1(\eta) : Y(\eta, k), Y_j(\eta), \lambda_j, j = 1, \cdots, L, \lambda_j \epsilon R^+ \\ &u_2(\xi) : X(\xi, k), X_j(\xi), \mu_j, j = 1, \cdots, M, \mu_j \epsilon R^+ \end{aligned} \quad (5.4)$$

Then the solution of equation (4.2.3) with $u_1(\eta), u_2(\xi) \epsilon L_2^1$ is given by,

$$\begin{aligned} \hat{S}(\xi, \eta, t) = \; & \textstyle\sum_{j=1}^{M} \sum_{r=1}^{L} \rho_{jr} X_j(\xi) Y_r(\eta) e^{i(\mu_j^2+\lambda_r^2)t} \\ & + \textstyle\int_{\mathbf{R}^2} dk d\ell \rho(k,\ell) X(\xi,k) Y(\eta,\ell) e^{-i(k^2+\ell^2)t} \\ & + \textstyle\int_{\mathbf{R}} dk \left[\sum_{j=1}^{M} \rho_j(k) e^{i(\mu_j^2-k^2)t} X_j(\xi) Y(\eta,k) \right. \\ & \left. + \textstyle\sum_{r=1}^{L} \bar{\rho}_r(k) e^{i(\lambda_r^2-k^2)t} X(\xi,k) Y_r(\eta) \right], \end{aligned} \quad (5.5)$$

where

$$\begin{aligned} \rho_{jr} &= \int_{R^2} d\xi d\eta \hat{S}(\xi,\eta,0) X_j^*(\xi) Y_r^*(\eta), \\ \rho(k,\ell) &= \int_{R^2} d\xi d\eta \hat{S}(\xi,\eta,0) X^*(\xi,k) Y^*(\eta,\ell), \\ \rho_j(k) &= \int_{R^2} d\xi d\eta \hat{S}(\xi,\eta,0) X_j^*(\xi) Y^*(\eta,k), \\ \bar{\rho}_r(k) &= \int_{R^2} d\xi d\eta \hat{S}(\xi,\eta,0) X^*(\xi,k) Y_r^*(\eta). \end{aligned} \quad (5.6)$$

The derivation of the above results can be found in [20]. The stationary phase method implies the following behavior of $\hat{S}(\xi, \eta, t)$ as $t \to \infty$,

$$\hat{S}(\xi,\eta,t) \sim \sum_{j=1,r=1}^{M,L} \rho_{rj} X_j(\xi) Y_r(\eta) e^{i(\mu_j^2+\lambda_r^2)t}, \quad t \to \infty. \quad (5.7)$$

Thus, it is quite interesting that for any initial-boundary conditions, the scattering data become degenerate as $t \to \infty$. It turns out that for degenerate scattering data, the RH problem (3.35) can be solved in closed form (see [20] for details). The asymptotic value of q can be calculated in closed form, provided that X_j, Y_r can be found in closed form. For example, if u_1, u_2 are reflectionless then X_j, Y_r satisfy a linear system and hence q can be found in closed form;

this solution is called in [20] an (L, M) breather. From the above it follows that if at least one of the two boundaries does not give rise to bound states of the Schrödinger operator, then every initial condition $q_1(\xi, \eta, 0), q_2(\xi, \eta, 0)$ will disperse away. If bound states do exist the asymptotic behavior is essentially determined by these bound states; the initial condition only fixes the constant ρ_{jr}.

5.2 Time-Dependent Boundaries

If u_1, u_2 are time-dependent then separation of variables implies that the solution of equation (4.2.3) is intimately related to the analysis of the time-dependent Schrödinger equation

$$i\Psi_t + \Psi_{xx} + u(x,t) + \Psi = 0, \quad u \epsilon R. \tag{5.8}$$

We recall that this equation plays an important role in the integration of the Kadomtsev-Petviashvili (KP) equation where t is replaced by y. However now we demand u to be decreasing in x only as opposed to the case of KP where u is decreasing in both x and y. A completeness result is not available for the above equation yet, so we cannot repeat the above analysis here. However, we can still use the ideas of 5.1 to derive certain exact solutions. If a completeness result analogous to equation (5.1) exists for equation (5.8) then these solutions will be generic, i.e. they will determine the long t behavior of (5.8). In analogy with equation (5.3) we find,

$$\varphi_n + \sum_{j=1}^{N} \frac{c_n c_j^*}{p_n + p_j^*} e^{-(p_n+p_j^*)(x-i(p_n-p_j^*)t)} \varphi_j = c_n e^{-p_n(x-ip_nt)}; \tag{5.9a}$$

$$u = -2\partial_x \sum_{n=1}^{N} c_n^* e^{-p_n^*(x+ip_n^*t)} \varphi_n. \tag{5.9b}$$

Indeed, one may verify directly, using the approach of direct linearization [19], that if u is defined by (5.9b), $u \epsilon R$, and if φ_n satisfy (5.9a), then φ_n also solves

$$i\varphi_{n_t} + \varphi_{n_{xx}} + u\varphi_n = 0. \tag{5.10}$$

Hence as before if

$$\hat{s}(\xi, \eta, t) = \sum_{j=1, r=1}^{M,L} \rho_{jr} X_j(\xi, t) Y_r(\eta, t), \tag{5.11}$$

and if u_1, u_2 are reflectionless, the scattering data is degenerate and hence q can be found in closed form.

Suppose u_1, u_2 are reflectionless, then X_j, Y_j satisfy linear algebraic equations: If $u_1(\eta, t)$ is given by

$$u_1(\eta, t) = -2\partial_\eta \sum_{j=1}^{L} \ell_j^* e^{-\lambda_j^*(\eta + i\lambda_j^* t)} Y_j(\eta, t), \tag{5.12}$$

then the Y_j's solve the linear algebraic system

$$Y_r + \sum_{j=1}^{L}(C^\eta)_{rj}Y_j = \ell_r e^{-\lambda_r(\eta - i\lambda_r^2 t)}, \quad (C^\eta)_{rj} \doteqdot \frac{\ell_r \ell_j^*}{\lambda_r + \lambda_j^*} e^{-(\lambda_r+\lambda_j^*)\eta + i(\lambda_r^2 - \lambda_j^{*2})t}. \quad (5.13)$$

Similarly if $u_2(\xi, t)$ is given by

$$u_2(\xi, t) = -2\partial_\xi \sum_{j=1}^{M} m_j^* e^{-\mu_j^*(\xi + i\mu_j^* t)} X_j(\xi, t), \quad (5.14)$$

then

$$X_r + \sum_{j=1}^{M}(C^\xi)_{rj}X_j = m_r e^{-\mu_r(\xi - i\mu_r^2 t)}, (C^\xi)_{rj} = \frac{m_r m_j^*}{\mu_r + \mu_j^*} e^{-(\mu_r+\mu_j^*)\xi + i(\mu_r^2 - \mu_j^{*2})t}. \quad (5.15)$$

In the above equations $\ell_i, m_i, \lambda_i, \mu_i \epsilon \boldsymbol{C}$ and $\lambda_{i_R}, \mu_{i_R} \epsilon R^+$. Using (5.11), where X_j, Y_j are given by (5.13), (5.15) and solving the inverse problem in closed form we find that q is given by,

$$q = 2 \sum_{i=1,j=1}^{M,L} X_i(\xi, t) Y_j(\eta, t) Z_{ij}(\xi, \eta, t). \quad (5.16)$$

Z_{ij} solves

$$Z_{ij} - \varepsilon \sum_{r=1}^{M} A_{ir} Z_{rj} = \rho_{ij}, \quad (5.17)$$

and the matrix A is defined by

$$A \doteqdot \rho(I + C^\eta)^{-1} \left[(I + C^\xi)^{-1}\rho^*\right]^T, \quad (5.18)$$

where superscript T denotes the transpose of a matrix, and the matrix ρ can be found from initial data

$$\rho_{ij} = \int_{R^2} d\xi d\eta \hat{S}(\xi, \eta, 0) X_i^*(\xi, 0) Y_j^*(\eta, 0). \quad (5.19)$$

The above solution is called in [20] an (M, L) dromion. In the special case that u_1, u_2 are time independent, then $\lambda_{jI} = \mu_{jI} = 0$ and the above solution degenerates to an (M, L) breather.

In addition to the exchange of energy between the mean flow and the surface waves, the localized lumps on the surface can also exchange energy among themselves. The complete investigation of the asymptotic behavior of the (L, M) dromion is given in [38]. Several other exact solutions of the DS equation are also analyzed in [38]. It is shown in [39] that these dromions can also exhibit fusion.

6 Inverse Spectral Method for Singular $\bar{\partial}$ Problems

We shall now discuss briefly some cases leading to singular $\bar{\partial}$ problems.

6.1 Nonlinear Evolutions Associated with Scattering Problems for the Schrodinger Operator in $\mathbf{R}$

The scattering problem at 0 energy for $\Delta - q$ in $\mathbf{R}^2$ leads formally to a hierarchy of nonlinear evolution equations [2]. The $\bar{\partial}$ method leads naturally to a formal derivation of the relation between the evolutions and the $\bar{\partial}$ scattering data: We stress the word formal since these problems engender considerable analytic difficulties, as we shall describe.

We let $x = x_1 + ix_2 \in \mathbf{C}$, $L = \frac{\partial^2}{\partial x \partial \bar{x}} - q(x)$, and we consider solutions $\phi(x, z)$ of $L\phi = 0$ modeled after exponential solutions of the unpurturbed Laplacean $\frac{\partial^2}{\partial x \partial \bar{x}}$. In other words, we require

$$\phi_+(x, z) = e^{ixz}\mu_+(x, z), \phi_-(x, z) = e^{i\bar{x}z}\mu_-(x, z),$$

with $\lim_{x\to\infty} \mu_\pm(x, z) = 1$. Them $\mu_\pm(x, z)$ satisfy the "spectral" problems

$$\begin{aligned} \partial_{\bar{x}}(\partial_x + iz)\mu_+ &= q\mu_+, \\ \partial_x(\partial_{\bar{x}} + iz)\mu_- &= q\mu_-. \end{aligned} \tag{6.1}$$

Moreover, it can be shown that as $x \to \infty$ we have

$$\mu_+(x, z) = 1 + \frac{-i}{xz}\alpha_+(z) + \frac{i}{\bar{x}\bar{z}}e^{-i(xz+\bar{x}\bar{z})}T_+(z) + O\left(\frac{1}{|x|^2}\right),$$

where

$$\alpha_+ = \int_{\mathbf{C}} q(x)\mu(x, z)dx, \quad T_+(z) = \int_{\mathbf{C}} e^{i(xz+\bar{x}\bar{z})}q(x)\mu_+(x, z)dx.$$

A similar relation holds for μ_-.

A direct formal computation relates the $\bar{\partial}$ data in the spectral parameter z to the "scattering" data $T(z)$ as follows.

$$\partial_{\bar{z}}\mu_+(x, z) = \frac{\pi}{z}e^{-i(xz+\bar{x}\bar{z})}T_+(z)\overline{\mu_+(x, z)}. \tag{6.2}$$

This relation is formally sufficient to solve the inverse problem recoving μ, and hence q.

If we consider linear evolutions of T_+, such as

$$T_+(z, t) = e^{\{(-iz)^k - (i\bar{z})^k\}t}T_+(z) \quad k = \pm 1, \pm 2, \pm 3, \ldots \tag{6.3}$$

we obtain a hierarchy generalizing KdV. In particular, $k = 3$ leads to the Veselov-Novikov equation

$$q_t = -\partial_{\bar{x}}^3 q + 3\partial_{\bar{x}}(q\partial_x^{-1}\partial_{\bar{x}}q) + 3\partial_x(q\partial_{\bar{x}}^{-1}\partial_x q)$$

(observe that $\partial_{\bar{x}}^{-1}\partial_x f = \frac{1}{\pi}\int_{\mathbf{C}} \frac{1}{(x-y)^2} f(y)dy$ is the Beurling transform). We remark that this hierarchy can also be obtained by a certain reduction of the Davey-Stewartson hierarchy.

All of this would lead us to declare these problems solved. Unfortunately, we face several analytical difficulties: the first involves the existence of μ_+, and the second involves the solvability of a very singular $\bar{\partial}$ problem (6.2). These issues have been addressed (for small q), and solved by Tian-Yue Tsai [32] following a complicated renormalization scheme developed previously for the Benjanin-Ono scattering problem [31].

We now describe some of these issues: To solve (6.1) with the proper behavior, we convert equation (6.1) to the equation

$$\mu_+ = 1 + G_z^+ * (q\mu_+),$$

where

$$G_z^+ = -\int_{\mathbf{C}} \frac{e^{i(x\bar{\xi}+\bar{x}\xi)}}{\xi(\bar{\xi}+z)} d\xi.$$

Unfortunately, estimates on G_z^+ grow like $\ln\left(\frac{1}{|z|}\right)$ for $|z|$ small, precluding direct use of a fixed point theorem. This difficulty is bypassed by defining

$$G_z^{+,0}(x) = G_z^+(x) - \ell_0(z),$$

where $\ell_0(z) = -\int \frac{\chi(y)}{y(\bar{y}+z)}dy$ and χ is smooth, $\chi = \begin{bmatrix} 1 & \text{for} & |y| \le \frac{1}{2} \\ 0 & \text{for} & |y| > 1 \end{bmatrix}$.

For G_z^{+-}, one can prove uniform estimates in z permitting the construction of "eigenfunctions" $\mu_+^0(x,z) = (I - G_z^{+0}q)^{-1}1$. The desired solutions for (6.1) are obtained via

$$\mu_+(x,z) = \frac{1}{1-\ell_0(z)\alpha_+^0(z)}\mu_+^0(x,z),$$

where $\alpha_+^0 = \int q\mu_+^0 dx$ and

$$T_+(z) = \frac{T_+^0(z)}{1-\ell_0(z)\alpha_+^0(z)}.$$

The expression $1 - \ell_0(z)\alpha_+^0(z)$ should be thought of as a determinant of $(I - G_z q)$

It was shown in [32] that, generically, there exists a curve $\Gamma \subseteq \mathbf{C}$, depending on q such that $1 - \ell_0(z)\alpha_0(z) = 0$ on Γ.

As a result, we see that the inverse problem (6.2) is a singular $\bar{\partial}$ problem. Moreover, any formal series approach for reconstruction of q from T is bound to fail since no smallness assumption can be made to guarantee convergence.

The direct approach followin in [31], [32] consists in finding the $\bar{\partial}$ relations verified by μ_+^0. Unfortunately, these relations lead to the analysis of related "eigenfunctions" and to a complicated set of identities permitting the solution of the inverse problem (for small potentials).

A more direct and natural approach would be to define $T(z)$ as a distribution and to develop the theory of equation (6.2), under the assumption that near a curve Γ the distribution is like a principal value across the curve.

6.2 The scattering problem for the Benjamin-Ono equation

As mentioned previously, the complicated renormalization in the Schrodinger scattering problem followed the pattern developed in [31] for the Benjamin-Ono which exhibits similar analytic difficulties.

Here we consider the spectral theory of $\frac{1}{i}\frac{d}{dx} - V(q)$ where $V(q) = P^+(qP^+f) - P^-(qP^-f)$ and

$$P^{\pm}f = \pm\frac{1}{2\pi}\int_0^{\pm\infty} e^{ix\xi}\hat{f}(\xi)d\xi.$$

As before, we are led to consider

$$\mu^+ = 1 + G_z^+ g\mu^+, \quad G_z^+(x) = \frac{1}{2\pi}\int_0^{\infty}\frac{e^{ix\xi}}{\xi - z}d\xi,$$

where G_z^+ exhibits a logarithmic blowup as $|z| \to 0$. The same method leads to $\mu^0(x,z)$ and to

$$\mu_+(x,z) = \frac{\mu^0(x,z)}{1 - \ell_0(z)\alpha_0(z)}, \quad T_+(z) = \int q\mu e^{ix\xi}dx.$$

Unlike the two dimensional Schrodinger, it is shown that for real q there are no blowups at zeros of the denominator.

For q small, it was shown formally in [30] that the following curious identity must hold.

$$\int_{\mathbf{R}} q dx = -\int_{-\infty}^{\infty}\frac{|T(\xi)|^2}{\xi}d\xi + \sum_{\text{bound states}} 2\pi,$$

indicating that any formula expressing T as a power series (around 0) in function space, or expressing q as a linearizing series in T, as obtained by Anderson and Taflin, is purely formal. The various attempts to renormalize the series of [30] have led to the complicated nonlinear substitutions obtained in [31]. Here again, a more direct approach to the problem allowing for vanishing of $\det(I - G_z^+ q)$ would be desirable.

Acknowledgements

The authors were partially supported by the National Science Foundation under Grant Numbers DMS-8803471, DMS-8916968 and by the Air Force Office of Scientific Research under Grant Number 87-0310.

References

[1] C.S. Gardner, J.M. Greene, M.D. Kruskal, and R.M. Miura, Phys. Rev. Lett. **19** (1967) 1095; Comm. Pure Appl. Math. **27** (1974) 97.

[2] A.P. Veselov and S.P. Novikov, Soviet Math. Dokl. **30**, (1984) 558-591.

[3] V.E. Zakharov and P.B. Shabat, Funct. Anal. Appl. **8** (1974) 226; Funct. Anal. Appl. **13** (1979) 166; V.E. Zakharov and A.V. Mikhailov, Sov. Phys. JETP **47** (1978) 1071.

[4] S.V. Manakov, Physica D **3** (1981) 420.

[5] A.S. Fokas and M.J. Ablowitz, Stud. Appl. Math. **69** (1983) 211.

[6] R. Beals and R.R. Coifman, Comm. Pure and Appl. Math. **37**, (1984) 39-90; Inverse Problems **3** (1987) 577-593.

[7] M.J. Ablowitz, D. BarYaacov, and A.S. Fokas, Stud. Appl. Math. **69** (1983) 135.

[8] A.S. Fokas, Phys. Rev. Lett. **51**, No. 1 (1983) 3; A.S. Fokas and M.J. Ablowitz, J. Math. Phys. **25**, No. 8 (1984) 2494.

[9] A.S. Fokas and M.J. Ablowitz, Phys. Rev. Lett. **51**, No. 1, 3 (1983); A.S. Fokas and M.J. Ablowitz, The Inverse Scattering Transform for 2 + 1 Problems, in Nonlinear Phenomena, K.B. Wolf ed., Lecture Notes in Physics **189** Springer-Verlag (1984).

[10] R. Beals and R.R. Coifman, Proc. Symp. Pure Math. **43**, Amer. Math. Soc. Providence, (1985) 45.

[11] R. Beals and R.R. Coifman, Physica D **18**, (1986) 242-249.

[12] R. Beals and R.R. Coifman, The Spectral Problem for the Davey-Stewartson and Ishimori Hierarchies, Proc. Conf. on Nonlinear Evolution Equations: Integrability and Spectral Methods, Como, University of Manchester (1988).

[13] R. Beals and R. Coifman, Linear Spectral Problem, Nonlinear Equations, and the $\bar{\partial}$ Method, Inverse Problem, **5**, 87-130 (1989).

[14] M.V. Wickerhauser, Comm. Math. Phys. **108** (1987) 67.

[15] L.Y. Sung and A.S. Fokas, SIAM J. Math. Anal. **22**, No. 5, 1303-13131 (1991); Comm. Pure and Appl. Math., Vol. XLIV, (1991) 535-571.

[16] A.S. Fokas and L.Y. Sung, On the Solvability of the N Wave, the Davey-Stewartson and the Kadomtsev-Petviashvili Equations, Clarkson University, INS #176, preprint (1991).

[17] P.G. Grinevich and S.P. Novikov, 2D Inverse Scattering Problem for Negative Energy and Generalized-Analytic Functions, Func. Anal. Appl. **22** (1988) 23-33.

[18] P.G. Grinevich and R.G. Novikov, Dokl. Akad. Nauk SSSR **286** (1986) 19-22.

[19] A.S. Fokas and M.J. Ablowitz, Phys. Rev. Lett. **47** (1981) 1096; P. Santini, M.J. Ablowitz and A.S. Fokas, J. Math. Phys. **25**, 2614 (1985).
G.R.W. Quispel and H.W. Capel, Phys. Lett. **88A** (1981) 371; ibid **85A** (1981) 248; ibid Physica **110A** (1981) 41; F.W. Nijhoff, J. Van der Linden, G.R.W. Quispel, H.W. Capel and J. Velthuizen, Physica A (1982).

[20] A.S. Fokas and P.M. Santini, Physica D **44** (1990) 99; Phys. Rev. Lett. **63** (1989) 1329.

[21] M.J. Ablowitz and J. Villaroel, On the KP Equation and Associated Constraints, preprint (1990).

[22] Y.L. Rodin, Generalized Analytic Functions on Riemann Surfaces, Lect. Notes in Mathematics, Springer Verlag (1987).

[23] L.Y. Sung, The Inverse Scattering Method for DS (preprint) (1991).

[24] D.J. Kaup, Physica D **1** (1980) 45-67.

[25] L.P. Niznik, Ukranian Math. J. **24** (1972), 110-114 [in Russian].
[26] L.P. Niznik, Soviet Phys. Dokl. **25** (1980) 707.
[27] H. Segur, AIP Conference Proc. **88** (1982) 211.
[28] X. Zhou, Inverse Scattering Transform for the Time Dependent Schrodinger Equation with Application to the KPI Equation, Comm. Math. Phys., **128**, 551-564 (1990).
[29] A.S. Fokas and M.J. Ablowitz, Stud. Appl. Math. **68** (1983) 1-10.
[30] R.L. Anderson and E. Taflin, Lett. Math. Phys. **9**, (1985) 299-311.
[31] R. Coifman and V. Wickerhauser, Inverse Problems (1990), 825-801.
[32] Tian-Yue Tsai, Yale Dissertation, 1989.
[33] M. Boiti, J. Leon, L. Martina, F. Pempinelli, Phys. Lett. A, **132** 432 (1988).
[34] A.S. Fokas and V.E. Zakharov, The Dressing Method for Nonlocal Riemann-Hilbert Problems, to appear in the J. of Nonlinear Science (1992).
[35] M. Boiti, F. Pempinelli, A.K. Pogrebkov, and M.C. Polivanov, Inverse Problems **7** (1991) 43.
[36] A. Arkadiev, A.K. Pogrebkov, and M.C. Polivanov, Physica D **36**, (1989) 189-197.
[37] J.M. Ghidaglia and J.G. Saut, Nonlinearity **3**, (1990) 475-506.
[38] P.M. Santini, Physica D **41**, (1990) 26-54.
[39] J. Hientarita and R. Hirota, Phys. Lett. A **145** (1990) 239.

Dispersion Relations for Nonlinear Waves and the Schottky Problem

B. Dubrovin

Department of Mechanics and Mathematics,
Moscow State University, Moscow 119899, Russia

An approach to the Schottky problem of specification of the periods of holomorphic differentials on Riemann surfaces (or, equivalently, specification of Jacobians among all principally polarized Abelian varieties) based on the theory of Kadomtsev - Petviashvili equation, is discussed.

Introduction. Dispersion relations for linear and nonlinear waves

One of the first exercises in a course of PDE is finding particular solutions. For linear PDE the simplest solutions can be found immediately using the well-known properties of the exponential. For example, for the linear wave (or Helmholtz) equation

$$u_{tt} - u_{xx} + m^2 u = 0 \tag{0.1}$$

one can try to find a solution of the form

$$u(x,t) = Ae^{i(kx+\omega t)}. \tag{0.2}$$

Here A, k, ω are unknown parameters. After substitution in the equation one obtains a constraint for the parameters ω, k

$$\omega^2 - k^2 = m^2 \tag{0.3}$$

and no constraints for the amplitude A because of the linearity of the equation. The solution (0.2) is called plane wave, or one-phase solution of (0.1). The parameters A, k, ω are the amplitude, the wave number† and the frequency of the plane wave. The equation (0.3) thus is the ***dispersion relation*** for the plane waves. The solution is $\frac{2\pi}{k}$-periodic in x and $\frac{2\pi}{\omega}$-periodic in t for real ω, k. The solution (0.2) is a complex one; to obtain a real solution one can take the real part of (0.2).

Multiphase quasi-periodic solutions of (0.1) are linear superpositions of plane waves

$$u(x,t) = \sum_s A_s e^{i(k_s x + \omega_s t)}, \tag{0.4a}$$

$$A_s \text{ arbitrary}, \ \omega_s^2 - k_s^2 = m^2. \tag{0.4b}$$

Considering infinite sums (or integrals over s) for real k_s, ω_s one obtains the general

† In more standard terminology the wave number is $-k$.

solution of the Cauchy problem for the equation (0.1) for appropriate functional classes of initial data.

Nonlinear analogues of simple waves can be constructed for a wide class of PDE. A feature of them is that now the amplitude is involved in the dispersion relation for the nonlinear waves. To see it let us consider a nonlinear wave equation

$$u_{tt} - u_{xx} + V'(u) = 0. \tag{0.5}$$

Let us assume the potential $V(u)$ to satisfy the following condition: the equation

$$V(u) = E \tag{0.6a}$$

has two solutions

$$u_- = u_-(E) < u_+ = u_+(E) \tag{0.6b}$$

for some interval $E_{\min} < E < E_{\max}$. Then nonlinear simple waves have the form

$$u(x,t) = U(kx + \omega t + \phi_0; E) \tag{0.7a}$$

for arbitrary phase shift ϕ_0, where the function $U = U(\phi; E)$ has the form

$$\int_{u_-(E)}^{U} \frac{du}{\sqrt{2(E - V(u))}} = I(E)\phi, \tag{0.7b}$$

$$I(E) = \frac{1}{\pi} \int_{u_-(E)}^{u_+(E)} \frac{du}{\sqrt{2(E - V(u))}} \tag{0.7c}$$

and the parameters ω, k, E satisfy

$$\omega^2 - k^2 = I^{-2}(E). \tag{0.7d}$$

The function $U(\phi; E)$ plays the role of the exponential ($U = \frac{\sqrt{2E}}{m} \cos\phi$ for the linear case). It is 2π-periodic in ϕ. So, again, the parameters ω, k have the sense of the frequency and wave number of the nonlinear waves (0.7). The shape of the wave is determined by the amplitude parameter E. The constraint (0.7d) for ω, k, E is nothing but the nonlinear dispersion relation for the frequency, wave number and amplitude of the nonlinear waves.

One can try to look for multiphase oscillating solutions of a nonlinear PDE of the form (in the spatially one-dimensional case)

$$u(x,t) = U(k_1 x + \omega_1 t + \phi_{10}, \ldots, k_m x + \omega_m t + \phi_{m0}; \mathbf{A}) \tag{0.8}$$

where the function $U(\phi_1, \ldots, \phi_m; \mathbf{A})$ is 2π-periodic in ϕ_1, ..., ϕ_m. The vector of parameters **A** plays the role of "amplitudes". It turns out that existence of such multiphase solutions for sufficiently big m (probably, for $m > 2$; see [16] for examples of 2-phase solutions of a nonintegrable equation) is a feature of integrable evolutionary equations (though this statement is still to be proved). For the KdV equation these are the famous finite-gap (or algebraic-geometrical) solutions that were constructed in the papers of 1974 - 1976 by S.Novikov and B.Dubrovin [1, 3-5], P.Lax [2], A.Its and V.Matveev [6], H.McKean and P.van Moerbeke [8]. On this basis a program of constructing and investigating multiphase solutions of nonlinear integrable systems was developed (see surveys [7, 9-12]). An extremely important step in the development of this program was taken by I. Krichever [13-14, 9]. He found a general approach to construct algebraic-geometrical solutions of spatially 2-dimensional integrable systems such as the Kadomtsev - Petviashvili (KP) equation

$$\frac{3}{4}u_{yy} = \partial_x(u_t - \frac{1}{4}(6uu_x - u_{xxx})). \tag{0.9}$$

Multiphase solutions of this equation have the form

$$u(x,y,t) = U(k_1x + l_1y + \omega_1t + \phi_{10}, \dots, k_mx + l_my + \omega_mt + \phi_{m0}; \mathbf{A}) \tag{0.10}$$

for arbitrary phase shifts ϕ_{10}, ..., ϕ_{m0}, where, as above, the function $U(\phi_1, \dots, \phi_m; \mathbf{A})$ is 2π-periodic in each ϕ_1, ..., ϕ_m, and $\mathbf{A}$ is a vector of amplitude parameters. This function can be expressed via multidimensional theta-functions. Theta-functions are defined by a multiple Fourier series

$$\theta(\phi|\tau) = \sum_{-\infty<n_1,\dots,n_m<\infty} \exp(\pi i \sum_{p,q=1}^{m} \tau_{pq}n_pn_q + \sum_{p=1}^{m} in_p\phi_p), \tag{0.11}$$

$$\phi = (\phi_1, \dots, \phi_m),\ \tau = (\tau_{pq})_{1\le p,q\le m}.$$

Parameters of the theta-function form a *period matrix*, i.e. a symmetric $m \times m$ complex matrix $\tau = (\tau_{pq})$ with positive definite imaginary part. This is 2π-periodic in ϕ_1, ..., ϕ_m. It also posesses the quasi-periodicity property

$$\theta(\phi + N\tau|\tau) = \exp(-\pi i < N\tau, N > -i < N, \phi >)\theta(\phi|\tau) \tag{0.12}$$

for any integer vector $N = (N_1, \dots, N_m)$, where the brackets $< \, , \, >$ mean the Euclidean inner product

$$< N, \phi >= \sum_1^m N_s\phi_s.$$

The Krichever's solutions of KP have the form (0.10) where

$$U = U(\phi; \tau) = -2\partial_k^2 \log \theta(\phi|\tau) + c, \tag{0.13}$$

$$\partial_k = \sum_{p=1}^{m} k_p \frac{\partial}{\partial \phi_p},$$

c is an arbitrary constant. (This can be taken out by the transformation

$$\omega_p \mapsto \omega_p - \frac{3}{2}ck_p,\ p = 1, \dots, m$$

$$u \mapsto u - c.)$$

So the period matrix τ can be considered as the amplitude of the multiphase solutions of KP. The main object of our investigation will be dispersion relations for these multiphase solutions

$$F_{KP}(k,l,\omega,\tau)=0$$

where $F_{KP}(k,l,\omega,\tau)$ is a vector-valued analytic function. Explicit form of this system of equations will be given below. It turns out that these dispersion relations for $m\geq 4$ constrain also the period matrix τ. These constraints will give a solution of the classical Schottky problem (see Section 2 below) exactly specifying period matrices of holomorphic differentials on Riemann surfaces.

1. Dispersion relations for multiphase solutions of KP. Novikov's conjecture

To obtain dispersion relations for multiphase solutions (0.10), (0.13) let us substitute the theta-functional formula (0.11) to the KP equation (the constant c is assumed to equal zero). After substitution one obtains

$$\partial_x^2[(\theta_{xxxx}\theta-4\theta_{xxx}\theta_x-3\theta_{xx}^2-4\theta_x\theta_t+4\theta_{xt}\theta-3\theta_{yy}\theta+3\theta_y^2)/\theta^2]=0 \qquad (1.1)$$

where

$$\theta=\theta(kx+ly+\omega t+\phi_0|\tau)$$

ϕ_0 is an arbitrary complex vector. If the theta function is indecomposable (see below) then the expression in the square brackets equals a constant. Let us denote this integration constant by $-8d$. To reduce the equality

$$\theta_{xxxx}\theta-4\theta_{xxx}\theta_x-3\theta_{xx}^2-4\theta_x\theta_t+4\theta_{xt}\theta-3\theta_{yy}\theta+3\theta_y^2+8d\theta^2=0$$

to a finite number of dispersion relations for k, l, ω, and τ, let us introduce the theta-functions of the second order

$$\hat\theta[p](\phi|\tau)=\sum_{-\infty<n_1,\dots,n_m<\infty}\exp(2\pi i\sum_{q,r=1}^{m}\tau_{qr}(n_q+\frac{p_q}{2})(n_r+\frac{p_r}{2})$$

$$+\sum_{q=1}^{m}i(n_q+\frac{p_q}{2})\phi_q). \qquad (1.2)$$

Here $p\in \mathbf{Z}_2^m$, i.e. it is an arbitrary m-vector with the coordinates being equal to 0 or 1. We have 2^m such theta-functions of the second order. Values of these theta-functions and of their derivatives in the origin $\phi=0$ are called ***theta-constants***. They are functions only on τ. For brevity let us omit the arguments of the theta-constants:

$$\hat\theta[p]\equiv\hat\theta[p](0|\tau),$$

$$\hat\theta_{ij}[p]\equiv\frac{\partial^2}{\partial\phi_i\partial\phi_j}\hat\theta[p](0|\tau),$$

$$\hat\theta_{ijqr}[p]\equiv\frac{\partial^4}{\partial\phi_i\partial\phi_j\partial\phi_q\partial\phi_r}\hat\theta[p](0|\tau).$$

(The theta-functions (1.2) are even functions of ϕ, so only derivatives of even order in the origin could be nonzero.)

Theorem 1.[15] *Dispersion relations for the multiphase solutions (0.10), (0.13) have the form*

$$\partial_k^4\hat{\theta}[p]+\partial_k\partial_\omega\hat{\theta}[p]+\frac{3}{4}\partial_l^2\hat{\theta}[p]+d\hat{\theta}[p]=0 \tag{1.3}$$

for arbitrary $p\in\mathbf{Z}_2^m$.

Here

$$\partial_k^4\hat{\theta}[p]=\sum_{i,j,q,r}k_ik_jk_qk_r\hat{\theta}_{ijqr}[p]$$

$$\partial_k\partial_\omega\hat{\theta}[p]=\sum_{i,j}k_i\omega_j\hat{\theta}_{ij}[p]$$

$$\partial_l^2\hat{\theta}[p]=\sum_{i,j}l_il_j\hat{\theta}_{ij}[p].$$

The dispersion relations (1.3) are written in the form of a system of algebraic equations for the coordinates of the vectors k, l, and ω and for an auxilary unknown variable d with the coefficients depending on the period matrix τ. For $m=1,\ 2,\ 3$ for generic matrix τ one can solve the dispersion relations in the form

$$k=k(\tau),\ l=l(\tau),\ \omega=\omega(\tau)$$

(in fact, one obtains a one-parameter family of solutions, see [15, 10, 17] and Section 3 below). This parametrization of 2-phase solutions of KP was used in [18] for constructing physicaly realistic models of nonlinear waves on shallow water.

For $m>3$ the dispersion relations are an overdetermined system of algebraic equations for k, l, ω, d. Compatibility conditions of these overdetermined equations constrain the "amplitude" τ. It was conjectured by S.Novikov in 1980 that these constraints exactly specify periods of Riemann surfaces providing a solution of the classical Schottky problem. We are coming now to the formulation of this problem.

2. Periods of Riemann surfaces. Schottky problem

A challenge of the theory of functions of the XIXth century is the problem of moduli of Riemann surfaces (still far from having been solved). Intuitively, the problem is to obtain a complete "list" of pairwise inequivalent Riemann surfaces of a given genus g. For $g=0$ the "list" consists only of one point: the Riemann sphere ($=\mathbf{CP}^1$) since any Riemann surface of genus 0 is equivalent (bi-holomorphic) to the Riemann sphere. For $g=1$ one obtains a 1-parameter family of Riemann surfaces (elliptic curves). If an elliptic curve is represented in the Weierstrass canonical form

$$y^2=4x^3-g_2x-g_3 \tag{2.1}$$

(where g_2 and g_3 are complex numbers) then the combination

$$J = \frac{g_2^3}{g_2^3 - 27g_3^2} \tag{2.2}$$

depends only on the equivalence class of the curve. So it can serve as the parameter of equivalence classes of elliptic curves (2.1) (in fact, any elliptic curve can be represented in the Weierstrass form (2.1)). Another choice of the parameter is the *period* of the elliptic curve

$$\tau = \oint_b \frac{dx}{y} : \oint_a \frac{dx}{y} \tag{2.3}$$

where a and b are basic cycles on the elliptic curve (i.e., on the torus) oriented in such a way that the intersection number $a \circ b = 1$. This is a complex number with positive imaginary part $\mathrm{Im}\tau > 0$. Ambiguity in the choice of the basis a, b provides the following transformation of the period τ

$$\tau \mapsto \frac{A\tau + B}{C\tau + D} \tag{2.4}$$

$$A,\ B,\ C,\ D \text{ are integers, } \det\begin{pmatrix} A & B \\ C & D \end{pmatrix} = 1.$$

Therefore the family of all elliptic curves can be represented as a quotient of the upper half plane $\mathbf{H}$ over the action (2.4) of the modular group $\mathbf{M}_1 = SL(2, \mathbf{Z})/(\pm 1)$.

For higher genera $g > 1$ the moduli space of Riemann surfaces has the complex dimension $3g-3$ (see, e.g. [19]). Periods of a Riemann surface R of any genus $g > 0$ are natural parameters uniquely specifying the class of bi-holomorphic equivalence of the surface. These periods are defined as follows.

Let us fix a symplectic basis $a_1, \ldots, a_g, b_1, \ldots, b_g \in H_1(R, \mathbf{Z})$ of cycles on the surface R. That means that the intersection numbers of these cycles have the following canonical form

$$a_i \circ a_j = b_i \circ b_j = 0,\ a_i \circ b_j = \delta_{ij}.$$

Here δ_{ij} is the Kronecker delta. The basis of cycles a_i, b_j uniquely specifies a basis of holomorphic differentials (Abelian differentials of the first kind) on the surface R $\Omega_1, \ldots, \Omega_g$ such that

$$\oint_{a_i} \Omega_j = \delta_{ij}. \tag{2.5}$$

The period matrix $\tau = (\tau_{ij})$ of the surface R (with respect to the symplectic basis a_i, b_j) has the form

$$\tau_{ij} = \oint_{b_i} \Omega_j,\quad i,\ j = 1, \ldots, g. \tag{2.6}$$

It is a symmetric matrix with positive definite imaginary part. For $g = 1$ it coincides with (2.3).

A change of the symplectic basis a_i, b_j implies the following transformation of the period matrix

$$\tau \mapsto \tilde{\tau} = (A\tau + B)(C\tau + D)^{-1} \tag{2.7a}$$

$$\begin{pmatrix} A & B \\ C & D \end{pmatrix} \in Sp(g,\mathbf{Z})$$

$$= \{ \begin{pmatrix} A & B \\ C & D \end{pmatrix} \mid \begin{pmatrix} A^{\mathrm{T}} & C^{\mathrm{T}} \\ B^{\mathrm{T}} & D^{\mathrm{T}} \end{pmatrix} \begin{pmatrix} 0 & 1 \\ -1 & 0 \end{pmatrix} \begin{pmatrix} A & B \\ C & D \end{pmatrix} = \begin{pmatrix} 0 & 1 \\ -1 & 0 \end{pmatrix} \}. \qquad (2.7b)$$

According to the classical Torelli's theorem (see [19]) the class of (bi-holomorphic) equivalence of a Riemann surface R is uniquely determined by the class of equivalence (2.7) of the period matrix of the surface.

Remark. Using periods of a Riemann surface R as the parameters of the theta-function (0.11) (here $m = g$) one obtains the *theta-function of the Riemann surface R*. This is a very important special function associated with a Riemann surface both in algebro-geometrical calculations (see, e.g. [20]) and in application to nonlinear equations (see the next section). The above Novikov's conjecture means that *only* theta-functions of Riemann surfaces occur in the multiphase solutions of nonlinear equations (particularly, of KP). The advantage of KP (and, more generally, of any 2+1 integrable system) is that the theta-function of *arbitrary* Riemann surface gives a multiphase solution of the equation. Multiphase solutions of a given 1+1 integrable system (like KdV) are expressed via more particular classes of theta-functions of Riemann surfaces represented as coverings of the Riemann sphere with fixed number of sheets. So dispersion relations of none of the 1+1 integrable systems can be used for specification of period matrices of *all* the family of Riemann surfaces.

Let us consider the family of *all* $g \times g$ symmetric matrices τ with positive definite imaginary part. They form the *Siegel upper half plane* $\mathbf{H}_g$. The Siegel modular group $\mathbf{M}_g = Sp(g,\mathbf{Z})/(\pm 1)$ acts on the Siegel upper half plane by the transformations (2.7) (see [21]). Theta-functions $\theta(\phi|\tau)$ and $\theta(\tilde{\phi}|\tilde{\tau})$, where $\tilde{\phi} = \phi(C\tau + D)^{-1}$, for equivalent matrices τ and $\tilde{\tau}$ coincide up to a shift of the argument and multiplication by exponential of a quadratic form of ϕ (see [21] for the explicit formula of the transformation law).

A matrix $\tau \in \mathbf{H}_g$ is called *decomposable* if it is equivalent to a block-diagonal matrix

$$\tilde{\tau} = \begin{pmatrix} \tau' & 0 \\ 0 & \tau'' \end{pmatrix}.$$

The correspondent theta-function is factorized to a product of theta-functions with less than g arguments. The period matrices of a Riemann surface always is indecomposable [19].

We are ready now to formulate the Schottky problem [31]. The periods of Riemann surfaces of a given genus g determine the *period map*

$$\text{Moduli space of Riemann surfaces of genus } g \to \mathbf{H}_g/\mathbf{M}_g.$$

The Torelli theorem provides this map to be injective (in fact, being an analytic embedding of the complex varieties: see [19]). For $g = 1, 2, 3$ the image of the period map is an open dense subset in $\mathbf{H}_g/\mathbf{M}_g$ (the completion is empty for $g = 1$ and coincides with the sublocus of decomposable matrices τ for $m = 2, 3$). For $g > 3$ the dimension $3g - 3$ of the moduli space of Riemann surfaces is less than the dimension $g(g+1)/2$ of the Siegel upper half plane. The Schottky problem reads: to specify the image of the period map for $g > 3$, i.e. to find a system of $g(g+1)/2 - (3g-3)$ equations for unknowns $\tau_{ij} \in \mathbf{H}_g$ specifying periods of Riemann surfaces.

The above Novikov's conjecture can be reformulated as follows: *the dispersion relations (1.3) for multiphase solutions of KP as equations for τ for $g = m > 3$ exactly specify periods of Riemann surfaces.*

The main motivation for this conjecture was the Krichever's construction of algebro-geometric solutions of the KP equation given in the next section.

3. Krichever's multiphase solutions of KP

Let us come back in more details to the multiphase solutions of KP. In fact in [13] it was constructed a family of the multiphase solutions (0.10), (0.13) where the period matrix τ is just the period matrix of a Riemann surface and components of the wave numbers and frequency vectors are certain Abelian integrals on the Riemann surface. More precisely, let R be a Riemann surface of genus g with a marked point $\infty \in R$, and with a local parameter z on R near this point such that $z(\infty) = 0$, and with a marked symplectic basis of cycles a_i, b_j. Then for the Krichever's multiphase solutions (0.10), (0.13) $m = g$, τ is the period matrix of the surface R,

$$k_j = \oint_{b_j} \eta^{(1)} \tag{3.1a}$$

$$l_j = \oint_{b_j} \eta^{(2)} \tag{3.1b}$$

$$\omega_j = \oint_{b_j} \eta^{(3)} \tag{3.1c}$$

where $\eta^{(q)}$ are the normalized

$$\oint_{a_j} \eta^{(q)} = 0, \; j = 1, \ldots, g,$$

Abelian differentials of the second kind with a pole only at ∞ with the principal parts

$$\eta^{(q)} = d(z^{-q}) + \text{regular terms}, \; q = 1, \; 2, \; 3.$$

Theorem 2. [13] *For an arbitrary Riemann surface R of genus g with a marked point ∞ and with a marked local parameter z near this point and with a marked symplectic basis a_i, b_j the formulae (0.10), (0.13), (2.6), (3.1) with $m = g$ determine a multiphase solution of the KP equation.*

Remark. We consider here complex multiphase solutions of KP being meromorphic functions of complex variables ϕ_1, ..., ϕ_m. One should impose certain reality constraints for the data R, ∞, z and for the phase shift ϕ_0 to obtain real smooth multiphase solutions of the two real modifications of KP: the equations KP1 (coinciding with (0.9)) and KP2 (this can be obtained from (0.9) by the substitution $y \mapsto iy$). The reality constraints were obtained in [10, 22] (for periodic multiphase solutions also in [23]). In [23] it was proved that the multiphase double periodic (in x, y) solutions of the KP2 form a dense subset in the space of all double periodic solutions of this equation.

It turns out that the solution does not depend on a choice of the symplectic basis of cycles. A change of the local parameter

$$z \mapsto a_1 z + a_2 z^2 + a_3 z^3 + \dots$$

implies the following transformation of the solution

$$x \mapsto a_1 x + 2a_2 y + 3a_3 t$$

$$y \mapsto a_1^2 y + 3a_1 a_2 t$$

$$t \mapsto a_1^3 t$$

$$u \mapsto u a_1^{-2} + 2(a_2^2 - a_1 a_3) a_1^{-2}$$

(so u transforms like a projective connection on the Riemann surface).

4. KP equation and Schottky problem

Let us come back to the Novikov's conjecture (see the end of Section 2 above). The system (1.3) has trivial solutions when the theta-function is a decomposable one. To get rid of these solutions let us impose the following *nondegeneracy condition* for the theta-function: the matrix of theta-constants

$$(\hat{\theta}_{11}[p], \hat{\theta}_{12}[p], \dots, \hat{\theta}_{mm}[p], \hat{\theta}[p])$$

(the matrix has $\frac{m(m+1)}{2} + 1$ columns and 2^m lines enumerated by the vectors $p \in \mathbb{Z}_m$) has maximal rank $= \frac{m(m+1)}{2} + 1$. The nondegeneracy condition holds for period matrices of Riemann surfaces [17]. We will consider solutions of the system (1.3) only satisfying the nondegeneracy condition.

The system of dispersion relations is invariant with respect to the action of the group of changes of local parameter z

$$k \mapsto \lambda k \tag{4.1a}$$

$$l \mapsto \pm(\lambda^2 l + 2\alpha\lambda k) \tag{4.1b}$$

$$\omega \mapsto \lambda^3 \omega + 3\lambda^2 \alpha l + 3\lambda\alpha^2 k \tag{4.1c}$$

$$d \mapsto \lambda^4 d, \quad \tau \mapsto \tau \tag{4.1d}$$

and also with respect to the following action of the Siegel modular group (2.7)

$$\tau \mapsto (A\tau + B)(C\tau + D)^{-1} \tag{4.2a}$$

$$k \mapsto kM^{-1}, \text{ where } M = C\tau + D, \tag{4.2b}$$

$$l \mapsto lM^{-1} \tag{4.2c}$$

$$\omega \mapsto \omega M^{-1} + \frac{1}{3}\{k,k\}kM^{-1}, \text{ where } \{x,y\} = xM^{-1}Cy^{\mathrm{T}} \tag{4.2d}$$

$$d \mapsto d + \frac{3}{8}\{l,l\} - \frac{1}{2}\{k,\omega\} - \frac{3}{4}\{k,k\}^2. \tag{4.2e}$$

Hence eliminating the variables k, l, ω, d from the dispersion relations we obtain a system of equations (for $m > 3$) for the matrix τ being invariant with respect to the action of the modular group (i.e. if the system (1.3) is compatible for a period matrix τ then it will be compatible for any matrix $\tilde{\tau}$ being equivalent (2.7) to τ). So the compatibility conditions of the system (1.3) specify a sublocus

$$X_m \subset \mathbf{H}_m/\mathbf{M}_m.$$

Due to the Theorem 2 this sublocus contains period matrices of Riemann surfaces of genus $g = m$.

The first test of the Novikov's conjecture was done in [17]: it was shown that the irreducible component of X_m containing period matrices of Riemann surfaces consists only of period matrices of Riemann surfaces. Even this statement sounds surprising. Indeed, from the construction of the Section 3 it follows that for a given τ = period matrix of a Riemann surface the system (1.3) is more than compatible: it has a one-parameter family of solutions due to ambiguity in the choice of the marked point ∞ on the Riemann surface. So the crucial point to prove the Novikov's conjecture is to prove that from compatibility of (1.3) it follows that (1.3) has a one-parameter family of solutions $k = k(\tau,\epsilon)$, $l = l(\tau,\epsilon)$, $\omega = \omega(\tau,\epsilon)$. If we identify the parameter ϵ with the z-coordinate of the displaced marked point $\infty \mapsto \infty'$, $\epsilon = z(\infty')$, then the coefficients of the expansion

$$k(\tau,\epsilon) = k + \frac{\epsilon}{2}k^{(2)} + \frac{\epsilon^2}{2}k^{(3)} + \dots$$

are periods of the normalized Abelian differentials of the second kind with poles at ∞,

$$k_i^{(q)} = c_q \oint_{b_i} \eta^{(q)}, \; i = 1,\dots,g$$

(for certain constants c_q). One has $k^{(2)} = l$, $k^{(3)} = \omega$, other vectors $k^{(q)}$ are the frequencies of the multiphase solutions of the same form (0.10), (0.13) to the q-th equation of the KP hierarchy. Thus one needs to prove that any theta-function solution (0.10), (0.13) of KP can be extended to a solution of all the KP hierarchy.

The final step in proof of Novikov's conjecture was obtained by T.Shiota who proved that X_m has no extra irreducible components:

Theorem 3.[24] *An indecomposable theta-function gives a multiphase solution (0.10), (0.13) of the KP equation iff it is the theta-function of a Riemann surface.*

Because of limits of the paper we have no possibility to discuss here this remarkable theorem. Another proof of the Novikov's conjecture was obtained by E.Arbarello and C.De Concini [25].

Conclusion

Investigating of dispertion relations of other integrable differential equations [26-28, 32-38] turned out to be fruitful for the algebraic geometry of Abelian varieties (e.g., Prym varieties [28]) as well as for the theory and applications of integrable systems. An approach to calculating parameters of *real* multiphase solutions of KP based on the classical theory of Schottky uniformization of Riemann surfaces and Burnside series was proposed in [29]. Recently it was found [30] that, periods of Riemann surfaces with fixed number of sheets and fixed ramification at infinity as functions of moduli of these Riemann surfaces, are themselves solutions of certain integrable systems arising in topological field theory. But these beautiful new developments, probably, should be a part of the next decade of history of the soliton theory.

References

1. S.Novikov, The periodic problem for the Korteweg - de Vries equation, *Funct. Anal. Appl.* **8** (1974) 236 - 246.
2. P.Lax, Periodic solutions of the Korteweg - de Vries equation, *Lect. in Appl. Math.* **15** (1974) 65 - 96.
3. B.Dubrovin, S.Novikov, Periodic and conditionally periodic analogues of multisoliton solutions of the Korteweg - de Vries equation, *Sov. Phys. JETP* **40** (1974) 1058 - 1063.
4. B.Dubrovin, S.Novikov, The periodic problem for the Korteweg - de Vries and Sturm - Liouville equations. Their connection with algebraic geometry, *Sov. Math. Dokl.* **15** (1974) 1597 - 1601.
5. B.Dubrovin, Inverse problem for periodic finite-zone potentials, *Funct. Anal. Appl.* **9** (1975) 61 - 62; Periodic problem for the Korteweg - de Vries equation in the class of finite-zone potentials, *Ibid.*, 215 - 223.
6. A.Its and V.Matveev, Hill's operators with a finite number of lacunae, *Funct. Anal. Appl.* **9** (1975) 65 - 66; Hill's operators with a finite number of lacunae and multisoliton solutions of the KdV equation, *Theor. Math. Phys.* **23** (1975) 343 - 355.
7. B.Dubrovin, V.Matveev, and S.Novikov, Nonlinear equations of Korteweg - de Vries type, finite-zone linear operators, and Abelian varieties, *Russ. Math. Surv.* **31:1** (1976) 59 - 146.
8. H.McKean and P.van Moerbeke, The spectrum of Hill's equation, *Invent. Math.* **30** (1975) 217 - 274.
9. I.Krichever, Methods of algebraic geometry in the theory of nonlinear equations, *Russ. Math. Surv.* **32:6** (1977) 185 - 213.
10. B.Dubrovin, Theta-functions and nonlinear equations, *Russ. Math. Surv.* **36:2** (1981) 11 - 92.

11. E.Belokolos, A.Bobenko, V.Matveev, and V.Enol'skii, Algebraic-geometric principles of superposition of finite-zone solutions of integrable non-linear equations, *Russ. Math. Surv.* **41:2** (1986) 1 - 49.

12. B.Dubrovin, I.Krichever, and S.Novikov, Integrable Systems. I. Encyclopedia of Mathematical Sciences 4, 173 - 280. Springer-Verlag, 1990.

13. I.Krichever, An algebraic-geometric construction of the Zakharov - Shabat equations and their periodic solutions, *Sov. Math. Dokl.* **17** (1976) 394 - 397.

14. I.Krichever, Integration of nonlinear equations by the methods of algebraic geometry, *Funct. Anal. Appl.* **11** (1977) 12 - 26.

15. B.Dubrovin, On a conjecture of Novikov in the theory of theta-functions and nonlinear equations of Korteweg - de Vries and Kadomtsev - Petviashvili type, *Sov. Math. Dokl.* **21** (1980) 469 - 472.

16. A.Nakamura, A direct method of calculating periodic wave solutions to nonlinear wave equations. I. Exact two-periodic wave solutions, *J. Phys. Soc. Japan* **47** (1979) 1701 - 1705; A direct method of calculating periodic wave solutions to nonlinear wave equations. II. Exact one- and two-periodic wave solutions of the coupled bilinear equations, *Ibid.* **48** (1980) 1365 - 1370.

17. B.Dubrovin, The Kadomtsev - Petviashvili equation and relations between periods of holomorphic differentials on Riemann surfaces, *Math. USSR Izvestia* **19** (1982) 285 - 296.

18. H. Segur and A. Finkel, An analytical model of periodic waves in shallow water, *Stud. Appl. Math.* **73** (1985) 183-220.
J. Hammack, N. Scheffner, and H. Segur, Two-dimensional periodic waves in shallow water, *J. Fluid Mech.* **209** (1989) 567-589; A note on the generation and narrowness of periodic rip currents, *J. Geophysical Research* **96:C3** (1991) 4909-4914.

19. Ph.Griffiths and J.Harris, Principles of Algebraic Geometry, New York, Wiley, 1978.

20. J.Fay, Theta-functions on Riemann Surfaces, *Lecture Notes in Mathematics* **352** (1973), Springer-Verlag.

21. C.Siegel, Topics in Complex Functions Theory, Chichester, Wiley, 1988.

22. B.Dubrovin and S.Natanzon, Real theta-function solutions of the Kadomtsev - Petviashvili equation, *Math. USSR Izvestia* **32** (1988) 269 - 288.

23. I.Krichever, Spectral theory of two-dimensional periodic operators and its applications, *Russ. Math. Surv.* **44:2** (1989) 145 - 225.

24. T.Shiota, Characterization of Jacobian varieties in terms of soliton equations, *Inv. Math.* **83** (1986) 333-382.

25. E. Arbarello and C. De Concini, Another proof of a conjecture of S.P. Novikov on periods of abelian integrals on Riemann surface, *Duke Math. J.* **54** (1987) 163 - 178.

26. B.Dubrovin and S.Natanzon, Real 2-zone solutions of the sine-gordon equation, *Funct. Anal. Appl.* **16** (1982) 21 - 33.

27. E.Horozov, Neumann's problem and equations that define matrices of periods of hyperelliptic Riemann surfaces, *C.R. Acad. Bulgare Sci.* **37** (1984) 277 - 280.

28. I.Taimanov, Effectivization of theta-function formulas for two-dimensional Schrödinger potential operators that are finite-gap at a certain energy level, *Sov. Math. Dokl.* **32** (1985) 843 - 846; On an analogue of the Novikov conjecture in a problem of Riemann - Schottky type for Prym varieties, *Sov. Math. Dokl.* **35** (1987) 420 - 424.

29. A.Bobenko, Uniformization and finite-gap integration, Preprint LOMI P-10-86, Leningrad, 1986 (in Russian); Schottky uniformization and finite-gap integration, *Sov. Math. Dokl.* **36** (1988) 38 - 42; Uniformization of Riemann surfaces and effectivization of theta-functional formulae, Preprint 257 (1990), Technische Universität Berlin, to be published in: E.Belokolos, A.Bobenko, V.Enol'skii, and A.Its, Algebraic-geometrical approach to nonlinear integrable equations, Springer-Verlag.

30. B.Dubrovin, Differential geometry of moduli spaces and its applications to soliton equations and to topological conformal field theory, Preprint n.117, Scuola Normale Superiore, November 1991, 31 pp.; Hamiltonian formalism of Whitham-type hierarchies and topological Landau - Ginsburg models, *Comm. Math. Phys.* **145** (1992) 195-207; Integrable systems in topological field theory, *Nucl. Phys. B379* (1992) 627-689; Integrable systems and classification of 2-dimensional topological field theories, Preprint SISSA 162/92/FM, September 1992, 42 pp.

31. F.Schottky, Über die Moduln der Thetafunktionen, *Acta Math.* **27** (1903), 235-288.

32. J.Zagrodzinski, Dispersion equations and a comparison of different quasi-periodic solutions of the sine-gordon equation, *J. Phys.* **A15** (1982) 3109-3118.

33. E.Belokolos and V.Enol'skii, Generalized Lamb ansatz, *Theor. Math. Phys.* **53** (1982) 1120-1127.

34. E.Arbarello and C.De Concini, On a set of equations characterizing Riemann matrices, *Ann. Math.* **120** (1984) 119 - 140.

35. G.Welters, Polarized abelian varieties and the heat equation, *Compos. Math.* **49** (1983) 173 - 194; A characterization of non-hyperelliptic Jacobi varieties, *Invent. Math.* **74** (1983) 437 - 440; A criterion for Jacobi varieties, *Ann. Math.* **120** (1984) 497 - 504.

36. A. Beauville and O. Debarre, Une relation entre deux approaches du problème de Schottky, *Invent. Math.* **86** (1986) 195 - 207.

37. G.van der Geer, The Schottky problem, In: *Proceedings of Arbeitstagung Bonn 1984*, Springer-Verlag, 1985.

38. M.Mulase, Cohomological structure in soliton equations and Jacobian varieties, *J. Diff. Geom.* **19** (1984) 403 - 430.

The Isomonodromy Method and the Painlevé Equations

*A.S. Fokas and A.R. Its**

Department of Mathematics and Computer Science
and The Institute for Nonlinear Studies, Clarkson University,
Potsdam, NY 13699-5815, USA
*On leave of absence from Leningrad University, Leningrad, USSR

1 Introduction

Although the six Painlevé transcendents, PI-PVI, were introduced at the turn of the century by Painlevé and his school from strictly mathematical considerations, they have recently appeared in a wide range of physical applications (see for example [1]-[13]). It is becoming increasingly evident, that these equations play in nonlinear physics the same role that certain classical special functions (such as Airy function, Bessel function, etc.) play in linear physics. In particular, just like their linear counterparts, they describe certain transitional and self-similar processes. One of the most important developments in the theory of nonlinear ODE's, has been the discovery that the Cauchy problems for the Painlevé equations can be linearized. The relevant method is called the inverse monodromy (in analogy with inverse spectral) or the isomonodromy (in analogy with isospectral) method and is the extension of the inverse spectral method from PDE's to ODE's. This method, which can be thought of as a nonlinear analogue of the Laplace's method for solving linear ODE's, allows one to investigate the asymptotic behavior of the solutions of Painlevé transcendents and to obtain connection formula.

The isomonodromy method was introduced in [14] and [15], where it was realized that solving a Cauchy problem for a given Painlevé equation is essentially equivalent to solving an inverse problem for an associated isomonodromy linear equation. This inverse problem can be formulated in terms of monodromy data which can be calculated from initial data. It was shown in [16] that the inverse problem can be formulated as a matrix, singular, discontinuous Riemann-Hilbert (RH) problem defined on a complicated contour. Hence techniques from RH theory can be employed to study the solvability of certain nonlinear ODE's. PII, a special case of PIII, PIV, and PV were formally studied in [14], [16] and [17]. A rigorous methodology for implementing the isomonodromy method was developed in [18] and has been applied to PI-PV [18], [19]. Using this method it can be shown that the Painlevé equations in general admit global meromorphic in t solutions. Actually, the proof of the existence of meromorphic solutions is quite more transparent than the original proof of Painlevé . Also, using this method it is possible to find those monodromy data (and hence those initial data) for which the solution is free from poles. For examples in the case of PII,

$$y_{tt} = 2y^3 + ty + \theta, \tag{1.1}$$

for real t, one finds the constraints

$$s_2 = \bar{s_2} \text{ and } |s_1 - \bar{s_3}| < 2, \quad |Re\theta| < \frac{1}{2}, \tag{1.2}$$

where the monodromy data s_1, s_2, s_3 are related via

$$s_1 + s_2 + s_3 + s_1 s_2 s_3 = -2i \sin \theta\pi. \tag{1.3}$$

The special cases

$$s_1 = \bar{s_3}, \quad \theta = 0 \tag{1.4}$$

and

$$s_1 = s_3 = i\alpha, \quad |\alpha| < 1, \quad \theta = 0, \tag{1.5}$$

were suggested in [20] where it is formally shown that if the monodromy data satisfy (1.4) or (1.5) then the solution of PII for real and large t has no poles. It is also noted in [20] that the case (1.4) corresponds to y being purely imaginary.

The first results about the evaluation of the connection formulae for the Painlevé transcendents appeared in [21]-[27], and have been summarized in monograph [20]; by now such asymptotic results have been obtained for most of the Painlevé equations [28]-[40]. Furthermore, for PII, Boutroux type asymptotics (where the asymptotics are expressed in terms of elliptic functions) have been parameterized via the monodromy data [41], [42]. Rigorous aspects of some of these results remain open; in particular for some of these results it is necessary to assume an apriori information about the solutions. However, we expect that these difficulties will be soon overcome. Our expectation is based on two facts: (a) There exists now a rigorous methodology for studying the Cauchy problem of the Painlevé transcendents [19], [18]. (b) A rigorous methodology has been developed for studying certain asymptotic questions without assuming an apriori information [43]. The combination of (a) and of a proper generalization of (b) should provide a general methodology for making the asymptotic results of the Painlevé equations rigorous (for certain special cases rigorous results already exist [23], [26], [27], [35], [40], [44]).

A renewed interest in Painlevé equations has recently occurred because of their appearance in the so called matrix model of two dimensional quantum gravity [11]-[13]. The understanding of certain physical questions has led to the mathematical understanding of new type of asymptotic limit, namely to the limit from a "discrete Painlevé equation" to a continuous one [55], [57].

In this article we review some of these developments: In §2 we use PII to illustrate the rigorous methodology developed in [18]. In §3 we use PII to illustrate the efficiency of the asymptotic analysis of Painlevé equations developed in [20,25-40] on the basis of the isomonodromy method; we give exhaustive results for purely imaginary and purely real solutions. In §4 we review the occurrence of continuous and discrete Painlevé equations in the 2D quantum gravity and indicate how the isomonodromy method can be effective for obtaining physically significant results.

We conclude this introduction with some remarks:

1. A historical perspective on Painlevé equations can be found in [14] and [17].

2. The Painlevé equations possess particular solutions which are either rational or can be expressed in terms of certain classical transcendental functions [45]. Such solutions can be obtained by an appropriate use of Schlessinger transformations. A systematic investigation of these transformations is given in [46].

3. Special cases of Painlevé equations can be investigated rigorously by linear integral equations of the Gel'fand-Levitan-Marchenko type [23], [47], [48].

4. It is possible to obtain certain asymptotic results about the Painlevé equations without the isomonodromy method [50].

5. Painlevé equations appear as similarity reductions of nonlinear PDE's solvable by the inverse spectral method [49].

6. Action-angle variables for PII are discussed in [51].

7. An approach for the rigorous justification of the asymptotic results obtained via the isomonodromy method is proposed in [58].

8. Certain results concerning particular solutions for the Painlevé equations can be obtained by means of conventional methods of the analytical theory of differential equations (see the reviews [59], [60]).

2 The Cauchy Problem for PII

The PII equation (1.1) can be written as the compatibility condition of the following system of equations:

$$Y_z = AY, \tag{2.1a}$$

$$Y_t = BY, \tag{2.1b}$$

where

$$A = -i(4z^2 + t + 2y^2)\sigma_3 + (4zy - \frac{\theta}{z})\sigma_1 - 2y_t\sigma_2, \quad B = -iz\sigma_3 + y\sigma_1, \tag{2.2}$$

the Pauli matrices $\sigma_j, j = 1,2,3$ are defined by

$$\sigma_1 = \begin{pmatrix} 0 & 1 \\ 1 & 0 \end{pmatrix}, \quad \sigma_2 = \begin{pmatrix} 0 & -i \\ i & 0 \end{pmatrix}, \quad \sigma_3 = \begin{pmatrix} 1 & 0 \\ 0 & -1 \end{pmatrix}, \tag{2.3}$$

and $Y(z,t)$ is a 2×2 matrix valued function in $\boldsymbol{C} \times \boldsymbol{C}$.

The method of [19] consists of investigating the following:

1. The direct problem: This involves characterizing a sectionally meromorphic solution $Y(z,t)$ of equation (2.1a). This solution has certain jumps across certain contours in the complex z-plane. These jumps, called monodromy data, are determined in terms of Stokes matrices (denoted by G) and connection matrices (denoted by E). By using appropriate symmetry and consistency conditions it is possible to eliminate the connection matrices and to establish the monodromy data as a two-dimensional algebraic variety.

2. The inverse problem: This includes reconstructing $Y(z,t)$ in terms of the monodromy data. This yields to a novel RH problem. For example, in connection with PII one needs to reconstruct a sectionally holomorphic function $m(z;t)$ with the following properties: (a) $m(z;t)$ has certain jumps on the contour defined by $Re(\frac{4iz^3}{3}+itz)=0$. (b) $m(z;t)$ tends to the identity for large z off the contour, but it oscillates on the contour. (c) $m(z;t)\sim \hat{m}_0(z;t)z^{\theta\sigma_3}$ as $z\to 0$, where $\hat{m}_0$ is analytic at $z=0$, $\sigma_3=diag(1,-1)$ and θ is a constant parameter. To study a RH problem of this type we first study a RH problem which is formulated on a new contour, obtained from the original one by: (a) inserting a circle around the origin; (b) performing a small clockwise rotation. The new RH problem is analytic both at the origin and at infinity and hence can be studied by the method of [54]; in particular is equivalent to a certain Fredholm integral equation. Having established the solution of the new RH problem it is straightforward to establish the solution of the original one. The associated Fredholm integral equation depends analytically on t, and since it is solvable at $t=0$ (which follows from the direct problem), it has solutions meromorphic in t [54].

3. Proving that the inverse problem solves the direct problem.

4. Proving a vanishing lemma, i.e. finding certain constraints on the monodromy data, such that if t is on Stokes lines, the homogeneous RH problem has only the zero solution.

2.1 The Direct Problem

The essence of the direct problem is to establish the analytic structure of Y with respect to z, in the entire complex z-plane. It should be pointed out that, in contrast to the analogous problem in the inverse scattering transform this task here is straightforward: Equation (2.1a) is a linear ODE in z, therefore its analytic structure is completely determined by its singular points. Equation (2.1a) has a regular singular point at the origin (if $\theta\neq 0$) and an irregular singular point at infinity.

(i) Analysis near $z=0$. It is well known that if the coefficient matrix of a linear ODE has a regular singular point at $z=0$, then the solution in the

neighborhood of $z = 0$ can be obtained via a convergent power series. In this particular case

$$Y_0(z) = \hat{Y}_0(z) z^{\theta\sigma_3}, \quad \text{as} \quad z \to 0, \quad \theta \neq \frac{2n+1}{2}, n \epsilon \mathbf{Z}, \tag{2.4a}$$

where $\hat{Y}_0(z)$ is holomorphic at $z = 0$. The dominant behavior near $z = 0$ is characterized by $Y_{0_z} \sim -\theta\sigma_1 Y_0/z$; thus $\hat{Y}_{0_z}(z) \sim -\frac{\theta}{z}(\sigma_1 \hat{Y}_0(z) + \hat{Y}_0(z)\sigma_3)$ and

$$\sigma_1 \hat{Y}_0(0) + \hat{Y}_0(0)\sigma_3 = 0. \tag{2.4b}$$

(ii) Analysis near $z = \infty$. The solution of equation (2.1a), for large z, possesses a formal expansion of the form $Y \sim \tilde{Y}$, $\tilde{Y} = \hat{Y}_\infty exp\left[-i(\frac{4z^3}{3} + tz)\sigma_3\right]$, where $\hat{Y}_\infty$ is a formal power series. However, because $z = \infty$ is an irregular singular point, the actual asymptotic behavior of Y changes form in certain sectors of the complex z-plane. These sectors are determined by $Re[i\frac{4}{3}z^3 + itz] = 0$; thus for large z the boundaries of the sectors, Σ_j, are asymptotic to the rays $argz = \frac{j\pi}{3}$, $0 \leq j \leq 7$. Let S_j be the sector containing the boundaries Σ_j, i.e. if $z \epsilon S_1, 0 \leq argz < \frac{\pi}{3}$, etc. Then, according to the Stokes phenomenon, if $Y \sim \tilde{Y}$ as $z \to \infty$ in S_1, $Y \sim \tilde{Y} G_1 G_2 \cdots G_j$, as $z \to \infty$ in S_{j+1}, $1 \leq j \leq 6$. The matrices G_j, $1 \leq j \leq 6$ are triangular and are called Stokes multipliers. Alternatively, for the formulation of the RH problem it is more convenient to introduce different solutions Y_j, $1 \leq j \leq 7$ such that Y_j is asymptotic to $\tilde{Y}$ in S_j. Then $Y_{j+1} = Y_j G_j$, $1 \leq j \leq 6$; also it can be shown [52] that $Y_1(z) = Y_7(ze^{2i\pi})$. Thus the nonsingular matrices Y_j satisfy:

$$Y_j(z) \sim \hat{Y}_\infty(z) e^{Q(z)}, \quad \text{as} \quad z \to \infty, z \epsilon S_j, Q = -i(\frac{4}{3}z^3 + tz)\sigma_3, \tag{2.5}$$

where $\hat{Y}_\infty(z)$ is piecewise holomorphic (relative to the contours of Figure 2.1) at $z = \infty$, with asymptotic expansion of the form $\hat{Y}_\infty(z) \sim I + 0(1/z)$ as $z \to \infty$. They are related by

$$Y_{j+1}(z) = Y_j(z) G_j, 1 \leq j \leq 5; \;\; Y_1(z) = Y_6(ze^{2i\pi}) G_6, \tag{2.6}$$

where the Stokes multipliers are given by

$$G_1 = \begin{pmatrix} 1 & 0 \\ s_1 & 1 \end{pmatrix}, \; G_2 = \begin{pmatrix} 1 & s_2 \\ 0 & 1 \end{pmatrix}, \; G_3 = \begin{pmatrix} 1 & 0 \\ s_3 & 1 \end{pmatrix}$$
$$G_4 = \begin{pmatrix} 1 & \hat{s}_1 \\ 0 & 1 \end{pmatrix}, \; G_5 = \begin{pmatrix} 1 & 0 \\ \hat{s}_2 & 1 \end{pmatrix}, \; G_6 = \begin{pmatrix} 1 & \hat{s}_3 \\ 0 & 1 \end{pmatrix}. \tag{2.7}$$

We note that although $Y_1 \sim \tilde{Y}$ as $z \to \infty$ on Σ_1, $Y_1 \sim \tilde{Y} G_1^{-1}$ as $z \to \infty$ on Σ_2, similarly $Y_2 \sim \tilde{Y} G_2^{-1}$ as $z \to \infty$ on Σ_3, etc.

(iii) Connection between Y_0 and Y_1. Since both Y_0 and Y_1 satisfy (2.1a) they are related by a constant matrix,

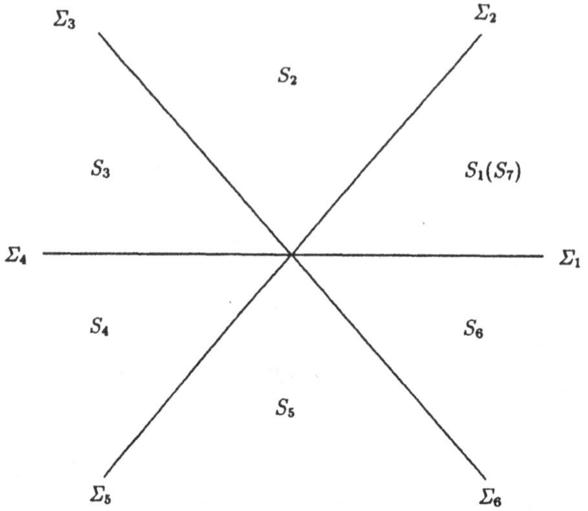

Figure 2.1

$$Y_1 = Y_0 E_0, \quad E_0 = \begin{pmatrix} \alpha & \beta \\ \gamma & \delta \end{pmatrix}, \quad det E_0 = 1. \tag{2.8}$$

The condition on the determinant follows from the fact that we have normalized Y_j, Y_0 to have unit determinant.

(iv) Symmetry Relations. For a complex matrix function f we denote

$$f^{\sigma_1}(ze^{i\pi}) = \sigma_1 f(z) \sigma_1. \tag{2.9}$$

Equations (2.2) imply that the A and B of equations (2.1) satisfy

$$A^{\sigma_1} = -A, \quad B^{\sigma_1} = B. \tag{2.10}$$

Hence

$$Y_z^{\sigma_1} = A^{\sigma_1} Y^{\sigma_1}, \quad Y_t^{\sigma_1} = B^{\sigma_1} Y^{\sigma_1}. \tag{2.11}$$

Equations (2.11) and the fact that Y, Y^{σ_1} have the same asymptotics at $z = \infty$ imply $Y = Y^{\sigma_1}$,

$$Y_{j+3}(ze^{i\pi}) = \sigma_1 Y_j(z) \sigma_1, \quad j = 1, 2, 3. \tag{2.12}$$

Equations (2.12) imply

$$G_{j+3} = \sigma_1 G_j \sigma_1, \quad j = 1, 2, 3, \quad \text{i.e. } \hat{s_1} = s_1, \quad \hat{s_2} = s_2, \quad \hat{s_3} = s_3. \tag{2.13}$$

Furthermore equations (2.12) also imply (for an appropriate choice of $Y_0(z)$) the important relationship

$$-E_0^{-1}\sigma_3 e^{-i\pi\theta\sigma_3}E_0 = G_1G_2G_3\sigma_1. \tag{2.14}$$

Indeed, the equations

$$Y_4(z) = Y_1(z)G_1G_2G_3 \text{ and } Y_4(ze^{i\pi}) = \sigma_1 Y_1(z)\sigma_1,$$

yield,

$$\sigma_1 Y_1(z)\sigma_1 = Y_1(ze^{i\pi})G_1G_2G_3.$$

Using $Y_1 = Y_0E_0$, and $Y_0 = \hat{Y}_0 z^{\theta\sigma_3}$, this equation becomes

$$z^{-\theta\sigma_3}\hat{Y}_0^{-1}(-z)\sigma_1\hat{Y}_0(z)z^{\theta\sigma_3} = e^{i\pi\theta\sigma_3}E_0G_1G_2G_3\sigma_1E_0^{-1}. \tag{2.15}$$

Let $F \doteqdot e^{i\pi\theta\sigma_3}E_0G_1G_2G_3\sigma_1E_0^{-1}$; using equation (2.4b), it follows from the limit of (2.15) as $z \to 0$ that $(F)_{Diag} = -\sigma_3$. By equation (2.4a), $\theta \notin \frac{1}{2} + \boldsymbol{Z}$. If $\theta = 0$, or if $\theta \notin \boldsymbol{Z}$, then $F = F_{Diag} = -\sigma_3$, and equation (2.14) follows. If $\theta \epsilon \boldsymbol{Z}/\{0\}$, then F in general is upper (lower) triangular for $Re\theta > 0$ ($Re\theta < 0$); however, it follows from equations (2.4a) and (2.8) that it is possible to choose a $Y_0(z)$ such that the corresponding F is also diagonal.

We note that, since $\sigma_1G_1G_2G_3\sigma_1 = G_4G_5G_6$, the square of equation (2.14) yields

$$E_0^{-1}e^{2i\pi\theta\sigma_3}E_0\Pi_{j=1}^{6}G_j = I. \tag{2.16}$$

This equation is a consistency condition (see [14]).The trace of equation (2.14) implies

$$s_1 + s_2 + s_3 + s_1s_2s_3 = -2i\sin\theta\pi. \tag{2.17}$$

(v) The Monodromy Data. In previous investigations the components of the connection matrix E_0 were taken as part of the monodromy data. This is not necessary since E_0 can be determined from (s_1, s_2, s_3). Here we call monodromy data the set (s_1, s_2, s_3) defined on the algebraic variety (2.17). Since $\det G_1G_2G_3\sigma_1 = -1$ and $trG_1G_2G_3\sigma_1 = -2i\sin\pi\theta$, it follows that $G_1G_2G_3\sigma_1$ has eigenvalues $-e^{i\pi\theta}$ and $e^{-i\pi\theta}$. Also $\theta \notin \frac{1}{2} + \boldsymbol{Z}$, thus these two eigenvalues are unequal and $G_1G_2G_3\sigma_1$ is diagonalized to $-\sigma_3e^{i\pi\theta\sigma_3}$. Therefore there exists a matrix E_0, with $\det E_0 = 1$, such that equation (2.14) is valid. For the inverse problem we will define a RH problem for the matrices Y_j defined in equations (2.5) and (2.6), where Y_1 satisfies $Y_1 = \hat{Y}_0(z)z^{i\theta\sigma_3}E_0$, $\hat{Y}_0(z)$ is analytic at $z = 0$, and E_0 is any matrix obtained from equation (2.14) ($\det E_0 = 1$). We note that the equation for Y_1 is well defined. Indeed, if $\hat{E}_0$ is another solution of (2.14), then $E_0\hat{E}_0^{-1}$ is diagonal and Y_1 satisfies a similar equation with $\hat{Y}_0(z)$ replaced by $\hat{Y}_0(z)E_0\hat{E}_0^{-1}$. Therefore, the different choices of E_0 do not affect $Y_1, \cdots, Y_6$.

If y evolves according to PII, then the monodromy data (s_1, s_2, s_3) are time independent.

2.2 The Inverse Problem

For PII we define $\Phi_j, j = 0, 1, \cdots, 6$ by

$$Y_j \doteqdot \Phi_j e^Q, \quad j = 1, \cdots, 6, \quad Y_0 \doteqdot \Phi_0 e^Q z^{\theta\sigma_3}, \quad Q \doteqdot -i(\frac{4z^3}{3} + tz)\sigma_3.$$

Then we obtain the RH

$$\Phi_+ = \Phi_- e^Q V e^{-Q} \tag{2.18}$$

where the jumps are given by

$$\Phi_2 = \Phi_1 e^Q G_1 e^{-Q}, \quad \Phi_2 = \Phi_3 e^Q G_2^{-1} e^{-Q}, \quad \Phi_4 = \Phi_3 e^Q G_3 e^{-Q},$$

$$\Phi_4 = \Phi_5 e^Q G_4^{-1} e^{-Q}, \quad \Phi_6 = \Phi_5 e^Q G_5 e^{-Q}, \quad \Phi_6 = \Phi_1 e^Q G_6^{-1} e^{-Q}.$$

$$\Phi_0 = \Phi_1 e^Q f_1^{-1} e^{-Q}, \quad \Phi_2 = \Phi_0 e^Q f_2 e^{-Q}, \quad \Phi_0 = \Phi_3 e^Q f_3^{-1} e^{-Q},$$

$$\Phi_4 = \Phi_0 e^Q f_4 e^{-Q}, \quad \Phi_0 = \Phi_5 e^Q f_5^{-1} e^{-Q}, \quad \Phi_6 = \Phi_0 e^Q f_6 e^{-Q},$$

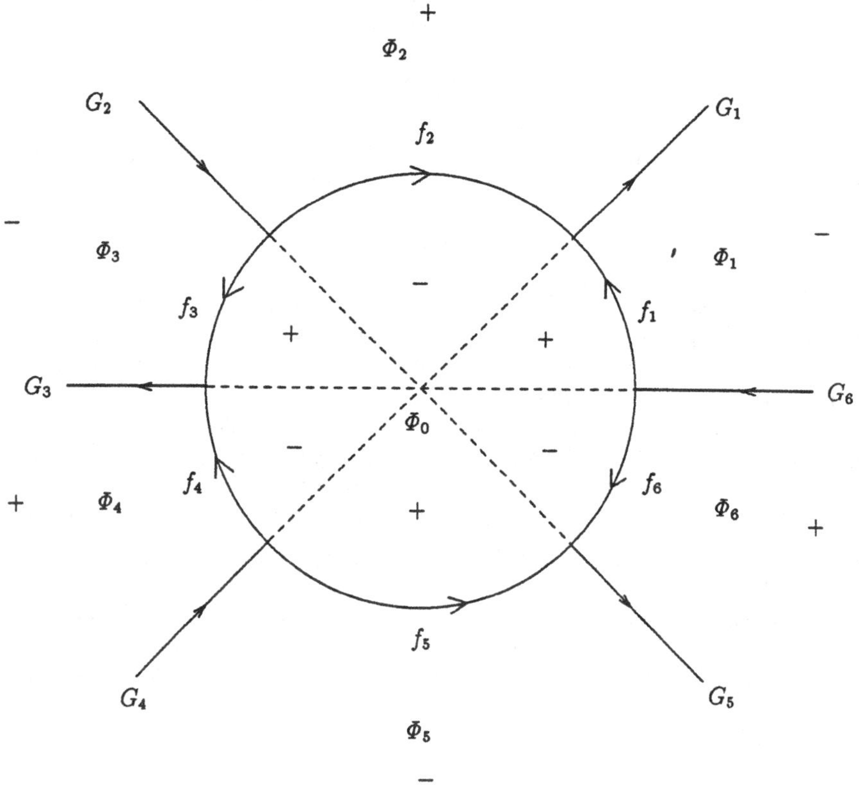

Figure 2.2

where

$$f_j = z^{\theta\sigma_3} E_0 G_1 \cdots G_{j-1}, \quad j = 1, \cdots, 6, \quad G_0 \doteq I.$$

The function V is smooth (in fact analytic) away from the intersections. At these intersections the product of the jump matrices equals the identity. Roughly speaking, these products represent the "smoothness" of V at the intersection points. Furthermore, it is proven [19] that the solvability of this RH problem is equivalent to the solvability of a "rotated" RH problem. Using this fact and the global smoothness of V it is then possible to use the results of [54] and establish the equivalence of this RH problem to a Fredholm integral equations.

2.3 The Inverse Solves the Direct

Recall that $Y = \Phi e^Q z^{\theta\sigma_3}$ inside the circle and $Y = \Phi e^Q$ outside the circle. Introducing the notation

$$f^{\sigma_1}(e^{i\pi} z) = \sigma_1 f(z) \sigma_1,$$

the equations $G_{j+3} = \sigma_1 G_j \sigma_1$, $j = 1, 2, 3$ can be written in the compact form $G = G^{\sigma_1}$. This implies

$$\Phi = \Phi^{\sigma_1}, \quad z \text{ outside the circle.} \tag{2.19}$$

The symmetry condition (3.19) plays a crucial role in establishing that the solution of the inverse problem also solves the direct problem. Let us first consider equation (2.2): Since Y and Y_t have the same jumps, $Y_t Y^{-1}$ is holomorphic and tends to $O(z)$ for large z, thus it follows that

$$\Phi_t - iz\Phi\sigma_3 = (B_0 + zB_1)\Phi. \tag{2.20}$$

The large z asymptotics of (2.20) yields

$$B_1 = -i\sigma_3, \quad B_0 = i[\sigma_3, \Phi_{-1}]. \tag{2.21}$$

However, we need to show that $B_0 = y\sigma_1$ for some y. This can be achieved by using (2.19):

$$\Phi_{-1} = -\Phi_{-1}^{\sigma} \Longrightarrow B_0^{\sigma_1} = i[\sigma_3^{\sigma_1}, \Phi_{-1}^{\sigma_1}] = i[\sigma_3, \Phi_{-1}] = B_0,$$

which implies

$$B_0 = i[\sigma_3, \Phi_{-1}] = y\sigma_1, \quad \text{for some } y. \tag{2.22}$$

We now consider equation (2.1).

$$\Phi_z - i(4z^2 + t)\Phi\sigma_3 + \theta z^{-1}\Phi\sigma_3 = (z^{-1}A_{-1} + A_0 + zA_1 + z^2 A_2)\Phi, \tag{2.23}$$

where z is inside the circle. Hence

$$A_{-1} = \theta\Phi(0,t)\sigma_3\Phi^{-1}(0,t). \tag{2.24}$$

However, we need to show that $A_{-1} = -\theta\sigma_1$. This can be achieved by using the symmetry relation (2.14). This relation implies (for z inside the circle),

$$\Phi^\sigma = -i\Phi e^Q \sigma_2 e^{-Q} \Longrightarrow \Phi^\sigma(0,t) = -i\Phi(0,t)\sigma_2. \tag{2.25}$$

Equations (2.24), (2.25) yield

$$A_{-1} = -\theta\sigma_1.$$

To determine A_0, A_1, A_2 we use

$$\Phi_z - i(4z^2+t)\Phi\sigma_3 = (z^{-1}A_{-1} + A_0 + zA_1 + z^2A_2)\Phi, \tag{2.26}$$

for z outside the circle. The large z asymptotics of (2.26) yields

$$A_2 = -4i\sigma_3, A_1 = 4i[\sigma_3, \Phi_{-1}] = 4y\sigma_1, A_0 = -it\sigma_3 - 4y\sigma_1\Phi_{-1} + 4i[\sigma_3, \Phi_{-2}]. \tag{2.27}$$

Equation (2.20) implies

$$\Phi_{-1_t} = -i[\sigma_3, \Phi_{-2}] + y\sigma_1\Phi_{-1}. \tag{2.28}$$

Using equations (2.28) and (2.22) we find

$$(\Phi_{-1})_{\text{off-diag}} = \frac{y}{2}\sigma_2, \quad (\Phi_{-1_t})_{\text{diag}} = \frac{iy^2}{2}\sigma_2. \tag{2.29}$$

Using the relations (2.29) in (2.27c) we obtain

$$A_0 = -it\sigma_3 - 4\Phi_{-1_t} = -it\sigma_3 - 2y_t\sigma_2 - 2iy^2\sigma_3. \tag{2.30}$$

2.4 A Vanishing Lemma

To prove the solvability of the RH problem is equivalent to proving that the homogeneous RH has only the vanishing solution.

We denote by $f^\dagger(z) \doteqdot f^*(\bar{z})$ the Schwarz reflection of a matrix function f. Consider the RH problem $\varphi^+ = \varphi^- V$ on the contour Σ containing the real axis. Let $V\Lambda\boldsymbol{R}$ and $V\Lambda\Sigma\backslash\boldsymbol{R}$ denote V on the real axis and on the rest of Σ respectively. It is shown in [18] that if V satisfies

$$V\Lambda\Sigma\backslash\boldsymbol{R} \text{ is Schwarz reflection invariant} \tag{2.31a}$$

and

$$ReV\backslash\boldsymbol{R} > 0, \tag{2.31b}$$

then every vanishing solution to the above RH problem must be zero. This result is obtained as follows: Let $H \doteqdot \varphi\varphi^\dagger$, where φ is a vanishing solution. Then equation (2.31a) implies that H is holomorphic on $\boldsymbol{C}\backslash\boldsymbol{R}$, and on $\boldsymbol{R}$

$$H^+ = \varphi^+(\varphi^-)^\dagger = \varphi^- V(\varphi^-)^*.$$

Since $H = 0(\frac{1}{z^2})$ as $z \to \infty$,

$$\int_{\mathbf{R}} H^+ = \int_{\mathbf{R}} \varphi^- V(\varphi^-)^* = 0.$$

Then equation (2.31b) implies that $\varphi^- \equiv 0$.

Using the above result it is possible to obtain vanishing lemmas for all Painlevé equations. The only difference here is that the relevant solutions may possess singularities at $z = 0$ and $z = \infty$. We therefore also impose appropriate constants on θ, to ensure that such singularity for H^+ are integrable.

It is interesting that a direct application of the above result to the original RH problems for Painlevé equations fails. However, it is possible to reduce the number of contours, and then to apply directly vanishing lemmas. It is also possible to obtain vanishing lemmas for the original RH problems provided one uses $H = \varphi h \varphi^\dagger$, instead of $H = \varphi\varphi^\dagger$, where h is an appropriately chosen piecewise constant matrix.

For PII we find: Assume that the matrix function Y, holomorphic on $\mathbf{C}/\Sigma$, has the following properties:

(i) Possesses the jumps $G_j, 1 \le j \le 6$ on Σ given by (2.6).

(ii) Behaves near $z = 0$ according to equation (2.4) with $|Re\theta| < \frac{1}{2}$.

(iii) The analytic continuation of Y behaves near $z = \infty$ according to (2.5) ($\hat{Y}_\infty \sim 0(\frac{1}{z})$).

Then $Y(t) \equiv 0$,

$$\text{for } t\epsilon\mathbf{R} \text{ if } s_2 = \bar{s}_2, \quad |s_1 - \bar{s}_3| < 2, \tag{2.32a}$$

$$\text{for } t\epsilon e^{\frac{2\pi i}{3}}\mathbf{R} \text{ if } s_3 = \bar{s}_3, \quad |s_2 - \bar{s}_1| < 2, \tag{2.32b}$$

$$\text{for } t\epsilon e^{\frac{-2\pi i}{3}}\mathbf{R} \text{ if } s_1 = \bar{s}_1, \quad |s_3 - \bar{s}_2| < 2. \tag{2.32c}$$

Proof.

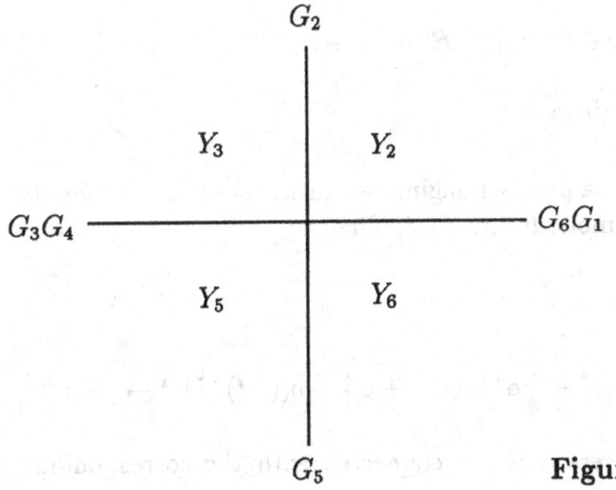

Figure 2.3

As it is illustrated in Figure 2.3, the conditions (2.31) yield

$$G_2 = G_5^*, \quad ReG_3G_4 > 0, \quad ReG_6G_1 > 0. \tag{2.33}$$

Equation $G_2 = G_5^*$ implies $s_2 = \bar{s}_2$, while equations (2.33b) and (2.33c) imply $|s_1 - \bar{s}_3| < 2$. The conditions (ii) and (iii) in the lemma guarantee that the singularities of H^* at $z = 0, \infty$ are integrable. Hence the Vanishing Lemma follows.

The cases (2.32b) and (2.32c) are obtained in a similar way after performing suitable rotations of the $\boldsymbol{C}$-plane.

Combining the results of 2.2-2.3 we have:

Theorem: The Cauchy problem for PII admits global in t meromorphic solution. This solution can be found by solving the RH problem $\Phi_+ = \Phi_- e^Q V e^{-Q}$ (see Figure 2.2), which is uniquely specified in terms of the MD (s_1, s_2, s_3) defined on the algebraic variety V_{II} given by equation (2.17). (E_0 is any solution of equation (2.14).) Having obtained Φ, y can be found from

$$y = 2(\Phi_{-1})_{12}; \quad \Phi = I + \frac{\Phi_{-1}}{z} + O(\frac{1}{z^2}), \quad z \to \infty.$$

For each $t_0 \epsilon \boldsymbol{C}$, there exists an analytic variety V_{II}^0, $dim V_{II}^0 < 2$, such that the monodromy transformation is a bijection, $\boldsymbol{C}^2 \to V_{II}/V_{II}^0$, $(y(t_0), y'(t_0))$ $(s_1, s_2, s_3) \epsilon V_{II}/V_{II}^0$, and $(s_1, s_2, s_3) \epsilon V_{II}^0$ iff the inverse monodromic transformation does not exist at t_0.

3 Connection Formulae for PII

We consider for simplicity the case that the constant θ in PII equation (1.1) is zero. Then the consistence condition implies $s_2 = -(s_1 + s_3)/(1 + s_1 s_3)$. Thus the quantities $s_1 = s_1(t, y, y_t)$ and $s_3 = s_3(t, y, y_t)$ can be taken as the independent monodromy data. It can be easily seen that if y is real or imaginary then

$$y \epsilon \boldsymbol{R} : s_3 = -\bar{s}_1; y \epsilon i \boldsymbol{R} : s_3 = \bar{s}_1,$$

where bar denotes complex conjugate.

Proposition 3.1 [35] Let $y(t)$ be a purely imaginary solution of (1.1) with $\theta = 0$, and let $s = s_1$ be the relevant monodromy data. Then

1. For any s:

$$y(t) = i(-t)^{-1/4} \alpha \sin\left\{\frac{2}{3}(-t)^{\frac{3}{2}} + \frac{3}{4}\alpha^2 \ln(-t) + \varphi\right\} + o((-t)^{-\frac{1}{4}}), t \to -\infty$$

where the asymptotic parameters α, φ are connected with the corresponding monodromy parameter s through the explicit formulae

$$\alpha^2 = \frac{1}{\pi}\ln(1+|s|^2),$$

$$\varphi = \frac{3}{2}\alpha^2\ln 2 + \frac{\pi}{4} - arg\Gamma(\frac{i\alpha^2}{2}) - args.$$

2. For $Res \neq 0$

$$y(t) = \sigma i\sqrt{\frac{t}{2}} + \sigma i(2t)^{-1/4}\rho\cos\left\{\frac{2\sqrt{2}}{3}t^{3/2} - \frac{3}{2}\rho^2\ln t - \psi\right\} +$$

$$+o(t^{-1/4}),\quad t\to-\infty$$

where

$$\rho^2 = \tfrac{1}{\pi}\ln\tfrac{1+|s|^2}{2|Res|},$$
$$\psi = -\tfrac{\pi}{4} + \tfrac{7}{2}\rho^2\ln 2 - arg\Gamma(i\rho^2) - arg(1-s^2)$$
$$\sigma = signRes.$$

3. For $Res = 0$

$$y(t) = \frac{s}{2\sqrt{\pi}}t^{-1/4}e^{-\frac{2}{3}t^{3/2}}(1+o(1)),\quad t\to+\infty.$$

Proposition 3.2 Let $y(t)$ be a real solution of (1.1) with $\theta = 0$, and let $s = s_1$ be the relevant monodromy data. Then

1. $|s| < 1, Ims = 0$ [22], [47], [53]

$$y(t) = \alpha(-t)^{-1/4}\cos\{\frac{2}{3}(-t)^{3/2} - \frac{3}{4}\alpha^2\ln(-t) + \varphi\} + o(t^{-1/4}),\quad t\to-\infty$$

where

$$\alpha^2 = -\frac{1}{\pi}\log(1-|s|^2),$$

$$\varphi = -\frac{3}{2}\alpha^2\ln 2 - \frac{\pi}{4} - \arg\Gamma(-i\alpha^2/2) + \frac{\pi}{2}sign\ \ s$$

and

$$y(t) = \frac{s}{2\sqrt{\pi}}t^{-1/4}e^{-\frac{2}{3}t^{3/2}}(1+o(1)),\quad t\to+\infty.$$

2. $|s| < 1, Ims \neq 0$ [29], [20]

$$y(t) = \alpha(-t)^{-1/4}\cos\left\{\frac{2}{3}(-t)^{3/2} - \frac{3}{4}\alpha^2\ln(-t) + \varphi\right\} + o(t^{-1/4}),\quad t\to-\infty$$

where

$$\alpha > 0, \quad \alpha^2 = -\frac{1}{\pi}\log(1-|s|^2),$$

$$\varphi = -\frac{3}{2}\alpha^2 \ln 2 + \frac{\pi}{4} - arg\Gamma(-i\alpha^2/2) - args.$$

$y(t)$ has infinitely many poles t_n^+ in the neighborhood of $+\infty$ such that

$$(t_n^+)^{3/2} = \frac{3}{\sqrt{2}}\pi n - \beta \log n + \psi + o(1), \quad n \to \infty, \qquad (*)$$

where

$$\beta = \frac{3}{2\sqrt{2}}\frac{1}{\pi}\log\left|\frac{2Im\ s}{1-|s|^2}\right|,$$

$$\psi = -\beta \log 24\pi - \frac{3}{2\sqrt{2}} arg\Gamma\left(\frac{1}{2} - i\frac{2\sqrt{2}}{3}\beta\right) - \frac{3}{2\sqrt{2}} arg\frac{1-s^2}{1-|s|^2}.$$

3. $|s| > 1$, $Im\ s = 0$ [29], [20].

$y(t)$ has infinitely many poles t_n^- in the neighborhood of $-\infty$ such that

$$(-t_n^-)^{3/2} = 3\pi n - \alpha \log n + \varphi + o(1), \quad n \to \infty$$

$$\alpha = \frac{3}{4\pi}\log(|s|^2 - 1), \qquad (**)$$

$$\varphi = -\alpha \log 24\pi + \frac{3}{2} arg\Gamma\left(\frac{1}{2} + i\frac{2\alpha}{3}\right) - \frac{3\pi}{4}(1 - (-1)^{n+n_0}) + \frac{3}{2} args - \frac{3\pi}{4}$$

and

$$y(t) = \frac{s}{2\sqrt{\pi}} t^{-1/4} e^{\frac{-2}{3}t^{3/2}}(1 + o(1)), \quad t \to +\infty.$$

4. $|s| > 1$, $Im\ s \neq 0$ [29], [33].

$y(t)$ has infinitely many poles t_n^+, t_n^- in the neighborhood of both $+\infty$ and $-\infty$. The sequences t_n^+ and t_n^- are described by the formulae (*) and (**) respectively.

5. $s = \pm 1, s_2 = 0$ [23].

$$y(t) = \pm\left[(-\frac{t}{2})^{1/2} - \frac{1}{2^{7/2}}(-t)^{-5/2} + O((-t)^{-11/2})\right], \quad t \to -\infty$$

and

$$y(t) = \pm\frac{1}{2\sqrt{\pi}} t^{-1/4} e^{-\frac{2}{3}t^{3/2}}(1 + o(1)), \quad t \to +\infty.$$

6. $s = \pm 1, s_2 \neq 0$ [33].

$$y(t) = \pm \left[(-\frac{t}{2})^{1/2} - \frac{1}{2^{7/2}}(-t)^{-5/2} + O((-t)^{-11/2}) \right]$$

$$\pm \frac{is_2}{2\sqrt{2\pi}}(-t)^{-1/4} \exp\left\{ -\frac{2\sqrt{2}}{3}(-t)^{3/2} \right\} (1 + o(1)), \quad t \to -\infty$$

and $y(t)$ has infinitely many pdes t_n^+ in the neighborhood of $+\infty$ such that

$$(t_n^+)^{3/2} = \frac{3}{\sqrt{2}}\pi n - \beta \log n + \psi + o(1), \quad n \to \infty$$

$$\beta = \frac{3}{2\sqrt{2}} \frac{1}{\pi} \log |s_2|$$

$$\psi = -\beta \log 24\pi - \frac{3}{2\sqrt{2}} arg \Gamma(\frac{1}{2} - i\frac{2\sqrt{2}}{3}\beta) - \frac{3}{2\sqrt{2}} arctg s_2.$$

Remarks

1. In the real case, the manifold M of the monodromy data can be represented as

$$M = \{(s_1, s_3), s_3 = -\bar{s}_1, \quad |s_1| \neq 1\} \cup \{s_3 = -s_1 = \pm 1, s_2 \epsilon i\boldsymbol{R}\}.$$

Therefore, the cases 1-6 actually cover all possible types of the asymptotic behavior for the real solutions of (1.1), $\theta = 0$.

2. By eliminating the monodromy data from the above equalities one can obtain explicit connection formulae between the asymptotic parameters (α, φ) and (β, θ).

3. The general complex case with $\theta \neq 0$ has been studied in [37], where the exhaustive information about the behavior of the second Painlevé transcendents on the rays $\arg t = \frac{\pi k}{3}$ has been obtained.

We now indicate how some of these results can be obtained. Assume that

$$y = O((-t)^{-1/4}), \quad y_t = O((-t)^{1/4}), \quad t \to -\infty \tag{3.1}$$

Putting $\lambda = (-t)^{-1/2} z$ one can rewrite the system (2.2), $\theta = 0$ in the form

$$\frac{dY}{d\lambda} = \tau A_o(\lambda, \tau) Y, \quad \tau = (-t)^{3/2}, \tag{3.2}$$

where

$$A_0 = -(4i\lambda^2 - i - 2iy^2 t^{-1})\sigma_3 + 4y\lambda(-t)^{-1/2}\sigma_1 + \frac{2y_t}{t}\sigma_2.$$

In (3.2) τ is a large parameter and for the matrix $A_0(\lambda,\tau)$, equation (3.1) implies the estimate

$$A_0(\lambda,\tau) = O(1), \quad \tau \to +\infty,$$

where λ is fixed. That means, that equation (3.2) can be investigated by the WKB-method.

Diagonalizing the matrix A_0 we find

$$T^{-1}A_0T = -\mu\sigma_3; \mu = \sqrt{-detA_0},$$

$$T = \frac{1}{8i(\lambda^2-\frac{1}{4})}\left[(4i\lambda^2 - i - 2i\frac{y^2}{t} + \mu)\sigma_0 + \frac{4i\lambda y}{(-t)^{1/2}}\sigma_2 - \frac{2iy_t}{t}\sigma_1\right], \quad \sigma_0 = I. \tag{3.3}$$

The turning points λ_T are given by

$$\mu(\lambda_T) = 4i\sqrt{(\lambda^2-\frac{1}{4})^2 + O(\tau^{-1})} = 0,$$

i.e. there exist two turning points of second order,

$$\lambda_\pm = \pm\frac{1}{2}.$$

The WKB-solutions of (3.2) can be represented by

$$Y^{WKB} \cong Texp\{-\tau\sigma_3\int_{\lambda_\pm}^{\lambda}\mu(\eta)d\eta\},$$

or

$$Y^{WKB} \cong \left[I + O\left(\frac{1}{\sqrt{\tau}(\lambda^2-\frac{1}{4})}\right)\right] exp\{-\tau\sigma_3\int_{\lambda_\pm}^{\lambda}\mu(\eta)d\eta\}. \tag{3.4}$$

The Stokes lines of the WKB solution are given by $Re\int_{\lambda_\pm}^{\lambda}\mu(\eta)d\eta = 0$. Therefore WKB solutions about λ_+ and λ_- can be used for the evaluation of the canonical solution Y_1, Y_2, Y_6 and Y_3, Y_4, Y_5 respectively (see Figure 3.1). Hence,

$$Y_k(\lambda) = Y_k^{WKB}(\lambda)e^{\sigma_3\delta_\infty}, \quad k = 1,2,6.$$

$$\delta_\infty \doteqdot \lim_{\lambda\to\infty}\left[\tau\int_{\lambda_+}^{\lambda}\mu(\eta)d\eta - \frac{4i}{3}\tau\lambda^3 + \tau\lambda\right]. \tag{3.5}$$

In the neighborhood of the turning point λ_+, the WKB asymptotic (3.5) breaks down. In this neighborhood we introduce the new variable

$$\xi = 2\sqrt{2\tau}(\lambda - \frac{1}{2})e^{-i\pi/4}. \tag{3.6}$$

Assuming $\xi = O(1)$ we will put in equation (3.2) the Taylor series expansion of the matrix $A_0(\lambda,\tau)$ near $\lambda_+ = \frac{1}{2}$ and we cancel terms quadratic in $(\lambda - \frac{1}{2})$ or $\tau^{-1/2}$. As a result we find the equation

$$\frac{dY}{d\xi} = \begin{pmatrix} \frac{\xi}{2} & v \\ w & -\frac{\xi}{2} \end{pmatrix} Y, \tag{3.7}$$

where

$$\begin{aligned} v &= \sqrt{2}(-t)^{\frac{3}{4}} e^{\frac{-i\pi}{4}} \left(\tfrac{i}{2}(-t)^{-\frac{1}{2}} y + \tfrac{y_t}{2t}\right), \\ w &= \sqrt{2}(-t)^{\frac{3}{4}} e^{\frac{-i\pi}{4}} \left(\tfrac{i}{2}(-t)^{-\frac{1}{2}} y - \tfrac{y_t}{2t}\right). \end{aligned} \tag{3.8}$$

Equation (3.7) can be solved in terms of parabolic cylinder functions $D_\nu(\xi)$ with

$$\nu = -vw - 1 = O(1). \tag{3.9}$$

Thus we get the following representation for the solutions of equation (3.2) in the neighborhood of the turning point t_+ (for the technical details see [20]):

$$\begin{gathered} \overset{TP}{Y}_k(\lambda) = (I + o(1)) \, \overset{o}{Y}_k(\xi), \quad k = 1,2,6 \\ \tau \to \infty, \quad arg(\lambda - \tfrac{1}{2}) = \begin{cases} -\frac{\pi}{2}, & k = 6 \\ 0, & k = 1 \\ \frac{\pi}{2}, & k = 2 \end{cases}, \\ |\lambda - \tfrac{1}{2}| \le \tau^{-\frac{1}{2}+\varepsilon}, \quad 0 < \varepsilon < \tfrac{1}{6}. \end{gathered} \tag{3.10}$$

In (3.10), $\overset{o}{Y}_k$ are the canonical solutions of the parabolic cylinder equation

$$y_{\xi\xi} + (\nu + \frac{1}{2} - \frac{\xi^2}{4})y = 0,$$

written in the matrix form and fixed by the asymptotics

$$\overset{o}{Y}_k(\xi) = \left(I + O(\xi^{-1})\right) exp\left\{\sigma_3 \left(\frac{\xi^2}{4} - (1+\nu)\ln\xi\right)\right\}, \quad \xi \to \infty, \tag{3.11}$$

where

$$arg\xi = -\frac{3\pi}{4}, \quad k = 6; \quad -\frac{\pi}{4}, \quad k = 1; \quad \frac{\pi}{4}, \quad k = 2.$$

In the matching domain $(\lambda - 1/2) \approx \tau^{-1/2+\varepsilon}$, we can match the WKB-solutions with the turning-point solutions:

$$\overset{TP}{Y}_k(\lambda) = Y^{WKB}(\lambda) e^{\sigma_3 \delta_+} (1 + o(1))$$

$$\delta_+ = \tau \int_{\lambda_+}^{\lambda} \mu(\eta) d\eta + \frac{\xi^2}{4} - (1+\nu)\ln\xi. \tag{3.12}$$ [1]

---

[1] Note that nonvanishing terms of δ_+ as $\tau \to \infty$ do not depend on λ (see formula (3.15)).

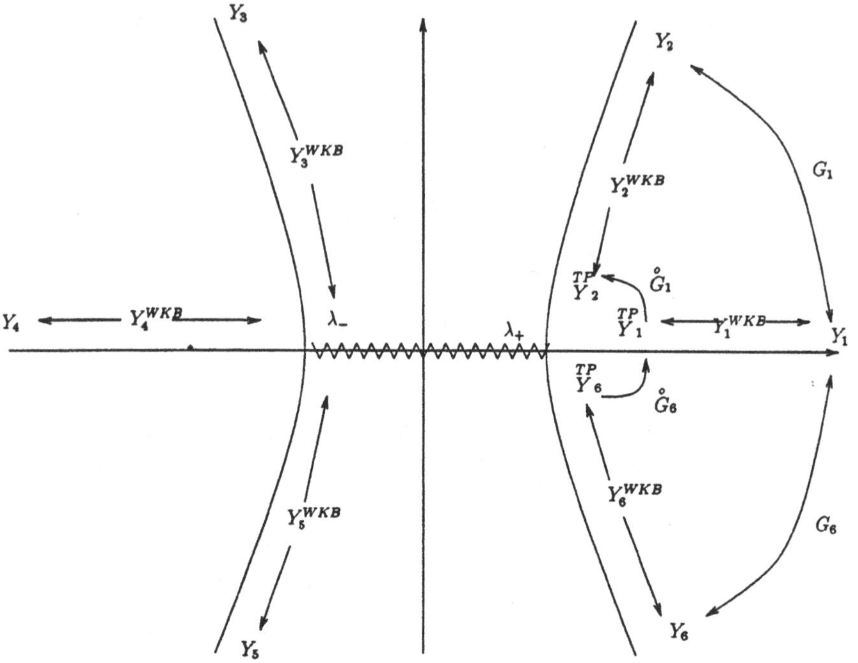

Figure 3.1

Using this formula and equalities (3.5), (3.10), we obtain for the Stokes matrices G_k, $k = 1,6$ the following asymptotic representations (cp. Figure 3.1):

$$G_k \cong e^{\gamma\sigma_3} \overset{o}{G}_k e^{-\gamma\sigma_3}, \quad \tau \to \infty, \quad k = 1, 6; \gamma = \delta_+ - \delta_\infty, \tag{3.13}$$

where $\overset{o}{G}_k \doteqdot \overset{o}{Y}{}_k^{-1} \overset{o}{Y}_{k+1}$ $(= \overset{TP}{Y}{}_k^{-1} \overset{TP}{Y}_{k+1})$ are the Stokes matrices of the parabolic cylinder equation (3.7), which can be described by the classical formulae [61]:

$$\overset{o}{G}_1 = \begin{pmatrix} 1 & 0 \\ g_1 & 1 \end{pmatrix}, \quad \overset{o}{G}_6 = \begin{pmatrix} 1 & \hat{g}_3 \\ 0 & 1 \end{pmatrix},$$

$$g_1 = \frac{i\sqrt{2\pi}}{v\Gamma(1+\nu)}, \quad \hat{g}_3 = \frac{-\sqrt{2\pi}}{w\Gamma(-1-\nu)} e^{-i(1+\nu)\pi}. \tag{3.14}$$

The principal point of the above scheme is the possibility of evaluating the nonvanishing terms in δ_+ and δ_∞ as $\tau \to \infty$ in explicit form, so for γ we obtain Gathering together the formulae (3.13), (3.14), (3.15) and taking into account the equality $\hat{s}_3 = s_3$ we have the explicit asymptotic representations for the functions s_1 and s_3:

$$
\begin{aligned}
s_1 &= \frac{i\sqrt{2\pi}}{v\Gamma(1+\nu)} exp\left\{\frac{2i}{3}\tau + (1+\nu)\ln 8\tau - i(1+\nu)\frac{\pi}{2}\right\} + o(1), \\
s_3 &= -\frac{\sqrt{2\pi}}{w\Gamma(-1-\nu)} exp\left\{-\frac{2i}{3}\tau - (1+\nu)\ln 8\tau - i(1+\nu)\frac{\pi}{2}\right\} + o(1), \quad \tau \to \infty,
\end{aligned}
\tag{3.16}
$$

where quantities v, w and ν are defined in (3.8) and (3.9).

The equalities (3.16) give the asymptotic solution of the direct monodromy problem for the equation (2.2) under the asymptotic assumption (3.1). Multiplying term by term the equalities in (3.16) we find

$$
-i(1+\nu) = ivw = \frac{1}{2\pi}\ln(1 + s_1 s_3) + o(1). \tag{3.17}
$$

That means that estimates (3.1) are matched with the isomonodromic conditions

$$
s_1 = const, \quad s_3 = const
$$

only if

$$
-s_1 s_3 < 1.
$$

In particular, this inequality is satisfied for any pure imaginary solution of equation (1.1). Solving equation (3.16) with respect to $y = -\bar{y}$ we obtain the asymptotics given in Proposition 3.1.

4 Painlevé Equations and the Matrix Model

A renewed interest in Painlevé equations has recently occurred because of their appearance in the so called matrix model of two dimensional quantum gravity [11]-[13]. The understanding of certain physical questions has led to the mathematical understanding of a new type of asymptotic limit, namely to the limit from a "discrete Painlevé equation" to a continuous one [55], [57].

Let us consider the orthogonal polynomials $Q_n(\lambda)$ with respect to the measure $\exp(-U(\lambda))d\lambda$:

$$
\delta_{nm} = \int_{-\infty}^{\infty} Q_n(\lambda) Q_m(\lambda) e^{-U(\lambda)} d\lambda, \quad U(\lambda) = \sum_{j=1}^{N} t_j \lambda^{2j}, \quad t_N > 0. \tag{4.1}
$$

It is well known that $Q_n(\lambda)$ satisfy the linear recurrence relations

$$
(L \cdot Q)_n \equiv \sum L_{nm} Q_m = \lambda Q_n, \quad L_{nm} = \frac{1}{2} w_m^{\frac{1}{2}} \delta_{n+1,m} + \frac{1}{2} w_n^{\frac{1}{2}} \delta_{n-1,m}. \tag{4.2}
$$

The coefficients w_n are functions of t_j, $w_n \equiv w_n(t_1, ..., t_N)$, and satisfy [11], [12], [62] the nonlinear recurrence relation

$$
n = \frac{1}{2} w_n^{\frac{1}{2}} (U'(L))_{n,n-1}, \quad n > 0, \tag{4.3}
$$

supplemented by the initial conditions

$$w_0 = 0, w_1 = \frac{4\int_{-\infty}^{\infty}\lambda^2 e^{-U(\lambda)}d\lambda}{\int_{-\infty}^{\infty}e^{-U(\lambda)}d\lambda}, \quad w_j = \frac{4h_j}{h_{j-1}}, j = 1, ..., N-1, \tag{4.4}$$

where $1/\sqrt{h_j}$ is the coefficient of the term λ^j in the expression of Q_j. It has also been shown [63] that $w_n(t_1, ..., t_N)$, as function of both n and $\{t_j\}$, satisfy the hierarchy of the Volterra equations:

$$\partial_{t_j} \ln w_n = L^{2j}_{n-1,n-1} - L^{2j}_{n,n}. \tag{4.5}$$

Let us denote as $\tau_n(t_1, ..., t_N)$ the corresponding τ-function. During the last two years this τ-function has been the object of investigation of a large number of papers. The reason is the works [11], [12] where the following two facts were discovered: (a) Let $t_j = \beta q_j, q_1 = \frac{1}{2}$ and consider the following double-scaling limit of equation (4.3):

$$\beta = C_1 h^{-5}, \quad \frac{n}{\beta} = C_2 + C_1^{-1} h^4 x, \quad h \to 0. \tag{4.6}$$

It turns out that it is possible to choose $C_2 = C_2(q_2, ..., q_N)$ in such a way that $\tau_n \to Z(x)$ where $Z(x)$ is the partition function for the two-dimensional quantum gravity. The variable x plays the role of the renormalized string coupling.

(b) Assume that under the limit (4.6) the following ansatz is valid

$$w_n \cong \rho(1 - 2h^2 u(x)). \tag{4.7}$$

Then, it is possible to choose $\rho = \rho(q_2, ..., q_N)$ in such a way that the ansatz (4.7) maps the equations (4.3) into the PI equation for the function $u(x)$:

$$u_{xx} = 6u^2 + x \tag{4.8}$$

(note, that $Z_{xx} = u$). Furthermore, the limiting equation (4.8) is the same for all of generic values of $q_2, ..., q_N$.

In connection with these exciting results the following analytical problems arise:

(a) To give the rigorous confirmation of the ansatz (4.7). (b) To calculate the parameters of the limiting solution $u(x)$.

It turns out that using the asymptotic analysis of the isomonodromy method one can obtain complete answers to both these questions: Because of space limitations we only give the main results. The details can be found in [55]-[57]. Regarding the first question, the answer is negative. Actually, this has already been known indirectly form [64]. Nevertheless, if one replaces in equation (4.4)

$$\int_{-\infty}^{\infty} d\lambda \text{ by } s_1 \int d\lambda - s_2 \int d\lambda,$$

where the integrals are along the lines $\arg\lambda = \pm\frac{\pi}{N}$, then the answer becomes positive. Let's suppose that such modification has been accomplished. Then, the limiting solution $u(x)$ of PI equation can be uniquely characterized by the following asymptotic behavior on the ray $\arg x = \pi - \frac{2\pi}{5}$:

$$u(x) = e^{-\frac{i\pi}{5}}\sqrt{\frac{|x|}{6}} + \gamma_0 |x|^{-\frac{1}{8}} e^{-\frac{8i}{5}(\frac{3}{2})^{1/4}|x|^{5/4}} + o(|x|)^{-\frac{1}{8}}$$

$$\gamma_0 = \frac{i}{\sqrt{8\pi}} e^{\frac{\pi i}{20}} (\frac{2}{3})^{\frac{1}{8}} \frac{p}{1+p}, \quad p = -\frac{s_2}{s_1}, \quad p \neq -1.$$

In [11] it was shown that $u(x)$ belongs to the one-parameter family of solutions of PI characterized by the following asymptotic behavior as $x \to -\infty$:

$$u(x) = \sqrt{\frac{-x}{6}} + \sum_{l=1}^{\infty} C_l (-x)^{\frac{1}{2}-\frac{5l}{2}} + \gamma(-x)^{-\frac{1}{8}} e^{-\frac{8}{5}(\frac{3}{2})^{1/4}(-x)^{5/4}} (1 + o(1)).$$

One of the main results of [55]-[57] is the calculation of the parameter γ,

$$\gamma = -\frac{i}{2} \frac{1}{\sqrt{8\pi}} (\frac{2}{3})^{\frac{1}{8}} \frac{1-p}{1+p}.$$

This parameter describes the nonperturbative effect. To obtain these results we made essential use of the asymptotic analysis of the PI equaiton developed in [36].

The isomonodromy approach to the general string equations that come from the matrix model in the continuous limit was introduced in [65], [66]. The nonperturbative effect for $m = 3$ has been calculated in [67]. An interesting approach to the string equation using algebraic geometric methods is used in [66], [68], [69].

Acknowledgements

This work was partially supported by the National Science Foundation under Grant Number DMS-8803471 and by the Air Force Office of Scientific Research under Grant Number 87-0310.

References

[1] E. Barouch, B.M. Mcoy and T.T. Wu, Phys. Rev. Lett. **31** 1409 (1973); T.T. Wu, B.M. McCoy, C.A. Tracy and E. Barouch, Phys. Rev. B **13** 316 (1976).

[2] M. Jimbo, T. Miwa, Y. Mori and M. Sato, Physica **1D** 80 (1980); M. Jimbo and T. Miwa, Proc. Jpn. Acad. **56A** 405 (1980).

[3] J.W. Miles, Proc. R. Soc. London A **361** 277 (1978).

[4] B.M. McCoy, J.H.H. Perk and R. Schrock, Nucl. Phys. B (1983); Correlation functions of the transverse Ising chain at the critical field for large temporal and spatial separations.
[5] B.M. McCoy, C.A. Tracy and T.T. Wu, J. Math. Phys. **18** 1058 (1977).
[6] B.M. McCoy and T.T. Wu, Nucl. Phys. **B180** (FS2) 89 (1981).
[7] D.B. Creamer, H.B. Thacker and D. Wilkinson, Phys. Rev. D **23** 3081 (1981).
[8] J.H.H. Perk, H.W. Capel, G.R.W. Quispel and F.W. Nijhoff, Finite-temperature correlations for the Ising chain in a transvere field, Physica **123A** 1 (1984).
[9] V.I. Gromak and V.V. Tsegelnik, Teor. Mat. Fiz. **55** 189-196 (1983).
[10] V.E. Zakharov, E.A. Kuznetsov and S.L. Musher, IETP Lett. **41** 125-127 (1985).
[11] D.J. Gross and A.A. Migdal, Nonperturbative 2D Quantum Gravity, preprint, Princeton University (1989).
[12] M. Douglas and S. Shenker, Strings in Less than One Dimension, Rutgers preprint RU-89-34.
[13] E. Brezin, E. Marinari, and G. Parisi, A Non-Perturbative Ambiguity Free Solution of a String Model, preprint (1990).
[14] H. Flaschka and A.C. Newell, Commun. Math. Phys. **76** 67 (1980).
[15] K. Ueno, Proc. Jpn. Acad. **56A** 97, 103 (1980; M. Jimbo, T. Miwa and K. Ueno, Physica **2D** 306 (1981); M. Jimbo and T. Miwa, Physica **2D** 407 (1981); **4D** 47 (1981); M. Jimbo, Prog. Theor. Phys. **61** 359 (1979).
[16] A.S. Fokas and M.J. Ablowitz, Comm. Math. Phys. **91** 381 (1983).
[17] A.S. Fokas, U. Mugan and M.J. Ablowitz, Physica **30D** 247 (1988).
[18] A.S. Fokas and X. Zhou, On the Solvability of Painleve II and IV, to appear in Comm. Math. Phys.
[19] A.S. Fokas, U. Mugan, and X. Zhou, On the Solvability of PI, III, and V, Clarkson University, preprint INS #180 (1991).
[20] A.R. Its and V.Yu Novokshenov, The Isomonodromic Deformation Method in the Theory of Painleve Equations, Lect. Notes in Math., A. Dodd and B. Eckamm ed., Springer Verlag, 1191 (1986).
[21] B.M. McCoy, C.A.Tracy, and T.T. Wu, J. Math. Phys. **18**, no. 5, 1058 (1977).
[22] M.J. Ablowitz, and H. Segur, Stud. Appl. Math. **57**, 13 (1977).
[23] S.P. Hastings, and J.B. McLeod, Arch. rational Mech. Anal. **73**, 31 (1980).
[24] M. Jimbo, Publ. RIMS Kyoto University, **18**, 1137 (1982).
[25] A.R. Its, and V.E. Petrov, Sov. Math. Dokl. **26**, no. 1, 244 (1982).
[26] V.Yu Novokshenov, Funct. Anal. Appl. **18** (1984).
[27] V.Yu Novokshenov, and B.I. Suleimanov, Usphekhi Mat. Nauk **39**, no. 4, 114 (Russian) (1984).
[28] A.V. Kitaev, Teor. Math. Phys. **64**, no. 3, 878 (1985).
[29] A.A. Kapaev, and V.Yu Novokshenov, Sov. Phys. Dokl. **31**, 719 (1986).
[30] B.M. McCoy, and Sh. Tang, Physica D **19**, 42 (1986).
[31] B. M. McCoy, and Sh. Tang, Physica D **20**, 187 (1986).
[32] V.Yu Novokshenov, On the Asymptotics of the General Real-Valued Solution to the Third Painleve Equation, Dokl. Akad. Nauk SSSR, **283**, 5, 1161 (1985).
[33] A.A. Kapaev, *Singular Solutions of the Painleve II Equations*, Lect. Notes in Math, Springer-Verlag, **1191**, 261 (1986).
[34] B.I. Suleimanov, *On Asymptotics of Regular Solutions for a Special Kind of Painleve V Equation*, Lect. Notes in Math., Springer Verlag, **1191**, 230 (1986).
[35] A.R. Its, and A.A. Kapaev, Math. USSR Izvestiya, **31**, no. 1, 193 (1988).
[36] A.A. Kapaev, Diff. Eq. **24**, 1107 (1988).

[37] A.A. Kapaev, Theoret. Math. Phys. **77**, 323 (1988).
[38] A.V. Kitaev, *The Isomonodromic Deformation Method for "Degenerate" Third Painleve Equation*, in: "Problems in Quantum Field Theory and Statistical Physics. **7**, Zap. Nauch. Semin. LOMI, **161**, P.P. Kulish and V.N. Popov, ed., Nauka, Leningrad, 45. (Russian) (1987).
[39] A.V. Kitaev, A.V., *Asymptotic Description of the Fourth Painleve Equation Solutions on the Stokes Rays Analogies*, in: "Problems in Quantum Field Theory and Statistical Physics **8**, Zap. Nauch. Semin. LOMI, **169**, P.P. Kulish and V.N. Popov, ed., Nauka, Leningrad, 84 (1988). (Russian)
[40] A.V. Kitaev, Mat. Sb., **134**, 421; 1989 English transl. in Math. USSR Sbornik, **62**,421 (1987).
[41] V.Yu Novokshenov, Dokl. Akad. Nauk SSSR, **311**, 288; 1990 English transl. in Sov. Math. Dokl (1990).
[42] A.A. Kapaev, *Irregular Singularity of the Painleve Function of the Second Type and Nonlinear Stokes Phenomenon*, in: Zap. Nauch. Semin. LOMI, Nauka, Leningrad, to appear (1990). (Russian).
[43] P. Deift and X. Zhou, A Steepest Descent Method for Oscillatory Riemann-Hilbert Problems, to appear in Annals of Math.
[44] A.R. Its, and V.Yu Novokshenov, Funt. Anal. and Appl. **22**, 190 (1988).
[45] A.S. Fokas and M.J. Ablowitz, J. Math. Phys. **23** 2033 (1982).
[46] U. Mugan and A.S. Fokas, Schlessinger Transformations for the First Five Painleve Transcendents, preprint (1991).
[47] P.A. Clarkson, and J.B. McLeod, Arch. Rational Mech. Anal. **103**, no. 2, 97 (1988).
[48] P.A. Clarkson, and J.B. McLeod, *Connection Formulae for the Second Painleve Transcendent*, Proc. of the Conf. Differential Equations and Mathematical Physics", U. of Birmingham, Alabama, USA (1986).
[49] M.J. Albowitz, A. Ramani, and H. Segur, Lett. Nuovo Cimento **23** 333 (1978); J. Math. Phys. **21** 716, 1006 (1980).
[50] N. Hoshi, M.D. Kruskal, Phys. Lett. A **130**, no. 3, 129 (1988).
[51] H. Flaschka and A.C. Newell, The Inverse Monodromy Transform is a Canonical Transformation, "Math. Studies", **61**, 65-91 (1982).
[52] G.D. Birkhoff, Trans. AMS **436** (1909); **199** (1910).
[53] B.I. Suleimanov, The Connection Between the Asymptotics at the Different Infinites for the Solutions of the Second Painleve Equaiton, Dif. Urav. **23**, 5, 834-842 (1987) (Russian).
[54] X. Zhou, SIAM J. Math. Anal., **20**, No. 4, 966-986 (1989).
[55] A.S. Fokas, A.R. Its, and A.V. Kitaev, Discrete Painleve Equations and Their Appearance in Quantum Gravity, to appear in Comm. Math. Phys. (1991).
[56] A.V. Kitaev, Calculation of Nonperturbation Parameter in Matrix Model Φ^4, Zap. Nauch. Semin. LOMI, **187**, 12, 31-40 (1991).
[57] A.S. Fokas, A.R. Its and A.V. Kitaev, The Isomonodromy Approach to 2D Quantum Gravity, preprint (1991).
[58] A.V. Kitaev, The Justification of the Asymptotic Formulae Obtained by the Isomonodromic Deformation Method, in: "Mathematical Questions of the Wave Propagation Theory" 19, Zap. Nauch. Semin. LOMI, 179, (Russian) (1989).
[59] N.P. Erugin, The Theory of the Movable Singularities of the Second Order Equations, Diff. Uraun. **12** 579 (1976).

[60] L.A. Bordag, Painleve Equaitons and Their Connection with Nonlinear Evolution Equations, preprint JINR, E5-80-477 (1980).
[61] Erdelyi, et al, *Higher Transcendental Functions*, **2**, McGraw-Hill (1953).
[62] C. Itzykson and J.B. Zuber, The Planar Approximation II, J. Math. Phys. **21** (3) (1980).
[63] M. Kac and P. von Moerbeke, Advances in Math. **16**, 160-164 (1975).
[64] F. David, Loop Equations and Non-Perturbative Effects in Two-Dimensional Quantum Gravity, MPLA **5**, 1019-1029 (1990).
[65] G. Moore, Geometry of the String Equations, Commun. Math. Phys. **133**, 261-304 (1990).
[66] G. Moore, Matrix Models of 2D Gravity and Isomonodromic Deformation, YCTP-P17-90, RU-90-53.
[67] A.A. Kapaev, Weak Nonlinear Solutions of the P_1^2 Equation, Zap Nauch. Semin. LOMI, **187**, Differential Geometry, Li Groups, and Mechanics, **12**, p. 88 (1991).
[68] S.P. Novikov, Quantization of Finite-Gap Potentials and Nonlinear Quasiclassical Approximation in Nonperturbative String Theory, Funkt. Analiz i Ego Prilozh., **24**, 4, 43-53 (1990).
[69] I.M. Krichever, On Heisenberg Relations for the Ordinary Linear Differential Operators, ETH preprint (1990).

The Cauchy Problem for Doubly Periodic Solutions of KP-II Equation

I. Krichever

Landau Institute for Theoretical Physics,
Russian Academy of Science, Moscow, Russia

1 Introduction

Since the middle of seventies algebraic geometry has become a very powerful tool in various problems of mathematical and theoretical physics. In the theory of integrable equations the algebraic-geometrical methods provide a construction of the periodic and quasi-periodic solutions which can be written exactly in terms of the theta-functions of auxiliary Riemann surfaces.

Roughly speaking, the algebraic-geometrical construction gives a map from a set of algebraic-geometrical data to a space of solutions of integrable non-linear partial differential equations

$$\{algebraic - geometrical\ data\} \longmapsto \{solutions\ of\ NLPDE\}$$

In a generic case the space of algebraic-geometrical data is a union for all g of the spaces

$$\tilde{M}_{g,N} = \{\Gamma_g, P_\alpha, k_\alpha^{-1}(Q), \gamma_1, \ldots, \gamma_g\},\ \alpha = 1, \ldots, N,$$

where Γ_g is an algebraic curve of genus g with fixed local coordinates $k_\alpha^{-1}(Q)$, $k_\alpha^{-1}(P_\alpha) = 0$, in neighbourhoods of N punctures P_α, and $\gamma_1, \ldots, \gamma_g$ are points of Γ_g in general position. (It is to be mentioned that $\tilde{M}_{g,N}$ are "universal" data. For the given non-linear integrable equation the corresponding subset of data has to be specified.)

All the integrable equations which are considered in the soliton theory can be represented as compatibility conditions of auxiliary linear problems. For two-dimensional integrable systems one of the general types of such representations has the form

$$[\partial_y - L, \partial_t - A] = 0, \tag{1.1}$$

where L and A are differential operators

$$L = \sum_{i=0}^{n} u_i(x,y,t)\partial_x^i\ ,\ A = \sum_{i=0}^{m} v_i(x,y,t)\partial_x^i \tag{1.2}$$

with scalar or matrix coefficients.

The most important example of these equations is the Kadomtsev-Petviashvili (KP) equation

$$\frac{3}{4}\sigma^2 u_{yy} + (u_t - \frac{3}{2}uu_x + \frac{1}{4}u_{xxx})_x = 0, \tag{1.3}$$

which is equivalent to (1.1), where

$$L = \sigma^{-1}(-\partial_x^2 + u(x,y,t)) \ , \ A = -\partial_x^3 + \frac{3}{2}u\partial_x - w(x,y,t). \tag{1.4}$$

The two real forms of this equation are called KP-1 (for $\sigma^2 = -1$) and KP-2 (for $\sigma^2 = 1$).

The algebraic-geometrical construction of the solutions of integrable equations is based on the concept of the Baker-Akhiezer functions which are defined by their very specific analytical properties on auxiliary Riemann surfaces. For example, the Baker-Akhiezer function in the case of the KP equation is defined for each set of data $\tilde{M}_{g,1}$ as a function $\psi(x,y,t,Q), Q \in \Gamma$ such that:

(i) the function ψ (as a function of the variable Q which is a point of Γ) is meromorphic everywhere except for the point P_1 and it has at most simple poles at the points $\gamma_1, \ldots, \gamma_g$ (if all of them are distinct).

(ii) in the neighbourhood of the point P_1 the function ψ has the form

$$\psi(x,y,t,Q) = e^{ikx+\sigma^{-1}k^2y+ik^3t}(1 + \sum_{s=1}^{\infty} \xi_s(x,y,t)k^{-s}) \ , \ k = k(Q). \tag{1.5}$$

These properties uniquely define $\psi(x,y,t,Q)$

Let us give a brief sketch of the proof that from the uniqueness of the Baker-Akhiezer function it follows that ψ is a common solution of the equations

$$(\partial_y - L)\psi(x,y,t,Q) = 0, \tag{1.6}$$

$$(\partial_t - A)\psi(x,y,t,Q) = 0, \tag{1.7}$$

where L and A have the form (1.4). The coefficients of L, A equal

$$u = 2i\xi_{1,x} \ , \ w = -\frac{3}{2}u_x + \frac{3}{2}iu\xi_1 + 3\xi_{2,x}. \tag{1.8}$$

For any formal series of the form (1.5) from (1.8) it follows that

$$(\partial_y - L)\psi = O(k^{-1})e^{ikx+\sigma^{-1}k^2y+ik^3t}, \tag{1.9}$$

$$(\partial_t - A)\psi = O(k^{-1})e^{ikx+\sigma^{-1}k^2y+ik^3t}. \tag{1.10}$$

It turns out, that if (1.5) is not a formal series but the expansion of the Baker-Akhiezer function the congruences (1.9,1.10) imply (1.5,1.6). Indeed, consider the left hand sides of (1.9,1.10). They define functions on Γ with the same analytical properties outside P_1 as ψ and have the form (1.9,1.10) near this point. Hence, they are equal to zero (because ψ is defined uniquely).

Corollary . The operators L, A of the form (1.4) with the coefficients (1.8), satisfy the operator equation (1.1). Therefore, $u(x,y,t)$ is a solution of the KP equation.

The Baker-Akhiezer function $\psi(x,y,t)$ can be exactly written in terms of Abelian differentials and Riemann theta-functions. From the corresponding formulae it follows [1] that the algebraic-geometrical solutions of the KP equation have the form

$$u(x,y,t) = 2\partial_x^2 \ln \theta(Ux + Vy + Wt + \Phi|B) + const. \tag{1.11}$$

Here $\theta(z_1, \dots, z_g)$ is the Riemann theta-function which is defined by the matrix $B = B(\Gamma)$ of the b-periods of the normalized holomorphic differentials on Γ. The vectors $2\pi iU$, $2\pi iV$, $2\pi iW$ are the vectors of b-periods of the normalized Abelian differentials of the second kind with the only poles at P_1 of orders 2,3,4, respectively. The vector Φ corresponds to the set of points γ_j and can be considered in (1.11) as an arbitrary vector.

This construction was proposed in [1,2] and was developed in different ways for various types of integrable equations (see, for example, the reviews [3-6]). It is by definition a sort of "inverse" transform: from "spectral" (algebraic-geometrical) data to "solutions". A direct transform: from "solutions" to "spectral" data is an unavoidable step for the solution of the Cauchy problem for doubly periodic initial conditions for (2+1)-integrable equations. At the same time only a "direct" transform can give an answer to the following question: "how many solutions can be obtained with the help of the algebraic-geometrical construction ?".

The answer on this question in lower dimensions is as follows. For finite dimensional integrable systems (0+1-systems) a typical Lax representation has the form:

$$\partial_t L(t,\lambda) = [M(t,\lambda), L(t,\lambda)], \tag{1.12}$$

where L and M are finite dimensional matrices depending on the "spectral parameter" λ as rational or algebraic functions. In that case *all* solutions of (1.12) can be expressed in terms of the Riemann theta-functions.

For one-dimensional evolution equations (1+1-systems) the algebraic-geometrical solutions are *dense* in the space of periodic solutions of Lax equations (though it is not proved rigorously in all generic cases).

The appearance of Reimann surfaces in the periodic theory of the Lax equations looks nowadays more or less obvious (although in the classical Floque spectral theory of linear periodic ordinary differential opeartors it did not play any essential role). The Lax equations are equations which have the representation of the form

$$L_t = [A, L], \tag{1.13}$$

where L, A are operators of the form (1.2).

For any complex number E the space $\mathcal{L}(E)$ of solutions of the equation

$$Ly = Ey \tag{1.14}$$

is finite dimensional. The monodromy operator

$$\hat{T} : y(x) \longmapsto y(x+T), \tag{1.15}$$

where T is a period of L preserves $\mathcal{L}(E)$. Hence, it defines a finite dimensional operator

$$\hat{T}(E) : \mathcal{L}(E) \longmapsto \mathcal{L}(E). \tag{1.16}$$

The solutions of (1.14) which are eigenfunctions of the monodromy operator (1.15) :

$$L\psi_i(x,E) = E\psi_i(x,E), \ \psi_i(x+T,E) = w_i\psi_i(x,E) \tag{1.17}$$

are called the Bloch solutions. The eigenvalues $w_i = w_i(E)$ of the monodromy operator are the roots of the characteristic equation

$$\det(w1 - \mathcal{L}(E)) = R(w,E) = 0. \tag{1.18}$$

The Bloch solutions which are multivalued functions of the variable E become single-valued function on the Riemann surface Γ which is defined with the help of the characteristic equation (1.18). In a generic case the Riemann surface Γ has infinite genus. The operators for which it has finite genus are called finite-gap (or algebraic-geometrical) operators. The analytical properties of the Baker-Akhiezer functions are the natural generalization of the analytical properties of the Bloch solutions which were found in the cases of the KdV equation, the sine-Gordon equation and some other one-dimensional integrable systems in the works of Novikov, Dubrovin, Matveev and Its (see review [7,8]; independently and a little bit later they were found also in the works of Lax, MacKean, Trubowitz, van Moerbeke).

In the two-dimensional case the situation with the periodic Cauchy problem turns out to be considerably more complicated than in lower dimensions. For example, as it was shown in [9], the periodic problem for the KP-1 equation is not integrable even formally. In this paper we are going to present the results of [10], from which it follows that the periodic problem for the KP-2 equation is integrable and that any smooth periodic solution of this equation can be approximated by the finite-gap (algebraic-geometrical) solutions.

The crucial point in the solution of the Cauchy problem for periodic initial conditions for two-dimensional integrable systems is a construction of the Floque spectral theory for two-dimensional periodic linear differential operators of the form

$$\partial_y - \partial^n + \sum_{i=0}^{n-2} u_i(x,y)\partial^i . \tag{1.19}$$

For the case of the KP equation ($n = 2$) such theory is presented in the next chapters. (It has to be mentioned that in [10] the spectral theory "on one energy level" for periodic two-dimensional Shrödinger operator was constructed. The investigation of the geometry of the global Bloch variety for such operators were completed in[11].) To begin with we shall give in the second chapter the spectral interpretation of the set of data $\tilde{M}_{g,1}$ for periodic algebraic-geometrical solutions of the KP equation. In the third chapter the perturbation theory for formal Bloch solutions for the operator (1.19) ($n = 2$) is presented. It turns out that the series of this perturbation theory converge and define a global Riemann surface of Bloch functions. The corresponding Riemann surface is a "integral" for the KP equation. The poles of the Bloch functions are a sort of "angle" variables. In the fifth chapter the sketch of the proof of the theorem of approximation is given. This complets the main part of this paper but not the main part of the periodic theory for the KP equation.

As usual, any construction of a set of exact solutions for partial differential equation is in some sense a begining of the story. The next step is the construction of their perturbation. It is necessary for a solution of perturbed equations, or for a solution of the same equation but with more compicated initial conditions. The perturbation theory of finite-gap solutions $u_0(x,y,t)$ of the KP-2 equation was developed in [10,12]. It contains the construction of the biortogonal basis for the linearized KP-2 equation

$$\frac{3}{4}v_{yy} + \partial_x(v_t - \frac{3}{2}u_0 v_x - \frac{3}{2}u_{0x} v + \frac{1}{4}v_{xxx}) = 0 \tag{1.20}$$

and its adjoint equation

$$\frac{3}{4}\Phi_{yy} + (-\Phi_t - \frac{3}{2}u_0\Phi_x + \frac{1}{4}\Phi_{xxx})_X = 0. \tag{1.21}$$

Knowing this basis, it is easy to write down an asymptotic solution of the form

$$u(x,y,t) = u_o(x,y,t) + \sum_{i=1}^{\infty} \varepsilon^i u_i(x,y,t) \tag{1.22}$$

both for the KP-2 equation itself and for its perturbations (ε is a small parameter). By analogy with the multiphase non-linear WKB-method (the Whitham method, see [13,14]) in the one-dimensional case, even the requirement of uniform boundeness of the first term of the series (1.22) leads to the fact that the parameters $I_1, \ldots, I_N$ of a finite-gap solution must depend on the "slow" variables $X = \varepsilon x$, $Y = \varepsilon y$, $T = \varepsilon t$. Equations that describe the slow modulation $I_k(X,Y,T)$ are called *Whitham equations*. For the two-dimensional systems they were obtained for the first time in [12], where the construction of their exact solution was also proposed. In the last chapter we present the recent application ([15,16]) of the Whitham theory to the theory of the tological Landau-Ginzburg models and to the theory of the matrix models for the non-perturbative two-dimensional gravity. In partiqular, the definition of the τ-function for the Whitham equations is given.

2 The spectral theory of algebraic-geometrical periodic non-stationary Schrödinger operators

The solutions $\psi(x,y,w_1,w_2)$ of the non-stationary Schrödinger equation

$$(\sigma\partial_y - \partial_x^2 + u(x,y))\psi(x,y,w_1,w_2) = 0 \tag{2.1}$$

with a periodic potential $u(x,y) = u(x+a_1,y) = u(x,y+a_2)$ are called the Bloch solutions, if they are eigenfunctions for the monodromy operators, i.e.

$$\psi(x+a_1,y,w_1,w_2) = w_1\psi(x,y,w_1,w_2), \tag{2.2a}$$

$$\psi(x,y+a_2,w_1,w_2) = w_2\psi(x,y,w_1,w_2). \tag{2.2b}$$

The Bloch functions will always be assumed to be normalized so that $\psi(0,0,w_1,w_2) = 1$. The set of pairs $Q = (w_1,w_2)$, for which there exists such a solution is called the Floque set and will be denoted by Γ. The multivalued functions $p(Q)$ and $E(Q)$ such that

$$w_1 = e^{ipa_1} \ , \ w_2 = e^{iEa_2}$$

are called quasi-momentum and quasi-energy, respectively.

The gauge transform $\psi \to e^{h(y)}\psi$, where $\partial_y h(y)$ is a periodic function, transfer the solutions of (2.1) into solutions of the same equation but with another potential $\tilde{u} = u - \sigma\partial_y h$. Consequently, the spectral sets correspondingto the potentials u and $\tilde{u}$ are isomorphic. Therefore, in what follows we restrict ourselves to the case of periodic potentials such that

$$\int_0^{a_1} u(x,y)dx = 0.$$

For the "free" operator with zero potential $u_0 = 0$, the Floque set is parametrized by the points of the complex plane of the variable k

$$w_1^0 = e^{ika_1} \ , \ w_2^0 = e^{-\sigma^{-1}k^2 a_2} \tag{2.3}$$

and the Bloch solutions have the form

$$\psi(x,y,k) = e^{ikx-\sigma^{-1}k^2 y}. \tag{2.4}$$

In the next chapters it would be shown that if $Re\ \sigma \neq 0$, then the Floque set of the operator (2.1) with a smooth periodic potential is isomorphic to a Riemann surface Γ (which in a generic case has infinite genus).

In this chapter we prove that for periodic algebraic-geometrical operators which were constructed in the previous chapter the auxiliary Riemann surface coincides with the Floque set and the Baker-Akhiezer functions of such operators coincide with the Bloch solutions.

First of all, let us specify the set of algebraic-geometrical data which give periodic potentials $u(x,y)$ of the equation (2.1). For any smooth algebraic curve Γ with a puncture P_1 and a local parameter $k^{-1}(Q)$ near P_1 the differentials $dp(Q)$ and $dE(Q)$ can be defined as meromorphic differentials on Γ which are holomorphic outside P_1 and which have the form

$$dp = dk(1+O(k^{-2}) \ , \ dE = i\sigma^{-1}dk^2(1+O(k^{-3})) \tag{2.5}$$

$$\mathrm{Im} U_\gamma = 0 \ , \ U_\gamma = \frac{1}{2\pi}\oint_\gamma dp, \tag{2.6a}$$

$$\mathrm{Im} V_\gamma = 0 \ , \ V_\gamma = \frac{1}{2\pi} \oint_\gamma dE. \tag{2.6b}$$

If for any cycle γ the corresponding periods U_γ, V_γ have the form

$$U_\gamma = \frac{n_\gamma}{a_1} \ , \ V_\gamma = \frac{m_\gamma}{a_2} \ , \ n_\gamma, m_\gamma \text{ are integers,} \tag{2.7}$$

then the potentials $u(x,y)$ corresponding to such data (Γ, P_1, k^{-1}) have periods a_1, a_2. Indeed, from (2.7) it follows that the functions

$$w_1(Q) = \exp(ia_1 \int^Q dp), \ w_2(Q) = \exp(ia_2 \int^Q dE) \tag{2.8}$$

are well defined on Γ. As a corollary of (2.5) and the uniqueness of the Baker-Akhiezer function we have

$$\psi(x + a_1, y, Q) = w_1(Q)\psi(x,y,Q), \tag{2.9a}$$

$$\psi(x, y + a_2, Q) = w_2(Q)\psi(x,y,Q) \tag{2.9b}$$

(the left and right hand sides of (2.9) have the same analytical properties on Γ). From (2.9) it follows that $u(x,y)$ which is defined with the help of the formula (1.8) is periodic. The equalities (2.9) mean that the evaluation of the Baker-Akhiezer function at any point Q is a Bloch solution of the corresponding equation (2.1). Let us prove that any Bloch solution of (2.1) corresponds to some point of Γ (these two statements imply that

$$Q \in \Gamma \longmapsto (w_1(Q), w_2(Q)) \in C^2 \tag{2.10}$$

is an isomorphism between Γ and the Floque set).

Let us introduce a notion of the dual Baker-Akhiezer function $\psi^+(x,y,t,Q)$. For any given set of g points in general position $\gamma_1, \ldots, \gamma_g$ the dual set of points $\gamma_1^+, \ldots, \gamma_g^+$ is defined as a unique set of points such that the union $\gamma_1, \ldots, \gamma_g$ and $\gamma_1^+, \ldots, \gamma_g^+$ are zeros of a meromorphic differential $d\Omega$ on Γ with the only pole of order 2 at P_1. The dual Baker-Akhiezer function $\psi^+(x,y,t,Q)$ is by definition a function which is meromorphic on Γ outside P_1 and has at most simple poles at the points $\gamma_1^+, \ldots, \gamma_g^+$. In the neighbourhood of P_1 the function ψ^+ has the form

$$\psi^+(x,y,t,Q) = e^{-(ikx + \sigma^{-1}k^2 y + ik^3 t)}(1 + \sum_{s=1}^{\infty} \xi_s^+(x,y,t)k^{-s}) \ , \ k = k(Q). \tag{2.11}$$

Lemma 2.1 . For the coefficients ξ_1 and ξ_1^+ of the expansions (1.8) and (2.11) the following equality holds

$$\xi_1(x,y,t) + \xi_1^+(x,y,t) = 0. \tag{2.12}$$

Proof . As it follows from the definitions of the Baker-Akhiezer function and its dual the differential

$$\hat{d\Omega}(x,y,t,Q)=\psi(x,y,t,Q)\psi^+(x,y,t,Q)d\Omega \tag{2.13}$$

is holomorphic outside P_1 where it has a pole of the second order. Therefore, the residue of $\hat{d\Omega}$ at P_1 equals zero. Since it is equal to the left hand side of (2.12) the lemma is proved.

Corollary . The dual Baker-Akhiezer function satisfies the equation

$$(-\sigma\partial_y-\partial_x^2+u(x,y))\psi^+(x,y,Q)=0\ . \tag{2.14}$$

It can be proved in the same way as for ψ that from the definition of ψ^+ it follows that ψ^+ is a solution of the equation of the form (2.14) with potential $v(x,y)=-2i\xi^+_{1,x}$. From (2.12) it follows that $u=v$ and (2.14) is proved.

Let w be any complex number and let $Q_n=Q_n(w)$ be points on Γ such that

$$w_1(Q)=w\ . \tag{2.15}$$

The following statement was proved in [17] using the simple contour integrals on Γ.

Theorem 2.1 . Any smooth function $f(x)$ such that $f(x+a_1)=wf(x)$ can be expanded in the following series

$$f(x)=\sum_n\frac{<\psi_n^+f>_x}{<\psi_n^+\psi_n>_x}\psi_n(x,y)\ , \tag{2.16}$$

where $\psi_n=\psi_n(x,y)=\psi(x,y,Q_n)$, $\psi_n^+=\psi_n^+(x,y)=\psi^+(x,y,Q_n)$; $<\ldots>_x$ denotes the mean value with respect to x.

Consider any Bloch solution $\psi=\psi(x,y,w,w_2)$ of (2.1). It can be expanded in the series (2.16). From (2.1) and (2.14) we have that

$$\partial_y<\psi_n^+\psi(x,y,w,w_2)>_x=0. \tag{2.17}$$

The equality of $<\psi_n^+\psi>_x$ for y and $y+a_2$ implies

$$<\psi_n^+\psi(x,y,w,w_2)>_x=w_2^{-1}(Q_n)w_2<\psi_n^+\psi(x,y,w,w_2)>_x\ . \tag{2.18}$$

Hence, $<\psi_n^+\psi>_x=0$ if $w_2\neq w_2(Q_n)$. For generic w the values $w_{2n}=w_2(Q_n)$ are different. Therefore, only one term in (2.16) is non-zero. Hence,

$$\psi(x,y,w,w_2)=\psi_n(x,y) \tag{2.19}$$

for a certain n and the pair w,w_2 corresponds to the point Q_n. Everything under consideration is analytical with respect to w. Consequently, (2.19) is true for all w. The isomorphism between Γ and the Floque set is proved. (That's why we use the same notations for the auxiliary Riemann surface in algebraic-geometrical data and for the Floque set of (2.1).)

As it follows from the formula (1.11) a generic algebraic-geometrical solution of the KP equation is a quasi-periodic and merophorphic function of all the variables x, y, t. We have specified the subset of data corresponding to periodic in x, y. At the end of this section we present the conditions for algebraic-geometrical data which lead to real and smooth solution of the KP-2 equation.

Finite-gap solutions of the KP-2 equation are real and non-singular if and only if their data $(\Gamma, P_1, k^{-1}, \gamma_s)$ satisfy the following conditions: There is an anti-holomorpgic involution τ on the curve Γ that has $g+1$ fixed ovals (such curves are called *M-curves*) ;each fixed oval of τ containes on of the points $P_1, \gamma_1, \ldots, \gamma_g$; the local parameter k^{-1} in a neighbourhood of P_1 must be chosen so that $k(\tau(Q)) = --\bar{k}(Q)$.

3 The perturbation theory for formal Bloch solutions

The main purpose of this chapter is to construct the perturbation theory for formal Bloch solutions of the equation (2.1) with a potential $u_0 + \delta u$ which enables us to express these solutions in terms of the basis $\psi_n(x,y)$ of the Bloch solutions for "unperturbed" equation (2.1) with the potential u_0.

Let us call a point Q_0 on Γ "non-resonant" if for any $Q_n \neq Q_0$ such that $w_1(Q_n) = w_{10} = w_1(Q_0)$ the following condition holds

$$w_{2n} = w_2(Q_n) \neq w_{20} = w_2(Q_0) \,. \tag{3.1}$$

For such Q_0 we can easily construct a formal Bloch solution of the equation

$$(\sigma\partial_y - \partial_x^2 + u_0 + \delta u)\tilde{\psi} = 0 \tag{3.2}$$

as a formal series

$$\tilde{\psi} = \sum_{s=0}^{\infty} \tilde{\phi}_s(x, y, , Q_0) \,, \quad \tilde{\phi}_0(x, y, , Q_0) = \psi(x, y, , Q_0) = \psi_0. \tag{3.3}$$

This series describes a "perturbation" of the Bloch solution ψ_0 of the non-perturbed equation.

Lemma 3.1 . If (3.1) is satisfied, then there exists a unique formal series

$$F(y, Q_0) = \sum_{s=1}^{\infty} F_s(y, Q_0) \tag{3.4}$$

such that the equation

$$(\sigma\partial_y - \partial_x^2 + u_0 + \delta u)\Psi(x, y, Q_0) = F(y, Q_0)\Psi(x, y, Q_0) \tag{3.5}$$

has a formal solution of the form

$$\Psi(x, y, Q_0) = \sum_{s=0}^{\infty} \phi_s(x, y, , Q_0) \,, \quad \phi_0 = \psi_0, \tag{3.6}$$

satisfying the conditions

$$< \psi_0^+ \Psi >_x = < \psi_0^+ \psi_0 >_x, \tag{3.7}$$

$$\Psi(x + a_1, y, Q_0) = w_1 \Psi(x, y, Q_0), \tag{3.8}$$

$$\Psi(x, y + a_2, Q_0) = w_{20} \Psi(x, y, Q_0). \tag{3.9}$$

(The corresponding solution is unique and is given by the recursion formulae (3.11-3.14).

Proof . The equation (3.5) is equivalent to the system of the equations

$$(\sigma \partial_y - \partial_x^2 + u_0)\phi_s = \sum_{i=1}^{n-1} F_i \phi_{s-i} - \delta u \phi_{s-1}. \tag{3.10}$$

The function ϕ_s can be represented in the form

$$\phi_s = \sum_n c_n^s(y, Q_0) \psi_n(x, y). \tag{3.11}$$

From (3.7) it follows that

$$c_0^s = 0, \ s > 0. \tag{3.11a}$$

Substitution of (3.11) in (3.10) and the expansion in ψ_n of the right hand side of (3.10) give the formulae for the derivatives $\partial_y c_n^s$. The initial condition for c_n^s can be found with the help of the condition (3.9) which implies that

$$w_{2n} c_n^s(y + a_2) = w_{20} c_n^s(y). \tag{3.12}$$

The final formula for c_n^s is as follows

$$c_n^s(y, Q_0) = \sigma^{-1} \frac{w_{2n}}{w_{20} - w_{2n}} \int_y^{y+a_2} \left(\sum_{i=1}^{s-1} F_i c_n^{s-i} - \frac{< \psi_n^+ \delta u \phi_{s-1} >_x}{< \psi_n^+ \psi_n >_x}\right) dy'. \tag{3.13}$$

The equality (3.11a) implies that the coefficient before ψ_0 in the expansion of the right hand side of (3.10) is equal to zero. Therefore,

$$F_s(y, Q_0) = \frac{< \psi_0^+ \delta u \phi_{s-1} >_x}{< \psi_0^+ \psi_0 >_x}, \tag{3.14}$$

and the Lemma is proved.

From (3.5, 3.8, 3.9) it follows that the formula

$$\tilde{\psi}(x, y, Q_0) = \frac{\Psi(x, y, Q_0)}{\Psi(0, 0, Q_0)} e^{-\sigma^{-1} \int_0^y F(y', Q_0) dy'} \tag{3.15}$$

defines a formal Bloch solution of the equation (3.2):

$$\tilde{\psi}(x + a_1, y, Q) = w_1 \psi(x, y, Q), \tag{3.16}$$

$$\tilde{\psi}(x, y + a_2, Q) = \tilde{w}_{20}\psi(x, y, Q), \tag{3.17}$$

where the corresponding Bloch multiplier is equal to

$$\tilde{w}_{20} = w_{20}e^{-\sigma^{-1}\int_0^{a_2} F(y',Q_0)dy'} \tag{3.18}$$

In the stationary case, when u does not depend on y, the preceding formulae turn out into the usual formulae of the perturbation theory of eigenfunctions corresponding to simple eigenvalues. The condition (3.1) is an analogue of simplicity of an eigenvalue of an operator. In cases when it is violated, it is necessary to proceed along the same lines as in the perturbation theory of multiple eigenvalues.

As the set of indices corresponding to the resonances we can take an arbitrary set of interges $I \in Z$ such that

$$w_{2\alpha} \neq w_{2n}, \ \alpha \in I, \ n \notin I$$

(up to the end of this section, integral indices belonging to I will be denoted by Greek letters, and all the others - by Latin).

Lemma 3.2 . There are unique formal series

$$F_\beta^\alpha(y, w_1) = \sum_{s=1}^{\infty} F_{\beta,s}^\alpha(y, w_1) \tag{3.19}$$

such that the equations

$$(\sigma\partial_y - \partial_x^2 + u_0 + \delta u)\Psi^\alpha(x, y, w_1) = \sum_\beta F_\beta^\alpha(y, w_1)\Psi^\beta(x, y, w_1) \tag{3.20}$$

have unique formal Bloch solutions of the form

$$\Psi^\alpha(x, y, w_1) = \sum_{s=0}^{\infty} \phi_s^\alpha(x, y, w_1) \ , \ \phi_0^\alpha = \psi_\alpha = \psi(x, y, Q_\alpha). \tag{3.21}$$

$$\Psi^\alpha(x + a_1, y, w_1) = w_1\Psi^\alpha(x, y, w_1), \tag{3.22}$$

$$\Psi^\alpha(x, y + a_2, w_1) = w_{2\alpha}\Psi^\alpha(x, y, w_1). \tag{3.23}$$

such that

$$< \psi_\beta^+\Psi^\alpha >_x = \delta_{\alpha,\beta} < \psi_\beta^+\psi_\alpha >_x . \tag{3.24}$$

The proof of the lemma is completely analogous to the proof of the lemma 3.1. The corresponding formulae for F_β^α and Ψ^α are the matrix generalization of the formulae (3.13,3.14) that's why we shall skip them (see details in [10]).

Let us define the matrix $T = T_\beta^\alpha(y, w_1)$ by the equation

$$\sigma T_y + TF = 0, \ T(0, w_1) = 1. \tag{3.25}$$

The functions

$$\hat{\Psi}^\alpha(x,y,w_1) = \sum_\beta T^\alpha_\beta(y,w_1)\Psi^\beta(x,y,w_1) \tag{3.26}$$

are solutions of (3.2). Under the translation by the period in x they are multiplied by w_1, while under the translation by the period in y they are transformed as follows

$$\hat{\Psi}^\alpha(x,y+a_2,w_1) = \sum_\beta \hat{T}^\alpha_\beta(w_1)w_{2\beta}\hat{\Psi}^\beta(x,y,w_1), \tag{3.27}$$

where

$$\hat{T}^\alpha_\beta(w_1) = T^\alpha_\beta(a_2,w_1).$$

It is natural to call a finite set of formal solutions $\hat{\Psi}^\alpha$ quasi-Bloch, since it remains invariant under the translation by the periods in x and y.

The characteristic equation

$$R(w_1,\tilde{w}_2) = \det(\tilde{w}_2\delta_{\alpha,\beta} - \hat{T}^\alpha_\beta(w_1)w_{2,\beta}) = 0 \tag{3.28}$$

is an analogue of the "secular equation" in the ordinary perturbation theory of multiple eigenvalues.

Let $h_\alpha(w_1,\tilde{w}_2)$ be an eigenvector of the matrix $\hat{T}^\alpha_\beta(w_1)w_{2,\beta}$ normalized so that

$$\sum_\alpha h_\alpha(\tilde{Q})\hat{\Psi}^\alpha(0,0,w_1) = 1 \ , \ \tilde{Q} = (w_1,\tilde{w}_2) \tag{3.29}$$

then

$$\tilde{\psi}(x,y,\tilde{Q}) = \sum_\alpha h_\alpha(\tilde{Q})\hat{\Psi}^\alpha(x,y,w_1) \tag{3.30}$$

is a formal Bloch solution of (3.2) with multipliers w_1 and $\tilde{w}_2$, normalized in the standart way. The last statement means that the Bloch solutions are defined (locally) on the Riemann surface (3.28).

4 The structure of the Riemann surface of Bloch functions

Any equation of the form (2.1) can be formaly considered as the perturbation of the "free" equation

$$(\sigma\partial_y - \partial_x^2)\psi = 0.$$

Therefore, formal Bloch solutions for (2.1) can be written with the help of the formulae of the previous section, in which δu must be replaced by $u(x,y)$.

The Bloch multiplicators for the free operator are given by the formulae (2.3). By definition a point k of the complex plane is "resonant" if there exists a point $k' \neq k$ such that

$$w_i(k) = w_i(k') \ , \ i = 1,2. \tag{4.1}$$

From (2.3) it follows that

$$k - k' = \frac{2\pi N}{a_1} , \tag{4.2a}$$

$$k^2 - (k')^2 = \frac{\sigma 2\pi i M}{a_2} , \tag{4.2b}$$

where N and M are intergers. Hence, all resonant points have the form

$$k = k_{N,M} = \frac{\pi N}{a_1} - \frac{\sigma i M a_1}{N a_2}, \ N \neq 0, \ k' = k_{-N,-M}. \tag{4.3}$$

For any $k_0 \neq k_{N,M}$ we can use the perturbation series corresponding to the non-resonant case. In these formulae $\psi_n(x,y)$ and $\psi_n^+(x,y)$ are equal to

$$\psi_n = \psi(x, y, k_n) , \ \psi_n^+ = \psi^+(x, y, k_n) , \tag{4.4}$$

where now ψ and ψ^+ are solutions of the free equation and its adjoint

$$\psi = e^{ikx - \sigma^{-1} k^2 y} , \ \psi^+ = e^{-ikx + \sigma^{-1} k^2 y}, \tag{4.5}$$

$$k_n = k_0 + \frac{2\pi n}{a_1}. \tag{4.6}$$

For sufficiently small $u(x,y)$ it is not too hard to show that the series of the perturbation theory converge outside some neighbourhood of the resonant points (4.3) and determine there a function $\tilde{\psi}(x, y, k_0)$ which is analytical in k_0. This is true for any σ. The principle distinction between the cases $Re\ \sigma = 0$ and $Re\ \sigma \neq 0$ is revealed under an attempt to extend $\tilde{\psi}$ to a "resonant" domain. In the case $Re\ \sigma = 0$ the resonant points are dense on the real axis. In the case $Re\ \sigma \neq 0$ there is only a finite number of the resonant points (4.3) in any finite domain of the complex plane. The discreteness of the resonant points in the last case is crucial for the extention of $\tilde{\psi}$ to a "resonant" domain (and for the following proof of the approximation theorem).

To begin with we shall give here the explanation of the structure of "global" Riemann surface of Bloch solutions in the case of small u. Let us consider some neighbourhoods $R_{N,M}$ and $R_{-N,-M}$ of a resonant pair of the points $k_{N,M}$ and $k_{-N,-M}$, respectively. The function $w_1(k)$ identifies them with some neighbourhood $\hat{R}_{N,M}$ of the point $w_1(k_{N,M}) = w_1(k_{-N,-M})$ on the complex plane of the variable w_1. The series (3.6, 3.11) of the non-resonant perturbation theory diverge in $R_{N,M}$ and $R_{-N,-M}$, but it turns out that the series of the lemma 2.2 converge in $\hat{R}_{N,M}$ and in this domain define quasi-Bloch solutions of (2.1) which are analytical in w_1. The chracteristic equation (3.28) defines two-sheet covering $\tilde{R}_{N,M}$ over $\hat{R}_{N,M}$ on which the Bloch solutions of (2.1) are defined. The boundary of $\tilde{R}_{N,M}$ can be naturaly identified with the boundaries of $R_{N,M}$ and $R_{-N,-M}$. Hence, the structure (local) of the Riemann surface Γ of the Bloch functions looks as follows. Let us cut out $R_{N,M}$ and $R_{-N,-M}$ from the complex plane and glue instead of them a corresponding piece of the Riemann surface $\tilde{R}_{N,M}$. From the topological point of view this surgery is a gluing of a "handle" between two resonant points.

The remarkable thing is that this perturbation approach works even when $u(x, y)$ is not small. Of course, in that case the estimations of the perturbation theory series are much more complicated. It turns out that the perturbation series for the non-resonant case converge outside some central finite domain R_0 and outside $R_{N,M}$ for $k_{N,M} \notin R_0$. Outside R_0 we again have toperform a surgery of the previous type ("glue" handles between $k_{N,M}$ and $k_{-N,-M}$ for $k_{N,M} \notin R_0$). In the central domain R_0 we have to glue some finite genus piece of the corresponding Riemann surface $\tilde{R}_0$ instead of disc R_0. As a result we shall obtain the global Riemann surface Γ of the Bloch solutions of the equation (2.1) with $Re\ \sigma \neq 0$.

In the case of real and smooth $u(x, y)$ for $\sigma = 1$ the final form of the Floque set can be represented in the following form ([11]). Let us fix some finite or infinite subset S of integer pairs $(N > 0, M)$. The set of pairs of complex numbers $\pi = \{p_{s,1}, p_{s,2}\}$ where $s \in S$ would be called "admissible", if

$$Re\ p_{s,i} = \frac{\pi N}{a_1}\ ,\ |p_{s,i} - k_s| = o(|k_s|^{-1}),\ i = 1, 2, \qquad ((4.7)$$

and the intervals $[p_{s,1}, p_{s,2}]$ do not intersect. (Here k_s are resonant points (4.3), $s = (N, M)$.)

Let us define the Riemann surface $\Gamma(\pi)$ for any admissible set π. It is obtained from the complex plane of the variable k by cutting it along the intervals $[p_{s,1}, p_{s,2}]$ and $[-\bar{p}_{s,1}, -\bar{p}_{s,2}]$ and by sewing after that the left side of the first cut with the right side of the second cut and vice versa. (After this surgery for any cut $[p_{s,1}, p_{s,2}]$ there corresponds a nontrivial cycle a_s on $\Gamma(\pi)$.

Theorem 4.1 ([10]).*For any real periodic potential $u(x, y)$ which can be analytically extended into some neighbourhood of the real values x, y, the Bloch solutions of the equation (2.1) with $\sigma = 1$ are parametrized by points Q of the Riemann surface $\Gamma(\pi)$ corresponding to some admissible set π. The function $\psi(x, y, Q)$ which is normalized by the condition $\psi(0, 0, Q) = 1$ is meromorphic on Γ and has a simple pole γ_s on each cycle a_s. If the addmissible set π contains only a finite number of pairs, then $\Gamma(\pi)$ has finite genus and is compactified by only one point P_1 ($k = \infty$), in the neighbourhood of which the Bloch function ψ has the form (1.5).*

The potentials u for which $\Gamma(\pi)$ has finite genus are called finite-gap and as it follows from the last statement of the theorem they coincide with the algebraic-geometrical potentials.

Remark. As it follows from the statements which were formulated in the introduction and at the end of the second section, for any finite set of pairs $p_{s,1}, p_{s,2}$ and for any set of points γ_s belonging to the intervals $[p_{s,1}, p_{s,2}]$ there corresponds a smooth real solution of the KP-2 equation. The first part of the conditions (4.7) are necessary for a periodicity of the corresponding solution in x. In generic case it is quasi-periodic in y. For solutions which are periodic in y also only half of $p_{s,1}, p_{s,2}$ are independent.

5 The approximation theorem

Let us consider the first order term $\phi_1(x, y, k_0)$ of the non-resonant perturbation theory for the formal Bloch solutions of the equation (2.1). (Again as in the previous chapter we consider (2.1) as a perturbation of the "free" equation with $u_0 = 0$.) From (3.13) it follows that ϕ_1 may have poles at the points $k_{N,M}$. The residue of ϕ_1 at $k_{N,M}$ is proportional to the Fourier coefficient of u

$$u^{N,M} = << \psi^+_{N,M} u \psi_{N,M} >> = << u \exp(\frac{2\pi i N}{a_1} + \frac{2\pi i M}{a_2} >> . \quad (5.1)$$

(Here and below $\psi_{N,M} = \psi(x, y, k_{N,M})$, $\psi^+_{N,M} = \psi^+(x, y, k_{N,M})$ and ψ, ψ^+ are given by the formulae (4.5);$<< ... >>$ denotes the mean value with respect to x, y.)

Therefore, if u is a trigonometrical polynomial, $u^{N,M} = 0, | N | + | M |> G$, then there are no resonances for such N, M in the first order of the perturbation theory. Of course, singularities will appear in the second order term. One can try to find the correction u_2 of the second order for the potential which cancels the singularity in the second order of the perturbation theory. After that a singularity in the next order will appear. Again one can try to cancel the corresponding singularity with the help of a correction potential. This can be done in all orders of the perturbation theory. This is the idea. Its exact realization looks as follows.

Lemma 5.1 . Let u_1 be a trigonometrical polynimial, $u_1^{N,M} = 0, | N | + | M |> G$. Then there exists a unique formal series

$$U(x, y) = \sum_{s=1}^{\infty} u_s(x, y) \quad (5.2)$$

in which for $s > 1$

$$u_s^{N,M} = 0 \ , \ | N | + | M |< G, \quad (5.3)$$

and such that for any $k_0 \neq k_{N,M}, | N | + | M |< G$ there exists a unique formal series

$$F(y, k_0) = \sum_{s=1}^{\infty} F_s(y, k_0) \quad (5.4)$$

for which the equation

$$(\sigma \partial_y - \partial_x^2 + U(x, y))\Psi(x, y, k_0) = F(y, k_0)\Psi(x, y, k_0) \quad (5.5)$$

has a formal solution of the form

$$\Psi(x, y, k_0) = \sum_{s=0}^{\infty} \phi_s(x, y, , k_0) \ , \ \phi_0 = \psi_0. \quad (5.6)$$

satisfying the conditions

$$< \psi_0^+ \Psi >_x = 1 \ , \quad (5.7)$$

$$\Psi(x+a_1,y,k_0)=w_1(k_0)\Psi(x,y,k_0), \tag{5.8a}$$
$$\Psi(x,y+a_2,k_0)=w_2(k_0)\Psi(x,y,k_0). \tag{5.8b}$$

The equation (5.5) is equivalent to the system of the equations

$$(\sigma\partial_y-\partial_x^2)\phi_s=\sum_{i=1}^{n-1}(F_i-u_i)\phi_{s-1}. \tag{5.9}$$

For $k_0\neq k_{N,M}$ the tems of the series (5.4,5.6) are given by formulae completely analogous to (3.11, 3.13, 3.14)

$$F_s=\sum_{i=1}^{s}<\psi_0^+u_i\phi_{s-i}>_x, \tag{5.10}$$

$$\phi_s=\sum_{n\neq 0}c_n^s(y,k_0)\psi_n(x,y),\ \psi_n=\psi(x,y,k_0+\frac{2\pi n}{a_1}. \tag{5.11}$$

$$c_n^s(y,k_0)=\sigma^{-1}\frac{w_{2n}}{w_{20}-w_{2n}}\int_y^{y+a_2}(\sum_{i=1}^{s-1}(F_ic_n^{s-i}-<\psi_n^+u_i\phi_s>_x)dy'. \tag{5.12}$$

Suppose that the terms u_i, $i<s$, of (5.2) are constructed so that ϕ_i have no poles at $k_{N,M}$, $|N|+|M|>G$. Hence, $\phi_i, i<s$ can be defined at these points by continuity. The next term u_s can be found from the condition that ϕ_s has no poles at these points. This condition defines the Fourier coefficients of u_s with the indices N,M such that $|N|+|M|>G$

$$u_s^{N,M}=<<\sum_{i=1}^{s-1}F_ic_N^{s-i}-\psi_{N,M}^+u_i\phi_{s-i}>>. \tag{5.13}$$

The formulae (5.13), (5.3) define all Fourier coefficients of u_s.

Theorem 5.1 . Each smooth periodic potencial $u(x,y)$ of the equation (2.1) with $Re\ \sigma\neq 0$ analytically extendable to a neighbourhood of real x,y can be approximated by finite-gap potentials uniformly with any number of derivatives.

Let u_1^G be the trigonometrical polynomial such that

$$u_1^{N,M}=0\ ,|N|+|M|>G. \tag{5.14a}$$
$$u_1^{N,M}=u^{N,M},\ |N|+|M|<G. \tag{5.14b}$$

By Lemma 5.1 to the potential u_1^G there corresponds the unique formal series $U^G(x,y)$. It turns out ([10]) that for $G>G_0$ there exists a constant G_0 depending on u) such that the series U^G converges and defines a smooth periodic function also analytically extendable in some neighbourhood of real x,y. At the same time

$$|U^G(x,y)-u_1^G|<Ae^{(-const\ G)}. \tag{5.15}$$

Moreover, it turns out that the formal series (5.6) also converges and define for

k_0, $| k_0 |> R(G)$ an analytical function which does not vanish for any k_0 in this region. At the infinity the function

$$\Psi(x,y,k_0)e^{-ikx+\sigma^{-1}k^2y} \tag{5.16}$$

is bounded. Therefore, $U^G(x,y)$ is a finite-gap potential.From (5.14, 5.15) we have

$$| U^G(x,y) - u |< Be^{(-const\ G)}. \tag{5.17}$$

The estimation (5.17) is valid in some neighbourhood of the real values x, y which proves that $U^G \to u$, $G \to \infty$ with any number of the derivatives.

The potentials $U^G(x,y)$ are finite-gap. Hence, they correspond to some finite genus curve of the form $\Gamma(\pi)$ and to some divisor γ_s. As it was explained above any set of such data defines a smooth real solution of the KP-2 equation $U^G(x,y,t)$, $U^G(x,y) = U^G(x,y,t=0)$. It turns out that for any finite interval in t these solutions uniformly converge and define a solution of the KP-2 equation $u(x,y,t)$, $u(x,y) = u(x,y,t=0)$.

6 The Whitham equations and their applications

The non-linear WKB (or Whitham) method can be applied to any non-linear equation which has a set of exact solutions of the form

$$u_0(x,y,t) = u_0(Ux + Vy + Wt + \Phi | I_1, \ldots, I_N), \tag{6.1}$$

where $u_0(z_1, \ldots | I_k)$ is a periodic function of the variables z_i depending on parameters (I_k). The vectors U, V, W are also functions of the same parameters : $U = U(I), V = V(I), W = W(I)$.

In the framework of the non-linear WBK-method asymptotic solutions of the form

$$u(x,y,t) = u_0(\varepsilon^{-1}S(X,Y,T)|I(X,Y,T)) + \varepsilon u_1 + \ldots \tag{6.2}$$

are constructed for the perturbed or non-perturbed initial equation. Here $X = \varepsilon x, Y = \varepsilon y, T = \varepsilon t$ are "slow" variables. If the vector $S(X,Y,T)$ is defined from the relations

$$\partial_X S = U(I(X,Y,T)) = U(X,Y,T),$$
$$\partial_Y S = V(X,Y,T), \quad \partial_T S = W(X,Y,T), \tag{6.3}$$

then the main term u_0 in the expansion (6.2) satisfies the initial equation up to the first order in ε. After that all the other terms of (4.2) are defined from the non-homogeneous linear equations. They can be easily solved if a full set of solutions for a homogeneous linear equation are known.

The asymptotic solutions of the form (6.2) can be constructed with an arbitrary dependence of the parameters I_k on slow variables. In this case the expansion (6.2) will be valid on a scale of order 1. The right hand side of the non-homogeneous linear equation for u_1 contains the first derivatives of the parameters I_k. Therefore, the choice of the dependence of I_k on slow variables can be used for the cancellation of the "secular" term in u_1.

In [12] it was shown that the necessary conditions of such cancelation can be represented in the following form. Let us introduce meromorphic differentials $d\Omega_i$ with the only pole at P_1 of the form

$$d\Omega_i = d(k^i + O(k^{-1})) \tag{6.4}$$

and such that all their periods on Γ are *real*. (In our previous notations: $dp = d\Omega_1$, $dE = d\Omega_2$).

The integrals Ω_i of these differentials are multivalued functions on the bundle M_g^* over $\hat{M}_{g,1}$

$$M_g^* \; = \; (\Gamma, P_1, k^{-1}, Q \in \Gamma) \longmapsto \hat{M}_{g,1} = (\Gamma, P_1, k^{-1}). \tag{6.5}$$

If I_k are local coordinates on $\hat{M}_{g,1}$ and (λ, I_k) are local coordinates on M_g^*, then for any dependence of I_k on the variables X, Y, T the integrals $p, E, \Omega = \Omega_3$ become functions of the variables X, Y, T: $p = p(\lambda, X, Y, T)$, $E = E(\lambda, X, Y, T)$, $\Omega = \Omega(\lambda, X, Y, T)$.

Theorem [12]. *The necessary conditions for the existence of the asymptotic solutions of the equation*

$$\frac{3}{4}u_{yy} + (u_t - \frac{3}{2}uu_x + \frac{1}{4}u_{xxx})_x + \varepsilon K[u] \; = 0. \tag{6.6}$$

which has the form (6.2) with uniformly bounded first-order term are equivalent to the equation

$$\frac{\partial p}{\partial \lambda}(\frac{\partial E}{\partial T} - \frac{\partial \Omega}{\partial Y}) - \frac{\partial E}{\partial \lambda}(\frac{\partial p}{\partial T} - \frac{\partial \Omega}{\partial X}) + \frac{\partial \Omega}{\partial \lambda}(\frac{\partial p}{\partial Y} - \frac{\partial E}{\partial X}) \; = \; \frac{< \psi^+ K \psi > }{< \psi^+ \psi} \frac{\partial p}{\partial \lambda}. \tag{6.7}$$

Here $K[u]$ is an arbitrary differential polynomial.

Important remark. The equation (6.7) does not depend on the choice of the local coordinate λ along Γ.

The Whitham equations (in the case $K = 0$) for the whole hierachy can be written for any pair of times $i, j > 1$

$$\frac{\partial p}{\partial \lambda}(\partial_i \Omega_j - \partial_j \Omega_i) - \frac{\partial \Omega_j}{\partial \lambda}(\partial_i p - \partial_x \Omega_i) + \frac{\partial \Omega_i}{\partial \lambda}(\partial_j p - \partial_x \Omega_j) \; = \; 0. \tag{6.8}$$

(we preserve here the same notation t_i for "slow" variables εt_i).

Let us consider now n-th reduction of the KP hierarchy which is the hierarchy of Lax equation. The Lax equations are equations on the coefficients of differential operator

$$L = \partial_1^n + u_{n-2}\partial_1^{n-2} + \ldots + u_0, \tag{6.9}$$

which have the form

$$\partial_i L = [[L^{i/n}]_+, L], i = 1, 2, ... \tag{6.10}$$

The subset $M_g(n)$ of algebraic geometrical data $\hat{M}_{g,1}$ which give the solutions of (6.10) is the following set of data: { curve Γ with puncture P_1 is such that there exists a function $E(Q)$ on Γ with the only pole at P_1 of order n; the local parameter in the neighbourhood of P_1 is $E^{-1/n}$}.

The Whitham equations (6.8) for the choice $\lambda(Q) = E(Q)$ take the form

$$\partial_i p(t, E) \ = \ \partial_x \Omega_i(t, E). \tag{6.11}$$

(for the KdV equation such form of the Whitham equations were obtained for the first time in [13]).

In [12] the construction of exact solutions for the Whitham equations was proposed. Only some particular cases of this construction would be considered below in details. To begin with we shall consider the simplest zero genus case.

The hierarchy of the dispersionless Lax equations describes solutions of (6.10) which are slow functions of all the variables. They can be written as a system of evolution equation on the coefficients of a polynomial $E(p)$

$$E(p) = p^n + u_{n-2}p^{n-2} + ... + u_0. \tag{6.12}$$

The analogues of (6.10) are the equations which are a particular case of the Whitham equations

$$\partial_i E = \frac{d\Omega_i}{dp}\partial_x E - \frac{dE}{dp}\partial_x \Omega_i, \tag{6.13}$$

where $\Omega_i(p)$ are the polynomials

$$\Omega_i(p) = [K^i(p)]_+, \ K^n(p) = E(p), \tag{6.14}$$

$[\ldots]_+$ denotes the non-negative part of a Laurent series.

Let $p(E)$ be the inverse (multivalued) function for (6.12)

$$p(E) = K + O(K^{-1}), \ \ K^n = E. \tag{6.15}$$

Then $\Omega_i(p) = \Omega_i(p(E)) = \Omega_i(E)$ can be also considered as multivalued function of the variable E (or K)

$$\Omega_i(E) = K^i + \sum_{j=1}^{\infty} \chi_{i,j} K^{-j}. \tag{6.16}$$

The equations (6.13) are equivalent to the equations

$$\partial_i p(E) = \partial_x \Omega_i(E). \tag{6.17}$$

which is genus zero partiqular case of representation (6.11).

In this chapter we consider mainly only one special case of the construction [12]. Let us define the formal series

$$S_+(p) = \sum_{i=1}^{\infty} t_i \Omega_i(p) = \sum_{i=1}^{\infty} t_i K^i + O(K^{-1}) \tag{6.18}$$

(if only a finite number of t_i is not equal to zero, then $S_+(p)$ is a polynomial). The coefficients of S_+ are linear functions of t_i and polynomials on u_i. We introduce the dependence of u_j on the variables t_i with the help of the following equations:

$$\frac{dS_+}{dp}(q_s) = 0, \ \ s = 1, \ldots, n-1, \tag{6.19}$$

where q_s are zeros of the polynomial

$$\frac{dE}{dp}(q_s) \ = \ 0. \tag{6.20}$$

The relations (6.19) define $u_i(t_1, t_2, ...)$ as implicit functions. They do not depend on the variables $t_n, t_{2n}, t_{3n}, ...$, because $\Omega_{pn} = E^p, \ , p = 1, 2, 3, \ldots$.

It [12] it was proved that if (6.19) is fulfilled, then

$$\partial_i S_+(E) = \Omega_i(E). \tag{6.21}$$

The compartibility conditions of (6.21) imply (6.17) (because $p = \Omega_1, x = t_1$).

Let us define a function

$$F(t_1, t_2, ...) = -\frac{1}{2} \operatorname{res}(S dS_+), \quad S = \sum_{i=1}^{\infty} t_i K^i. \tag{6.22}$$

From the following formulae it follows that the function

$$\tau_K(t_1, t_2, ...) = \exp F(t_1, t_2, ...) \tag{6.23}$$

is an analoge of the τ-function for usual Lax equations.

Remark . The functions F, τ do not depend on $t_n, t_{2n}, \ldots$.

In [15] it was proved, that

$$\partial_i F = -\operatorname{res}(K^i dS_+). \tag{6.24}$$

The function F contains all the information about Ω_i. For example, the coefficients $\chi_{i,j}$ of the expansions (6.16) equal:

$$\partial_i \partial_j F = -i\chi_{j,i} = -j\chi_{i,j}. \tag{6.25}$$

The function F satisfies the truncated version of the Virasoro constraints which were obtained in[18,19] for the partition function of two-dimensional gravity models. The dispersionless analogue of these constraints have the form ([15]):

$$\sum_{i=n+1}^{\infty} it_i\partial_{i-n}F + \frac{1}{2}\sum_{j=1}^{n-1} t_j t_{n-j} = 0, \tag{6.26}$$

$$\sum_{i=1}^{\infty} it_i\partial_i F = 0, \tag{6.27}$$

$$\sum_{i=1}^{\infty} it_i\partial_{i+m}F + \frac{1}{2}\sum_{j=1}^{m-1} \partial_j F\partial_{m-j}F = 0, m = n, 2n, 3n, ... \tag{6.28}$$

In [20] the ring of primary fields for perturbed A_{n-1} Landau-Ginzburg models was found. This ring is completely defined with the help the corresponding superpotential $W(p, t_1, \dots, t_{n-1})$. The relation of these results to the dispersionless Lax equation is given by the following theorem ([15]).

Theorem 6.1 . Let $E(p, t_1, t_2, ...t_{n-1}, t_{n+1}, ...)$ be the solution of the dispersionless Lax hierarchy which was constructed above. Then the superpotential of the perturbed A_{n-1} topological minimal model is equal to

$$W(p, t_1, ...t_{n-1}) = \frac{1}{n}E(p, t_1, \frac{t_1}{2}, ..., \frac{t_{n-1}}{n-1}, \frac{1}{n+1}, 0, 0, ...). \tag{6.29}$$

The partition function of this model is equal to

$$F = F(t_1, \frac{t_2}{2}, ..., \frac{t_{n-1}}{n-1}, \frac{1}{n+1}, 0, 0, ...), \tag{6.30}$$

where $F(t_1, t_2, ...)$ is given by the formula (6.22). (We would like to mention here the recent paper [21] where this results were generalized for the Landau-Ginzburg models on Riemann surfaces.)

At the end of this section we consider the loop equations for hermitean matrix models. (The review of the latest works on the application of the loop equations to matrix models and $2d$ quantum gravity can be found in [22].) The hermitean matrix model is defined by the partition function

$$Z_N = \int DM \exp(-\mathrm{Tr}(V(M)), \tag{6.31}$$

where M is $N \times N$ hermitean matrix,

$$V(K) = \sum_{i=0}^{\infty} \tilde{t}_i K^i. \tag{6.32}$$

The Wilson loop correlator is by definition

$$\mathcal{W}(K) = \langle \mathrm{tr}\frac{1}{K-M}\rangle = -\sum_{i=0}^{\infty} K^{-i-1}\partial_i \log Z_N. \tag{6.33}$$

The loop equations are derived from the invariance of the integral (6.31) with respect to the infinitesimal shift of M and have the form

$$[\sum_{i=1}^{\infty} i\tilde{t}_i K^{i-1})\mathcal{W}(K)]_- = \mathcal{W}^2(K) + \frac{\delta}{\delta V}\mathcal{W}(K). \tag{6.34}$$

$[\ldots]_-$ denotes the negative part of Laurent series. The equation (6.34) has to be supplemented with the condition

$$\mathcal{W}(K) = \frac{N}{K} + O(K^{-2}). \tag{6.35}$$

The leading term of $1/N^{-2}$ expansion of a solution of (6.34) should be a solution of the truncated equation

$$(\sum_{i=1}^{\infty} i\tilde{t}_i K^{i-1})\mathcal{W}_0(K) = \mathcal{W}_0^2(K). \tag{6.36}$$

Below we consider only the "even" case $\tilde{t}_{2i+1} = 0$.

Let us consider the solution of the dispersionless KdV equation (the $n = 2$ case of the dispersionless Lax equations) which was constructed above.

From (6.24,6.26-6.28) it follows that after the redefinition of "times"

$$\tilde{t}_{2i} = \frac{2i+1}{2i} t_{2i+1}, \quad N = -\frac{1}{2} t_1, \tag{6.37}$$

the solution of the dispersionless KdV hierarchy gives the the solution of (6.36) with the condition (6.35) with the help of the formula:

$$\mathcal{W}_0 = -\frac{d}{dE}(t_1 K + S_-(K)), \; S_- = S - S_+. \tag{6.38}$$

This solution coincides with, so-called,"one-cut" solution [22]. It turns out that the Whitham equations (or more exactly, their special solutions) on the moduli space of hyperelliptic curves give "multi-cut" solutions of the equation (6.36).

Let us consider the Whitham equations for the genus g algebraic-geometrical solutions of the Lax equations (6.10). They are equations on the moduli space $M_g(n)$. In full analogy with the genus zero case their particular solutions can be obtained with the help of the following defining relations

$$dS_+(q_s) = 0, \; s = 1, \ldots, n + 2g - 1, \tag{6.39}$$

where q_s are zeros of the defferential $dE(q_s) = 0$.

Example. $n = 2$ In this case the space $M_g(2)$ is the space of sets of distinct points $E_1, \ldots, E_{2g+1}$. The corresponding curves are hyperelliptic curves which are defined by the equation

$$y^2 = \prod_{i=1}^{2g+1}(E - E_i) = R(E). \tag{6.40}$$

The differentials $d\Omega_i$ has a'priori the form:

$$d\Omega_{2i+1} = \frac{Q_i(E)}{\sqrt{R(E)}} dE = \frac{2i+1}{2} \frac{E^{g+i} + \ldots}{\sqrt{R}} dE. \tag{6.41}$$

The coefficients of the polynomial $Q_i(E)$ are uniquely defined from the normalizing conditions for $d\Omega_i$, which are equivalemt to a system of linear equations. They become the functions of E_i and can be expressed through complete hyperelliptic integrals. Therefore, the polynomial Q_i is also a function of the variables E_i, i.e.

$$Q_i(E) = Q_i(E|E_1, \ldots E_{2g+1}). \tag{6.42}$$

The defining relations (6.39) are equivalent to a set of non-linear transcendent (but not differential) equations

$$\sum_{i=1}^{\infty} t_i Q_i(E_m|E_1, E_2, \ldots, E_{2g+1}) = 0, m = 1, 2, \ldots, 2g+1. \tag{6.43}$$

They define E_m as functions of t_i, which are the solution of the Whitham equations on $M_g(n)$.

Let $F(t_1, \ldots)$ be the function given by the same formula (6.22) where S_+ corresponds to the solution of the Whitham equations on $M_g(n)$. Then all the relations (6.24,6.25) and the constraints (6.26-6.28) would be fulfilled as well. Hence, for $n = 2$, the formulae (6.38) after redefinition of "times" (6.37) provides "multi-cut solutions of the equation (6.36).

Important remark. These solutions are not analytical functions of t_i, because the normalizing conditions for Ω_i are real equations. Locally, analytical solutions of (6.36) can be obtained if we consider the Whitham-type equations on the Teichmüller space, which covers $M_g(n)$. Corresponding equations have the same form (6.11), but now normalizing conditions which define Ω_i, should be chosen in the form:

$$\oint_{a_i} d\Omega_i = 0, \tag{6.44}$$

where a_i, b_i is a canonical basis of cycles on Γ. After that all the previous statements will be fulfilled.

References

[1] I. Krichever, Soviet Doklady 227:2 (1976),291.
[2] I. Krichever, Funk. Anal. i Pril. 11, No 13 (1977) , 15.
[3] I.Krichever, Uspekhi Mat. Nauk 32, No 6 (1977), 180.
[4] B. Dubrovin, Uspekhi Mat. Nauk 36, No 2 (1981),11-80.
[5] I.Krichever, S. Novikov, Uspekhi Mat. Nauk, 35, No 6 (1980).
[6] B. Dubrovin, I .Krichever, S. Novikov, Integrable systems, VINITY AN USSR, 1985.

[7] B. Dubrovin, V. Matveev, S. Novikov, Uspekhi Mat. Nauk 31, No 1 (1976), 55-136.
[8] V. Zakharov, S. Manakov, S. Novikov, L. Pitaevski, Soliton theory, Moscow, Nauka, 1980.
[9] V.E. Zakharov, E.I. Shul'man, Doklady Akad Nauk SSSR, 223 (1985), 1325.
[10] I. Krichever, Uspekhi Mat. Nauk, 44, No 2 (1989), 121.
[11] D. Gieseker, H. Knörrer, E. Trubowitz, Contemporary Mathematics, 116 (1991), 19.
[12] I. Krichever, Funk. Anal. i Pril. 22(3) (1988) , 37-52 .
[13] H. Flashka, M. Forest, L.McLaughlin, Comm. Pure and Appl. Math. 33(6)
[14] S. Yu. Dobrokhotov, V. P. Maslov , Soviet Scientific Reviews, Math. Phys.Rev. OPA Amsterdam 3 (1982) , 221-280.
[15] I. Krichever, Topological minimal models and dispersionless Lax equations, preprint ISI Turin (to appear in Comm. Math. Phys.) (1991).
[16] I. Krichever, Whitham theory for integrable systems and topological quantum field theories (in press; to appear in procidings of the Cargese summer school in "New symmetry principles in quantum theory, 1991).
[17] I. Krichever, Funk. Anal.i Pril. 20, No 3 (1986), 42.
[18] M. Fukuma, H. Kawai , Continuum Schwinger-Dyson equations and universal structures in two-dimensional quantum gravity , preprint Tokyo University UT-562 , May 1990 .
[19] M. Fukuma, H. Kawai, Infinite dimansional Grassmanian structure of two-dimensional quantum gravity , preprint Tokyo University UT-572 , November 1990.
[20] E. Verlinder, H. Verlinder, A solution of two-dimensional topological quantum gravity, preprint IASSNS-HEP-90/40, PUPT-1176 (1990).
[21] B. Dubrovin, Hamiltonian formalism of Whitham-type hierarchies and topogical Landau-Ginsburg models (in press).
[22] Yu. Makeenko, Loop equations in matrix models and in 2D quantum gravity, Submitted in Mod. Phys. Lett.A .

Integrable Singular Integral Evolution Equations

P.M. Santini

Dipartimento di Fisica, Università di Roma "La Sapienza", I-00185 Roma, Italy
Istituto Nazionale di Fisica Nucleare, Sezione di Roma, I-00185 Roma, Italy

1. INTRODUCTION

Integro-differential evolution equations (IDEE's) appear in several areas of applied Science. For instance, in a fluid dynamical context, equation

$$u_t + auu_x + \int_{-\infty}^{\infty} dx' K(x-x') u_{x'}(x',t) = 0, \quad K(x) := \int_{-\infty}^{\infty} \frac{dk}{2\pi} c(k) e^{ikx} \tag{1.1}$$

is a simple mathematical model combining the breaking term uu_x of shallow water theory with a linear dispersion $\omega(k) = kc(k)$ [1]. As observed in [2],[3], in a fluid of total depth D characterized by a thin termocline located at depth d and in a long wave regime, the dispersion takes the form [4]

$$c(k) = c_0\left(1 - \frac{kd}{2}\left(coth(kD) - \frac{1}{kD}\right)\right) \tag{1.2}$$

and gives rise to the IDEE

$$u_t + \left(c_0 + \frac{d}{2D}\right)u_x + auu_x + \frac{d}{4D} P \int_{-\infty}^{\infty} dx' coth\left(\frac{\pi}{2D}(x'-x)\right) u_{x'x'}(x',t) = 0, \tag{1.3}$$

where the symbol $P\int$ denotes principal value integrals.

In the shallow water limit $D \to 0$, $c(k) \sim c_0 - \frac{dD}{6}k^2$, equation (1.3) reduces to the celebrated Korteweg - de Vries (KdV) equation [5]

$$u_t + c_0 u_x + auu_x + \frac{dD}{6} u_{xxx} = 0, \tag{1.4}$$

while, in the deep water limit $D \to \infty$, $c(k) = -\frac{d}{2}|k|$, it reduces to the Benjamin-Ono (BO) equation [6]:

$$u_t + \left(c_0 + \frac{d}{2D}\right)u_x + auu_x + \frac{d}{2\pi} P \int_{-\infty}^{\infty} dx' (x'-x)^{-1} u_{x'x'}(x',t) = 0. \tag{1.5}$$

Staying intermediate between the KdV and BO equations, (1.3) has been named intermediate long wave (ILW) equation.

We recall that the KdV equation is a prototype example of partial differential equation (PDE) integrable by the inverse scattering (or spectral) transform (IST), a powerful technique for investigating and solving the initial value problem for classes of nonlinear evolution equations (NEE's) [7],[8]. Other well-known examples of integrable PDE's are the nonlinear Schrodinger (NLS) [9]

$$iu_t + u_{xx} - 2\epsilon|u|^2u = 0, \tag{1.6a}$$

the modified Korteweg-de Vries (MKdV) [10]

$$u_t - u_{xxx} + 6\epsilon u^2 u_x = 0 \tag{1.6b}$$

and the Kadomtsev- Petviashvili (KP) [11], [12]

$$(u_t + u_{xx} + 6uu_x)_x + 3\sigma^2 u_{yy} = 0 \tag{1.6c}$$

equations.

IDEE's entered the field of integrable systems with the discovery of the existence of a number of soliton solutions and conserved quantities for equations (1.3) [2], [3] and (1.5) [6], [13], [14]. It was subsequently shown that equations (1.3) and (1.5) share with integrable PDE's all the basic integrability properties, like the existence of N-soliton solutions [15], [16], [17], of infinitely many constants of motion [18], [19], of Backlund transformations (BT) [19] and of novel IST schemes [18], [19].

Instead of differential equations, the spectral problem associated with (1.3) turns out to be the Riemann - Hilbert (RH) boundary value problem [19], [20]

$$i\psi_x^+(x,z) + u(x)\psi^+(x,z) = z\psi^-(x,z) \tag{1.7}$$

in the strip $0 < Im\ x < \eta$ of the complex x-plane, where the width η of the strip is proportional to the depth D of the fluid. The relevance of RH problems for the solution of PDE's and discrete systems was already known [21], but only at the level of inverse problems in the dual space k. Therefore the IST schemes associated with IDEE's (unlike those of PDE's) exhibit an interesting symmetry, since their direct and inverse transforms are both characterized by RH boundary value problems (see §3.1,2).

Other RH problems in x space have been introduced to generate and solve new integrable IDEE's which reduce, in the "shallow water limit" $\eta \to 0$, to well-known integrable PDE's. Integro-differential analogues of KdV, MKdV, NLS and KP equations have been derived and studied (see §3.2,3).

IDEE's share with integrable PDE's also several interesting algebraic - geometrical properties: they are bi-Hamiltonian systems with ∞ - many constants of motion in involution, they possess a recursion operator which maps symmetries into symmetries and whoose adjoint maps gradients of conserved quantities into gradients of conserved quantities [22]-[27] (see §4.2).

We finally remark that, in the philosophy of a unification of the algebraic properties of integrable systems, it is possible to establish a deep (and surprising) connection between IDEE's and discrete systems [24]. This is due to the fact that IDEE's associated with RH problems in the strip $0 < Im\ x < \eta$ are generated by rational combinations of the shift operator $\mathcal{E} = exp(i\eta\frac{\partial}{\partial x})$, while discrete systems on a lattice are generated by suitable combinations of the lattice shift $(\tilde{\mathcal{E}}f)(n) = f(n+1)$.

This article is organized as follows. In §2.1 we study scalar RH problems in the strip, as a necessary background for all the results. In §2.2 we use the above RH problems as "change of variables" to generate and solve singular integral evolution equations. §3 is dedicated to IDEE's integrable by IST. In §3.1 we present the IST formalism of the ILW and BO equations; in §3.2 we review the IST formalism of a singular integral version of the AKNS class; in §3.3 we list other relevant examples of IDEE's integrable by IST. §4 is dedicated to algebraic and geometrical properties of IDEE's. In §4.1 we present the zero curvature representation of the ILW_n class in terms of pseudo-differential operators; in §4.2 we apply the theory of recursion operators and bi-Hamiltonian structures to the nonlocal AKNS and ILW hierarchies; in §4.3 we establish an algebraic equivalence between relevant examples of singular integral evolution equations and discrete systems.

2. ELEMENTARY RH PROBLEMS AND ASSOCIATED IDEE'S

2.1 Riemann - Hilbert problems in the strip and their limits

The following two RH boundary - value problems play a central role in this article. Given a function $v(x)$, Holder continuous for $x \in \mathcal{R}$ and localized, determine a function $\psi(x)$, analytic in the strip $0 < Im\ x < \eta$ and such that its boundary values $\psi^{\pm}(x)$, defined by

$$\psi^{+}(x) := \lim_{y \to 0^{+}} \psi(x+iy), \qquad x \in \mathcal{R}, \tag{2.1a}$$

$$\psi^{-}(x) := \lim_{y \to \eta^{-}} \psi(x+iy), \qquad x \in \mathcal{R}, \tag{2.1b}$$

satisfy the equation

$$\psi^{+}(x) - \psi^{-}(x) = v(x), \qquad x \in \mathcal{R}, \tag{2.2}$$

or the equation

$$\psi^{+}(x) + \psi^{-}(x) = v(x), \qquad x \in \mathcal{R}. \tag{2.3}$$

Remark 2.1 $\psi^{\pm}(x)$ are, by construction, related by the periodicity condition

$$\psi^{-}(x) = \psi^{+}(x+i\eta), \qquad x \in \mathcal{R}. \tag{2.4}$$

Remark 2.2 $\psi^{-}(x)$ is defined for $x \in \mathcal{R}$ and is analytically extendible downstairs in the strip $-\eta < Im\ x < 0$.

A straightforward way to solve the linear problems (2.2) and (2.3) consists in using Fourier Transform (FT). If

$$\psi^{+}(x) = \int_{-\infty}^{\infty} \frac{dk}{2\pi} e^{ikx} \hat{\psi}(k) \ + \ \alpha, \qquad \alpha = const., \tag{2.5a}$$

then

$$\psi^{-}(x) = \int_{-\infty}^{\infty} \frac{dk}{2\pi} e^{ikx} \hat{\psi}(k) e^{-\eta k} \ + \ \alpha \tag{2.5b}$$

and equation (2.2) implies that

$$\hat{\psi}(k) = \hat{v}(k)(1-e^{-\eta k})^{-1}, \tag{2.6}$$

where $\hat{v}(k)$ is the FT of $v(x)$. Using the convolution theorem it follows that

$$\psi^{+}(x) + \psi^{-}(x) = P\int_{-\infty}^{\infty} \frac{dk}{2\pi} e^{ikx} coth(\frac{\eta k}{2}) \hat{v}(k) \ + \ 2\alpha = -i(\mathcal{T}v)(x) + 2\alpha, \tag{2.7}$$

where $\mathcal{T}$ is the singular integral operator defined by

$$(\mathcal{T}f)(x) := \frac{1}{\eta} P \int_{-\infty}^{\infty} dx' coth(\frac{\pi}{\eta}(x'-x)) f(x'). \tag{2.8}$$

From equations (2.2) and (2.7) it follows that

$$\psi^{\pm} = \frac{1}{2i\eta} \int_{-\infty}^{\infty} dx' coth(\frac{\pi}{\eta}(x'-(x \pm i0))) v(x') + \alpha = \frac{1}{2}(\pm 1 - i\mathcal{T}) v(x) + \alpha, \quad x \in \mathcal{R}, \tag{2.9a}$$

$$\psi(x) = \frac{1}{2i\eta} \int_{-\infty}^{\infty} dx' coth(\frac{\pi}{\eta}(x'-x))v(x') + \alpha, \qquad 0 < Im\ x < \eta \tag{2.9b}$$

is the solution of the RH problem (2.2). Furthermore

$$i(\psi^+ + \psi^-) = \mathcal{T}(\psi^+ - \psi^-) + 2\alpha. \tag{2.10}$$

The arbitrary constant α can be specified imposing boundary conditions; if, for instance, $\psi(x) \to 1$ as $x \to +\infty$, then $\alpha = 1 + \frac{1}{2i\eta} \int_{-\infty}^{\infty} dx v(x)$.

If equation (2.3) holds, an analogous procedure yields

$$\hat{\psi}(k) = \hat{v}(k)(1 + e^{-\eta k})^{-1}, \tag{2.11}$$

$$\psi^+(x) - \psi^-(x) = \int_{-\infty}^{\infty} \frac{dk}{2\pi} e^{ikx} tgh(\frac{\eta k}{2})\hat{v}(k) = -i(\mathcal{D}v)(x), \tag{2.12}$$

where the operator $\mathcal{D}$, inverse of $-\mathcal{T}$, is defined by the convolution integral

$$(\mathcal{D}f)(x) := \frac{1}{\eta} P \int_{-\infty}^{\infty} dx' csch(\frac{\pi}{\eta}(x'-x))f(x'), \qquad x \in \mathcal{R}. \tag{2.13}$$

Therefore

$$\psi^{\pm}(x) = \pm\frac{1}{2i\eta} \int_{-\infty}^{\infty} dx' csch(\frac{\pi}{\eta}(x' - (x \pm i0)))v(x') = \frac{1}{2}(1 \mp i\mathcal{D})v(x), \quad x \in \mathcal{R}, \tag{2.14a}$$

$$\psi(x) = \frac{1}{2i\eta} \int_{-\infty}^{\infty} dx' csch(\frac{\pi}{\eta}(x'-x))v(x'), \qquad 0 < Im\ x < \eta \tag{2.14b}$$

is the solution of the RH problem (2.3). Furthermore

$$i(\psi^+ - \psi^-) = \mathcal{D}(\psi^+ + \psi^-), \quad x \in \mathcal{R}. \tag{2.15}$$

Remark 2.3 The solution (2.9b) of the RH problem (2.2) is well - defined outside the strip and is periodic:

$$\psi(x + ih\eta) = \psi(x), \qquad h \in \mathcal{Z}, \qquad 0 < Im\ x < \eta. \tag{2.16}$$

The solution (2.14b) of the RH problem (2.3) is also defined outside the strip and is antiperiodic:

$$\psi(x + ih\eta) = (-1)^h \psi(x), \qquad h \in \mathcal{Z}, \qquad 0 < Im\ x < \eta. \tag{2.17}$$

Remark 2.4 The periodicity condition (2.4) suggests the introduction of the shift operator

$$\mathcal{E} := e^{i\eta \frac{\partial}{\partial x}} \tag{2.18}$$

such that $\psi^-(x) = (\mathcal{E}\psi^+)(x) = \psi^+(x + i\eta)$. In terms of $\mathcal{E}$ the RH problems (2.2) and (2.3) read, respectively, $(1 - \mathcal{E})\psi^+ = v$ and $(1 + \mathcal{E})\psi^+ = v$. Therefore the integral operators $\mathcal{D}$ and $\mathcal{T}$ admit the following rational representations

$$\mathcal{D} = i(1 - \mathcal{E})(1 + \mathcal{E})^{-1}, \quad \mathcal{T} = i(1 + \mathcal{E})(1 - \mathcal{E})^{-1} \tag{2.19}$$

in terms of $\mathcal{E}$.

If $\eta \to \infty$, the strips $0 < Im\, x < \eta$ and $-\eta < Im\, x < 0$ become the upper and lower half x - plane respectively, and $\psi^{\pm}$ become the boundary values of a function $\psi(x)$ analytic in the upper and lower half x - plane. With these analyticity properties the solution of the RH problems (2.2) and (2.3) can be obtained directly, or taking the $\eta \to \infty$ limit of equations (2.9b) and (2.14b), and is given, respectively, by

$$\psi^{\pm}(x) = \frac{1}{2\pi i}\int_{-\infty}^{\infty} dx' \frac{v(x')}{x' - (x \pm i0)} + \alpha = \frac{1}{2}(\pm 1 - i\mathcal{H})v(x) + \alpha, \qquad x \in \mathcal{R}, \tag{2.20}$$

and by

$$\psi^{\pm}(x) = \pm\frac{1}{2\pi i}\int_{-\infty}^{\infty} dx' \frac{v(x')}{x' - (x \pm i0)} = \frac{1}{2}(1 \mp i\mathcal{H})v(x), \qquad x \in \mathcal{R}, \tag{2.21}$$

where $\mathcal{H}$ is the Hilbert transform

$$(\mathcal{H}v)(x) := \frac{1}{\pi} P \int_{-\infty}^{\infty} dx' \frac{v(x')}{x' - x} \tag{2.22}$$

Both equations (2.10) and (2.15) reduce, up to the constant α, to the Plemelj formula

$$\mathcal{H}(\psi^+ - \psi^-) = i(\psi^+ + \psi^-). \tag{2.23}$$

If $\eta \to 0$, the strip degenerates into a straight line, moreover

$$\psi^-(x) = \psi^+(x + i\eta) = \phi(x) + i\eta\phi_x(x) + \mathcal{O}(\eta^2), \quad \psi^+(x) = \phi(x),$$

and the RH problem (2.2) reduces to the trivial differential equation $\phi_x = g$, $v(x) = -i\eta g(x) + \mathcal{O}(\eta^2)$, whose solution can also be obtained taking the $\eta \to 0$ limit of $(2.9a)_+$:

$$\phi(x) = \frac{1}{2}\left(\int_{-\infty}^{x} - \int_{x}^{\infty} dx' g(x') + \alpha.\right.$$

Remark 2.5 Equations (2.10) and (2.15) are a useful tool to obtain several identities for $\mathcal{T}$, $\mathcal{D}$ and $\mathcal{H}$. Indeed, if $\psi^{\pm}$ satisfy (2.10) (resp. (2.15)) it is clear that also $f(\psi^{\pm})$ do so; replacing for instance $\psi^{\pm}$ in equations (2.10) by ${\psi^{\pm}}^2 = \frac{1}{4}((\pm 1 - i\mathcal{T})v)^2$, we obtain the well-known identity

$$2\mathcal{T}v\mathcal{T}v = (\mathcal{T}v)^2 - v^2 - c, \qquad c = \left(\frac{1}{\eta}\int_{\mathcal{R}} dx v(x)\right)^2, \tag{2.24a}$$

where the constant c is fixed from large x asymptotics. Analogously, from equations (2.15) and (2.14a) we get

$$2v\mathcal{D}v = \mathcal{D}(v^2) - \mathcal{D}((\mathcal{D}v)^2), \;\Rightarrow\; 2\mathcal{T}v\mathcal{D}v = (\mathcal{D}v)^2 - v^2. \tag{2.24b}$$

Using ${\psi^{\pm}}^3$ and $e^{-2\psi^{\pm}}$ we obtain,

$$\mathcal{T}(v^3) + (\mathcal{T}v)^3 = 3v^2\mathcal{T}v + 3\mathcal{T}(v(\mathcal{T}v)^2), \tag{2.24c}$$

$$\mathcal{D}(v^3) + (\mathcal{D}v)^3 = 3v^2\mathcal{D}v + 3\mathcal{D}(v(\mathcal{D}v)^2) \tag{2.24d}$$

and

$$\mathcal{D}(e^{-v}\cos(\mathcal{D}v)) = -e^{-v}\sin(\mathcal{D}v). \tag{2.24e}$$

In the $\eta \to \infty$ limit we replace in equations (2.24) $\mathcal{T}$ and $\mathcal{D}$ by $\mathcal{H}$, obtaining identities for the Hilbert transform. These identities are all related; for instance (2.24b) follows from (2.24a), while (2.24b) implies (2.24a) in the special case $\int_{\mathcal{R}} dx v(x) = 0$ (why?). For future convenience, we summarize other useful formulas for the operators $\mathcal{T}$ and $\mathcal{D}$:

$$\mathcal{D}\mathcal{T} = \mathcal{T}\mathcal{D} = -1, \tag{2.25a}$$

$$(\mathcal{D}f)(x) \to o, \quad (\mathcal{T}f)(x) \to \mp\frac{1}{\eta}\int_{\mathcal{R}} dx' f(x'), \quad x \to \pm\infty, \tag{2.25b}$$

$$\mathcal{D},\ \mathcal{T} \to \mathcal{H}, \quad \eta \to \infty; \quad \mathcal{H}^2 = -1, \tag{2.25c}$$

$$\eta\mathcal{T} \sim -(\int_{-\infty}^{x} - \int_{x}^{\infty})dx' + \frac{1}{6}\eta^2\frac{\partial}{\partial x} + \mathcal{O}(\eta^4), \qquad \frac{1}{\eta}\mathcal{D} \sim \frac{1}{2}\frac{\partial}{\partial x} + \mathcal{O}(\eta^2), \qquad \eta \sim 0. \tag{2.25d}$$

2.2 Singular integral evolution equations associated with (2.2) and (2.3)

It is possible to associate with the elementary RH problems (2.2) and (2.3) integrable evolution equations. We first consider the RH problem (2.2) and its solution (2.9a). If $v(x) \in \mathcal{R}$ and $\alpha = ia$, $a \in \mathcal{R}$ in (2.9a), it follows that

$$\overline{\psi^+} = -\psi^-, \tag{2.26}$$

where $^-$ denotes complex conjugation. If we postulate nonlinear (but integrable) time evolutions for $\psi^\pm$ compatible with (2.26), like

$$\psi_t^\pm = \beta_n(\psi^\pm)^n, \quad \beta_n \in \mathcal{R}\ for\ n\ odd; \quad \beta_n = ib_n,\ b_n \in \mathcal{R}\ for\ n\ even, \tag{2.27}$$

the compatibility between (2.2) and (2.27) gives rise to singular integral evolution equations for v; the first few are:

$$v_t = v, \qquad n = 1, \quad \beta_1 = 1, \tag{2.28a}$$

$$v_t = v\mathcal{T}v - 2av, \quad a_t + a^2 = -\frac{1}{4\eta^2}(\int_{\mathcal{R}} dx' v(x'))^2, \quad a = a(t), \quad n = 2, \quad \beta_2 = i, \tag{2.28b}$$

$$v_t = v^3 - 3v(\mathcal{T}v)^2, \qquad n = 3, \quad \beta_3 = 4. \tag{2.28c}$$

Analogously, the first few equations arising from the compatibility between (2.3) and

$$\psi_t^\pm = \beta_n(\psi^\pm)^n, \quad \beta_n \in \mathcal{R}, \quad n \in \mathcal{N}, \tag{2.29a}$$

or

$$\psi_t^\pm = \frac{1}{2}e^{-2\psi^\pm}, \tag{2.29b}$$

are

$$v_t = v^2 - (\mathcal{D}v)^2 = -2\mathcal{T}v\mathcal{D}v, \quad n = 2, \quad \beta_2 = 2, \tag{2.30a}$$

$$v_t = v^3 - 3v(\mathcal{D}v)^2, \qquad n = 3, \quad \beta_3 = 4 \tag{2.30b}$$

or

$$v_t = e^{-v}cos(\mathcal{D}v). \tag{2.30c}$$

The derivation of equations (2.28) and (2.30) is straightforward; for instance, to get equation (2.28), replace in (2.27) equation (2.9a) and use the identities (2.24).

The solution of the initial value problem for equations (2.28) is given by

$$v(x,t) = 2Re\psi^+(x,t), \tag{2.31}$$

where ψ^+ is the solution of equations (2.27) for the initial condition

$$\psi^+(x,0) = \frac{1}{2}(1 - iT)v(x,0) + ia(0). \tag{2.32a}$$

The solution of the initial value problem for equations (2.30) is given by (2.31) and (2.29), for the initial condition

$$\psi^+(x,0) = \frac{1}{2}(1 - i\mathcal{D})v(x,0). \tag{2.32b}$$

The above ideas can be generalized to generate integrable IDEE's depending on derivatives with respect to more variables. For instance the following integrable evolutions

$$\psi_t^\pm + (\psi_x^\pm + \beta(\psi^\pm)^2)_x = 0, \quad \textit{potential Burgers equation,} \tag{2.33a}$$

$$\psi_t^\pm + (\psi_{xx}^\pm + \beta(\psi^\pm)^2)_x = 0, \quad \textit{potential KdV equation,} \tag{2.33b}$$

$$(\psi_t^\pm + \psi_{xx}^\pm + \beta(\psi^\pm)^2)_x + 3\sigma^2\psi_{yy}^\pm = 0, \quad \textit{potential KP equation,} \tag{2.33c}$$

give rise, through (2.2) and $\beta = i$, to the integrable evolution equations

$$v_t + v_{xx} + (vTv)_x = 0, \tag{2.34a}$$

$$v_t + v_{xxx} + (vTv)_x = 0, \tag{2.34b}$$

$$(v_t + v_{xxx} + vTv)_x + 3\sigma^2 v_{yy} = 0. \tag{2.34c}$$

They also give rise, through (2.3) and $\beta = 1$, to the integrable evolution equations

$$v_t + v_{xx} + \frac{1}{2}((\mathcal{D}v)^2 - v^2)_x = v_t + v_{xx} + (Tv\mathcal{D}v)_x = 0, \tag{2.35a}$$

$$v_t + v_{xxx} + (Tv\mathcal{D}v)_x = 0, \tag{2.35b}$$

$$(v_t + v_{xxx} + Tv\mathcal{D}v)_x + 3\sigma^2 v_{yy} = 0. \tag{2.35c}$$

If $\eta \to \infty$ it is sufficient to replace T and $\mathcal{D}$ by $\mathcal{H}$ in all the above equations to get new integrable systems; in this case equation (2.28b) admits the reduction $a = 0$ to $v_t = v\mathcal{H}v$. It is on such equation (which describes the motion of vorticity for an inviscid fluid [28]) that the basic ideas associated with this linearization technique were introduced [28]; the technique was generalized in [29] to solve other interesting equations, like the $\eta \to \infty$ limits of equations (2.28) and (2.34). Equation (2.34b) arises in diffusive models of population and was first solved in [31] using a dependent variable transformation and splitting into upper and lower functions. As observed in [29], other nonlocal operators corresponding to different geometries (see f.i. [30]) could also be included in this approach.

3. RH SPECTRAL PROBLEMS AND IDEE's INTEGRABLE VIA IST

In §2.2 we have seen that the elementary RH boundary value problems (2.2) and (2.3) can be used to generate and solve singular integral evolution equations. In this section we construct examples of integrable IDEE's with a richer mathematical structure; they are associated with RH spectral problems.

3.1 The ILW and BO hierarchies

Our first example is given by the following scalar, differential RH problem

$$L_1\psi^+ := i\psi_x^+ + u\psi^+ = z\psi^-, \quad u(x) \in \mathcal{R}, \tag{3.1}$$

in the strip $0 < Im\ x < \eta$. The hierarchy of IDEE's associated with it can be obtained seeking compatible time evolutions in the form

$$i\psi_t^\pm = M^\pm\psi^\pm := \sum_{j=0}^{m} \alpha_j^\pm \frac{\partial^j \psi^\pm}{\partial x^j}. \tag{3.2}$$

The first nontrivial examples, corresponding to $m = 2$ and $m = 3$, are given by the ILW equation in dimensionless form

$$u_t = au_x + 2uu_x + \mathcal{T}u_{xx}, \qquad \alpha_2^\pm = 1 \tag{3.3}$$

and by

$$u_t = [\frac{1}{2}(\mathcal{T}^2 - 1)q_{xx} + q^3 + \frac{3}{2}(q\mathcal{T}q_x + \mathcal{T}qq_x)]_x. \tag{3.4}$$

The derivation of such equations is interesting; for instance, differentiating (3.1) with respect to t and multiplying (3.2) by $-z$, we obtain

$$\frac{\partial}{\partial t}(z\psi^-) = \alpha_2^+\psi_{xxx}^+ + (\alpha_1^+ + \alpha_{2\ x}^+ - iu\alpha_2^+)\psi_{xx}^+ + (\alpha_0^+ + \alpha_{1\ x}^+ - iu\alpha_1^+)\psi_x^+ + (\alpha_{0_x}^+ + u_t - iu\alpha_0^+)\psi^+$$

and

$$z\psi_t^- = \alpha_2^-\psi_{xxx}^+ + (\alpha_1^- - iu\alpha_2^-)\psi_{xx}^+ + (\alpha_0^- - iu\alpha_1^- - 2iu_x\alpha_2^-)\psi_x^+ + (-iu\alpha_0^- + -iu_x\alpha_1^- - i\alpha_2^- u_{xx})\psi^+.$$

Equating to zero the coefficients of the derivatives of ψ^+ we generate three RH problems of the type (2.2) for the coefficients $\alpha_j^\pm$:

$$\alpha_2^+ - \alpha_2^- = 0, \quad \Rightarrow \quad \alpha_2^\pm = \alpha_2 = const.,$$

$$\alpha_1^+ - \alpha_1^- = 0, \quad \Rightarrow \quad \alpha_1^\pm = \alpha_1 = const. = ia,$$

$$\alpha_0^+ - \alpha_0^- = -2i\alpha_2 u_x, \quad \Rightarrow \quad \alpha_0^\pm = -i\alpha_2(\pm 1 - i\mathcal{T})u_x + a.$$

The coefficient of ψ^+ is just equation (3.3).

In the limit $\eta \to \infty$ equations (3.3) and (3.4) reduce, respectively, to the BO equation in dimensionless form

$$u_t = 2uu_x + \mathcal{H}u_{xx} \tag{3.5}$$

and to

$$u_t = (-\frac{1}{2}q_{xx} + q^3 + \frac{3}{2}q\mathcal{H}q_x + \mathcal{H}qq_x)_x. \tag{3.6}$$

The solution of the initial value problem for equation (3.3) (respectively (3.5)) has been obtained in [20] (respectively in [32]); it requires the study of the direct and inverse problems associated with equation (3.1), interpreted as a RH spectral problem in the strip of the complex x plane (respectively in the x plane). In the following we present a short review of these two results.

3.1.1 The IST formalism for the ILW equation [20]

Direct problem. For the eigenfunction $\mu^+ = \psi^+ e^{-i\xi x}$, $z = \xi e^{-\eta\xi}$ the RH problem takes the more convenient form

$$i\mu_x^+ + \xi(\mu^+ - \mu^-) + u\mu^+ = 0, \quad u \in \mathcal{R}, \quad x \in \mathcal{R}, \tag{3.7}$$

and the spectral parameter ξ is expressed in terms of the new parameter λ through the transformation

$$\xi = f(\lambda) = \lambda(1 - e^{-\eta\lambda})^{-1}. \tag{3.8}$$

The inverse of (3.8) is multivalued and the principal branch is a deformed strip of the λ plane including the real axis and intersecting the imaginary axis at $\lambda = \pm\frac{2\pi}{\eta}i$. The line $Im\ \lambda = 0$ corresponds to the half line $\xi > 0$ of the ξ plane. The + solutions of (3.7), satisfying the asymptotics

$$M \to 1, \quad \tilde{M} \to e^{i\lambda x + \lambda\eta}, \quad x \to -\infty, \tag{3.9a}$$

$$\tilde{N} \to 1, \quad N \to e^{i\lambda x + \lambda\eta}, \quad x \to +\infty, \tag{3.9b}$$

are characterized by the Fredholm integral equations

$$\begin{pmatrix} M \\ \tilde{M} \end{pmatrix}(x,\lambda) = \begin{pmatrix} 1 \\ e^{i\lambda x + \lambda\frac{\eta}{2}} \end{pmatrix} + \int_{-\infty}^{\infty} G_+(x-y,\xi)u(y)\begin{pmatrix} M \\ \tilde{M} \end{pmatrix}(y,\lambda), \tag{3.10a}$$

$$\begin{pmatrix} N \\ \tilde{N} \end{pmatrix}(x,\lambda) = \begin{pmatrix} e^{i\lambda x + \lambda\frac{\eta}{2}} \\ 1 \end{pmatrix} + \int_{-\infty}^{\infty} G_-(x-y,\xi)u(y)\begin{pmatrix} N \\ \tilde{N} \end{pmatrix}(y,\lambda), \tag{3.10b}$$

where

$$G_\pm(x) := \int_{C_\pm} \frac{dp}{2\pi} e^{ipx}(p - \xi(1 - e^{-\eta p}))^{-1} \tag{3.10c}$$

and $C_\pm$ are the contours $Im\ p = \mp i0$. The eigenfunctions are related through the scattering equation

$$M(x,\lambda) = a(\lambda)\tilde{N}(x,\lambda) + \theta(\xi)b(\lambda)N(x,\lambda), \tag{3.11}$$

where θ is the Heaviside function and

$$a(\lambda) := 1 + \frac{1}{2i(\xi(\lambda) - 1/\eta)} \int_{-\infty}^{\infty} dy u(y)M(y,\lambda), \tag{3.12a}$$

$$b(\lambda) := \frac{1}{2i(\xi(-\lambda) - \frac{1}{\eta})} \int_{-\infty}^{\infty} u(y)M(y,\lambda)e^{-\lambda y - \lambda\frac{\eta}{2}}. \tag{3.12b}$$

Fredholm theory implies that M and N are, respectively, + and − functions in the ξ plane; $a(\lambda)$ is also a + function.

The bound states correspond to those $\tilde{\lambda}_j = ik_j$, $0 < k_j < \frac{2\pi}{\eta}$, $j = 1,..,\nu$ for which

$$a_j := a(\tilde{\lambda}_j) = 0, \quad M_j(x) := M(x,\lambda_j) = b(\tilde{\lambda}_j)N(x,\tilde{\lambda}_j) =: b_jN_j(x). \tag{3.13}$$

Inverse problem. Equation (3.11), the analyticity properties of the eigenfunctions and the symmetry condition

$$N(x,\lambda) = \tilde{N}(x,-\lambda)e^{i\lambda x + \lambda\frac{\eta}{2}} \tag{3.14}$$

define the RH problem in the ξ plane

$$\frac{M(x,\lambda)}{a(\lambda)} = \tilde{N}(\lambda) + \theta(\xi)\rho(\lambda)\tilde{N}(-\lambda)e^{i\lambda x + \lambda\eta}, \quad \rho(\lambda) := \frac{b(\lambda)}{a(\lambda)}, \quad \xi \in \mathcal{R} \tag{3.15}$$

with poles (the zeroes of $a(\lambda)$), whose solution is expressed in terms of the following integral equation

$$\tilde{N}(\xi)-\frac{1}{2\pi i}\int_0^\infty \frac{d\xi'(\lambda')}{2\pi i}\frac{\rho(\lambda')\tilde{N}(-\lambda')e^{i\lambda' x+\lambda'\eta}}{\xi'-(\xi-i0)}=1+i\sum_{j=1}^{\nu}\frac{C_jN_j}{\xi-\xi_j},\quad \xi\in\mathcal{R} \tag{3.16a}$$

$$\xi_j:=\xi(\tilde{\lambda}_j),\qquad C_j:=-i\frac{b(\lambda)}{a_\xi(\lambda)}|_{\lambda=\tilde{\lambda}_j} \tag{3.16b}$$

Finally $u(x)=u^+(x)-u^-(x)=u^+(x)+\overline{u^+(x)}$ is recovered through the $\xi\to\infty$ limit of (3.16a):

$$u^+(x)=\frac{1}{2\pi i}\int_0^\infty d\xi\rho(\lambda)N(\lambda)-i\sum_{j=1}^{\nu}C_jN_j. \tag{3.17}$$

Equation (3.17) defines $u(x)$ in terms of the spectral data, whose linear evolution is described by

$$\tilde{\lambda}_j(t)=\tilde{\lambda}_j(0),\qquad C_j(t)=C_j(0)e^{i\tilde{\lambda}_j[\tilde{\lambda}_j coth(\tilde{\lambda}_j\frac{\eta}{2})-2/\eta]t}, \tag{3.18a}$$

$$\rho(\lambda,t)=\rho(\lambda,0)e^{i\lambda(\lambda coth(\lambda\frac{\eta}{2})-2/\eta)t}. \tag{3.18b}$$

The 1-soliton solution obtains for $\rho=0$, $\nu=1$ and reads

$$u(x,t)=\frac{k_1 sin(k_1\eta/2)}{cos(k_1\eta/2)+cosh(k_1(x-x_1(t)))},\quad x_1(t)=\frac{1}{k_1}ln\frac{c_1}{k_1}+[-k_1cot(\frac{k_1\eta}{2})+\frac{2}{\eta}]t.$$

3.1.2 The IST formalism for the BO equation [32]

Direct problem. In this case equation (3.7) is interpreted as a differential RH problem in the complex x plane and $\mu^\pm$ represent the boundaries of functions analytic in the upper (+) and lower (-) half plane. The + solutions of (3.7), satisfying the asymptotics

$$m\to 1,\quad \bar{m}\to e^{i\lambda x},\quad x\to-\infty,$$

$$\bar{n}\to 1,\quad n\to e^{i\lambda x},\quad x\to\infty,$$

are characterized by the integral equations

$$\begin{pmatrix} m\\ \bar{m}\end{pmatrix}(x,\lambda)=\begin{pmatrix} 1\\ e^{i\lambda x}\end{pmatrix}+\int_{-\infty}^{\infty}dyg_+(x-y,\lambda)u(y)\begin{pmatrix} m\\ \bar{m}\end{pmatrix}(y,\lambda), \tag{3.19a}$$

$$\begin{pmatrix} n\\ \bar{n}\end{pmatrix}(x,\lambda)=\begin{pmatrix} e^{i\lambda x}\\ 1\end{pmatrix}+\int_{-\infty}^{\infty}dyg_-(x-y,\lambda)u(y)\begin{pmatrix} n\\ \bar{n}\end{pmatrix}(y,\lambda), \tag{3.19b}$$

where

$$g_\pm(x,\lambda)=\int_0^\infty\frac{dp}{2\pi}e^{ipx}(p-(\lambda\pm i0))^{-1}. \tag{3.19c}$$

The eigenfunctions are related through the scattering equation

$$m(x,\lambda)=\bar{n}(x,\lambda)+\theta(\lambda)\beta(\lambda)n(x,\lambda),\quad \beta(\lambda):=i\int_{-\infty}^{\infty}dyu(y)m(y,\lambda)e^{-i\lambda y},\quad \lambda\in\mathcal{R}. \tag{3.20}$$

Unlike the ILW case, the bound states $\lambda_j < 0, j = 1,..,\nu$ are associated now with the homogeneous solutions ϕ_j of equations (3.19):

$$\phi_j(x) = \int_{-\infty}^{\infty} dy g(x-y,\lambda_j)u(y)\phi_j(y) \tag{3.21}$$

Since the kernels of equation (3.19a) and (3.19b) are + and − functions, respectively, in λ, then the eigenfunctions m and $\tilde{n}$ are also + and − functions, apart from poles, corresponding to the homogeneous solutions, i.e.:

$$m(x,\lambda) = 1 + \sum_{j=1}^{\nu} \frac{c_j\phi_j(x)}{\lambda-\lambda_j} + m_+(x,\lambda), \quad \tilde{n}(x,\lambda) = 1 + \sum_{j=1}^{\nu} \frac{\tilde{c}_j\phi_j(x)}{\lambda-\lambda_j} + \tilde{n}_-(x,\lambda) \tag{3.22}$$

where m_+ and $\tilde{n}_-$ are + and − functions in λ, respectively.

It turns out that equations (3.20), (3.22), toghether with the symmetry condition

$$\frac{\partial}{\partial\lambda}(n(x,\lambda)e^{-i\lambda x}) = f(\lambda,t)e^{-i\lambda x}\tilde{n}(x,\lambda), \quad f(\lambda) := -\frac{1}{2\pi\lambda}\int_{-\infty}^{\infty} dy u(y) n(y,\lambda), \tag{3.23a}$$

and with equations

$$c_j = \tilde{c}_j = -i, \quad j = 1,..,\nu, \tag{3.23b}$$

$$\tilde{n} - i\phi_j/(\lambda-\lambda_j) \rightarrow (x+\gamma_j)\phi_j, \quad \lambda \rightarrow \lambda_j, \tag{3.23c}$$

define a *nonlocal* RH problem in the complex λ plane, equivalent to the following linear integral equations

$$n(x,\lambda) = \int_0^{\infty} \frac{dl}{2\pi}\beta(l)h(x,\lambda,l)n(x,l) + \sum_{j=1}^{\nu}\phi_j(x)h(x,\lambda,\lambda_j) = v(x,\lambda), \tag{3.24a}$$

$$(x+\gamma_j)\phi_j(x) - \frac{1}{2\pi i}\int_0^{\infty} d\lambda \frac{\beta(\lambda)n(x,\lambda)}{\lambda-\lambda_j} + i\sum_{s=1,\neq j}^{\nu}\frac{\phi_j(x)}{\lambda_j-\lambda_s} = 1, \tag{3.24b}$$

where

$$v(x,\lambda) := \int_0^{\lambda} dl[f(l)e^{i(\lambda-l)x} + \frac{e^{i\lambda x}}{l\ ln(l)}] \tag{3.25a}$$

$$h(x,\lambda,l) := e^{i(\lambda-l)x}\int_{\infty}^{x} dp v(p,\lambda)e^{-i(\lambda-l)p}, \quad l > 0, \tag{3.25b}$$

$$h(x,\lambda,\lambda_j) := e^{i(\lambda-l)x}\int_{\alpha}^{x} dp v(p,\lambda)e^{-i(\lambda-\lambda_j)p} +$$

$$e^{i(\lambda-\lambda_j)(x-\alpha)}\int_0^{\lambda} dl(\frac{f(l)e^{i\alpha(\lambda-l)}}{\lambda_j-l} + \frac{1}{\lambda_j l\ ln(l)}). \tag{3.25c}$$

The potential $u(x) = u^+(x) - u^-(x) = u^+(x) + \overline{u^+(x)}$ is reconstructed in terms of the scattering data through

$$u^+(x) = \int_0^\infty \frac{dl}{2\pi i}\beta(l)n(x,l) + i\sum_{j=1}^{\nu}\phi_j(x), \tag{3.26}$$

and the scattering data evolve according to

$$\lambda_j(t) = \lambda_j(0), \quad \gamma_j(t) = 2\lambda_j t + \gamma_j(0), \tag{3.27a}$$

$$\beta(\lambda,t) = \beta(\lambda,0)e^{i\lambda^2 t}, \quad f(\lambda,t) = f(\lambda,0)e^{i\lambda^2 t}. \tag{3.27b}$$

The 1-soliton solution obtains for $\beta = 0,\ \nu = 1$ and reads

$$u(x,t) = \frac{2\gamma_{1_I}(0)}{(x + 2\lambda_1 t + \gamma_{1_R}(0))^2 + \gamma_{1_I}(0)^2}, \quad \gamma_{1_I} = Im\ \gamma_1, \quad \gamma_{1_R} = Re\ \gamma_1.$$

Remark 3.1 Nonlocal RH problems of this type are peculiar of inverse problems in multidimensional spaces [12].

3.1.3 Connection between the two IST formalisms [33]

The IST formalism of ILW is conceptually similar to that of the KdV equation and the $\eta \to 0$ limit of it is straightforward [19]. On the other hand the IST formalism of BO is similar to that of KPI, therefore the $\eta \to \infty$ limit of the IST formalism of ILW provides an interesting link between two different types of IST formalisms, appropriate to one and two space dimensions. Such a limit has been considered in detail in [33]. Here we summarize few aspects of it.

i) The information associated with the symmetry condition (3.14), which follows from

$$G_\pm(x,\lambda) = G_\pm(x,-\lambda)e^{i\lambda x}, \tag{3.28}$$

is apparently lost in the $\eta \to \infty$ limit, which leads to an identity. To restore such information we use the noncommutativity of the operations $\eta \to \infty$ and $\partial/\partial\lambda$. Indeed if we first take the λ derivative of equation (3.28) and of equation (3.10b), and then we take the $\eta \to \infty$ limit of them, we obtain

$$g_{\pm_\lambda}(x,\lambda) = ixg_\pm(x,\lambda) - \frac{1}{2\pi\lambda} \tag{3.29}$$

and equation (3.23a).

ii) For large η each of the ILW eigenvalues $\tilde\lambda_j$ tends to the upper edge of the principal branch:

$$\tilde\lambda_j = \frac{2\pi i}{\eta}\left(1 + \frac{1}{\eta\lambda_j} + \frac{1}{(\eta\lambda_j)^2} + \mathcal{O}(\eta^{-3})\right); \tag{3.30}$$

at the same time $M_j = \mathcal{O}(\eta)$ and $\phi_j = -2\pi \lim_{\eta\to\infty}(M_j/\eta)$ is the solution of the homogeneous equation (3.21), corresponding to the eigenvalue λ_j. That's how to each zeroe $\tilde\lambda_j$ of a it corresponds, in the limit $\eta \to \infty$, an eigenvalue λ_j of the Fredholm equation (3.19).

3.2 A singular integral extension of the AKNS and KdV hierarchies

The second relevant example of RH spectral problem in the strip $0 < Im\ x < \eta$ is given by the following 2x2 matrix equation introduced in [34]:

$$\psi^-(x,z) = U(x,z)\psi^+(x,z), \quad -1 < z < 1, \quad x \in \mathcal{R}, \tag{3.31a}$$

$$U(x,z) := z\sigma_3 + Q(x), \tag{3.31b}$$

where σ_3 :=diag$(1,-1)$ and z is the spectral parameter. For the solvability of (3.31) it is important that $det\ U$ be x-independent, namely that

$$det\ Q - 1 = tr(\sigma_3 Q) = 0 \quad \Rightarrow \quad Q = \begin{pmatrix} r & q_{12} \\ q_{21} & r \end{pmatrix}, \quad r := \sqrt{1 + q_{12}q_{21}}. \tag{3.32}$$

In this case $det\ U = 1 - z^2$ and i) U is invertible for $z \neq \pm 1$ for every $x \in \mathcal{R}$ (a necessary condition for the solvability of (3.31)). ii) The total index $J = \frac{1}{2\pi}[arg(det\ U(x))]_{-\infty}^{\infty}$ is zero; therefore the two partial indeces of the RH problem are generically zero [36]. This fact garanties the existence and uniqueness of a bounded fundamental matrix solution of (3.31).

It turns out that the constraint (3.32), introduced here as a requirement for the solvability of (3.31), is compatible with the evolutions associated with (3.31). Such equations can be obtained seeking evolutions compatible with (3.31) in the form [34]

$$\psi_t^{\pm} = B^{\pm}(x,t;z)\psi^{\pm} := \sum_{j=0}^{m} z^j B_j^{\pm}(x,t)\psi^{\pm}, \quad B_m^{\pm} = \frac{a_m}{2}\sigma_3, \tag{3.33}$$

The first few examples are

$$\begin{pmatrix} q_{12} \\ q_{21} \end{pmatrix}_t = ic_1 \begin{pmatrix} r\mathcal{D}q_{12} \\ r\mathcal{D}q_{21} \end{pmatrix}, \quad r := \sqrt{1 + q_{12}q_{21}}, \quad m = 1, \tag{3.34a}$$

$$\begin{pmatrix} q_{12} \\ q_{21} \end{pmatrix}_t = c_2 \begin{pmatrix} r\mathcal{D}r\mathcal{D}q_{12} - \frac{1}{2}q_{12}\mathcal{T}(q_{12}\mathcal{D}q_{21} + q_{21}\mathcal{D}q_{12}) \\ -r\mathcal{D}r\mathcal{D}q_{21} + \frac{1}{2}q_{21}\mathcal{T}(q_{12}\mathcal{D}q_{21} + q_{21}\mathcal{D}q_{12}) \end{pmatrix}, \quad m = 2, \tag{3.34b}$$

$$\begin{pmatrix} q_{12} \\ q_{21} \end{pmatrix}_t = ic_3 \begin{pmatrix} r\mathcal{D}r\mathcal{D}r\mathcal{D}q_{12} - \frac{r}{2}\mathcal{D}q_{12}\mathcal{T}(q_{12}\mathcal{D}q_{21} + q_{21}\mathcal{D}q_{12}) + \frac{1}{2}q_{12}\mathcal{T}(q_{12}\mathcal{D}r\mathcal{D}q_{21} - q_{21}\mathcal{D}r\mathcal{D}q_{12}) \\ r\mathcal{D}r\mathcal{D}r\mathcal{D}q_{21} - \frac{r}{2}\mathcal{D}q_{21}\mathcal{T}(q_{12}\mathcal{D}q_{21} + q_{21}\mathcal{D}q_{12}) + \frac{1}{2}q_{21}\mathcal{T}(q_{21}\mathcal{D}r\mathcal{D}q_{12} - q_{12}\mathcal{D}r\mathcal{D}q_{21}) \end{pmatrix},$$
$$m = 3. \tag{3.34c}$$

If $q_{12} = \epsilon q_{21} = \epsilon q$, equations (3.34a) and (3.34c) become, respectively,

$$q_t = c\sqrt{1 + \epsilon q^2}\mathcal{D} \quad \Rightarrow \quad \theta_t = c\mathcal{D}sin\theta, \quad \mathcal{T}\theta_t = -csin\theta, \quad q = isin\theta, \tag{3.35}$$

$$q_t = cr\mathcal{D}(r\mathcal{D}r\mathcal{D}q - \epsilon q\mathcal{T}q\mathcal{D}q), \quad r := \sqrt{1 + \epsilon q^2}, \tag{3.36}$$

while, if $q_{12} = \epsilon q_{\bar{21}} = -i\epsilon\bar{q}$, equation (3.34b) reduces to

$$iq_t + c[r\mathcal{D}r\mathcal{D}q - \epsilon q\mathcal{T}(Re(\bar{q}\mathcal{D}q))] = 0, \quad r = \sqrt{1 + \epsilon|q|^2}. \tag{3.37}$$

Equations (3.34a) and (3.34c) admit also the reduction $q_{21} = q$, $q_{12} = 1$ to, respectively,

$$q_t = c\sqrt{1 + q}\mathcal{D}q \quad \Rightarrow \quad u_t = \frac{c}{2}\mathcal{D}u^2, \quad q = u^2 - 1, \tag{3.38}$$

$$q_t = cr(\mathcal{D}r\mathcal{D}r\mathcal{D}q + \frac{1}{2}\mathcal{D}(q^2) + \frac{1}{2}q\mathcal{D}q), \qquad r = \sqrt{1+q}. \tag{3.39}$$

In the $\eta \to \infty$ limit, equation (3.31) becomes a RH problem in the complex x plane and the associated IDEE's obtain from equations (3.34)- (3.39) replacing the operators $\mathcal{T}$ and $\mathcal{D}$ by the Hilbert transform $\mathcal{H}$. In particular, equations (3.35) and (3.38) reduce to the so-called Sine-Hilbert equation

$$\mathcal{H}\theta_t = -c\sin\theta, \tag{3.40}$$

and to

$$u_t = \frac{c}{2}\mathcal{H}(u^2) \quad \Rightarrow \quad \mathcal{H}u_t = -\frac{c}{2}u^2. \tag{3.41}$$

In the $\eta \to 0$ limit, equation (3.31) reduces to the generalized Zakharov- Shabat spectral problem [37] and then the associated evolution equations (3.34)-(3.37) and (3.38,39) reduce to equations of the AKNS and KdV hierarchies respectively. More precisely, equations (3.37), (3.36) and (3.39) go, respectively, to the NLS, MKdV and KdV equations. Both equations (3.35) and (3.38) reduce to the linear wave equation.

The spectral problem (3.31), the general form of the associated evolution equations and conservation laws were first obtained in [34]; the analyticity reduction (3.32) and soliton solutions in [35]. A BT was derived in [38] and reads

$$(r + q_{12}(\mathcal{D}q'_{21}))\mathcal{D}q'_{21} + q_{12}q'_{21} - iz_0 q'_{21} + \frac{1}{4}q_{21} = 0,$$

where (q_{12}, q_{21}) and (q'_{12}, q'_{21}) are different solutions of the same equation of the hierarchy. Its remarkable solution is

$$q'_{21} = -\frac{i}{4}\frac{q_{12}\varphi^2 + 2r\varphi + q_{21}}{q_{12}\varphi + r + z_0}, \qquad \varphi = \frac{\psi_2}{\psi_1},$$

where $(\psi_1, \psi_2)^T$ is a vector solution of (3.31).

The solution of the initial value problem for the above IDEE's (respectively for the $\eta \to \infty$ limits of them) has been considered in [39] (respectively in [40]) and here we briefly review the two results.

3.2.1 The IST formalism for the nonlocal AKNS hierarchy [39]

Direct problem. It is convenient to parametrize z in terms of a new spectral parameter k through the transformation

$$z = z(k) = tgh(\eta k). \tag{3.42}$$

Its inverse is multivalued and the principal branch is the strip $-\frac{\pi}{2\eta} < Im\ k < \frac{\pi}{2\eta}$ of the complex k plane. The real line $Im\ k = 0$ corresponds to the segment $-1 < z < 1$ of the complex z plane. It is also convenient to introduce the eigenfunction

$$\mu^+(x,z) = \psi^+(x,z)e^{i\zeta x}, \qquad e^{\eta\zeta} := I + z\sigma_3. \tag{3.43}$$

The + matrix solutions (3.43), satisfying the asymptotics

$$\mu_L(x,z) \to I, \quad x \to -\infty; \qquad \mu_R(x,z) \to I, \quad x \to \infty$$

are characterized by the integral equations

$$\mu_L(x,z) = I - \frac{e^{-\eta\zeta}}{2i\eta}\int_{-\infty}^{\infty} dy[coth(\frac{\pi}{\eta}(y-(x+i0))) - 1]e^{i\zeta(y-x)}(Q(y)-I)\mu_L(y,z)e^{-i\zeta(y-x)}, \tag{3.44a}$$

$$\mu_R(x,z) = I - \frac{e^{-\eta\zeta}}{2i\eta}\int_{-\infty}^{\infty} dy[coth(\frac{\pi}{\eta}(y-(x+i0))) + 1]e^{i\zeta(y-x)}(Q(y)-I)\mu_L(y,z)e^{-i\zeta(y-x)}, \tag{3.44b}$$

and are related through the scattering equation

$$\mu_L(x,z) = \mu_R(x,z)e^{-i\zeta x}S(z)e^{i\zeta x}, \quad -1 < z < 1, \tag{3.45a}$$

where the scattering matrix S is given by

$$S(z) := I - \frac{e^{-\eta\zeta}}{i\eta}\int_{-\infty}^{\infty} dx e^{i\zeta x}(Q(x)-I)\mu_L(x,z)e^{-i\zeta x}.$$

Furthermore equations (3.32) imply that $det\ S = 1$.

It is possible to show that $(\mu_{L_{11}},\mu_{L_{21}})^T$, $(\mu_{R_{12}},\mu_{R_{22}})^T$ and $S_{11}(k)$ are analytically extendable upstairs, in the strip $0 < Im\ k < \frac{\pi}{\eta}$, while $(\mu_{L_{12}},\mu_{L_{22}})^T$, $(\mu_{R_{11}},\mu_{R_{21}})^T$ and $S_{22}(k)$ are analytically extendable downstairs, in the strip $-\frac{\pi}{\eta} < Im\ k < 0$.

The bound states correspond to those $k_j,\ j = 1,...,n$ and $\bar{k}_j,\ j = 1,..,\bar{n}$ such that

$$S_{11}(k_j) = 0,\ \ j = 1,..,n; \quad S_{22}(\bar{k}_j) = 0,\ \ j = 1,..,\bar{n}, \tag{3.46a}$$

$$\begin{pmatrix}\mu_{L11}\\ \mu_{L21}\end{pmatrix}(k_j) = S_{21}(k_j)e^{2ik_jx}\begin{pmatrix}\mu_{R12}\\ \mu_{R22}\end{pmatrix}(k_j),\ \ j = 1,..,n, \tag{3.46b}$$

$$\begin{pmatrix}\mu_{L12}\\ \mu_{L22}\end{pmatrix}(\bar{k}_j) = S_{12}(\bar{k}_j)e^{-2i\bar{k}_jx}\begin{pmatrix}\mu_{R11}\\ \mu_{R21}\end{pmatrix}(\bar{k}_j),\ \ j = 1,..,n, \tag{3.46c}$$

Inverse problem. The analyticity informations of the direct problem suggest to rewrite equation (3.45a) in the following form:

$$\Phi_+(k) = \Phi_-(k)e^{-i\zeta x}\Pi(k)e^{i\zeta x}, \quad k\in\mathcal{R}, \tag{3.47a}$$

$$\Phi_+(k) = \begin{pmatrix}\mu_{L_{11}}/S_{11} & \mu_{R_{12}}\\ \mu_{L_{21}}/S_{11} & \mu_{R_{22}}\end{pmatrix}, \qquad \Phi_-(k) = \begin{pmatrix}\mu_{R_{11}} & \mu_{L_{12}}/S_{22}\\ \mu_{R_{21}} & \mu_{L_{22}}/S_{22}\end{pmatrix} \tag{3.47b}$$

$$\Pi(k) := \begin{pmatrix}1-r_{12}r_{21} & -r_{12}\\ r_{21} & 1\end{pmatrix}, \qquad r_{ij} := \frac{S_{ij}}{S_{jj}},\ i\neq j, \tag{3.47c}$$

Furthermore we have the periodicity condition

$$\Phi_+(k+\frac{i\pi}{\eta}) = \Phi_-(k), \quad S_{11}(k+\frac{i\pi}{\eta}) = 1/S_{22}(k). \tag{3.48}$$

Equation (3.47),(3.48) define a matrix RH problem in the strip of the complex k plane with poles (the zeroes of S_{11} and S_{22}). Notice the beautiful symmetry between direct and inverse problems: they are both RH problems in strips!

The solution of (3.47), (3.48) is given by the integral equations

$$\begin{pmatrix}\mu_{L_{11}}\\ \mu_{L_{21}}\end{pmatrix}(z) = \begin{pmatrix}1\\ 0\end{pmatrix} + \frac{1}{2\pi i}\int_{-1}^{1}\frac{dz'}{z'-(z-i0)}r_{21}(z')e^{2ik'x}\begin{pmatrix}\mu_{R_{12}}\\ \mu_{R_{22}}\end{pmatrix}(z') +$$

$$\sum_{j=1}^{n} \frac{c_j}{z-z_j} e^{2ik_j x} \begin{pmatrix} \mu_{R_{12}} \\ \mu_{R_{22}} \end{pmatrix}(z_j), \tag{3.49a}$$

$$\begin{pmatrix} \mu_{R_{12}} \\ \mu_{R_{22}} \end{pmatrix}(z) = \begin{pmatrix} 0 \\ 1 \end{pmatrix} - \frac{1}{2\pi i} \int_{-1}^{1} \frac{dz'}{z' - (z + i0)} r_{12}(z') e^{-2ik' x} \begin{pmatrix} \mu_{L_{11}} \\ \mu_{L_{21}} \end{pmatrix}(z') +$$

$$\sum_{j=1}^{\tilde{n}} \frac{\tilde{c}_j}{z - \tilde{z}_j} e^{-2ik_j \frac{\nu}{\eta} x} \begin{pmatrix} \mu_{L_{11}} \\ \mu_{L_{21}} \end{pmatrix}(z_j), \tag{3.49b}$$

where $c_j := (S_{21}/S_{11_z})|_{z=z_j}$, $\quad \tilde{c}_j := (S_{12}/S_{22_z})|_{z=\tilde{z}_j}$.

The reconstruction of $Q(x)$ is acheived noticing that

$$Q(x) - I = M^-(x)\sigma_3 - \sigma_3 M^+(x), \quad M^+(x) \lim_{|z|\to\infty} z(\mu_R - I), \quad M^-(x) = M^+(x + i\eta). \tag{3.50}$$

The time evolution of the spectral data is finally given by

$$z_j(t) = z_j(0), \quad \tilde{z}_j(t) = \tilde{z}_j(0), \quad c_j(t) = c_j(0) e^{-2\beta(z_j)t}, \quad \tilde{c}_j(t) = \tilde{c}_j(0) e^{2\beta(\tilde{z}_j)t}, \tag{3.51a}$$

$$r_{21}(z,t) = r_{21}(z,0) e^{-\beta(z)t}, \quad r_{12}(z,t) = r_{12}(z,0) e^{\beta(z)t}, \tag{3.51b}$$

where $\beta(z) = \lim_{|x|\to\infty} \sigma_3 B^{\pm}(x,t;z)$, and the 1-soliton solution, corresponding to $r_{12} = r_{21} = 0$ and $\nu = \tilde{\nu} = 1$, reads

$$q_{12} = (-\tilde{c}_1 e^{-\gamma_1})[cosh(\eta k_1) e^{-\theta_-(x,\tilde{k}_1)} + cosh(\eta \tilde{k}_1) e^{-\theta_+(x,k_1)}]/d(x),$$

$$q_{21} = (c_1 e^{-\gamma_1})[cosh(\eta k_1) e^{\theta_-(x,\tilde{k}_1)} + cosh(\eta \tilde{k}_1) e^{\theta_+(x,k_1)}]/d(x),$$

$$\theta_\pm(x,k) := 2ik_1 x - \eta k_1 \pm \gamma_1, \quad d(x) := cosh(\eta(k_1 - \tilde{k}_1)) + cosh(\theta_+(x,k_1) - \theta_-(x,\tilde{k}_1)),$$

$$e^{2\gamma_1} := \frac{c_1 \tilde{c}_1}{(z_1 - \tilde{z}_1)^2}, \quad z_1 = tgh(\eta k_1), \quad \tilde{z}_1 = tgh(\eta \tilde{k}_1), \quad 0 < Im\, k_1 < \frac{\pi}{2\eta}, \quad -\frac{\pi}{2\eta} < Im\, \tilde{k}_1 < 0.$$

Remark 3.2 We stress the beautiful symmetry between direct and inverse transforms: they are both RH problems in strips! This is a nonlinear counterpart of the symmetry between direct and inverse FT. If $\eta \to 0$ such a symmetry is broken, since the direct problem becomes an ODE (the generalized Zakharov - Shabat spectral problem [37]) and the inverse problem a RH problem in the k plane.

3.2.2 The IST formalism for SH hierarchy [40]

Direct problem. In the $\eta \to \infty$ limit of (3.31) the eigenfunctions become the boundary values of functions analytic in the upper (+) and lower (-) half x plane. Those satisfying the asymptotics $\mu^{\pm} \to I + z\sigma_3$, $x \to \pm\infty$ are characterized by the Fredholm integral equations

$$\mu^+(x,z) + (I + z\sigma_3)^{-1} \frac{1}{2}(1 - i\mathcal{H})((Q - I)\mu^+)(x,z) = I \tag{3.52a}$$

$$\mu^-(x,z) - \frac{1}{2}(1 + i\mathcal{H})((Q - I)\mu^+)(x,z)(I + z\sigma_3)^{-1} = I \tag{3.52b}$$

These Fredholm equations can have homogeneous matrix solutions $\phi_j^{\pm(j)}(x)$, $j \in \mathcal{Z}$ corresponding to the eigenvalues z_j; they are singular matrices:

$$\begin{pmatrix} \phi_{12}^{\pm(j)} \\ \phi_{22}^{\pm(j)} \end{pmatrix}(x) = \alpha_j^{\pm} \begin{pmatrix} \phi_{11}^{\pm(j)} \\ \phi_{21}^{\pm(j)} \end{pmatrix}(x) =: \alpha_j^{\pm} \Pi^{\pm(j)}(x), \quad j \in \mathcal{Z}. \tag{3.53}$$

From Fredholm theory it follows that $\mu^\pm$ are holomorphic in z, apart from the poles z_j which generically cluster at ± 1:

$$\mu^\pm(x,z) = I + \sum_{j\in\mathbb{Z}} \frac{1}{z - z_j}\phi^{\pm(j)}(x); \qquad z \to \pm 1 \quad as \quad j \to \pm\infty. \tag{3.54}$$

Inverse Problem. The vector solutions $\Pi^{\pm(j)}$ of equations (3.31) satisfy the infinite dimensional algebraic system

$$(x - \gamma_j^\pm)\Pi^{\pm(j)}(x) = c_j[\begin{pmatrix} 1 \\ -1/\alpha_j^\pm \end{pmatrix} + \sum_{s\in\mathbb{Z}\neq j} \frac{\alpha_j^\pm - \alpha_s^\pm}{\alpha_j^\pm(z_j - z_s)}\Pi^{\pm(s)}(x)] \tag{3.55}$$

in terms of the spectral data $\{z_j, c_j, \alpha_j^\pm, \gamma_j^\pm\}$, where:

$$c_j := \lim_{|x|\to\infty} x\phi_{11}^{\pm(j)}, \quad \gamma_j^\pm := \sum_{s\in\mathbb{Z}\neq j} \frac{\alpha_j^\pm - \alpha_s^\pm}{\alpha_j^\pm(z_j - z_s)}c_s + \lim_{|x|\to\infty}(\frac{1 - x\phi_{11}^{\pm(j)}}{c_j}) \tag{3.56a}$$

and

$$(1 - z_j)\alpha_j^- = (1 + z_j)\alpha_j^+, \quad \gamma_j^+ - \gamma_j^- = \frac{2c_j}{1 - z_j^2}. \tag{3.56b}$$

Matrix $Q(x)$ is finally reconstructed using equation

$$Q(x) - I = \sum_{j\in\mathbb{Z}} (\phi^{-(j)}(x)\sigma_3 - \sigma_3\phi^{+(j)}(x)) \tag{3.57}$$

which follows from equation (3.31) for large z. The scattering data evolve linearly:

$$z_j(t) = z_j(0), \quad c_j(t) = c_j(0), \quad \alpha_j^\pm(t) = \alpha_j^\pm(0)e^{a_m z_j{}^m t}, \tag{3.58a}$$

$$\gamma_j^\pm(t) = \gamma_j^\pm(0) - na_m c_j z_j{}^{m-1}t, \quad j \in \mathbb{Z} \tag{3.58b}$$

and the 1-soliton solution obtains choosing only one pole z_0:

$$q_{12} = -\frac{2c_0\alpha_0^+(t)}{1 - z_0}\frac{x + \gamma_0^+(t) - c_0/(1 + z_0)}{(x - \gamma_0^+(t))(x + \gamma_0^-(t))},$$

$$q_{21} = -\frac{2c_0}{\alpha_0^+(t)(1 + z_0)}\frac{x + \gamma_0^+(t) - c_0/(1 - z_0)}{(x - \gamma_0^+(t))(x + \gamma_0^-(t))}.$$

Remark 3.3 For the simplest equation of this hierarchy (the SH equation (3.40)) it is possible to obtain large classes of solutions through a particularly simple direct method, based on the change of variables (a RH problem!) [41]

$$\theta = i(ln\ f^+ - ln\ f^-), \quad x \in \mathcal{R}, \tag{3.59a}$$

which transforms equation (3.40) into the equation

$$f^+[f_t^+ - \frac{1}{2i}(f^- - f^+)] = -f^-[f_t^+ - \frac{1}{2i}(f^- - f^+)] \tag{3.59b}$$

satisfied by

$$f_t^\pm = \frac{1}{2i}(f^- - f^+) \pm ipf^\pm. \tag{3.59c}$$

If p does not depend on $f^{\pm}$, equations (3.59) linearize the SH equation and provide large classes of solutions, of rational and periodic type, investigated in great detail [41]. However this technique should not provide the linearization of the general Cauchy problem since, for an arbitrary initial condition, one expects a complicated functional dependence of p on $f^{\pm}$. The technique illustrated above has been succesfully applied to other IDEE's [42].

3.3 Other IDEE's integrable by the IST

In this section we list other interesting examples of integrable IDEE's known in literatute.

i) We begin with the IDEE

$$u_t + \mathcal{T}u_{xx} + u_x(c_1 + c_2 e^{iu} + i\mathcal{T}u_x) = 0 \tag{3.60}$$

which is called modified intermediate long wave (MILW) equation because it was derived in [43] from the complete Baklund transformation (BT) of the ILW equation, in perfect analogy with the derivation of MKdV from KdV; in addition its $\eta \to 0$ limit reduces to the MKdV equation (1.6b).

A BT, an infinite number of conserved quantities and an associated IST scheme for the MILW equation (3.60) have been derived in [44].

ii) The system of IDEE's

$$u_{1_t} = (2u_0 - u_{1_x} - 2i\mathcal{T}u_{1_x} - u_1^2)_x \tag{3.61a}$$

$$u_{0_t} = (u_0 - u_{1_x} - i\mathcal{T}u_{1_x})_x + u_1(u_0 - u_{1_x} - i\mathcal{T}u_{1_x})_x - u_1 u_{0_x} - 2u_0 u_{1_x} \tag{3.61b}$$

is the first nontrivial member of a hierarchy (called ILW_2 hierarchy) of IDEE's associated with the RH problem [27], [45], [46]

$$L_2\psi^+ = \lambda\psi^-, \quad L_2 := \frac{\partial^2}{\partial x^2} + u_1\frac{\partial}{\partial x} + u_0. \tag{3.62}$$

The Miura-type transformation

$$\left(\frac{\partial}{\partial x} + q_1\right)\left(\frac{\partial}{\partial x} + q_2\right) = \frac{\partial^2}{\partial x^2} + u_0\frac{\partial}{\partial x} + u_1 \tag{3.63}$$

maps the first equation (3.61) of the ILW_2 hierarchy to the first equation [45]

$$q_{1_t} = -(-q_{2_x} + i\mathcal{T}(q_1 + q_2)_x + q_1^2)_x, \tag{3.64a}$$

$$q_{2_t} = -(q_{1_x} + i\mathcal{T}(q_1 + q_2)_x + q_2^2)_x \tag{3.64b}$$

of a "modified" class of IDEE's (called the $MILW_2$ hierarchy), whose spectral prblem is conveniently written in matrix form

$$L\psi = \hat{\Lambda}\psi, \qquad L := I\frac{\partial}{\partial x} + \begin{pmatrix} q_1 & 0 \\ 0 & q_2 \end{pmatrix}, \quad \hat{\Lambda} := \begin{pmatrix} 0 & \lambda\mathcal{E} \\ 1 & 0 \end{pmatrix}. \tag{3.65}$$

iii) In the Lax pair (3.1), (3.2) of the ILW hierarchy, the operators $M^{\pm}$ are polynomial expansions in $\frac{\partial}{\partial x}$, but nothing prevents us from rewriting $M^{\pm}$ in terms of the operator $\lambda\mathcal{E}$, obtaining:

$$i\psi_x = (-u + \lambda\mathcal{E})\psi, \quad \psi = \psi^+, \quad \mathcal{E}\psi = \psi^-, \tag{3.66a}$$

$$i\psi_t = \sum_j b_j(\lambda\mathcal{E})^j\psi; \tag{3.66b}$$

for instance, for the ILW equation:

$$i\psi_t = ((\lambda\mathcal{E})^2 + \lambda(\mathcal{E}u + u\mathcal{E}) + u^2 + i(\mathcal{T}u_x))\psi. \tag{3.67}$$

Expansion (3.66b) becomes obviously convenient if we allow for negative powers of $\lambda\mathcal{E}$; for instance, the compatibility between (3.66a) and

$$i\psi_t = \frac{i}{\lambda}e^{i[\theta(x)-\theta(x-i\eta)]}(\lambda\mathcal{E})^{-1}\psi, \qquad \theta_x = u, \tag{3.68a}$$

gives rise to the following intermediate Toda lattice (ITL_1) [45]

$$\theta_{xt} = e^{i[\theta(x+i\eta)-\theta(x)]} - e^{i[\theta(x)-\theta(x-i\eta)]}. \tag{3.68}$$

Analogously, the compatibility between the RH problem (3.65) of ILW_2 and

$$\psi_t = e^{-\Theta}\hat{\Lambda}^{-1}e^{\Theta}\psi, \qquad \Theta_x = \begin{pmatrix} \theta_{1_x} & 0 \\ 0 & \theta_{2_x} \end{pmatrix} = \begin{pmatrix} q_1 & 0 \\ 0 & q_2 \end{pmatrix}$$

gives rise to the two-component ITL_2 [45]

$$\theta_{1_{xt}} = e^{\theta_1(x)-\theta_2(x)} - e^{\theta_2(x+i\eta)-\theta_1(x)}, \tag{3.69a}$$

$$\theta_{2_{xt}} = e^{\theta_2(x)-\theta_1(x-i\eta)} - e^{\theta_1(x)-\theta_2(x)}. \tag{3.69b}$$

Remark 3.4. Although the spectral problems of this paper depend polynomially on the shift operator $\mathcal{E}$, in the associated evolution equations derived so far $\mathcal{E}$ appears only in the special rational combinations $\mathcal{T}$ and $\mathcal{D}$. For equations (3.68), (3.69), which are the first exceptions, the initial condition must be analytically extendable into the x plane, unlike all the previous IDEE's for which the initial datum must be well behaved only for $x \in \mathcal{R}$ (for instance a Schwartz function).
Remark 3.5. Since, by construction, the three linear problems (3.66a), (3.67) and (3.68a) are compatible, it follows that the ILW equation is a symmetry of the ITL (3.68b) and viceversa. The same is true for the equations (3.61) and (3.69).

The bilinear formalisms and the solitonic τ functions associated with the ILW_n, $MILW_n$ and ITL_n hierarchies have been recently constructed in [47]. Furthermore it has been shown that they are reductions of the so-called infinite dimensional Toda lattice hierarchy.

iv) The last example of IDEE in 1+1 dimensions is [48]

$$(c - 2q_x)q_{xt} + (1 - 2q_t)q_{xx} - \mathcal{T}q_{txx} = 0 \tag{3.70}$$

Periodic and soliton solutions, an infinite number of conserved quantities and an associated IST scheme are given in [49].

v) We finally remark that 2+1 dimensional extensions of the above IDEE's are also possible. The following 2+1 dimensional extension of ILW [50]

$$u_t = \mathcal{T}u_{yy} + 2uu_y \tag{3.71}$$

is associated with the differential RH problem

$$i\psi_y^+ + u\psi^+ = \lambda\psi^-, \quad \psi^-(x,y) = \psi^+(x+i\eta, y). \tag{3.72}$$

Equation (3.71) is applicable in Fluid dynamics and reduces, in the shallow water limit, to the KP equation. Solitons and finite gap solutions of (3.71) were derived in [50]; its bi-Hamiltonian structures was obtained in [51].

4. ALGEBRAIC PROPERTIES AND OPERATOR STRUCTURES OF IDEE'S

4.1 Pseudo-differential operators and zero curvature representation [27]

It was shown in [27] that the dressing method can be extended to pseudo- differential operators in order to construct and solve integrable IDEE's. We have the following *Proposition 1.* The hierarchy of IDEE's associated with the RH spectral problem

$$L_n\psi^+ = \lambda\psi^-, \quad L_n := \sum_{j=0}^{n} u_j \frac{\partial^j}{\partial x^j}, \quad u_N = 1, \tag{4.1}$$

(called the ILW_n hierarchy) admits the following representation

$$L_{n_t} = M_{n,m}^- L_n - L_n M_{n,m}^+, \quad n \in \mathcal{N} \tag{4.2a}$$

where the operators $M_{n,m}^{\pm}$ are defined by

$$M_{n,m}^{\pm} := (K^{\pm}\frac{\partial^m}{\partial x^m}(K^{\pm})^{-1})_+ \tag{4.2b}$$

in terms of the pseudo - differential operators

$$K^{\pm} = \sum_{j\geq N}^{\infty} K_j^{\pm}(x,t)\partial_x^{-j}, \quad \partial_x := \frac{\partial}{\partial x}, \tag{4.3}$$

characterized by the RH problem

$$K^-\partial_x^n = L_n K^+, \qquad K_j^-(x) = K_j^+(x+i\eta). \tag{4.4}$$

In formula (4.2b) $()_+$ is the positive part of (the expansion of) a pseudo-differential operator.

As we have seen in the previous section, the construction of IDEE's can be acheived either using expansions in derivatives or translation operators and, sometimes, this second representation is more convenient, like in examples (3.68) and (3.69). A version of the Dressing method based on a formal expansion in terms of the "noncommutative spectral parameter" $\lambda\mathcal{E}$ has been recently developed to obtain the zero curvature representation of IDEE's like the ILW_n, the $MILW_n$ and the ITL_n hierarchies [45], [46].

4.2 Recursion and bi-Hamiltonian operators, symmetries, conservation laws and BT's of IDEE's

The basic property of an integrable evolutionary system

$$q_t = k(q) \tag{4.5}$$

is the existence of a "sufficient" number of symmetries $\sigma = \sigma(q)$, i.e. of flows $q_t = \sigma$

commuting with the flow (4.5). This implies that

$$\frac{\partial \sigma}{\partial t} + \sigma_f[k] - k_f[\sigma] = 0, \tag{4.6}$$

where by $\sigma_f[k]$ we denote the Fréchet derivative of σ in the direction k, i.e.

$$\sigma_f[k] := \frac{\partial}{\partial \epsilon}\sigma(u + \epsilon k)|_{\epsilon=0}. \tag{4.7}$$

The algebraic properties of integrable systems are conveniently described by introducing the so-called recursion operator Φ, which satisfies the following properties.
1) Φ generates hierarchies of evolution equations associated with a given spectral problem (first shown in [37]); 2) a suitable extension of Φ generates BT's [52]; 3) the adjoint Φ^* of Φ is the "squared eigenfunction" operator [37]; 4) Φ generates ∞-many commuting symmetries [53],[54]; 5) it is factorizable in terms of two Hamiltonian operators, then the associated equations are bi-Hamiltonian [55]; 6) Φ^* generates ∞-many gradients of constants of motion in involution [55].

4.2.1 The nonlocal AKNS hierarchy

As illustrative example of such a theory we consider the nonlocal AKNS hierarchy, which is conveniently represented in the following form:

$$\begin{pmatrix} q_{12} \\ q_{21} \end{pmatrix}_t = \mathbf{k}^{(n)} := c_n \Phi^n \sigma^{(0)} \tag{4.8}$$

where $\sigma^{(0)}$ is the "starting symmmetry"

$$\sigma^{(0)} := \begin{pmatrix} q_{12} \\ -q_{21} \end{pmatrix} \tag{4.9a}$$

and Φ is the recursion operator

$$\Phi := i \begin{pmatrix} r\mathcal{D} - \frac{1}{2}q_{12}\mathcal{T}\frac{1}{r}q_{21} & -\frac{1}{2}q_{12}\mathcal{T}\frac{1}{r}q_{21} \\ \frac{1}{2}q_{21}\mathcal{T}\frac{1}{r}q_{21} & -r\mathcal{D} + \frac{1}{2}q_{21}\mathcal{T}\frac{1}{r}q_{12} \end{pmatrix}, \quad r := \sqrt{1 + q_{12}q_{21}} \tag{4.9b}$$

Equations (3.34a), (3.34b) and (3.34c) correspond to $n = 1$, 2 and 3 respectively (with $c_n = ic$, $c \in \mathcal{R}$).

The recursion operator Φ and the starting symmetry $\sigma^{(0)}$ satisfy the following properties.
1) Φ is a hereditary or Nijenhuis operator, i.e.

$$\Phi_f[\Phi\mathbf{g}]\tilde{\mathbf{g}} - \Phi\Phi_f[\mathbf{g}]\tilde{\mathbf{g}} \tag{4.10}$$

is symmetric with respect to the two-dimensional vectors $\mathbf{g}$ and $\tilde{\mathbf{g}}$.
2) Φ is a strong symmetry for $\sigma^{(0)}$, i.e. the Lie derivative of Φ along $\sigma^{(0)}$ is zero:

$$\Phi_f[\sigma^{(0)}] + \Phi\sigma_f^{(0)} - \sigma_f^{(0)}\Phi = 0 \tag{4.11}$$

3) Φ admits the factorization $\Theta_2 = \Phi\Theta_1$ in terms of the operators

$$\Theta_1 := r\begin{pmatrix} 0 & 1 \\ -1 & 0 \end{pmatrix}, \quad \Theta_2 := \begin{pmatrix} \frac{1}{2}q_{12}\mathcal{T}q_{12} & r\mathcal{D}r - \frac{1}{2}q_{12}\mathcal{T}q_{21} \\ r\mathcal{D}r - \frac{1}{2}q_{21}\mathcal{T}q_{12} & \frac{1}{2}q_{21}\mathcal{T}q_{21} \end{pmatrix}. \tag{4.12}$$

It turns out that $\Theta = \Theta_2 + \alpha\Theta_1$ is a Hamiltonian operator $\forall$ constants α, i.e.

a) Θ is skew-symmetric $\Theta^* = -\Theta$ with respect to the symmetric bilinear form

$$(\mathbf{f},\mathbf{g}) := \int_{-\infty}^{\infty} dx(f_1 g_1 + f_2 g_2); \tag{4.13}$$

where f_1, f_2 are the components of vector $\mathbf{f}$.
b) Θ satisfies the Jacobi identity with respect to the bracket $\{a,b,c\} := (a, \Theta_f[\Theta b]c)$.
4) The adjoint Φ^* of Φ with respect to the bracket (4.13) satisfies the following well-coupling condition

$$\Phi\Theta_1 = \Theta_1\Phi^*. \tag{4.14}$$

5) The vector $\gamma^{(0)} := \Theta_1^{-1}\sigma^{(0)} = \frac{1}{r}\binom{q_{21}}{q_{12}}$ is a gradient, i.e. the operator γ_f is symmetric:

$$\gamma_{\mathbf{f}}^{(0)} = (\gamma_{\mathbf{f}}^{(0)})^*. \tag{4.15}$$

The above results are crucial in the proof of the following proposition, which summarizes all the relevant properties of equations (4.8).
Proposition 2 We consider the compatible pair of Hamiltonian operators (4.12b), the recursion operator (4.9b) and we define the quantities

$$\sigma^{(m)} := \Phi^m\sigma^{(0)}, \quad \gamma^{(m)} := \Phi^m\gamma^{(0)}, \quad m \in \mathcal{N}; \tag{4.16}$$

then the following hold.
1) $\sigma^{(m)}$ are symmetries of each equation of the hierarchy (4.8), i.e.

$$\sigma^{(m)}{}_f[\mathbf{k}^{(n)}] = \mathbf{k}^{(n)}{}_f[\sigma^{(m)}], \quad n, m \in \mathcal{N} \tag{4.17}$$

2) $\gamma^{(m)}$ are gradients of conserved quantities for (4.8), i.e.

$$\gamma^{(m)}{}_f = (\gamma^{(m)})^*, \tag{4.18a}$$

$$\gamma^{(m)}{}_f[\mathbf{k}^{(n)}] + (\mathbf{k}^{(n)})^*[\gamma^{(m)}] = 0. \tag{4.18b}$$

3) Equations (4.8) are bi-Hamiltonian systems, i.e. they can be written in the following two Hamiltonian forms:

$$\begin{pmatrix} q_{12} \\ q_{21} \end{pmatrix}_t = c_n\Theta_2\gamma^{(n-1)} = c_n\Theta_1\gamma^{(n)}. \tag{4.19}$$

4) The potentials I_n, defined by

$$\gamma^{(n)} = grad\ I_n, \quad (I_n)_f[\mathbf{g}] =: (grad\ I_n, \mathbf{g}), \tag{4.20}$$

are constants of motion for equations (4.8).
5) These constants of motion are in involution with respect to the Poisson brackets

$$\{I_n, I_m\}_j := (\gamma^{(n)}, \Theta_j\gamma^{(m)}), \qquad j = 1, 2. \tag{4.21}$$

The recursion operator (4.9b) was derived in [34], toghether with a recursive procedure to generate constants of motion; BT's were obtained in [38].

Proposition 2 for the nonlocal AKNS hierarchy is a straightforward application of the theory for recursion operators of integrable 1+1 dimensional systems, developed at the end of the 70th in [54]-[56]. The extension of such a theory to integrable PDE's in more than 1+1 dimensions (like KP) and to IDEE's like ILW, BO and their 2+1 dimensional extension (3.71)) is not straightforward, due to the lack of a representation of the type

(4.8). To bypass this difficulty an alternative approach, the mastersymmetry approach, was developed [25],[57] and used to prove the existence of symmmetries and constants of motion of the BO [25] and of the KP [58] equations. We recall that a mastersymmetry of (4.5) is a function τ which maps symmetries of (4.5) into symmetries of (4.5) via the transformation

$$\sigma' = [\tau, \sigma]_f = \tau_f[\sigma] - \sigma_f[\tau]. \tag{4.22}$$

The generalization of the recursion operator theory to multidimensional systems and to IDEE's was made, respectively, in [59] and in [22], [23] (see also [60] and [61] for an alternative approach).

In order to extend the recursion operator theory to such cases it is important to replace the "local" representation of the hierarchy of integrable systems

$$q_t = k^{(n)}(q), \quad n \in \mathcal{N} \tag{4.23}$$

by the "bilocal" representation

$$\delta(\mathbf{x} - \mathbf{x}')q_t' = c_n\delta(\mathbf{x} - \mathbf{x}')K^{(n)}(q, q'), \quad \Rightarrow \quad k^{(n)}(q) = K^{(n)}(q, q), \tag{4.24a}$$

$$K^{(n)}(q, q') := \Phi^n \hat{K}^{(n)} \cdot 1, \tag{4.24b}$$

where $\mathbf{x}, \mathbf{x}' \in \mathcal{R}^\nu$, δ is the Dirac function, $q = q(\mathbf{x}, t)$, $q' = q(\mathbf{x}', t)$, $K^{(n)}$ belongs to a suitable space of polynomials in q, q' and the recursion operator Φ and the starting operator $\hat{K}^{(0)}$ are operator valued functions on such a space. For a review of other results associated with a bilocal formalism, see [62].

4.2.2 The ILW hierarchy [23]

For the ILW hierarchy, $\nu = 1$ and

$$\Phi := q_+ - iTq_-, \qquad \hat{K}^{(0)} := q_-, \tag{4.25}$$

where

$$q_\pm := q(x) \pm q(x') + i(\partial_x \mp \partial_{x'}), \tag{4.26a}$$

$$(Tg)(x, x') := \frac{1}{2\eta} \int_{-\infty}^{\infty} dx' coth(\frac{\pi}{2\eta}(\xi - (x - x')))G(\xi, x - x'), \quad g(x, x') = G(x + x', x - x'). \tag{4.26b}$$

The first bilocal functions $K^{(n)}$ read

$$K^{(0)} = \hat{K}^{(0)} \cdot 1 = q - q', \tag{4.27a}$$

$$K^{(1)} = \Phi\hat{K}^{(0)} \cdot 1 = i(q_x + q'_{x'}) + Tq_x - (Tq_x)' + (q + q')(q - q') - i(q - q')(Tq - (Tq)'), \tag{4.27b}$$

where $(Tq)' = (Tq)(x')$. If $c_n = -i2^{-n}$, equations (4.24) give: $n = 0:\ q_t = 0; n = 1:\ q_t = q_x$; $n = 2:$ *equ.*(3.3); $n = 3:$ *equ.*(3.4).

The representation (4.24) suggests that the new theory has to make essential use of a formalism for generalized functions (distributions) with bilocal coefficients (the $K^{(n)}(q, q')$'s), generated by the bilocal operators Φ and $\hat{K}^{(0)}$.

The d-derivative. We introduce a new operation, a directional derivative (called d - derivative), which generalizes the usual Fréchet derivative to the space of bilocal operators, in the following way. The d-derivative $(q_\pm)_d$ of the bilocal operators $q_\pm$ in a direction characterized by the function $f = f(x, x')$ is given by

$$(q_\pm)_d[f] := f_\pm, \tag{4.28a}$$

where $f_\pm$ are integral operators defined by

$$(f_\pm g)(x,x') := \int_{\mathcal{R}} dx''(f(x,x'')g(x'',x') - g(x,x'')f(x'',x')). \tag{4.28b}$$

Since Φ and $\hat{K}^{(0)}$ are expressed in terms of $q_\pm$, their d-derivative is well defined:

$$\Phi_d[f] = f_+ - if_-T, \qquad \hat{K}_d^{(0)}[f] = f_-. \tag{4.29}$$

The connection between d-derivative and Fréchet derivative is given by the projective formula

$$(K(q,q'))_d[\delta(x-x')g(x,x')] = (K(q,q'))_f[g(x,x)] = \partial_\epsilon K(q+\epsilon g(x,x), q' + \epsilon g(x',x'))|_{\epsilon=0}. \tag{4.30}$$

The Lie algebra of the starting symmetry. It turns out that the starting operator $\hat{K}^{(0)}$, acting on functions $h = h(x-x')$, forms a Lie algebra; more precisely:

$$[\hat{K}^{(0)} \cdot h, \hat{K}^{(0)} \cdot \tilde{h}]_d = -\hat{K}^{(0)} \cdot [h, \tilde{h}]_I, \tag{4.31}$$

where

$$[A,B]_d := A_d[B] - B_d[A], \tag{4.32a}$$

$$[h,\tilde{h}]_I := \int_{\mathcal{R}} dx''[h(x-x'')\tilde{h}(x''-x') - \tilde{h}(x-x'')h(x''-x')]. \tag{4.32b}$$

The introduction of the d-derivative allows to extend the results of Proposition 2:
Proposition 3 With respect to the d-derivative:
i) Φ is hereditary, i.e.:

$$\Phi_d[\Phi f]g - \Phi\Phi_d[f]g \tag{4.33}$$

is symmetric with respect to f and g.
ii) Φ is a strong symmetry for $\hat{K}^{(0)} \cdot h$, i.e.

$$\Phi_d[\hat{K}^{(0)} \cdot h] + \Phi(\hat{K}^{(0)} \cdot h)_d - (\hat{K}^{(0)} \cdot h)_d \Phi = 0 \tag{4.34}$$

iii) Φ admits a factorization $\Theta_2 = \Phi\Theta_1$ in terms of the operators

$$\Theta_1 := q_-, \qquad \Theta_2 := (q_+ - iq_-T)q_-; \tag{4.35}$$

$\Theta = \Theta_2 + \alpha\Theta_1$ is a Hamiltonian operator $\forall\alpha$, i.e.:
a) Θ is skew-symmetric: $\Theta^* = -\Theta$ with respect to the extended bilinear form

$$< f,g > := \int_{\mathcal{R}^2} dx dx' f(x,x')g(x',x) \tag{4.36}$$

b) Θ satisfies the Jacobi identity with respect to the bracket $< a, \Theta_d[\Theta b]c >$.
iv) The adjoint $\Phi^* = q^+ - iq_-T$ of Φ, with respect to the bracket (4.36), satisfies the well-coupling condition $\Phi\Theta_1 = \Theta_1\Phi^*$.

The recursion operator Φ and its adjoint Φ^* generate a sequence of bilocal functions

$$\sigma^{(m)}(q,q') := \Phi^m \hat{K}^{(0)} \cdot 1, \quad \gamma^{(m)}(q,q') := \Phi^{*m} \cdot 1; \tag{4.37}$$

Their characterization with respect to the evolution equations leads to the introduction of the new notions of "extended symmetry" and "extended gradient". Indeed Proposition 3 implies the following

Proposition 4.

i) $\sigma^{(m)}(q,q')$ are extended symmetries for the evolution equations (4.24), i.e.

$$(\sigma^{(m)}(q,q'))_f[k^{(n)}(q)] = (\delta K^{(n)}(q,q'))_d[\sigma^{(m)}(q,q')], \tag{4.38}$$

where

$$(\delta K^{(n)})_d = \sum_{l=0}^{n}(-2i)^l \binom{n}{l}(\Phi^{n-l}\hat{K}^{(0)}\cdot\delta^{(l)}(x-x'))_d; \tag{4.39}$$

ii) $\gamma^{(m)}(q,q')$ are extended gradients, i.e.

$$(\gamma^{(m)}(q,q'))_d = (\gamma^{(m)}(q,q'))_d^*; \tag{4.40}$$

iii) $\gamma^{(m)}(q,q')$ are extended covariants for equations (4.24), i.e.

$$(\gamma^{(m)}(q,q'))_d[K^{(n)}] + (\delta K^{(n)}(q,q'))_d^*[\gamma^{(m)}(q,q')] = 0. \tag{4.41}$$

Remark 4.1 Equation (4.39) is obtained applying the d-derivative to $\delta\Phi^n\hat{K}^{(0)}\cdot 1$, after commuting δ to the right. The first three operators $(\delta K^{(n)})_d$ read:

$$(\delta K^{(0)})_d = 0, \quad (\delta K^{(1)})_d = 2i(\partial_x + \partial_{x'}),$$

$$(\delta K^{(2)})_d = 4i\{T(\partial_x+\partial_{x'})^2 + (\partial_x+\partial_{x'})(q+q') + i[Tq_x - (Tq_x)' - (q-q')T(\partial_x+\partial_{x'})]\}.$$

The usefulness of extended symmetries and extended covariants follows from the following:

Proposition 5.

i) If $\sigma^{(m)}(q,q')$ is an extended symmetry of equation (4.24), then

a) equations $\sigma^{(m)}(q,q')=0$ are BT for equation (4.24), where q and q' are now interpreted as two different solutions of (4.24).

b) $\sigma^{(m)}(q) := \sigma^{(m)}(q,q)$ are symmetries of (4.24), i.e. $\sigma_f^{(m)}[k^{(n)}] = k_f^{(n)}[\sigma^{(m)}]$.

ii) If $\gamma^{(m)}(q,q')$ are extended gradients, then $\gamma^{(m)}(q) := \gamma^{(m)}(q,q)$ is a gradient, i.e. $\gamma_f^{(m)} = (\gamma_f)^+$, where $^+$ is the adjoint operation with respect to the bracket

$$(f,g) := \int_R dx f(x)g(x). \tag{4.42}$$

iii) If $\gamma^{(m)}(q,q')$ are extended conserved covariants, then $\gamma^{(m)}(q)$ are conserved covariants, i.e. $\gamma_f^{(m)}(q)[k^{(n)}] + (k_f^{(n)})^+[\gamma^{(m)}(q)] = 0$.

iv) The potentials I_m, defined by $\gamma^{(m)} =: grad\ I_m$, $\quad (I_m)_f[g] = (\gamma^{(m)}, g)$, are constants of motion for (4.24).

Remark 4.2 Equations (4.27) are the first two BT's of the ILW hierarchy; they are different from the BT derived in [19],[43].

The Hamiltonian (actually bi-Hamiltonian) nature of the ILW hierarchy follows from the following

Proposition 6.

i) The equations of ILW hierarchy are bi-Hamiltonian, i.e. they can be written in the following forms

$$\delta q_t' = \delta\Theta_1\gamma^{(n)}(q,q') = \delta\Theta_2\gamma^{(n-1)}(q,q'); \tag{4.43a}$$

ii) the constants of motion are in involution with respect to the brackets

$$\{I_n, I_m\}_j := <\gamma^{(n)}(q,q'), \Theta_j \gamma^{(m)}(q,q')>, \quad j=1,2. \tag{4.43b}$$

Remark 4.3 The first Hamiltonian operator $\Theta_1 = q_-$ commutes with δ and reduces to the local Hamiltonian operator $i\partial_x$; therefore equations (4.43) have their local counterpart:

$$q_t = i\partial_x \gamma^{(n)}(q), \qquad \{I_n, I_m\}_1 := (\gamma^{(n)}(q), \partial_x \gamma^{(m)}(q))).$$

The second Hamiltonian operator **does not** commute with δ and a local counterpart of equations (4.43) is not available for $j = 2$. If $\eta \to \infty$, we obtain the corresponding theory of the BO equation [22].

Remark 4.4 Bilocal recursion operators allow a simple and elegant characterization of mastersymmetries [59], [22]. For instance, in the BO case, we have that

$$[\delta K^{(n)}(q,q'), \tau^{(m)}(q,q')]_d = 4inK^{(n+m-1)}(q,q'), \quad \tau^{(m)}(q,q') := \Phi^m \hat{K}^{(0)} \cdot (x+x').$$

Therefore $\tau^{(m)}(q) := \tau^{(m)}(q,q)$ are mastersymmetries of degree 1, since

$$[k^{(n)}(q), \tau^{(m)}(q)]_f = 4ik^{(n+m-1)}(q).$$

Remark 4.5 The recursion operator (4.25) was derived in [23] "expanding" in powers of ∂_x. Expanding in powers of $\mathcal{E}$ one obtains a different recursion operator and Hamiltonian structures, but all the results of this sections remain true [23].

The involutivity of conservation laws for the ILW and BO equations was first obtained in [25] and in [26]. This result was extended in [27] to the ILW_n hierarchies. The recursion operator approach in a bilocal setting was applied to the BO and ILW hierarchies in [22] and [23]. Connections of IDEE's with some realization of the so-called Yang-baxter algebra have been established [61].

4.3 Algebraic equivalence of IDEE's and discrete systems [24]

In this section we investigate the algebraic equivalence [24] of protoype examples of discrete evolutionary systems, like the Discrete Chiral Field (DCF) [63]:

$$((G(n))^{-1}G_t(n+\hat{1}))_t = (G(n+1))^{-1}G_t(n+1) - (G(n))^{-1}G_t(n) \tag{4.44a}$$

and the Discrete Sine-Gordon (DSG) [63,64]:

$$\phi_t(n+1) - \phi_t(n) = c[sin\phi(n) + sin\phi(n+1)] \tag{4.44b}$$

with prototype examples of singular integral evolution equations, like the equations of the nonlocal AKNS hierarchy. We will show that equations (4.44) and equations (4.8), (4.9) are just different concrete realisations of the same *abstract* and *elementary* algebraic structures, associated with the following spectral problem:

$$\mathcal{E}^{-1}\psi = (Q + \lambda A)\psi, \tag{4.45}$$

where $\mathcal{E}$ is the translation operator defined on a suitable function space, $\mathcal{E}^{-1}$ is its inverse, Q and A take values in an associative algebra with unit element I, endowed with a trace form $< X, Y >$, and λ is the spectral parameter. The following two different concrete realisations of the translation operator $\mathcal{E}$ will be considered:

$$(\mathcal{E}f)(x) = f(x + i\eta), \quad x \in \mathcal{R}, \quad \eta > 0, \tag{4.46a}$$

$$(\mathcal{E}f)(n) = f(n+1), \quad n \in \mathcal{Z}. \tag{4.46b}$$

The class of bi-hamiltonian systems associated with (4.45) can be obtained by standard means in the form:

$$f(\Phi)Q_t = g(\Phi)K^{(0)} \tag{4.47}$$

where $f(z)$, $g(z)$ are analytic functions of their argument; the recursion operator Φ, defined by

$$\Phi = (Q_L\mathcal{E} - Q_R)(A_L\mathcal{E} - A_R)^{-1}, \tag{4.48}$$

is the factorization $\Phi = \Theta_2\Theta_1{}^{-1}$ of the two Hamiltonian operators

$$\Theta_1 := Q_L A_R - Q_R A_L, \qquad \Theta_2 = (Q_L\mathcal{E} - Q_R)(A_L\mathcal{E} - A_R)^{-1}(Q_L A_R - Q_R A_L) \tag{4.49}$$

and the starting vector field (starting symmetry) is given by:

$$K^{(0)} := (Q_L - Q_R)B, \qquad (A_L - A_R)B = 0. \tag{4.50}$$

In formulas (4.48)-(4.50) the subscripts R and L denote right and left multiplication, i.e.:

$$X_R F := FX; \quad X_L F := XF$$

The whole analysis so far developed holds of course whatever be the concrete realization of the translation operator $\mathcal{E}$ and the particular choice of the constant element A. In the following sections, starting from the abstract results derived above, we will investigate the reduction induced by the choice $A = \sigma_3$ (resp. $A = I$) and, correspondingly, we consider the realisation of the translation operator given by (4.46a) (resp. (4.46b)).

4.3.1 Reduction to integral evolution equations.

We restrict considerations to the algebra $\mathcal{A} := C^\infty(\mathbf{R}, gl(2,\mathbf{C}))$ of 2x2 matrix-valued smooth functions of the real variables x and take, as mentioned before, $A = \sigma_3$, i.e. the Pauli matrix $\sigma_3 = diag(1,-1)$; the trace form on $\mathcal{A}$ is given by $< X, Y >= tr \int_{-\infty}^{+\infty} dx XY$.

To get the nonlocal AKNS hierarchy, we perform a canonical reduction (reduction by restriction)[65] of the abstract structures (4.30-32) on the image of Θ_1 ($Im\Theta_1$).

We denote the field matrix Q as:

$$Q = \begin{pmatrix} q_{11} & q_{12} \\ q_{21} & q_{22} \end{pmatrix}, \quad Q \to I \ as \ |x| \to \infty \tag{4.51}$$

It turns out that $Im\Theta_1 := \{K : K = \Theta_1\gamma, \gamma \in \mathcal{A}\}$, is the set of vector fields of the form:

$$K = \begin{pmatrix} K_{11} & K_{12} \\ K_{21} & K_{22} \end{pmatrix}, \quad K_{11} = K_{22} = \frac{q_{21}K_{12} - q_{12}K_{21}}{q_{11} + q_{22}}. \tag{4.52}$$

The submanifolds tangent to $Im\Theta_1$ (technically speaking, the leaves of the characteristic distribution $Im\Theta_1$) are readily derived from (4.52) by integration, obtaining the constraints $q_{11} = q_{22} + c_1$; $q_{11}^2 - q_{12}q_{21} + c_1 q_{11} + c_2 = 0$. The boundary condition (4.51) selects the special leaf S, defined by: $c_1 = 0$, $c_2 = -1$ on which:

$$q_{11} = q_{22} = \sqrt{1 + q_{12}q_{21}} =: r \tag{4.53}$$

which is just the reduction (3.32), obtained now through geometric considerations.

On the other hand, the elements β of the cotangent bundle T^*S are defined by the duality condition:

$$< \gamma, K >= tr \int_{\mathcal{R}} dx \gamma K := \int_{\mathcal{R}} dx(\beta_1 K_{12} + \beta_2 K_{21}), \qquad \gamma = \begin{pmatrix} \gamma_{11} & \gamma_{12} \\ \gamma_{21} & \gamma_{22} \end{pmatrix} \tag{4.54a}$$

which implies:

$$\beta_1 = \gamma_{21} + \frac{q_{21}(\gamma_{11} + \gamma_{22})}{2r}, \qquad \beta_2 = \gamma_{12} + \frac{q_{12}(\gamma_{11} + \gamma_{22})}{2r}. \tag{4.54b}$$

Formulas (4.53), (4.54b) allow us to rewrite the equation $K = \Theta_1 \gamma$ in the following vector form:

$$\begin{pmatrix} K_1 \\ K_2 \end{pmatrix} = 2rJ \begin{pmatrix} \beta_1 \\ \beta_2 \end{pmatrix}, \qquad J = \begin{pmatrix} 0 & -1 \\ 1 & 0 \end{pmatrix}. \tag{4.55}$$

We emphasize the fact that, in our vector reduction, the first Hamiltonian operator is given by $2\sqrt{1 + q_{12}q_{21}}\,J$, which is remarkably "as Hamiltonian as" the canonical one J!

In order to obtain the reduction by restriction of the operators Φ, Θ_j, $j = 1,2$ we reduce equations $K' = \Phi K$ and $K = \Theta_2 \gamma$. It turns out that:

$$\begin{pmatrix} K_1' \\ K_2' \end{pmatrix} = \begin{pmatrix} r\frac{\mathcal{E}-1}{\mathcal{E}+1} - q_{12}\frac{\mathcal{E}+1}{\mathcal{E}-1}\frac{1}{2r}q_{21} & -q_{12}\frac{\mathcal{E}+1}{\mathcal{E}-1}\frac{1}{2r}q_{12} \\ q_{21}\frac{\mathcal{E}+1}{\mathcal{E}-1}\frac{1}{2r}q_{21} & -r\frac{\mathcal{E}-1}{\mathcal{E}+1} + q_{21}\frac{\mathcal{E}+1}{\mathcal{E}-1}\frac{1}{2r}q \end{pmatrix} \begin{pmatrix} K_1 \\ K_2 \end{pmatrix}, \tag{4.56a}$$

$$\begin{pmatrix} K_1 \\ K_2 \end{pmatrix} = \begin{pmatrix} -q_{12}\frac{\mathcal{E}+1}{\mathcal{E}-1}q_{12} & r\frac{\mathcal{E}-1}{\mathcal{E}+1}2r + q_{12}\frac{\mathcal{E}+1}{\mathcal{E}-1}q_{21} \\ -r\frac{\mathcal{E}-1}{\mathcal{E}+1}2r + q_{21}\frac{\mathcal{E}+1}{\mathcal{E}-1}q_{12} & -q_{21}\frac{\mathcal{E}+1}{\mathcal{E}-1}q_{21} \end{pmatrix} \begin{pmatrix} \beta_1 \\ \beta_2 \end{pmatrix}. \tag{4.56b}$$

Introducing the concrete realisation (4.46a) of the translation operator $\mathcal{E}$, equations (4.56a) and (4.56b) are the recursion operator (4.9b) and the Hamiltonian operators (4.12) of the nonlocal AKNS hierarchy.

4.3.2 Reduction to discrete systems

Following the scheme outlined in the previous subsection, we list now the results associated with the canonical reduction induced by the choice $A = I$, for the concrete realisation (4.46b) of the translation operator. Our algebra is now the algebra of 2x2 matrix-valued sequences and shall be denoted as $\mathcal{B}$. The trace form on $\mathcal{B}$ is then given by: $< X, Y >:= tr \sum_{n=-\infty}^{+\infty} X(n)Y(n)$. We choose the boundary condition:

$$Q(n) \to \sigma_3 \quad (|n| \to \infty) \tag{4.57}$$

and, motivated by (4.51), we choose the integral leaf S on which $Q^2 = I$. It follows that

$$Q = \begin{pmatrix} \rho & q_{12} \\ q_{21} & -\rho \end{pmatrix}, \qquad \rho := \sqrt{1 - q_{12}q_{21}}, \tag{4.58}$$

and thus:

$$K = \begin{pmatrix} K_{11} & K_q \\ K_r & -K_{11} \end{pmatrix}, \qquad K_{11} = -K_{22} = -\frac{q_{12}K_{21} + q_{21}K_{12}}{2\rho} \tag{4.59}$$

$$\begin{pmatrix} K'_{12} \\ K'_{21} \end{pmatrix} := \mathcal{N}|_S \begin{pmatrix} K_{12} \\ K_{21} \end{pmatrix} = \begin{pmatrix} \rho\frac{\mathcal{E}+1}{\mathcal{E}-1} + \frac{q_{12}}{2}\frac{\mathcal{E}+1}{\mathcal{E}-1}\frac{q_{21}}{\rho} & \frac{q_{12}}{2}\frac{\mathcal{E}+1}{\mathcal{E}-1}\frac{q_{12}}{\rho} \\ -\frac{q_{21}}{2}\frac{\mathcal{E}+1}{\mathcal{E}-1}\frac{q_{21}}{\rho} & -\rho\frac{\mathcal{E}+1}{\mathcal{E}-1} - \frac{q_{21}}{2}\frac{\mathcal{E}+1}{\mathcal{E}-1}\frac{q_{12}}{\rho} \end{pmatrix} \begin{pmatrix} K_1 \\ K_2 \end{pmatrix}, \tag{4.60}$$

$$\theta_2|_S = \begin{pmatrix} q_{12}\frac{\mathcal{E}+1}{\mathcal{E}-1}q_{12} & -2\rho\frac{\mathcal{E}+1}{\mathcal{E}-1}\rho - q_{12}\frac{\mathcal{E}+1}{\mathcal{E}-1}q_{21} \\ -2\rho\frac{\mathcal{E}+1}{\mathcal{E}-1}\rho - q_{21}\frac{\mathcal{E}+1}{\mathcal{E}-1}q_{12} & q_{21}\frac{\mathcal{E}+1}{\mathcal{E}-1}q_{21} \end{pmatrix}. \tag{4.61}$$

Choosing $f(z) = 1, g(z) = z$ in the general formula (4.47) we get the equation:

$$\begin{pmatrix} q_{12} \\ q_{21} \end{pmatrix}_t = 2\rho\frac{\mathcal{E}+1}{\mathcal{E}-1}\begin{pmatrix} q_{12} \\ q_{21} \end{pmatrix} \tag{4.62}$$

where now:

$$(\frac{\mathcal{E}+1}{\mathcal{E}-1}f)(n) = f(n) + 2\sum_{j=n+1}^{\infty} f(j) \tag{4.63a}$$

$$(\frac{\mathcal{E}-1}{\mathcal{E}+1}f)(n) = -f(n) + 2\sum_{j=n+1}^{\infty} (-1)^{n+1-j} f(j) \tag{4.63b}$$

Performing the further reduction $q_{12} = q_{21} = sin\phi$, yielding $\rho = cos\phi$, we finally obtain the DSG equation (4.44b). If $f(z) = 1 - z, g(z) = 0$ we obtain the DCF equation

$$(\rho\frac{\mathcal{E}+1}{\mathcal{E}-1} - 1)q_{12_t} - q_{12}\frac{\mathcal{E}+1}{\mathcal{E}-1}\rho_t = 0, \quad -(\rho\frac{\mathcal{E}+1}{\mathcal{E}-1} + 1)q_{21_t} + q_{21}\frac{\mathcal{E}+1}{\mathcal{E}-1}\rho_t = 0 \tag{4.64}$$

and if we keep the matrix notation, we can recast (4.64) in the more compact form:

$$Q_t = [Q, \frac{\mathcal{E}+1}{\mathcal{E}-1}Q_t] \tag{4.65}$$

whence, finally, its standard form (4.44a) is recovered through the substitution $Q(n) = G^{-1}(n)G(n+1)$.

REFERENCES

1. G. B. Witham, Proc. R. Soc. **A 299**,6 (1967); *Linear and nonlinear waves*, Wiley, New York, 1974.
2. R. I. Joseph, J. Phys. **A 101**, 1225 (1977).
   T. Kubota, D. R. Ko and D. Dodds, H. Hydronaut. **12**, 157 (1978).
3. R. I. Joseph and R. Egri, J. Phys **A 11**, L97 (1978).
4. O. M. Phillips, *The dynamics of the upper ocean*, Cambridge Univ. Press, Cambridge, 1966.
5. D. J. Korteweg and G. De Vries, Philos. Mag. Ser. **5**, 39, pp422 (1876).
6. T. B. Benjamin, J. Fluid Mech. **29**, 559 (1967).
   H. Ono, J. Phys. Soc. Jap. **39**, 1082 (1975).
7. C. S. Gardner, J. M. Green, M. D. Kruskal and R. M. Miura, Phys. Rev. Lett. **19**, 1095 (1967); Comm. Pure Appl. Math. **27**, 97 (1974).
8. M. J. Ablowitz and H. Segur, *Solitons and the Inverse Scattering Transform*, Vol.4, SIAM, Philadelphia, PA, 1981.
   S. P. Novikov, S. V. Manakov, L. P. Pitaeskii and S. V. Zakharov, *Theory of Solitons, The Inverse Scattering Method*, Contemporary Soviet Mathematics, Consultant Bureau, New York and London, 1984.
   F. Calogero and A. Degasperis, *Spectral Transform and Solitons*, North - Holland, Amsterdam, 1982.

A. C. Newell, *Solitons in Mathematics and Physics*, Vol.45, SIAM, Phyladelphia, PA, 1985.
B. G. Konopelchenko, *Nonlinear integrable equations*, Lecture Notes in Physics, Springer Verlag, Berlin, 1987.
9. V. E. Zakharov and A. B. Shabat, Sov. Phys. JETP **34**, 62 (1972).
10. M. Wadati, J. Phys. Soc. Jap. **32**, 1681 (1972).
11. B. B. Kadomtsev and V. I. Petviashvili, Sov. Phys. Doklady **15**, 539 (1970).
12. S. V. Manakov, Physica **3D**, 420 (1981).
A. S. Fokas and M. J. Ablowitz, Stud. Appl. Math. **69**, 211 (1983).
M. J. Ablowitz, D. Bar Yaacov and A. S. Fokas, Stud. Appl. Math. **69**, 135 (1983).
13. R. I. Joseph, J. Math. Phys. **18**, 2251 (1977).
14. J. D. Meiss and N. R. Pereira, Phys. Fluids **21**, 700 (1978).
15. Y. Matsuno, J. Phys. A **12**, 619 (1979); *Bilinear transformation method*, Academic Press, New York, 1984.
16. J. Satsuma and Y. Ishimori, J. Phys. Soc. Jap. **46**, 681 (1979).
17. H. H. Chen and Y. C. Lee, Phys. Rev. Lett. **43**, 264 (1979).
H. H. Chen, Y. C. Lee and N. R. Pereira, Phys. Fluids **22**, 187 (1979).
18. T. L. Bock and M. D. Kruskal, Phys. Lett. **74A**, 173 (1979).
19. J. Satsuma, M. J. Ablowitz and Y. Kodama, Phys. Lett. **73 A**, 283 (1979).
Y. Kodama, J. Satsuma and M. J. Ablowitz, Phys. Rev. Lett. **46**, 687 (1981).
20. Y. Kodama, M. J. Ablowitz and J. Satsuma, J. Math. Phys. **23**, 564 (1982).
21. V. E. Zakharov and A. B. Shabat, Funct. Anal. Appl. **13**, 166 (1979).
22. A. S. Fokas and P. M. Santini, J. Math. Phys. **29**, 604 (1988).
23. P. M. Santini, Inverse Problems **5**, 203 (1989).
24. O. Ragnisco and P. M. Santini, Inverse Problems **6**, 441 (1990).
25. A. S. Fokas and B. Fuchssteiner, Phys. Lett. **86A**, 341, (1981).
26. B. Kuppershmidt, Libertas Math. **1**, 125 (1981).
27. D. R. Lebedev and A. O. Radul, Comm. Math. Phys. **91**, 543 (1983).
28. P. Costantin, P. D. Lax and A. Majda, Comm. Pure Appl. Math. **38**, 715 (1985).
29. M. J. Ablowitz, A. S. Fokas and M. D. Kruskal, Phys. Lett. **120A**, 215 (1987).
30. M. J. Ablowitz, A. S. Fokas, J. Satsuma and H. Segur, J. Phys. A **15**, 781 (1982).
31. J. Satsuma, J. Phys. Soc. Jap. **50**, 1423 (1981).
32. A. S. Fokas and M. J. Ablowitz, Stud. Appl. Math. **68**, 1 (1983).
33. P. M. Santini, M. J. Ablowitz and A. S. Fokas, J. Math. Phys. **25**, 892 (1984).
34. A. Degasperis and P. M. Santini, Phys. Lett. **98A**, 240 (1983).
35. A. Degasperis, P. M. Santini and M. J. Ablowitz, J. Math. Phys. **26**, 2469 (1985).
36. I. Gohberg and M. G. Krein, Usp. Mat. Nauk. **13**, 2 (1958).
37. M. J. Ablowitz, D. J. Kaup, A. C. Newell and H. Segur, Stud. Appl. Math. **53**, 249 (1974).
38. A. Degasperis, in *Nonlinear evolution equations: integrability and spectral methods*, ed. A. Degasperis, A. P. Fordy and M. Lakshmanan, Manchester Univ. Press, 1990.
39. A. Degasperis, P. M. Santini and M. J. Ablowitz, The initial value problem for a nonlocal generalization of the AKNS class, Preprint Dip. Fisica, Univ. Roma I, 1992.
40. P. M. Santini, M. J. Ablowitz and A. S. Fokas, J. Math. Phys. **28**, 2310 (1987).
41. Y. Matsuno, Phys. Lett. **119A**, 229 (1986); **120A**, 187 (1987); J. Phys. A **20**, 3587 (1987).
42. Y. Matsuno, J. Math. Phys **32**, 120 (1991).
43. J. Satsuma and M. J. Ablowitz, in *Nonlinear partial differential equations in engineering and applied science*, ed. R. L. Stenberg, A. J. Kalinoski and J. S. Papadakis, M. Dekker, New York, 1980.
44. J. Satsuma, Thiab R. Taha and M. J. Ablowitz, J. Math. Phys **25**, 900 (1984).
45. A. Degasperis, D. Lebedev, M. Olshanetsky, S. Pakuliak, A. Perelomov and P. M. Santini, Comm. Math. Phys. **141** 133 (1991).

46. A. Degasperis, D. Lebedev, M. Olshanetsky, S. Pakuliak, A. Perelomov and P. M. Santini, Generalized intermediate long wave hierarchy in zero curvature representation with noncommutative spectral parameter, Preprint MPI 63, Max Planck Inst., 1991.
47. D. Lebedev, A. Orlov, S. Pakuliak and A. Zabrodin, Nonlocal integrable equations as reductions of the Toda hierarchy, Preprint Bonn Univ., ISSN 0172 (1991).
48. Y. Matsuno, J. Math. Phys. **29**, 49 (1980); **30**, 241 (1989).
49. Y. Matsuno, J. Math. Phys. **31**, 2904 (1990).
50. V. B. Matveev and M. A. Sall, Dokl. Akad. Nauk. SSSR **261**, 533 (1981).
A. I. Bobenko, V. B. Matveev and M. A. Sall, Sov. Phys. Dokl. **27**, 610 (1982).
51. P. M. Santini, Inverse Problems **6**, 99 (1990).
52. F. Calogero, Lett. Nuovo Cimento **14**, 433 (1975).
53. P. J. Olver, J. Math. Phys. **18**, 1212 (1977).
54. B. Fuchssteiner, Nonlinear Anal. **3**, 849 (1979).
A. S. Fokas and B. Fuchssteiner, Lett. Nuovo Cimento **28**, 299 (1980).
B. Fuchssteiner and A. S. Fokas, Physica **4D**, 47 (1981).
55. F. Magri, J. Math. Phys. **19**, 1156 (1978); in *Nonlinear evolution equations and dynamical systems*, ed. M. Boiti, F. Pempinelli and G. Soliani, Lecture Notes in Physics, Vol. 120, Berlin, Springer 1980.
56. I. M. Gel'fand and I. Dorfman,Funct. Anal. Appl. **13**, 13 (1979); **14**, 71 (1980).
57. B. Fuchssteiner, Progr. Theor. Phys. **70**, 150 (1983).
58. W. Oevel and B. Fuchssteiner, Phys. Lett. **88A**, 323 (1982).
59. A. S. Fokas and P. M. Santini, Stud. Appl. Math. **75**, 179 (1986).
P. M. Santini and A. S. Fokas, Comm. Math. Phys. **115**, 375 (1988).
A. S. Fokas and P. M. Santini, Comm. Math. Phys. **116**, 449 (1988).
60. F. Magri, C. Morosi and G. Tondo, Comm. Math. Phys. **115**, 1 (1988).
61. C. Morosi and G. Tondo, Comm. Math. Phys. **122**, 91 (1989).
62. B. G. Konopelchenko, Inverse Problems **4**, 795, 1988.
63. M. Bruschi, D. Levi and O. Ragnisco, Lett. Nuovo Cimento **33**, 284 (1982).
64. S. J. Orfanidis, Phys. Rev. **D18**, 3828 (1978).
65. F. Magri and C. Morosi, Geometrical characterization of integrable Hamiltonian systems through the theory of Poisson-Nijenhuis manifolds, Quaderno S19, Dip. di Matematica, Univ. di Milano, 1984.
F. Magri, C. Morosi and O. Ragnisco, Comm. Math. Phys. **99**, 115 (1985).

Part II

Asymptotic Results

Long-Time Asymptotics for Integrable Nonlinear Wave Equations

P.A. Deift[1], *A.R. Its*[2], *and X. Zhou*[3]

[1]Courant Institute, New York University, New York, NY 10012, USA
[2]Department of Mathematics, (Leningrad University and) Clarkson University, Potsdam, NY 13676, USA
[3]Department of Mathematics, Yale University, Box 2155 Yale Station, New Haven, CT 06520, USA

Introduction

We recall the remarkable discovery of Gardner, Green, Kruskal and Miura [9]. Consider the solution of the Korteweg de Vries (KdV) equation

$$q_t - 6qq_x + q_{xxx} = 0 \tag{0.1}$$

with initial data

$$q(x,0) = q_0(x) \in \mathcal{S}(\boldsymbol{R}). \tag{0.2}$$

Let S denote the scattering map that assigns to a Schrödinger operator $H = -d^2/dx^2 + q(x)$ its reflection coefficient, r,

$$S : q \mapsto r(z;q). \tag{0.3}$$

Then (for simplicity we ignore questions of discrete spectrum), the KdV equation is solved by the following inverse scattering formula,

$$q(x,t) = S^{-1}(r(\cdot;q_0)e^{8i(\cdot)^3t}). \tag{0.4}$$

This discovery was followed by a flood of papers showing that a wide variety of "integrable" nonlinear wave equations could also be solved by such an inverse scattering method. The exciting and fundamental question then arose whether the rather abstract formalism of inverse scattering theory could be used to describe in detail the long-time behavior of the associated non-linear wave equations.

It was recognized early on that the maps $S : q \mapsto r$ and $S^{-1} : r \mapsto q$ had many features of a nonlinear Fourier-type correspondence. For example, the decay and regularity properties of q correspond to the regularity and decay properties of r, respectively. When r is small, the inverse map S^{-1} reduces to

$$q(x,t) \sim \text{const.} \int_{\boldsymbol{R}} r(z;q_0)e^{8iz^3t+2izx}dz, \quad |r| \ll 1, \tag{0.5}$$

from which the Fourier-type correspondence is particularly clear. Evaluating the

RHS by the method of stationary phase/steepest descent, we obtain a formula for $q(x,t)$ as $t \to \infty$ in the case $|r| << 1$. It turns out a posteriori, that the formula is not quite correct (a logarithmic phase term is missing–cf. (3.30) and the discussion in [8]), but if the view of S as a nonlinear Fourier map holds true, then one may hope to find some "nonlinear" version of the method of stationary phase/steepest descent which computes the asymptotics of the solution exactly, not only in the case $|r| << 1$, but also in the fully nonlinear case when $|r|$ is no longer small.

Significant work on the long-time behavior of nonlinear wave equations solvable by the inverse scattering method was first carried out by Shabat [20], Manakov [16] and by Ablowitz and Newell [1] in 1973. The decisive step was taken in 1976 when Zakharov and Manakov [27] were able to write down precise formulae, depending explicitly on the initial data for the leading asymptotics of the nonlinear Schrödinger (NLS) equation in the physically interesting region $x = O(t)$. A complete description of the leading asymptotics of the solution of the Cauchy problem for KdV, with connection formulae between different asymptotic regions, was presented by Ablowitz and Segur [21], but without precise information on the phase. In a later development [22], Ablowitz and Segur used a modification of the method of [27] to derive the leading asymptotics for the solution of the MKdV, KdV and sine-Gordon equations, including full information on the phase. The asymptotic formulae of Zakharov/Manakov type were rigorously justified and extended to all orders by Buslaev and Sukhanov [6 I-IV] in the case of the KdV equation, and by Novokshenov [17] in the case of NLS. Also, Novokshenov [18], [19] and Sukhanov [24–26] extended the method to other equations.

The method of Zakharov and Manakov as well as the elegant approach to the calculation of the modulus of the solution proposed by Ablowitz and Segur [21] are represented in Section 1 of this paper in the case of NLS equation. A distinguishing feature of these schemes is the use of an a priori ansatz for the symptotic form of the solution. To avoid the utilization of any a priori information about the asymptotics one needs to develop a direct asymptotic analysis of the inverse scattering transform and this in turn reduces to the asymptotic analysis of an oscillatory matrix Riemann-Hilbert problem. The first step in this direction was made back in 1973 by Manakov [16]. In [16] the explicit formula for he asymptotics of the modulus of the solution of NLS equation was obtained without any a priori conjectures regarding its form. In 1981, Its [12] further developed Manakov's approach and proposed a method which made it possible to derive the complete asymptotics for the solution of the NLS equation directly from the corresponding Riemann–Hilbert problem. We present this scheme, as well as Manakov's original approach, in Section 2. The basic idea is the reduction of the original Riemann–Hilbert problem to a model isomonodromy problem that can be solved explicitly in terms of special functions. In the other cases - MKdV [13], sine-Gordon [14], Landau-Lifshitz Equation [5], NLS with non-zero boundary conditions [15] - the method of Its reduced the original problems to model Riemann–Hilbert problems, whose explicit solutions, unfortunately could not be found. These Riemann–Hilbert problems appear as the isomonodromy problems associated with the classical Painlevé transcendents, for which the long time behavior can be computed independently via the Isomonodromic Deformation Method (see [13] and [10]). In other words, except for the simplest case of NLS, the method of Its still uses external information, but now from the theory of the solutions of ordinary differential equations.

What emerges from the developments in [12] and [13] is the following. In realizing one's hope for a nonlinear stationary phase/steepest descent method, the classical analysis of an oscillatory integral at the points of stationary phase must be replaced by the analysis of an (explicitly solvable) Riemann-Hilbert problem "localized" at the points of stationary phase.

Recently, Deift and Zhou [8] developed a steepest descent method for oscillatory Riemann-Hilbert problems. The method is computationally systematic and yields, with rigorous error estimates, the long-time asymptotics of a general class of integrable systems. A key ingredient in the method is to deform the given Riemann-Hilbert problem to an equivalent Riemann-Hilbert problem on an augmented contour adapted to the directions of steepest descent of the associated phase factor ($e^{8iz^3t+2izx}$ for MKdV). The jump matrix v_t for the deformed Riemann-Hilbert problem converges in $L^1 \cap L^2 \cap L^\infty(dz)$ to the identity as $t \to \infty$ away from any neighborhood of the stationary points ($\pm z_0 = \pm\sqrt{\frac{-x}{12t}}$ for MKdV). The problem then reduces to a Riemann-Hilbert problem with nontrivial jumps only in a small neighborhood of the stationary phase points. After scaling at each stationary point, one again obtains a Riemann-Hilbert problem of isomonodromy type which can be solved explicitly as in [12] and [13] above.

An essential technical tool in the method is the analysis of Riemann-Hilbert problems on self-intersecting contours as developed in the context of inverse scattering problems in the 80's by [3], [4], [28], [29] and [7]. The estimates for error terms are based on operator estimates which are controlled in turn by $L^1 \cap L^2 \cap L^\infty$ estimates of the jump matrix. Such $L^1 \cap L^2 \cap L^\infty$ estimates are possible, after certain scaling operations on the deformed contours, but not on the original contours.

In Section 3 of this chapter, we illustrate the method of [8] by calculating the long-time asymptotics of the defocusing NLS and the MKdV equations. We hope that our exposition will be useful not only to the reader concerned with rigorous results, but also to the reader whose main interest lies in a systematic and simple derivation of the final result.

The results in this chapter concern solutions without solitons. The analysis can be extended to include solitons, should they exist, in one of two ways. Firstly, the problem can be solved without solitons, and then the solitons can be added in by Darboux transformations, which are purely algebraic. Calculations of this type were first performed by Shabat in 1973 in his D.S. thesis for the case of KdV, and by Segur in 1976 [23] for the case of NLS, and then by Novokshenov in 1980 [17], where the complete description of asymptotics for NLS is presented. Alternatively the solitons can be analyzed directly together with the continuous spectrum, by using the steepest descent method of Section 3.

1. Approaches based on a priori information on the form of the asymptotics. Ablowitz-Segur and Zakharov-Manakov methods.

We begin by reminding the reader of certain well-known facts concerning the scattering transform for the NLS-equation. We will use these facts and definitions in all of the following sections of this chapter.

Within the framework of the IST-method the Cauchy problem for NLS-equation (defocusing case)

$$iy_t + y_{xx} - 2|y|^2 y = 0 \qquad (1.1)$$
$$y(x,0) = y_0(x) \in S(\boldsymbol{R})$$

is associated with the scattering problem for the linear matrix equation

$$\psi_x = -iz\sigma_3\psi + \begin{pmatrix} 0 & iy \\ -i\bar{y} & 0 \end{pmatrix}\psi. \qquad (1.2)$$

For Im $z \neq 0$, let $\psi(z,x,t)$ be the (unique) solution of (1.2) normalized at $x \to +\infty$,

$$\psi e^{izx\sigma_3} \to I \text{ as } x \to +\infty \qquad (1.3)$$

and for which

$$\sup_{-\infty<x<\infty} \|\psi\, e^{izx\sigma_3}\| < \infty.$$

Then for each x and t

$$m(z,x,t) \equiv \psi(z,x,t)e^{izx\sigma_3}$$

is analytic in z for Im $z \neq 0$ and provides (see for example [**11**]) the (unique) solution of the associated Riemann–Hilbert problem

$$m_+(z) = m_-(z)v_{x,t} \quad z \in \boldsymbol{R} \qquad (1.4)$$
$$m(z) \to I \ \text{ as } z \to \infty,$$

where

$$m_\pm(z) = \lim_{\epsilon \to 0+} m(z \pm i\epsilon)$$
$$v_{x,t}(z) = e^{-i(2tz^2+xz)\sigma_3}v(z)e^{i(2tz^2+xz)\sigma_3} \equiv e^{-i(2tz^2+xz)\mathrm{ad}\sigma_3}v(z)$$
$$v(z) = \begin{pmatrix} 1-|r(z)|^2 & -\bar{r}(z) \\ r(z) & 1 \end{pmatrix}$$

Here $r(z)$ lies in Schwartz space and is the reflection coefficient (scattering data) corresponding to the initial data $y_0(x)$. The solution $y(x,t)$ of (1.1) can be obtained from the solution $m(z,x,t)$ of (1.4) through the relation

$$y(x,t) = 2\lim_{z\to\infty}(zm_{12}(z)). \qquad (1.5)$$

The alternative (and more traditional) definition of the scattering data, which we will use in this and the next sections, is as follows. Let $\psi^\pm(z,x,t)$ be the standard Jost solutions of the equation (1.2) normalized by the conditions

$$\psi^\pm \to e^{-izx\sigma_3}, \quad x \to \pm\infty, \quad z \in \boldsymbol{R}.$$

The scattering data corresponding to the potential $y(x,t)$ is introduced as the entries of the scattering (more precisely transition) matrix

$$S(z,t) = [\psi^+]^{-1}\psi^- \equiv \begin{pmatrix} a(z,t) & \bar{b}(z,t) \\ b(z,t) & \bar{a}(z,t) \end{pmatrix} \quad z \in \boldsymbol{R} \qquad (1.6)$$

and has the following properties:

a) $a(z,t) = a(z),\ b(z,t) = b(z)e^{4iz^2t}$

b) $b \in S(\boldsymbol{R})$

c) $|a(z)|^2 - |b(z)|^2 = 1$

d) $a(z) = \exp\{\frac{1}{2\pi i}\int_{-\infty}^{\infty}\frac{\log(1+|b(z')|^2)}{z'-z-i0}dz'\}$

Here $a(z)$, $b(z)$ are the scattering data corresponding to the initial potential $y_0(x)$. The reflection coefficient is defined as the ratio b/a and coincides with the function $r(z)$ in (1.4),

$$r(z) = b(z)/a(z). \tag{1.7}$$

Equation (1.1) has infinitely many conservation laws, i.e. first integrals of motion. These integrals can be described as follows (see, for example, [11]):

$$I_n = \int_{-\infty}^{\infty} \sigma_n(x)dx \tag{1.8}$$

$$\sigma_1 = (i/2)|y|^2, \sigma_2 = -(1/4)y_x\bar{y}, \tag{1.9}$$

$$2i\sigma_{n+1} = -\sigma_{nx} + \frac{\bar{y}_x}{\bar{y}}\sigma_n + \sum_{\substack{k+l=n \\ k,l=1,\cdots,n-1}} \sigma_k\sigma_l, \quad n \geq 2. \tag{1.10}$$

One of the principal results provided by the IST-method is the derivation of the expression for constants of motion (1.9–1.10) through the scattering data (trace formulae):

$$\begin{aligned} I_n &= \frac{i}{\pi}\int_{-\infty}^{\infty} \log|a(z)|z^{n-1}dz \qquad (1.11)\\ &= -\frac{i}{2\pi}\int_{-\infty}^{\infty} \log(1-|r(z)|^2)z^{n-1}dz \ \ (\text{note that } |r|<1)\ n = 1,2,\cdots. \end{aligned}$$

These trace formulae form the basis of the asymptotic approach of M. Ablowitz and H. Segur for the calculation of the modulus of the solution.

The starting point for the Ablowitz-Segur method, as well as for Zakharov-Manakov method, is an a priori asymptotic ansatz for the solution of the Cauchy problem (1.1). Using quite general arguments (see, for example, [21]) it is natural to assume that

$$y(x,t) \cong t^{-\frac{1}{2}}(\alpha + \sum_{n=1}^{\infty}\sum_{k=0}^{2n}\frac{(\log t)^k}{t^n}\alpha_{nk})e^{\frac{ix^2}{4t} - i\nu\log t} \tag{1.12}$$
$$\text{as } t \to +\infty,$$

where α, α_{nk} and ν are (complex) functions of the slow variable x/t. The amplitude function $\alpha(x/t)$ plays the major role; all of the subsequent coefficients α_{nk} and the functional parameter ν can be found explicitly in terms of α via the simple substitution of (1.12) into equation (1.1). For example,

$$\nu = 2|\alpha|^2. \tag{1.13}$$

Thus, to determine the asymptotic behavior of the solution $y(x,t)$ we have to connect the amplitude function α with the initial data y_0. But by the inverse method, knowledge of y_0 is equivalent to knowledge of $r(z)$ – the reflection coefficient corresponding to y_0 – and so we can specify the connection problem as the computation of α explicitly in terms of r. The observation of M. Ablowitz and H. Segur is that to determine the modulus of $\alpha(x/t)$ it is enough to take into account the trace formulae (1.11). Indeed, substituting (1.12) into the recurrence relations (1.10), one easily proves that

$$\sigma_n = \frac{i}{2t}\eta^{n-1}|\alpha(\eta)|^2 + O(\frac{\log t}{t^2})$$

where

$$\eta = -\frac{x}{4t}.$$

Therefore, the integrals (1.9) can be rewritten as

$$\begin{aligned} I_n &= \int_{-\infty}^{\infty} \sigma_n(x,t)dx = \lim_{t\to\infty}\int_{-\infty}^{\infty}\sigma_n(x,t)dx \\ &= 2i\int_{-\infty}^{\infty}\eta^{n-1}|\alpha(\eta)|^2 d\eta \end{aligned} \tag{1.14}$$

On the other hand, we have the expression (1.11) for I_n. Comparing (1.11) and (1.14) we observe that both the functions $2|\alpha(\eta)|^2$ and $1/\pi \log|a(\eta)|$ solve exactly the same moment problem, and so (modulo same technical conditions)

$$|\alpha(\eta)|^2 = 1/2\pi \log|a(\eta)| = -1/4\pi\log(1-|r(\eta)|^2) \tag{1.15}$$

and the problem of the calculation of the amplitude of the solution $y(x,t)$ is solved. This result was first obtained by Manakov in 1973 [**16**], using another method, which we present in Section 3.

The derivation of the phase of the solution of the Cauchy problem (1.1) turns out to be much more complicated. The formula for the phase was obtained by Zakharov and Manakov in 1976 by the method described below and based again on the a priori information (1.12) for the solution.

The basic idea of the Zakharov-Manakov approach can be described as follows. Substituting the master term

$$\hat{y}(x,t) = t^{-\frac{1}{2}}\alpha(x/t)\exp\{ix^2/4t - i\nu(x/t)\log t\} \tag{1.16}$$

of the series (1.12) into the linear problem (1.2), one can try to calculate the corresponding scattering data $\hat{r}(z,t;\alpha)$. One then chooses the functional parameter α in such a way that

$$\hat{r}(z,t;\alpha) \to r(z;y_0)e^{4itz^2} \text{ as } t\to +\infty \tag{1.17}$$

and hence

$$\begin{aligned} y(x,t) = \hat{y}(x,t) + S^{-1}(r(\cdot;y_0)e^{4i(\cdot)^3 t}) \\ - S^{-1}(\hat{r}) = \hat{y} + \text{ small} \end{aligned}$$

provides the asymptotics for y.

It turns out that the reflection coefficient $\hat{r}$ can be calculated asymptotically as $t \to \infty$ in closed form. The crucial observation is that the system (1.2) with $y = \hat{y}$ can be analyzed by the classical WKB-method. Indeed, substituting

$$y = \hat{y}, \quad \psi = e^{ix^2/8t\sigma_3}\phi, \quad \eta = -\frac{x}{4t}$$

into (1.2), we obtain

$$\frac{d\phi}{d\eta} = 4itA\phi \tag{1.18}$$

where

$$A \equiv A(z, \eta, t) = \begin{pmatrix} z - \eta, & -t^{-i\nu-1/2}\alpha(\eta) \\ t^{i\nu-1/2}\bar{\alpha}(\eta) & -z + \eta \end{pmatrix} \tag{1.19}$$

is bounded as $t \to +\infty$ and does not contain rapid oscillations. Applying to (1.18) the standard form of the matrix WKB-ansatz,

$$\begin{aligned} &\phi \cong T \exp(4it\sigma_3 \int^{\eta} \sqrt{-det\ A(z, \eta', t)} d\eta'), \\ &T^{-1}AT = \sqrt{-det\ A}\sigma_3, \end{aligned} \tag{1.20}$$

we come to the following representation for the matrix Jost solutions $\psi^{\pm}$ of the system (1.2)

$$\begin{aligned} \psi^{\pm} = &(I + \frac{1}{2(z-\eta)}\begin{pmatrix} 0 & \hat{y} \\ \bar{\hat{y}} & 0 \end{pmatrix} + O(\frac{\log t}{t(z-\eta)})) \\ &\cdot \exp\{-izx\sigma_3 - i\sigma_3 \int_{\mp\infty}^{\eta} \frac{\nu(\eta')}{z - \eta'} d\eta'\}. \end{aligned} \tag{1.21}$$

In the neighborhood of the turning point $\eta = z$, the representations (1.21) break down. Introducing in this region the new variable

$$\zeta = 2\sqrt{2t}(\eta - z)$$

and replacing $\alpha(\eta)$ by its value at the point $\eta = z$, we obtain the following representation for equation (1.18) near the turning point

$$\frac{d}{d\zeta}\phi^{T_p} = \begin{pmatrix} -i\zeta/2 & \beta \\ \bar{\beta} & i\zeta/2 \end{pmatrix} \phi^{T_p}, \quad \beta = -i\sqrt{2}t^{-i\nu}\alpha(z).$$

This equation is one of the matrix forms of the equation for the parabolic cylinder functions. Therefore, one can derive an explicit expression for ϕ^{T_p} in terms of parabolic cylinder functions (see formula (2.19) of the next section with $z \to \zeta$).

In the matching domain, $\eta - z \sim t^{-\frac{1}{2}+\epsilon}$, $\epsilon > 0$, the solution $\psi^{T_p} \equiv \exp(\frac{ix^2}{8t}\sigma_3)\phi^{T_p}$ must match the solutions ψ^+ and ψ^- as $\zeta \to -\infty$ and $\zeta \to +\infty$ respectively. This means that we can calculate the asymptotics of the matrices

$$N_{\pm}(z, t) = \psi^{T_p - 1}\psi^{\pm} \tag{1.22}$$

as $t \to +\infty$. In turn, this provides us with the asymptotics for the scattering matrix $\hat{S}(z,t)$ corresponding to the potential (1.16). Indeed,

$$\hat{S} = N_{+}^{-1} N_{-} \tag{1.23}$$

and taking into account the known asymptotic formulae for the parabolic cylinder functions and the relation

$$\int_{\mp\infty}^{\eta} \frac{\nu(\eta')}{z-\eta'} d\eta' \cong -\nu(z) \log|z-\eta| + \int_{\mp\infty}^{z} \log|\eta' - z| d\nu(\eta'), \quad \eta \sim z,$$

we have from (1.21)–(1.23) that

$$\begin{aligned} \hat{r}(z,t) e^{-4itz^2} &= \frac{\sqrt{\pi} e^{-\nu(z)\pi/2}}{\Gamma(-i\nu(z))\alpha(z)} \exp\{-i\pi/4 - 3i\nu(z)\log 2 \\ &- 2i \int_{-\infty}^{z} \log|z-\eta| d\nu(\eta)\} + o(1), \quad t \to \infty. \end{aligned} \tag{1.24}$$

Formula (1.24) solves asymptotically the direct scattering problem for the potential $\hat{y}$. In accordance of the general idea of the method, as represented by the relation (1.17), we obtain immediately from (1.24) the desired connection between $\alpha(\eta)$ and $r(z)$:

$$\begin{aligned} |\alpha(\eta)|^2 &= \nu(\eta)/2 = -\frac{1}{4\pi} \log(1 - |r(\eta)|^2), \\ \arg \alpha(\eta) &= -3\nu(\eta) \log 2 - \pi/4 + \arg \Gamma(i\nu(\eta)) \\ &- \arg r(\eta) + 1/\pi \int_{-\infty}^{\eta} \log|\eta - \eta'| d \log(1 - |r(\eta')|^2), \\ \eta &= -x/4t. \end{aligned}$$

2. Manakov's Approach and the "Isomonodromic" Riemann-Hilbert Problem

Now we come back to the original formulation of the problem and we try to solve asymptotically the Riemann–Hilbert problem (1.4). The variable x will be treated as a parameter, and we will not make any use of the Lax operator (1.2). This means that we will not need any a priori information about the solution $y(x,t)$.

Let us define the following piecewise analytic matrix function

$$\hat{m}(z) = \tag{2.1}$$
$$\begin{pmatrix} 1 & -\frac{1}{2\pi i} \int_{-\infty}^{\infty} \frac{\bar{r}(z')\delta_{+}(z')\delta_{-}(z')}{z'-z} e^{-2it\theta(z')} dz' \\ \frac{1}{2\pi i} \int_{-\infty}^{\infty} \frac{r(z')\delta_{+}^{-1}(z')\delta_{-}^{-1}(z')}{z'-z} e^{2it\theta(z')} dz' & 1 \end{pmatrix}$$
$$e^{\sigma_3 \log \delta(z)}$$

where

$$\theta(z) = 2z^2 - 4z_0 z,$$
$$\delta(z) = \exp\{\frac{1}{2\pi i}\int_{-\infty}^{z_0} \frac{\log(1-|r(z')|^2)}{z'-z}dz'\}, \tag{2.2}$$
$$z_0 = -x/4t.$$

Observe that for sufficiently smooth functions $f(z)$,

$$\frac{1}{2\pi i}\int_{-\infty}^{\infty} \frac{f(z')e^{\pm 2it\theta(z')}}{z'-z-i0}dz' = \begin{cases} f(z)e^{\pm 2it\theta(z)}, & \pm\theta'(z) > 0 \\ 0, & \pm\theta'(z) < 0 \end{cases}$$
$$+ O(\frac{1}{\sqrt{t}(z-z_0)}), \qquad \text{as } t \to +\infty,$$

and taking into account the relation

$$\delta_+(z) = \begin{cases} \delta_-(z)(1-|r(z)|^2), & z < z_0 \\ \delta_-(z) \equiv \delta(z), & z > z_0, \end{cases} \tag{2.3}$$

one easily proves that

$$[m_-(z)]^{-1} m_+(z) = v_{x,t}(z) + O(\frac{1}{\sqrt{t}(z-z_0)}) \tag{2.4}$$

for $z \in \boldsymbol{R}$, $z \neq z_0$. This means that formula (2.1) provides us with the asymptotic solution of the Riemann–Hilbert problem (1.4).

In slightly different form, formula (2.1) for the asymptotic solution of the Riemann–Hilbert problem (1.4) was proposed by Manakov in 1973, [16], (Manakov was not dealing with the Riemann–Hilbert problem itself, but with the corresponding singular integral equations). The above reformulation of Manakov's results was presented in [12]. We will call the function $\hat{m}(z)$ *Manakov's ansatz*. It should be emphasized that this term is different and has nothing in common with the a priori information (1.12) for the solution $y(x,t)$.

A direct consequence of the estimate (2.4) is the asymptotic equality

$$m(z) = (I + O(t^{-1/2}))\hat{m}(z), \qquad Im\ z \neq 0 \tag{2.5}$$

for the exact solution $m(z)$ of the Riemann–Hilbert problem (1.4). This fact enables us to calculate the asymptotics of the modulus of the solution $y(x,t)$ of the Cauchy problem (1.1). Indeed, representing function $\psi(z) \equiv m\exp(-ixz\sigma_3)$ as $\psi(z) = (I + m_1/z + \dots)\exp(-iz\sigma_3 x)$, substituting this into the linear equation (1.2) and taking into account condition (1.3), we get the relation (compare with (1.5)):

$$\lim_{z\to\infty} z(m_{11}(z) - 1) = -i/2\int_x^{\infty} |y(x')|^2 dx'. \tag{2.6}$$

At the same time, from (2.1), (2.5), we obtain

$$\lim_{z\to\infty} z(m_{11}(z) - 1) = \lim_{z\to\infty} z(\delta(z) - 1) + o(1)$$

$$= \frac{i}{2\pi} \int_{-\infty}^{-x/4t} \log(1 - |r(z)|^2) dz + o(1) \tag{2.7}$$

$$t \to +\infty$$

Comparing (2.7) and (2.6) one obtains the asymptotic formula

$$|y(x,t)|^2 \cong -\frac{1}{4\pi t} \log(1 - |r(-x/4t)|^2), \tag{2.8}$$

which we have already obtained in the previous section. We note again that the above method to obtain formula (2.8) was historically the first one.

It is evident from the above calculation that the approximation to the explicit solution $m(z)$ of the Riemann–Hilbert problem (1.4) given by Manakov's ansatz $\hat{m}(z)$ is too crude for the calculation of the asymptotics of the function $y(x,t)$ itself. The technical reason is that function $\hat{m}(z)$ does not satisfy the conjugation relation (1.4) in the neighborhood of the stationary point $z = z_0$ (see (2.4)).

To improve the asymptotic formula (2.5) notice that in the neighborhood of the stationary phase point $z = z_0$, the function $\delta(z)$ appearing in the formula (2.1) for $\hat{m}$ can be represented as

$$\delta_\pm(z) \cong (z - z_0)_\pm^{i\nu} e^{i/2\pi \int_{-\infty}^{z_0} \log|z - z_0| d \log(1 - |r(z)|^2)} \tag{2.9}$$

where

$$\nu = -1/2\pi \log(1 - |r(z_0)|^2) \tag{2.10}$$

and $(z - z_0)_\pm^{i\nu}$ denotes the boundary values of the corresponding multivalued function defined on the z-plane with the cut along $(-\infty, z]$. Equality (2.9) means that if we replace in the formula (2.1)

$$\begin{aligned} &\delta(z) \to (z - z_0)^{i\nu} \\ &r(z) \to r(z_0)e^{-2i\gamma} \equiv r_0, \\ &\gamma = 1/2\pi \int_{-\infty}^{z_0} \log|z - z_0| \; d\log(1 - |r(z)|^2) \end{aligned} \tag{2.11}$$

we obtain a function $\hat{m}^0(z)$, which models the singularities of the function $\hat{m}(z)$ at $z = z_0$. Furthermore, the function $\hat{m}^0(z)$ satisfies relation (2.4) with a matrix $v^0_{x,t}(z)$ that differs from $v_{x,t}$ by replacing

$$r(z) \to r_0.$$

This means that the initial Riemann–Hilbert problem is reduced to the following one (the notation m^{iso} will be explained in the end of this Section (see Remark 1)):

$$m_+^{\text{iso}}(z) = m_-^{\text{iso}}(z) v^0_{x,t}(z), \tag{2.12}$$

$$m^{\text{iso}}(z) \sim z^{i\nu\sigma_3}, z \to \infty.$$

The new asymptotic solution $m^{as}(x)$ of the Riemann–Hilbert problem (1.4) can be represented as

$$m^{as} = m^{iso}(\hat{m}^0)^{-1}\hat{m}.$$

Unlike $\hat{m}$, the function $m^{as}(z)$ satisfies relation (1.4) with an error that is small uniformly on $z \in \mathbb{R}$. This fact leads to the asymptotic representation for the explicit solution $m(z)$,

$$m(z) = (I + o(t^{-1/2}))m^{as}(z), \quad Im\ z \neq 0 \tag{2.13}$$

which is sharper than (2.5), and leads in turn to the asymptotic relation

$$\begin{gathered} y(x,t) = y^{iso}(x,t) + o(t^{-1/2}) \\ t \to \infty, x/t = O(1) \\ y^{iso} = 2 \lim_{z\to\infty}(z m^{iso}_{12}(z) z^{-i\nu}) \end{gathered} \tag{2.14}$$

for the solution of the Cauchy problem (1.1).

The crucial observation is that the model Riemann–Hilbert problem (2.12) can be solved explicitly in terms of parabolic cylinder functions ([I]). Indeed, let us perform the following scaling transformation:

$$m^0(z) = (8t)^{i\nu/2\sigma_3} e^{-(2itz_0^2+i\gamma)\mathrm{ad}\sigma_3} m^{iso}(z/\sqrt{8t} + z_0) z^{-i\nu\sigma_3}. \tag{2.15}$$

In terms of the function $m^0(z)$, the Riemann–Hilbert problem (2.12) can be rewritten as

$$\begin{aligned} &m^0_+(z) = \overset{o}{m}_-(z) e^{-iz^2/4\mathrm{ad}\ \sigma_3} z_-^{i\nu\sigma_3} v(z_0) z_+^{-i\nu\sigma_3}, \\ &m^0(z) \to I. \end{aligned} \tag{2.16}$$

Introducing

$$\psi^0(z) = m^0(z) z^{i\nu\sigma_3} e^{-iz^2/4\sigma_3},$$

we can represent the Riemann–Hilbert problem (2.16) as

$$\begin{aligned} &\psi^0_+ = \psi^0_- v(z_0) \\ &\psi^0(z) = (I + m^0_1/z + \dots) z^{i\nu\sigma_3} e^{-iz^2/4\sigma_3}, \quad z \to \infty. \end{aligned} \tag{2.17}$$

The crucial characteristic of this Riemann–Hilbert problem is that the conjugation matrix does not depend on z, so the logarithmic derivative $\partial_z \psi^0 \cdot (\psi^0)^{-1}$ has no jump across the real axis. Taking into account the asymptotic condition (2.17) and using Liouville's theorem, one concludes that $\partial_z\psi^0 \cdot (\psi^0)^{-1}$ must be the polynomial

$$-iz/2\sigma_3 + \frac{i}{2}[\sigma_3, m^0_1].$$

In other words, $\psi^0(z)$ should satisfy the equation

$$\frac{d\psi^0}{dz} = A(x)\psi^0 \equiv \begin{pmatrix} -iz/2 & \beta \\ \bar{\beta} & iz/2 \end{pmatrix} \psi^0(z), \quad \beta = i(m^0)_{12} \tag{2.18}$$

which can be solved (compare with the previous section) in terms of the parabolic cylinder functions. Taking into account the asymptotic condition (2.17), and the known asymptotic formulae for the parabolic cylinder functions, we obtain

$$\psi^0(z) = e^{\pi|\beta|^2/4}$$

$$\times \begin{pmatrix} D_{i|\beta|^2}(e^{i\pi/4}z) & \beta e^{-3\pi i/4} D_{-1-i|\beta|^2}(e^{-i\pi/4}z) \\ \bar{\beta} e^{3\pi i/4} D_{-1+i|\beta|^2}(e^{i\pi/4}z) & D_{-i|\beta|^2}(e^{-i\pi/4}z) \end{pmatrix} S(z) \tag{2.19}$$

where

$$S(z) = \begin{cases} S_+, & \operatorname{Im} z > 0 \\ S_- & \operatorname{Im} z < 0, \end{cases}$$

$$S_+ = \begin{pmatrix} 1 & 0 \\ \frac{1}{\beta}\frac{\sqrt{2\pi}}{\Gamma(-i|\beta|^2)} e^{-\pi|\beta|^2/2+i\pi/4} & 1 \end{pmatrix}, \quad S_- = \sigma_1 \bar{S}_+ \sigma_1 .$$

In particular, formula (2.19) implies that

$$\begin{aligned} v(z_0) &= S_-^{-1} S_+ \\ &= \begin{pmatrix} e^{-2\pi|\beta|^2} & -\frac{1}{\beta}\frac{\sqrt{2\pi}}{\Gamma(i|\beta|^2)} e^{-\pi|\beta|^2/2-i\pi/4} \\ 1/\beta \frac{2\pi}{\Gamma(-i|\beta|^2)} e^{-\pi|\beta|^2/2+i\pi/4} & 1 \end{pmatrix} \end{aligned}$$

or

$$\begin{aligned} |\beta|^2 &= |(m_1^0)_{12}|^2 = -\frac{1}{2\pi}\log(1 - |r(z_0)|^2 = \nu \\ \arg\beta &= \frac{\pi}{2} + \arg(m_1^0)_{12} = \frac{\pi}{4} + \arg\Gamma(i\nu) - \arg r(z_0). \end{aligned} \tag{2.20}$$

Equalities (2.20) yield the asymptotics for the solution of the Cauchy problem (1.1). In fact, from (2.14), (2.15) and (2.20) we have

$$\begin{aligned} y(x,t) &= \frac{1}{\sqrt{2t}}(8t)^{-i\nu} e^{4itz_0^2+2i\gamma} \cdot (m_1^0)_{12} + o(t^{-1/2}) \\ &\equiv y_{as}(x,t) + o(t^{-1/2}), \end{aligned} \tag{2.21}$$

where

$$\begin{aligned} y_{as} &= t^{-1/2}\alpha(z_0)\exp\left(\frac{ix^2}{4t} - i\nu(z_0)\log t\right), \\ |\alpha(z_0)|^2 &= \nu(z_0)/2 = -\frac{1}{4\pi}\log(1 - |r(z_0)|^2), \\ \arg\alpha(z_0) &= -3\nu\log 2 - \pi/4 + \arg\Gamma(i\nu) - \arg r(z_0) \\ &+ \frac{1}{\pi}\int_{-\infty}^{z_0} \log|z - z_0|\, d\log(1 - |r(z)|^2)\, , \\ z_0 &= -\frac{x}{4t}. \end{aligned} \tag{2.22}$$

Remark 1. It is possible to perform the scaling transformation (2.15) after the derivation of the z-equation (2.18). In this case, the matrix A will depend on the parameters x, t, y^{iso}:

$$A(z) \equiv A(z; x, t, y^{iso}). \tag{2.23}$$

The matrices $S_\pm$ are the monodromy data – Stokes matrices – for equation (2.18).

The fact that they do not depend on the rapid variables x, t (we keep $\eta = -x/4t$ fixed) means that the RHS in (2.21) describes the isomonodromy deformations of the coefficient matrix (2.23). This is the reason why we refer to the solution of the Riemann–Hilbert problem (2.12) as $m^{\text{iso}}(z)$.

3. A Nonlinear Steepest Descent Method

The two examples we consider are the defocusing NLS equation and the MKdV equation

$$y_t - 6y^2 y_x + y_{xxx} = 0 \tag{3.1}$$

with real initial data

$$y(\cdot, 0) = y_0 \in \mathcal{S}(\boldsymbol{R}).$$

In both cases the associated linear Lax operators are self-adjoint with no discrete spectrum.

The Riemann-Hilbert problem for NLS is given by (1.4). The Rie- mann-Hilbert problem for MKdV (see [**13**]), has the same form as (1.4), but now

$$v_{x,t}(z) = e^{-i(4tz^3 + xz)\operatorname{ad}\sigma_3} v(z). \tag{3.2}$$

We also have the symmetry condition

$$r(z) = \overline{-r(-z)} \tag{3.3}$$

and

$$\sup_{z \in \boldsymbol{R}} |r(z)| < 1.$$

The connection with the solution $y(x,t)$ of the Cauchy problem in both cases is given by the same formula (1.5).

In both examples, the oscillatory character of the Riemann-Hilbert problem is clear. The inverse scattering problem for many 1+1-dimensional integrable systems may be formulated as oscillatory Rie- mann-Hilbert problems of similar type. In the classical method of steepest descent for oscillatory integrals, one deforms contours in order to obtain decay for the integrand away from the stationary phase points. In the steepest descent method for oscillatory Riemann-Hilbert problems that follows, one deforms contours in the spirit of the classical case in order to obtain a Riemann–Hilbert problem with a jump matrix that decays to the identity away from stationary phase points.

We now describe in more detail what it means to "deform a Riemann-Hilbert problem".

Let Γ be an oriented contour in the Riemann sphere $\overline{\boldsymbol{C}}$ with non-degenerate jump matrix v (i.e. $\det v \neq 0$) on Γ. The associated Riemann–Hilbert problem is to find a non-degnerate matrix function m, holomorphic in $\overline{\boldsymbol{C}} \setminus \Gamma$, satisfying

$$\begin{aligned} m_+(z) &= m_-(z)v(z), \qquad z \in \Gamma, \\ m(z) &\to I, \qquad z \to \infty \end{aligned} \tag{3.4}$$

where $m_\pm$ denote the boundary values of m on the left and right of Γ respectively,

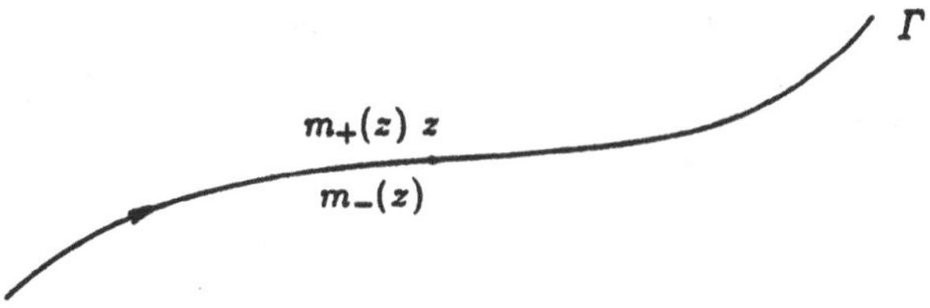

Figure 3.1

In general, the contour Γ may be deformed at will within the domain of holomorphy and non-degeneracy of v in $\overline{\boldsymbol{C}}$. For example, if v is holomorphic and non-degenerate in the shaded region Ω of Figure 3.2a

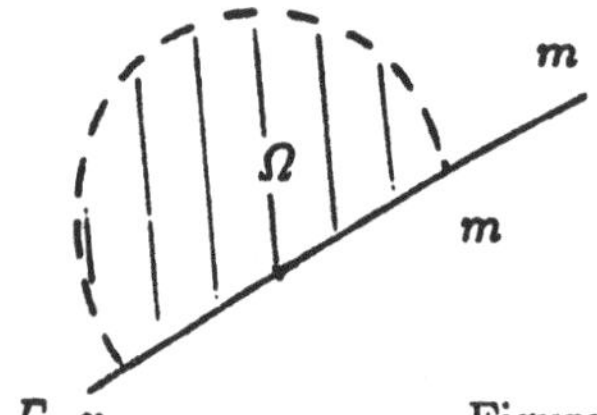

Figure 3.2a

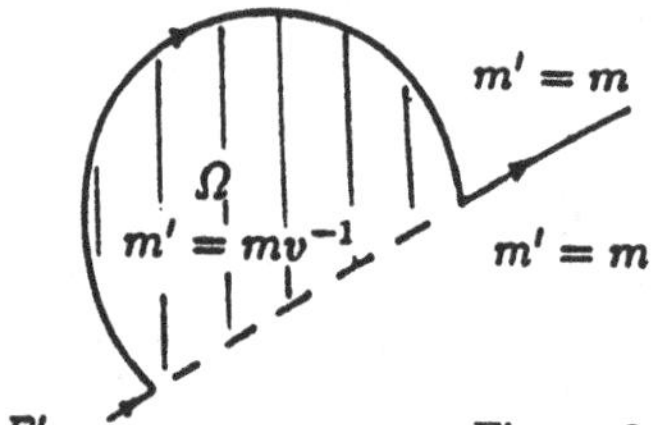

Figure 3.2b

then the contour Γ may be deformed to Γ' in Figure 3.2b with associated jump matrix v, which is just the analytic continuation of the original jump matrix v on Γ. The solution m' of the Riemann-Hilbert problem associated with Γ' agrees with m away from Ω, and for $z \in \Omega$, $m'(z) = m(z)v(z)^{-1}$. Note that for z in Ω, $m'(z)$ is the continuation of m across Γ from the right. We also consider deformations of a more complicated type, in which we assume that v has a factorization $v = b_-^{-1} v_1 b_+$ on $\Gamma \cap \overline{(\Omega_+ \cup \Omega_-)}$.

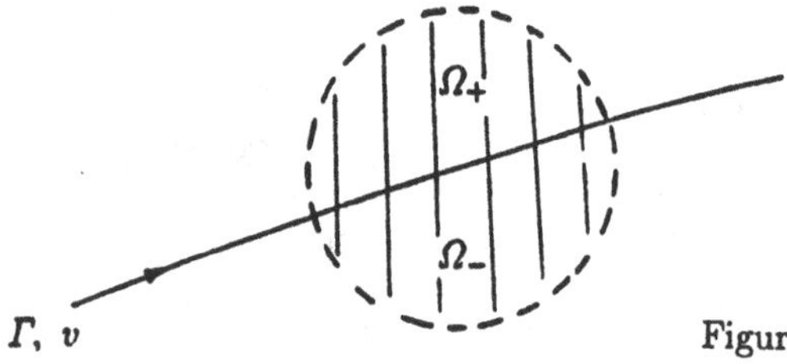

Figure 3.3a

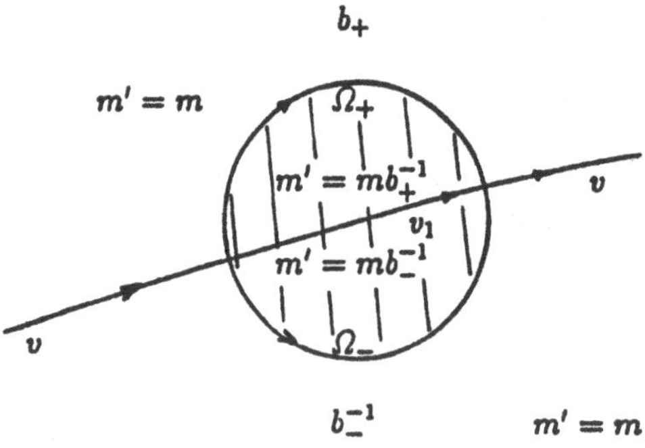

Figure 3.3b

with $b_\pm$ holomorphic and non-degenerate in $\Omega_\pm$ respectively. Here the deformed contour Γ' consists of the original contour Γ together with two semicircles and the jump matrices are as indicated in Figure 3.3b.

We now begin the analysis of defocusing NLS. Let $\theta = 2z^2 + \frac{x}{t}z$ as in Section 2, with stationary phase point $z_0 = -x/4t$. For simplicity, we restrict ourselves here to the physically interesting region $|z_0| \le M$, for some fixed constant M. The matrix v admits the following triangular factorizations

$$v = \begin{pmatrix}1 & \bar{r}\\ 0 & 1\end{pmatrix}\begin{pmatrix}1 & 0\\ r & 1\end{pmatrix}$$

$$= \begin{pmatrix}1 & 0\\ \frac{r}{1-|r|^2} & 1\end{pmatrix}\begin{pmatrix}1-|r|^2 & 0\\ 0 & \frac{1}{1-|r|^2}\end{pmatrix}\begin{pmatrix}1 & \frac{\bar{r}}{(1-|r|^2)}\\ 0 & 1\end{pmatrix}. \tag{3.5}$$

Consider the function $\delta(z)$ introduced in (2.2) that solves the scalar Riemann–Hilbert problem (2.3) and whose significance will become clear below. One checks that $\delta(z)$ and $\delta(z)^{-1}$ are uniformly bounded in z and for $|z_0| \le M$.

The function $\tilde{m} = m\delta^{-\sigma_3}$ satisfies a Riemann–Hilbert problem across $\Gamma = \mathbb{R}$ with jump matrix

$$\tilde{v}_{x,t}(z) = e^{-it\theta\,\mathrm{ad}\sigma_3}(\delta_-^{\sigma_3} v \delta_+^{-\sigma_3})$$

$$= \begin{cases} e^{-it\theta\,\mathrm{ad}\sigma_3}\begin{pmatrix}1 & 0\\ \frac{r\delta_-^{-2}}{1-|r|^2} & 1\end{pmatrix}\begin{pmatrix}1 & \frac{-\bar{r}\delta_+^2}{1-|r|^2}\\ 0 & 1\end{pmatrix}, & z < z_0\\ e^{-it\theta\,\mathrm{ad}\sigma_3}\begin{pmatrix}1 & -\bar{r}\delta^2\\ 0 & 1\end{pmatrix}\begin{pmatrix}1 & 0\\ r\delta^{-2} & 1\end{pmatrix}, & z > z_0. \end{cases} \tag{3.6}$$

Note that as δ^{σ_3} is diagonal, we can replace m by $\tilde{m}$ in (1.5).

Having made the above definitions, we now describe the strategy. Suppose that the coefficients

$$\frac{r}{1-|r|^2},\ \frac{\bar{r}}{1-|r|^2},\ r,\ \bar{r} \tag{3.7}$$

can be replaced by some rational functions

$$[\frac{r}{1-|r|^2}],\ [\frac{\bar{r}}{1-|r|^2}],\ [r],\ [\bar{r}] \tag{3.8}$$

respectively. Then if the poles of these functions are appropriately placed, the Riemann-Hilbert problem on $\mathbb{R}$ can be deformed to the contour Σ.

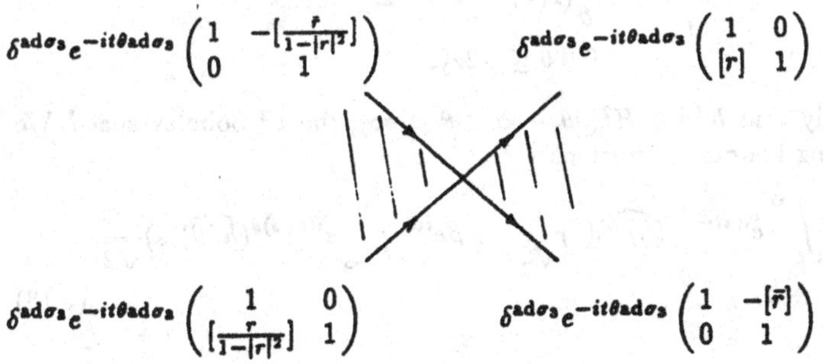

Figure 3.4, Σ

Using the signature table for Re$i\theta$,

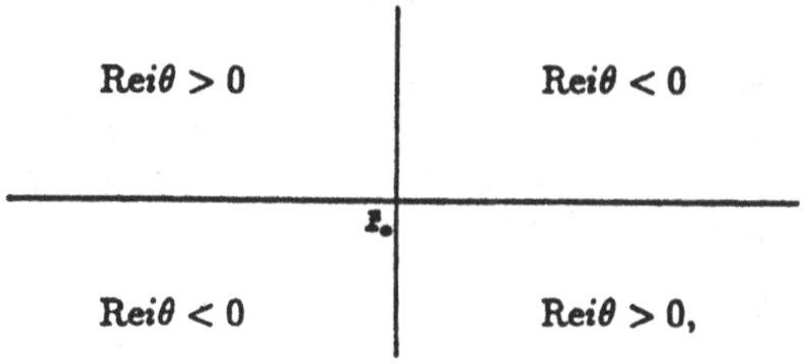

Figure 3.5. z-plane.

the reader now sees that the scalar factorization (2.3) for δ, and the triangular factorizations (3.6) for v, *were chosen specifically* to ensure that the jump matrices for the deformed problem Σ converge rapidly to the identity away from any neighborhood of z_0, as $t \to \infty$.

To verify that the coefficients (3.7) can be replaced by the rational functions (3.8) with well-controlled errors, we proceed as follows. Expand $(i+z)^{10}r(z)$ in a fifth order Taylor series around z_0,

$$(i+z)^{10}r = \mu_0 + \mu_1(z-z_0) + \cdots + \mu_5(z-z_0)^5 + (i+z)^{10}h$$

and set

$$[r] = \frac{\mu_0 + \mu_1(z-z_0) + \cdots + \mu_5(z-z_0)^5}{(i+z)^{10}}. \tag{3.9}$$

Also, set

$$\beta = \frac{(z-z_0)^2}{(z+i)^4}. \tag{3.10}$$

Observe that

$$\frac{h}{\beta} = \frac{r-[r]}{\beta} = \frac{(z-z_0)^4}{(z+i)^6}g(z;z_0), \quad z \geq z_0, \tag{3.11}$$

where

$$|g| + |\frac{\partial g}{\partial z}| + |\frac{\partial^2 g}{\partial z^2}| \leq C(M). \tag{3.12}$$

Since $\theta(z) = 2(z-z_0)^2 - 2z_0^2$ is one-to-one from (z_0,∞) to $(-2z_0^2,\infty)$, we can consider h/β as a function of θ:

$$\begin{aligned}\frac{h}{\beta}(\theta) &\equiv \frac{h}{\beta}(z(\theta)), \quad \theta > -2z_0^2,\\ &\equiv 0, \quad \theta \leq -2z_0^2.\end{aligned}$$

One checks easily that $h/\beta \in H^2(d\theta, -\infty < \theta < \infty)$, the L^2 Sobolev space. We have the following Fourier expressions,

$$\begin{aligned}e^{2it\theta}h &= \beta\int_t^{\infty} e^{i(2t-s)\theta}(\widehat{h/\beta})(s)\frac{ds}{\sqrt{2\pi}} + \beta e^{it\theta}\int_{-\infty}^{t} e^{i(t-s)\theta}(\widehat{h/\beta})(s)\frac{ds}{\sqrt{2\pi}}\\ &\equiv h_I + h_{II},\end{aligned} \tag{3.13}$$

where

$$(\widehat{h/\beta})(s) = \int_{-\infty}^{\infty} e^{is\theta}(h/\beta)(\theta)\frac{d\theta}{\sqrt{2\pi}}. \tag{3.14}$$

By Plancherel

$$\int_{-\infty}^{\infty}(1+s^2)^2|(\widehat{h/\beta})(s)|^2 ds < \infty \tag{3.15}$$

and hence

$$|h_I(z)| = \frac{const.}{|z+i|^2 t^{3/2}}. \tag{3.16}$$

On the other hand, $h_{II}(z)$ has an analytic continuation to the line $z_0 + e^{i\pi/4}\boldsymbol{R}_+$, where it satisfies the estimate

$$|h_{II}(z)| \le \frac{const.}{|z+i|^2 t}, \tag{3.17}$$

again by (3.15). Hence

$$\|h_I\|_{L^1\cap L^2\cap L^\infty(z_0+e^{i\pi/4}\boldsymbol{R})} = O(t^{-3/2}) \tag{3.18}$$

and

$$\|h_{II}\|_{L^1\cap L^2\cap L^\infty}(z_0,\infty) = O(t^{-1}) \tag{3.19}$$

Arbitrarily high order of decay in (3.18) and (3.19) can be obtained by using a higher order Taylor expansion in (3.9) at z_0.

Now deform the Riemann–Hilbert problem for $\tilde{v}$ on $\boldsymbol{R}$ to the contour

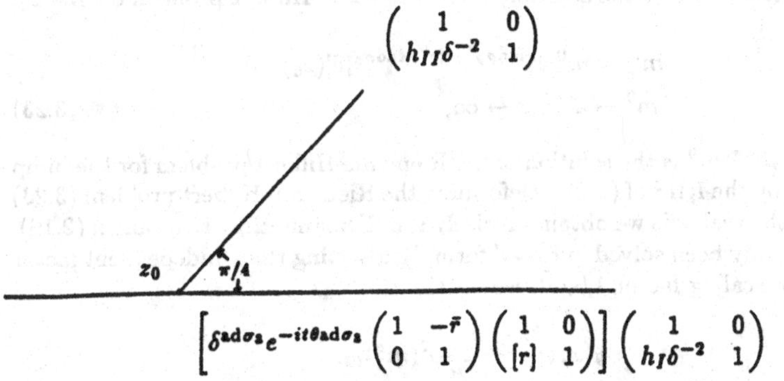

Figure 3.6

It turns out (see **[DZ 2]**; see also the discussion at the end of this section) that the error estimates (3.18) and (3.19) are sufficient to ensure that the contributions of $h_{II}\delta^{-2}$ on $z_0 + e^{i\pi 4}\boldsymbol{R}_+$ and $h_1\delta^{-2}$ on (z_0,∞) are negligible for the leading asymptotics as $t \to \infty$. Repeating the above arguments for the remaining functions in (3.7), we see that they all can be replaced in the Riemann–Hilbert problem $\tilde{v}$ by the appropriate rational functions in (3.8), with effective error control. Deforming the contour as above, we arrive at the Riemann–Hilbert on Σ in Figure 3.4.

Define the scaling operator $N : L^2(\Sigma) \to L^2(\Sigma - z_0)$,

$$f(z) \mapsto Nf(z) = f\left(\frac{z}{\sqrt{8t}} + z_0\right). \tag{3.20}$$

Denote the jump matrix in Figure 3.4 by $\delta^{\mathrm{ad}\sigma_3}e^{-it\theta\,\mathrm{ad}\sigma_3}[\tilde{v}]$. A straightforward computation shows that as $t \to \infty$

$$N\delta^{\operatorname{ad}\sigma_3} e^{-it\theta\operatorname{ad}\sigma_3}[\tilde{v}] \to \phi^{\operatorname{ad}\sigma_3} z^{\nu i \operatorname{ad}\sigma_3} e^{-iz^2/4\operatorname{ad}\sigma_3}[\tilde{v}](z_0), \tag{3.21}$$

where

$$\phi = (8t)^{\frac{-\nu i}{2}} e^{2itz_0^2} e^{\frac{i}{2\pi}\int_{-\infty}^{z_0} \log(z_0-\zeta)\,d\log(1-|r(\zeta)|^2)} \tag{3.22}$$

and $[\tilde{v}](z_0)$ is defined by the figure

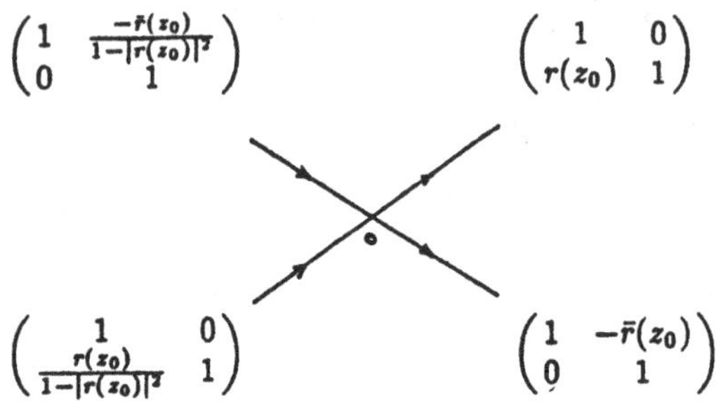

Figure 3.7

It follows from the exponential decay of $e^{-iz^2/4\operatorname{ad}\sigma_3}[\tilde{v}](z_0)$, that the asymptotic formula in (3.21) has an $L^1 \cap L^2 \cap L^\infty(\Sigma - z_0)$ error of order $(\log t)/t^{1/2}$. Since ϕ is independent of z, m^0 is the solution of the Riemann–Hilbert problem on $\Sigma - z_0$,

$$\begin{aligned} m_+^0 &= m_-^0 z^{\nu i \operatorname{ad}\sigma_3} e^{-iz^2/4\operatorname{ad}\sigma_3}[\tilde{v}](z_0), \\ m^0 &\to I \text{ as } z \to \infty, \end{aligned} \tag{3.23}$$

if and only if $\phi^{\operatorname{ad}\sigma_3} m^0$ is the solution of the Riemann–Hilbert problem for the jump matrix given by the RHS of (3.21). Deforming the Riemann–Hilbert problem (3.23) on $\Sigma - z_0$ to the real axis we obtain precisely the Riemann–Hilbert problem (2.16), which has already been solved in closed form. Reinserting the z-independent factor $\phi^{\operatorname{ad}\sigma_3}$ and the scaling factor $1/\sqrt{8t}$, we obtain

$$y(x,t) \sim \frac{1}{i\sqrt{2t}} \phi^2 (m_1^\circ)_{12}.$$

Careful bookkeeping of the error terms that arise in the above method, yields the following result.

THEOREM 3.1. *Let $y(x,t)$ be the solution of the Cauchy problem (1.1). Then as $t \to \infty$*

$$y(x,t) = y_{as}(x,t) + O\left(\frac{\log t}{t}\right) \tag{3.24}$$

in the region $|\frac{x}{4t}| \le M$, where $y_{as}(x,t)$ is given by (2.22).

We now consider the MKdV equation. Here $\theta = 4z^3 + \frac{x}{t}z$ with two stationary phase points $\pm z_0 = \pm\sqrt{\frac{-x}{12t}}$. As before we restrict ourselves to the physically interesting region, here described by $M^{-1} \le z_0 \le M$ for any fixed constant $M > 1$. In particular, this implies $x < 0$.

In this case, the signature table for $\mathrm{Re}\, i\theta$ consists of six regions

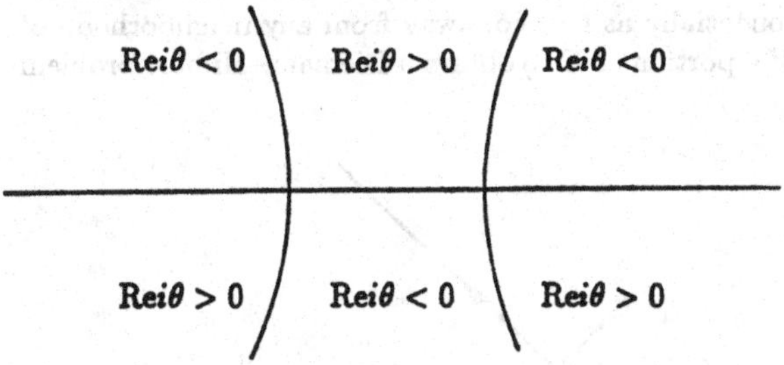

Figure 3.8

Here we solve the scalar Riemann–Hilbert problem

$$\delta_+ = \begin{cases} \delta_-(1-|r|^2), & |z| < z_0, \\ \delta_- = \delta, & |z| > z_0, \end{cases} \tag{3.25}$$

with explicit solution

$$\delta(z) = \left(\frac{z-z_0}{z+z_0}\right)^{i\nu} e^{\chi(z)}, \quad \chi(z) = \frac{1}{2\pi i}\int_{-z_0}^{z_0} \log\left(\frac{1-|r(\xi)|^2}{1-|r(z_0)|^2}\right)\frac{d\xi}{\xi - z}. \tag{3.26}$$

The analog of (3.6) is

$$\tilde{v} \equiv \delta_-^{\sigma_3}\, v \delta_+^{-\sigma_3} = \begin{cases} \begin{pmatrix} 1 & 0 \\ \frac{r\delta_-^{-2}}{1-|r|^2} & 1 \end{pmatrix}\begin{pmatrix} 1 & \frac{-\bar{r}\delta_+^2}{1-|r|^2} \\ 0 & 1 \end{pmatrix}, |z| < z_0 \\ \begin{pmatrix} 1 & -\bar{r}\delta^2 \\ 0 & 1 \end{pmatrix}\begin{pmatrix} 1 & 0 \\ r\delta^{-2} & 1 \end{pmatrix}, \quad |z| > z_0. \end{cases} \tag{3.27}$$

As before, on each interval $(-\infty, -z_0), (-z_0, z_0)$ and (z_0, ∞), we replace the coefficients r etc., by rational functions $[r]$ etc., which agree with the original coefficients at the stationary phase points $\pm z_0$. We deform the contour $\boldsymbol{R}$ to

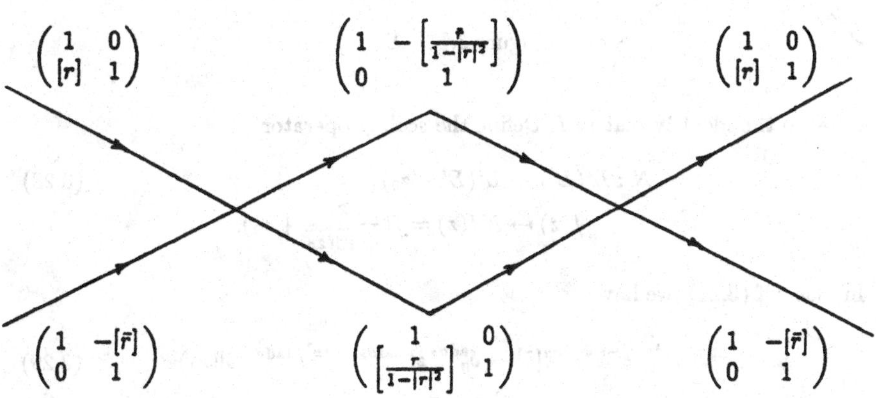

Figure 3.9 Σ, $[\tilde{v}]$

(The conjugating factors $\delta^{\operatorname{ad}\sigma_3} e^{-it\theta \operatorname{ad}\sigma_3}$ are omitted from the Figure in the interests of brevity.) The jump matrix $\delta^{\operatorname{ad}\sigma_3} e^{-it\theta\operatorname{ad}\sigma_3}[\tilde{v}]$ for the Riemann–Hilbert problem converges to I exponentially as $t \to \infty$, away from any neighborhood of $\pm z_0$. We can therefore cut off a portion of Σ to obtain a Riemann–Hilbert problem on two crosses $\Sigma_A \cup \Sigma_B$

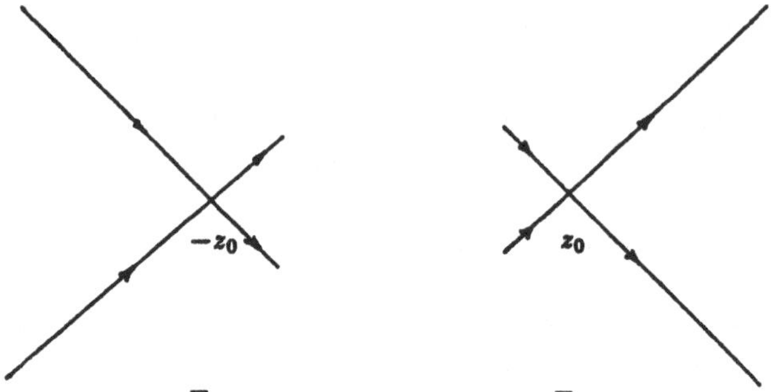

Σ_A Σ_B Figure 3.10

separated by a distance of order M^{-1}. As $t \to \infty$ the interaction between the Riemann–Hilbert problem on Σ_A and on Σ_B goes to zero faster than the leading order of the solution and the contribution of $\Sigma_A \cup \Sigma_B$ to $y(x,t)$ is simply the sum of the contributions from the two Riemann–Hilbert problems separately, $y \sim y_A + y_B$. Symmetry implies that $y_A = \bar{y}_B$, and hence it is enough to compute y_B.

We first extend Σ_B to a full cross by setting the jump matrix on the dotted lines in the figure

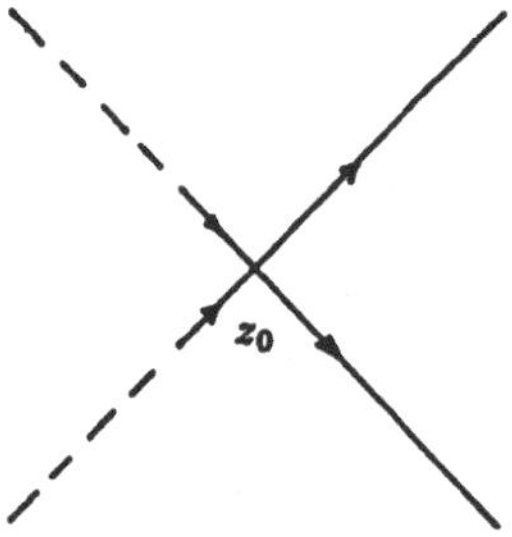

Figure 3.11. Σ'

equal to the identity matrix I. Define the scaling operator

$$\begin{aligned} N : L^2(\Sigma') &\to L^2(\Sigma' - z_0) \\ f(z) \mapsto Nf(z) &= f\left(\frac{z}{\sqrt{48tz_0}} + z_0\right). \end{aligned} \tag{3.28}$$

In place of (3.21), we have

$$N\delta^{\operatorname{ad}\sigma_3} e^{-it\theta\operatorname{ad}\sigma_3}[\tilde{v}] \to \delta_0^{\operatorname{ad}\sigma_3} z^{\nu i \operatorname{ad}\sigma_3} e^{-iz^2/4 \operatorname{ad}\sigma_3}[\tilde{v}](z_0) \tag{3.29}$$

where $\delta_0 = (192tz_0^3)^{-i\nu/2}(2z_0)^{\nu i} e^{8itz_0^3 \chi(z_0)}$ and $[\tilde{v}](z_0)$ appears in Figure 3.7. Thus the calculation of the long-time behavior of the MKdV equation reduces to the *same* explicitly solvable isomonodromy problem as in NLS. Set

$$\begin{aligned} y_{as} = 2Re(y_B)_{as} &= 2Re\left(\frac{1}{\sqrt{48tz_0}}\delta_0^{-2}(2i\beta_{21})\right) \\ &= \left(\frac{\nu}{3tz_0}\right)^{\frac{1}{2}} \cos(16tz_0^3 - \nu\log(192tz_0^3) + \varphi(z_0)) \end{aligned} \tag{3.30}$$

where

$$\varphi(z_0) = \arg\Gamma(i\nu) - \frac{\pi}{4} - \arg r(z_0) + \frac{1}{\pi}\int_{-z_0}^{z_0} \log|\zeta - z_0|.$$

Once again careful bookkeeping yields the following result.

THEOREM 3.2. *Let $y(x,t)$ be the solution of the Cauchy problem (3.1). Then as $t \to \infty$*

$$y(x,t) = y_{as}(x,t) + O\left(\frac{\log t}{t}\right) \tag{3.32}$$

in the region $M^{-1} \leq \sqrt{\frac{-x}{12t}} \leq M$.

A complete description of the long-time behavior of the MKdV equation in all regions of x-space, can be found in [8].

Finally, we will indicate how to use the estimates on the jump matrices obtained above, to obtain error estimates on the asymptotic solutions.

By way of illustration, in the case of NLS, we show how to use the estimates (3.18) and (3.19) to bound the contribution of h_I and h_{II} to the asymptotic solution. Other error estimates are similar.

For an oriented countour Γ with a factored $n \times n$ jump matrix $v = b_-^{-1}b_+$, the Riemann-Hilbert problem on Γ

$$\begin{aligned} m_+ &= m_- v \quad \text{on } \Gamma \\ m &\to I \quad \text{as } i \to \infty \end{aligned}$$

is solved as follows (see [3]). Set

$$w_\pm = \pm(b_\pm - I)$$

and consider the equation

$$\begin{aligned} C_w : M_n(\boldsymbol{C}) + L^2(\Sigma, M_n(\boldsymbol{C})) &\to L^2(\Sigma, M_n(\boldsymbol{C})) \\ f \mapsto C_w f &= C_+(fw_-) + C_-(fw_+) \end{aligned} \tag{3.33}$$

where $C_\pm$ denotes the Cauchy operator

$$(C_\pm f)(z) = \int_\Gamma \frac{f(\zeta)}{\zeta - z_\pm}\frac{d\zeta}{2\pi i} \tag{3.34}$$

where $z_\pm$ indicates a (non-tangential) limit from $\pm$ sides of Γ respectively.

Let $\mu \in I + L^2(\Sigma, M_n(\boldsymbol{C}))$ be the solution of

$$(1 - C_w)\mu = I; \tag{3.35}$$

then

$$m(z) = I + \int_\Gamma \frac{\mu(\zeta)(w_+(\zeta) + w_-(\zeta))}{\zeta - z} \frac{d\zeta}{2\pi i}, \quad z \in \boldsymbol{C} \setminus \Gamma \tag{3.36}$$

The asymptotics (1.5) of m needed for NLS can be read off from the limit

$$\begin{aligned} \lim_{z\to\infty} z\, m(z) &= -\frac{1}{2\pi i} \int \mu(\zeta)(w_+(\zeta) + w_-(\zeta)) d\zeta \\ &= -\frac{1}{2\pi i} \int ((1 - C_w)^{-1} I)(w_+ + w_-) d\zeta. \end{aligned} \tag{3.37}$$

As an example, let us illustrate how to control the error when we replace r by $[r]$ in the case of NLS. For the Riemann–Hilbert problem in Figure 3.6, the error in the jump matrix is controlled by (3.18) and (3.19). In general, let us assume that two sets of data $w_\pm$ and $w'_\pm$ differ by

$$\|w_\pm - w'_\pm\|_{L^1\cap L^2\cap L^\infty} = O(t^{-1})$$

and $\|w_\pm\|_{L^1\cap L^2\cap L^\infty} = O(1)$. We also assume that

$$\|(1 - C_w)^{-1}\|_{L^1\to L^2} = O(1). \tag{3.38}$$

The estimate (3.38) indeed holds for NLS (and MKdV) (see [8]). Thus the following estimates are easily derived,

$$\|C_w I - C_{w'} I\|_{L^2} = O(\|w - w'\|_{L^2}) = O(t^{-1}) \tag{3.39}$$

$$\|C_w - C_{w'}\|_{L^2\to L^2} = O(\|w - w'\|_{L^\infty}) = O(t^{-1}) \tag{3.40}$$

$$\|(1 - C_w)^{-1} - (1 - C_{w'})^{-1}\|_{L^2\to L^2} \overset{\text{by (3.38)}}{=} O(\|C_w - C_{w'}\|) = O(t^{-1}). \tag{3.41}$$

We finally obtain the uniform estimate

$$\begin{aligned} &\int ((1 - C_w)^{-1} I)(w_+ + w_-) \\ &\quad = \int ((1 - C_w)^{-1} C_w I)(w_+ + w_-) + \int (w_+ + w_-) \\ &\quad = \int ((1 - C_{w'})^{-1} C_{w'} I)(w'_+ + w'_-) \\ &\quad + \int (w'_+ + w'_-) + O(t^{-1}). \end{aligned} \tag{3.42}$$

If w and w' correspond to r and $[r]$ respectively, this shows that we can replace r by $[r]$ in the asymptotic solution of the inverse problem with a controlled error.

Acknowledgements. The research of the first and third authors was supported in part by NSF Grants DMS-9001857 and DMS-9196033 respectively.

REFERENCES

[1] M.J. Ablowitz and A.C. Newell. The decay of the continuous spectrum for solutions of the Korteweg de Vries equation, J. Math. Phys. 14, 1277-1284 (1973).

[2] M.J. Ablowitz and H. Segur. Asymptotic solutions of the Korteweg de Vries equation. Stud. Appl. Math. 57, No. 1, 13-14 (1977).

[3] B. Beals and R. Coifman, Scattering and inverse scattering for first order systems. Comm. Pure Appl. Math. 37, 39-90, (1984).

[4] R. Beals, P. Deift and C. Tomei. Direct and Inverse Scattering on the Line, AMS, Math. Surveys and Monographs No. 28 (1988).

[5] R.F. Bikbaev, A.R. Its, The asymptotics as $t \to \infty$ of the solution of the Cauchy problem for the Landau–Lifshitz equation. Teoretich. i Mat. Fiz., 76, No. 1, (1988).

[6] V.S. Buslaev and V.V. Sukhanov:

1. Asymptotic behavior of solutions of the Korteweg de Vries equation, Proc. Sci. Seminar LOMI, 120, 32-50 (1982) (Russian); Jour. Sov. Math. 34, 1905-1920 (1986)(English).

2-4. On the asymptotic behavior as $t \to \infty$ of the solutions of the equation $\psi_{xx} + u(x,t)\psi + (\lambda/4)\psi = 0$ with potential u satisfying the Korteweg de Vries equation, I,II,III.

I. Prob. Math. Phys., ed. M. Birman, 10, 70-102 (1982) (Russian); Sel. Math. Soc. 4, No. 3, 225-248 (1985) (English).

II. Proc. Sci. Seminar LOMI, 138, 8-32 (1984)(Russian); Jour. Sov. Math. 32, 426-446 (1986)(English).

III. Prob. Math. Phys., Ed. M. Birman 11, 78-113 (1986) (Russian).

[7] P. Deift and X. Zhou, Direct and Inverse Scattering on the Line with arbitrary Singularities. Comm. Pure Appl. Math. 44, 485-533 (1991).

[8] P. Deift and X. Zhou, A Steepest Descent method for oscillatory Riemann-Hilbert Problems. Ann. of Math. To appear.

[9] C.S. Gardner, J.M. Greene, M.D. Kruskel and R.M. Miura, Method for Solving the Korteweg de Vries Equation, Phys. Rev. Lett., 19, 1967, 1095-1097.

[10] A.S. Fokas and A.R. Its, The Isomonodromy Method and the Painlevé Equations, in this book.

[11] L.D. Faddeev and L.A. Takhtajan. Hamiltonian Methods in the Theory of Solitons, Springer–Verlag, Berlin, Heidelberg, 1987.

[12] A.R. Its. Asymptotics of solutions of the nonlinear Schrödinger equation and isomonodromic deformations of systems of linear differential equations, Sov. Math. Dokl., 24, No. 3, 452-456 (1981).

[13] A.R. Its and V.Yu. Novokshenov, The isomonodromic method in the theory of Painlevé equations, Lecture Notes in Math., 1191, Springer-Verlag, Berlin, Heidelberg, 1986.

[14] A.R. Its and V.E. Petrov, "Isomonodromic" solutions of the sine-Gordon equation and its time asymptotics of the rapidly decreasing solutions, Soviet Math. Dokl., 26, No. 1, 244–247 (1982).

[15] A.R. Its and A.F. Ustinov, The time asymptotics of the solution of the Cauchy problem for the nonlinear Schrödinger equation with finite density boundary conditions, Dokl. Acad. Nauk SSSR, 291, No. 1, 91 (1986).

[16] S.V. Manakov, Nonlinear Fraunhofer diffraction, Zh. Eksp. Teor. Fiz., 65, 1392-1398 (1973)(Russian): Sov. Phys. -JETP, 38, No.4, 693-696 (1974)(English).

[17] V.Yu. Novokshenov, Asymptotics as $t \to \infty$ of the solution of the Cauchy problem for the nonlinear Schrödinger equation, Sov. Math. Dokl. 21, No. 2, 529-533 (1980).

[18] V.Yu. Novokshenov, Asymptotics as $t \to \infty$ of the solution of the Cauchy problem for the two dimensional Toda lattice, Izv. Acad. Nauk SSSR, 48, No. 2, 372–410 (Russian) (1984).

[19] V.Yu. Novokshenov, Asymptotic behavior as $t \to \infty$ of the solutions of the Cauchy problem for the nonlinear differential–difference Schrödinger equation, Diff. Equ. 21 (Russian), No. 11, 1915–1926 (1985).

[20] A.B. Shabat. On the Korteweg de Vries equation. Sov. Math. Dokl., 14, 1266 (1973).

[21] H. Segur and M.J. Ablowitz, Asymptotic solutions and conservation laws for the nonlinear Schrödinger equation, Part I, J. Math. Phys., 17, 710–713, (1976).

[22] H. Segur and M. J. Ablowitz, Asymptotic solutions of nonlinear evolution equations and a Painlevé transcendent, Physica 3D, 1+2, 165–184 (1981).

[23] H. Segur, Asymptotic solutions and conservation laws for the nonlinear Schrödinger equation, Part II, J. Math. Phys. 17, 714–716. (1976).

[24] V.V. Sukhanov, Asymptotic behavior of solutions of the Cauchy problem for a system of the KdV type for large time, Sov. Phys. Dokl., 28 (4), 325–327, (1983).

[25] V.V. Sukhanov, On the asymptotic behavior of solutions of the nonlinear string equations at large time, Zap. Nauch. Semin. Lomi, 161, 122–138, (1987)

[26] V.V. Sukhanov, Large time aymptotics for principal chiral field, Teoretich. i Mat. Fiz., 84, No. 1, 23–27, (1990)

[27] V.E. Zakharov and S.V. Manakov, Asymptotic behavior of nonlinear wave systems integrated by the inverse method, Zh. Eksp. Teor. Fiz., 71, 203-215 (1976)(Russian); Sov. Phys. -JETP, 44, No. 1, 106-112 (1976)(English).

[28] X. Zhou, The Riemann-Hilbert problem and inverse scattering, SIAM J. Math. Anal., 20, No. 4, 966-986 (1989).

[29] X. Zhou, Direct and Inverse Scattering Transforms with Arbitrary Spectral Singularities, Comm. Pure Appl. Math. 42, 895-938 (1989).

The Generation and Propagation of Oscillations in Dispersive Initial Value Problems and Their Limiting Behavior

P.D. Lax[1], *C.D. Levermore*[2], *and S. Venakides*[3]

[1]Courant Institute, New York University, New York, NY 10012, USA

[2]Mathematics Department, University of Arizona, Tucson, AZ 85721, USA

[3]Mathematics Department, Duke University, Durham, NC 27706, USA

We review a variety of equations describing physical systems in which dissipative or diffusive mechanisms are absent, but which undergo dispersive processes. We shall investigate the limiting behavior of such a system when the parameter in the dispersive term tends to zero. Numerical experiments reveal, and theory confirms by proof, fascinating patterns: smoothness in some regions coexist with regular oscillations in other regions, and irregular oscillations in yet other parts. The limit exists in the weak, i.e. average sense, and can be described with great precision. No general theory is given — since none exists — but several success stories are presented. Cases where the limiting behavior has been analyzed and understood are completely integrable, which makes them explicitly solvable. The strategy that has been used is to study these explicit solutions, complicated though they are, and to trace within their structure the passage to zero of the small parameter. In this way, not only the weak limit, but the microstructure of the oscillations can be understood. Less is known about nonintegrable cases. While they sometimes share properties exhibited by the integrable cases, new phenomena arise that have yet to be completely understood.

1. Introduction

Most equations describing the behavior of physical systems contain parameters. Examples abound: the speed of light, the mass of the electron, Planck's constant, Reynolds's number, etc. Of particular interest is the behavior of solutions when such a parameter approaches a critical value, usually zero or infinity.

In this review article we look at a variety of equations describing physical systems in which dissipative or diffusive mechanisms are absent, but which undergo dispersive processes. We shall investigate the limiting behavior of such systems when the parameter in the dispersive term tends to zero. Numerical experiments reveal, and theory confirms by proof, fascinating patterns: smoothness in some regions coexist with regular oscillations in other regions, and irregular oscillations in yet other parts. In spite of this complexity, the limit exists in the weak, i.e. average, sense and can be described with great precision. We shall give no general theory — since none exists — but present several examples where the limiting behavior has been analyzed and understood.

All these cases are completely integrable, which makes them explicitly solvable. The strategy that has been used successfully is to study these explicit solutions, complicated though they are, and to trace within their structure the passage to zero of the small parameter. To leading order, the result is an explicit formula for the weak limit, in terms of density of states characterized as the solution of a variational problem. Most properties of the weak limit can be deduced from this variational formula. In addition, a more refined analysis yields the small scale oscillations in great detail. The weak limit then can be regarded as the result of a Whitham averaging of these oscillations.

A variety of mathematical tools come into play. Exact solutions of completely integrable systems, criteria for positiveness of quadratic form, variational conditions for constrained minima, function theoretic methods for solving variational problems, hyperelliptic integrals, and Riemann θ functions.

The organization of this review is as follows: in Section 2 we present the structure of the KdV hierarchy. Section 3 contains the explicit solution of the initial value problem for the KdV hierarchy. In Section 4 we derive the leading order asymptotics for KdV in terms of weak convergence, and Section 5 is devoted to the higher order theory that analyzes the microstructure. Section 6 describes recent results on the focusing and defocusing cubic Schrödinger equation and their associated hierarchies, including the modified KdV equation. Section 7 describes completely integrable discrete analogues of small dispersion limits, namely continuum limits, in particular the limits of the Kac-van Moerbeke system and the Toda chain. In Section 8 we present numerical evidence that nonintegrable discrete systems such as dispersive difference schemes and the vibrations of chains governed by other spring laws, nevertheless sometimes exhibit weak limits in the small dispersion limit. While there is comparatively little theory yet, these weak limits, and the approach to them, seem to share some of the properties that can be rigorously proved in the integrable cases.

2. The Structure of the KdV Hierarchy

The Korteweg-deVries (KdV) equation is the prototypical integrable nonlinear wave equation; it was the first to be solved and remains the best understood. We will use it as our primary example in this paper. Its initial value problem for $u = u(x,t)$ is

$$\partial_t u + 6u\partial_x u + \epsilon^2 \partial_{xxx} u = 0\,, \tag{2.1a}$$

$$u(x,0) = v(x)\,. \tag{2.1b}$$

This was first solved by Gardner, Greene, Kruskal and Miura [16] for initial data $v(x)$ that decays sufficiently rapidly as $|x| \to \infty$. Their critical observation was that the KdV is the solvability condition for the linear system

$$Lf = \lambda f\,, \qquad \partial_t f = Bf\,, \tag{2.2}$$

in which λ is an eigenvalue, and where the operators L and B are given by

$$Lf \equiv -\epsilon^2 \partial_{xx} f - uf\,, \tag{2.3a}$$
$$Bf \equiv -4\epsilon^2 \partial_{xxx} f - 3(u\partial_x + \partial_x u)f\,. \tag{2.3b}$$

Restated, the local existence of solutions $f = f(\lambda, x, t)$ of the system (2.2) for every constant λ is equivalent to

$$\partial_t L = [B, L] \equiv BL - LB\,, \tag{2.4}$$

which is equivalent to the KdV equation (2.1a). This observation enables the IVP for the KdV equation to be solved for a variety of initial data, including periodic data [6], [24], [37], [38] and data with differing limits at $\pm\infty$ [4], [5].

The operators L and B are referred to as the Lax pair for the KdV equation [31]. Formally, equation (2.4) insures that the spectrum of the Schrödinger operator L is independent of time. This structure, as well as the general solution strategy outlined below, is shared by all integrable nonlinear wave equations.

The strategy for solving (2.1) that was introduced in [16] is the so-called inverse scattering method. Given $u(x,0)$, the asymptotics of the eigenfunctions $f(\lambda, x, 0)$ as $|x| \to \infty$, referred to as the scattering data, can be calculated in principle. The evolution of the scattering data is obtained explicitly from the second equation of (2.2) since the terms containing u drop out when $|x|$ is large. The potential $u(x,t)$ is then obtained from the knowledge of the large $|x|$ asymptotics of $f(\lambda, x, t)$ using the inverse theory of scattering potentials. This theory was developed in the 40's and 50's by Gelfand, Levitan, Marchenko, Faddeev and others within the program of determining nuclear potentials from the data of scattering experiments.

More specifically, the L^2 spectrum of the Schrödinger operator (2.3a) consists of the nonnegative semi-axis $\lambda \geq 0$ along with a finite set (possibly empty) of negative simple eigenvalues $\lambda_1, \cdots, \lambda_N$. The asymptotic behavior of an eigenfunction $f = f(\lambda, x)$ corresponding to a $\lambda = k^2 > 0$ in the continuous spectrum is given by

$$f(\lambda, x) \sim \begin{cases} T(k)\exp\left(\dfrac{-ikx}{\epsilon}\right), & \text{for } x \to -\infty, \\ \exp\left(\dfrac{-ikx}{\epsilon}\right) + R(k)\exp\left(\dfrac{ikx}{\epsilon}\right), & \text{for } x \to +\infty. \end{cases} \tag{2.5}$$

The asymptotic behavior of a real unit normalized eigenfunction $f = f_j(x)$ corresponding to a discrete eigenvalue $\lambda_j = -\eta_j^2 < 0$ is given by

$$f_j(x) \sim \exp\left(\frac{-\eta_j x + \chi_j}{\epsilon}\right), \qquad \text{for } x \to +\infty\,. \tag{2.6}$$

The inverse theory prescribes that the fundamental scattering data consist of the reflection coefficient $R(k)$, the eigenvalues λ_j (bound states in quantum mechanics terminology) and the norming exponents χ_j. The transmission coefficient $T(k)$, as well as all other asymptotic information, can be computed in terms of this fundamental set.

By (2.4), the eigenvalues η_j are independent of time as u evolves according to (2.1). The time dependence of the remaining scattering data inferred from the linear system (2.2) is

$$\chi_j(t) = \chi_j(0) + 4\eta_j^3 t\,, \qquad R(k,t) = R(k,0)\exp\left(i\frac{8k^3t}{\epsilon}\right). \tag{2.7}$$

Hence, given $R(k,0)$, η_j, and $\chi_j(0)$ computed from the initial data $u(x,0)$, the solution $u(x,t)$ of the KdV equation (2.1) is then determined by inverse scattering from the $R(k,t)$, η_j, and $\chi_j(t)$ given by (2.7).

The KdV equation can be recast in the Hamiltonian form

$$\partial_t u = \partial_x \frac{\delta H}{\delta u}\,, \qquad H = \int_{-\infty}^{\infty} \tfrac{1}{2}\epsilon^2|\partial_x u|^2 - u^3\,dx\,. \tag{2.8}$$

The associated Poisson bracket of any two functionals F_1 and F_2 is defined as

$$\{F_1,F_2\} \equiv \int_{-\infty}^{\infty} \frac{\delta F_1}{\delta u}\partial_x \frac{\delta F_2}{\delta u}\,dx\,, \tag{2.9}$$

and the evolution of any functional F under the KdV flow (2.8) is then

$$\frac{dF}{dt} = \{F,H\}\,. \tag{2.10}$$

Thus, a functional F is conserved under the KdV flow (2.8) if and only if it Poisson commutes with the KdV Hamiltonian (i.e. if and only if $\{F,H\}=0$).

The complete integrability of the KdV equation yields the existence of an infinite family of conserved functionals H_m that are in involution (i.e. that Poisson commute) [15], [56]; thus

$$0 = \{H_m,H_n\} \equiv \int_{-\infty}^{\infty} \frac{\delta H_m}{\delta u}\partial_x \frac{\delta H_n}{\delta u}\,dx\,. \tag{2.11}$$

Denote their corresponding densities as ρ_m, so that

$$H_m = \int_{-\infty}^{\infty} \rho_m\,dx\,, \qquad \text{for } m = -1,0,1,\cdots.$$

The first three conserved densities are simply

$$\rho_{-1} = -u\,, \qquad \rho_0 = u^2\,, \qquad \rho_1 = \tfrac{1}{2}\epsilon^2|\partial_x u|^2 - u^3\,; \tag{2.12}$$

the last being just the density of the KdV Hamiltonian H defined in (2.8), so that $H_1 = H$.

All of the H_m except for $m = -1$ are Hamiltonians that generate flows which commute with the KdV flow (2.8), the so-called KdV hierarchy [31]. Letting t_m denote the time variable associated with the mth flow, its evolution is then given by

$$\partial_{t_m} u = \partial_x \frac{\delta H_m}{\delta u}\,, \qquad \text{where } m = 0, 1, \cdots. \tag{2.13}$$

Every H_n is conserved by each of these flows; their densities satisfy the local conservation laws

$$\partial_{t_m} \rho_{n-1} + \partial_x \mu_{m,n} = 0\,, \qquad \text{for } m, n = 0, 1, \cdots. \tag{2.14}$$

Here the $\mu_{m,n}$ is the flux for the $(n-1)^{th}$ conserved density under the mth flow. Note that for $n = 0$ in (2.14) recovers the flows (2.13).

Since all these flows commute with each other, they may be solved simultaneously for $u = u(x, \mathbf{t})$, where $\mathbf{t} = (t_0, t_1, \cdots)$ such that all but finitely many t_m are zero. Associated with each $\mathbf{t}$ is the odd polynomial $p(\cdot, \mathbf{t})$ defined by

$$p(\eta, \mathbf{t}) = \sum_{m=0}^{\infty} t_m 4^m \eta^{2m+1}\,. \tag{2.15}$$

The simultaneous evolution of the scattering data is then given by

$$\chi_j(\mathbf{t}) = \chi_j(0) + p(\eta_j, \mathbf{t})\,, \qquad R(k, \mathbf{t}) = R(k, 0) \exp\left(\frac{-2p(ik, \mathbf{t})}{\epsilon}\right)\,, \tag{2.16}$$

and $u(x, \mathbf{t})$ is determined by inverse scattering.

The problem of the zero dispersion limit for the KdV can then be expanded to consider the solution $u_\epsilon(x, \mathbf{t})$ of the whole KdV hierarchy (2.13) for some initial data $v(x)$ that is independent of ϵ, and examine the limiting behavior of the associated conserved densities ρ_n^ϵ and fluxes $\mu_{m,n}^\epsilon$ as ϵ tends to zero.

3. The Asymptotic Scattering Data for the KdV

For simplicity, we only present the formulas for the case when the initial data $v(x)$ is a single positive bump with a maximum value of η_{max}^2 as depicted in Fig. 3.1. The scattering data for $v(x)$ can then be computed asymptotically for small ϵ by the semiclassical (WKB) method. The discrete eigenvalues are packed in the range of the potential $-\eta_{max}^2 < \lambda < 0$. In terms of the transformed spectral variable $\eta = (-\lambda)^{1/2}$, as was first proved by Weyl, the asymptotic density of eigenvalues is given by the formula

$$\text{density of eigenvalues over } (0, \eta_{max}) \sim \frac{1}{\pi\epsilon}\varphi(\eta)\,,$$

where

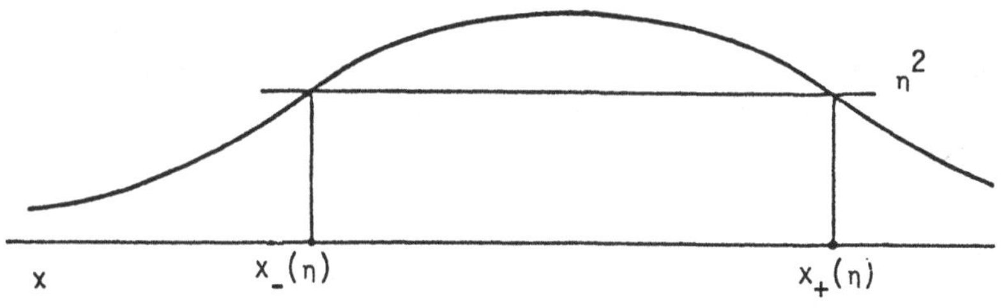

Fig. 3.1. The initial data $v(x)$ considered here.

$$\varphi(\eta) = \int_{x_-(\eta)}^{x_+(\eta)} \frac{\eta}{\sqrt{v(x) - \eta^2}}\, dx\,, \tag{3.1}$$

and $x_-(\eta) < x_+(\eta)$ are defined by $v(x_\pm) = \eta^2$ (see Fig. 3.1). The total number of eigenvalues $N = N_\epsilon$ is given asymptotically by

$$N_\epsilon \sim \frac{1}{\pi\epsilon} \int_0^{\eta_{max}} \varphi(\eta)\, d\eta\,. \tag{3.2}$$

The norming exponent obtained from the WKB analysis at $\mathbf{t} = 0$ is given by

$$\chi_j \sim \chi(\eta_j)\,, \qquad \chi(\eta) = \eta\, x_+(\eta) + \int_{x_+(\eta)}^{\infty} \left(\eta - \sqrt{\eta^2 - v(x)}\right) dx\,, \tag{3.3}$$

and the reflection coefficient is found to be zero to all orders of the expansion.

Based on the above calculation, we choose to neglect the scattering data related to the continuous spectrum. More precisely, we replace the initial data $v(x)$ with the reflectionless potential $v_\epsilon(x)$ corresponding to the WKB scattering data given by (3.1) and (3.3). This device sidesteps one aspect of the important question concerning the limiting behavior of the inverse scattering machinery, however is justified a posteriori by the result that v_ϵ converges strongly to v.

The KdV solution corresponding to this reflectionless initial data is given in terms of the Kay-Moses [26] determinant $\tau_\epsilon(x,\mathbf{t})$ as

$$u_\epsilon(x,\mathbf{t}) = 2\epsilon^2 \partial_{xx} \log \tau_\epsilon(x,\mathbf{t})\,. \tag{3.4}$$

We write the expansion of the Kay-Moses determinant in a slightly unfamiliar form by replacing summation by integration over an atomic measure. Thus

$$\tau_\epsilon(x,\mathbf{t}) = 1 + \sum_{k=1}^{N_\epsilon} \frac{1}{k!\pi^k} \tau_\epsilon^{(k)}(x,\mathbf{t})\,, \tag{3.5a}$$

(3.5b)

$$\tau_\epsilon^{(k)}(x,\mathbf{t}) = \int\cdots\int \exp\left(\frac{2\sum_{j=1}^k a(\xi_j, x, \mathbf{t})}{\epsilon}\right) \frac{\prod\limits_{\substack{i,j=1\\ i\neq j}}^{k} |\xi_i - \xi_j|}{\prod\limits_{i,j=1}^{k} |\xi_i + \xi_j|}\, d\nu(\xi_1)\cdots d\nu(\xi_k)\,,$$

where

$$a(\eta, x, t) = -\eta x + p(\eta, t) + \chi(\eta), \tag{3.6}$$

the atomic measure $d\nu(\xi)$ is given by

$$d\nu(\xi) = \pi\epsilon \sum_{i=1}^{N_\epsilon} \delta(\xi - \eta_i)\, d\xi, \tag{3.7}$$

and the η_i corresponds to the ith eigenvalue.

We recast $\tau_\epsilon^{(k)}$ in the more transparent form

$$\tau_\epsilon^{(k)}(x,t) = \int \cdots \int \exp\left(\frac{2(a_\epsilon, \psi_k) + (L\psi_k, \psi_k)}{\pi\epsilon^2}\right) d\nu(\xi_1) \cdots d\nu(\xi_k), \tag{3.8}$$

where the distribution ψ_k is defined by

$$\psi_k(\eta; \xi_1, \cdots, \xi_k) \equiv \pi\epsilon \sum_{i=1}^{k} \delta(\eta - \xi_i), \tag{3.9}$$

and where we have introduced the function a_ϵ and operator L by

$$a_\epsilon(\eta, x, t) \equiv a(\eta, x, t) - \frac{\epsilon}{2} \log(2\eta), \tag{3.10a}$$

$$L\psi(\eta) \equiv \frac{1}{\pi} \int_0^{\eta_{max}} \log\left|\frac{\eta - \xi}{\eta + \xi}\right| \psi(\xi)\, d\xi. \tag{3.10b}$$

The kernel of L is defined to be zero on the diagonal and (3.10a) incorporates the diagonal term of the denominator of (3.5b). Here the formal inner product is defined by

$$(a, \psi) = \int_0^{\eta_{max}} a(\eta)\psi(\eta)\, d\eta. \tag{3.11}$$

It is possible to write the function τ_ϵ in real exponential form because it is a sum of positive terms.

At this point we make the important observation [25] that for each ϵ the function

$$(x, t) \mapsto \log \tau_\epsilon(x, t) \qquad \text{is convex.} \tag{3.12}$$

Combining the Kay-Moses formula (3.4) with the local conservation laws (2.14), induction on n shows that the associated conserved densities and fluxes are given by

$$\rho_{n-1}^\epsilon(x, t) = 2\epsilon^2 \partial_{x t_n} \log \tau_\epsilon(x, t), \tag{3.13a}$$

$$\mu_{m,n}^\epsilon(x, t) = -2\epsilon^2 \partial_{t_m t_n} \log \tau_\epsilon(x, t). \tag{3.13b}$$

The Kay-Moses formula (3.4) is recovered from (3.13a) upon setting $n = 0$ by using (2.12) and the fact that $\partial_{t_0} = -\partial_x$.

4. Leading Order Theory for the KdV: The Weak Limit

The leading order theory, developed by Lax and Levermore in [33] and [34] establishes the existence of the limit

$$\lim_{\epsilon \to 0} \epsilon^2 \log \tau_\epsilon(x, \mathbf{t}) = Q(x, \mathbf{t}), \tag{4.1}$$

where the limit is uniform over compact subsets of $(x, \mathbf{t})$. Inserting this limit into (3.13) produces the weak limits

$$\lim_{\epsilon \to 0} \rho_{n-1}^{\epsilon}(x, \mathbf{t}) = 2\,\partial_{x t_n} Q(x, \mathbf{t}), \qquad \lim_{\epsilon \to 0} \mu_{m,n}^{\epsilon}(x, \mathbf{t}) = -2\,\partial_{t_m t_n} Q(x, \mathbf{t}), \tag{4.2}$$

for the conserved densities and fluxes associated with the solution $u_\epsilon(x, \mathbf{t})$. In particular, this gives the weak limit of $u_\epsilon(x, \mathbf{t})$, henceforth denoted $u(x, \mathbf{t})$, as

$$u(x, \mathbf{t}) \equiv \lim_{\epsilon \to 0} u_\epsilon(x, \mathbf{t}) = 2\,\partial_{xx} Q(x, \mathbf{t}). \tag{4.3}$$

Moreover, by analyzing the expansion (3.5) using an argument in the spirit of steepest-descents, they showed that the leading contribution to $\epsilon^2 \log \tau_\epsilon(x, \mathbf{t})$ comes from the largest term in that sum. As is suggested by the form of (3.8), they then characterized the limit $Q(x, \mathbf{t})$ of these terms by the maximization problem

$$\begin{aligned} Q(x, \mathbf{t}) &= \frac{2}{\pi} \max\left\{(a, \psi) + \tfrac{1}{2}(L\psi, \psi) \,:\, \psi \in \mathcal{A}\right\}, \\ \mathcal{A} &\equiv \left\{\psi \in L^1([0, \eta_{max}]) \,:\, 0 \le \psi \le \phi\right\}, \end{aligned} \tag{4.4}$$

the maximum being attained within the admissible set $\mathcal{A} \subset L^1$. Observe that this maximization is possed over nonnegative L^1 functions, not over the atomic distributions ψ_k of (3.9). The functional maximizer $\psi^*(x, \mathbf{t})$ is the continuum limit of the multivariate maximizers ψ_k as $\epsilon \to 0$. The variational problem is strictly concave and therefore has a unique solution that depends continuously on $(x, \mathbf{t})$ in the weak topology of measures. This in turn implies that $Q(x, \mathbf{t})$ is continuously differentiable with

$$\partial_x Q(x, \mathbf{t}) = -\frac{2}{\pi}(\eta, \psi^*(x, \mathbf{t})), \qquad \partial_{\mathbf{t}} Q(x, \mathbf{t}) = \frac{2}{\pi}(p_{\mathbf{t}}(\eta), \psi^*(x, \mathbf{t})), \tag{4.5}$$

where $p_{\mathbf{t}}(\eta) \equiv \partial_{\mathbf{t}} p(x, \mathbf{t})$ is seen to be independent of $\mathbf{t}$ from (3.6). The convexity of $(x, \mathbf{t}) \mapsto Q(x, \mathbf{t})$ is manifest in (4.4) where it is given as a supremum of linear functions of $(x, \mathbf{t})$. From the convexity (3.12) and regularity of the approximating functions $\log \tau_\epsilon$, it follows [25] that

$$\lim_{\epsilon \to 0} \epsilon^2 \partial_x \log \tau_\epsilon(x, \mathbf{t}) = \partial_x Q(x, \mathbf{t}), \qquad \lim_{\epsilon \to 0} \epsilon^2 \partial_{\mathbf{t}} \log \tau_\epsilon(x, \mathbf{t}) = \partial_{\mathbf{t}} Q(x, \mathbf{t}), \tag{4.6}$$

where again the limits are uniform over compact subsets of $(x, \mathbf{t})$.

The maximization problem (4.4) is attacked analytically by solving its variational conditions given by

$$
(4.7)\qquad \begin{aligned} &L\psi(\eta) + a(\eta, x, t) \le 0\,, && \text{when } \psi(\eta, x, t) = 0\,,\\ &L\psi(\eta) + a(\eta, x, t) = 0\,, && \text{when } 0 < \psi(\eta, x, t) < \varphi(\eta)\,,\\ &L\psi(\eta) + a(\eta, x, t) \ge 0\,, && \text{when } \psi(\eta, x, t) = \varphi(\eta)\,. \end{aligned}
$$

Let I be the interior of the set of (η, x, t) in which equality holds; let $\bar{I}$ be its closure. Directly differentiating (4.7) while using formula (3.6) for a leads to

$$
(4.8)\qquad \begin{aligned} &L\psi_x(\eta) = \eta\,, && L\psi_t(\eta) = -p_t(\eta)\,, && \text{when } (\eta, x, t) \in I,\\ &\psi_x(\eta) = 0\,, && \psi_t(\eta) = 0\,, && \text{when } (\eta, x, t) \notin \bar{I}. \end{aligned}
$$

These differentiated variational conditions do not have any explicit dependence on (x, t), but rather their dynamics, as well as all their memory of the initial data, is contained in the set I. Consequently, these conditions have more general validity than the variational conditions (4.7). Indeed, the latter vary (although in a sense insignificantly) when the initial data are considered in different classes (such as periodic or tending to different limits as $x \to \pm\infty$), while the differentiated conditions remain unchanged.

Making the Ansatz that at fixed (x, t) the set $I(x, t)$, defined as

$$
(4.9)\qquad I(x,t) \equiv \{\eta \in [0, \eta_{max}) : (\eta, x, t) \in I\}\,,
$$

consists of a finite union of disjoint open intervals, one can uniquely determine the functions $\psi_x(\eta)$ and $\psi_t(\eta)$ in terms of the endpoints of these intervals. More precisely, $I(x, t)$ is assumed to take the form

$$
I(x,t) = [0, \beta_{2g+1}) \cup (\beta_{2g}, \beta_{2g-1}) \cup \cdots \cup (\beta_2, \beta_1)\,,
$$

for some nonnegative integer g, where the β_i and g depend on (x, t) and

$$
0 \le \beta_{2g+1} < \beta_{2g} < \cdots < \beta_2 < \beta_1 \le \eta_{max}\,.
$$

Indeed, if any function $\psi(\eta)$ supported in $[0, \eta_{max}]$ is extended to be an odd function over $\mathbf{R}$ that vanishes when $|\eta| \ge \eta_{max}$, with the corresponding symmetric extension of $I(x, t)$, then the η-derivative of L is the Hilbert transform:

$$
(4.10)\qquad \frac{d}{d\eta} L\psi(\eta) = \frac{1}{\pi}\int_{-\infty}^{\infty} \frac{\psi(\mu)}{\eta - \mu}\, d\mu \equiv H\psi(\eta)\,.
$$

Applying this observation to ψ_x and ψ_t allows the differentiated variational conditions (4.8) to be transformed into Riemann-Hilbert problems through which they can be solved explicitly for ψ_x and ψ_t in terms of hyperelliptic functions involving the radical

$$
(4.11)\qquad R(\eta, x, t) = \left(\prod_{i=1}^{2g+1} (\beta_i^2 - \eta^2)\right)^{1/2}.
$$

Note that the extended $I(x, t)$ is the set over which $R(\eta)$ is real-valued and

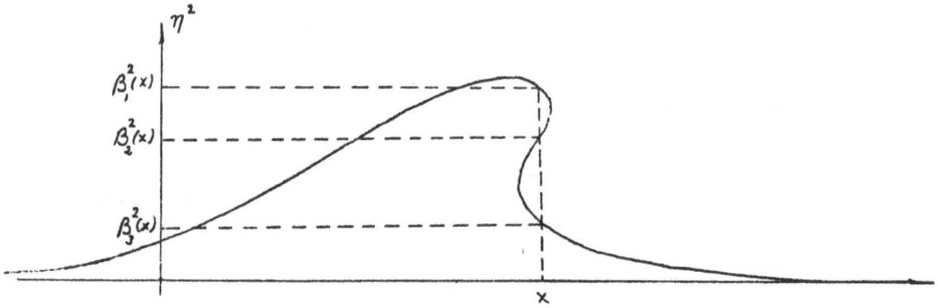

Fig. 4.1. A folded curve at some $t \neq 0$.

g is just the genus of the associated Riemann surface. The dependence of the β_i on $(x, \mathbf{t})$ is then derived from the compatibility constraint $\partial_{\mathbf{t}}\psi_x = \partial_x\psi_{\mathbf{t}}$. This reduces to the first order system of hyperbolic equations in the Riemann invariant (diagonal) form

$$\partial_{t_k}\beta_i + S_{ki}(\beta_1, \cdots, \beta_{2g+1})\,\partial_x\beta_i = 0\,, \qquad \text{for } i = 1, \cdots, 2g+1. \tag{4.12}$$

When $g = 1$ these are identical to Whitham's modulation equations for periodic conidal waves [53], hence (4.12) are referred to as Whitham equations. They were derived by Lax and Levermore [33] via the above argument, and simultaneously by Flaschka, Forrest and McLaughlin [13] as the modulation equations for g-phase quasi-periodic waves. The reconciliation of these two viewpoints was made manifest in the higher order Lax-Levermore theory of Venakides [50] that showed u_ϵ to be a g-phase quasi-periodic wave at small scales; this is dicussed in the next section. The advantage of the Lax-Levermore-Venakides approach is that it is global, allowing the initial value problem to be solved in the limit.

The global picture of the solution is as follows. At $\mathbf{t} = 0$ one finds that $g = 0$ and the sole β_1 is given by $\beta_1(x) = \sqrt{v(x)}$. For $\mathbf{t}$ near 0 the β_1 evolves by (4.12), which for t_1 is simply

$$\partial_{t_1}\beta_1^2 + 6\beta_1^2\partial_x\beta_1^2 = 0\,, \tag{4.13}$$

until the time at which the solution developes an infinite derivative. After this time the solution curve folds over as shown in Fig. 4.1. The dynamics of this fold is not given by the multivalued solution of the characteristic equation for (4.13), but rather by the $g = 1$ Whitham equations with the different levels of the fold given by $\beta_3 < \beta_2 < \beta_1$. This curve can evolve further, developing more folds over $(x, \mathbf{t})$ that are governed in a similar way by the Whitham equations (4.12) for higher g as is illustrated in Fig. 4.2. The genus $g(x, \mathbf{t})$ is a piecewise constant that gives the number of folds over $(x, \mathbf{t})$, each new fold adding two new levels. The weak limits are strong wherever $g = 0$. In particular this is true initially, so that our modified data v_ϵ converges strongly to the original data v.

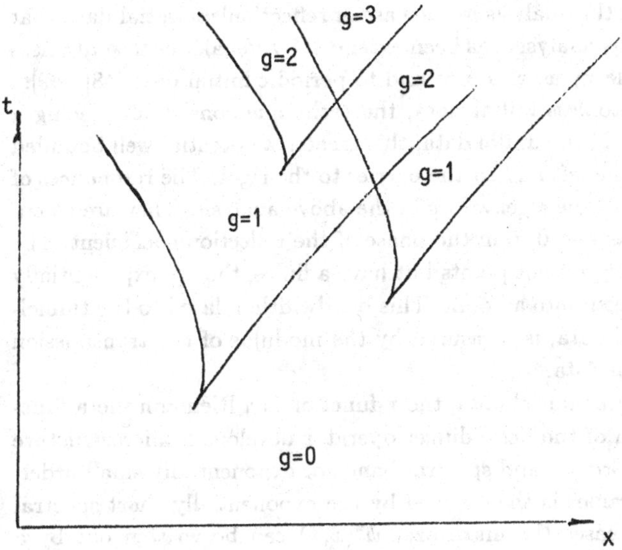

Fig. 4.2. Looking down at folds over the (x, t_1) plane.

The limits are not strong wherever $g > 0$, indicating the presence of limiting oscillations. An explicit formula for the maximizer $\psi^*(x, \mathbf{t})$ can be written down in terms of this evolving curve [34].

The intrinsic structure of the Whitham equations (4.12) has been the object of much recent study. Their Hamiltonian structure was found by Novikov and Dubrovin [7]. Levermore [35] showed they are strictly hyperbolic and genuinely nonlinear so long as the Riemann invariants remain distinct. He derived some compatibility equations to show that the first derivatives of the Riemann invariants satisfy Riccatti equations; these are used to show that their solutions can only develop an infinite derivative where the Riemann invariants decrease or come together. Bikbaev and Novokshenov [1] had also found part of this structure. Serre [42] used these same compatibility equations to construct a rich class of entropies for equations of Whitham type. But by far the most striking fact about these equations, first observed by Tsarev [45], is that they can be solved by a hodograph method, even though the number of unknowns exceeds the number of independent variables. This has been put into a algebro-geometric setting by Krichever [30], and has been exploited by Potemin [41], Tian [44], Gurevich and Krylov [19] and Wright [54] to obtain many new solutions, some of which had been studied numerically much earlier by Gurevich and Pitaevskii [20]. Further discussion of this exciting topic would take us far afield from the focus of this paper; we refer the reader to [8] and [10] for more complete discussions.

We have described the analysis in the case of reflectionless initial data that decay as $x \to \pm\infty$. This analysis has been extended by Venakides to solitonless initial data [46] that decay as $x \to \pm\infty$ and to periodic initial data [48], [49].

In the case of solitonless initial data, the technique consists of placing a barrier to the far right of the initial data; this creates a potential well bounded by the initial data to the left and by the barrier to the right. The resonances of this well play the role of the eigenvalues in the above analysis. They are calculated asymptotically as $\epsilon \to 0$ from the phase of the reflection coefficient. The fact that the resonances are not points but have a finite, though exponentially small, bandwidth is taken into account. This bandwidth, related to the tunneling through the initial data, is measured by the modulus of the transmission coefficient of the initial data.

In the case of periodic initial data, the τ-function is a Riemann theta function and the spectrum of the Schrödinger operator develops a microstructure with spectral gaps of order ϵ and spectral bands of exponentially small order. The role of the eigenvalues is then played by the exponentially short spectral bands. In both these cases the maximizer $\psi^*(x, \mathbf{t})$ can be written out by a time-dependent curve which evolves according to the evolution equation of Lax and Levermore.

Finally, McLaughlin and Strain [39] have recently shown how to numerically construct the weak limits by directly solving the maximization problem (4.4) for any given $\mathbf{t}$. Their method does not require the solution of an initial value problem, thus taking full advantage of the parametric nature of the $\mathbf{t}$ dependence.

5. Higher Order Theory for the KdV: Microstructure

One criticism of the original Lax-Levermore theory is that it does not reveal the nature of the small-scale oscillations that lead to the weak limit. This difficulty has been examined by Venakides through two different approaches. The most recent approach [50] makes a sharper calculation of the Kay-Moses determinant $\tau_\epsilon(x,t)$ of the Lax-Levermore theory. The Kay-Moses determinant (3.5) is an example of a so-called τ-function [43] for KdV depending on the linear phases $-\eta_j x + 4\eta_j^3 t + \chi(\eta_j)$ for $j = 1, \cdots, N_\epsilon$. It can be thought of as the functional vehicle that linearizes KdV. The higher-order theory shows that the τ-function factors as

$$(5.1) \qquad \tau_\epsilon(x,t) \sim \exp\left(\frac{Q(x,t)}{\epsilon^2}\right) \bar{\tau}_\epsilon\left(x, t; \frac{x}{\epsilon}, \frac{t}{\epsilon}\right),$$

where $Q(x,t)$ is given by (4.4). Inserting this into (3.4) and using (4.3) leads to

$$(5.2) \qquad u_\epsilon(x,t) = u(x,t) + 2\epsilon^2 \partial_{xx} \log \bar{\tau}_\epsilon\left(x, t; \frac{x}{\epsilon}, \frac{t}{\epsilon}\right).$$

The higher order theory shows that in the (x,t) region where $g \geq 1$ (g is the number of gaps in the support of the maximizing ψ of the leading order theory) the second term represents a g-phase quasiperiodic zero-mean oscillation on the small scales. Indeed, on these scales, $\bar{\tau}_\epsilon$ is a function of g phases $(\kappa_i\, x+\omega_i\, t+\delta_i)/\epsilon$ for $i = 1,\cdots,g$, and can be easily transformed to a multiple of the classical Riemann theta-function representing such KdV solutions. The theta function corresponds to the hyperelliptic curve arising in the leading order theory (4.11). The wave numbers κ_i and frequencies ω_i depend on the slow scales (x,t) as do the weak limits. Thus, while the global flow is on a degenerate torus of genus $N_\epsilon = O(\epsilon^{-1})$, the local flow effectively occurs on a torus of genus g. In contrast to modulation theory this is not assumed apriori but is a conclusion of the theory.

We give a flavor of the calculation. The key idea of the higher order theory is to quantize the admissible functions ψ in the variational problem (4.4) as

$$\frac{1}{\pi\epsilon}\int_c \psi(\eta)\, d\eta = \text{integer}, \tag{5.3}$$

where c is a topological component of the support of ψ. This reflects the fact that ψ has emerged as a continuum approximation of the atomic measure (3.9) whose mass is quantized. The condition is of a higher order in the sense that only an order ϵ correction of ψ^* is required for it to be satisfied.

In the spirit of the method of steepest descents, write τ_ϵ as

$$\tau_\epsilon \sim \sum_\psi K(\epsilon,\psi)\exp\left(\frac{2(a,\psi)+(L\psi,\psi)}{\pi\epsilon^2}\right), \tag{5.4}$$

where ψ ranges over all integrable functions bounded by $0 \leq \psi(\eta) \leq \varphi(\eta)$ which are local maxima of the functional $2(a,\psi)+(L\psi,\psi)$ and are constrained by the quantum condition. Clearly, all contributions which are not negligible should be perturbations of the leading order nonquantized maximizer $\psi^*(\eta)$. We pose the Ansatz that the coefficients $K(\varepsilon,\psi)$ are independent of ψ to leading order for those ψ's which are near ψ^*.

We then can then write τ_ϵ as

$$\tau_\epsilon \sim K(\epsilon)\sum_\psi \exp\left(\frac{2(a,\psi)+(L\psi,\psi)}{\pi\epsilon^2}\right). \tag{5.5}$$

Although a rigorous proof of the Ansatz still eludes us, it follows from the study of Zhang and Venakides [58] that the Ansatz yields the correct answer, including phase shifts, in the particular case of a quasiperiodic finite zoned nonmodulated global solution of (2.1a).

Setting $\psi(\eta) = \psi^*(\eta) + \epsilon\bar{\psi}(\eta) + \cdots$ and performing the maximization with respect to $\bar{\psi}$ we obtain the variational conditions

$$
(5.6)\quad \begin{cases} L\bar{\psi}(\eta) = \sum_{i=0}^{g} \frac{1}{\pi} c_i \mathbf{1}_i(\eta), & \text{when } \eta \in \operatorname{supp} \psi^*, \\ \bar{\psi}(\eta) = 0, & \text{when } \eta \notin \operatorname{supp} \psi^*, \\ \frac{1}{\pi} \int_{\beta_{2i}}^{\beta_{2i-1}} \bar{\psi}(\eta)\, d\eta = m_i - \frac{1}{\pi\epsilon} \int_{\beta_{2i}}^{\beta_{2i-1}} \psi^*(\eta)\, d\eta, & \text{for } i = 1, \cdots, g, \end{cases}
$$

where $\mathbf{1}_i(\eta)$ is the characteristic function of the interval $(\beta_{2i}, \beta_{2i-1})$ and c_i a Lagrange multiplier.

As in the leading order analysis, the solutions are obtained by reduction to a Riemann-Hilbert problem. They form a vector space over the multi-integers $(m_1, \cdots, m_g)$ and essentially coincide with the holomorphic differentials corresponding to the hyperelliptic curve (4.11) in the variable η^2. Upon substitution into (5.5) we obtain a multiple of the theta function corresponding to this hyperelliptic curve.

Remark 1. The calculation can be carried out under a weaker Ansatz which allows the logarithm of $K(\varepsilon, \psi)$ to depend on $(m_1, \cdots, m_g)$ linearly. This affects only the phase shifts of the theta function. A precise calculation of the phase shifts seems to be beyond our method at present and may require the inclusion of the higher order WKB contribution to the scattering data including the neglected reflection coefficient.

Remark 2. The higher order calculation is valid when (x, t) is not a point of discontinuity of the local genus $g(x, t)$.

Remark 3. The local solution has a rich mathematical structure through its connection to the hyperelliptic curve (4.11). The meromorphic differentials corresponding to the hyperelliptic curve present themselves in the leading order theory, while the holomorphic differentials and the period matrix arise naturally in the higher order theory.

We finish this section by a graphic explanation of how the global flow, with number of degrees of freedom of order $O(\epsilon^{-1})$, effectively reduces to a flow on a torus of genus g near a point (x, t). This was obtained by Venakides in [47] and [48] in his first approach to the microstructure problem. The original multi-soliton approximation is a degenerate case of a periodic solution (period $\to$ infinity). The spectral bands collapse to the bound states $\lambda_1, \cdots, \lambda_{N_\epsilon}$, $N_\epsilon = O(\epsilon^{-1})$, separated by microgaps with lengths of order ϵ. The dynamical variables are the Dirichlet eigenvalues $\mu_1, \cdots, \mu_{N_\epsilon}$. Exactly one of these lies in each spectral microgap. The Dirichlet eigenvalues depend on x and t and satisfy a set of N_ϵ equations involving the related Abel sums. The continuum limit of this system of N_ϵ equations in N_ϵ unknowns is shown to be *exactly the variational condition of the leading order theory.*

A finer analysis of the discrete system reveals more structure. In an effective gap of the local solution, i.e. an interval $(\beta_{2i+1}^2, \beta_{2i}^2)$, it is proven that each bound state (represented by ▪ in the Fig. 5.1) has a Dirichlet eigenvalue (rep-

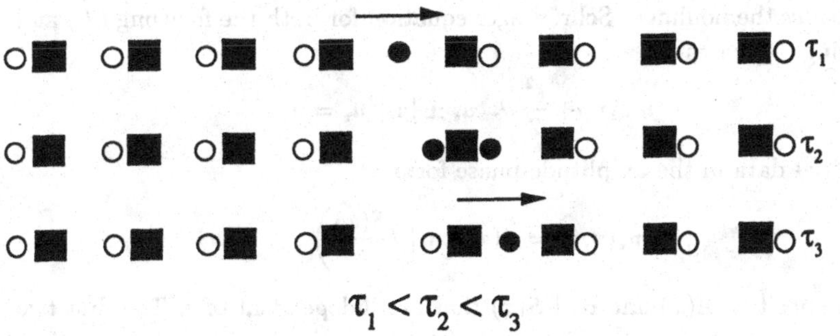

Fig. 5.1. The effective gap dynamics.

resented by o in the figure) at an exponentially small distance from it. One Dirichlet eigenvalue (represented by •) is *left free to move* (see Fig. 5.1). As it approaches the endpoint of the microgap it dislodges the Dirichlet eigenvalue from the endpoint of the adjacent gap and lets it move freely. Effectively, the free Dirichlet eigenvalue passes from one microgap to another. There is only one Dirichlet eigenvalue moving in the effective gap. Making the Ansatz that the Dirichlet eigenvalues in the effective spectral *bands* are located, on the average, in the midpoint of their microgaps (this can be proven before breaktime), it can be shown that the g free Dirichlet eigenvalues in the effective gaps obey the dynamical system of equations of the Dirichlet eigenvalues of a quasiperiodic problem with the local effective band/gap structure.

6. The Nonlinear Schrödinger Hierarchies

Other classes of integrable problems that have been studied are associated with the Lax operators

$$L \equiv \begin{pmatrix} -i\epsilon\,\partial_x & \pm i\bar{u} \\ iu & i\epsilon\,\partial_x \end{pmatrix}. \tag{6.1}$$

These were introduced by Zakharov and Shabat [57] to solve the focusing (+) and defocusing (−) nonlinear Schrödinger (NLS) equations by their respective inverse scattering transforms. Note that L given in (6.1) is selfadjoint only for the defocusing case.

At this time we have a better understanding of small dispersion limits for those integrable systems associated with selfadjoint Lax operators than for those associated with nonselfadjoint ones. In fact, at first it was thought that the selfadjointness of the Lax operator might be essential to establish a small dispersive limit. However, recently some insights have been gained that cast new light into the richness of these problems.

Consider the nonlinear Schrödinger equation for both the focusing (+) and defocusing (−) cases

$$i\epsilon\,\partial_t u_\epsilon + \frac{\epsilon^2}{2}\,\partial_{xx}u_\epsilon \pm |u_\epsilon|^2 u_\epsilon = 0\,, \tag{6.2}$$

with initial data in the amplitude-phase form

$$u_\epsilon(x,0) = A(x)\exp\left(i\frac{S(x)}{\epsilon}\right)\,, \tag{6.3}$$

for some positive $A(x)$ and real $S(x)$ that are independent of ϵ. The first two conserved quantities, usually identified with mass and momentum conservation, have densities

$$\rho_\epsilon = |u_\epsilon|^2\,, \qquad \mu_\epsilon = -i\frac{\epsilon}{2}\left(\bar{u}_\epsilon\partial_x u_\epsilon - u_\epsilon\partial_x\bar{u}_\epsilon\right); \tag{6.4}$$

these determine the wave function u up to a constant phase. Their evolution, derived from (6.2) is governed by the local conservation laws

$$\partial_t\rho_\epsilon + \partial_x\mu_\epsilon = 0\,, \tag{6.5a}$$

$$\partial_t\mu_\epsilon + \partial_x\left(\frac{\mu_\epsilon^2}{\rho_\epsilon} \mp \frac{\rho_\epsilon^2}{2}\right) = \frac{\epsilon^2}{4}\,\partial_x\left(\rho_\epsilon\partial_{xx}\log\rho_\epsilon\right). \tag{6.5b}$$

These form a closed system for ρ_ϵ and μ_ϵ subject to the initial conditions

$$\rho_\epsilon(x,0) = |A(x)|^2\,, \qquad \mu_\epsilon(x,0) = |A(x)|^2\partial_x S(x)\,. \tag{6.6}$$

The left side of (6.5) has the form of the barotropic Euler equations of fluid dynamics with a "pressure" given by $p = \pm\frac{1}{2}\rho^2$ (a γ-law gas with $\gamma = 2$); in the focusing case (−) the pressure decreases with increasing mass density. In that case the solutions therefore develop mass concentrations, the so-called focusing phenomena, which are regularized by the dispersive term on the right side of the momentum equation (6.5b) once the mass density gradients become large enough.

Taking the formal zero dispersion limit (semiclassical limit) of (6.5) by setting ϵ to zero gives

$$\partial_t\rho + \partial_x\mu = 0\,, \qquad \partial_t\mu + \partial_x\left(\frac{\mu^2}{\rho} \mp \frac{\rho^2}{2}\right) = 0\,. \tag{6.7}$$

As was pointed out in [25], this system is hyperbolic for the defocusing case, while it is elliptic for the focusing case.

For the defocusing case the solutions of (6.7) will evolve until the development of an infinite derivative as the "fluid" attempts to form a shock. After that time the dispersive term cannot be neglected and oscillations develope as in KdV. In fact, the whole defocusing hierarchy was treated in a similar fashion to that of KdV by Jin, Levermore and McLaughlin [25].

For the focusing case the Cauchy-problem for (6.7) is ill-posed for all but analytic initial data. This indicates that the semiclassical limit, if it even ex-

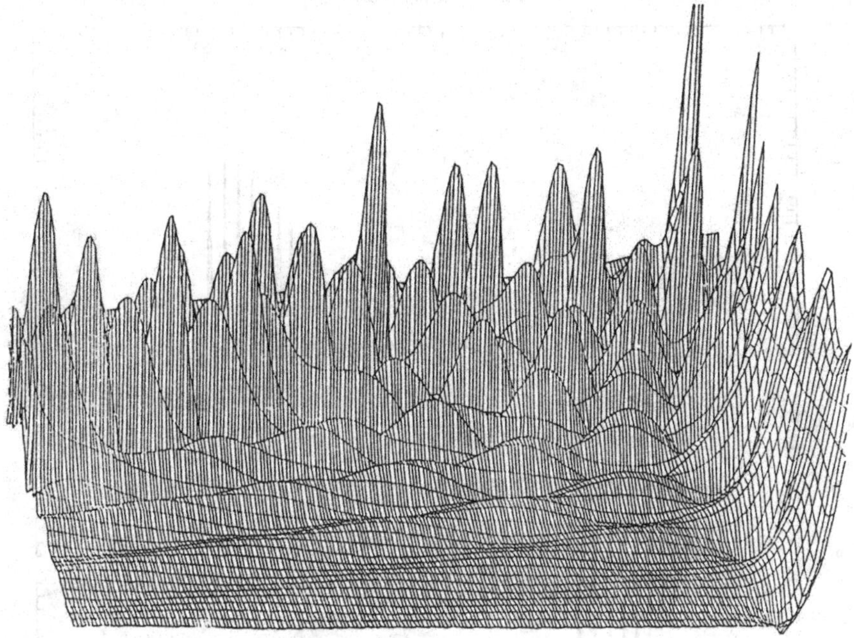

Fig. 6.1. Focusing NLS $|u|^2$ concentrations.

ists, is not governed by a simple modulational dynamics. This observation is consistent with the numerical experiment shown in Fig. 6.1; it shows a sea of intermittent concentrations rather than the coherent wavetrains seen for the defocusing case.

In contrast, consider the modified KdV equations

$$\partial_t u_\epsilon \pm \tfrac{3}{2} u_\epsilon^2 \partial_x u_\epsilon + \epsilon^2 \tfrac{1}{4} \partial_{xxx} u_\epsilon = 0 \,, \tag{6.8}$$

or, more generally, their complex forms

$$\partial_t u_\epsilon \pm \tfrac{3}{2} |u_\epsilon|^2 \partial_x u_\epsilon + \epsilon^2 \tfrac{1}{4} \partial_{xxx} u_\epsilon = 0 \,. \tag{6.9}$$

These are the third members of the focusing (+) and defocusing (−) NLS hierarchies [14], and hence share the same respective Lax operator. They, along with all the other odd flows in those hierarchies, commute with complex conjugation, and therefore preserve the reality of the initial data.

A similar analysis to that which lead to the formal zero dispersion limit for the NLS equations (6.7) leads here to

$$\partial_t \rho + \partial_x \left(\pm \tfrac{3}{4} \rho^2 - \tfrac{3}{4} \frac{\mu^2}{\rho} \right) = 0 \,, \qquad \partial_t \mu + \partial_x \left(\pm \tfrac{3}{2} \rho \mu - \tfrac{3}{4} \frac{\mu^3}{\rho^2} \right) = 0 \,. \tag{6.10}$$

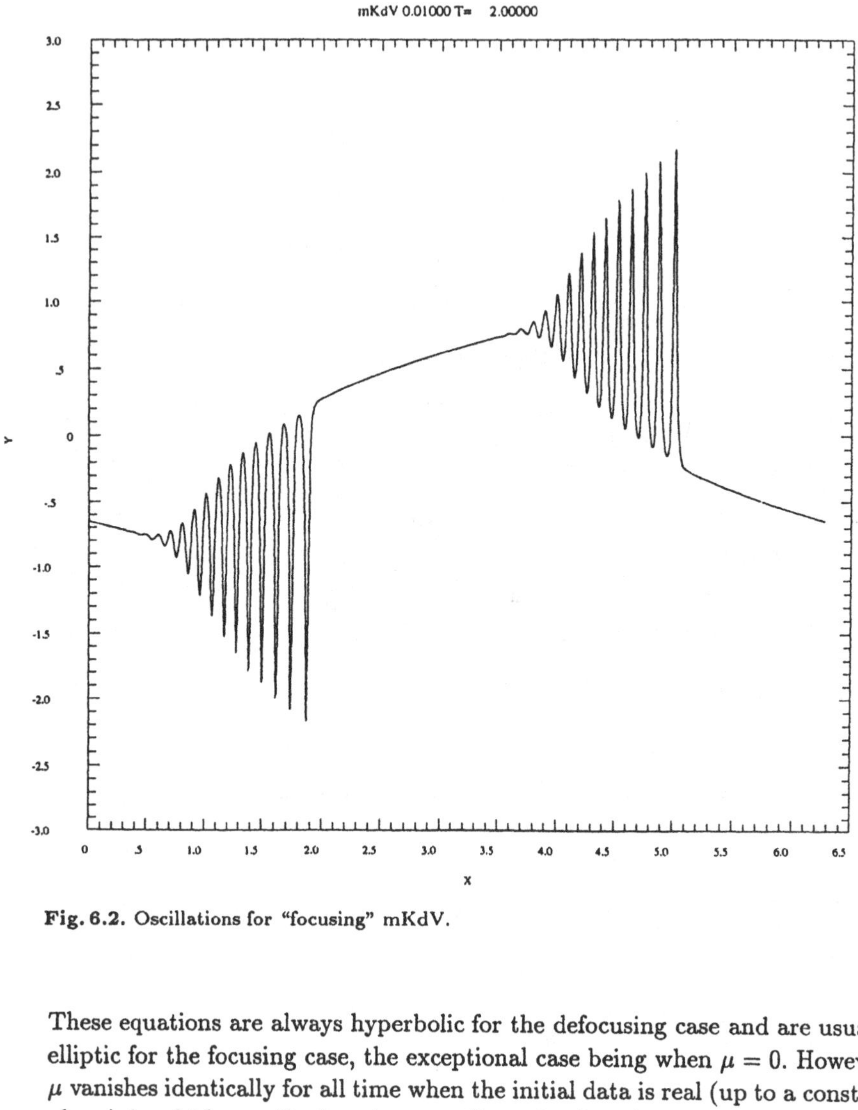

Fig. 6.2. Oscillations for "focusing" mKdV.

These equations are always hyperbolic for the defocusing case and are usually elliptic for the focusing case, the exceptional case being when $\mu = 0$. However, μ vanishes identically for all time when the initial data is real (up to a constant phase), in which case the focusing zero dispersion limit becomes hyperbolic too.

Recently, the zero dispersion limit for all the odd flows of the focusing NLS hierarchy with real initial data was worked out [9] and the hyperbolicity of the Whitham equations was established. Correspondingly, all the odd conserved densities vanish identically for such data. The limit is indicated by the numerical experiment [9] seen in Fig. 6.2, which clearly shows coherent modulated oscillations that evolved from an initial sine wave. Later, this run clearly showed the development of modulated multiphase solutions. Thus, the zero dispersion limit for flows corresponding to the nonselfadjoint Lax operator in (6.1) does not exist as a simple modulation theory for the even flows, but does for the odd flows. Whether this kind of behavior occurs for other nonselfadjoint Lax operators remains to be seen. In this case at least, symmetries play an important role in singling out which flows have limits. Hopefully, by working out other examples, we will be able to identify what mathematical structures are critical to the existence of small dispersion limits.

Finally, Bronski and McLaughlin [3] have recently undertaken a numerical study of the semiclassical limit for the focusing NLS equation in the weakly nonlinear case. In this way they try to understand the failure of the above theory as manifestation of modulational instability. They also point out an application of these theories to the problem of optical shocking in fibers.

7. Integrable Discrete Problems

It is well known to those interested in solving numerically nonlinear hyperbolic partial differential equations that dispersive difference schemes give rise to oscillations on mesh scale near incipient discontinuities. The passage to the continuum limit as the mesh size tends to zero is analogous to the zero dispersion limit in the partial differential equations discussed in the previous sections. We shall present here two examples that have been studied so far.

7.1 A Discretization of the Hopf Equation

The simpler of the two examples, investigated by Goodman and Lax [17], is an approximation to the Hopf (inviscid Burgers) initial-value problem

$$\partial_t u + u\partial_x u = 0\,, \tag{7.1a}$$

$$u(x,0) = v(x)\,. \tag{7.1b}$$

Space is discretized, the spatial derivative is replaced by a *centered* difference quotient, and the initial data is discretized to obtain

$$\frac{d}{dt}u_k + u_k\frac{u_{k+1}-u_{k-1}}{2\Delta} = 0\,, \tag{7.2a}$$

$$u_k(0) = v(k\Delta)\,. \tag{7.2b}$$

Here Δ is mesh size, and $u_k(t)$ an approximation to $u(k\Delta, t)$. We consider the continuum limit, which is to let the mesh size Δ tends to zero for fixed $v(x)$.

In order to see the dispersive nature of the differnce scheme (7.2), approximate the difference quotient by its the Taylor expansion as

$$\frac{u_{k+1}-u_{k-1}}{2\Delta} \simeq \partial_x u + \tfrac{1}{6}\Delta^2\partial_{xxx}u\,.$$

Setting this into (7.2) we obtain

$$\partial_t u + u\partial_x u + \tfrac{1}{6}\Delta^2 u\partial_{xxx}u \simeq 0\,. \tag{7.3}$$

This equation is very much like the KdV equation, as long as the factor u does not change sign. Note that in (7.3) the mesh size Δ takes the place of the parameter ϵ in the KdV equation (2.1a).

The system (7.2) is completely integrable. To show this we observe first that (7.2) is in conservation form, and therefore $\sum u_k$ is a conserved quantity; to avoid convergence questions we take the periodic case and extend the sum over a period. Dividing (7.2) by u_k we get another system in conservation form, from which we deduce that $\sum \log u_k = \log \prod u_k$ is a conserved quantity. From these two conservation laws we deduce that if the u_k are initially positive, they remain positive for all t. This allows us to introduce as new variables

$$a_k = \text{const.}\sqrt{u_k}$$

Setting $u_k = ca_k^2$ into (7.2) we get, after division by $2a_k$, and setting $c = 4\Delta$ that

$$\frac{d}{dt}a_k + a_k\left(a_{k+1}^2 - a_{k-1}^2\right) = 0\,. \tag{7.4}$$

Now define T to be the matrix of right translation acting on finite periodic sequences; it is

$$T = \begin{pmatrix} 0 & 0 & \cdots & 0 & 1 \\ 1 & 0 & 0 & \ddots & 0 \\ 0 & 1 & 0 & \ddots & \vdots \\ \vdots & \ddots & \ddots & \ddots & 0 \\ 0 & \cdots & 0 & 1 & 0 \end{pmatrix}.$$

Letting A be the diagonal matrix with the vector $(a_1, \dots, a_N)$ on the diagonal, define the operators L and B by

$$L \equiv TA + AT^{-1}\,, \qquad B \equiv (TA)^2 - (AT^{-1})^2\,.$$

The matrix L is symmetric with respect to the ℓ^2 scalar product, and B is antisymmetric. A quick calculation shows that the equation

$$\frac{d}{dt}L = [B, L]\,, \tag{7.5}$$

when written out componentwise, is just the system of equations (7.4). This system has been studied by Kac and van Moerbeke [27] and shown to be completely intetgrable. Moser [40] has shown how to reduce this system to the Flaschka form [12] of the Toda lattice.

Solutions of the system (7.2) show the same behavior as solutions of KdV. There is a time t_{crit} during which the initial value problem for the first order equation (7.1) has a smooth solution; for $t < t_{crit}$, the solution of the initial value problems for (7.2) tends, as Δ tends to zero, to the smooth solution of (7.1). For $t > t_{crit}$, however, solutions of (7.2) develop oscillations. Figures 7.1

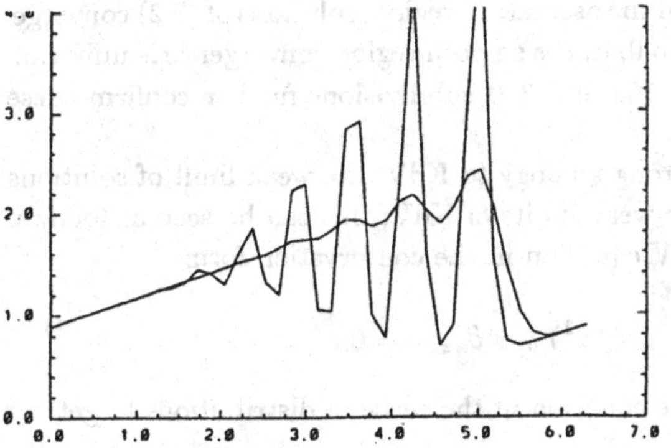

Fig. 7.1a. Computed with 40 mesh points.

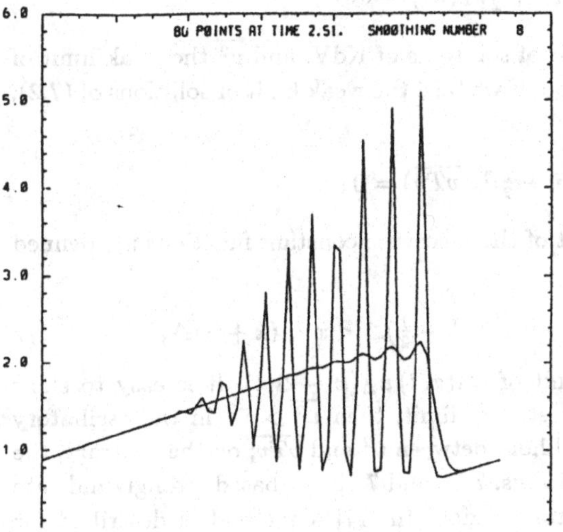

Fig. 7.1b. Computed with 80 mesh points.

depict the solution of (7.2) as function of x, linearly interpolated from the points $x = k\Delta$, at time $t = 2.51$, with initial values

$$v(x) = 1.5 + \cos x\,.$$

For these initial values, breakdown occurs at $t_{crit} = 1$. In Fig. 7.1a the x period of length 2π has been subdivided into 40 equal parts, in Fig. 7.1b into 80 equal parts. The noteworthy features of these figures are:

(1) There is a distinct oscillatory region, the same in both figures, outside of which the solution is smooth.

(2) The number of oscillations in Fig. 7.1b is twice the number of oscillations in Fig. 7.1a. This confirms the surmise that the oscillations are at a wavelength proportional to Δ.

(3) The average value of $u(x)$, computed as a moving average, is well nigh indistinguishable in Figs. 7.1a and 7.1b. This suggests fairly

convincingly that, in the oscillatory region, solutions of (7.2) converge weakly as Δ tends to 0; in the smooth region convergence is uniform. Calculations performed with 160 and 320 subdivisions further confirm these conclusions.

Although (7.2) bears a strong analogy to KdV, the weak limit of solutions of (7.2) are different from the weak limits of KdV; this can be seen as follows. First write the (rescaled) KdV equation in the conservation form

$$\partial_t u + \tfrac{1}{2}\partial_x(u^2) + \epsilon^2 \partial_{xxx} u = 0\,,$$

and then take the limit of this equation in the sense of distributions to get

$$\partial_t \bar{u} + \tfrac{1}{2}\partial_x\left(\overline{u^2}\right) = 0\,, \tag{7.6}$$

where $\bar{u}$ denotes the weak limit of solutions of KdV, and $\overline{u^2}$ the weak limit of their squares. On the other hand, if we take the weak limit of solutions of (7.2), we get

$$\partial_t \bar{v} + \tfrac{1}{2}\partial_x\left(\overline{vTv}\right) = 0\,, \tag{7.7}$$

where $\bar{v}$ dentoes the weak limit of the piecewise constant functions u_Δ defined as

$$u_\Delta(x,t) = u_k(t)\,, \qquad \text{for } (k-\tfrac{1}{2})\Delta < x < (k+\tfrac{1}{2})\Delta\,,$$

and $\overline{vTv}$ denotes the weak limit of $u_\Delta(x,t)u_\Delta(x-\Delta,t)$. It is easy to show that if $\bar{u}$ is the weak, but not strong, limit, then $\overline{u^2} > \bar{u}^2$ in the oscillatory region. No such inequality need hold between $\bar{v}^2$ and $\overline{vTv}$; on the contrary, the numerical calculation on which Figs. 7.1a and 7.1b are based strongly indicate that $\bar{v}^2 > \overline{vTv}$ in the oscillatory region. In [17] a method is described for constructing solutions of (7.2) that are modulated binary oscillations; for these solutions $\bar{v}^2 > \overline{vTv}$. The upshot of these considerations is that $\bar{u}$ and $\bar{v}$ are different in the oscillatory region.

The system (7.2) is, as we have shown, completely integrable; therefore there is an explicit formula for its solutions. Letting $\Delta \to 0$ in these formulas it ought to be possible to prove the experimentally observed phenomena by the techniques described in the previous sections; so far this hasn't been carried out.

7.2 A Discrete Lagrangian Fluid Dynamics and the Toda Chain

We now turn to a discrete system obtained from the equations of one-dimensional compressible gas dynamics; when written in terms of the Lagrangian coordinate x, these are

$$\begin{aligned} \partial_t u + \partial_x p &= 0\,, \\ \partial_t V - \partial_x u &= 0\,. \end{aligned} \tag{7.8}$$

Here u is the velocity, V the specific volume, and the pressure is given as a function of V by an equation of state $p = p(V)$; so we are dealing with an isentropic fluid. If $p(V)$ is decreasing then (7.8) is hyperbolic i.e. a wave equation. The first equation in (7.8) is conservation of momentum, while the second is conservation of mass. Consider the spatial semi-discretization of (7.8) given by

$$\text{(7.9a)} \qquad \frac{d}{dt}u_k + \frac{p_{k+1/2} - p_{k-1/2}}{\Delta} = 0\,,$$

$$\text{(7.9b)} \qquad \frac{d}{dt}V_{k+1/2} - \frac{u_{k+1} - u_k}{\Delta} = 0\,,$$

where $p_{k\pm\frac{1}{2}} = p(V_{k\pm\frac{1}{2}})$. Notice that by staggering the points where u and V are located, we can conveniently center the difference quotients that replace the derivatives in (7.9). This suggests, in analogy with (7.2), that (7.9), too, is a dispersive system. That this is indeed so can be seen by rewriting (7.9) in terms of discrete Eulerian coordinates X_k, defined by

$$\text{(7.10)} \qquad \frac{d}{dt}X_k = u_k\,,$$

the initial values to be chosen so that at $t = 0$

$$\text{(7.11)} \qquad \frac{X_{k+1} - X_k}{\Delta} = V_{k+1/2}\,.$$

Using (7.9b) and (7.10) we deduce that

$$\frac{d}{dt}\frac{X_{k+1} - X_k}{\Delta} = \frac{d}{dt}V_{k+1/2}\,;$$

therefore if (7.11) holds for $t = 0$, it holds for all t. Setting (7.11) into (7.9a) and into the equation of state yields

$$\text{(7.12)} \qquad \frac{d^2}{dt^2}X_k + \frac{p(\frac{X_{k+1}-X_k}{\Delta}) - p(\frac{X_k - X_{k-1}}{\Delta})}{\Delta} = 0\,.$$

This can be regarded as a discretization of the quasilinear equation

$$\text{(7.13)} \qquad \partial_{tt}X + \partial_x p(\partial_x X) = 0\,.$$

Since V is the reciprocal of density, p is a decreasing function of V. Therefore (7.13) is hyperbolic, i.e. a wave equation.

Alternatively, (7.12) can be regarded as Newton's law for a laterally vibrating chain of particles of mass Δ. The position of the kth particle is X_k. Two adjacent particles are connected by a spring, the same for all pairs; the force exerted by the spring when its ends are separated by a distance d is const. $-p(d/\Delta)$.

Holian and Straub [22] took various force laws $p(\cdot)$, associated with the names of Lennard-Jones, Morse, and Toda. They set $\Delta = 1$ in equation (7.12) and investigated the limit $t \to \infty$ for the shock initial data

$$X_k(0) = k\Delta\,, \quad \frac{d}{dt}X_k(0) = \begin{cases} -a & \text{for} \quad k > 0\,, \\ 0 & \text{for} \quad k = 0\,, \\ a & \text{for} \quad k < 0\,; \end{cases} \tag{7.14}$$

this is equivalent, after rescaling, to the $\Delta \to 0$ limit. Holian and Straub not only confirmed an early finding of von Neumann that the shock wave reflected from the origin with large oscillations behind the shock [52] (see next section), but discovered by numerical experimentation a striking new phenomenon: the existence of a critical shock strength a_{crit} with the following property:

i) For $a < a_{crit}$, the solution of the von Neumann shock problem (7.12), (7.14) is oscillatory in the regions

$$-s_1 t < x < -s_2 t\,, \qquad s_2 t < x < s_1 t\,, \tag{7.15}$$

while it tends to zero in the region

$$-s_2 t < x < s_2 t\,, \tag{7.16}$$

as t tends to ∞. Here s_1 is the shock speed, and s_2 a second kind of speed; both s_1 and s_2 are functions of a.

ii) For $a > a_{crit}$, the solution is oscillatory everywhere behind the shock, i.e. for $|x| < s_1 t$. Yet the quality of the oscillations is different in the inner region (7.16): it is periodic in time and binary in space, i.e. neighboring particles are moving in opposite directions.

Holian, Flaschka and McLaughlin [21] exploited complete integrability to analyze the von Neumann shock problem for the Toda chain (also referred to as the Toda lattice), which has the force law $p(V) = \exp(-V)$. They derived a formula for the shock speed $s_1(a)$, and for the critical speed a_{crit}. The reason for the occurrence of a critical speed is this: if you write the equation for the Toda chain over the whole axis in commutator form

$$\frac{d}{dt}L = [B, L]\,,$$

the continuous spectrum of L is the union of the intervals $[-a-1, -a+1]$ and $[a-1, a+1]$; for $a < a_{crit}$ L has a continuous spectrum consisting of a single interval, for $a > a_{crit}$, the continuous spectrum of L consists of two disjoint intervals. The shock problem is equivalent to the initial boundary value problem, for the seminfinite lattice ($k = 0, 1, \dots$) in which the zeroth particle is driven at a constant velocity $2a$ and the remaining particles are initially at rest. This has been investigated by Kaup [29]. The shock problem is obtained from the IBV problem by first using a coordinate frame fixed on the forcing zeroth particle and then reflecting the system about the origin.

In a recent article [51], Venakides, Deift and Oba studied the supercritical von Neumann shock problem for the Toda chain by analyzing the long time behavior of the τ-function of the initial value problem. They obtained the oscillatory structure and the speeds s_1 and s_2 *without assuming the generation*

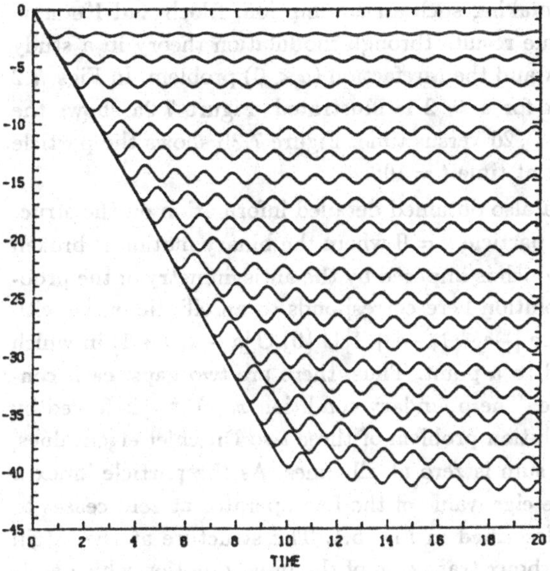

Fig. 7.2a. Particle positions for the supercritical Toda shock with $a = 2$.

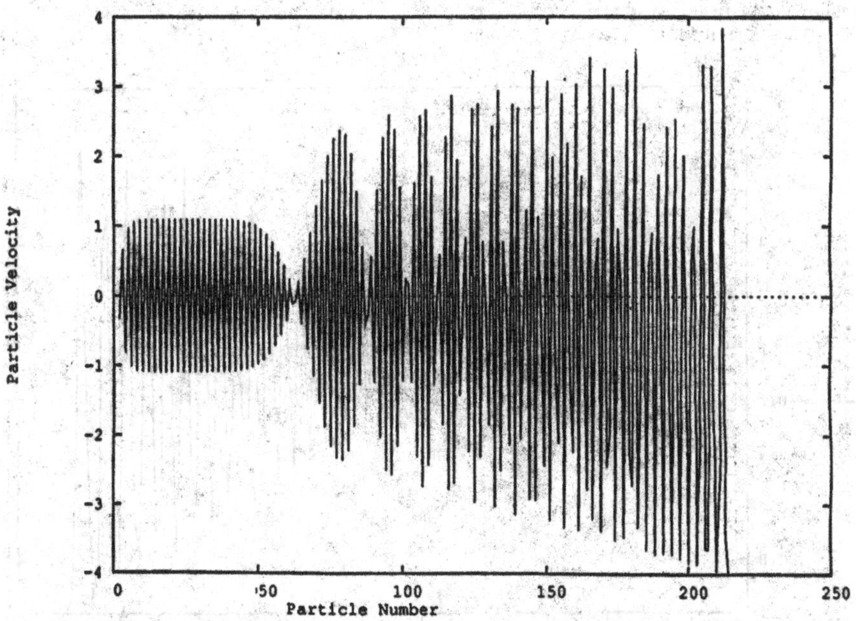

Fig. 7.2b. Particle velocities for the supercritical Toda shock with $a = 2$.

of oscillations a priori. By making such an assumption, Bloch and Kodama subsequently derived the same results through modulation theory in a study [2] that treats both the shock and the rarefaction ($a < 0$) problem. In Figs. 7.2 the supercritical Toda shock for $a = 2$ is illustrated. Figure 7.2a shows the position of particles $k = 1, \cdots, 20$ versus time. Figure 7.2b shows the particle velocity versus particle index at time $t = 100$.

Venakides, Deift and Oba also obtained detailed information on the structure of the solution near the particle $k = 0$ where the binary motion is broken by the condition $x_0(t) \equiv 0$, which is imposed by the antisymmetry of the problem. They found that the solution here corresponds to an elliptic curve with the degenerate band structure $[-a-1, -a+1] \cup \{0\} \cup [a-1, a+1]$ in which the middle band is collapsed to a point. Thus, there are two gaps, each containing a Dirichlet eigenvalue. The boundary condition $x_0(t) \equiv 0$ is forced by the antisymmetry of the evolution problem of these two Dirichlet eigenvalues, which guarantees that their sum is zero at all times. As the particle index k increases, the presence of the eigenvalue of the Lax operator at zero ceases to be felt, by the mechanism described in Fig. 5.1. The structure at the origin may also be derived by a Darboux transform of the binary motion which adds an eigenvalue at the origin.

The critical and subcritical cases were tackled by Kamvissis [28]. Figure 7.3 shows the subcritical Toda shock for $a = .5$; it differs from Fig. 7.2b in that the oscillations die away behind the shock. The resulting characteristic shape resembles a martini glass.

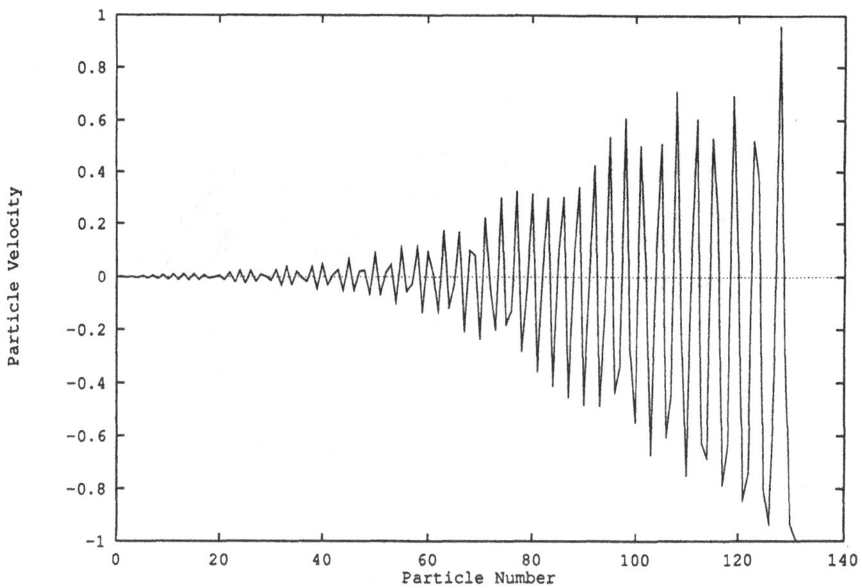

Fig. 7.3. Particle velocities for the subcritical Toda shock with $a = .5$ (the martini glass).

The Toda chain is completely integrable; therefore there is an explicit formula for the solution of the initial value problem. The limiting behavior as $\Delta \to 0$ has been carried out so far only for the special initial values (7.14) and finite perturbations of it, see [51] and [28]; but it should be possible to handle more general initial data.

8. Nonintegrable Discrete Problems

The study of zero dispersion limits for nonintegrable problems has proven a difficult road to travel. Without the machinery of inverse scattering, theoretical results have been limited. Moreover, careful numerical studies of these limits in the setting of partial differential equations are hampered by the fact that in order to resolve a term like $\epsilon^2 \partial_{xxx} u$ for a tiny value of ϵ requires a much tinier value of Δx. This is because the oscillations generated by the breaking process will have wavelengths of order at least ϵ that need to be resolved with many mesh points. However, this latter difficulty is circumvented if the problem is set as the continuum limit of a numerical scheme. In that case the smallest scale that needs to be resolved, the mesh scale $\Delta \equiv \Delta x$, is also the dispersion parameter, and it is therefore much easier to produce convincing numerical experiments. We shall present here two examples that have been studied so far.

8.1 Another Discretization of the Hopf Equation

Recently, a numerical and theoretical study has been carried out by Levermore and Liu [36] on a semi-discrete approximation to the Hopf initial-value problem

$$\partial_t u + u\,\partial_x u = 0\,, \qquad u(x,0) = v(x)\,. \tag{8.1}$$

They examined the dispersive center difference scheme

$$\frac{d}{dt}u_j + \tfrac{1}{3}\left(u_{j-1} + u_j + u_{j+1}\right)\frac{u_{j+1} - u_{j-1}}{2\Delta} = 0\,, \qquad u_j(0) = v(x_j)\,, \tag{8.2}$$

where $\Delta \equiv x_{j+1} - x_j$ is the mesh scale. While the origin of this scheme is somewhat clouded, it has a long and distinguished history; for example, it was used by Zabusky and Kruskal [55] to treat the nonlinear term when they discovered the remarkable interaction properties of solitons for the Korteweg-de Vries equation.

The historical appeal of this scheme derives from the fact that it possesses the two local conservation forms

$$\begin{aligned} &\frac{d}{dt}u_j + \frac{f_{j+\frac{1}{2}} - f_{j-\frac{1}{2}}}{\Delta} = 0\,, \qquad f_{j+\frac{1}{2}} = \tfrac{1}{6}\left(u_j^2 + u_j u_{j+1} + u_{j+1}^2\right), \\ &\frac{d}{dt}u_j^2 + \frac{g_{j+\frac{1}{2}} - g_{j-\frac{1}{2}}}{\Delta} = 0\,, \qquad g_{j+\frac{1}{2}} = \tfrac{1}{3}\left(u_j^2 u_{j+1} + u_j u_{j+1}^2\right). \end{aligned} \tag{8.3}$$

Without the richness of conserved quantities found in the integable schemes, the focus of their theory was narrowed to concentrate on modulated period-two spatial oscillations. The envelope of such oscillations is described by the averaged quantities

$$(8.4) \qquad v_{j+\frac{1}{2}} \equiv \tfrac{1}{2}(u_j + u_{j+1}), \qquad w_{j+\frac{1}{2}} \equiv \tfrac{1}{2}(u_j^2 + u_{j+1}^2),$$

which will then vary smoothly in the continuum limit. The fluxes in (8.3) can be expressed in terms of these quantites as

$$(8.5) \qquad \begin{aligned} f_{j+\frac{1}{2}} &= \tfrac{1}{3}v_{j+\frac{1}{2}}^2 + \tfrac{1}{6}w_{j+\frac{1}{2}}, \\ g_{j+\frac{1}{2}} &= \tfrac{4}{3}v_{j+\frac{1}{2}}^3 - \tfrac{2}{3}v_{j+\frac{1}{2}}w_{j+\frac{1}{2}}. \end{aligned}$$

Averaging the conservation forms (8.3) over adjacent spatial points yields

$$(8.6) \qquad \begin{aligned} \frac{d}{dt}v_{j+\frac{1}{2}} + \frac{f_{j+\frac{3}{2}} - f_{j-\frac{1}{2}}}{2\Delta} &= 0, \\ \frac{d}{dt}w_{j+\frac{1}{2}} + \frac{g_{j+\frac{3}{2}} - g_{j-\frac{1}{2}}}{2\Delta} &= 0. \end{aligned}$$

Passing to the continuum limit of vanishing Δ in (8.6) while employing the expressions for the fluxes given in (8.5) leads to the 2×2 system

$$(8.7) \qquad \begin{aligned} \partial_t v + \partial_x\left(\tfrac{1}{3}v^2 + \tfrac{1}{6}w\right) &= 0, \\ \partial_t w + \partial_x\left(\tfrac{4}{3}v^3 - \tfrac{2}{3}vw\right) &= 0. \end{aligned}$$

Indeed, (8.6) is precisely the central difference approximation of (8.7). These are the modulation equations for period-two oscillations.

Levermore and Liu noted that while $v^2 \leq w$ is a an invariant region for classical solutions of the modulation equations, a fact consistent with the origins (8.4) of v and w, the system remains hyperbolic only so long as $w \leq 10v^2$. Furthermore, unlike all the integable cases discussed previously, this hyperbolic region is not invariant under the evolution. This is illustrated in Figs. 8.1a and 8.1b, which show the breakdown of an envelope of modulated period-two oscillations upon the crossing of this hyperbolic threshold. Here the period-two oscillations originated from the singularity that develops from the initial data $v(x) = .3\sin(x)$. Wherever the modulated period-two oscillations have broken up, the dynamics is no longer slowly varying in time and the spatial mesh size was such that the weak limit, if it even exists, could not be resolved. However, they where able to prove, using standard techniques from numerical analysis, that smoothy modulated period-two oscillations within the hyperbolic regime are stable.

The onset of period-two oscillations was studied through weakly nonlinear matched asymptotic analysis. The connecting layer between the regions of period-two and no oscillations was described by a self-similar solution of a mKdV eqution satisfying the second Painlevé equation. This was consistent with the numerical experiments [36].

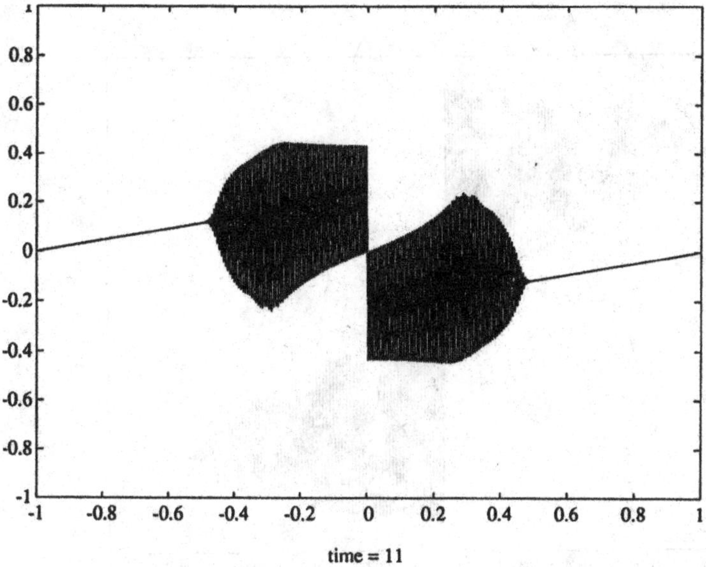

Fig. 8.1a. Modulated period-two oscillations before leaving the hyperbolic regime.

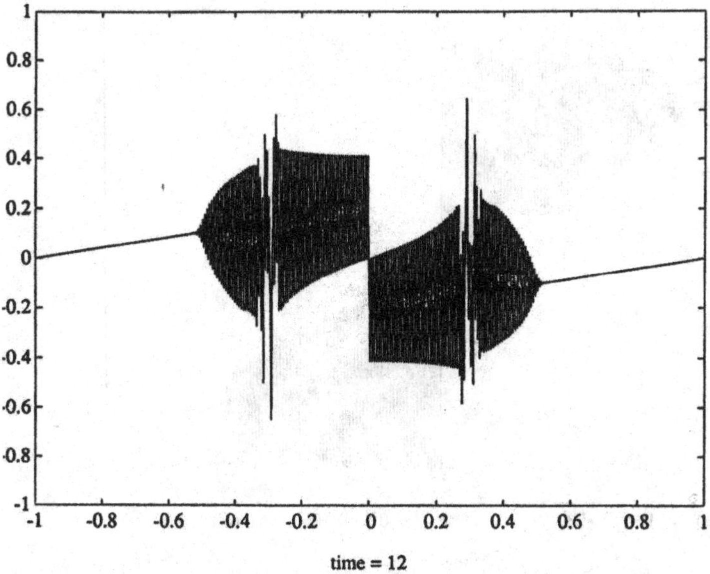

Fig. 8.1b. Breakdown of period-two oscillations after leaving the hyperbolic regime.

Fig. 8.1b. Breakdown of period-two oscillations after leaving the hyperbolic regime.

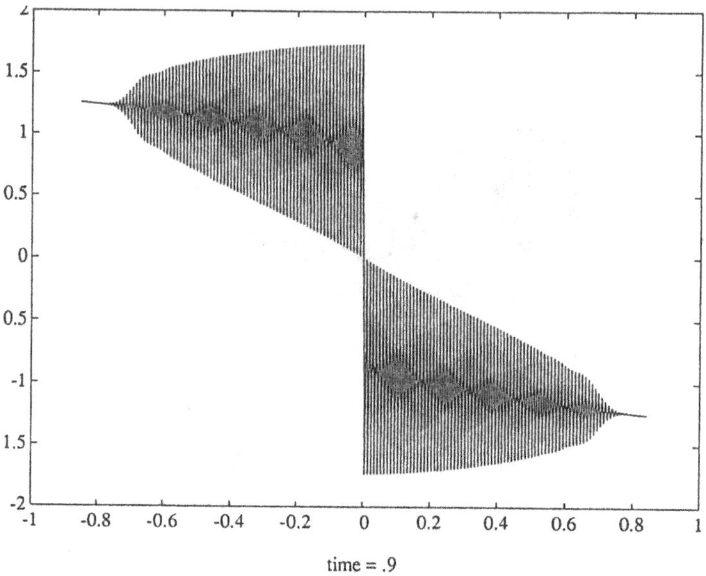

Fig. 8.2a. The evolution on a symmetric mesh.

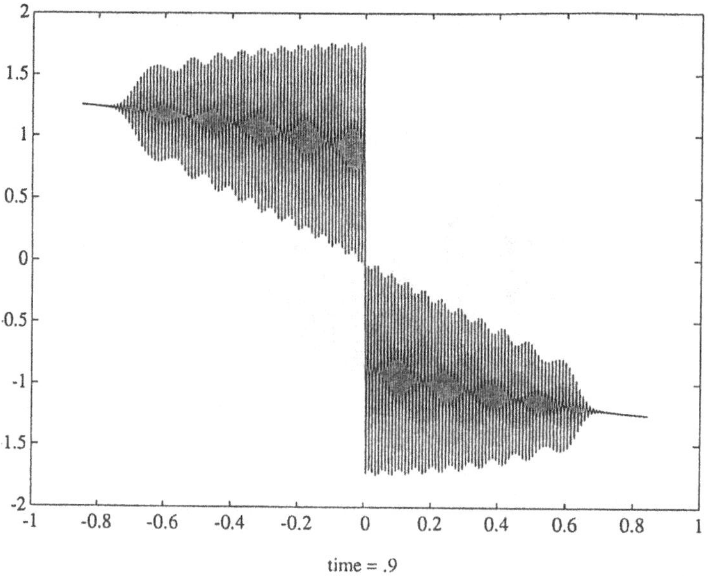

Fig. 8.2b. The evolution on a mesh shifted by $\frac{1}{4}\Delta$.

They also discovered another new phenomenon that had not been observed in the integable systems, namely that when the envelope of modulated period-two solutions develops a discontinuous contact solution, it is then unstable to shifts in the underlying mesh by a fraction of the cell size Δ. Figures 8.2a and 8.2b show the envelopes of modulated period-two oscillations that arise from the same initial data $v(x) = -x^{1/3}$, and the same Δ, but the mesh of Fig. 8.2a has $x = 0$ as a mesh point whereas that of Fig. 8.2b is shifted by $\frac{1}{4}\Delta$. This sensativity to the mesh placement shows that no limiting dynamics exists for this problem that is independent of details at the smallest scales.

8.2 More Discrete Fluid Dynamics and Spring Chains

We now return to the discretization of fluid dynamics (7.12), presented as a nonlinear spring chain in the form

$$\frac{d^2}{dt^2}X_k + \frac{p(\frac{X_{k+1}-X_k}{\Delta}) - p(\frac{X_k-X_{k-1}}{\Delta})}{\Delta x} = 0\,. \tag{8.8}$$

but here with an eye toward discussing nonintegrable choices for the force law $p(V)$. Before presenting the known results and conjectures concerning the vibrations of this nonlinear spring chain, we give a brief history.

In 1944, von Neumann [52], in connection with his work at Los Alamos, suggested solving the initial value problem for the compressible gas dynamics equations (7.8) numerically by replacing the space and time derivatives with central differences. He staggered the variables u and V in time as well as space, arriving at the completely centered system

$$\frac{u_k^{n+\frac{1}{2}} - u_k^{n-\frac{1}{2}}}{\Delta t} + \frac{p_{k+\frac{1}{2}}^n - p_{k-\frac{1}{2}}^n}{\Delta x} = 0\,, \tag{8.9a}$$

$$\frac{V_{k+\frac{1}{2}}^{n+1} - V_{k+\frac{1}{2}}^n}{\Delta t} - \frac{u_{k+1}^{n+\frac{1}{2}} - u_k^{n+\frac{1}{2}}}{\Delta} = 0\,. \tag{8.9b}$$

At that time von Neumann made no explicit provisions to accommodate shock waves. The initial data represent gas in a tube with one of its ends sealed off; initially the gas is set in motion with constant velocity a in the direction of the sealed end. Calculations carried out on punched card equipment at the Aberdeen Proving Ground showed a shock wave emerging from the sealed end, followed by rapid oscillations. At the time von Neumann interpreted the oscillations in velocity as heat energy created by the irreversible action of the shock; he conjectured that as Δx and Δt tend to zero, the oscillatory approximate solutions generated by the difference scheme tend weakly to the exact discontinuous solution of the equations of gas dynamics. We quote [52]:

"In the mathematical terminology, the surmise means that the quasi-molecular kinetic solutions converge to the hydrodynamical one but in the weak sense .... A mathematical proof of this surmise would be most

important, but it seems to be very difficult, even in the simplest special cases. The procedure to be followed will therefore be a different one: We shall test the surmise experimentally by carrying out the necessary computations ⋯ ."

It was pointed out in [32] and [23] that whereas von Neumann was right in conjecturing weak convergence, he was wrong to surmise that the weak limits would be exact solutions of the gas dynamical equations.

The study of the spring chain (8.8) was taken up in the early fifties by Fermi, Pasta and Ulam [11]; they were however not interested in the limit $\Delta \to 0$, but the limit $t \to \infty$. They took sinusoidal initial data, and in their numerical experiments observed almost periodic behavior. The study of the von Neumann shock problem was resumed in 1978, again at Los Alamos, by the aforementioned works of Holian and Straub [22] and Holian, Flaschka and McLaughlin [21].

Hou and Lax [23] turned to the general initial value problem for chains. Figures 8.3a and 8.3b depict the velocity component u of the solution of gas dynamics discretization (7.9) with the equation of state $p(V) = V^{-\gamma}$, $\gamma = 1.4$, as function of x, linearly interpolated from the points $x = k\Delta$, at time $t = 0.7$, with initial values

$$u(x,0) = 2 + 0.5\sin 2\pi x\,, \qquad V(x) = 1\,.$$

For these initial values, breakdown occurs at $t_{crit} \simeq 0.41$, In Fig. 8.3a, $\Delta = 0.01$, in Fig. 8.3b, $\Delta = 0.005$. The phenomena displayed by these figures are similar to the ones found in Figs. 7.1:

(1) There are two distinct oscillatory regions, the same in both figures, separating two smooth regions.
(2) Oscillations are on the mesh scale.
(3) The average value of u is the same in both figures.

Calculations with initial values of a rarefaction wave show uniform convergence to the rarefaction waves.

Hou and Lax also experimented with the von Neumann scheme (8.9), using various initial data, including Riemann data. Figures 8.4a and 8.4b show the velocity component u of the solution of a Riemann initial value problem whose discontinuous solution is a shock propagating with speed 1, with $u = 1$ behind the shock, and $u = 0.666$ in front of the shock. In Fig. 8.4a, $\Delta x = 0.005$ and $\Delta t = 0.0005$; in Fig. 8.4b, $\Delta x = 0.0025$ and $\Delta t = 0.00025$. These figures suggest strongly that

(1) u converges uniformly to $u = 1$ in some interval $0 < x < s_2$, where the approximate value of s_2 is $\simeq 0.5$.
(2) There is an oscillatory region $s_2 < x < s_1$, $s_1 \simeq 1.06$; the wave length of the oscillations is $O(\Delta)$.
(3) In the oscillatory region the averaged value of u is a decreasing function of x.

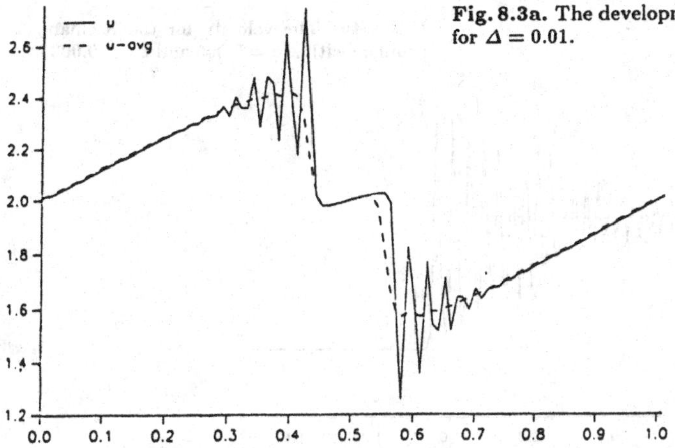

Fig. 8.3a. The development of oscillations for $\Delta = 0.01$.

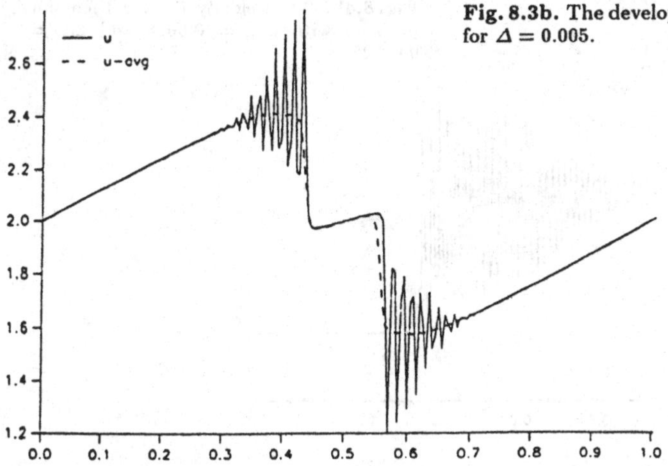

Fig. 8.3b. The development of oscillations for $\Delta = 0.005$.

These calculations were made with a CFL number ($\Delta t/\Delta x$) of .1. Further calculations using different CFL numbers produce the same shock speeds s_1 and s_2, and also apparently the same averaged quantities. Although strong limits are independent of the choice of relation between Δx and Δt, this is by no means expected of weak limits, and needs to be understood.

Figure 8.5 depicts the velocity component u of the solution of the von Neumann shock problem for the spring chain with the force law $p(V) = V^{-\gamma}$, $\gamma = 1.4$. Observe that the region behind the shock consists of two regions; in the inner region the oscillation is binary, i.e. neighboring particles are moving in opposite direction, just as in the case of the Toda chain depicted in Fig. 7.2b.

The numerical evidence presented in this section, and an abundance not reported here, indicate very strongly that the main features of the zero dispersion limit — oscillatory behavior and weak convergence — are true for systems

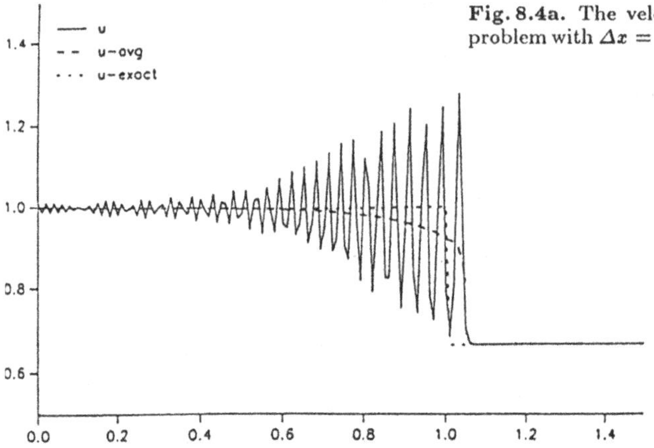

Fig. 8.4a. The velocity for the Riemann problem with $\Delta x = 0.005$ and $\Delta t = 0.0005$.

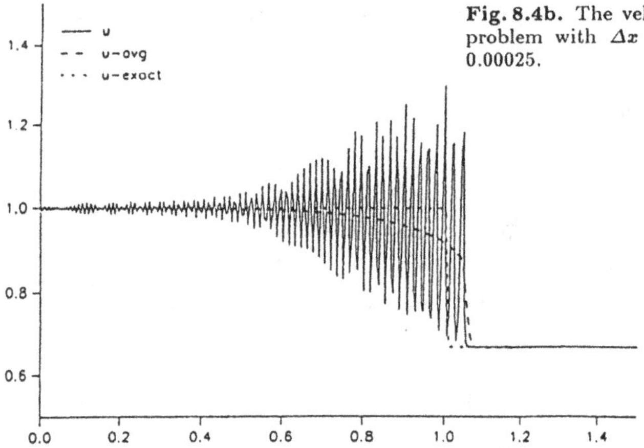

Fig. 8.4b. The velocity for the Riemann problem with $\Delta x = 0.0025$ and $\Delta t = 0.00025$.

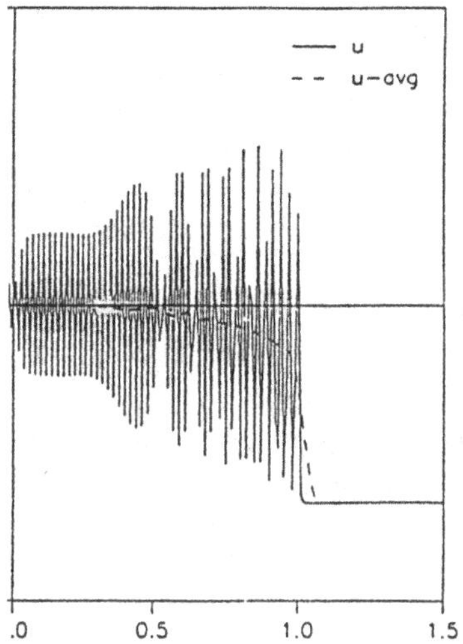

Fig. 8.5. The velocity for the von Neumann shock problem for the γ-law chain with $\gamma = 1.4$.

that are not completely integrable but resemble in form integrable ones, such as chains of particles. So far there are no results of any generality, nor are there methods for tackling the nonintegrable case; an interesting example for short range forces has been worked out by Greenberg [18].

Acknowledgements. The authors are grateful to a number of colleagues for their helpful comments during the preparation of this article; in particular N. Ercolani, S. Jin and T. Zhang. They also gladly acknowledge the support provided to them: P.D.L. by the Department of Energy under contract DE-FG02-88ER25053; C.D.L. from the NSF under grant DMS-8914420; S.V. from the NSF under grant DMS-9103386 and from the ARO under grant DAAL03-91-G0178.

References

[1] R. Bikbaev and V. Novokshonov, *Self-similar solutions of the Whitham Equations and the Korteweg-de Vries Equation with Finite-gap Boundary Condition*, in *Nonlinear and Turbulent Processes in Physics*, V.G. Bar'yakhtov, V.M. Chernousenko, N.S. Erokhin, and V.E. Zakharov eds. Proc. Third Internm Workshop, 1987 Kiev **1**, 32–, World Scientific, Singapore (1988).

[2] A.M. Bloch and Y. Kodama, *Dispersive Regularization of the Whitham Equations for the Toda Lattice*, SIAM J. Appl. Math. **54(4)**, (to appear 1992).

[3] J.C. Bronski and D.W. McLaughlin, *Semiclassical Behavior in the NLS Equation: Optical Shocks–Focusing Instabilities*, in *Singular Limits of Dispersive Waves*, N. Ercolani, I. Gabitov, D. Levermore and D. Serre eds., NATO ARW series, Plenum, New York (to appear 1992).

[4] V.S. Buslaev and V.N. Fomin, *An Inverse Scattering Problem for the One-Dimensional Schrödinger Equation on the Entire Axis*, Vestnik Leningrad Univ. **17**, 56–64 (in Russian) (1962).

[5] A. Cohen and T. Kappeler, *Scattering and Inverse Scattering for Steplike Potentials in the Schrödinger Equation*, Indiana U. Math. J. **34(1)**, 127–180 (1985).

[6] B.A. Dubrovin, V.B. Matveev, and S.P. Novikov, *Nonlinear Equations of Korteweg-de Vries Type, Finite Zoned Linear Operators, and Abelian Varieties*, Uspekhi Mat. Nauk **31**, 55–136 (1976).

[7] B.A. Dubrovin and S.P. Novikov, *The Hamiltonian Formalism of One-Dimensional Systems of Hydrodynamic Type and the Bogolyubov-Whitham Averaging Method*, Sov. Math. Doklady **27**, 665–669 (1983).

[8] B.A. Dubrovin and S.P. Novikov, *Hydrodynamics of Weakly Deformed Soliton Lattices: Differential Geometry and Hamiltonian Theory*, Russ. Math. Surveys **44**, 35–124 (1989).

[9] N. Ercolani, S. Jin, D. Levermore and W. MacEvoy, *The Zero Dispersion Limit of the NLS/mKdV Hierarchy for the Nonselfadjoint ZS Operator*, preprint (1992).

[10] N. Ercolani, I. Gabitov, D. Levermore and D. Serre eds., *Singular Limits of Dispersive Waves*, Proceedings of the NATO Advanced Research Workshop at Lyon during 8-12 July 1991, Plenum, New York (to appear 1992).

[11] E. Fermi, J. Pasta and S. Ulam, *Studies in Nonlinear Problems I*, in *Nonlinear Wave Motion*, A.C. Newell ed., Lectures in Applied Math. **15**, 143–196, Amer. Math. Soc., Providence (1974).

[12] H. Flaschka, *On the Toda Lattice II*, Prog. Theor. Phys. **51**, 703–716 (1974).

[13] H. Flaschka, M.G. Forest and D.W. McLaughlin, *Multiphase Averaging and the Inverse Spectral Solutions of the Korteweg-de Vries Equation*, Comm. Pure Appl. Math. **33**, 739–784 (1980).

[14] H. Flaschka, A.C. Newell and T. Ratiu, *Kac-Moody Lie Algebras and Soliton Equations II: Lax Equations Associated with* $A_1^{(1)}$, Physica D **9**, 300–323 (1983).

[15] C.S. Gardner, *The Korteweg-de Vries Equation and Generalizations VI: The Korteweg-de Vries Equation as a Hamiltonian System*, J. Math. Phys. **12**, 1548–1551 (1971).
[16] C.S. Gardner, J.M. Greene, M.D. Kruskal and R.M. Miura, *Method for Solving the Korteweg-deVries Equation*, Phys. Rev. Lett. **19**, 1095–1097 (1967).
[17] J. Goodman and P.D. Lax, *On dispersive difference schemes* I, Comm. Pure Appl. Math. **41**, 591–613 (1988).
[18] J.M. Greenberg, *The Shock Generation Problem for a Discrete Gas with Short Range Repulsive Force*, Comm. Pure Appl. Math., (to appear 1992).
[19] A.V. Gurevich and A.L. Krylov, *Dissipationless Shock Waves in Media with Positive Dispersion*, Sov. Phys. JETP **65**, 944–953 (1987).
A.V. Gurevich, A.L. Krylov and G.A. El', *Riemann Wave Breaking in Dispersive Hydrodynamics*, JETP Lett. **54**, 102–107 (1991).
[20] A.V. Gurevich and L.P. Pitaevski, *Nonstationary Structure of a Collisionless Shock Wave*, Sov. Phys. JETP **38**, 291–297 (1974).
[21] B.L. Holian, H. Flaschka and D.W. McLaughlin, *Shock Waves in the Toda Lattice: Analysis*, Phys. Rev. A **24**, 2595–2623 (1981).
[22] B.L. Holian and G.K. Straub, *Molecular Dynamics of Shock Waves in One-Dimensional Chains*, Phys. Rev. B **18**, 1593–1608 (1978).
[23] T. Hou and P.D. Lax, *Dispersive Approximations in Fluid Dynamics*, Comm. Pure Appl. Math. **44**, 1–40 (1991).
[24] A.R. Its and V.B. Matveev, *The Schrödinger Operators with Finite-Gap Spectrum and N-Soliton Solutions of the Korteweg-de Vries Equation*, Theo. Math. Phys. **23**, 51–68 (in Russian) (1975).
[25] S. Jin, C.D. Levermore and D.W. McLaughlin, *The Semiclassical Limit for the Defocusing Nonlinear Schrödinger Hierarchy*, preprint (1991).
S. Jin, C.D. Levermore and D.W. McLaughlin, *The Behavior of Solutions of the NLS Equation in the Semiclassical Limit*, in *Singular Limits of Dispersive Waves*, N. Ercolani, I. Gabitov, D. Levermore and D. Serre eds., NATO ARW series, Plenum, New York (to appear 1992).
[26] I. Kay and H.E. Moses, *Reflectionless Transmission through Dielectrics and Scattering Potentials*, J. Appl. Phys. **27**, 1503–1508 (1956).
[27] M. Kac and P. van Moerbeke, *On an Explicitly Solvable System of Nonlinear Differential Equations Related to the Toda Lattice*, Adv. in Math. **16**, 160–169 (1975).
[28] S. Kamvissis, *On the Long Time Behavior of the Doubly Infinite Toda Chain under Shock Initial Data*, Dissertation, NYU (1991).
[29] D.J. Kaup, *Nonlinear Schrödinger Solitons in the Forced Toda Lattice*, Physica D **25**, 361–368 (1987).
[30] I.M. Krichever, *The Method of Averaging for Two-Dimensional "Integrable" Equations*, Funct. Anal. Appl. **22**, 200–213 (1988).
[31] P.D. Lax, *Integrals of Nonlinear Equations of Evolution and Solitary Waves*, Comm. Pure Appl. Math. **21**, 467–490 (1968).
[32] P.D. Lax, *On Dispersive Difference Schemes*, Physica D **18**, North-Holland, Amsterdam, 250–254 (1986).
[33] P.D. Lax and C.D. Levermore, *The Zero Dispersion Limit of the Korteweg-deVries Equation*, Proc. Nat. Acad. Sci. USA **76**(8), 3602–3606 (1979).
[34] P.D. Lax and C.D. Levermore, *The Small Dispersion Limit of the Korteweg-deVries Equation I, II, III*, Comm. Pure Appl. Math. **36**, 253–290, 571–593, 809–829 (1983).
[35] C.D. Levermore, *The Hyperbolic Nature of the Zero Dispersion KdV Limit*, Comm. P.D.E. **13**(4), 495–514 (1988).
[36] C.D. Levermore and J.-G. Liu, *Large Oscillations Arising in a Dispersive Numerical Scheme*, Comm. Pure Appl Math. (submitted 1992).
C.D. Levermore and J.-G. Liu, *Oscillations Arising in Numerical Experiments*, in *Singular Limits of Dispersive Waves*, N. Ercolani, I. Gabitov, D. Levermore and D. Serre eds., NATO ARW series, Plenum, New York (to appear 1992).
[37] H.P. McKean and E. Trubowitz, *Hill's Operator and Hyperelliptic Function Theory in the Presence of Infinitely Many Branch Points*, Comm. Pure Appl. Math. **29**, 143–226 (1976).
[38] H.P. McKean and P. Van Moerbeke, *The Spectrum of Hill's Equation*, Invent. Math. **30**, 217–274 (1975).
[39] D.W. McLaughlin and J. Strain, *Calculating the Weak Limit of KdV by Quadratic Programming*, in *Singular Limits of Dispersive Waves*, N. Ercolani, I. Gabitov, D. Levermore and D. Serre eds., NATO ARW series, Plenum, New York (to appear 1992).

[40] J. Moser, *Three Integrable Hamiltonian Systems Connected with Isospectral Deformations*, Adv. in Math. **16**, 197–220 (1975)

[41] G.V. Potemin, *Algebro-geometric Construction of Self-Similar Solutions of the Whitham Equations*, Russian Math. Surveys **43(5)**, 252–253 (1988).

[42] D. Serre, *Systèmes Hyperboliques Riches de Lois de Conservation*, in *Nonlinear PDEs and Their Applications* H. Brezis and J.-L. Lions eds., Collège de France Seminar **XII** 1989, (to appear 1992).
D. Serre, *Systèmes d'EDO Invariants sous l'Action de Systèmes Hyperboliques d'EDP*, Annales de l'Institut Fourier **39**, 953–968 (1989).
D. Serre, *Richness and the Classification of Quasilinear Hyperbolic Systems*, in *Multidimensional Hyperbolic Problems* J. Glimm and A. Majda eds., IMA volume in Mathematics and its Applications **29** 315–333, Springer-Verlag, New York (1991).

[43] Siegel and Wilson, *Loop Groups and Equations of KdV Type*, IHES Publ. Math. 61, 5–65 (1985).

[44] F.R. Tian, *Oscillations of the Zero Dispersion Limit of the Koteweg-de Vries Equation*, Comm. Pure Appl. Math. (to appear 1993).
F.R. Tian, *On the Initial Value Problem of the Whitham Averaged System*, in *Singular Limits of Dispersive Waves*, N. Ercolani, I. Gabitov, D. Levermore and D. Serre eds., NATO ARW series, Plenum, New York (to appear 1992).

[45] S.P. Tsarev, *On Poisson Brackets and One-Dimensional Systems of Hydrodynamic Type*, Sov. Math. Doklady **31**, 488–491 (1985).

[46] S. Venakides, *The Zero Dispersion Limit of the Korteweg-de Vries Equation with Nontrivial Reflection Coefficient*, Comm. Pure Appl. Math. **38**, 125–155 (1985).

[47] S. Venakides, *The Generation of Modulated Wavetrains in the Solution of the Korteweg-de Vries Equation*, Comm. Pure Appl. Math. **38**, 883–909 (1985).

[48] S. Venakides, *The Zero Dispersion Limit of the Korteweg-de Vries Equation with Periodic Initial Data*, AMS Trans. **301**, 189–225 (1987).

[49] S. Venakides, *The Continuum Limit of Theta Functions*, Comm. Pure Appl. Math. **42**, 711–728 (1989).

[50] S. Venakides, *Higher Order Lax-Levermore Theory*, Comm. Pure Appl. Math. **43**, 335–362 (1990).

[51] S. Venakides, P. Deift and R. Oba, *The Toda Shock Problem*, Comm. Pure Appl. Math. **44**, 1171–1242 (1991).

[52] J. von Neumann, *Proposal and Analysis of a New Numerical Method in the Treatment of Hydrodynamical Shock Problems*, Collected Works **VI**, Pergamon Press, New York (1961).

[53] G.B. Whitham, *Non-Linear Dispersive Waves*, Proc. Royal Soc. London Ser. A **283**, 238–261 (1965).
G.B. Whitham, *Linear and Nonlinear Waves*, J. Wiley, New York (1974).

[54] O.C. Wright, *Korteweg-de Vries Zero Dispersion Limit: A Restricted Initial Value Problem*, Ph.D. Dissertation, Princeton University (1991).

[55] N.J. Zabusky and M.D. Kruskal, *Interaction of "Solitons" in a collisionless Plasma and the Recurrence of Initial States*, Phys. Rev. Lett. **15**, 240–243 (1965).

[56] V.E Zakharov and L. Faddeev, *The Korteweg-de Vries Equation: A Completely Integrable Hamiltonian System*, Funct. Anal. Appl. **5**, 280–287 (1971).

[57] V.E. Zakharov and A.B. Shabat, *Exact Theory of Two-dimensional Self-focusing and One-dimensional Self-modulation of Waves in Nonlinear Media*, Sov. Phys. JETP **34**, 62–69 (1973).

[58] T. Zhang and S. Venakides, *Periodic Limit of Inverse Scattering*, Comm. Pure Appl. Math. (to appear 1993).

Differential Geometry and Hydrodynamics of Soliton Lattices

S.P. Novikov

Landau Institute of Theoretical Physics,
Academy of Sciences, Moscow, Russia

Introduction

I am going to discuss here some results in the soliton theory of a Moscow group during the last years. The group of people who worked here include B.A. Dubrovin, I.M. Krichever, S.P. T̂sarev (and the present author). More details may be found in the survey article [1]. Modern needs in the large new classes of hydrodynamic type systems appear in connection with very interesting asymptotic method – so called "nonlinear analog of WKB-method", method of the slow modulations of parameters or "Whitham method" in the theory of solitons (see for example the book [2], chapter 4). This method is based on the large family of exact solutions $\varphi_0(x,t;u)$ periodic or quasiperiodic in x and t, of the form which we call "soliton lattice":

$$\varphi_0(x,t;u) = F(\eta_0 + Ux + Vt; u^1, \ldots, u^N). \tag{0.1}$$

Here $F(\eta_1, \ldots, \eta_m; u^1, \ldots, u^N)$ is the function , which is 2π–periodic in each variable η_j and depends on the N parameters u^p. All quantities (a_j, b_k, u^p) are constants and $\varphi_0(x,t;u)$ satisfies some nonlinear P.D. equation

$$\varphi_t(x,t) = K(\varphi, \varphi_x, \ldots, \varphi^{(s)}), \tag{0.2}$$

describing the propagation of solitons or some other nonlinear waves.

We have to have therefore some N–parametric family of invariant m–tori in the functional space of quasiperiodic functions. There are many such classical examples for $m = 1$. For example in the case (0.2) is exactly the famous KDV–equation and for $m = 1$ everybody knows the family of "knoidal waves" in elliptic functions

$$\varphi_0(x,t;u^1,u^2,u^3) = -2\partial_x^2 log\Theta(kx + \omega t + \eta_0) + c$$

Here we have $u^1 = c, u^2 = k, u^3 = \omega$. We write KDV in the form (0.3)

$$\varphi_t - 6\varphi\varphi_x + \varphi_{xxx} = 0. \tag{0.3}$$

The algebro–geometric solutions (or finite-gap solutions) of KDV discovered in [3] and investigated by many authors present us the analogous families for any $m > 1, N = 2m + 1$ (see details in [2,4]) The analogs of these solutions are known for all systems, integrable by the famous "Inverse scattering transform"(IST): SG (Sine-Gordon), NS (Nonlinear Schrodinger), KP (Kadomtsev-Petviashvili) and others. Consider now the function

$$\varphi_0\left(\frac{S(X,T)}{\epsilon}; u(X,T)\right). \tag{0.4}$$

Here $X = \epsilon x, T = \epsilon t, S_X = U, S_T = V$. In 1965 Whitham observed (in some cases— $m = 1$,the system (0.2) is a nondegenerate Lagrangian system or it is exactly the KDV)— that the function (0.4) satisfies equation (0.2J) plus something small ($\epsilon \to 0$) if the "slow functions" $u^p(X, T)$ satisfy some "hydrodynamic type" equation:

$$u^p{}_T = v^p{}_q(u) u^q{}_X. \qquad (0.5)$$

A lot of authors developed this idea since 1965 (see the papers [5-10]; more complete list of references may be found in [1]). Serious investigations for $m > 1$ started after the discovery of "finite-gap solutions" (1974–1975) — see the papers [11-13].

The activity of our group in this area started in 1982–1983.Its investigations were mainly concentrated on the next problems:

Problem 1: Suppose the original system (0.2) is hamiltonian corresponding to some local field-theoretical hamiltonian formalism. Is the associated hydrodynamic type system (or the "averaged system") (0.5) also hamiltonian? How to construct explicitly its coefficients using the hamiltonian formalism of the original system (0.2)?

Problem 2: Which kind of boundary problems for the equations of hydrodynamic type (0.5) appear from the "Nonlinear WKB" method? Is it possible to investigate them directly inside of the theory of the 1st–order systems as a new kind of hydrodynamics?

Problem 3: Suppose the original system (0.2) is completely integrable by the IST–method (Inv. Scatt. Transform).Is the averaged system (0.5) completely integrable ? How to construct its solutions explicitly?

Problem 1 was solved by the present author and B.A.Dubrovin ([14]). This solution leads to the discovery of some deep and natural differentially–geometric structure for the hamiltonian hydrodynamic type equations (0.5), which does not exist for the generic hydrodynamic type equations (0.5) and was not observed during the 100 years of their investigations starting from Riemann. After that the Problem 3 (on the level of local differential geometry) was solved by S.P.Tsarev in his thesis (see [15,16]. The effective global algebro–geometric construction of some important solutions was found later by Krichever [17].

Problem 2 was investigated numerically by Avilov, Krichever and the present author ([18,19]), for $m = 1$ only. Very interesting classes of "multivalued functions" appear in the hydrodynamics of soliton lattices, but nothing rigorous was proven here.

In Appendix we discuss some beautiful restrictions on the initial data for the KDV-hierarchy which lead to rapidly oscillating functions and therefore to the "Nonlinear WKB". They have appeared recently in the so-called matrix models in the form of the "string equation".

We shall use the notation H.T. and P.B.H.T. for hydrodynamic type and Poisson brackets of hydrodynamic type respectively. Also P.B. denotes Poisson brackets, SH denotes semi-Hamiltonian, AFG denotes averaged finite gap solutions.

1. Differential Geometry and Hydrodynamic Type Poisson Brackets

1.1 Basic definitions

It will be most convenient to explain these ideas starting from the general geometric definitions. Let $X = (x_1, \ldots, x_n)$,$n =$ "dimension of the space" and $x_1 = x$ for $n = 1$. Let M be some N–dimensional manifold, where N is the "number of components" and $u = (u^1, \ldots, u^N)$ are some local coordinates on M. The "hydrodynamic structure" on M is by definition some collection of tensor fields numerated by the same indices as the X–space coordinates $v_q^{p,\alpha}(u)$.Definitions D1–D4 give us the base of geometric theory.

D1: The H.T. system is defined by the tensor fields $v_q^{p,\alpha}$ in the form

$$u_t^p = v_q^{p,\alpha}(u)u_\alpha^q, \tag{1.1}$$

for the fields $u(x)$ belonging to some functional space of mappings $X \to M$.

D2: The Riemann invariants for the system (1.1) are such coordinates $u^1, \dots, u^N$ (if they exist) that all matrix fields $v_q^{p,\alpha}(u)$ are diagonal

$$v_q^{p,\alpha} = v^{p,\alpha}\delta_q^p. \tag{1.2}$$

For $(N = 2, n = 1)$ the Riemann invariants always exist for hyperbolic systems (i.e. all individual matrices $v_q^{p,\alpha}(u_0)$ are equivalent to diagonal ones).

D3: The hydrodynamic quantities are such local functionals on the space of fields $u(x)$,whose densities do not depend of the derivatives:

$$J = \int j(u(x))dx. \tag{1.3}$$

The most important new definition of the paper [14] is the next one.

D4: The local homogeneous hydrodynamic type Poisson bracket is defined by the formula:

$$\begin{aligned}\{u^p(x), u^q(y)\} = g^{pq,\alpha}(u(x))\partial_\alpha\delta(x-y)+ \\ +b_r^{pq,\alpha}(u(x))\partial_\alpha u^r(x)\delta(x-y)\end{aligned}. \tag{1.4}$$

for some functions $g^{pq,\alpha}(u), b_r^{pq,\alpha}(u)$ in the local coordinates (u).

The H.T. hamiltonian (1.3) generates the H.T. system (1.1) using the Poisson bracket (1.4). If $H = \int h(x)dx$ we have

$$\begin{aligned} u_t^p &= g^{pk}u_x^l h_{;k;l} \\ v_q^p &= g^{pk}\nabla_k(\partial_q h) \end{aligned} \tag{1.5}$$

The P.B. operation $(1,3)$ has to be bilinear and skew symmetric; it has to satisfy Leibnitz and Jacobi identities. As a consequence of these requirements our quantities $g^{pq,\alpha}(u)$ should transform as symmetric tensors on M for each α and $b_k^{pq,\alpha}$ should transform as $g^{pr,\alpha}\Gamma_{rk}^{q,\alpha}(u)$ for some Christoffell symbols Γ on M under the local changes of coordinates $u(w)$. The connections $\Gamma_{rk}^{q,\alpha}$ are compatible with the "metric" $g^{pq,\alpha}$ (for $det(g^{pq,\alpha}) \neq 0$), if *torsion* and *curvature* are equal to zero.

There exist therefore some new coordinates (w) such that

$$\begin{aligned} g^{pq,\alpha} &= \text{const} \\ \Gamma_{rk}^{q,\alpha} &= 0 \end{aligned} \tag{1.6}$$

(for one value of α only!). For $n = 1$ it gives us the complete invariant for the local classification of P.B.T.H. with $(detg \neq 0)$–SIGNATURE of the metric g.

So the canonical form of the nondegenerate P.B.H.T. is the generalized Gardner-Zakharov-Faddeev's δ'-bracket for $n = 1$.

For $n > 1$ we may kill Christoffell symbols for one value of $\alpha = 1$ only; after that (as the present author, Dubrovin and Mokhov proved in [20,21]) all other "metrics" will be linear (may be nonhomogeneous) functions in that variables $(u^1, \dots, u^N)$, all other $b_r^{pq,\alpha}$ will be equal to **const** for $\alpha \neq 1$. Here we should have $detg^{pq,1} \neq 0$.

Such structures are invariant under the affine transformations of coordinates

$$u = Aw + u_0$$

1.2 Special P.B.H.T. Lie algebras.

In the case when all metrics may be linear in u we have

$$\begin{aligned} g^{pq,\alpha} &= g_0^{pq,\alpha} + C_k^{pq,\alpha} u^k \\ b_k^{pq,\alpha} &= \text{const}, g_0^{pq,\alpha} = \text{const} \\ C_k^{pq,\alpha} &= b_k^{pq,\alpha} + b_k^{qp,\alpha} \end{aligned} \tag{1.7}$$

The linear (homogeneous) part of such P.B.H.T. determines some very interesting class of infinite–dimensional Lie algebras ("hydrodynamic algebras"): for two vector-functions $f(x)$ and $g(x)$ with N components f_p, g_q we may define the commutator in the "local translation-invariant first–order Lie algebra" or hydrodynamic algebra

$$[f,g]_r = [(\partial_\alpha f_p)g_q - (\partial_\alpha g_p)f_q]b_r^{pq,\alpha}. \tag{1.8}$$

For $n = 1$ these algebras were investigated in [22]. The hamiltonian formalism of classical liquid (including the magnetohydrodynamics and superfluid systems ad the references to surveys of Khalatnikov, Dzyaloshinsky and other physicists especially from Landau Institute) in connection with some special Lie algebras was discussed in [23].

It is useful to introduce new algebra B as a multiplication in the N–space M with basis $e^1, \ldots, e^N$

$$e^p \circ e^q = b_k^{pq} e^k. \tag{1.9}$$

For the functions $f(x) = f_p(x)e^p$ and $g(x) = g_q e^q$ we write (1.8) in the form

$$f' \circ g - g' \circ f$$

using the multiplication $\circ$ in algebra B –see (1.9). In that case ($n = 1$) our formula (1.8) determines correctly the Lie algebra L_B if and only if for any three elements of the algebra B the next identities are true: if $L_a(b) = a \circ b, R_b(a) = a \circ b$ we should have

$$\begin{aligned} [L_a, L_b] &= 0 \\ [R_b, R_c] &= R_{b\circ c - c\circ b} \end{aligned}$$

The constant part of P.B. (1.7) $g_0^{pq,\alpha}$ determines some 2–cocycle on the Lie algebra L_B for $n = 1$

$$\gamma[f,g] = \int f'_p g_0^{pq} g_q(x) dx. \tag{1.10}$$

In general , local translational–invariant 2-cocycles of the type $\tau = 0, 1, 2, 3$, on the algebra L_B may be defined by the formula

$$\begin{aligned} \gamma[f,g] &= -\gamma[g,f] = \\ &= \int \gamma_0^{pq} f_p^{(\tau)}(x) g_q(x) dx \end{aligned} \tag{1.11}$$

We have $\tau = 1$ in the case (1.10). We may have nontrivial cocycles for ($\tau = 0, 1, 2, 3$) only. The classical Gelfand–Fuchs cocycle for $B = R, \tau = 3$ generates the well–known Virasoro algebra– extension of the algebra of vector fields on the circle L_R). Formula (1.11) determines some cocycles of the type $\tau = 3$ for commutative algebra B if the symmetric form γ_0^{pq} determines scalar product $<,>$ on B with the next properties

$$< ab, c > = < a, cb > \tag{1.12}$$

If $1 \in B$ and the form is nondegenerate we come to the classical Frobenius algebras. In that case we always have

$$\gamma_0^{pq} = C_k^{pq} u_0^k \tag{1.13}$$

for some point u^0. The case $\tau = 1$ is also very important. Symmetric form in that case is such that (1.12) is also true but the algebra B may be noncommutative.

The corresponding cocycle of the type ($\tau = 1$) is cohomologous to zero if (1.13) is true. Interesting examples presents us P.B.H.T. for the classical gas dynamics (for $n = 1$). We have here $N = 3$—the densities of momenta, of mass and of entropy are our basic fields. Let $f(x) = p(x)e^1 + \rho(x)e^2 + s(x)e^3, e^1 = e, e^2 = a, e^3 = b$ For the multiplication in algebra B we have (we omit the symbol $\circ$)

$$e^2 = e, ea = a, eb = b, be = ae = a^2 = b^2 = ab = ba = 0. \tag{1.14}$$

The "metric" $g^{pq}(u)$ here has the rank 2. This P.B.H.T. does not satisfy the nondegeneracy condition

$$detg^{pq} = 0$$

But there are nontrivial cocycles for $\tau = 1$; g_0^{pq} should be such that

$$< e, B >= 0$$

Therefore g_0^{pq} is some arbitrary two-by-two symmetric matrix concentrated on the subspace $(2, 3)$. The perturbed P.B.H.T. has nondegenerate metric

$$g^{pq}(u) = g_0^{pq} + C_r^{pq} u^r$$

which has zero curvature.

The case $\tau = 2$ may also be interesting. The form here and the corresponding scalar product should be such that

$$< a, b > = - < b, a >$$
$$< ac, b > = - < a, bc >$$

Any cocycle of the type τ generates the extension of P.B. where the extended P.B. is equal to original P.B.H.T. plus $(g_0^{pq}\delta^{(\tau)}(x - y))$

The theory of extensions for Lie algebras L_B was constructed in [22,23] As Zelmanov proved in [24] the algebra B (over C)with $N > 1$ contains some ideal $I \subset B$ such that $I^2 = 0$. Therefore all our algebras B may be constructed from the factor–algebras A of B by I and the $A--$modules$I$ using the cocycles of the work [22]. There was some nice example of the algebra B in the paper [25]; some of their algebraic properties were discussed by S.Gelfand.

1.3 Liouville structures.

Suppose $n = 1$.

D.6: The coordinates $u^1, \ldots, u^N$ are Liouville for the P.B.H.T. (1.7) if there exist some tensor fields $\gamma^{pq,\alpha}(u)$ such that

$$g^{pq,\alpha} = \gamma^{pq,\alpha} + \gamma^{qp,\alpha} \tag{1.15}$$

$$b_r^{pq,\alpha} = \frac{\partial \gamma^{pq,\alpha}}{\partial u^r}. \tag{1.16}$$

Example: for the brackets (1.6) linear in u the same coordinates u are Liouville

$$\gamma^{pq,\alpha} = b_r^{pq,\alpha} u^r + \text{const} \tag{1.17}$$

The Liouville structure is covariant under the affine transformations only.

D.7: We call the coordinates **strongly Liouville** for P.B.H.T. if its restriction to any group $(u^{i_1}, \ldots, u^{i_k})$ of coordinates (after affine transformation) correctly determines some P.B.H.T. with Liouville structure. As the present author and Dubrovin observed in [1] they appear from the "Nonlinear WKB" for the hydrodynamic of soliton lattices. For $n = 2$ any Liouville structure is strong , but for $N > 2$ it is not so. The P.B. of the classical gas dynamic is strongly Liouville in the coordinates (p, ρ, s). A nice example comes from a relativistic liquid. Its energy–momentum tensor $((t_{ij}) = T; i, j = 0, 1)$ is symmetric and

$$t_{00} = \varepsilon, t_{01} = p, trT = t_{00} - t_{11} = q$$

The H.T. equations have the form

$$\partial_i T^i_j = 0 \tag{1.18}$$

The eigenvalues of T in the Minkovsky metric are the density of inner energy ε_0 and pressure P. We have $q = \varepsilon_0 - P$. With state equation $F(\varepsilon_0, P) = 0$ we obtain the complete system. The P.B.H.T. for it is such that $H = \int t_{00} dx$ and

$$\gamma^{ij} = \begin{pmatrix} t_{01} & t_{00} \\ t_{11} & t_{10} \end{pmatrix}$$

There is discussion in [1] about some interesting generalizations of P.B.H.T. : higher order analogs, nonhomogeneous brackets, discrete analog (which leads to the linearized Yang–Baxter equations with very interesting additional structure).

2. Weakly Deformed Soliton Lattices and Their Hamiltonian Hydrodynamics

Suppose the original system (0.2) is hamiltonian corresponding to some local field-theoretical P.B. $\{,\}_0$ and hamiltonian H

$$H = \int h(\varphi, \varphi_x, \ldots) dx \tag{2.1}$$

$$\{\varphi(x), \varphi(y)\}_0 = \sum_{k=0}^{L} B_k(\varphi(x), \ldots, \varphi^{(n_k)}(x)) \tag{2.2}$$

$$\varphi_t = \{\varphi(x), H\}_0 \tag{2.3}$$

Consider any family of the "soliton lattices" (0.1) which is nondegenerate

$$U = (k_1, \ldots, k_m), \mathrm{rk} \left(\frac{\partial k_i}{\partial u^j}\right) = m. \tag{2.4}$$

It means that the first m parameters may be k_p and $N \geq m$. Suppose also that there are at least N independent local commuting integrals

$$\begin{gathered} I_p = \int P_p(\varphi, \varphi_x, \ldots, \varphi^{(l_p)}) dx \\ \{I_p, I_q\}_0 = 0 \end{gathered} \tag{2.5}$$

For the densities we have some finite sum

$$\{P_p, P_q\}_0 = \sum_k A_k^{pq}(\varphi(x), \ldots, \varphi^{(l_k)}(x)) \delta^{(k)}(x - y). \tag{2.6}$$

From the commutativity (2.5) we deduce that $\int A_0^{pq} = 0$ for any field φ . It means that

$$A_0^{pq}(\varphi, \ldots) = \partial_x Q^{pq}(\varphi, \ldots). \tag{2.7}$$

Let us introduce the **physical coordinates** u**, metric and Christoffel symbols** by the formulas:

$$g^{pq}(u) = \bar{A}_1^{pq} \\ =(2\pi)^{-m} \int_{T^m} A_1^{pq}(F, F', \ldots) d^m\eta. \tag{2.8}$$

$$\gamma^{pq}(u) = \bar{Q}^{pq} \\ =(2\pi)^{-m} \int_{T^m} Q^{pq}(F, F', \ldots) d^m\eta. \tag{2.9}$$

$$b_r^{pq} = \frac{\partial \gamma^{pq}}{\partial u^r}, u^q = \bar{P}_q. \tag{2.10}$$

Here we have

$$F' = k_j \frac{\partial F}{\partial \eta_j}, \ldots$$

and F is from (0.1)

Our theorem states that that the metric (2.9) has zero curvature and the corresponding P.B.H.T. in physical coordinates (2.8) has the strongly Liouville form (2.10). The averaged H.T. equation is the hamiltonian H.T. system in the bracket (2.8)$-$(2.10) .Its hamiltonian is H.T. quantity whose density exactly coincides with the averaged density of energy

$$\varepsilon = \bar{h}(u), \\ H = \int \varepsilon dx \tag{2.11}$$

All averaged densities $\bar{P}_q = u^q$ determine the H.T. conservative quantities

$$I_p = \int u^p dx$$

The averaged P.B.H.T. of them are equal to zero. This procedure may be considered as an alternative definition of the averaged system. Our works were dedicated to the studying of that procedure. Some other people investigated this question — in which cases this equations lead to the nonlinear WKB-type approximation (see for example [11])? Our goal was to investigate the formal properties of the resulting H.T. systems (including their hamiltonian formalism , exact integrability and also the boundary conditions for them natural in the physical problems).

The special case of nondegenerate Lagrangian system was investigated by Whitham in 1965 for $m = 1$. The formal properties for any $m > 1$ are the same here. In that case we don't need the local integrals as above. For example it is important for some nonintegrable systems like the perturbed NS

$$i\psi_t = \psi_{xx} + V(|\psi|^2)\psi.$$

Here $m = 2, N = 4$;but we have generically here only three conservative quantities. In the integrable case we have more, of course.

The Lagrangian averaging leads to the H.T. systems in the form (2.12):

$$v_t^j = \partial_x \left(\frac{\delta H}{\delta w^j(x)} \right) \\ w_t^j = \partial_x \left(\frac{\delta H}{\delta v^j(x)} \right) \tag{2.12}$$

Here we have $j = 1,\ldots,m, N = 2m, (u) = (v,w)$. In that case the metric has a constant form

$$g^{pq} = \begin{pmatrix} 0 & 1 \\ 1 & 0 \end{pmatrix}. \tag{2.13}$$

Its signature is (m,m). In the variables $(v,y), y = \int w dx$ the system (2.12) will be Lagrangian. This form exactly corresponds to the classical "Clebsch variables" in hydrodynamics. A lot of people considered the hamiltonian formalism of H.T. systems in the form (2.12) only (see for example [26]) .

It is natural to suppose that for the hamiltonian system (0.1) the subspace (0.1) presents some finite-dimensional system, integrable in the strict sense of Liouville or some family of such systems depending from $N - 2m$ parameters. Suppose $J_1,\ldots,J_m$ are the action variables. In [14] we formulated a theorem that the quantities

$$u^j = k_j, u^{m+j} = J_j, j = 1,\ldots,m$$

have the averaged P.B. of the form (2.13). This observation for $m = 1$ was made in the work of Hayes [9]. Up to now this theorem was proven for the integrable systems only.

Motivated by the last applications of the action variables the present author, Veselov and Dubrovin investigated them and constructed a theory of "algebro-geometric" finite dimensional P.B. for the systems, integrable by the method of Riemann surfaces (see [27,28]; these ideas were initiated in some calculations of the papers [29,30])

3. Integrability of the Hamiltomian H.T. Systems

3.1 Riemann Invariants for the Hamiltonian H.T. systems. Integrability and Differential Geometry.

The Riemann invariants are known for some classical systems of gas dynamics more than 100 years. Their existence for $N > 2$ is the true sign for some degeneracy in the system. For $(N = 2, n = 1)$ they always exist . The transformation to the inverse functions $x(u^1,u^2), t(u^1,u^2)$ linearizes the H.T. system. This is well-known "hodograph transformation". Is there any generalization of the "hodograph" for $N > 2$? This problem was solved by Tsarev in his thesis [12] in the process of investigating the diagonal hamiltonian H.T. systems. The conjecture about the integrability of that class was formulated by the present author and posed to Tsarev as a problem.

Tsarev's main theorem states that such systems are integrable in the strict sense of Liouville at least in the class of monotonic functions. He found also a natural analog of the "hodograph" method for the construction of exact solutions.

Let us consider the diagonal hamiltonian H.T. system

$$u^p_t = v^p(u)u^p_x, v^p \neq v^q. \tag{3.1}$$

corresponding to some hamiltonian

$$H = \int h(u)dx$$

and P.B.H.T. (1.4) for $n = 1$ with nondegenerate metric $g^{pq}(u)$. It is easy to prove that the metric will be also diagonal

$$g^{pq} = g^p(u)\delta^p_q. \tag{3.2}$$

in the same coordinates (but not constant).

The classification of the orthogonal coordinates in the euclidean space is the classical problem. E. Cartan proved that they depend on the $N(N-1)$ independent functions of 2 variables.

As Tsarev observed, all diagonal H.T. systems which are hamiltonian in the same P.B.H.T. commute with each other. We have:

$$\begin{aligned} u_t^p &= w^p(u)u_x^p, H_2 = \int h_2(u)dx \\ \Gamma_{ki}^k &= \left(\frac{\partial_i w^k}{w^i - w^k}\right) = \left(\frac{\partial_i v^k}{v^i - v^k}\right) \cdot \\ 2\Gamma_{ki}^k &= \partial_i \log |g_k| \end{aligned} \tag{3.3}$$

All collections of the systems (3.3) depend on N functions of 1 variable. They may be found "in principle" from the equation (3.3) on the level of local differential geometry. The next consequence will be important:

$$\partial_i \left(\frac{\partial_j w^k}{w^j - w^k}\right) = \partial_j \left(\frac{\partial_i w^k}{w^i - w^k}\right). \tag{3.4}$$

Definition. We shall call the diagonal H.T. system a **semihamiltonian** (SH) one if (3.4) is valid for its coefficients.

The class of SH systems is larger than the class of the diagonal hamiltonian ones. For ($N = 2$) all H.T. systems are diagonalizable but the hamiltonian subclass depends on 3 functions only. SH systems have the most important property of the hamiltonian H.T. systems: They are deeply connected with some diagonal metric, determined from the equation (3.3),but may have nonzero curvature if the SH system is really nonhamiltonian).

The integration process works for all SH systems: We have a large collection of commuting SH systems from (3.3) . Each system of that class generates some exact solution. Consider the functions $u^p(x,t)$ obtained from the equation (3.5):

$$v^p(u(x,t))t + x = w^p(u). \tag{3.5}$$

They satisfy to the original equation (3.1) . It gives locally the complete solution of (3.1) .

There is also a construction of a large family of conserved quantities for the SH systems. It was known for $N = 2$ many years. For the hamiltonian systems the commuting flows and the conserved quantities are in a natural one-to-one correspondence. It is unclear now— is there any hamiltonian explanation for the SH class?

3.2 Soliton lattices for the integrable systems.

The applications of these methods to any concrete integrable system is nontrivial problem. Consider for example KDV in the form (0.3) . Even the simplest case $m = 1, N = 3$ leads to the highly nontrivial "Whitham system" after averaging . Whitham observed that it admits the Riemann invariants (r_j). Its analytical form is:

$$(r_j)_T = v^j(r)(r_j)_X$$

where

$$-3v^j = r_1 + r_2 + r_3 - 2f_j$$

$$f_1 = (r_2 - r_1)\left(\frac{K}{K-E}\right), f_2 = (r_3 - r_1)\left(\frac{(1-s^2)K}{E-(1-s^2)K}\right), \tag{3.6}$$

$$f_3 = (r_1 - r_3)\frac{(1-s^2)K}{E}$$

and $K(s), E(s)$ are the standard "complete elliptic integrals",

$$s^2 = \frac{r_2 - r_1}{r_3 - r_1}, v_1 \le v_2 \le v_3, r_1 \le r_2 \le r_3. \tag{3.7}$$

On the boundary our family degenerates to a constant solution of KDV for $r_2 = r_1$ and to soliton for $r_3 = r_2$. The result about the existence of the Riemann invariants was generalized in 1980 ([12]) to the families of finite-gap solutions $m \ge 1, N = 2m+1$: they are exactly the branching points of Riemann surfaces Γ and the endpoints of the spectrum of corresponding Schrodinger operator with periodic (quasiperiodic) potential on the line:

$$(r_j)_T = v^j(r)(r_j)_X, j = 1, \ldots, 2m+1. \tag{3.8}$$

The same results are valid for the systems SG and NS. The family of finite -gap solutions has a form for KDV:

$$\varphi(x,t;u) = -2\partial_x^2 \log \Theta(Ux + Vt + \eta_0 | B) + C. \tag{3.9}$$

Here we have the Riemann surface

$$\begin{gathered} \Gamma : \mu^2 = \prod (\lambda - r_j), (r_j real) \\ U = (k_1, \ldots, k_m), V = (\omega_1, \ldots, \omega_m), B = (b_{ij}) \\ \oint_{b_i} \Omega_j = b_{ij}, \oint_{a_j} \Omega_k = 2\pi i \delta_{jk}, \oint_{b_j} dp(\lambda) = k_j, \oint_{b_j} dq(\lambda) = \omega_j \end{gathered} \tag{3.10}$$

Here ω_j are the 1st kind differentials and dp, dq are the second kind ones on the Riemann surface Γ normalized as in (3.11) :

$$\begin{gathered} \oint_{a_j} dp = \oint_{a_j} dq = 0 \\ dp = dk + \text{regular}, dq = d(k^2) + \text{regular} \\ z = k^{-1} = \lambda^{-\frac{1}{2}} \to 0, a_j = ([r_{2j-1}, r_{2j}]; \pm) \end{gathered} \tag{3.11}$$

The multivalued functions $p(\lambda)$, $q(\lambda)$ are the "quasi-momentum" and "quasi-energy", a_k, b_k–canonical basis of cycles:

$$a_k \circ a_j = b_k \circ b_j = 0, a_k \circ b_j = \delta_{kj}. \tag{3.12}$$

It is well-known that the Kruskal integrals exactly coincide with the asymptotic coefficients

$$\begin{gathered} p(\lambda) = \sqrt{\lambda} + \sum_{s=0}^{\infty} I_s \frac{1}{(2\sqrt{\lambda})^{2s+1}} \\ I_s = \int P_s(\varphi, \ldots) dx \end{gathered} \tag{3.13}$$

Here $P_0 = \varphi, P_1 = \frac{\varphi^2}{2}, P_2 = \frac{\varphi_x^2}{2} - \varphi^3, \ldots$.

As it was observed in [12], the averaged H.T. system has the nice algebraic form

$$\partial_T p(\lambda) = \partial_X q(\lambda). \tag{3.14}$$

The generalization of the form (3.14) to KP was found by Krichever [17] who was able also to develop a "Nonlinear WKB"–method for KP (this is more difficult because KP is a non-local evolution system). The averaged densities $\bar{P}_s$ generate the conserved quantities

$$I_s = \int u^s dx, s = 1, 2, \ldots.$$

Any linear combination of them is also a conserved quantity

$$I = \sum_j c_j I_j$$

So we have a huge family of conserved H.T. quantities after averaging

$$\sum_p u^p(r_1, \ldots, r_{2m+1})$$

Any independent group of $(2m+1)$ of such quantities determines the "physical coordinates"

$$w^p = \sum_{j\geq 0} c_j^p u^j(r_1, \ldots, r_{2m+1}), \tag{3.15}$$

such that the H.T.P.B. has a strongly Liouville form. This property gives us some characterization of the averaged P.B. for the integrable KDV-type systems.

All quantities v^p determine some exact solutions by the Tsarev's procedure (above). We call these solutions "the averaged finite- gap solutions" (AFG).

For the basic quantities $u^q = \bar{P}_q$ the AFG–solutions are self-similar as it was observed by Krichever. There is a nice formula:

$$v^j(r) = \left(\frac{dp}{dq}\right)_{\lambda = r_j}. \tag{3.16}$$

The AFG-solution generated by the averaged Kruskal integral $\int u^q dx$ using the hamiltonian formalism and (3.5) may be written in the form:

$$\begin{aligned} r_p(X,T) &= T^{\gamma} f_p(XT^{-1-\gamma}) \\ \gamma &= \frac{1}{q-2}, q = 3, 4, \ldots \end{aligned} \tag{3.17}$$

(details see in [1]). For $q = 4, m = 1$ computations were done in [31]. We obtain from (3.5)

$$\begin{aligned} w^p &= \frac{1}{35}[(3v^p - a)f_p + f] \\ f &= 5a^3 - 12ab + c \\ a = \sum_{p=1}^{3} r_p, b &= \sum_{p<q} r_p r_q, c = r_1 r_2 r_3 \\ f_p &= \frac{\partial f}{\partial r^p}. \end{aligned} \tag{3.18}$$

The system (3.18) is nondegenerate in the domain $\Delta = (z_-, z_+)$:

$$z_- = -\sqrt{2}, z_+ = \frac{+\sqrt{10}}{27}, z = xt^{-\frac{3}{2}}$$

Consider the curve $r(x,t) = t^{\frac{1}{2}}l(z)$ determined outside of Δ by the equation

$$x + 6rt = (r)^3$$

It will be by definition the C^0–continuation of the 3-valued curve (r_1, r_2, r_3) at any moment t equal to constant.

We have the "boundary condition":

$$\begin{aligned} r_2(x_-) = r_1(x_-), r_3(x_-) = r(x_-) \\ r_3(x_+) = r_2(x_+), r(x_+) = r(x_+) \end{aligned} . \tag{3.19}$$

So the complete C^0–curve $r(z)$ is such that

$$r(z) = (r_1, r_2, r_3), z \in \Delta. \tag{3.20}$$

The "dispersive shock wave" $r(x,t) = t^{\frac{1}{2}}l(z)$ which was found numerically by Gurevitch and Pitaevsky in 1973 and which we call "GP"-solution belongs to C^1-class in the points (z_-, z_+) and may have singularity in some point $z_0, l_2(z_0) = 0$. After proving that in fact the functions (3.18) – (3.20) are C^1 we shall come to the result that our formulas present the "dispersive shock wave" exactly (and there is no singularity in the point $r_2 = 0$). This important difference in the definition of the functional classes has been missed in [31] as the present author pointed out to Krichever and Potemin.

Consider the meromorphic form $ds(\lambda)$ on the surface Γ such that:

$$\oint_{a_j} ds = 0, ds = \sum_{p \geq 1} c_p d(k^{2p-1}) + (regular)$$

All collection of the "averaged finite-gap" solutions may be obtained from the next equation of Krichever:

$$\begin{aligned} (Xdp + Tdq - ds)_{\lambda = r_p} = 0 \\ p = 1, \ldots, 2m+1 \end{aligned} . \tag{3.21}$$

Here we have $c_4 = 0, c_p = 0 (p \neq 4)$ in the GP-case above. More general class of the forms ds with some jumps along the curves on the surface Γ leads to the general solution of the averaged H.T. system of hydrodynamics of soliton lattices for KDV.

3.3 Time evolution of the multivalued functions. Numerical investigations.

Hydrodynamics of soliton lattices based on the Whitham type equations cannot be realised on the ordinary spaces of one-valued functions in the physically interesting cases. We should define the time evolution for the multivalued functions. (The same conclusion may be deduced from the results of Lax , Levermore and Venakides [32,33].) We constructed in the works [18,19] the class of special 1- and 3-valued functions for the investigation of stability of the "dispersive shock wave" which should be realized also as the asymptotics $t \to \infty$.

Consider now 2 classes of multivalued functions $r(x)$, one-valued for $|x| \to \infty$:

Class 1: $r(x) \to A_{\pm}, |x| \to \infty$

Class 2: $r(x) \sim x^{\frac{1}{3}}, |x| \to \infty$

We suppose that there is one and only one finite interval Δ for each function $r(x)$ such that it is one-valued outside Δ and 3-valued inside $\Delta = [x_-, x_+]$. The branches r_p of r should be such that

$$\begin{gathered} r_1 < r_2 < r_3(inside\Delta) \\ r_1(x_-) = r_2(x_-) = r_- < r_3, r_2(x_+) = r_3(x_+) = r_+ > r_1 \end{gathered} \tag{3.22}$$

The graph of the curve $r(x)$ should be C^1-smooth. The most important requirements are the asymptotics near the boundary of Δ :

$$0 > x - x_- = [a_+ + b_+(r - r_+)]f(1 - s^2) + O(r - r^+)^3 \tag{3.23}$$

$$0 < x - x_- = [a_- + b_-(r - r_-)](r - r_-)^2 + O(r - r_-)^3 \tag{3.24}$$

The parameter s^2 was defined in (3.7) and

$$f(u) = u \log\left(\frac{16}{u} + \frac{1}{2}\right), u = 1 - s^2. \tag{3.25}$$

Time evolution was realized numerically [18] for the functional classes above. Our conclusion is that for the C^1-small perturbations of the GP-solutions (in the both classes 1 and 2) the time dynamics is correctly defined for all $t > t_0$ and has self-similar GP-asymptotics for $t \to \infty$. For the class 1 this result may be rigorously deduced from the IST-method for KDV (see[34]). For the important class 2 connected with the "dispersive shock wave" it is probably impossible now.

The influence of small viscosity in the same classes was investigated in [19].More details and discussion may be found in the survey [1].

Appendix

The situation of dispersive shock wave appeared recently in the so-called "quantum 2D-gravity", as you may find in the preprints of Douglas, Seiberg and Shenker, Molinari and Parisi (1990). After the results of Gross and Migdal, Brezin and Kazakov, Douglas and Shenker (the end of 1989) we know that the so-called renormalization group for the "matrix models" in some special continuum limit (for the large order of matrices) may be described as a KDV-type hierarchy:

$$\frac{\partial L}{\partial t_j} = [A_j, L], j \leq l - 1, t_0 = x. \tag{A.1}$$

The critical points, corresponding to concrete conformal theories coupled with gravity should satisfy to the next equation (string equation)

$$[A_l, L] = 1. \tag{A.2}$$

(and L should have very special asymptotic for $|x| \to \infty$; if $L = -\partial^2 + u$ we have here exactly the KDV hierarchy and $u \sim x^{\frac{1}{l+1}}$.

For $l = 1$ the equation $(A.2)$ coincides with the classical "Painleve-1" equation (but it never was written before in the form (A.2)). For $l = 2$ its asymptotic for the large x is exactly like in the case of the "dispersive shock wave" (see the last part 3.3 of the paragraph 3). The asymptotic behavior of solution for $t \to \infty$ is written in (3.18)-(3.20).

The present author and Krichever developed recently useful asymptotic methods (nonlinear WKB and linear WKB, using Riemann surfaces and Lax-type pairs) for the studying the equation (A.2), written in the form:

$$[A, L] = \varepsilon 1. \tag{A.3}$$

(see [36,37]).

Very recently the present author investigated the equation (A.3) and found very interesting formula for the physically important "string solution". In the simplest non-trivial case (order of A equal 3, order of L equal 2, where the equation (A.3) is exactly the Painleve-1) the potential $u(x)$ has the form:

$$\begin{aligned} u &= 2\wp(\frac{\omega}{2}\|g_2; g_3) \\ 2g_3' &= \epsilon u \\ g_2 &= -\epsilon x, x \to -\infty \end{aligned} \qquad (A.4)$$

The quantity 2ω here is the real period. The basic periods are complex-adjoint. This equation may be obviously integrated by 1 quadrature. Our main conjecture in [36] was not true; the ψ-function is in fact multivalued on the Riemann surface. After the investigation I found that the conjecture has to be replaced by the formula (A.4). The analogous result is true for the operators of the higher order also. This work is still not finished: I still do not know – is this formula exact or only asymptotic for $x \to \infty$?

References

1. Dubrovin B.A., Novikov S.P., UMN (Russian Math. Surveys) **44** (6) 29–98 (1989).
2. Novikov S.P., Manakov S.V., Pitaevsky L.P., Zakharov V.E., Theory of solitons. Plenum Press, New York (1984).
3. Novikov S.P., Funk. Anal. Appl. **8** (3) 54–66 (1974).
4. Dubrovin B.A., Matveev V.B., Novikov S.P., UMN, **31** (1) 55–136 (1976).
5. Whitham C.B., a)J. Fluid. Mech., **22** (2) 273–283 (1965). b)Proc. Royal Society London, A139 283–291 (1965).
6. Luke J.C. Proc. Royal Soc. London, A292 (1430) 410–412 (1966).
7. Maslov V.P., Theor. Math. Phys., **1** (3) 378–383 (1969).
8. Ablowitz M.J., Benney D.J., Studies Appl. Math. **49** (3) 225–238 (1970).
9. Hayes W.D., Proc. Royal Soc. London, **332** 199–221 (1973).
10. Gurevitch A.V., Pitaevsky L.P., a)JETP **65** (2) 590–604 (1973). b) Letters of JETP, **17** (5) 268–271 (1973).
11. Dobrokhotov S.Yu., Maslov V.P., Soviet Sci. Rev. (ser. C, Math. Phys. Rev.) **3** 1–150 (1982).
12. Flaschka H., Forest M.G., Mclaughlin D.W., Comm. Pure Appl. Math. **33** (6) 739–784 (1980).
13. Flaschka H., Mclaughlin D.W., Studies Appl. Math., **68** (1) 11–59 (1983).
14. Dubrovin B.A., Novikov S.P., DAN SSSR (= Soviet Math. Dokl.) **270** (4) 781–785 (1983).
15. Tsarev S.P., DAN SSSR, **282** (3) 534–537 (1985).

16. Novikov S.P., UMN, **40** (4) 79–89 (1985).

17. Krichever I.M., Funk. Anal. Appl., **44** (2) 37–52 (1989).

18. Avilov V.V., Novikov S.P., DAN SSSR, **294** (2) 325–329 (1987).

19. Avilov V.V., Krichever I.M., Novikov S.P., DAN SSSR, **295** (2) 345–349 (1987).

20. Dubrovin B.A., Novikov S.P., DAN SSSR, **279** (2) 294–297 (1984).

21. Mokhov O.I., Funk. Anal. Appl., **22** (4) 92–93 (1988).

22. Balinsky A.A., Novikov S.P., DAN SSSR **283** (5) 1036–1039 (1985).

23. Novikov S.P., UMN, **37** (5) 3–49 (1982).

24. Zelmanov E.I., DAN SSSR, **292** (6) (1987).

25. Gelfand I.M., Dorfman I.A., Funk. Anal. Appl., **13** (4) 13–30 (1979).

26. Ercolani N., Forest M.G., Mclaughlin D.W., Montgomery R., Duke Math. J., **55** (4) 949–983 (1987).

27. Veselov A.P., Novikov S.P., a) DAN SSSR **266** (3) (1982). b) Trans. of Steklov Math. Inst. **165** 49–61 (1984).

28. Dubrovin B.A., Novikov S.P., DAN SSSR, **267** (6) 1295–1300 (1982).

29. Flaschka H., Mclaughlin D.W., Progr. Theor. Phys., **55** (6) 438–456 (1976).

30. Alber S.I., Preprint of the Institut of the Chemical Physics ,Chernogolovka, AN SSSR, (1976).

31. Potemin G.V., UMN, **43** (5) 211–212 (1988).

32. Lax P.D., Levermore C.D., Comm. Pure Appl. Math., **36** 253–290, 571–593, 809–830 (1983).

33. Venakides S., a) AMS Transactions, **301** 189–226 b) Comm. Pure Appl. Math., **38** 125–155 (1985).

34. Bikbaev R.V., Novokshenov V.Yu., Proc. of the third International Workshop, Kiev, **1** 32–35 (1988).

35. Gurevitch A.V., Pitaevsky L.P., JETP, **93** (3) 871–880 (1987).

36. Novikov S.P., a) Funk. Anal. Appl., **24** (4) (1990). b) Progr. in Theor. Phys. (to appear in 1991).

37. Krichever I.M., Preprint of the Forshungsinstitut fur Math. , ETH, Zurich (June 1990).

Part III

Algebraic Aspects

Bi-Hamiltonian Structures and Integrability

A.S. Fokas[1] *and I.M. Gel'fand*[2]

[1]Department of Mathematics and Computer Science and
The Institute for Nonlinear Studies, Clarkson University,
Potsdam, NY 13699-5815, USA

[2]Department of Mathematics, Rutgers University,
New Brunswick, NJ 08903, USA

1 Introduction

The Hamiltonian approach to integrability of nonlinear evolution equations in $1+1$, i.e. in one spatial and one temporal dimension, has its origin in the works of Zakharov and Faddeev [1] and Gardner [2] who interpreted the Korteweg-deVries (KdV) equation

$$u_t = u_{xxx} + 6uu_x \tag{1.1}$$

as a completely integrable Hamiltonian system in an infinite dimensional phase space (the relevant Hamiltonian operator is ∂_x). Furthermore, it was shown in [1], that the inverse spectral method yields action-angle variables for this equation. Hierarchies of infinitely many commuting vector fields and constants of motion in involution for the KdV equation, were constructed by Lax [3]-[4] and by Gel'fand and Dikii [5]-[7], using the equation of the resolvent of the Schrödinger operator (which is related to the equation for the squared eigenfunctions). The approach of Lax is based on what became later known as the Lenard scheme (see below), while Gel'fand and Dikii obtained the constants of motion explicitly in terms of fractional powers of the Schrödinger operator, using what we shall call the resolvent scheme.

The hierarchies of vector fields associated with the nonlinear Schrödinger , the modified KdV, and the sine-Gordon equations were given in [8] using the squared eigenfunction equation of the Dirac operator. The squared eigenfunction operator was interpreted in [9] as an operator generating higher symmetries and was given the name recursion operator.

Important developments in the Hamiltonian theory took place in 1978: Magri [10] realized that integrable Hamiltonian systems have additional structure, namely they are bi-Hamiltonian systems, i.e. they are Hamiltonian with respect to two different compatible Hamiltonian operators (this was implicit in [5]-[7]). Adler [11], proposed a scheme for deriving such Hamiltonian operators starting from a given Lax operator. Gel'fand and Dikii [12] proved that the Adler scheme indeed produces Hamiltonian operators. This scheme is often referred to in the literature as the Adler-Gel'fand-Dikii (AGD) scheme.

The squared eigenfunction operator can be factorized in terms of two compatible Hamiltonian operators. It was shown in [13] and [14] that this operator possesses the Nijenhuis property (this property, for finite dimensional systems was introduced in 1951 [15], and for infinite dimensional systems in [16]). Nijenhuis operators are nonlocal; the problem of working with nonlocal operators was bypassed in [17] by working with Nijenhuis relations instead of Nijenhuis operators. Also in [17], a coordinate free formulation was given, where starting from two compatible Hamiltonian operators, hierarchies of commuting vector fields and of constants of motion in involution were generated. This scheme, which is called the Lenard scheme, has the disadvantage that it constructs explicitly the gradients of the conserved quantities, rather than the conserved quantities themselves. The explicit construction of the constants of motion for several $1+1$ integrable systems has been achieved using the resolvent scheme [18].

The understanding of the central role played by compatible Hamiltonian operators for equations in $1 + 1$, motivated a search for such operators for equations in $2 + 1$. However, in this direction several negative results appeared in the literature. In particular Zakharov and Konopelchenko [19] proved that bi-Hamiltonian structures of a certain type (naturally motivated by the results in $1 + 1$) did not exist in multidimensions; a similar result was proven for the Benjamin-Ono equation [20]. (It should be noted that the Benjamin-Ono equation has more similarities with the Kadomtsev-Petviashvili (KP) equation, than with the KdV equation). These results cast some doubt to whether the bi-Hamiltonian approach could be extended to equations in $2+1$. In 1986 it was shown by Fokas and Santini [21]-[23] that equations in $2 + 1$, and in particular the KP equation

$$u_t = u_{xxx} + 6uu_x + 3\alpha^2(\partial_x)^{-1}u_{yy}, \quad \alpha^2 = \pm 1, \tag{1.2}$$

admit a bi-Hamiltonian formulation. The bi-Hamiltonian structures found in [21]-[23] are of a novel type not found in $1+1$, and thus the result of [19] is not violated. These structures we made more transparent in [24] by defining them in terms of only two spatial variables (in [21]-[23] they were described in kernel language and thus were defined in terms of three spatial variables). It has been recently shown [25] that bi-Hamiltonian operators in multidimensions can be generated by a proper generalization of the AGD scheme. In $1 + 1$, since the bi-Hamiltonian theory is built over a ring of functions, Hamiltonian operators act on functions. The main novelty is $2 + 1$, is that the bi-Hamiltonian theory is built over a ring of pseudodifferential operators, and therefore Hamiltonian operators now act on operators instead of functions. (In this sense, Hamiltonian operators in $2+1$ should be called **operands**, i.e. operators acting in the space of operators.)

In this article we present some of the above results in a unified way. In §2 we present a new Hamiltonian formalism based on the notion of a graded Lie algebra (see also [31] and [42]). In [17] the Hamiltonian formulation was based on the introduction of a certain complex, but its explicit construction beyond the space of 1-forms was left open. Here we show how this construction can be carried out. In §3 we present two methods for generating bi-Hamiltonian structures. Both methods start with a Lax operator (this operator defines the time-independent part of a Lax pair); one uses the square eigenfunction equation and the other is the AGD scheme. Several examples, for equations in $1+1$ and $2+1$ are used to illustrate these methods. In §4 we discuss the Lenard scheme for equations in $1+1$ and $2+1$.

One of the important recent applications of bi-Hamiltonian structures associated with equations in $1+1$, has been the derivation of certain Virasoro algebras. A short review of this development is given in §5.

The bi-Hamiltonian approach reviewed here is closely related to the r-matrix approach developed by Faddeev and his school, and to the group theoretic approach of Reyman and Semenov-Tian-Shansky (see [32] and reference therein). However, due to space limitations we cannot elaborate on these connections.

We should point out that there exist several other schemes for obtaining hierarchies of commuting vector fields. They include the master symmetry approach [20], the Sato-Segal-Wilson [33]-[34] construction, the dressing method, and the application of the inverse spectral method.

For the sake of brevity of presentation we cannot describe several important Hamiltonian structures, which have appeared in connection with integrable equations. They include Kirillov-Kostant type structures and structures arising in hydrodynamics. These structures, together with the notion of the Dirac structure (which unifies Hamiltonian and symplectic structures) are discussed in [35]. Other important developments of the bi-Hamiltonian theory not discussed here include: (a) The implementation of the Liouville integration of the stationary KdV hierarchy [26] (the theta function solutions obtained in [27]-[30] and the associated Abel transform arise naturally in this process), and; (b) the investigation of the spectral theory associated with bi-Hamiltonian systems in classical mechanics [31], [40].

We conclude this introduction by presenting different bi-Hamiltonian formulations of the KP equation. We first recall that the KdV equation (1.1) can be written as

$$u_t = \partial_x \frac{\delta f_2}{\delta u} = \left(\partial_x^3 + 2u\partial_x + 2\partial_x u\right) \frac{\delta f_1}{\delta u}, \tag{1.3}$$

where $\tilde{f}_i = \int f_i dx$, $i = 1, 2$, $f_1 = \frac{u^2}{2}$, $f_2 = -\frac{u_x^2}{2} + u^3$, and $\frac{\delta}{\delta u}$ denotes the usual variational derivative. The operators $\theta_1 = \partial_x$ and $\theta_2 = \partial_x^3 + 2u\partial_x + 2\partial_x u$ are Hamiltonian operators; the associated Poisson brackets are

$$\left\{\tilde{f}_1, \tilde{f}_2\right\}_\theta = \int (\theta d\tilde{f}_1)(d\tilde{f}_2)dx = \int \left(\theta \frac{\delta f_1}{\delta u}\right)\left(\frac{\delta f_2}{\delta u}\right) dx.$$

Using θ_1 and θ_2, the Lenard scheme implies a hierarchy of conserved gradients, $\frac{\delta f_j}{\delta u}$, $j = 1, 2, 3, \ldots$.

It will be shown in this article that the KP equation (1.2) can be written as

$$u_t = \partial_x(u_{xx} + 3u^2 + 3\alpha^2 \partial_x^{-2} u_{yy}) = \Theta(u - 2\partial_y) \tag{1.4}$$

where the action of the operand Θ on an operator B is defined by

$$\Theta B \doteqdot B_{xxx} + [u_x, B]_+ + 2[u + \alpha\partial_y, B_x]_+ + \left[u + \alpha\partial_y, (\partial_x^{-1}[u + \alpha\partial_y, B])\right], \quad (1.5)$$

and $[,]$, $[,]_+$ denote commutator and anticommutator respectively. Furthermore,

$$u_{xx} + 3u^2 + 3\alpha^2\partial_x^{-2}u_{yy}, \quad u - 2\partial_y,$$

are gradients and Θ is a Hamiltonian operand.

It turns out that the KP equation (1.2) can also be written as

$$u_t = \Theta_1(u_{xx} + 3u^2 + 3\alpha^2(\partial_x^{-2}u_{yy})) = -\Theta_2(3(\partial_x^{-1}u_y) + 9u_x + 18u\partial_x + 24\partial_x^3), \quad (1.6)$$

where the action of the Hamiltonian operands Θ_1 and Θ_2 on the operator B are defined by

$$\Theta_1 B = [\partial_x, B], \quad \Theta_2 B = \alpha B_y + [u + \partial_x^2, B]. \quad (1.7)$$

We recall that equations in $1+1$ which are integrable by the inverse spectral method, are formally bi-Hamiltonian systems. However, some of these equations are reductions of certain larger systems, and in the process of reduction local Hamiltonian operators give rise to nonlocal ones. Equations (1.4) and (1.6) indicate that equations in $2+1$ are formally quartic-Hamiltonian systems; the operands ∂_x, Θ_1 and Θ_2 are local, while the operand Θ is nonlocal because it arises in the process of a certain reduction (see §3).

The Lenard scheme for equations in $2+1$ yields hierarchies of commuting flows. However, in the general case, instead of the notion of a conserved functional, there is the notion of the conservation of a certain equivalence class (see §3). The question of obtaining conventional conservation laws in this formalism will be discussed in [41].

Integrable equations in $2+1$ are nonlocal. This means that one is now forced to work with a ring involving the basic dynamical variable $u(x,y,t)$, its x and y derivations, and primitives with respect to ∂_x. We assume that ∂_x^{-1} is skew symmetric; for example for u decaying sufficiently fast as $|x| \to \infty$, $\partial_x^{-1} = \frac{1}{2}(\int_{-\infty}^{x} - \int_{x}^{\infty})$. One can pass from the formal to the rigorous theory provided that the action of ∂_x^{-1} is well defined. For example for decaying functions it must be shown that ∂_x^{-1} acts always on decaying function. Establishing this fact can be quite complicated. For example if u is decaying, $\partial_x^{-1}u$ is well defined but it is not decaying any more, thus $\partial^{-2}u$ is not well defined unless $\int_{-\infty}^{\infty} u(x,y)dx = 0$.

Introducing the formal operator $\Psi = \partial_x^{-1}\Theta$, it follows that the KP equation (1.4) can be written as $u_t = \partial_x\Psi(u - 2\partial_y)$. It can be shown [24] that members of the KP hierarchy have similar form, namely,

$$u_t = \partial_x \sum_{j=0}^{n} c_{n_j}\Psi^{n-j}\partial_y^j, \quad (1.8)$$

where c_{n_j} are appropriate constants, chosen by the requirement that the RHS of equation (1.8) is a function and not an operator. Denoting by $\hat{\Psi}(x,y,y')$ the kernel of the operand $\Psi(x,y)$, it can be shown [21]-[23] that equation (1.8) reduces to

$$u_t = \partial_x(\hat{\Psi}^n 1)(x,y,y')\,|_{y'=y}\,. \tag{1.9}$$

We note that the representation (1.8) has the advantage of involving only the variables x and y, but it has the disadvantage of involving a sum of n terms. It has been recently shown [39] that the above representations are equivalent to the elegant representation

$$u_t = Res_{\partial_y}\partial_x\Psi^n\partial_y^{-1}. \tag{1.10}$$

The case $n=2$ corresponds to the KP. Indeed if Θ is defined by equation (1.5),

$$\Theta\partial_y^{-1} = 2u_x\partial_y^{-1} + (-u_{xy} + \alpha(\partial_x^{-1}u_{yy}))\partial_y^{-2} + O(\partial_y^{-3}),$$

and

$$Res_{\partial_y}\partial_x\Psi^2\partial_y^{-1} = Res_{\partial_y}\Theta\partial_x^{-1}\Theta\partial_y^{-1} = \text{ the RHS of KP.}$$

2 Basic Concepts

We begin with a graded Lie algebra $\mathcal{L}$ with elements of grades $-1, 0$, and 1, $\mathcal{L} = \mathcal{L}^{-1} \oplus \mathcal{L}^0 \oplus \mathcal{L}^1$. $\mathcal{L}^1$ is one dimensional with generator d. To describe the linear space $\mathcal{L}^{-1}$, it is convenient to use the auxiliary linear space V which is isomorphic to $\mathcal{L}^{-1}$, $i: V \to \mathcal{L}^{-1}$, if $v \epsilon V$ then $i_v \epsilon \mathcal{L}^{-1}$. We use the notations [,] and $[,]_+$ to denote a commutator and an anticommuter respectively. Since the anticommutator of i_{v_1} and i_{v_2} is of degree -2, it follows that $[i_{v_1}, i_{v_2}]_+ = 0$. Similarly, $[d,d]_+ = 0$, which we write as $d^2 = 0$. The anticommutor of i_v and d, which is of degree 0 is denoted by L_v, i.e.

$$L_v = [i_v, d]_+; \quad i_v \epsilon \mathcal{L}^{-1}, \quad d \epsilon \mathcal{L}^1, \quad L_v \epsilon \mathcal{L}^0. \tag{2.1}$$

We call L_v the Lie derivative in the direction v. Equation (2.1) defines a map from $\mathcal{L}^{-1}$ to $\mathcal{L}^0$. In what follows we assume that there exists an isomorphism between i_v and L_v. There exist important cases that $\mathcal{L}^1$ is not one dimensional, see for example [31], [42], and for such cases this assertion is not valid; however our aim here is to discuss the simplest possible case. The commutator of L_{v_1} and L_{v_2} is of degree 0 thus it equals L_{v_3}. Because V is isomorphic to $\mathcal{L}^0$, V inherits from $\mathcal{L}^0$ the structure of a Lie algebra, i.e. $v_3 = [v_1, v_2]$, where this commutator is defined via

$$[L_{v_1}, L_{v_2}] = L_{[v_1,v_2]}. \tag{2.2}$$

The elements of V equipped with the above commutator are called vector fields. The commutator of L_{v_1} and i_{v_2} is of degree -1, thus it equals i_{v_3}; actually

$$[L_{v_1}, i_{v_2}] = i_{[v_1,v_2]}. \tag{2.3}$$

Indeed, taking the anticommutator of the RHS of this equation with d it follows after some simple calculations that

$$[[L_{v_1}, i_{v_2}], d]_+ = [L_{v_1}, L_{v_2}] = L_{[v_1,v_2]},$$

thus $[i_{v_3}, d]_+ = L_{[v_1,v_2]}$, and hence $v_3 = [v_1, v_2]$.

We also consider a graded linear space Ω which is a left $\mathcal{L}$ module. Graded means that Ω is a direct sum of subspaces Ω^i of grade i, i.e. $\Omega = \oplus\Omega^i$. Left $\mathcal{L}$ module means that we have a left action of the Lie algebra $\mathcal{L}$ in such a way that this action is consistent with the grades of the elements of Ω, i.e.

$$i_v\Omega^i \subset \Omega^{i-1}, \quad d\Omega^i \subset \Omega^{i+1}; \quad i_v : \Omega^i \to \Omega^{i-1}, \quad d : \Omega^i \to \Omega^{i+1}. \tag{2.4}$$

If $\Omega^i = 0$ for $i < 0$, i.e. $i_v\Omega^0 = (0)$, then we call the above complex a deRham complex. Elements of Ω^0 are called functionals, and elements of Ω^q, $q \neq 0$, are called q forms. If $w\epsilon\Omega^q$, then $i_{v_q}i_{v_{q-1}}...i_{v_1}w\epsilon\Omega^0$ and is denoted by $w(v_1, v_2, ..., v_q)$. Since the i_v's anticommute, it follows that $w(v_1, ..., v_q)$ is skew symmetric.

Let $U(\mathcal{L})$ denote the enveloping algebra of $\mathcal{L}$. Then equation (2.1) implies that every element of the enveloping algebra is some polynomial of i_v and d. Polynomial identities involving i_v and d can easily be derived using equations (2.1) and (2.3). For example, equation (2.3) yields

$$i_{v_2}i_{v_1}d = di_{v_1}i_{v_2} + i_{v_1}di_{v_2} - i_{v_2}di_{v_1} - i_{[v_1,v_2]}. \tag{2.5}$$

Applying equation (2.5) to a 0-form w and using $i_v w = 0$ it follows that

$$(dw)(v) = L_v w, \quad w\epsilon\Omega^0. \tag{2.6}$$

Similarly, applying equation (2.5) to a 1-form w, using $di_{v_2}i_{v_1}w = 0$, and rewriting some of the terms of equation (2.5) in terms of L_{v_1}, and L_{v_2}, it follows that

$$(dw)(v_1, v_2) = L_{v_1}w(v_2) - L_{v_2}w(v_1) - w([v_1, v_2]), \quad w\epsilon\Omega^1. \tag{2.7}$$

This procedure yields formulae for $i_{v_q}i_{v_{q-1}}...i_{v_1}d$, and for $(dw)(v_1, v_2, ..., v_q)$, which coincide with the well known formulae of differential forms in classical mechanics.

We note that equation (2.5) provides an expression for $i_{[v_1,v_2]}$ in terms of combinations of i_{v_1}, i_{v_2}, and d. It is interesting that it is possible to build a Lie algebra structure starting from this equation and $[d, d]_+ = 0$, $[i_{v_1}, i_{v_2}]_d = 0$.

The equation $d^2 = 0$ can be used for the introduction of cohomologies: Let C^q be the subspace of Ω^q consisting of closed forms, i.e. forms w^q such that $dw^q = 0$; let E^q be the subspace of Ω^q consisting of exact forms, i.e. forms w^q such that $w^q = dw^{q-1}$. Clearly $E^q \subset C^q$; the quotient C^q/E^q is called the qth cohomology space.

It is possible to construct different modules. It would be interesting to give a systematic investigation of all different possibilities. However, our aim here is to clarify and unify the algebraic properties of integrable systems appearing in classical mechanics, in the so-called $1+1$ theory (i.e. evolution equations in our space variable), and the $2+1$ theory. So, in what follows we present the appropriate contructions for the $1+1$ and the $2+1$ theories. The construction associated with the finite dimensional case is straightforward.

2.1 The 1 + 1 Case

We first consider the case that the basic dynamic variable $u(x,t)$ is scalar. We do not specify the classes of functions to which u belongs. We only consider u as a symbol to which the operator ∂_x can be applied. We denote by $\mathcal{R}$ the ring of smooth functions of u and its x-derivatives; in most applications $\mathcal{R}$ is a polynomial ring over u, u_x, u_{xx}, etc. Let $r_i \epsilon \mathcal{R}$, $i=1,2$; we denote the Fréchet derivative of r_1 in the direction r_2 by

$$r_1' r_2 = \frac{\partial}{\partial \varepsilon} r_1(u+\varepsilon r_2)\,|_{\varepsilon=0}\,, \quad r_i \epsilon \mathcal{R}, \quad i=1,2. \tag{2.8a}$$

This equation implies,

$$r' = \frac{\partial r}{\partial u} + \frac{\partial r}{\partial u_x}\partial_x + \frac{\partial r}{\partial u_{xx}}\partial_x^2 + \ldots, \quad r' \epsilon \mathcal{R}\{\partial_x\}_+, \tag{2.8b}$$

where $\mathcal{R}\{\partial_x\}_+$ denotes the ring of differential operators with coefficients in $\mathcal{R}$, i.e. elements of this ring have the form $\sum_{j=0}^N r_j \partial_x^j$, $r_j \epsilon \mathcal{R}$.

The Lie derivative of v_2 along v_1 is defined by $L_{v_1} v_2 = [v_1, v_2]$, where

$$L_{v_1} v_2 = v_2' v_1 - v_1' v_2, \quad v_1, v_2 \epsilon \mathcal{R}.$$

The space V, i.e. the space of vector fields, consists of elements of $\mathcal{R}$ equipped with this commutator.

This important commutator, in a slightly different form, was first introduced in [5]. The authors of [5] considered vector fields as expressions of the form $v\frac{\partial}{\partial u} + v_x \frac{\partial}{\partial u_x} + \cdots$, equipped with the commutator

$$\left[v_1 \frac{\partial}{\partial u} + v_{1_x}\frac{\partial}{\partial u_x} + \cdots, v_2\frac{\partial}{\partial u} + v_{2_x}\frac{\partial}{\partial u_x} + \cdots\right] = v_{12}\frac{\partial}{\partial u} + v_{12_x}\frac{\partial}{\partial u_x} + \cdots$$

where $v_{12} = v_1 \frac{\partial v_2}{\partial u} + v_{1_x}\frac{\partial v_2}{\partial u_x} + \cdots - (1 \leftrightarrow 2)$. However, v_{12} is nothing but $v_2' v_1 - v_1' v_2$. For example, to the vector field $\frac{\partial}{\partial x}$ of [5] there corresponds the vector field u_x in the present theory; this vector field commutes with any other v.

To construct the Ω^0, the space of functionals, we introduce in $\mathcal{R}$ a relation of equivalence: $f_1 \sim f_2$ iff $f_1 - f_2$ is an "exact derivative", i.e. $f_1 - f_2 = g_x$ where g_x is the total derivation of g w.r.t. x, i.e. $g_x = g_u u_x + g_{u_x} u_{xx} + \ldots = g' u_x$. An equivalence class is called a functional; the set of all functionals belongs to the coset $\mathcal{R}/\partial\mathcal{R}$, where $\partial\mathcal{R}$ consists of all elements of $\mathcal{R}$ which are exact derivatives. The mapping assigning to each $f \epsilon \mathcal{R}$, its equivalence class $\tilde{f} \epsilon \mathcal{R}/\partial\mathcal{R}$, is called a formal integral and is denoted by $\tilde{f} = \int f dx$. It is easy to prove that $\int f_x g dx = -\int f g_x dx$, which we call integration by parts. If one passes from the formal to the rigorous theory (where the classes of functions to which u belongs are specified), all the relations with integrals will be preserved if these classes are such that $\int f_x dx = 0$ (e.g. functions decreasing sufficiently fast at infinity, periodic in x functions, etc).

We now define Ω^1, the space of 1-forms: Let $\xi \epsilon \mathcal{R}$, then $\xi(v) \epsilon \Omega^0$, where $\xi(v)$ is defined by

$$i_v \xi = \xi(v) = \int \xi v dx, \quad \xi, v \epsilon \mathcal{R}. \tag{2.10a}$$

Equation (2.10a) will also be denoted as

$$\xi(v) = \langle \xi, v \rangle. \tag{2.10b}$$

We define the Lie derivative L_v on the space Ω^0 of functionals $\tilde{f}$, by the formula

$$L_v \tilde{f} = \int f' v dx, \quad f, v \epsilon \mathcal{R}. \tag{2.11}$$

To prove that this action is well defined, and that it induces the structure of a left $\mathcal{L}$-module on Ω^0, we need to prove

$$L_v \tilde{f}_x = \partial_x (L_v \tilde{f}) \quad \text{and} \quad [L_{v_1}, L_{v_2}] \tilde{f} = L_{[v_1, v_2]} \tilde{f} \tag{2.12}$$

Equation (2.12a) can be written as $L_v L_{u_x} \tilde{f} = L_{u_x} L_v \tilde{f}$ and hence, since $[u_x, v] = 0$, it is a consequence of (2.12b). To prove (2.12b) we use the symmetry of the second Fréchet derivative, i.e. $(\tilde{g}'' v_1) v_2 = (\tilde{g}'' v_2) v_1$, where $r_1'' r_2$ is defined by

$$r_1'' r_2 = \left(\frac{\partial r_1}{\partial u} \right)' r_2 + \left(\frac{\partial r_1}{\partial u_x} \right)' r_2 \partial_x + \left(\frac{\partial r_1}{\partial u_{xx}} \right)' r_2 \partial_x^2 + \ldots . \tag{2.13}$$

Indeed,

$$[L_{v_1}, L_{v_2}] \tilde{f} = (\tilde{f}' v_2)' v_1 - (\tilde{f}' v_1)' v_2 = (\tilde{f}'' v_1) v_2 + \tilde{f}' v_2' v_1 - (\tilde{f}'' v_2) v_1 - \tilde{f}' v_1' v_2 =$$

$$= \tilde{f}' (v_2' v_1 - v_1' v_2) = \tilde{f}' [v_1, v_2] = L_{[v_1, v_2]} \tilde{f}.$$

Equation (2.11) together with $L_v = d i_v + i_v d$ can be used to define $d\tilde{f}$:

$$(d\tilde{f})(v) = L_v \tilde{f} = \int f' v dx = \int \frac{\delta f}{\delta u} v dx, \tag{2.14}$$

where $\frac{\delta}{\delta u}$ denotes the usual variation derivative, i.e. $\frac{\delta f}{\delta u} = \frac{\partial f}{\partial u} - \partial_x \frac{\partial f}{\partial u} + \partial_x^2 \frac{\partial f}{\partial u_{xx}} + \ldots$, and we have used integration by parts to transform $f' v$ to $\frac{\delta f}{\delta u} v$. Comparing equations (2.10a) and (2.14) shows that $d\tilde{f}$ is a 1-form, where $\xi = \frac{\delta f}{\delta u}$, i.e.

$$d\tilde{f} = \frac{\delta f}{\delta u}. \tag{2.15}$$

Having defined the Lie derivative of functionals, it is possible to define the Lie derivative of 1-forms. To define $L_v \xi$, the Lie derivative of a 1-form ξ, we use the equation

$$\langle L_v \xi, v_1 \rangle = L_v \langle \xi, v_1 \rangle - \langle \xi, L_v v_1 \rangle.$$

But since $\langle \xi, v_1 \rangle$ is a 0-form, $L_v\langle \xi, v_1\rangle = \langle \xi, v_1\rangle' v$; also $L_v v_1 = v_1' v - v' v_1$, thus

$$\langle L_v \xi, v_1 \rangle = \langle \xi' v, v_1 \rangle + \langle \xi, v' v_1 \rangle, \quad \xi \epsilon \Omega^1, \quad v \epsilon V, \tag{2.16a}$$

or

$$L_v \xi = \xi' v + (v')^* \xi, \tag{2.16b}$$

where $(v')^*$ denotes the adjoint of the operator v' with respect to $<,>$, i.e. $(v')^*$ is defined by the equation $\langle (v')^* \xi, v_1 \rangle = \langle \xi, v' v_1 \rangle$ (using equation (2.8b) and integration by parts, $(v')^*$ can be computed explicitly).

We now indicate how the construction of the entire complex can be achieved. To construct Ω^2, the space of 2-forms, we define $w \epsilon \Omega^2$ by

$$w(v_1, v_2) = \langle l v_1, v_2 \rangle - \langle l v_2, v_1 \rangle, w \epsilon \Omega^2, v_i \epsilon V, l \epsilon \mathcal{R}\{\partial_x\}_+. \tag{2.17}$$

We note that w is bilinear and skew-symmetric. Furthermore, Ω^2 contains the differentials of 1-forms. Indeed, using equation (2.7) we define $d\xi$ by

$$(d\xi)(v_1, v_2) = L_{v_1}\langle \xi, v_2 \rangle - L_{v_2}\langle \xi, v_1 \rangle - \langle \xi, L_{v_1} v_2 \rangle,$$

thus

$$(df)(v_1, v_2) = \langle \xi, v_2 \rangle' v_1 - \langle \xi, v_1 \rangle' v_2 - \langle \xi, v_2' v_1 - v_1' v_2 \rangle,$$

and

$$(d\xi)(v_1, v_2) = \langle \xi' v_1, v_2 \rangle - \langle \xi' v_2, v_1 \rangle, \xi \epsilon \Omega^1, v_i \epsilon V. \tag{2.18}$$

Hence $d\xi$ is a 2-form, with $l = \xi'$. We note that if $\xi = d\tilde{f}$, then $\xi = \frac{\delta f}{\delta u}$, thus ξ' is symmetric, and equation (2.18) implies $d\xi = 0$.

To construct Ω^3, the space of 3-forms, we define $w \epsilon \Omega^3$ by

$$\begin{aligned} w(v_1, v_2, v_3) = \sum_{j_1=0, j_2=0} & \left(\langle \alpha_{j, j_2} \left(\partial_x^{j_1} v_1\right) \left(\partial_x^{j_2} v_2\right), v_3 \rangle \right. \\ & \left. - \langle \alpha_{j_1 j_2} \left(\partial_x^{j_1} v_1\right) \left(\partial_x^{j_2} v_3\right), v_2 \rangle + c.p. \right), \\ & \alpha_{j_1 j_2} \epsilon \mathcal{R}, \quad v_i \epsilon V, \end{aligned} \tag{2.19}$$

where $c.p.$ denotes cyclic permutation w.r.t v_1, v_2, v_3. Defining dw by

$$(dw)(v_1, v_2, v_3) = L_{v_1} w(v_2, v_3) + w(v_1, [v_2, v_3]) + c.p., \quad w \epsilon \Omega^2, \quad v_i \epsilon V, \tag{2.20}$$

it follows that Ω^3 contains the differentials of Ω^2, where $\sum \alpha_{j_1 j_2} (\partial_x^{j_1} v_1)(\partial_x^{j_2} v_2) = (l' v_1) v_2$. We note that if $w = d\xi$, $\xi \epsilon \Omega^1$, then $l' = \xi''$, thus $(\xi'' v_1) v_2 = (\xi'' v_2) v_1$ and $dw = 0$.

In the general case, $w(v_1, v_2, ..., v_q)$ is constructed from skew symmetric combinations of $\sum \langle \alpha_{j_1 j_2 \ldots j_{q-1}} \, (\partial_x^{j_1} v_1) \ldots (\partial_x^{j_{q-1}} v_{q-1}), v_q \rangle$, where $\alpha_{j_1 j_2 \ldots j_{q-1}} \epsilon \mathcal{R}$.

Let θ be a linear mapping from Ω^1 to V. This mapping is called a Hamiltonian mapping, if it is skew symmetric with respect to $<,>$, i.e. if

$$\langle \theta\xi_1, \xi_2\rangle = -\langle \xi_1, \theta\xi_2\rangle, \quad \xi_i \epsilon \Omega^1, \tag{2.21a}$$

and if the Shouten bracket of θ with itself is zero, where the Shouten bracket is defined by

$$[\theta,\theta] : \Omega^1 \times \Omega^1 \times \Omega^1 \to \Omega^0, [\theta,\theta](\xi_1,\xi_2,\xi_3) = \langle \theta L_{\theta\xi_1}\xi_2, \xi_3\rangle + c.p.. \tag{2.21b}$$

We introduce in the space Ω^0 the bilinear operation

$$\{\tilde{f}_1, \tilde{f}_2\}_\theta = \langle \theta d\tilde{f}_1, d\tilde{f}_2\rangle, \tilde{f}_i \epsilon \Omega^0. \tag{2.22}$$

This operation is called the Poisson bracket associated with θ. It is easy to show that if θ is Hamiltonian, then there exists a homomorphism between the Poisson brackets and the Lie-algebra $\mathcal{L}$,

$$-\theta d\{\tilde{f}_1, \tilde{f}_2\}_\theta = [\theta d\tilde{f}_1, \theta d\tilde{f}_2]. \tag{2.23}$$

An element $v \epsilon V$ that has the representation $v = \theta d\tilde{f}$ will be called a Hamiltonian vector field with the Hamiltonian $\tilde{f} \epsilon \Omega^0$. An evolution equation for the dynamical variable $u(x,t)$ will be called a Hamiltonian equation if it can be written in the form

$$u_t = \theta d\tilde{f}, \quad \tilde{f} \epsilon \Omega^0. \tag{2.24}$$

Using equation (2.16a), it is straightforward to find the coordinate form of equation (2.21b):

$$-\langle \theta L_{\theta\xi_1}\xi_2, \xi_3\rangle = \langle L_{\theta\xi_1}\xi_2, \theta\xi_3\rangle = \langle \xi_2'\theta\xi_1, \theta\xi_3\rangle + \langle \xi_2, (\theta\xi_1)'\theta\xi_3\rangle =$$

$$= \langle \xi_2'\theta\xi_1, \theta\xi_3\rangle + \langle \xi_2, (\theta'\theta\xi_3)\xi_1\rangle + \langle \xi_2, \theta\xi_1'\theta\xi_3\rangle$$

$$= \langle \xi_2, (\theta'\theta\xi_3)\xi_1\rangle + \langle \xi_2'\theta\xi_1, \theta\xi_3\rangle - \langle \theta\xi_2, \xi_1'\theta\xi_3\rangle.$$

Thus

$$[\theta,\theta](\xi_1,\xi_2,\xi_3) = \langle (\theta'\theta\xi_1)\xi_2, \xi_2\rangle + c.p. \tag{2.25}$$

In the finite dimensional case this equation reduces to

$$[\theta,\theta]^{ijk} = \sum_l \frac{\partial \theta^{ij}}{\partial x^l}\theta^{lk} + c.p.,$$

which is the definition given by J. Shouten for the bracket of the bivector field θ^{ij} with itself.

The Nijenhuis torsion was introduced by A. Nijenhuis in 1951 as follows: Let $A = A_i^j$ be a (1,1) tensor field; associate with A its torsion $S_A = (S_A)_{ij}^k$ (which is a (1,2) tensor field) as follows

$$(S_A)_{ij}^k = \sum_l A_i^l \frac{\partial A_j^k}{\partial x^l} + A_l^k \frac{\partial A_i^l}{\partial x^j} - (i \leftrightarrow j).$$

The invariant definition of the Nijenhuis torsion is [17],

$$S_A(v_1,v_2) = [Av_1, Av_2] - A[v_1, Av_2] - A[Av_1, v_2] + A^2[v_1,v_2], v_i \epsilon V. \qquad (2.26)$$

Indeed, using $[Av_1, Av_2] = (Av_2)'(Av_1) - (1 \leftrightarrow 2) = A'(Av_1)v_2 + Av_2'(Av_1) - (1 \leftrightarrow 2)$, etc., it follows that

$$S_A(v_1,v_2) = A'(Av_1)v_2 + AA'(v_2)v_1 - (1 \leftrightarrow 2), \quad v_i \epsilon V. \qquad (2.27)$$

Sometimes one finds in the literature the association of the Hamiltonian mapping θ with the 2-form w defined by $w(\theta\xi_1, \xi_2) = \langle \theta\xi_1, \xi_2 \rangle$, $\xi_i \epsilon \Omega^1$. This equation is consistant with equation (2.17) if $l = \theta^{-1}$; however the operator θ is usually non invertible. If one can associate θ with a 2-form then θ is Hamiltonian iff w is closed, which then implies equation (2.23).

Example 1 The KdV equation can be written in the following forms

$$u_t = \partial_x d \int (-\frac{u_x^2}{2} + u^3)dx = (\partial_x^3 + 2u\partial_x + 2\partial_x u) d \int \frac{u^2}{2} dx. \qquad (2.28)$$

The operators $\theta_1 = \partial_x$ and $\theta_2 = \partial_x^3 + 2u\partial_x + 2\partial_x u$ are Hamiltonian operators (see §3), thus KdV is a bi-Hamiltonian system. The associated Poisson brackets are

$$\{\tilde{f}_1, \tilde{f}_2\}_\theta = \int (\theta d\tilde{f}_1) d\tilde{f}_2 dx = \int \left(\theta \frac{\delta f_1}{\delta u}\right) \frac{\delta f_2}{\delta u} dx.$$

The Poisson bracket with $\theta = \theta_1$ is the Gardner-Zakharov-Faddeev bracket, and that with $\theta = \theta_2$ is the Magri bracket. The Hamiltonians $\tilde{f}_1 = \int(-\frac{u_x^2}{2} + u^3)dx$ and $\tilde{f}_2 = \int \frac{u^2}{2} dx$ are constants of motion of the KdV. Actually KdV admits an infinite hierarchy of constants of motion, $I_1 = \tilde{f}_2$, $I_2 = \tilde{f}_1$, $I_3 = \int(\frac{u_{xx}^2}{2} + \frac{5}{2}u^2 u_{xx} + \frac{10}{4}u^4)dx$. The first two conserved quantities have a simple geometrical origin: We use the Hamiltonian operator ∂_x to construct the Hamiltonian vector fields v_1, v_2, v_3 corresponding to I_1, I_2, I_3: $v_1 = u_x$, $v_2 = u_{xxx} + 6uu_x$, $v_3 = u_{xxxxx} + 10uu_{xxx} + 20u_x u_{xx} + 30u^2 u_x$. These vector fields commute with the RHS of the KdV and thus they are symmetries of the KdV. The symmetries v_1 and v_2 are a consequence of the fact that the KdV is invariant under x and t translations. Indeed to the translation $x \to x + \alpha$, there corresponds the invariant vector field ∂_x, which is equivalent to $u_x \partial_u$, thus u_x is a symmetry of KdV. Similarly $t \to t + \beta$ implies ∂_t, which is equivalent to $u_t \partial_u$, thus $u_{xxx} + uu_x$ is a symmetry of KdV. The symmetry v_3 does not have an obvious geometrical origin.

The extension of the above formalism to the case that the basic dynamical variables are the set of scalars $u_1, ..., u_n$ is straightforward. Also the same formalism is valid if u is a matrix valued function provided that $\int$ is replaced by $\int trace$ and equation (2.10a) is replaced by $\int trace \xi^T v dx$ ($trace$ and T denote the usual trace and transpose operations as defined for matrices).

2.2 The 2 + 1 Case

We consider a ring $\mathcal{R}$ consisting of smooth functions of the basic dynamical variables $u(x,y,t)$, of its x and y derivatives, and of primitives (integrals) with respect to ∂_x of smooth functions of u and of $\frac{\partial_u^{\alpha+\beta}}{\partial_x^\alpha \partial_y^\beta}$. For example the expression $\partial_x^{-1}(u_x u_{yy})$ belongs to this ring. We assume that the operator ∂_x^{-1} is skew symmetric, for example for dynamical variables decaying sufficiently fast as $|x| \to \infty$, $\partial_x^{-1} = \frac{1}{2}(\int_{-\infty}^x - \int_x^\infty)$. The consideration of this ring rather than of a differential one is a necessary consequence of the fact that bi-Hamiltonian equations in $2+1$ are nonlocal equations. With the ring $\mathcal{R}$ we consider the ring $\mathcal{R}\{\partial_y\}$ of formal pseudodifferential operators in ∂_y with coefficients in the ring $\mathcal{R}$. The multiplication law in $\mathcal{R}\{\partial_y\}$ is uniquely defined by $\partial_y \cdot a = a\partial_y + (\partial_y a)$, $\partial_y^{-1} \cdot a = a\partial_y^{-1} - (\partial_y a)\partial_y^{-2} + (\partial_y^2 a)\partial_y^{-3} - \dots$, $a \epsilon \mathcal{R}$. Under this multiplication, $\mathcal{R}\{\partial_y\}$ becomes an associative ring. (Below we shall usually omit the $\cdot$.)

Vector fields, denoted by v, are elements of $\mathcal{R}\{\partial_y\}$ which are equipped with the commutator

$$L_{v_1} v_2 = v_2' v_1 - v_1' v_2; \quad v_1, v_2 \epsilon \mathcal{R}\{\partial_y\}. \tag{2.29}$$

To construct the Ω^0 we introduce in $\mathcal{R}\{\partial_y\}$ a relation of equivalence: $f_1 \sim f_2$ iff f_1 can be obtained from f_2 by cyclic permutation of basic variables in any term of it, or by overthrowing the x and y derivatives in a conventional way. This concept is related to cyclic cohomology. For example $u_{xx}u\partial_y^{-1} \sim \partial_y^{-1}u_{xx}u \sim u\partial_y^{-1}u_{xx} \sim -u_x\partial_y^{-1}u_x$. We denote the class of equivalence of f by $\uparrow f \downarrow$ and $\uparrow f \downarrow \epsilon \Omega^0$. By construction,

$$\uparrow f_x \downarrow = 0, \quad \uparrow f_y \downarrow = 0, \quad \uparrow f_1 f_2 \downarrow = \uparrow f_2 f_1 \downarrow, \quad f_i \epsilon \mathcal{R}\{\partial_y\}. \tag{2.30}$$

Having defined Ω^0, the rest of the construction is similar to the one presented in §2.1. For example

$$L_v \uparrow f \downarrow = \uparrow f'v \downarrow = \uparrow \frac{\delta f}{\delta u} v \downarrow; v, f, \frac{\delta f}{\delta u} \epsilon \mathcal{R}\{\partial_y\}, \tag{2.31}$$

where for the last equality we have used the operations of equivalence to express $\uparrow f'v \downarrow$ as a pseudo differential operator acting on v. For example if $\uparrow f \downarrow = \uparrow u\partial_y^{-1}u_{xx} \downarrow$, then $\uparrow f'v \downarrow = \uparrow v\partial_y^{-1}u_{xx} + u\partial_y^{-1}v_{xx} \downarrow = \uparrow (\partial_y^{-1}u_{xx} + u_{xx}\partial_y^{-1})v \downarrow$ and $\frac{\delta f}{\delta u}$ is the pseudodifferential operator $\partial_y^{-1}u_{xx} + u_{xx}\partial_y^{-1} = 2u_{xx}\partial_y^{-1} - u_{xxy}\partial_y^{-2} + \cdots$.

Similarly we define Ω^1 the space of 1-forms whose elements, denoted by ξ, are linear mappings from $\mathcal{V}$ to Ω^0 that can be represented in the form

$$\xi(v) = \uparrow \xi v \downarrow, \quad \xi, v \epsilon \mathcal{R}\{\partial_y\}. \tag{2.32}$$

As before Ω^1 contains the differentials of all the elements of Ω^0, where $d \uparrow f \downarrow$ is defined by

$$(d \uparrow f \downarrow)(v) = L_v \uparrow f \downarrow = \uparrow \frac{\delta f}{\delta u} v \downarrow. \tag{2.33}$$

3 Derivation of Hamiltonian Operators

In this section we review two methods for generating bi-Hamiltonian structures. Both methods start with a Lax operator L (this operator defines the time-independent part of a Lax pair). One method uses square eigenfunctions; the other method is the so called Adler-Gel'fand-Dikii (AGD) scheme. Both methods can be used for equations in $1+1$ as well as for equations in $2+1$.

3.1 The Squared Eigenfunction Method

3.1.1 The 1 + 1 Case Given the Lax operator L, one first rewrites the eigenvalue equation $Lv = 0$ in the form $\psi_x = w\psi$ and then one considers its "adjoint" equation,

$$\psi_x(x,t,\lambda) = w(x,t,\lambda)\psi(x,t,\lambda), \quad \hat{\psi}_x(x,t,\lambda) = -\hat{\psi}(x,t,\lambda)w(x,t,\lambda). \tag{3.1}$$

Multiplying equation (3.1a) from the right by $\hat{\psi}$, and equation (3.1b) from the left by ψ, it follows that the "square eigenfunction" equation is

$$\varphi_x = [w, \varphi], \quad \varphi = \psi\hat{\psi}. \tag{3.2}$$

This equation suggests that $\partial_x - [w, \cdot]$ can be used to obtain Hamiltonian structures.

Example 1 Consider the generalized Zakharov-Shabat operator

$$L = \partial_x - u(x,t) - \lambda a, \tag{3.3}$$

where $a = diag(a_1, ..., a_N)$, $a_i \neq a_j$, $i \neq j$, is considered given, the dynamic variable u is a full $N \times N$ matrix, and λ is the spectral parameter. In the usual Zakharov-Shabat case, u is off-diagonal, but here we take u full in order to avoid nonlocal operators arising in the process of reduction. Letting $w = u + \lambda a$ in equation (3.2) we find

$$(\varphi_x - [u, \varphi]) - \lambda[a, \varphi] = 0. \tag{3.4}$$

This suggests that

$$\theta_1 = [a, \cdot] \text{ and } \theta_2 = \partial_x - [u, \cdot] \tag{3.5}$$

are Hamiltonian operators. It can be easily verified that this is indeed the case.

Example 2 Consider the Schrödinger operator

$$L = \partial_x^2 + u(x,t) + \lambda. \tag{3.6}$$

Writing the equation $Lv = 0$ in matrix form we find

$$w = \begin{pmatrix} 0 & 1 \\ -u - \lambda & 0 \end{pmatrix}. \tag{3.7}$$

Letting $\varphi_{11} = a$, $\varphi_{12} = 2b$, $\varphi_{21} = c$, and $\varphi_{22} = d$, equation (3.2) yields

$$a_x = c - 2b\hat{u}, 2b_x = d - a, c_x = \hat{u}(a-d), d_x = 2\hat{u}b - c; \hat{u} = -(u+\lambda). \quad (3.8)$$

Equations (3.8a) and (3.8d) imply $d_x = -a_x$, or $d = -a + 2\gamma$, then equations (3.8b) and (3.8a) imply $a = \gamma - b_x$, $d = \gamma + b_x$, $c = 2b\hat{u} - b_{xx}$. Using the expression in equation (3.8c) we find

$$(\partial_x^3 + 2u\partial + 2\partial u)b + 4\lambda\partial_x b = 0, \quad (3.9)$$

which implies the two well known Hamiltonian operators associated with the KdV hierarchy.

3.1.2 The 2 + 1 Case Equation $Lv = 0$ can again be written in the form $\Psi_x = W\Psi$, but now W is **not** a function but an operator in ∂_y. We can still define the adjoint of this equation provided that we interpret Ψ as a pseudodifferential operator in ∂_y. Then the analogue of equation (3.2) is the **operator** equation

$$\Phi_x = [W, \Phi], \quad \Phi = \Psi\hat{\Psi}. \quad (3.10)$$

Example 3 The Lax operator associated with the KP can be taken in the form

$$L = \partial_x^2 + U(x,y,t), \quad U(x,y,t) = u(x,y) + \alpha\partial_y + \lambda, \quad (3.11)$$

where $\alpha = i$ and $\alpha = -1$ correspond to KPI and KPII respectively. Thus

$$W = \begin{pmatrix} 0 & 1 \\ -u - \alpha\partial_y - \lambda & 0 \end{pmatrix}. \quad (3.12)$$

Letting $\Phi_{11} = A$, $\Phi_{12} = 2B$, $\Phi_{21} = C$, $\Phi_{22} = D$ and using equation (3.10) we find

$$B_{xxx} + [u_x, B]_+ + 2[u + \alpha\partial_y, B_x]_+ + \left[u + \alpha\partial_y, (\partial_x^{-1}[u + \alpha\partial_y, B])\right] + 4\lambda B_x = 0, \quad (3.13)$$

where $[,]_+$ denotes anticommutator. We note that $\Phi \epsilon \mathcal{R}_\lambda\{\partial_y\}$, where $\mathcal{R}_\lambda$ denotes that the ring $\mathcal{R}$ is enlarged to include a smooth dependence on λ. Thus the operator

$$\Theta B = B_{xxx} + [u_x, B]_+ + 2[u + \alpha\partial_y, B_x]_+ + [u + \alpha\partial_y, (\partial_x^{-1}[u + \alpha\partial_y, B])],$$

acts in general on elements of $\mathcal{R}\{\partial_y\}$. If $\alpha = 0$, and if Θ acts on $\mathcal{R}$ instead of on $\mathcal{R}\{\partial_y\}$, one recovers the second Hamiltonian operator of the KdV. The appearance of the operator ∂_x^{-1} in Θ is the consequence of a certain reduction (see below).

Example 4 Using a simple exponential transformation, the operator L defined in equation (3.11) can be taken in the form $\partial_x^2 + u(x,y) + \alpha\partial_y + \lambda\partial_x$, thus

$$L = \alpha\partial_y + U(x,y,t), \;\; U(x,y,t) = u(x,y) + \partial_x^2 + \lambda\partial_x. \tag{3.14}$$

This equation implies

$$-\alpha\Psi_y = U\Psi, \;\; \alpha\hat{\Psi}_y = \hat{\Psi}U,$$

thus

$$-\alpha\Phi_y = [U, \Phi], \tag{3.15}$$

where $\Phi\epsilon\mathcal{R}_\lambda\{\partial_x\}$. This equation suggest the Hamiltonian operands defined by

$$\Theta_1 B = [\partial_x, B] \text{ and } \Theta_2 B = \alpha B_y + [u + \partial_x^2, B]. \tag{3.16}$$

3.2 The AGD Scheme

Let R be a given, possibly noncommutative, ring with a fixed derivation $\partial_0 : R \to R$. Let $R\{\partial_0\}$ denote the space of formal pseudodifferential operators with respect to ∂_0, i.e. elements of the form $\tilde{A} = \sum_{-\infty}^{N} A_i\partial_0^i$, $A_i\epsilon R$. For such an element we define $\tilde{A}_+ = \sum_0^N A_i\partial_0^i$, $\tilde{A}_- = \sum_{-\infty}^{-1} A_i\partial_0^i$, and use the usual multiplication rule with respect to ∂_0. Let $\varphi : R \to Q$, be a linear mapping from R to some linear space Q, such that

$$\varphi(AB) = \varphi(B\dot{A}), \;\; \varphi(\partial_0 A) = 0, \;\; A, B\epsilon R. \tag{3.17}$$

Denote the coefficient of ∂_0^{-1} in $\tilde{A}$ by $res_{\partial_0}\tilde{A}$, then it can be shown [11], [12] that

$$\varphi(res_{\partial_0}\tilde{A}\tilde{B}) = \varphi(res_{\partial_0}\tilde{B}\tilde{A}), \;\; \varphi(res_{\partial_0}\tilde{A}_-\tilde{B}\tilde{C}_+) + c.p. = \varphi(res_{\partial_0}\tilde{A}\tilde{B}\tilde{C}), \tag{3.18}$$

where $c.p.$ denote cyclic permutation w.r.t $\tilde{A}$, $\tilde{B}$, $\tilde{C}$.

3.2.1 The 1+1 Case Let $\mathcal{R}$, V, and Ω^1 be the ring, the space of vector fields and the space of 1-forms respectively introduced in 2.2.1 in connection with the $1+1$ theory. Given $\xi\epsilon\Omega^1$, define $X_\xi\epsilon\mathcal{R}\{\partial_x\}$ by

$$X_\xi = \sum_{i=0}^{n-1} \partial_x^{-i-1} \cdot \xi_i, \tag{3.19}$$

where $\xi_0, \cdots, \xi_{n-1}$ are the components of the vector ξ. Consider the fixed element $L\epsilon\mathcal{R}\{\partial_x\}$ given by

$$L = \partial_x^n + \sum_{i=0}^{n-1} u_i\partial_x^i, \tag{3.20}$$

where $u_0, \cdots, u_{n-1}$ are the basic dynamic variables. Then define the map

$$\mathcal{F} = (LX_\xi)_+L - L(X_\xi L)_+. \tag{3.21}$$

Note that $\mathcal{F}\epsilon\mathcal{R}\{\partial_x\}_+$, i.e. it belongs to the space of all differential with respect to ∂_x operators. Since $\mathcal{F}$ can also be written as $\mathcal{F} = (LX_\xi L)_- - (LX_\xi)_- L$, it follows that $\mathcal{F}$ is of order not greater than $n-1$ with respect to ∂_x. Therefore there exist a vector field v with comparts $v_0, \cdots, v_{n-1} \epsilon V$ such that

$$\mathcal{F} = \sum_0^{n-1} v_i \partial_x^i.$$

For the elements $v \epsilon V$ and $\xi \epsilon \Omega^1$ we have

$$(v, \xi) = \int trace \;\; v_i \xi_i dx = \int trace \;\; res_{\partial_x}(\mathcal{F} X_\xi) dx. \tag{3.22}$$

It was suggested by Adler [10] that the operator θ that associates with any $\xi \epsilon \Omega^1$, a $v \epsilon V$, according to the above procedure is a Hamiltonian operation. This was proved by Gel'fand and Dikii [11].

Example 5 Consider the operator L studied in example 1 (see equation (3.3)). In this case $X = \partial_x^{-1} \cdot \xi$. Then $(XL)_+ = (LX)_+ = \xi$. Thus

$$L(XL)_+ - (LX)_+ L = (\partial_x \xi - [u, \xi]) - \lambda[a, \xi].$$

Therefore the Lax operator L implies the two Hamiltonian operators (3.5).

3.2.2 The 2 + 1 Case Let $\mathcal{R}$, V and Ω^1 be the ring, the space of vector fields and the space of 1-forms respectively introduced in 2.2.2. Given $\xi \epsilon \Omega^1$, define X_ξ by

$$X_\xi = \sum_{i=0}^{n-1} \partial_x^{-i-1} \cdot \xi_i, \tag{3.23}$$

where $\xi_0, \cdots, \xi_{n-1}$ are the components of the vector ξ. Recall that $\xi \epsilon \mathcal{R}\{\partial_y\}$, thus $X_\xi \epsilon (\mathcal{R}\{\partial_y\})\{\partial_x\}$, i.e. it is a pseudodifferential operator in ∂_x whose coefficients are pseudodifferential operators in ∂_y. Consider the fixed element $L \epsilon (\mathcal{R}\{\partial_y\})\{\partial_x\}$ given by

$$L = \partial_x^n + \sum_{i=0}^{n-1} U_i \partial_x^i, \tag{3.24}$$

where $U_i \epsilon \mathcal{R}\{\partial_y\}$ are such elements that $U_i' v = v_i$, $v_0, \cdots, v_{n-1} \epsilon V$. For example $U_i = u_i + c$, where c denotes some constant pseudodifferential operator with respect to ∂_y. Define as before

$$\mathcal{F} = (LX_\xi)_+ L - L(X_\xi L)_+. \tag{3.25}$$

Note the $\mathcal{F} \epsilon (\mathcal{R}\{\partial_y\}_+)\{\partial_x\}$, and is of order not greater than $n-1$ with respect to ∂_x. Therefore, there exist a vector field v with components $v_0, \cdots, v_{n-1}$ such that $\mathcal{F} = \sum_{i=0}^{n-1} v_i \partial_x^i$. For the elements, $v \epsilon V$ and $\xi \epsilon \Omega^1$ we have

$$(v, \xi) = \uparrow \sum_{i=0}^{n-1} v_i \xi_i \downarrow = \uparrow res_{\partial_x}(\mathcal{F} X_\xi) \downarrow . \tag{3.26}$$

It was proved in [25] that the operator Θ that associates with any $\xi\epsilon\Omega^1$, a $v\epsilon V$, according to the above procedure is a Hamiltonian operand.

Example 6 Consider a generalization of the Lax operator associated with the KP equation,

$$L = \partial_x^2 + U_1\partial_x + U_0. \tag{3.27}$$

After some tedious calculations, equation (3.25) with $X_\xi = \partial_x^{-1} \cdot \xi_0 + \partial_x^{-2} \cdot \xi_1$, yields

$$v_0 = \xi_{0xxx} + [U_1, \xi_{0xx}] + (\xi_0 U_0)_x + U_0\xi_{0x} - 2\xi_{0x}U_{1x} - \xi_0 U_{1xx} - U_1\xi_0 U_{1x}$$

$$+U_1\xi_0 U_0 - U_0\xi_0 U_1 - U_1\xi_{0x}U_1 - \xi_{1xx} - U_1\xi_{1x} - [U_0, \xi_1], \tag{3.28}$$

$$v_1 = \xi_{0xx} - (\xi_0 U_1)_x - [U_0, \xi_0] - 2\xi_{1x} - [U_1, \xi_1].$$

The Lax operator of the KP given by equation (3.11) corresponds to the reduction $U_1 = 0$. This implies $\xi_1 = \frac{1}{2}(\xi_{0_x} - \partial_x^{-1}[U_0, \xi_0])$. Using this expression in (3.28) and letting $U_0 = u + \alpha\partial_y + \lambda$ we find

$$2v_0 = \xi_{0_{xxx}} + [u_x, \xi_0]_+ + 2[u + \alpha\partial_y, \xi_{0_x}]_+ + \Big[u + \alpha\partial_y, (\partial_x^{-1}[u + \alpha\partial_y, \xi_0])\Big] + 4\lambda\xi_{0_x}.$$

Example 7 Considering the operator L studied in example 4 (see equation (3.14)) and associating with it $X = \partial_y^{-1} \cdot \xi$, we find

$$v = \alpha\xi_y + [u + \alpha\partial_x^2, \xi] + \lambda[\partial_x, \xi],$$

which yields the two Hamiltonian operators defined in equation (3.16).

4 Bi-Hamiltonian Equations

Hamiltonian operators, such as those generated in §3, can be used, via the Lenard scheme, to produce integrable Hamiltonian evolution equations.

Proposition 4.1 (The Lenard Scheme.) Let Θ and Λ be two compatible Hamiltonian operators: $\Omega^1 \to V$, and let v_1 be a vector field ϵV such that

$$v_1 = \Theta\xi_0 = \Lambda\xi_1, \quad \xi_0, \xi_1 \epsilon \Omega^1, \quad d\xi_0 = d\xi_1 = 0. \tag{4.1}$$

Consider the hierarchy

$$v_i = \Theta\xi_{i-1} = \Lambda\xi_i, \quad i = 2, 3, ..., \tag{4.2}$$

and assume that if some $\xi\epsilon\Omega^1$ is closed on the image of Λ then $d\xi = 0$. Then,

(i) all ξ_i are closed, i.e. $d\xi_i = 0$,
(ii) all v_i commute in V,
(iii) all ξ_i commute.

Thus, if $\xi_i = d \downarrow f_i \uparrow$ then

$$\{f_i, f_j\}_\Theta = \{f_i, f_j\}_\Lambda = 0,$$

and $u_t = v_i, i = 1, 2, ...,$ is a completely integrable Hamiltonian system.

Example 1 Let $\Lambda = \partial_x$, $\theta = \partial_x^3 + 2u\partial_x + 2\partial_x u$, $\xi_0 = \frac{1}{2}$, $\xi_1 = u$; note that $\xi_0 = d\tilde{f}_0$ and $\xi_1 = d\tilde{f}_1$, where $\tilde{f}_0 = \int \frac{u}{2} dx$ and $\tilde{f}_1 = \int \frac{u^2}{2} dx$.

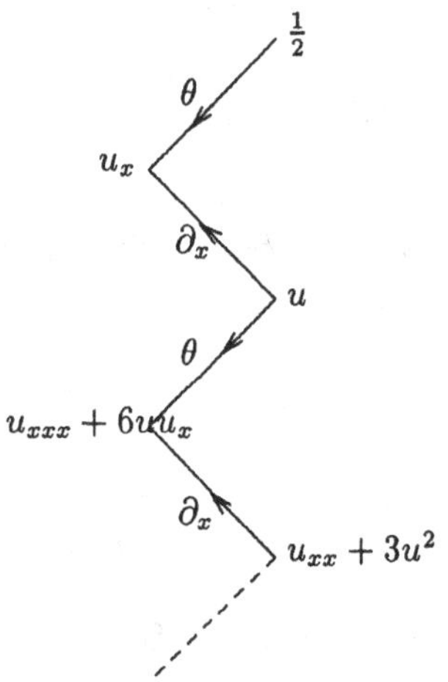

We remark again that the Lenard scheme constructs gradients of conserved quantities ($\frac{1}{2}$, u, $u_{xx} + 3u^2, ...$) and not conserved quantities ($\int \frac{u}{2} dx$, $\int \frac{u^2}{2} dx$, $\int (-\frac{u_x^2}{2} + u^3) dx, ...$).

The application of the Lenard scheme for equations in 2 + 1 will yield in general vector fields which are pseudodifferential operators. However, a suitable linear combination of the starting 1-forms yields vector fields which are functions

Example 2 Let $\Lambda = \partial_x$, Θ be defined by equation (1.5)

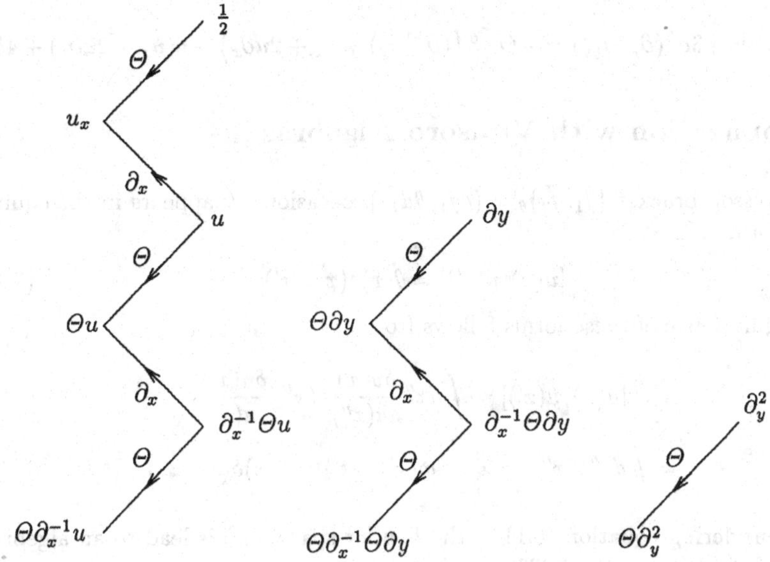

A simple computation shows that $\Theta u = u_{xxx}+6uu_x+2\alpha u_{xy}+4\alpha u_x\partial_y+\alpha^2\partial^{-1}u_{yy}$, and $\Theta\partial_y = \alpha u_{xy} + 2\alpha u_x\partial_y - \alpha^2\partial^{-1}u_{yy}$. Thus $\Theta u - 2\Theta\partial_y$ yields the RHS of the KP equation. Letting $\Psi = \partial_x^{-1}\Theta$ the RHS of the KP can be written as $\Theta(\Psi(\frac{1}{2}) - 2\partial_y)$. It is shown in [25] that the nth member of the KP hierarchy is $\Theta\sum_{k=0}^{n} 4^k \begin{pmatrix} n \\ k \end{pmatrix} \Psi^{n+1-k}\mathcal{R}^k$, where the operator $\mathcal{R}$ is defined by $\mathcal{R}\mathcal{F} = \mathcal{F}\partial_y$.

Example 3 Let $\Lambda B = [\partial_x, B]$, $\Theta B = \alpha B_y + [u + \partial_x^2, B]$. Then

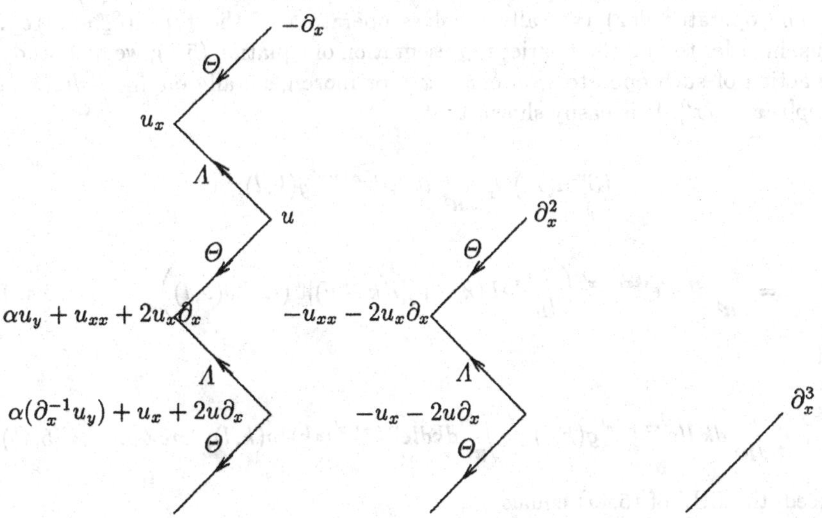

The first vector field of the KP hierarchy is $u_x = -\Theta\partial_x$, the second is $\alpha u_y = \Theta(u+\partial_x^2)$, the third is

$$u_{xxx}+6uu_x+3\alpha^2(\partial_x^{-1}u_{yy}) = -\Theta\left\{3\left((\partial_x^{-1}u_y)+u_x+2u\partial_x\right)+6(u_x+2u\partial_x)+4\partial_x^3\right\}.$$

5 Connection with Virasoro Algebras

The Poisson bracket $\{\tilde{f}_1,\tilde{f}_2\}_\theta = (d\tilde{f}_1, \theta d\tilde{f}_2)$, occasionally appears in the equivalent form

$$[u(x),u(x')] = \theta(x)\delta(x-x'). \tag{5.1}$$

The equivalence of these forms follows from

$$[u(x),u(x')] = \int dx'' \frac{\delta u(x)}{\delta u(x'')}\theta(x'')\frac{\delta u(x')}{\delta u(x'')}$$

$$= \int dx''\delta(x''-x)\theta(x'')\delta(x''-x') = \theta(x)\delta(x-x').$$

Considering equation (5.1) in the Fourier space, one is lead to an algebra, often of the Virasoro type [36]. Letting

$$u(x) = \int_{\boldsymbol{R}} dk e^{ikx}L(k), [u(x),u(x')] = i\int_{\boldsymbol{R}^2} dkdle^{ikx+ilx'}[L(k),L(l)], \tag{5.2}$$

the Fourier transform of equation (5.1) yields

$$[L(k),L(l)] = g(k,l), \tag{5.3}$$

where the concrete form of $g(k,l)$ depends on the concrete form of $\theta(x)$.

The operator $\theta(x)$ typically involves operators of the type $(\partial_x^m u(x))\,\partial_x^n$. Thus, in order to find the Fourier representation of equation (5.1), we first study the action of such operators on $\delta(x-x')$, or more generally on $\int_{\boldsymbol{R}^2} dkdlg(k,l) \times exp[ikx+ilx']$. It is easily shown that

$$(\partial_x^m u(x))\,\partial_x^n \int_{\boldsymbol{R}^2} dkdle^{ikx+ilx'}g(k,l)$$

$$= \int_{\boldsymbol{R}^2} dkdle^{ikx+ilx'}\left(\int_{\boldsymbol{R}} d\nu L(k-\nu)[i(k-\nu)]^m(i\nu)^n g(\nu,l)\right), \tag{5.4a}$$

and

$$\partial_x^n \int_{\boldsymbol{R}^2} dkdle^{ikx+ilx'}g(k,l) = \int_{\boldsymbol{R}^2} dkdle^{ikx+ilx'}(ik)^n g(k,l), \quad n\epsilon\boldsymbol{Z}_+. \tag{5.4b}$$

Indeed, the LHS of (5.4a) equals

$$\int_{\mathbf{R}^3} d\nu dl d\mu (i\mu)^m L(\mu)(i\nu)^n g(\nu,l) e^{i\nu x + ilx' + i\mu x},$$

which reduces to the RHS of (5.4a) by letting $\mu = k - \nu$.

Example 5.1 (KdV).

Let θ_1, θ_2 be given by equations

$$\theta_1 = \partial_x, \quad \theta_2 = \lambda\partial_x^3 + 2u\partial_x + u_x. \tag{5.5}$$

Then equations (5.4) imply

$$\begin{aligned}
(\theta_2 + \lambda\theta_1)(x)\delta(x - x') &= (\lambda\partial_x^3 + 2u\partial_x + u_x + \lambda\partial_x)\int_{\mathbf{R}^2} dk dl \delta(k+l) e^{ikx + ilx'} \\
= \int_{\mathbf{R}^2} dk dl e^{ikx+ilx'} \quad & [\lambda(ik)^3\delta(k+l) \\
& + \int_{\mathbf{R}} d\nu\, (2i\nu + i(k-\nu))\, L(k-\nu)\delta(\nu + l) + ik\lambda\delta(k+l)] \\
= \int_{\mathbf{R}^2} dk dl e^{ikx+ilx'} \quad & [\lambda(ik)^3\delta(k+l) + i(k-l)L(k+l) + ik\lambda\delta(k+l)] \\
= i\int_{\mathbf{R}^2} dk dl e^{ikx+ilx'} \quad & [L(k), L(l)]
\end{aligned}$$

Thus the linear combination of $\theta_2 + \lambda\theta_1$ yields

$$[L(k), L(l)] = (k - l)L(k + l) - \lambda k^3\delta(k + l) + \lambda k\delta(k + l). \tag{5.6}$$

In the physics literature $\lambda = -\frac{c}{12}$, where c is the central charge.

A general method of associating infinite dimensional Lie-algebras with Hamiltonian operators linear in the dynamic variables was presented in [43].

Example 5.2 (Boussinesq).

The Boussinesq equation

$$u_{tt} = \frac{1}{3}u_{xxxx} + \frac{4}{3}\left(u^2\right)_{xx}, \tag{5.7}$$

can be put into a bi-Hamiltonian form when it is written as a pair of first-order evolution equations

$$\begin{aligned}
u_t &= v_x \\
v_t &= \tfrac{1}{3}u_{xxx} + \tfrac{8}{3}u\, u_x\, .
\end{aligned}$$

The compatible pair of Hamiltonian operators is given by

$$\theta_1 = \begin{pmatrix} 0 & 1 \\ 1 & 0 \end{pmatrix}\partial_x, \tag{5.8}$$

and

$$\theta_2 = \begin{pmatrix} \partial_x^3 + 2u\partial_x + u_x & 3v\partial_x + 2v_x \\ 3v\partial_x + v_x & \frac{1}{3}\partial_x^5 + \frac{5}{3}(u\partial_x^3 + \partial_x^3 u) - (u_{xx}\partial_x + \partial_x u_{xx}) + \frac{16}{3}u\partial_x u \end{pmatrix}. \tag{5.9}$$

The algebra which corresponds to the Hamiltonian operator is the same as the Fateev-Zamolodchikov algebra [37].

It should be noted that equations (5.4) are valid even for negative integers provided that $\partial_x^{-1} f = \frac{1}{2}(\int_{-\infty}^{x} - \int_{x}^{\infty}) d\xi f(\xi)$ and the integrals in (5.4) are taken in the principal value sense. Hence nonlocal Hamiltonian structures can be used to generate nonlocal algebras.

Example 5.3 (Zakharov-Shabat).

The Zakharov-Shabat eigenvalue problem is associated with the Hamiltonian operator

$$\theta = \partial_x - Q^- \partial_x^{-1} Q^-, \quad Q^- = [Q, \cdot], \tag{5.10}$$

where Q is a 2×2 off-diagonal matrix. This θ gives rise to the following algebra [38]

$$\begin{aligned} [L_1(k), L_1(l)] &= -2 ⨍_{R} \tfrac{d\nu}{\nu} L_2(k-\nu) L_2(\nu + l) \\ [L_1(k), L_2(l)] &= k\,\delta(k+l) + 2 ⨍_{R} \tfrac{d\nu}{\nu} L_2(k-\nu) L_1(\nu + l). \\ [L_2(k), L_2(l)] &= -2 ⨍_{R} \tfrac{d\nu}{\nu} L_1(k-\nu) L_1(\nu + l), \end{aligned} \tag{5.11}$$

where $⨍$ denotes a principle value integral.

We recall that the two Hamiltonian operators associated with the KdV equation are generated from the linear eigenvalue equation $\psi_{xx} + (u + \lambda^2)\psi = 0$. Furthermore, the Hamiltonian operator θ given by equation (5.10) is associated with the linear eigenvalue equation $\psi_x = \lambda J \psi + Q\psi$, where $J = diag(1, -1)$ and Q is 2×2 off-diagonal matrix. Both of these eigenvalue problems are reductions of the eigenvalue equation $\psi_x = \lambda J \psi + u\psi$, where u is a full 2×2 matrix (see equation (3.3)). This eigenvalue equation gives rise to the Hamiltonian operator $\partial_x + [u, \cdot]$ (see equation (3.5)) which generates a Lie-algebra. It is quite remarkable, that this Lie-algebra, through the process of reduction, gives rise to the two interesting algebras given by equations (5.6) and (5.11).

Acknowledgements

This work was partially supported by the National Science Foundation under Grant Number DMS-8803471 and by the Air Force Office of Scientific Research under Grant Number 87-0310.

References

[1] V.E. Zakharov, L.D. Faddeev, Funct. Anal. Appl., **5**, 18-27 (1971).
[2] C.S. Gardner, J. Math. Phys., **12**, 1548-1551 (1971).
[3] P.D. Lax, Commun. Pure Appl. Math. **21**, 467-490 (1975).
[4] P.D. Lax, SIAM Review, **18**, 351-375 (1976).
[5] I.M. Gel'fand, L.A. Dikii, Uspeki Matem. Nauk, **30**:5, 67-100 (1975).
[6] I.M. Gel'fand and L.A. Dikii, Funct. Anal. and Appl., **10**, 18 (1976); Funct. Anal. and Appl., **10**, 13 (1976).
[7] I.M. Gel'fand, Y.I. Manin, and M.A. Shubin, Func. Anal. and Appl., **10**, 30 (1976).
[8] M.J. Ablowitz, D.J. Kaup, A.C. Newell, H. Segur, Stud. Appl. Math., **54**, 249-315 (1974).
[9] P.J. Olver, J. Math. Phys. **18**, 1212-1215 (1977).
[10] F. Magri, J. Math. Phys. **19**, 1156-1162 (1978).
[11] M. Adler, Invent. Math., **50**, 219-248 (1979).
[12] I.M. Gel'fand, Collected Works.
[13] I.M. Gel'fand, I.Ya. Dorfman, Funct. Anal. Appl., **13**:4, 13-30 (1979).
[14] A.S. Fokas, B. Fuchssteiner, Lett. in Nuovo Cimento. **28** 299-303; B (1980). Fuchssteiner, A.S. Fokas, Physica D, **4**, 47-66 (1981).
[15] see references in A.P. Stone, Can. J. Math. **25**, 903 (1973).
[16] B. Fuchssteiner, Nonl. Anal. Theory Meth. Appl., **3**, 849 (1979).
[17] I.M. Gel'fand, I.Ya. Dorfman, Funct. Anal. Appl., **14**:3, 71-74 (1980).
[18] L.A. Dickey, Soliton Equations and Hamiltonian Systems, World Scientific (1991).
[19] V.E. Zakharov, B.G. Konopelchenko, Commun. Math. Phys., **94**, 483 (1984).
[20] A.S. Fokas, B. Fuchssteiner, Phys. Lett. A, **86**, 341-345 (1981).
[21] A.S. Fokas and P.M. Santini, Stud. Appl. Math. **75**, 174 (1986).
[22] P.M. Santini, A.S. Fokas, Commun. Math. Phys. **115**, 375-419 (1988).
[23] A.S. Fokas, P.M. Santini, Commun. Math. Phys. **116**, 449-474 (1988).
[24] F. Magri, C. Morosi, G. Tondo, 1988, Commun. Math. Phys., **115**, 457-475 (1988).
[25] I.Ya. Dorfman and A.S. Fokas, Hamiltonian Theory over Non-Commutative Rings and Integrability in Multidimensions, INS #181, Clarkson University, (1991).
[26] I.M. Gel'fand and L.A. Dikii, Funct. Anal. and Appl., **1**, 8 (1979).
[27] S.P. Novikov, Funkts. Anal. Prilozhen., **8**, No. 3, 54-66 (1974).
[28] B.A. Dubrovin, Funkts. Anal. Prilozhen., **9**, No. 3, 41-52 (1975).
[29] A.R. Its and V.B. Matveev, Teor. Mat. Fiz., **23**, No. 1, 51-68 (1975).
[30] H. McKean and F.V. Moerbeke, "Sur le spectre de quelques operateurs et la variete de Jacobi," Sem. Bourbaki (1975-1976), p. 474.
[31] I.M. Gel'fand and I.S. Zakharevich, Webs, Veroneze Curves and BiHamiltonian Systems, Funct. Anal. and Appl. (1991).
[32] L.D. Faddeev and L.A. Takhtajan, Hamiltonian Methods in the Theory of Solitons, Springer Verlag (1987).
[33] M. Sato and Y. Sato, In Nonlinear PDE's in Applied Science, ed. by H. Fujita, P.D. Lax, and G. Strang, North Holland, Tokyo (1982).
[34] G. Segal and G. Wilson, Loop Groups and Equations of KdV-Type, Publ. Math., **63**, 1-64 (1985).

[35] I. Dorfman, Dirac Structures and Integrability of Nonlinear Evolution Equations, to appear.
[36] J.L. Gervais and A. Neveu, Nuc. Phys. **B199**, 59 (1982); **B209**, 125 (1982).
[37] V.A. Fateev and A.B. Zamolodchikov, Nucl. Phys. **B280**, 644 (1987).
[38] A.S. Fokas and Y. Nutku, Bi-Hamiltonian Structures and Virasoro Hierarchies, Preprint, Clarkson University, (1991).
[39] A.S. Fokas, G.Z. Tu, A Residue Representation for Integrable Equations in Multidimensions, preprint (1991).
[40] H. McKean, Bi-Hamiltonian Structures in Classical Mechanics, in this volume.
[41] A.S. Fokas and I.M. Gel'fand, The Resolvent Scheme for Integrable Multidimensional Equations (in preparation).
[42] I.M. Gel'fand, Y.L. Daletsky, and B. Tsygan, A Version of a Non-Commutative Differential Geometry, Docklady (1989).
[43] I.M. Gel'fnad and I.Y. Dorfman, Funkts. Anal. and Appl., **15**, 23-40 (1981).

On the Symmetries of Integrable Systems

P.G. Grinevich[1], *A.Yu. Orlov*[2], *and E.I. Schulman*[2]
[1]Landau Institute for Theoretical Physics, Kosygina 2,
Moscow GSP-1, 117940, Russia
[2]Oceanology Institute, Krasikova 23, 117218 Moscow, Russia

...the enormous usefulness of
mathematics in the natural sciences
is something bordering on the
mysterious and there is no rational
explanations for it.
Eugene Wigner [1].

Introduction

Symmetries play a very important role in mathematics as well as in physics. Many problems were solved by studying their symmetry properties. For example Gauss in his first and beloved work found how to construct a 17–sided polygon with compasses and ruler. The essential part of his method was studying the invariants of the equation $z^{17} - 1 = 0$. Other significant results based on the symmetry approach were obtained by Abel and Galois in the theory of algebraic equations. Abel proved that the solution of a general 5th degree algebraic equation cannot be expressed via radicals. Galois succeeded to elicit conditions of solvability of an algebraic equation in radicals using its symmetry group. The analog of Galois theory for ordinary differential equations was developed in our century.

The Noether theorem in Lagrangean mechanics and the Liouville theorem in Hamiltonian mechanics may be considered as other important examples of connections between symmetries and solvability . The last theorem states that any $2n$-dimensional Hamiltonian system possessing n first integrals in involution (equivalent to the existence of the n-dimensional Abel symmetry group) is completely integrable.

In contemporary physics an important role belongs to the so-called conformal theories. The string theory and the theory of second-order phase transitions are important examples. In the phase transitions theory the conformal invariance arises along with scale invariance at the critical point where the correlation radius length approaches infinity. There is an important difference between 2-dimensional and multidimensional theories, because in more than two dimensions conformal algebra has a finite number of generators while in two dimensions their number is infinite: any holomorphic function $z \to f(z)$ generates a local conformal transformation. Due to this infinite-dimensional group it is possible to classify second order phase transitions and calculate critical exponents [1].

The presence of an infinite - dimensional transformation group is a characteristic feature of solvable equations in the soliton theory. We will show that the corresponding algebra has a commutative part and a noncommutative one. The existence of commutative subalgebra is well known from the very beginning of the soliton theory. The noncommutative symmetries are much less known and the aim of this review is to outline a regular approach to this subject. Our interest in the symmetries of the integrable systems has several reasons. One of them is that the algebra of symmetries of solvable equations always contains conformal subalgebra [18], [19]. This proved to be important for studying connections between the soliton theory and the modern quantum field theory.

In the string theory the deformations of the complex structures of the Riemann surfaces play an important role. We will show that this action coincides with the action of the conformal symmetries on the quasiperiodic Kadomtsev-Petviashvili solutions. The τ-functions of such solutions coincide with the determinant of the $\bar{\partial}$ - operator on the appropriate bundles (see [21], [24]) - one of the central objects in the string theory. Another important example arises in the matrix model. The partition function of the matrix model in the double scale limit coincides with the Korteweg de-Vries τ-function, invariant with respect to some noncommutative symmetry.

The infinite number of commutative symmetries and correspondent motion invariants is equivalent to solvability. It may be used for testing wide classes of partial differential equations for solvability (see the example Mikhailov-Shabat review of an approach based on calculation of the commutative symmetry group in $1+1$ case and Zakharov's and one of the authors' review of an approach based on studying corresponding motion invariants for multidimensional case in this volume). We will not discuss these topics here.

1 Symmetries and exact solvability

1.1 Preliminary remarks

In the late 1960s, due to the pioneering work of Zabusky, Kruskal, Miura, Green and Gardner [2], a new method for exactly solving the nonlinear evolution equation (n.e.e.)

$$u_t + \frac{3}{2}uu_x - \frac{1}{4}u_{xxx} = 0 \qquad (1.1.1)$$

was discovered. This method was based on the fact that equation (1.1.1) may be represented as the compatibility condition for two linear equations $L\psi = E\psi$, $\psi_t = A\psi$, where ψ is some auxiliary function $L = -\partial_x^2 + u(x,t)$ is the Schrödinger operator and A is a third-order operator. The crucial property is that the spectrum of the L-operator is invariant under t-evolution. One may say that KdV defines isospectral deformations of the L-operator.

During several subsequent years the method used in [2] was generalized for other equations. As a result the inverse scattering transform (i.s.t.) method was created. It is difficult to overestimate the value of this discovery. The concept of soliton appeared and entered scientific thinking. A number of nontrivial finite-dimensional and infinite-dimensional integrable systems have been constructed. Essential results in the algebraic geometry and other areas of mathematics have been achieved. It is enough to mention the proof of Novikov hypothesis by Shiota. So A.C. Newell had good reason to entitle the last chapter of his book [3] as "Miracles of soliton mathematics".

We can also think about miracles of soliton physics, because the very equations, which are solvable by i.s.t. (they definitely have zero measure among possible equations) happen to coincide with universal ones, which describe the asymptotic behavior of various physical systems, and are also singled out of other physical equations by balance between nonlinearity and dispersion [3]. It is similar to phase transitions of second order where, at the parameter values corresponding to the transition, the universal behavior and an infinite number of internal symmetries (exact solvability) arise.

For example, the Korteweg-de-Vries equation (1.1.1) and its two-dimensional generalization, Kadomtsev-Petviashvili equation

$$(u_t + \frac{3}{2}uu_x - \frac{1}{4}u_{xxx})_x = 3\alpha^2 u_{yy},\ \alpha^2 = \pm 1, \tag{1.1.2}$$

describe one-dimensional and weakly two-dimensional waves of small amplitude in any isotropic weakly dispersive physical system. A number of examples of this kind of universality may be found in [4], see also F. Calogero's paper [5].

All autonomous equations solvable by i.s.t. may be considered as isospectral deformations of corresponding linear operators. This is equivalent to the existence of an infinite number of motion invariants and corresponding commutative symmetries which are also isospectral. For this reason it is natural to consider the solvable equation together with its isospectral symmetries as one commutative hierarchy [6], [3]. All equations participating in the hierarchy are symmetries to each other and commute. That is if $u(t)$, $t = (t_1, t_2, \ldots)$ is a field variable for hierarchy consisting of equations enumerated by integers n with the group parameters t_n:

$$\partial u/\partial t_n = K_n[u(t)], \qquad n = 1, 2, \ldots, \tag{1.1.3}$$

then for any m, n $\partial^2 u/\partial t_n \partial t_m = \partial^2 u/\partial t_m \partial t_n$. This means that equations solvable by i.s.t. possess an infinite-dimensional Abel algebra of symmetries.

There are rather new results that the symmetry algebra of solvable equations is wider and contains noncommutative symmetries. This fact is much less known. These symmetries depend explicitly on the "commutative times" t_n (usually t_1 is designated as x, like for KdV). The most obvious KdV symmetry of this type is related to conservation of the center-of-mass velocity.

$$\partial u/\partial \beta_{-2} = 3tu_x - 2 \qquad \text{(Galilean).} \tag{1.1.4}$$

Less trivial KdV symmetries may be obtained by the action of the recursion operator on the above-mentioned trivial symmetry (β_{2n} are group parameters of noncommutative symmetries), t_{2m+1} are the commutative ones, $t_1 = x$, $t_3 = t$,

$t_{2k+1} = 0$ as $k > 1$):

$$\partial u/\partial\beta_{2m} = (\partial/\partial x)\Lambda^{m+1}(3tu - 2x), \tag{1.1.5}$$

where $\Lambda = \frac{1}{4}\partial^2 - \partial^{-1}u\partial - \frac{1}{2}\partial^{-1}u_x$, $\partial = \partial/\partial_x$ is the recursion operator for KdV. For example for $m = 0, 1$ we have:

$$\partial u/\partial\beta_0 = 3tu_t + xu_x + 2u \qquad \text{(Scaling)}. \tag{1.1.6}$$

$$\partial u/\partial\beta_2 = 3tu_{t_5} + xu_{t_3} + u_{xx} - 2u^2 - \frac{1}{2}u_x\partial^{-1}u. \tag{1.1.7}$$

The formula (1.1.5) was pointed out for the first time in [10] without subsequent comments.

Ordinary higher KdVs are $\partial u/\partial t_{2k+1} = (\partial/\partial x)\Lambda^k u$, e.g.

$$u_{t_5} = (u^{IV} - 10uu_{xx} - 5u_x^2 + 10u^3)_x. \tag{1.1.8}$$

Symmetries of the type (1.1.5) are used for describing automodel asymptotic behavior of KdV-type equations [3] which satisfy their stationary versions [18], [19]. Subsequently some such symmetries were found for KP [11], [12] and several other equations [13]. Gauge equivalence between different solvable equations [14]-[16] is also closely related to symmetry. In [17] a two-dimensional analog of the recursion operator was constructed by treating the field variable as an integral operator. It makes it possible to treat Hamiltonian structure and symmetries of KP hierarchy similarly to the one-dimensional equations.

In [18] for 1+1 systems and in [19] for 2+1 systems it was shown that equations for noncommutative symmetries are themselves solvable by i.s.t. and a systematic approach was developed to calculate them, to construct their L-A pairs representation and to study their action on different soliton objects for a wide class of solvable hierarchies. This method was generalized to be applicable in periodic and quasiperiodic cases in [20]-[24].

1.2 Symmetries and Linearization

Consider for the simplicity the n.e.e. of the form (1.1.3) with some nonlinear operator $K[u(t)]$ The n.e.e. $u_\beta = T[u](t)$ is called a symmetry for (1.1.3) if $\partial^2 u/\partial\beta\partial t = \partial^2 u/\partial t\partial\beta$ or

$$K'_u[T] \stackrel{\text{def}}{=} (d/d\epsilon)K[u + \epsilon T]_{\epsilon=0} = \partial T/\partial t + T'_u[K] \ , \tag{1.2.1}$$

where $\stackrel{\text{def}}{=}$ means "by definition", and $K'_u[v]$ is called the Gateux derivative along the direction v. The (1.2.1) means that $v = T[u](t)$ satisfies the linearized version of (1.1.3) with some background $u(t)$, $dv/dt = K'_u[v]$. So all information about symmetries of nonlinear equation is hidden in its linearized version. It becomes obvious when taking into account that we consider continuous symmetries and the difference of two solutions with infinitesimally different group parameters β and $\beta + d\beta$ satisfies (1.2.1). This fundamental fact was discovered by Sophus Lie.

1.3 Exactly Solvable Equations. Baker-Akhiezer Functions

The most general statement about the equations solvable by i.s.t. is that they are compatibility conditions for some overdetermined system of linear equations on some auxiliary function Ψ , which is known in the soliton theory as the Baker-Akhiezer function and in the quantum scattering theory as the Jost function:

$$\mathcal{L}_1[u]\Psi = 0,\ \mathcal{L}_2[u]\Psi = 0 \tag{1.3.1}$$

where the operators $\mathcal{L}_1$ and $\mathcal{L}_2$ may be matrix differential or integrodifferential ones. The simplest compatibility condition for (1.3.1) is

$$[\mathcal{L}_1[u],\ \mathcal{L}_2[u]] = 0 \tag{1.3.2}$$

and for proper $\mathcal{L}_1$ and $\mathcal{L}_2$ this condition is a nonlinear system of equations on $u(t)$, e.g. of the form (1.1.3) and may be solved by (1.3.1) ([3], [8], [9]). Historically, the first regular method to find such operators in both 1+1 and 2+1 cases was the Zakharov - Shabat (Z-Sh) dressing method [25]. This method starts with the problem of decomposition of an integral operator

$$\hat{\mathcal{F}}\Psi(x) = \int_{-\infty}^{\infty} F(x,z)\Psi(z)dz,\ \hat{F} = 1 + \hat{\mathcal{F}}$$

by means of two Volterra operators :

$$K_-\hat{F} = K_+;\ \hat{K}_\pm = 1 + \hat{\mathcal{K}}_\pm,\ K_\pm(x,z) = 0 \ \text{ as } \ x \gtrless z. \tag{1.3.3}$$

If $\hat{F}$ satisfies the equations

$$\mathcal{L}_i^{(0)}\hat{F} + \hat{F}\mathcal{L}_i^{(0)+} = 0, \qquad i = 1,2 \tag{1.3.4}$$

and $\mathcal{L}_1^{(0)}, \mathcal{L}_2^{(0)}$ are differential commuting operators, then "dressed" operators

$$\mathcal{L}_i = K_\pm \mathcal{L}_i^{(0)} K_\pm^{-1}, \qquad i = 1,2 \tag{1.3.5}$$

automatically commute and are purely differential ones, giving a Z-Sh representation (1.3.2) for some nonlinear evolution equation. Solutions of the nonlinear equation (1.3.2) are expressed via values of $K(x,z)$ and a finite number of its derivatives on the diagonal $x = z$ [25]. E.g. for KP equation (1.1.2) the corresponding operators are

$$\mathcal{L}_1^{(0)} = i\alpha\partial_y - \partial_x^2,\ \mathcal{L}_2^{(0)} = \partial_t - \partial_x^3,\ \mathcal{L}_1 = i\alpha\partial_y - \partial_x^2 + u,$$

$$\mathcal{L}_2 = \partial_t - \partial_x^3 + \frac{3}{2}u\partial_x + \frac{3}{4}u_x + \frac{3}{4}i\alpha\partial_x^{-1}u_y. \tag{1.3.6}$$

Assuming in (1.3.6) that $\partial_y = 0$ we come to the KdV hierarchy operators.

The most general method for constructing and solving the n.e.e. is the $\bar{\partial}$ - method [26] - [29].

Let the matrix-valued function $\chi(\lambda, \bar{\lambda})$ satisfy the following $\bar{\partial}$ - equation, $\lambda \in \mathbb{C}$.

$$\bar{\partial}\chi(\lambda,\bar{\lambda}) = \frac{1}{2\pi i}\int \chi(\lambda',\bar{\lambda}')R(\lambda',\bar{\lambda}',\lambda,\bar{\lambda})d\lambda' d\bar{\lambda}', \qquad \bar{\partial} = \partial_x + i\partial_y. \qquad (1.3.7)$$

If the kernel $R(\lambda',\bar{\lambda}',\lambda,\bar{\lambda})$ depends on some additional variables t_n, $n = 1,2,\ldots,N$

$$\partial_{t_n}R(\lambda',\lambda) = I_n(\lambda',\bar{\lambda}')R(\lambda',\lambda) - R(\lambda',\lambda)I_n(\lambda,\bar{\lambda}) \qquad (1.3.8a)$$

and the matrices $I_n(\lambda,\bar{\lambda})$ do not depend on t_n, then the function $\mathcal{D}_n\chi$ also satisfies (1.3.7), where

$$\mathcal{D}_n = \partial_{t_n} - I_n(\lambda,\bar{\lambda}). \qquad (1.3.8b)$$

In the case of the KP equation $I_n = \lambda^n, N = \infty$, and (1.3.8a) means that

$$R(\lambda',\lambda,\mathbf{t}) = e^{\theta_0(\lambda',\mathbf{t})}R_0(\lambda',\lambda)e^{-\theta_0(\lambda,\mathbf{t})} \qquad (1.3.8c)$$

where

$$\theta_0 = \sum \lambda^n t_n. \qquad (1.3.9)$$

An important statement [29] is that if R satisfies (1.3.8a) and $N \geq 2$, then in the ring $\mathcal{M}$ of operators of the form

$$\mathcal{M} : M = \Sigma m_{nk}D_n^k,\ m_{nk} \in C(t_n) \qquad (1.3.10)$$

a special left ideal $\tilde{\mathcal{M}}$ is contained. $\tilde{\mathcal{M}}$ consists of operators $\tilde{M}$ such that $\tilde{M}\chi = 0$. It is natural to call operators $\tilde{M}$ dressed ones comparing with M. Operators $\tilde{M}$ determine the Z-Sh representations for corresponding nonlinear equations. If $R(\lambda',\lambda)$ is localized on some contour $\Gamma \in C$ we arrive at the nonlocal Riemann problem [29]. The analytic properties of $\chi(\lambda,\bar{\lambda})$ determine which approach is better, $\bar{\partial}$ or Riemann problem. For example for KP equation with vanishing at infinity in x it depends on the sign of the α^2 in (1.1.2), (1.2.6). For KP-I equation ($\alpha = 1$) the solutions are to be constructed by nonlocal Riemann problem [27], [28], [30]. The Baker-Akhiezer functions $\Psi(\lambda,\mathbf{t})$, $t_1 = ix$, $t_2 = y$, $t_3 = t$ satisfies to the following conditions :

1)$\Psi(\lambda,\mathbf{t})$ is holomorphic in $\lambda \in C$ everywhere but on the real axes $\mathrm{Im}\lambda = 0$

2) On the line $\mathrm{Im}\lambda = 0$ the function $\Psi(\lambda,\mathbf{t})$ has a jump while the boundary values Ψ_+ and Ψ_- satisfy (identically in t) the following nonlocal Riemann problem

$$\Psi_+(\lambda + i0,\mathbf{t}) = \Psi_-(\lambda - i0,\mathbf{t}) + \int_{-\infty}^{\infty}\Psi_-(\mu - i0,\mathbf{t})R(\mu,\lambda)d\mu \qquad (1.3.11a)$$

3) At the point $\lambda = \infty$ Ψ has an essential singularity of the form

$$\Psi(\lambda,\mathbf{t}) = \exp(\theta_0)(1 + O(1/\lambda)),\ \theta_0 = i\lambda x + \lambda^2 y + i\lambda^3 t + \ldots. \qquad (1.3.11b)$$

These conditions at almost arbitrary R unambiguously determine the Baker-Akhiezer function $\Psi(\lambda,\mathbf{t})$.

4) The conjugate Baker-Akhiezer function $\Psi^*(\lambda,\mathbf{t})$, which it is natural to consider as the differential 1-form in λ , satisfies the following equation

$$\Psi^*(\lambda + i0,\mathbf{t}) = \Psi^*(\lambda - i0,\mathbf{t}) + \int_{-\infty}^{\infty}\Psi^*(\mu - i0,\mathbf{t})R^+(\mu,\lambda)d\mu. \qquad (1.2.11c)$$

The functions R, R^+ satisfy the relation

$$R(\lambda,\mu) + R^+(\mu,\lambda) + \int_{-\infty}^{\infty} R(\lambda,\nu)R^+(\mu,\nu)d\nu = 0. \qquad (1.3.12)$$

As $\lambda \to \infty$ the conjugate function Ψ^* behaves as

$$\Psi^*(\lambda,t) = d\lambda \exp\{-(i\lambda x + \lambda^2 y + i\lambda^3 t + \ldots)\}(1 + O(1/\lambda)). \qquad (1.3.13)$$

The solutions for KP-II ($\alpha = i$) are constructed using Baker-Akhiezer functions [26] with the following properties ($t_1 = ix,\ t_2 = iy,\ t_3 = t$)
1) $\Psi(\lambda, \mathbf{t})$ and $\Psi^*(\lambda, \mathbf{t})$ are continuous in λ on the whole C^1 and satisfy

$$\bar{\partial}_\lambda \Psi(\lambda,\mathbf{t}) = T(\lambda)\Psi(\bar{\lambda},\mathbf{t}), \qquad (1.3.14a)$$

$$\bar{\partial}_\lambda \Psi^*(\lambda,\mathbf{t}) = T^+(\lambda)\Psi^*(\bar{\lambda},\mathbf{t}),\ T^+(\bar{\lambda}) = -T(\lambda), \qquad (1.3.14b)$$

(no assumptions about holomorphic properties of $T(\lambda)$ are made).
2) as $\lambda \to \infty$

$$\Psi \sim (1 + O(1/\lambda))\exp(\theta_0),\ \theta_0 = i\lambda x - i\lambda^2 y + i\lambda^3 t + \ldots \qquad (1.3.15a)$$

$$\Psi^* \sim d\lambda(1 + O(1/\lambda))\exp(-\theta_0). \qquad (1.3.15b)$$

To construct quasiperiodic (finite-gap) solutions of KP equations [31], consider a Riemann surface Γ of genus $g < \infty$ with a given point ∞, a local parameter $z = 1/\lambda$ in its neighborhood and a divisor $\gamma_1, \ldots, \gamma_g$. The Baker-Akhiezer function Ψ is uniquely determined by the conditions:
1) It is meromorphic everywhere but ∞ and has simple poles in $\gamma_1, \ldots, \gamma_g \in \Gamma$.
2) In the neighborhood of ∞ it has an essential singularity of the form (1.3.11b) for KP-I and (1.3.15a) for KP-II.

The conjugate Baker-Akhiezer function Ψ^* is a differential 1-form holomorphic everywhere but ∞ and having simple zeros at $\gamma_1, \ldots, \gamma_g$. At ∞ Ψ^* has the essential singularity of the form (1.3.13) for KP-I and (1.3.15b) for KP-II.

Remark. One may consider the potentials that vanish at infinity on the quasiperiodic finite gap background. This problem may be solved by using $\bar{\partial}$ - or nonlocal Riemann problem on the Riemann surface corresponding to background, see for example [42].

Along with the Baker-Akhieser function Ψ Baker-Akhiezer j-differentials can be considered, see for details [24]. They were introduced in [32], [33]. Then the conjugate Baker-Akhiezer differential will be a $1 - j$ form.

Remark. Tensor properties of the Baker-Akhiezer j-differentials are of no importance if we consider only isospectral Abel symmetries. But if we consider symmetry changing Riemann surface then the tensor properties play an important role. For example as we will see below the central charge in the Virasoro algebra which arises when calculating the action of the nonisospectral symmetries on the τ_j- function corresponding to the Baker-Akhiezer j-differential is $6j^2 - 6j + 1$, see sect. 4.

2 Symmetries and infinitesimal dressing

To construct symmetries it is important to combine two facts: i) symmetries of some equation satisfy its linearized version, see sect. 1.1 and ii) the infinitesimal dressing changes solutions infinitesimally. Therefore a general infinitesimal dressing contains full information about symmetries [18], [19].

2.1 Symmetries via Zakharov-Shabat dressing. Lagrangeans

Consider the dressing (1.3.3) - (1.3.5) in the case when $\mathcal{F} = \delta\mathcal{F}$, $\mathcal{K}_\pm = \delta\mathcal{K}_\pm$ are infinitesimal. We obtain

$$\delta\mathcal{L}_i = [\delta\mathcal{K}_\pm, \mathcal{L}_i],\ i = 1,2,\ \delta\mathcal{F} = \delta\mathcal{K}_+ - \delta\mathcal{K}_-, \tag{2.1.1}$$

so symmetries of (1.3.2) are represented as

$$[\mathcal{L}_1, \delta\mathcal{L}_2] + [\delta\mathcal{L}_1, \mathcal{L}_2] = 0$$

Consider in more detail the KP equation (1.1.2) with auxiliary linear operators (1.3.6). A very fruitful way to describe all the KP hierarchy was developed by Date, Miwa, Jimbo and Kashiwara in a series of papers reviewed in [34]. All the equations of the KP hierarchy are determined by a single pseudodifferential operator L which is $\partial = \partial/\partial x$ dressed by formal Volterra operator K (K_+ and K_- result in the same L)

$$K = 1 + \sum_1^\infty K_i(t)\partial^{-i} \tag{2.1.2}$$

$$L = K\partial K^{-1} = \partial + u_1\partial^{-1} + u_2\partial^{-2} + \dots \tag{2.1.3}$$
$$u_1 = -(K_1)_x.$$

Equations of the hierarchy are enumerated by two integers n, m and have the representation (1.3.2) in the form

$$[\partial/\partial t_n - B_n, \partial/\partial t_m - B_m] = 0,\ \text{where}\ \partial\Psi/\partial t_n = B_n\Psi, \qquad B_n = (L^n)_+, \tag{2.1.4}$$

and $(\ )_+$ is the projection on the differential part (for example $L_+ = \partial, K_+ = 1$.) One may prove that the Baker-Akhiezer function is an eigenfunction for the operator L ,

$$L\Psi = \lambda\Psi;\ \Psi = K\Psi_0;\ \Psi_0 = \exp\theta_0. \tag{2.1.5}$$

The infinitesimal "redressing" of L performed by means of variation of Volterra operator $\hat{K}$ leads to δL of the form (2.1.1) with $\delta\mathcal{K} = \delta K K^{-1}$. According to (2.1.4) and (1.3.4) we have $\partial F/\partial t_n = \hat{F}B_n^+ - B_nF$, $n \neq 1$. This equation has a special class of solutions, so-called operators with degenerate kernel,

$$\delta\mathcal{F}(x,z) = \sum_{i=1}^N \Psi(x,\mu_i)\Psi^*(z,\lambda_i). \tag{2.1.6}$$

Then according to (2.1.1)

$$\delta \mathcal{K}_{\pm}(*) = -\int_{\mp\infty}^{x} \delta \mathcal{F}(x,z)(*)dz, \tag{2.1.7}$$

$$\delta L = [L, \delta \mathcal{K}_{\pm}]; \ \delta B_n = [\partial/\partial t_n - B_n, \delta \mathcal{K}_{\pm}]. \tag{2.1.8}$$

Taking in (2.1.6) $N = 1$ we obtain

$$\mathcal{F} = \Psi(x, \lambda + \epsilon)\Psi^*(z, \lambda) = \tag{2.1.9a}$$

$$\sum (\partial_\lambda^n \Psi(x,\lambda))\Psi^*(z,\lambda)\epsilon^n/n! = \tag{2.1.9b}$$

$$\sum (\epsilon^n/n!)\Psi(x,\lambda))\Psi^*(z,\lambda)(\partial_\lambda + \theta'_\lambda)^n \cdot 1 \tag{2.1.9c}$$

Here $\theta = \ln \Psi = \theta_0 + \partial_x^{-1}\vartheta$, $\partial_x^{-1}\vartheta = \ln(\Psi\Psi_0^{-1})$, and we use $\partial_\lambda^n e^\theta = e^\theta(\partial_\lambda + \theta'_\lambda)^n$. Equation (2.1.9) may be considered as generating function creating all symmetries of KP. Really, according to (2.1.1) we may take $\delta L \sim \partial L/\partial \beta$ and obtain

$$\partial L/\partial \beta = [L, \delta \mathcal{K}_{\pm}] \ \text{ or } \ [L, M_{\pm}] = 0, \tag{2.1.10a}$$

$$\partial B_n/\partial \beta = [\partial t_n - B_n, \delta \mathcal{K}_{\pm}] \ \text{ or } \ [\partial t_n - B_n, M_{\pm}] = 0. \tag{2.1.10b}$$

Here the operator M has the form

$$M_{\pm} = \partial/\partial\beta + \Psi(x,\mu)\int_{\pm\infty}^{x} \Psi^*(z,\lambda)(*)dz. \tag{2.1.11}$$

If we decompose M and $\partial/\partial\beta$ in powers of λ and $\epsilon = \lambda - \mu$ then the coefficients are operators giving Z-Sh representations for all infinitesimal symmetries of KP equation.

2.2 Basis in the space of symmetries. Explicit calculations

We consider now in more detail symmetries of the KP equation given by $[\mathcal{L}_1, M] = 0$. Calculating $\mathcal{L}_1$ in terms of L_2^+ we obtain from (2.1.3), (2.1.2)

$$\mathcal{L}_1 = \partial_y - L_+^2 = \partial_y - (\partial^2 + 2u_1), \ u_1 = -(K_1)_x.$$

We see that $u = V_x$, where $V = -2K_1(t)$ is a potential for the KP equation which in terms of V is

$$V_{xt} + \frac{3}{2}V_x V_{xx} + \frac{1}{4}V_{xxx} = 3\alpha^2 V_{yy}. \tag{2.2.1}$$

From (2.1.10b) with $n = 2$ we obtain

$$\partial u/\partial\beta = \partial V_x/\partial\beta = -2\{\Psi(x,\mu)\Psi^*(x,\lambda)\}_x. \tag{2.2.2}$$

The usual commutative symmetries of KP arise if we put $\lambda = \mu$. Indeed, let us suppose in the vicinity of $\mu = \lambda = \infty$ that $\partial/\partial\beta = \sum \lambda^{-n-1}(\partial/\partial t_n)$. Expanding $\Psi\Psi^*(\lambda, x)$ into a power series we obtain the isospectral equations

$$\partial V/\partial t_n = -2 \ \mathrm{Res}_\lambda\{\lambda^n \Psi(x,\lambda)\Psi^*(x,\lambda)\}, \tag{2.2.3a}$$

$$\Psi(x,\mu)\Psi^*(z,\lambda) = 1 + \sum_{n=0}^{\infty} \lambda^{-n-1}\partial_x^{-1} u_{t_n}, \tag{2.2.3b}$$

where $u_{t_n} = K_n[u]$ are commutative equations from the KP hierarchy, which have the form $[\partial/\partial t_2 - B_2, \partial/\partial t_n - B_n] = 0$. Certainly, $\partial/\partial t_2 - B_2 = \mathcal{L}_1$, $\partial/\partial t_3 - B_3 = \mathcal{L}_3$ and at $n = 3$ we obtain (1.1.2) with operators (1.3.6). Both $[\mathcal{L}_1, M] = 0$ and $[\mathcal{L}_2, M] = 0$ give symmetries of the KP equation.

The commutation relation

$$[i\alpha\partial_y - \partial_x^2 - u, \partial/\partial\beta - \Psi(\mu)\partial^{-1}\Psi^*(\lambda)] = 0 \tag{2.2.4}$$

gives us the so-called "full KP hierarchy" [20] :

$$(i\alpha\partial_y - \partial_x^2 - u)\Psi = 0 \tag{2.2.5a}$$

$$(i\alpha\partial_y + \partial_x^2 + u)\Psi^* = 0 \tag{2.2.5b}$$

$$\partial u/\partial\beta = 2\partial/\partial x(\Psi\Psi^*) \tag{2.2.5c}$$

which describes all KP symmetries. Ψ and Ψ^* may be arbitrary solutions of (2.2.5a), (2.2.5b). If we consider u, Ψ, Ψ^* as independent field variables then this system appears to be a degenerate case of the Davey-Stewartson equation, see [8], page 289.

It can be written in Lagrangean form [20]

$$\delta S/\delta\Psi^* = 0, \ \delta S/\delta\Psi = 0, \ \delta S/\delta v = 0$$

where $u = v_x$ and the action is

$$S = \int (2\Psi^*(i\alpha\partial_y - \partial_x^2 - v_x)\Psi + v\partial^2 v/\partial x\partial\beta)dxdyd\beta.$$

To write down the basis of noncommutative symmetries we recall that $\theta_0' = \sum n t_n \lambda^{n-1}$ is linear in t_n and we have to select not only terms with different powers of λ but also with different powers of t_n. From (2.1.9) and (2.2.2) we find that all possible KP symmetries may be enumerated by integers n, m with group parameters β_{nm} and

$$\partial V/\partial\beta_{nm} = -2 \ \mathrm{Res}\{\lambda^n \Psi(x,\mu)\Psi^*(x,\lambda)(\partial_\lambda + \theta_\lambda')^n \cdot 1\}. \tag{2.2.6}$$

We will call these the (m, n) symmetries.

To find θ , or equally, ϑ, one may substitute $\Psi = \exp\theta$ into $\Psi_y = B_2\Psi$ to obtain $\theta_y = \theta_{xx} + \theta_x^2 + u$. Keeping in mind (1.3.9) we see that $\partial_x^{-1}\vartheta_y = \vartheta_x + 2\lambda\vartheta + \vartheta^2 + u$ and taking $\vartheta = \sum_{n\geq 1}\vartheta_n\lambda^{-n}$, $\vartheta_1 = -u/2$ as $\lambda \to \infty$ we find

$$2\vartheta_{n+1} = \partial_x^{-1}\vartheta_{ny} - \vartheta_{nx} + \sum_{k=1}^{n-1}\vartheta_k\vartheta_{n-k}.$$

The quantities $I_n = \int\int \vartheta_n dxdy$ are involutive motion invariants of the KP equation and Hamiltonians for the higher KPs, which may be written in the form $\partial u/\partial t_n = \partial_x(\delta I_n/\delta u)$. In the case of KP on the plane with rapidly vanishing initial data these motion invariants are rather constraints [50] while for periodic boundary conditions they are good motion invariants [51]. Any symmetry of the KP equation may be expressed explicitly in terms of right hand sides of higher KP equations and densities of their Hamiltonians [19]. For example one obtains

$$\partial u/\partial\beta_{02} = 9t^2 u_{t_4} + 12tyu_{t_3} + (4y^2 + 6xt)u_{t_2} + \\ +(4xy + 12t)u_x + 12t\partial^{-1}u_y + 8yu. \qquad (2.2.7)$$

Symmetries with parameters β_{nm} may be considered as basis in the algebra of KP symmetries, see also sect. 3. They have the same commutators as the differential operators $\lambda^n\partial_\lambda^m$. The KP symmetry with parameter β_{nm} has the Z-Sh representation

$$[\partial_y - B_2, \partial/\partial\beta_{nm} - M_{nm}] = 0 \qquad (2.2.8)$$

$$M_{nm} = [K\partial^n(\sum nt_n\partial^{n-1})K^{-1}]_+ \qquad (2.2.9)$$

It is interesting to extract from (2.2.8) local contact symmetries of KP, discovered in [12], [35]. They happen to correspond to (2.2.6) with $n = 1 - 2m$, $2 - 2m$ and $3 - 2m$ if putting $t_n = 0$ as $n \geq 4$.

According to the commutation rule one may use (2.2.7) to build an algebra of symmetries of any degree in t_n, since $[\partial_{02}, \partial_{n0}] = n(n-1)\partial_{n-2,0} + n\partial_{n-1,n}$ and so on.

2.3 Modified Zakharov-Manakov ring

The symmetries (2.2.4) - (2.2.7) may be obtained from the $\bar\partial$-method by adding new operators to the ring (1.3.8), (1.3.8a) [19], [20]. They are $D_{nm} = \partial_{nm} + \lambda^n D_\lambda^m$. Here $D_\lambda = \partial_\lambda + (\theta_0')_\lambda$ is the prolonged λ derivative, $\partial_{nm} = \partial/\partial\beta_{nm}$ and the following dependence of the kernel $R(\lambda,\lambda')$ on β_{nm} is assumed:

$$\partial_{nm}R(\lambda',\bar\lambda',\lambda,\bar\lambda) = (D_\lambda^{*\,m}(\lambda')^n - \lambda^n D_\lambda^m)R. \qquad (2.3.1)$$

Here $D_\lambda^* = -\partial_\lambda - (\theta_0')_\lambda$. Then in addition to operators $\tilde M$ of the left ideal of M (1.3.8a) we obtain

$$(D_{nm} - \tilde M_{nm})\chi = 0, \quad \tilde M_{nm} = M_{nm} \quad (\partial \to \partial + \lambda) \qquad (2.3.2)$$

The proof of (2.3.1) follows from the Stokes theorem and infinitesimal Riemann problem

$$\delta\Psi_+ = \delta\Psi_- + \beta_{nm}\lambda^n\partial^m\Psi/\partial\lambda^m$$

which is solved with the help of the Cauchy kernel :

$$\delta\Psi_+(\lambda) = \frac{1}{2\pi i}\oint \mu^n \frac{\partial^m\Psi}{\partial\mu^m}\omega(\lambda,\mu).$$

3 General quasiperiodic case [21] - [24]. Completeness

3.1 Action of the nonisospectral symmetries on the finite-gap solutions

We are going here to explain how the nonisospectral symmetries act on the finite-gap solutions. Special attention will be paid to the conformal $(m, 1)$ symmetries corresponding to the parameters β_{m1}. The reason is that only these $(m, 1)$ and Abel $(m, 0)$ symmetries preserve the finite-gap structure of solutions. The simple way to explain is the following. In the small amplitude limit the one-gap solution is a monochromatic wave $a\exp($ ikx $+i\omega(k)t + \phi)$. The action of Abel symmetries shifts the phase ϕ, the action of the $(m, 1)$ symmetries changes the wave vector k while the infinitesimal action of (m, n) symmetries with $n \geq 2$ transports the monochromatic wave to the modulated VKB-type wave $a(\epsilon x, \ldots)\exp(ik(\epsilon x, \ldots)x + \ldots)$.

Remark. The subalgebra of Abelian and conformal symmetries contains a part compatible with the KdV reduction.

We shall show that the action of $(m, 1)$ symmetries on the finite-gap solutions corresponds to the conformal algebra action on the Riemann surfaces [44]. Conformal algebra is the algebra of the vector fields $V_n = \lambda^{n+1}\partial_\lambda$ on the circle with the commutation relations

$$[V_n, V_m] = (m - n)V_{m+n}. \tag{3.1.1}$$

Now we consider how the algebra of the vector fields on the circle varies the structures of Riemann surfaces [44]. Let S be a small circle around ∞ on Γ and $U(S)$ be its small neighborhood such that $\infty \in U(S)$. Let Γ be covered by two regions Γ_+ and Γ_- such, that $U(S) = \Gamma_+ \cap \Gamma_-$ and $\infty \in \Gamma_-$. Γ may by treated as a result of gluing Γ_+ and Γ_-. We may vary the Riemann surface Γ by changing the gluing law: instead of $\gamma_+ \to \gamma_-, \gamma_- \in \Gamma_-, \gamma_+ \in \Gamma_+$ we glue $\gamma_+ \to \exp(\beta v)\gamma_-$, where $\exp(\beta v)$ is a shift along vector field v in $U(S), \beta$ is group time. Both Riemann surfaces Γ and Γ' (the new one) are constructed from the same regions Γ_+ and Γ_-. Then the unit maps $\Gamma_+ \to \Gamma_+$, $\Gamma_- \to \Gamma_-$ define a natural mapping $E : \Gamma \to \Gamma'$ with a jump on S. We assume the local parameter $z = 1/\lambda$ and the points $\gamma_1, \ldots, \gamma_g$ to be mapped by E. In the case of infinitesimal action ($\beta << 1$) holomorphic j-tensor field Δ' on the new surface Γ' can be treated as a field on the old surface Γ with a jump on S satisfying the following equation

$$\Delta'_+ - \Delta'_- = \beta L_v \Delta, \tag{3.1.2}$$

where Δ'_+ and Δ'_- are the boundary values of Δ' on S, L_v is the Lie derivative, Δ is the original field on Γ. Thus Δ' is a solution of the Riemann problem - a well-known object in the soliton theory.

Remark. From the geometrical point of view the map E determines a pair of connections on the following bundle: the base of this bundle is the moduli space of all Riemann surfaces with a marked point and a local parameter in it and the fibers of the bundle are these Riemann surfaces. One of these connections is correctly defined and nonsingular everywhere except the neighborhood of the marked points, the other one is defined and nonsingular in this neighborhood.

The difference between these connections is exactly the differential operator on the fiber L_v.

As we mentioned above we have to solve the Riemann problem to describe the conformal algebra action. This is performed using the appropriate Cauchy kernel, which solves the factorization problem (3.1.2).

It is worth noting that the Riemann problem plays a crucial role in the soliton theory. For example see [14], [47] for solutions construction and [45] for r-matrix method. When constructing symmetries, which are infinitesimal variations of solutions for 1+1 systems, the following linearized Riemann problem arises : $\delta\chi_+(\lambda) = \delta G(\lambda) + \delta\chi_-(\lambda)$ with some matrix $\delta G(\lambda)$ [18] on some contour γ . Its solutions perform infinitesimal dressing of operators of the underlying linear system [46],

$$\mathcal{L}\Psi \equiv (\partial_x - U(\lambda))\Psi = 0, \quad A_n\Psi \equiv (\partial_{t_n} - V_n(\lambda))\Psi = 0,$$

so that

$$\delta\mathcal{L} = [\mathcal{L}, \delta\chi_\pm], \ \delta A_n = [A_n, \delta\chi_\pm].$$

The factorization of this Riemann problem is performed using the usual Cauchy integral over γ :

$$\delta\chi_\pm = \oint_\gamma d\mu\delta G(\mu)/(\mu - \lambda).$$

To outline the scheme constructing of Z-Sh representations for any symmetry equation, consider for simplicity $U = \lambda U_0 + U_1$ and let $V_n(\lambda)$ be a polynomial in λ. Then for all n $V_n = (\lambda^n V)_+$, where $V = V_0 + \sum_1^\infty v_n\lambda^{-n}$, $[U_0, V_0] = 0$, V is a series in λ satisfying $[\mathcal{L}, V] = 0$, $\mathcal{L} = \partial_x - U(\lambda)$. Noncommutative conformal symmetries have the Z-Sh representation [18] $[\mathcal{L}, \partial/\partial\beta_{m1} - (\lambda^{m+1}.\hat{W})_+] = 0$, where $\hat{W} = \partial_\lambda - \sum_1^\infty w_n\lambda^{-n}$ satisfies $[\mathcal{L},, \hat{W}] = 0$. The factorization of matrix-valued series in λ, forming the so-called loop algebra is performed by integration with the Cauchy kernel over a small contour S surrounding $\lambda = \infty$,

$$V(\lambda)_+ = \oint_S d\mu V(\mu)/(\mu - \lambda), W_+ = \oint_S d\mu W(\mu)/(\mu - \lambda)$$

To calculate the variation of the Baker-Akhiezer function corresponding to the finite-gap solutions we need an appropriate Cauchy kernel $\omega(\lambda, \mu, t)$. It should have the following properties:

1) $\omega(\lambda, \mu, t)$ is a function (0-form) in λ and a 1-form in μ.
2) As $\lambda \to \mu$ $\quad \omega(\lambda, \mu, t) = \frac{1}{2\pi i}\frac{d\mu}{(\mu-\lambda)} +$ regular terms
3) $\omega(\lambda, \mu, t) = \frac{\exp(\theta_0)}{\lambda}O(1)d\mu$ as $\lambda \to \infty$
$\omega(\lambda, \mu, t) = \frac{\exp(-\theta_0)}{\mu}O(1)d\mu$ as $\mu \to \infty$
where θ_0 is the same as in the formulas (1.3.11a), (1.3.15b) respectively.
4) $\omega(\lambda, \mu, t)$ has the same analytical properties in λ as the Baker-Akhiezer function $\Psi(\lambda, t)$ and as $\Psi^*(\mu, t)$ in μ. For example, for the solutions decreasing at the infinity, $\omega(\lambda, \mu, t)$ obeys the nonlocal Riemann problem (1.3.11) or the $\bar{\partial}$ - problem (1.3.14a) by λ for all μ. For the finite-gap solutions $\omega(\lambda, \mu, t)$ is meromorphic as a function of λ on the corresponding Riemann surface and has simple poles

at the same points as $\Psi(\lambda, t)$ and is holomorphic in μ with simple zeroes in these points.

It is possible to prove the existence and uniqueness of such Cauchy kernels. But fortunately there exists a simple universal formula for $\omega(\lambda, \mu, t)$ [21] - [24] :

$$\omega(\lambda, \mu, x, y, \ldots) = \frac{1}{2\pi} \int_{\mp\infty}^{x} \Psi(\lambda, z, y, \ldots)\Psi^*(\mu, z, y, \ldots)dz \tag{3.1.3}$$

(a similar formula for systems with discrete variable x was obtained earlier by Krichever and Novikov [32]).

The limit of integration in (3.1.3) is chosen from the condition of the convergence, depending on mutual position of λ and μ .

In what follows we shall use the quasimomentum $p(\lambda)$. It is a multivalued function on the Riemann surface Γ such that $dp(\lambda)$ is an Abelian differential of the second kind with a second-order pole in p_∞ : $dp \sim d\lambda$, $z = 1/\lambda$ is a local parameter in p_∞ and the normalization condition for dp is:

$$\mathrm{Im} \oint_{a_i} dp = \mathrm{Im} \oint_{b_i} dp = 0. \tag{3.1.4}$$

In the case vanishing at infinity $p(\lambda) = \lambda$.

From (3.1.4) we see that $\mathrm{Im}p(\lambda)$ is a single-valued function in Γ. The $p(\lambda)$ is called quasimomentum because for the solutions purely periodic in x we have :

$$\Psi(\lambda, , x+T, y, \ldots) = \Psi(\lambda, x, y, \ldots)\exp(ip(\lambda)T). \tag{3.1.5}$$

(for additional information see the paper of Krichever in this book).

For a given $\mu \in \Gamma$ consider the following curve $\alpha(\mu)$:

$$\lambda \in \alpha(\mu) \text{ if } \mathrm{Im}\ p(\lambda) = \mathrm{Im}\ p(\mu). \tag{3.1.6}$$

From (3.1.5) we see that the limit of integration changes when λ crosses the curve $\alpha(\mu)$. For $\lambda \in \alpha(\mu)$ all the properties of $\omega(\lambda, \mu, t)$ can be easily deduced from (3.1.3). The case $\lambda \in \alpha(\mu)$ needs a more detailed consideration.

To prove that ω is the Cauchi kernel one has to begin with the orthonormality conditions: if $\lambda, \mu \in \alpha$ then on the curve α

$$\frac{1}{2\pi} \int_{-\infty}^{\infty} \Psi(\lambda, x)\Psi^*(\mu, x)dx = \delta(\lambda - \mu). \tag{3.1.7}$$

Really, $\partial\Psi/\partial t_n = B_n\Psi, \partial\Psi^*/\partial t_n = -B_n^+\Psi^*$ and

$$(\partial/\partial t_n) \int_{-\infty}^{\infty} \Psi\Psi^* dx = \int\{(B_n\Psi)\Psi^* - \Psi(B_n^+\Psi^*)\} = 0.$$

On the other hand for large t_n $\Psi\Psi^* \sim \exp\{i(p(\lambda^n) - p(\mu^n))t_n\}$. If $\lambda \neq \mu$ then for some n $\Psi\Psi^* \to 0$ as $t_n \to +\infty$ or $-\infty$. Thus the integral (3.1.7) is equal to 0 for $\lambda \neq \mu$. So we are to prove now that the integral (3.1.7) gives a δ-function with unit coefficient. This proof coincides with that in the discrete case and may be found in [31]. Going further we see that $\omega(\lambda, \mu)$ has the appropriate analytical properties (for example satisfies the $\bar\partial$ - problem in λ and the conjugate one in μ in the decreasing KP-II case) everywhere where $p(\lambda) \neq p(\mu)$. On the line $\alpha(\mu)$, ω may have a jump as a function of λ because we change the limit of integration. To prove that $\omega(\lambda, \mu)$ is continuous as $\lambda \neq \mu$ let us calculate the singular part of $\bar\partial_\lambda \omega$ on $\alpha(\mu)$. Consider an interval I in $\alpha(\mu)$ and a small domain $D \ni I$. Using Stokes theorem we have

$$\int_D \bar\partial_\lambda \omega d\lambda d\bar\lambda = \oint_{\partial D} \omega d\lambda.$$

Then deform the contour ∂D to go along both edges of I.

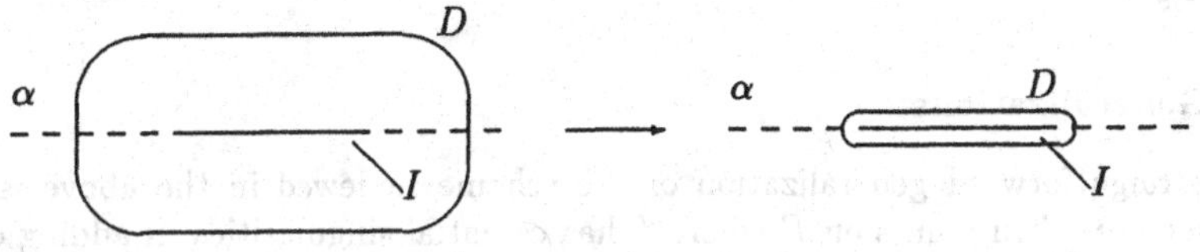

This gives (let ω_+, ω_- be the limit values of ω on edges of α)

$$\int_D \bar\partial_\lambda \omega d\lambda d\bar\lambda = \int_I (\omega_+ - \omega_-) d\lambda = \frac{1}{2\pi} \int_I \int_{-\infty}^{\infty} \Psi(\lambda)\Psi^*(\mu) dx d\lambda = \begin{cases} 0 \text{ for } \mu \notin I \\ 1 \text{ for } \mu \in I \end{cases}$$

We use here the jump of integration limit on α and (3.1.7).

Direct calculation of the variation of the Baker-Akhiezer function via a vector field v gives us the following answer:

$$\frac{\partial U}{\partial \beta} = \frac{1}{2\pi i} \oint_S (v\Psi(\lambda, t)\Psi^*(\lambda, t) \tag{3.1.8}$$

$$\frac{\partial \Psi(\lambda, t)}{\partial \beta} = \frac{1}{2\pi i} \oint_S (v\Psi(\mu, t))\omega(\lambda, \mu, t) \tag{3.1.9}$$

where S is a small circle surrounding p_∞ on Γ . The variations of Ψ by means of ω may be considered as a proper analytic form of infinitesimal Z-Sh dressing validating both in the finite-gap and in the rapidly decreasing cases. So we have proved that equations (3.1.8) coincide with the non-isospectral KP symmetries with the group parameters β_{n1}, e.g. linear in t_n, and for $t_n = 0, n > 3$ they appear to be master symmetries [13].

Remark. The action of the part of the symmetries (which in fact do not change the Riemann surface, was calculated for the first time in [32]. The problem of calculating other symmetries was posed there.

Invariant solutions of (3.1.5), (3.1.6) solve the isomonodromy problem [39]. The special role of $\omega(\lambda, \mu)$ in the symmetry theory may be explained also because symmetries may be considered as a result of small variations of the $\bar\partial$-problem kernel R . This small variation may be expanded via basis of symmetries to give (3.1.1).

Completeness. The fact that ω is a Green function of the $\bar\partial$-problem and any solution of perturbed $\bar\partial$-problem may be expressed through it proves the completeness of the obtained set of symmetries. It is worth noticing that for

1+1 equations, e.g. for KdV (1.1.1) symmetries of the type (1.1.5) have simple meaning [18]. They may be considered as small conformal transformations of the scattering data and it holds true in the periodic case too [23]. If the spectrum of the operator $L = -d_x^2 + u(x)$ consists of an infinite set of intervals $[E_0, E_1], [E_2, E_3], \ldots, E_0 < E_1 < \ldots$, then $\partial E_k/\partial\beta_{2m} = 2E_k^{m+1}$, $\beta_m = \beta_{m+1\ 1}$.

4 τ - functions, Virasoro algebra, Schlessinger transformations, Toda hierarchy, vertex operators and related topics

4.1 Generalizations

The straightforward generalization of the scheme reviewed in the above is to add some marked points on Γ where Ψ has essential singularities. If adding one point, so that the points p_0 and p_∞ are marked on Γ , we come to Veselov-Novikov equation theory [36] (symmetries are considered in [52]) and to Toda hierarchy theory which generalize the KP [7], hierarchy [21] - [24]. To obtain the 2-D Toda lattice [38]

$$\partial\theta_k/\partial_x\partial\beta = \exp(\theta_k - \theta_{k-1}) - \exp(\theta_{k+1} - \theta_k) \tag{4.1.1}$$

we have to treat the M-operator corresponding to (3.1.7), $M_k = \partial/\partial\beta - \Psi_k\partial^{-1}\Psi_k^*$, and discrete differential operator L_k, shifting variables, $L_k = (\partial - \theta'_{k-1})\exp(-\partial/\partial k)$, $\theta' = \partial\theta/\partial x$, see [7]. This condition $[M_k, L_k] = 0$ gives us Ψ_k , Ψ_k^* as eigenfunctions of L_k, L_k^+ respectively. In this case Ψ, Ψ^* and the kernel $\omega(\lambda,\mu)$ begin to depend on an additional discrete time t_0 [24]. The transformation from t_0 to t_0+1 is the so-called Schlessinger transformation. It arises in the isomonodromy problem [24], [41].

4.2 τ-function, vertex operators and symmetries

In studying various solvable equations a strange phenomena occurred. For almost all of them the solutions and corresponding Baker-Akhiezer functions of the hierarchy might be expressed in a closed form via one function of all commutative times. This function is called the τ-function, see [40], [53]. For example for KP $u = \partial_x^2 \ln\tau(\mathbf{t})$. The τ-function appeared to be an interesting object and study of the τ-function revealed intriguing connections between soliton theory and quantum field theory [40], [34]. In the context of symmetries, the Virasoro algebra arises [18]-[24]. This algebra plays a very important role in the quantum field theory of strings. The action of vector fields on the Riemann surfaces may be represented as Virasoro algebra action on τ_j where τ_j is the τ-function corresponding to the Baker-Akhieser j-differentials. This action was calculated in [48] for $j = 0$ and in [21]-[24] for all j.

$$\partial\tau_j(t_0,\mathbf{t})/\partial\beta_{m+1\ 1} = L_m^j\tau_j(t_0,\mathbf{t})$$

where L_m^j are the Virasoro algebra operators :

$$L^j_m = \sum(kt_k\partial_{k+m} + \frac{1}{2}(\partial_k\partial_{m-k})) + (t_0 - 2j + (j - \frac{1}{2})(m+1))\partial_m, \; m > 0,$$

$$L^j_0 = \sum(kt_k\partial_k + (t_0 - 2j)^2/2 + (j - \frac{1}{2})(t_0 - 2j),$$

$$L^j_m = \sum(kt_k\partial_{k+m} - m(t_0 - 2j + (j - \frac{1}{2})(m+1)t_{-m} +$$
$$+ k(m-k)t_{-k}t_{k-m}/2, \qquad m < 0.$$

(These formulas are known as Shugavara formulas).

The algebra of L^j_m and hence the action of symmetries on τ_j is a central extension of action on Ψ :

$$[L^j_m, L^j_{m'}] = (m' - m)L^j_{m+m'} + (6j^2 - 6j + 1)\frac{(m^3 - m)}{6}\delta_{m+m',0}.$$

More detailed information may be found in [24].

The Baker-Akhiezer functions may be expressed also in terms of τ-functions and vertex operators [34], [7]

$$\Psi(\lambda, t) = \tau^{-1}(t)X(\lambda, t)\tau(t)$$
$$\Psi^*(\lambda, t) = \tau^{-1}(t)X^*(\lambda, t)\tau(t)$$

where the vertex operators $X(\lambda, t)$, $X^*(\lambda, t)$ are

$$X(\lambda, t) = \exp\left\{\sum_1^\infty \lambda^m t_m\right\} \exp\left\{-\sum_1^\infty (n\lambda^n)^{-1}\partial/\partial t_n\right\}.$$

$$X^*(\lambda, t) = d\lambda \exp\left\{-\sum_1^\infty \lambda^m t_m\right\} \exp\left\{\sum_1^\infty (n\lambda^n)^{-1}\partial/\partial t_n\right\}.$$

The dynamics of Baker-Akhiezer functions Ψ, Ψ^* according to the linear problems underlying KP, (2.1.11) $\partial\Psi/\partial\beta = M\Psi$ leads to the τ-function dynamics [7], [19]:

$$\partial\tau/\partial\beta = X^*(\mu, t)X(\lambda, t)\tau(t, \beta).$$

References

[1] E. Wigner, Symmetries and reflections, Bloomington - London, 1967.

[2] N.J. Zabusky, M.D. Kruskal, Interaction of solitons in a collisionless plasma and the recurrence of initial states, Phys. Rev. Letters, 15, pp. 240-243, (1965); C.S. Gardner, J.M. Green, M.D. Kruskal, R.M. Miura, Comm. Pure. Appl. Math., 27, 97-133 (1974).

[3] A.C. Newell, Solitons in Mathematics and Physics, Conf. Bd. Math. Soc., 43, Soc. Indust. Appl. Math. (1985).

[4] V.E. Zakharov, E.I. Schulman, Integrability of Nonlinear Systems and Perturbation Theory, in "What is Integrability?", ed. by V.E. Zakharov, Springer Series in Nonlinear Dynamics pp. 185-250 (1991).
[5] F. Calogero, Why are Certain Nonlinear PDEs both Widely Applicable and Integrable?, in "What is Integrability?", ed. by V.E. Zakharov, Springer Series in Nonlinear Dynamics pp. 1-61 (1991).
[6] P.D. Lax, Comm. Pure Appl. Math., 21, pp. 467-490 (1968).
[7] A.Yu. Orlov, Symmetries for Unifying Different Systems into a Single Integrable Hierarchy, Preprint IINS/Oce-04/03, 1991.
[8] V.E. Zakharov, S.V. Manakov, S.P. Novikov, L.P. Pitaevsky, Soliton Theory, (Plenum, New-York, 1984).
[9] M.J. Ablowitz, H. Segur, Solitons and the Inverse Scattering Transform (SIAM, Philadelphia, 1981).
[10] N.Kh. Ibragimov, A.B. Shabat, Dokl. Acad. Nauk SSSR, 244, 1, pp. 56-61 (1979).
[11] H.H. Chen, Y.C. Lee, J.E., Lin. Physica 9D, 9, N3, pp. 439-445 (1983).
[12] F.J. Schwarz, J. Phys. Soc. Japan, 51, 8, p. 2387 (1982)
[13] B. Fuchssteiner, Progr. Theor. Phys., 70, pp. 1508-1522 (1983).; W. Oevel, B. Fuchssteiner, Phys. Lett. 88A, 323, (1982).
[14] V.E. Zakharov, A.B. Shabat, Funct. Anal. i ego Priloz. 13, 3, pp. 13-22 (1979).
[15] V.E. Zakharov, A.B. Mikhailov, ZhETF, 74, 6, pp. 1954-1973, (1978).
[16] V.E. Zakharov, L.A. Takhtajan, Teor. Mat. Fiz., 38, 1, pp. 26-35 (1979).
[17] A.S. Fokas, P. Santini, Communications in Math. Phys. 115, 116 (1988).
[18] A.Yu. Orlov, E.I. Schulman, Additional Symmetries of Integrable Systems and Conformal Algebra Representations. Preprint IA and E, N217, Novosibirsk, 1984. A.Yu. Orlov, E.I. Schulman, Teor. Mat. Fiz., 64, pp. 323-327 (1985).
[19] A.Yu. Orlov, E.I. Schulman, Additional Symmetries for 2-D Integrable Systems. Preprint IA and E, N277, Novosibirsk, 1985. A.Yu. Orlov, E.I. Schulman, Letters in Math. Phys., 12, pp. 171-179, (1986).
[20] A.Yu. Orlov, Vertex Operator, $\bar{\partial}$ - Problem, Symmetries, Variational Identities and Hamiltonian Formalism for 2+1 Integrable Systems. Proc. IV Int. Workshop on Nonlinear and Turb. Proc. Phys., Singapore, 1989, pp. 116-134.
[21] P.G. Grinevich, A.Yu. Orlov, Wilson τ-function and $\det \bar{\partial}$. In Proc. IV Int. Workshop on Nonlinear and Turb. Proc. Phys., Kiev, 1989, pp. 242-245.
[22] P.G. Grinevich, A.Yu. Orlov, Vector Fields Action on Riemann Surfaces and KP Theory. Krichever-Novikov Problem. ibid., pp. 246-249.
[23] P.G. Grinevich, A.Yu. Orlov, Higher Symmetries of KP Equation and the Virasoro Action on Riemann Surfaces. In "Nonlinear Eqs. and Dynamical Syst., ed. S. Carillo, O. Ragnisco, Springer, 1990, pp. 165-169.
[24] P.G. Grinevich, A.Yu. Orlov, In "Modern Problems of Quantum Field Theory", Springer-Verlag, 1989.
[25] V.E. Zakharov, A.V. Shabat, Funct. Anal. i ego Priloz., 8, pp. 43-53 (1974).
[26] A.S. Fokas, D.BarYaacov, M.J. Ablowitz, Stud. Appl. Math., 69, N2, pp. 135-143, (1983).
[27] A.S. Fokas, M.J. Ablowitz, Stud. Appl. Math., 69, N3, pp. 211-228, (1983).
[28] A.S. Fokas, M.J. Ablowitz, Phys. Rev. Letters, 51, 1, pp. 7-11, (1983).
[29] V.E. Zakharov, S.V. Manakov, Funct. Anal. i ego Priloz., 19, 2, pp. 11-25, (1985).
[30] S.V. Manakov, Physica P, 3, N1+2, pp. 420-427, (1981).
[31] I.M. Krichever, This volume. and Uspechy Math. Nauk, 44, 2, pp. 121-184, (1989); Uspechy Math. Nauk, 32(6), 183-208, (1977).
[32] I.M. Krichever, S.P. Novikov, Funct. Anal. i ego Priloz., 21, 2, 46 (1987).

[33] I.M. Krichever, S.P. Novikov, Funct. Anal. i ego Priloz., 23, 1, 24 (1989).
[34] E. Date, M. Kashiwara, M. Jimbo, T. Miwa, In Proc. RIMS Symposium, World Scientific, Singapore (1983).
[35] D. David, D. Levi, P. Winternitz, Phys. Lett. 118A, 390(1986).
[36] A.V. Veselov, S.P. Novikov, Dokl. Akad. Nauk SSSR,
[37] E. Schulman, L. Ryzhik, to appear.
[38] A.V. Mikhailov, Physica 3D, 73-117(1981).
[39] H. Flaschka, A.C. Newell, Comm. Math. Phys., 76, 65-116(1980); A.R. Its, Soviet Math. Dokl., 24(1981); M. Jimbo, T. Miwa, Physica 2D, 407 (1981).
[40] R. Hirota, In "Backlund transformations", R.M. Miura ed., Lect. Notes Math. 515 (1976).
[41] M. Sato, T. Miwa, M. Jimbo, Holonomic Quantum Fields I, II, III, IV, V Publ. RIMS, Kyoto Univ., 14, 223-267 (1978); 15, 201-278; 577-629; 871-972 (1979), 531-584 (1980); and also RIMS Preprint 246 (1978).
[42] P.G. Grinevich, Funct. Anal. i ego Priloz., 23, 4, 79-80 (1989).
[43] F. Calogero, A. Degasperis, Lett. Nuov. Cim., 22 (1978), 420; V.A. Belinskii, V.E. Zakharov, JETP, 12 (1978), 1953 (Russian); Maison, Phys Lett (1978).
[44] M. Shiffer, D.K. Spencer, Functional on Finite Riemann Surfaces, Moscow, 1957.
[45] M.A. Semenov-Tian-Shansky, Funct. Anal. i ego Priloz., 17, 4, 17-33 (1983).
[46] M.J. Ablowitz, D. Kaup, A.C. Newell, H. Segur, Stud. Appl. Math., 53, 249-315 (1974).
[47] I.M. Krichever, Funct. Anal. i ego Priloz. 12, 3, 20-31 (1978); Engl. transl. in Functional Anal. Appl. 12 (1978); I.M. Krichever, S.P. Novikov, Dokl. Akad. Nauk SSSR 247 (1979); Engl. transl. in Soviet. Math. Dokl, 20(1979).
[48] N. Kawamoto, Yu. Namikawa, A. Tsuchiya, Ya. Yamada, Comm. Math. Phys. 116, No 2, 247-308 (1988).
[49] F. Calogero, A. Degasperis, Spectral transform and solitons. North-Holland Publishing Company, 1982.
[50] H.H. Chen, J.E. Lin, Constraints in Kadomtsev-Petviashvili Equation. Preprint N82-112, Univ. of Maryland, 1981.
[51] A. Reiman, M. Semionov-Tian-Shamski, Notes of Sci. Seminars of LOMI, vol. 133, pp. 212-227 (1984).
[52] T. Shiota, Invent. Math., 83, 333-382 (1986).
[53] G. Segal, G. Wilson, Publ. Math. IHES 61, p. 5 (1985).

The n-Component KP Hierarchy and Representation Theory

V.G. Kac[1] *and J.W. van de Leur*[2]

[1]Department of Mathematics, MIT, Cambridge, MA 02139, USA
[2]Department of Mathematics, University of Utrecht, Utrecht, The Netherlands

§0. Introduction.

0.1. The remarkable link between the soliton theory and the group GL_∞ was discovered in the early 1980s by Sato [S] and developed, making use of the spinor formalism, by Date, Jimbo, Kashiwara and Miwa [DJKM1,2,3], [JM]. The basic object that they considered is the KP hierarchy of partial differential equations, which they study through a sequence of equivalent formulations that we describe below. The first formulation is a deformation (or Lax) equation of a formal pseudo-differential operator $L = \partial + u_1\partial^{-1} + u_2\partial^{-2} + \dots$, introduced in [S] and [W1]:

$$\frac{\partial L}{\partial x_n} = [B_n, L], \ n = 1, 2, \dots . \tag{0.1.1}$$

Here u_i are unknown functions in the indeterminates $x_1, x_2, \dots$, and $B_n = (L^n)_+$ stands for the differential part of L^n. The second formulation is given by the following zero curvature (or Zakharov-Shabat) equations:

$$\frac{\partial B_m}{\partial x_n} - \frac{\partial B_n}{\partial x_m} = [B_n, B_m], \ m, n = 1, 2, \dots . \tag{0.1.2}$$

These equations are compatibility conditions for the following linear system

$$Lw(x,z) = zw(x,z), \ \frac{\partial}{\partial x_n} w(x,z) = B_n w(x,z), \ n = 1, 2, \dots \tag{0.1.3}$$

on the wave function

$$w(x,z) = (1 + w_1(x)z^{-1} + w_2(x)z^{-2} + \dots)e^{x_1 z + x_2 z^2 + \dots}. \tag{0.1.4}$$

Provided that (0.1.2) holds, the system (0.1.3) has a unique solution of the form (0.1.4) up to multiplication by an element from $1 + z^{-1}\mathbb{C}[[z^{-1}]]$. Introduce the wave operator

$$P = 1 + w_1(x)\partial^{-1} + w_2(x)\partial^{-2} + \dots , \tag{0.1.5}$$

so that $w(x,z) = Pe^{x_1 z + x_2 z^2 + \dots}$. Then the existence of a solution of (0.1.3) is equivalent to the existence of a solution of the form (0.1.5) of the following Sato equation, which is the third formulation of the KP hierarchy [S], [W1]:

$$\frac{\partial P}{\partial x_k} = -(P \circ \partial \circ P^{-1})_- \circ P, \ k = 1, 2, \dots , \tag{0.1.6}$$

VK is supported in part by the NSF grant DMS-9103792.
JvdL is supported by a fellowship of the Royal Netherlands Academy of Arts and Sciences

where the formal pseudo-differential operators P and L are related by

$$L = P \circ \partial \circ P^{-1}. \tag{0.1.7}$$

Let $P^* = 1 + (-\partial)^{-1} \circ w_1 + (-\partial)^{-2} \circ w_2 + \dots$ be the formal adjoint of P and let

$$w^*(x,z) = (P^*)^{-1} e^{-x_1 z - x_2 z^2 - \dots}$$

be the adjoint wave function. Then the fourth formulation of the KP hierarchy is the following bilinear identity

$$\mathrm{Res}_{z=0} w(x,z) w^*(x',z) dz = 0 \text{ for any } x \text{ and } x'. \tag{0.1.8}$$

Next, this bilinear identity can be rewritten in terms of Hirota bilinear operators defined for an arbitary polynomial Q as follows:

$$Q(D) f(x) \cdot g(x) \overset{\text{def}}{=} Q(\frac{\partial}{\partial y})(f(x+y) g(x-y))|_{y=0}. \tag{0.1.9}$$

Towards this end, introduce the famous τ-function $\tau(x)$ by the formulas:

$$w(x,z) = \Gamma^+(z)\tau/\tau, \ w^*(x,z) = \Gamma^-(z)\tau/\tau. \tag{0.1.10}$$

Here $\Gamma^{\pm}(z)$ are the vertex operators defined by

$$\Gamma^{\pm}(z) = e^{\pm(x_1 z + x_2 z^2 + \dots)} e^{\mp(z^{-1}\tilde{\partial}/\partial x_1 + z^{-2}\tilde{\partial}/\partial x_2 + \dots)}, \tag{0.1.11}$$

where $\frac{\tilde{\partial}}{\partial x_j}$ stands for $\frac{1}{j}\frac{\partial}{\partial x_j}$. The τ-function exists and is uniquely determined by the wave function up to a constant factor. Substituting the τ-function in the bilinear identity (0.1.8) we obtain the fifth formulation of the KP hierarchy as the following system of Hirota bilinear equations:

$$\sum_{j=0}^{\infty} S_j(-2y) S_{j+1}(\tilde{D}) e^{\sum_{r=1}^{\infty} y_r D_r} \tau \cdot \tau = 0. \tag{0.1.12}$$

Here $y = (y_1, y_2, \dots)$ are arbitrary parameters and the elementary Schur polynomials S_j are defined by the generating series

$$\sum_{j \in \mathbb{Z}} S_j(x) z^j = \exp \sum_{k=1}^{\infty} x_k z^k. \tag{0.1.13}$$

The τ-function formulation of the KP hierarchy allows one to construct easily its N-soliton solutions. For that introduce the vertex operator [DJKM2,3]:

$$\Gamma(z_1, z_2) =: \Gamma^+(z_1)\Gamma^-(z_2) : \tag{0.1.14}$$

(where the sign of normal ordering : : means that partial derivatives are always moved to the right), and show using the bilinear identity (0.1.8) that if τ is a solution of (0.1.12), then $(1 + a\Gamma(z_1, z_2))\tau$, where $a, z_1, z_2 \in \mathbb{C}^\times$, is a solution as well. Since $\tau = 1$ is a solution, the function

$$f_N \equiv (1 + a_1 \Gamma(z_1^{(1)}, z_2^{(1)})) \dots (1 + a_N \Gamma(z_1^{(N)}, z_2^{(N)})) \cdot 1 \tag{0.1.15}$$

is a solution of (0.1.12) too. This is the τ-function of the N-soliton solution.

The first application of the KP hierarchy, as well as its name, comes from the fact that the simplest non-trivial Zakharov-Shabat equation, namely (0.1.2) with $m = 2$ and $n = 3$, is equivalent to the Kadomtsev-Petviashvili equation if we let $x_1 = x,\ x_2 = y,\ x_3 = t,\ u = 2u_1$:

(0.1.16)
$$\frac{3}{4}\frac{\partial^2 u}{\partial y^2} = \frac{\partial}{\partial x}\left(\frac{\partial u}{\partial t} - \frac{3}{2}u\frac{\partial u}{\partial x} - \frac{1}{4}\frac{\partial^3 u}{\partial x^3}\right).$$

Recall also that the celebrated KdV and Boussinesq equations are simple reductions of (0.1.16). Since the functions u and τ are related by

(0.1.17)
$$u = 2\frac{\partial^2}{\partial x^2}\log\tau,$$

the functions $2\frac{\partial^2}{\partial x^2}\log f_N$ are solutions of (0.1.16), called the N-soliton solutions.

0.2. The connection of the KP hierarchy to the representation theory of the group GL_∞ is achieved via the spinor formalism. Consider the Clifford algebra $C\ell$ on generators ψ_j^+ and ψ_j^- ($j \in \frac{1}{2}+\mathbb{Z}$) and the following defining relations (i.e. $\psi_i^\pm$ are free charged fermions):

(0.2.1)
$$\psi_i^+\psi_j^- + \psi_j^-\psi_i^+ = \delta_{i,-j},\quad \psi_i^\pm\psi_j^\pm + \psi_j^\pm\psi_i^\pm = 0.$$

The algebra $C\ell$ has a unique irreducible representation in a vector space F (resp. F^*) which is a left (resp. right) module admitting a non-zero vector $|0\rangle$ (resp. $\langle 0|$) satisfying

(0.2.2)
$$\psi_j^\pm|0\rangle = 0 \text{ (resp. } \langle 0|\psi_{-j}^\pm = 0) \text{ for } j > 0.$$

These representations are dual to each other with respect to the pairing

$$\langle\langle 0|a,\ b|0\rangle\rangle = \langle 0|ab|0\rangle$$

normalized by the condition $\langle 0|1|0\rangle = 1$.

The Lie algebra $g\ell_\infty$ embeds in $C\ell$ by letting

(0.2.3)
$$r(E_{ij}) = \psi_{-i}^+\psi_j^-.$$

Exponentiating gives a representation R of the group GL_∞ on F and F^*. Let for $n \in \mathbb{Z}$:

(0.2.4)
$$\alpha_n = \sum_{j\in\frac{1}{2}+\mathbb{Z}} \psi_{-j}^+\psi_{j+n}^- \text{ for } n \neq 0,\ \alpha_0 = \sum_{j>0}\psi_{-j}^+\psi_j^- - \sum_{j<0}\psi_j^-\psi_{-j}^+.$$

and consider the following operator on F:

(0.2.5)
$$H(x) = \sum_{n=1}^{\infty} x_n\alpha_n.$$

For a positive integer m let

$$\langle\pm m| = \langle 0|\psi_{\frac{1}{2}}^\pm\dots\psi_{m-\frac{1}{2}}^\pm \in F^* \text{ and } |\pm m\rangle = \psi_{-m+\frac{1}{2}}^\pm\cdots\psi_{-\frac{1}{2}}^\pm|0\rangle \in F.$$

Then the Fock space is realized on the vector space of polynomials $B = \mathbb{C}[x_1, x_2, \dots; Q, Q^{-1}]$ via the isomorphism $\sigma:\ F \xrightarrow{\sim} B$ defined by

(0.2.6)
$$\sigma(a|0\rangle) = \sum_{m\in\mathbb{Z}}\langle m|e^{H(x)}a|0\rangle Q^m.$$

This remarkable isomorphism is called the boson-fermion correspondence and goes back to the work of Skyrme [Sk] and many other physicists; this beautiful form of it is an important part of the work of Date, Jimbo, Kashiwara and Miwa [DJKM2,3], [JM].

Using that

(0.2.7) $$[\alpha_m, \alpha_n] = m\delta_{m,-n},$$

(i.e that the α_n are free bosons), it is not difficult to show that the isomorphism σ is characterized by the following two properties [KP2]:

(0.2.8) $$\sigma(|m\rangle) = Q^m, \ \sigma\alpha_n\sigma^{-1} = \frac{\partial}{\partial x_n} \text{ and } \sigma\alpha_{-n}\sigma^{-1} = nx_n \text{ if } n > 0.$$

Using (0.2.8), it is easy to recover the following well-known properties of the boson-fermion correspondence [DJKM2,3], [KP2]. Introduce the fermionic fields

$$\psi^{\pm}(z) = \sum_{j\in\frac{1}{2}+\mathbb{Z}} \psi_j^{\pm} z^{-j-1/2}.$$

Then one has:

(0.2.9) $$\sigma\psi^{\pm}(z)\sigma^{-1} = Q^{\pm 1} z^{\pm\alpha_0}\Gamma^{\pm}(z),$$

(0.2.10) $$\sigma\Big(\sum_{i,j\in\frac{1}{2}+\mathbb{Z}} r(E_{ij}) z_1^{i-\frac{1}{2}} z_2^{-j-\frac{1}{2}}\Big)\sigma^{-1} = \frac{1}{z_1 - z_2}\Gamma(z_1, z_2).$$

Hence $\Gamma(z_1, z_2)$ lies in a "completion" of the Lie algebra $g\ell_\infty$ acting on B via the boson-fermion correspondence. Therefore, the group GL_∞ and its "completion" act on B and Date, Jimbo, Kashiwara and Miwa show that all elements of the orbit $\mathcal{O} = GL_\infty \cdot 1$ and its completions satisfy the bilinear identity (0.1.12). Since $\Gamma(z_1, z_2)^2 = 0$ and $\Gamma(z_1, z_2)$ lies in a completion of $g\ell_\infty$, we see that $\exp a\Gamma(z_1, z_2) = 1 + a\Gamma(z_1, z_2)$ leaves a completion of the orbit $\mathcal{O}$ invariant, which explains why (0.1.15) are solutions of the KP hierarchy.

Since the orbit $GL_\infty|0\rangle$ (which is the image of $\mathcal{O}$ in the fermionic picture) can be naturally identified with the cone over a Grassmannian, we arrive at the remarkable discovery of Sato that solutions of the KP hierarchy are parameterized by an infinite-dimensional Grassmannian [S].

0.3. It was subsequently pointed out in [KP2] and [KR] that the bilinear equation (0.1.8) (in the bosonic picture) corresponds to the following remarkably simple equation on the τ-function in the fermionic picture:

(0.3.1) $$\sum_{k\in\frac{1}{2}+\mathbb{Z}} \psi_k^+\tau \otimes \psi_{-k}^-\tau = 0.$$

This is the fermionic formulation of the KP hierarchy. Since (0.3.1) is equivalent to

(0.3.2) $$\mathrm{Res}_{z=0}\psi^+(z)\tau \otimes \psi^-(z)\tau = 0,$$

it is clear from (0.1.10 and 11) that equations (0.1.8) and (0.3.2) are equivalent. Since $\tau = |0\rangle$ obviously satisfies (0.3.1) and $R\otimes R(GL_\infty)$ commutes with the operator $\sum_k \psi_k^+ \otimes \psi_{-k}^-$, we see why any element of $R(GL_\infty)|0\rangle$ satisfies (0.3.1). Thus, the most natural approach to the KP hierarchy is to start with the fermonic formulation (0.3.1), go over to the bilinear identity (0.1.8) and then to all other formulations (see [KP2], [KR], [K]). This approach was generalized in [KW].

0.4. Our basic idea is to start once again with the fermonic formulation of KP, but then use the n-component boson-fermion correspondence, also considered by Date, Jimbo, Kashiwara and Miwa [DJKM1,2], [JM]. This leads to a bilinear equation on a matrix wave function, which in turn leads to a deformation equation for a matrix formal pseudo-differential operator, to matrix Sato equations and to matrix Zakharov-Shabat type equations.

The corresponding linear problem has been already formulated in Sato's paper [S] and Date, Jimbo, Kashiwara and Miwa [DJKM1] have written the corresponding bilinear equation for the wave function, but the connection between these formulations remained somewhat obscure.

It is the aim of the present paper to give all formulations of the n-component KP hierarchy and clarify connections between them. The generalization to the n-component KP is important because it contains many of the most popular systems of soliton equations, like the Davey-Stewartson system (for $n = 2$), the 2-dimensional Toda lattice (for $n = 2$), the n-wave system (for $n \geq 3$). It also allows us to construct natural generalizations of the Davey-Stewartson and Toda lattice systems. Of course, the inclusion of all these systems in the n-component KP hierarchy allows us to construct their solutions by making use of vertex operators.

Hirota's direct method [H] requires some guesswork to introduce a new function (the τ-function) for which the equations in question take a bilinear form. The inclusion of the equations in the n-component KP hierarchy provides a systematic way of construction of the τ-functions, the corresponding bilinear equations and a large family of their solutions.

The difficulty of the τ-function approach lies in the fact that the hierarchy contains too many Hirota bilinear equations. To deal with this difficulty we introduce the notion of an energy of a Hirota bilinear equation. We observe that the most interesting equations are those of lowest energy. For example, in the $n = 1$ case the lowest energy ($= 4$) non-trivial equation is the classical KP equation in the Hirota bilinear form, in the $n = 2$ case the lowest energy ($= 2$) equations form the 2-dimensional Toda chain and the energy 2 and 3 equations form the Davey-Stewartson system in the bilinear form, and in the $n \geq 3$ case the lowest energy ($= 2$) bilinear equations form the n-wave system in the bilinear form.

There is a new phenomenon in the n-component case, which does not occur in the 1-component case: the τ-function and the wave function is a collection of functions $\{\tau_\alpha\}$ and $\{W_\alpha\}$ parameterized by the elements of the root lattice M of type A_{n-1}. The set $\operatorname{supp} \tau = \{\alpha \in M | \tau_\alpha \neq 0\}$ is called the support of the τ-function τ. We show that $\operatorname{supp} \tau$ is a convex polyhedron whose edges are parallel to roots; in particular, $\operatorname{supp} \tau$ is connected, which allows us to relate the behaviour of the n-component KP hierarchy at different points of the lattice M. It is interesting to note that the "matching conditions" which relate the functions W_α and W_β, $\alpha, \beta \in M$, involve elements from the subgroup of translations of the Weyl group [K, Chapter 6] of the loop group $GL(\mathbb{C}[z, z^{-1}])$ and are intimately related to the Bruhat decomposition of this loop group (see [PK]). We are planning to study this in a future publication.

The behaviour of solutions obtained via vertex operators in the n-component case is much more complicated than for the ordinary KP hiearchy. In particular, they are not necessarily multisoliton solutions (i.e. a collection of waves that preserve their form after interaction). For that reason we call them the multisolitary solutions. Some of the multisolitary solutions turn out to be the so called dromion solutions, that have become very popular recently [BLMP], [FS], [HH], [HMM]. This solutions decay exponentially in all directions (and they are not soliton solutions; in particular, they exist only for $n > 1$). It is a very interesting problem for which values of parameters the multisolitary solutions are soliton or dromion solutions.

Note also that the Krichever method for construction of the quasiperiodic solutions of the KP hierarchy as developed in [SW] and [Sh] applies to the n-component KP.

As shown in [S], [DJKM2], the m-th reduction of the KP hiearchy, i.e. the requirement that L^m is a differential operator, leads to the classical formulation of the celebrated KdV hierarchy for $m = 2$, Boussinesq for $m = 3$ and all the Gelfand-Dickey hierarchies for $m > 3$. The totality of τ-functions for the m-th reduced KP hierarchy turns out to be the orbit of the vacuum under the loop group of SL_m.

We define in a similar way the m-th reduction of the n-component KP and show that the totality of τ-functions is the orbit of the vacuum vector under the loop group of SL_{mn}. Even the case $m = 1$ turns out to be extremely interesting (it is trivial if $n = 1$), as it gives the $1+1$ n-wave system for $n \geq 3$ and the decoupled non-linear Schrödinger (or AKNS) system for $n = 2$. We note that the 1-reduced n-component KP, which we call the n-component NLS hierarchy, admits a natural generalization to the case of an arbitrary simple Lie group G (the n-component NLS corresponding to GL_n). These hierarchies which might be called the GNLS hierarchies, contain the systems studied by many authors [Di], [W1 and 2], [KW],... .

0.5. The paper is set out as follows. In §1 we explain the construction of the semi-infinite wedge representation F of the group GL_∞ and write down the equation of the GL_∞-orbit $\mathcal{O}$ of the vacuum $|0\rangle$ (Proposition 1.3). This equation is called the KP hierarchy in the fermionic picture. As usual, the Plücker map makes $\mathcal{O}$ a $\mathbb{C}^\times$-bundle over an infinite-dimensional Grassmannian. We describe the "support" of $\tau \in \mathcal{O}$ (Proposition 1.4).

In §2 we introduce the n-component bosonisation and write down the fermionic fields in terms of bosonic ones via vertex operators (Theorem 2.1). This allows us to transport the KP hierarchy from the fermionic picture to the bosonic one (2.3.3) and write down the n-component KP hierarchy as a system of Hirota bilinear equations (2.3.7). We describe the support of a τ-function in the bosonic picture (Proposition 2.3). At the end of the section we list all Hirota bilinear equations of lowest energy (2.4.3–9).

We start §3 with an exposition of the theory of matrix formal pseudo-differential operators, and prove the crucial Lemma 3.2. This allows us to reformulate the n-component KP hierarchy (2.3.3) in terms of formal pseudo-differential operators (see (3.3.4 and 12)). Using the crucial lemma we show that the bilinear equation (2.3.3) is equivalent to the Sato equation (3.4.2) and matching conditions (3.3.16) on the wave operators $P^+(\alpha)$. We show that Sato equation is the compatibility condition of Sato's linear problem (3.5.5) on the wave function (Proposition 3.5), and that compatibility of Sato equation implies the equivalent Lax and Zakharov-Shabat equations (Lemma 3.6). We prove that compatibility conditions completely determine the wave operators $P^+(\alpha)$ once one of them is given (Proposition 3.3). At the end of the section we write down explicitly the first Sato and Lax equations and relations between them.

In §4 we show that many well-known $2+1$ soliton equations are the simplest equations of the n-component KP hierarchy, and deduce from §3 expressions for their τ-functions and the corresponding Hirota bilinear equations.

Using vertex operators we write down in §5 the N-solitary solutions (5.1.11) of the n-component KP and hence of all its relatives. We discuss briefly the relation of this general solution to the known solutions to the relatives.

In §6 we discuss the m-reductions of the n-component KP hierarchy. They reduce the $2+1$ soliton equations to $1+1$ soliton equations. We show that at the group theoretic level it corresponds to a reduction from GL_∞ (or rather a completion of it) to the subgroup $SL_{mn}(\mathbb{C}[t,t^{-1}])$ (Proposition 6.1). We discuss in more detail the 1-reduced n-component KP, which is a generalization of the NLS system and which admits further generalization to any simple Lie group.

We would like to thank A.S. Fokas for asking one of us to write a paper for this volume. We are grateful to E. Medina for calling our attention to the paper [HMM]. The second named author would like to thank MIT for the kind hospitality.

§1. The semi-infinite wedge representation of the group GL_∞ and the KP hierarchy in the fermionic picture.

1.1. Consider the infinite complex matrix group

$GL_\infty = \{A = (a_{ij})_{i,j\in\mathbb{Z}+\frac{1}{2}} | A \text{ is invertible and all but a finite number of } a_{ij} - \delta_{ij} \text{ are } 0\}$
and its Lie algebra

$$gl_\infty = \{a = (a_{ij})_{i,j\in\mathbb{Z}+\frac{1}{2}} | \text{ all but a finite number of } a_{ij} \text{ are } 0\}$$

with bracket $[a,b] = ab - ba$. The Lie algebra gl_∞ has a basis consisting of matrices E_{ij}, $i,j \in \mathbb{Z}+\frac{1}{2}$, where E_{ij} is the matrix with a 1 on the (i,j)-th entry and zeros elsewhere.

Let $\mathbb{C}^\infty = \bigoplus_{j\in\mathbb{Z}+\frac{1}{2}} \mathbb{C}v_j$ be an infinite dimensional complex vector space with fixed basis $\{v_j\}_{j\in\mathbb{Z}+\frac{1}{2}}$. Both the group GL_∞ and its Lie algebra gl_∞ act linearly on $\mathbb{C}^\infty$ via the usual formula:

$$E_{ij}(v_k) = \delta_{jk} v_i.$$

The well-known semi-infinite wedge representation is constructed as follows [KP2]. The semi-infinite wedge space $F = \Lambda^{\frac{1}{2}\infty}\mathbb{C}^\infty$ is the vector space with a basis consisting of all semi-infinite monomials of the form $v_{i_1} \wedge v_{i_2} \wedge v_{i_3} \ldots$, where $i_1 > i_2 > i_3 > \ldots$ and $i_{\ell+1} = i_\ell - 1$ for $\ell >> 0$. We can now define representations R of GL_∞ and r of gl_∞ on F by

(1.1.1) $$R(A)(v_{i_1} \wedge v_{i_2} \wedge v_{i_3} \wedge \cdots) = Av_{i_1} \wedge Av_{i_2} \wedge Av_{i_3} \wedge \cdots,$$

(1.1.2) $$r(a)(v_{i_1} \wedge v_{i_2} \wedge v_{i_3} \wedge \cdots) = \sum_k v_{i_1} \wedge v_{i_2} \wedge \cdots \wedge v_{i_{k-1}} \wedge av_{i_k} \wedge v_{i_{k+1}} \wedge \cdots.$$

These equations are related by the usual formula:

$$\exp(r(a)) = R(\exp a) \text{ for } a \in gl_\infty.$$

1.2. The representation r of gl_∞ can be described in terms of a Clifford algebra. Define the wedging and contracting operators ψ_j^+ and ψ_j^- $(j \in \mathbb{Z} + \frac{1}{2})$ on F by

$$\psi_j^+(v_{i_1} \wedge v_{i_2} \wedge \cdots) = \begin{cases} 0 & \text{if } j = i_s \text{ for some } s \\ (-1)^s v_{i_1} \wedge v_{i_2} \cdots \wedge v_{i_s} \wedge v_{-j} \wedge v_{i_{s+1}} \wedge \cdots & \text{if } i_s > -j > i_{s+1} \end{cases}$$

$$\psi_j^-(v_{i_1} \wedge v_{i_2} \wedge \cdots) = \begin{cases} 0 & \text{if } j \neq i_s \text{ for all } s \\ (-1)^{s+1} v_{i_1} \wedge v_{i_2} \wedge \cdots \wedge v_{i_{s-1}} \wedge v_{i_{s+1}} \wedge \cdots & \text{if } j = i_s. \end{cases}$$

These operators satisfy the following relations $(i, j \in \mathbb{Z} + \frac{1}{2}, \lambda, \mu = +, -)$:

(1.2.1) $$\psi_i^\lambda \psi_j^\mu + \psi_j^\mu \psi_i^\lambda = \delta_{\lambda,-\mu}\delta_{i,-j},$$

hence they generate a Clifford algebra, which we denote by $C\ell$.

Introduce the following elements of F $(m \in \mathbb{Z})$:

$$|m\rangle = v_{m-\frac{1}{2}} \wedge v_{m-\frac{3}{2}} \wedge v_{m-\frac{5}{2}} \wedge \cdots.$$

It is clear that F is an irreducible $C\ell$-module such that

(1.2.2) $$\psi_j^\pm|0\rangle = 0 \text{ for } j > 0.$$

It is straightforward that the representation r is given by the following formula:

(1.2.3) $$r(E_{ij}) = \psi_{-i}^+ \psi_j^-.$$

Define the *charge decomposition*

(1.2.4) $$F = \bigoplus_{m \in \mathbb{Z}} F^{(m)}$$

by letting

(1.2.5) $$\text{charge}(v_{i_1} \wedge v_{i_2} \wedge \ldots) = m \text{ if } i_k + k = \tfrac{1}{2} + m \text{ for } k >> 0.$$

Note that

(1.2.6) $$\text{charge}(|m\rangle) = m \text{ and charge}(\psi_j^\pm) = \pm 1.$$

It is clear that the charge decomposition is invariant with respect to $r(g\ell_\infty)$ (and hence with respect to $R(GL_\infty)$). Moreover, it is easy to see that each $F^{(m)}$ is irreducible with respect to $g\ell_\infty$ (and GL_∞). Note that $|m\rangle$ is its highest weight vector, i.e.

$$r(E_{ij})|m\rangle = 0 \text{ for } i < j,$$
$$r(E_{ii})|m\rangle = 0 \text{ (resp. } = |m\rangle) \text{ if } i > m \text{ (resp. if } i < m).$$

1.3. The main object of our study is the GL_∞-orbit

$$\mathcal{O} = R(GL_\infty)|0\rangle \subset F^{(0)}$$

of the vacuum vector $|0\rangle$.

Proposition 1.3 ([KP2]). *A non-zero element τ of $F^{(0)}$ lies in $\mathcal{O}$ if and only if the following equation holds in $F \otimes F$:*

$$\sum_{k\in\mathbb{Z}+\frac{1}{2}} \psi_k^+\tau \otimes \psi_{-k}^-\tau = 0. \tag{1.3.1}$$

Proof. It is clear that $\sum_k \psi_k^+|0\rangle \otimes \psi_{-k}^-|0\rangle = 0$ and it is easy to see that the operator $\sum_k \psi_k^+ \otimes \psi_{-k}^- \in \operatorname{End}(F\otimes F)$ commutes with $R(g)\otimes R(g)$ for any $g \in GL_\infty$. It follows that $R(g)|0\rangle$ satisfies (1.3.1). For the proof of the converse statement (which is not important for our purposes) see [KP2] or [KR]. □

Equation (1.3.1) is called the *KP hierarchy in the fermionic picture.*

Note that any non-zero element τ from the orbit $\mathcal{O}$ is of the form:

$$\tau = u_{-\frac{1}{2}} \wedge u_{-\frac{3}{2}} \wedge u_{-\frac{5}{2}} \wedge \ldots, \text{where } u_j \in \mathbb{C}^\infty \text{ and } u_{-k} = v_{-k} \text{ for } k \gg 0. \tag{1.3.2}$$

This allows us to construct a canonical map $\varphi : \mathcal{O} \to \text{Gr}$ by $\varphi(\tau) = \sum_i \mathbb{C}u_{-i} \subset \mathbb{C}^\infty$, where Gr consists of the subspaces of $\mathbb{C}^\infty$ containing $\sum_{j=k}^{\infty} \mathbb{C}v_{-j-1/2}$ for $k \gg 0$ as a subspace of codimension k. It is clear that the map φ is surjective with fibers $\mathbb{C}^\times$.

1.4. Consider the free $\mathbb{Z}$-module $\tilde{L}$ with the basis $\{\delta_j\}_{j\in\frac{1}{2}+\mathbb{Z}}$, let $\tilde{\Delta}$ (resp. $\tilde{\Delta}_0$) $= \{\delta_i - \delta_j | i,j \in \frac{1}{2}+\mathbb{Z}$ (resp. $i,-j \in \frac{1}{2}+\mathbb{Z}_+$), $i \neq j\}$, and let $\tilde{M} \subset \tilde{L}$ (resp. $\tilde{M}_0 \subset \tilde{L}$) be the $\mathbb{Z}$-span of $\tilde{\Delta}$ (resp. $\tilde{\Delta}_0$). We define the weight of a semi-infinite monomial by

$$\text{weight}(\psi_{i_1}^+ \ldots \psi_{i_s}^+ \psi_{j_1}^- \ldots \psi_{j_t}^-|0\rangle) = \delta_{-i_1} + \ldots + \delta_{-i_s} - \delta_{j_1} - \ldots - \delta_{j_t}. \tag{1.4.1}$$

Note that weights of semi-infinite monomials from $F^{(0)}$ lie in $\tilde{M}_0$. Given $\tau \in F$ we denote by $f\text{supp}\,\tau$, and call it the *fermionic support of* τ, the set of weights of semi-infinite monomials that occur in τ with a non-zero coefficient.

Proposition 1.4. *If $\tau \in \mathcal{O}$, then $f\text{supp}\ \tau$ is the intersection of $\tilde{M}_0$ with a convex polyhedron with vertices in $\tilde{M}_0$ and edges in $\tilde{\Delta}_0$.*

Proof. According to the general result [PK, Lemma 4], the edges of the convex hull of $f\text{supp}\ \tau$ must be parallel to the elements of $\tilde{\Delta}_0$. But if the difference of weights of two semi-infinite monomials is a multiple of $\delta_i - \delta_j$, then it is clearly equal to $\pm(\delta_i - \delta_j)$. Hence edges of the convex hull of $f\text{supp}\ \tau$ are elements of $\tilde{\Delta}_0$, and the proposition follows. □

§2. The n-component bosonization and the KP hierarchy in the bosonic picture.

2.1. Using a bosonization one can rewrite (1.3.1) as a system of partial differential equations. There are however many different bosonizations. In this paper we focus on the n-component bosonizations, where $n = 1, 2, \ldots$.

For that purpose we relabel the basis vectors v_i and with them the corresponding fermionic operators (the wedging and contracting operators). This relabeling can be done in many different ways, see e.g. [TV], the simplest one is the following.

Fix $n \in \mathbb{N}$ and define for $j \in \mathbb{Z},\ 1 \le j \le n,\ k \in \mathbb{Z} + \frac{1}{2}$:

$$v_k^{(j)} = v_{nk - \frac{1}{2}(n-2j+1)},$$

and correspondingly:

$$\psi_k^{\pm(j)} = \psi^{\pm}_{nk \pm \frac{1}{2}(n-2j+1)}.$$

Notice that with this relabeling we have:

$$\psi_k^{\pm(j)}|0\rangle = 0 \text{ for } k > 0.$$

The charge decomposition (1.2.5) can be further decomposed into a sum of *partial charges* which are denoted by $\text{charge}_j,\ j = 1, \dots, n$, defined for a semi-infinite monomial $v \equiv v_{i_1} \wedge v_{i_2} \wedge \dots$ of weight $\sum_i a_i \delta_i$ by

$$\text{charge}_j(v) = \sum_{k \in \mathbb{Z}} a_{kn+j-1/2}, \tag{2.1.1}$$

which is equivalent to

$$\text{charge}_j\ \psi_k^{\pm(i)} = \pm\delta_{ij},\ \text{charge}_j\ |0\rangle = 0.$$

Another important decomposition is the *energy decomposition* defined by

$$\text{energy}\ |0\rangle = 0,\ \text{energy}\ \psi_k^{\pm(j)} = -k. \tag{2.1.2}$$

Note that energy is a non-negative number which can be calculated by

$$\text{energy}(v) = \sum_{k \in \frac{1}{2} + \mathbb{Z}} a_k([k/n] + \tfrac{1}{2}). \tag{2.1.3}$$

Introduce the fermionic fields ($z \in \mathbb{C}^\times$):

$$\psi^{\pm(j)}(z) \stackrel{\text{def}}{=} \sum_{k \in \mathbb{Z} + \frac{1}{2}} \psi_k^{\pm(j)} z^{-k-\frac{1}{2}}. \tag{2.1.4}$$

Next we introduce bosonic fields ($1 \le i, j \le n$):

$$\alpha^{(ij)}(z) \equiv \sum_{k \in \mathbb{Z}} \alpha_k^{(ij)} z^{-k-1} \stackrel{def}{=} :\psi^{+(i)}(z)\psi^{-(j)}(z):, \tag{2.1.5}$$

where : : stands for the *normal ordered product* defined in the usual way ($\lambda, \mu = +$ or $-$):

$$:\psi_k^{\lambda(i)}\psi_\ell^{\mu(j)}: := \begin{cases} \psi_k^{\lambda(i)}\psi_\ell^{\mu(j)} & \text{if } \ell > 0 \\ -\psi_\ell^{\mu(j)}\psi_k^{\lambda(i)} & \text{if } \ell < 0. \end{cases} \tag{2.1.6}$$

One checks (using e.g. the Wick formula) that the operators $\alpha_k^{(ij)}$ satisfy the commutation relations of the affine algebra $gl_n(\mathbb{C})^\wedge$ with central charge 1, i.e.:

$$[\alpha_p^{(ij)}, \alpha_q^{(k\ell)}] = \delta_{jk}\alpha_{p+q}^{(i\ell)} - \delta_{i\ell}\alpha_{p+q}^{(kj)} + p\delta_{i\ell}\delta_{jk}\delta_{p,-q}, \tag{2.1.7}$$

and that

(2.1.8) $$\alpha_k^{(ij)}|m\rangle = 0 \text{ if } k > 0 \text{ or } k = 0 \text{ and } i < j.$$

The operators $\alpha_k^{(i)} \equiv \alpha_k^{(ii)}$ satsify the canonical commutation relation of the associative oscillator algebra, which we denote by $\mathfrak{a}$:

(2.1.9) $$[\alpha_k^{(i)}, \alpha_\ell^{(j)}] = k\delta_{ij}\delta_{k,-\ell},$$

and one has

(2.1.10) $$\alpha_k^{(i)}|m\rangle = 0 \text{ for } k > 0.$$

It is easy to see that restricted to $g\ell_n(\mathbb{C})^\wedge$, $F^{(0)}$ is its basic highest weight representation (see [K, Chapter 12]). The $g\ell_n(\mathbb{C})^\wedge$-weight of a semi-infinite monomial v is as follows:

(2.1.11) $$\Lambda_0 + \sum_{j=1}^{n} \text{charge}_j(v)\delta_j - \text{energy}(v)\tilde{\delta}.$$

Here Λ_0 is the highest weight of the basic representation, $\{\delta_j\}$ is the standard basis of the weight lattice of $g\ell_n(\mathbb{C})$ and $\tilde{\delta}$ is the primitive imaginary root ([K, Chapter 7]).

In order to express the fermionic fields $\psi^{\pm(i)}(z)$ in terms of the bosonic fields $\alpha^{(ii)}(z)$, we need some additional operators Q_i, $i = 1, \dots, n$, on F. These operators are uniquely defined by the following conditions:

(2.1.12) $$Q_i|0\rangle = \psi_{-\frac{1}{2}}^{+(i)}|0\rangle, \; Q_i\psi_k^{\pm(j)} = (-1)^{\delta_{ij}+1}\psi_{k\mp\delta_{ij}}^{\pm(j)}Q_i.$$

They satisfy the following commutation relations:

(2.1.13) $$Q_iQ_j = -Q_jQ_i \text{ if } i \neq j, \; [\alpha_k^{(i)}, Q_j] = \delta_{ij}\delta_{k0}Q_j.$$

Theorem 2.1. *([DJKM1], [JM])*

(2.1.14) $$\psi^{\pm(i)}(z) = Q_i^{\pm 1} z^{\pm\alpha_0^{(i)}} \exp(\mp\sum_{k<0}\frac{1}{k}\alpha_k^{(i)}z^{-k})\exp(\mp\sum_{k>0}\frac{1}{k}\alpha_k^{(i)}z^{-k}).$$

Proof. See [TV].

The operators on the right-hand side of (2.1.14) are called vertex operators. They made their first appearance in string theory (cf. [FK]).

We shall use below the following notation

(2.1.15) $$|k_1, \dots, k_n\rangle = Q_1^{k_1} \dots Q_n^{k_n}|0\rangle.$$

Remark 2.1. One easily checks the following relations:

$$[\alpha_k^{(i)}, \psi_m^{\pm(j)}] = \pm\delta_{ij}\psi_{k+m}^{\pm(j)}.$$

They imply formula (2.1.14) for $\psi^{\pm(i)}(z)$ except for the first two factors, which require some additional analysis.

2.2. We can describe now the n-component boson-fermion correspondence. Let $\mathbb{C}[x]$ be the space of polynomials in indeterminates $x = \{x_k^{(i)}\}$, $k = 1, 2, \dots$, $i = 1, 2, \dots, n$. Let L be a lattice with a basis $\delta_1, \dots, \delta_n$ over $\mathbb{Z}$ and the symmetric bilinear form $(\delta_i|\delta_j) = \delta_{ij}$, where δ_{ij} is the Kronecker symbol. Let

$$(2.2.1)\qquad \varepsilon_{ij} = \begin{cases} -1 & \text{if } i > j \\ 1 & \text{if } i \le j. \end{cases}$$

Define a bimultiplicative function $\varepsilon:\ L \times L \to \{\pm 1\}$ by letting

$$(2.2.2)\qquad \varepsilon(\delta_i, \delta_j) = \varepsilon_{ij}.$$

Let $\delta = \delta_1 + \ldots + \delta_n$, $M = \{\gamma \in L | (\delta|\gamma) = 0\}$, $\Delta = \{\alpha_{ij} := \delta_i - \delta_j | i, j = 1, \ldots, n,\ i \neq j\}$. Of course M is the root lattice of $s\ell_n(\mathbb{C})$, the set Δ being the root system.

Consider the vector space $\mathbb{C}[L]$ with basis e^γ, $\gamma \in L$, and the following twisted group algebra product:

$$(2.2.3)\qquad e^\alpha e^\beta = \varepsilon(\alpha, \beta) e^{\alpha+\beta}.$$

Let $B = \mathbb{C}[x] \otimes_{\mathbb{C}} \mathbb{C}[L]$ be the tensor product of algebras. Then the n-component boson-fermion correspondence is the vector space isomorphism

$$(2.2.4)\qquad \sigma : F \xrightarrow{\sim} B,$$

given by

$$(2.2.5)\qquad \sigma(\alpha^{(i_1)}_{-m_1} \ldots \alpha^{(i_s)}_{-m_s} |k_1, \ldots, k_n\rangle) = m_1 \ldots m_s x^{(i_1)}_{m_1} \ldots x^{(i_s)}_{m_s} \otimes e^{k_1\delta_1 + \ldots + k_n\delta_n}.$$

The transported charge and energy then will be as follows:

$$(2.2.6)\qquad \mathrm{charge}(p(x) \otimes e^\gamma) = (\delta|\gamma),\ \mathrm{charge}_j(p(x) \otimes e^\gamma) = (\delta_j|\gamma),$$

$$(2.2.7)\qquad \mathrm{energy}(x^{(i_1)}_{m_1} \ldots x^{(i_s)}_{m_s} \otimes e^\gamma) = m_1 + \ldots + m_s + \tfrac{1}{2}(\gamma|\gamma).$$

We denote the transported charge decomposition by

$$B = \bigoplus_{m \in \mathbb{Z}} B^{(m)}.$$

The transported action of the operators $\alpha^{(i)}_m$ and Q_j looks as follows:

$$(2.2.8)\qquad \begin{cases} \sigma\alpha^{(j)}_{-m}\sigma^{-1}(p(x) \otimes e^\gamma) = m x^{(j)}_m p(x) \otimes e^\gamma, \text{ if } m > 0, \\ \sigma\alpha^{(j)}_m\sigma^{-1}(p(x) \otimes e^\gamma) = \frac{\partial p(x)}{\partial x_m} \otimes e^\gamma, \text{ if } m > 0, \\ \sigma\alpha^{(j)}_0\sigma^{-1}(p(x) \otimes e^\gamma) = (\delta_j|\gamma)p(x) \otimes e^\gamma, \\ \sigma Q_j \sigma^{-1}(p(x) \otimes e^\gamma) = \varepsilon(\delta_j, \gamma)p(x) \otimes e^{\gamma+\delta_j}. \end{cases}$$

2.3. Using the isomorphism σ we can reformulate the KP hierarchy (1.3.1) in the bosonic picture as a hierarchy of Hirota bilinear equations.

We start by observing that (1.3.1) can be rewritten as follows:

$$(2.3.1)\qquad \mathrm{Res}_{z=0}\, dz(\sum_{j=1}^n \psi^{+(j)}(z)\tau \otimes \psi^{-(j)}(z)\tau) = 0,\ \tau \in F^{(0)}.$$

Here and further $\mathrm{Res}_{z=0}\, dz \sum_j f_j z^j$ (where f_j are independent of z) stands for f_{-1}. Notice that for $\tau \in F^{(0)}$, $\sigma(\tau) = \sum_{\gamma \in M} \tau_\gamma(x) e^\gamma$. Here and further we write $\tau_\gamma(x)e^\gamma$ for $\tau_\gamma \otimes e^\gamma$. Using Theorem 2.1, equation (2.3.1) turns under $\sigma\otimes\sigma :\ F\otimes F \xrightarrow{\sim} \mathbb{C}[x', x'']\otimes(\mathbb{C}[L']\otimes\mathbb{C}[L''])$ into the following equation:

$$\text{(2.3.2)}\quad \begin{aligned}&\mathrm{Res}_{z=0}\, dz\Big(\sum_{j=1}^{n}\sum_{\alpha,\beta\in M}\varepsilon(\delta_j,\alpha-\beta)z^{(\delta_j|\alpha-\beta)}\\ &\times\exp\Big(\sum_{k=1}^{\infty}(x_k^{(j)'}-x_k^{(j)''})z^k\Big)\exp\Big(-\sum_{k=1}^{\infty}\Big(\frac{\partial}{\partial x_k^{(j)'}}-\frac{\partial}{\partial x_k^{(j)''}}\Big)\frac{z^{-k}}{k}\Big)\\ &\tau_\alpha(x')(e^{\alpha+\delta_j})'\tau_\beta(x'')(e^{\beta-\delta_j})''\Big)=0.\end{aligned}$$

Hence for all $\alpha,\beta\in L$ such that $(\alpha|\delta)=-(\beta|\delta)=1$ we have:

$$\text{(2.3.3)}\quad \begin{aligned}&\mathrm{Res}_{z=0}\Big(dz\sum_{j=1}^{n}\varepsilon(\delta_j,\alpha-\beta)z^{(\delta_j|\alpha-\beta-2\delta_j)}\\ &\times\exp\Big(\sum_{k=1}^{\infty}(x_k^{(j)'}-x_k^{(j)''})z^k\Big)\exp\Big(-\sum_{k=1}^{\infty}\Big(\frac{\partial}{\partial x_k^{(j)'}}-\frac{\partial}{\partial x_k^{(j)''}}\Big)\frac{z^{-k}}{k}\Big)\\ &\tau_{\alpha-\delta_j}(x')\tau_{\beta+\delta_j}(x'')\Big)=0.\end{aligned}$$

Now making the change of variables

$$x_k^{(j)}=\tfrac{1}{2}(x_k^{(j)'}+x_k^{(j)''}),\quad y_k^{(j)}=\tfrac{1}{2}(x_k^{(j)'}-x_n^{(j)''}),$$

(2.3.3) becomes

$$\text{(2.3.4)}\quad \begin{aligned}&\mathrm{Res}_{z=0}\Big(dz\sum_{j=1}^{n}\varepsilon(\delta_j,\alpha-\beta)z^{(\delta_j|\alpha-\beta-2\delta_j)}\\ &\times\exp\Big(\sum_{k=1}^{\infty}2y_k^{(j)}z^k\Big)\exp\Big(-\sum_{k=1}^{\infty}\frac{\partial}{\partial y_k^{(j)}}\frac{z^{-k}}{k}\Big)\tau_{\alpha-\delta_j}(x+y)\tau_{\beta+\delta_j}(x-y)\Big)=0.\end{aligned}$$

We can rewrite (2.3.4) using the elementary Schur polynomials defined by (0.1.13):

$$\text{(2.3.5)}\quad \sum_{j=1}^{n}\varepsilon(\delta_j,\alpha-\beta)\sum_{k=0}^{\infty}S_k(2y^{(j)})S_{k-1+(\delta_j|\alpha-\beta)}\Big(-\frac{\tilde\partial}{\partial y^{(j)}}\Big)\tau_{\alpha-\delta_j}(x+y)\tau_{\beta+\delta_j}(x-y)=0.$$

Here and further we use the notation

$$\frac{\tilde\partial}{\partial y}=\Big(\frac{\partial}{\partial y_1},\frac{1}{2}\frac{\partial}{\partial y_2},\frac{1}{3}\frac{\partial}{\partial y_3},\dots\Big)$$

Using Taylor's formula we can rewrite (2.3.5) once more:

$$\text{(2.3.6)}\quad \begin{aligned}&\sum_{j=1}^{n}\varepsilon(\delta_j,\alpha-\beta)\sum_{k=0}^{\infty}S_k(2y^{(j)})S_{k-1+(\delta_j|\alpha-\beta)}\Big(-\frac{\tilde\partial}{\partial u^{(j)}}\Big)\\ &\times e^{\sum_{j=1}^{n}\sum_{r=1}^{\infty}y_r^{(j)}\frac{\partial}{\partial u_r^{(j)}}}\tau_{\alpha-\delta_j}(x+u)\tau_{\beta+\delta_j}(x-u)|_{u=0}=0.\end{aligned}$$

This last equation can be written as the following generating series of Hirota bilinear equations:

$$\text{(2.3.7)}\quad \begin{aligned}&\sum_{j=1}^{n}\varepsilon(\delta_j,\alpha-\beta)\sum_{k=0}^{\infty}S_k(2y^{(j)})S_{k-1+(\delta_j|\alpha-\beta)}(-\widetilde{D^{(j)}})\\ &\times e^{\sum_{j=1}^{n}\sum_{r=1}^{\infty}y_r^{(j)}D_r^{(j)}}\tau_{\alpha-\delta_j}\cdot\tau_{\beta+\delta_j}=0\end{aligned}$$

for all $\alpha, \beta \in L$ such that $(\alpha|\delta) = -(\beta|\delta) = 1$. Hirota's dot notation used here and further is explained in Introduction (see (0.1.9)).

Equation (2.3.7) is known (see [DJKM1,2], [JM]) as the n-component KP hierarchy of Hirota bilinear equations. This equation still describes the group orbit: $\sigma(\mathcal{O}) = \sigma R \sigma^{-1}(GL_\infty) \cdot 1$.

Remark 2.3. Equation (2.3.7) is invariant under the transformations $\alpha \longmapsto \alpha+\gamma$, $\beta \longmapsto \beta+\gamma$, where $\gamma \in M$. Transformations of this type are called Schlessinger transformations.

Let $\tau = \sum_{\gamma \in L} \tau_\gamma(x) e^\gamma \in B$; the set $\text{supp}\, \tau \stackrel{\text{def}}{=} \{\gamma \in L |\ \tau_\gamma \neq 0\}$ is called the *support* of τ.

Proposition 2.3. *Let $\tau \in \mathbb{C}[[x]] \otimes \mathbb{C}[M]$ be a solution to the KP hierarchy (2.3.4). Then $\text{supp}\, \tau$ is the intersection of M with a convex polyhedron with vertices in M and edges parallel to elements of Δ.*

Proof. Consider the linear map $\overline{\sigma} : \tilde{L} \to L$ defined by $\overline{\sigma}(\delta_j) = \delta_{(j+1/2) \bmod n}$, where $a \bmod n$ stands for the element of the set $\{1, \dots, n\}$ congruent to $a \bmod n$. Then it is easy to see that for $\tau \in F$ we have:

$$\text{supp}\, \sigma(\tau) = \overline{\sigma}(f \text{supp}\, \tau).$$

Now Proposition 2.3 follows from Proposition 1.4. □

2.4. The indeterminates $y_k^{(j)}$ in (2.3.7) are free parameters, hence the coefficient of a monomial $y_{k_1}^{(j_1)} \dots y_{k_s}^{(j_s)}$ ($k_i \in \mathbb{N}$, $k_1 \leq k_2 \leq \dots$, $j_i \in \{1, \dots, n\}$) in equation (2.3.7) gives us a Hirota bilinear equation of the form

$$\sum_{i=1}^{n} \sum_{\alpha,\beta} Q_{k;\alpha,\beta}^{(j)}(D) \tau_{\alpha-\delta_i} \cdot \tau_{\beta+\delta_i} = 0, \tag{2.4.1}$$

where $Q_{k,\alpha,\beta}^{(j)}$ are polynomials in the $D_r^{(i)}$, $k = (k_1, \dots, k_s)$, $j = (j_1, \dots, j_s)$ and $\alpha, \beta \in L$ are such that $(\alpha|\delta) = -(\beta|\delta) = 1$. Each of these equations is a PDE in the indeterminates $x_k^{(j)}$ on functions τ_γ, $\gamma \in M$.

Recall that an expression $Q(D)\tau_\alpha \cdot \tau_\beta$ is identically zero if and only if $\alpha = \beta$ and $Q(D) = -Q(-D)$. The corresponding Hirota bilinear equation is then called *trivial* and can be disregarded.

Let us point out now that the energy decomposition (2.2.7) induces the following energy decomposition on the space of Hirota bilinear equations:

$$\text{energy}(Q_{k;\alpha,\beta}^{(j)}(D)\tau_{\alpha-\delta_i} \cdot \tau_{\beta+\delta_i}) = k_1 + \dots + k_s + \tfrac{1}{2}((\alpha|\alpha) + (\beta|\beta)) \tag{2.4.2}$$

It is clear that the energy of a nontrivial Hirota bilinear equation is at least 2.

Below we list the Hirota bilinear equations of lowest energy for each n.

$n = 1$. In this case we may drop the superscript in $D_k^{(1)}$ and the subscript in τ_α (which is 0). Each monomial $y_{k_1} \dots y_{k_s}$ gives a Hirota bilinear equation of the form

$$Q_k(D)\tau \cdot \tau = 0$$

of energy $k_1 + \dots + k_s + 1$. An easy calculation shows that the lowest energy of a non-trivial equation is 4, and that there is a unique non-trivial equation of energy 4, the classical KP equation in the Hirota bilinear form:

$$(D_1^4 - 4D_1D_3 + 3D_2^2)\tau \cdot \tau = 0. \tag{2.4.3}$$

$n \geq 2$. There is an equation of energy 2 for each unordered pair of distinct indices i and k (recall that $\alpha_{ik} = \delta_i - \delta_k$ are roots):

$$D_1^{(i)} D_1^{(k)} \tau_0 \cdot \tau_0 = 2\tau_{\alpha_{ik}} \tau_{\alpha_{ki}}. \tag{2.4.4}$$

Furthermore, for each ordered pair of distinct indices i and j there are three equations of energy 3:

$$(D_2^{(i)} + D_1^{(i)2})\tau_0 \cdot \tau_{\alpha_{ij}} = 0, \tag{2.4.5}$$

$$(D_2^{(j)} + D_1^{(j)2})\tau_{\alpha_{ij}} \cdot \tau_0 = 0, \tag{2.4.6}$$

$$D_1^{(i)} D_2^{(j)} \tau_0 \cdot \tau_0 + 2D_1^{(j)} \tau_{\alpha_{ij}} \cdot \tau_{\alpha_{ji}} = 0. \tag{2.4.7}$$

$n \geq 3$. There is an equation of energy 2 and an equation of energy 3 for each ordered triple of distinct indices i, j, k:

$$D_1^{(k)} \tau_0 \cdot \tau_{\alpha_{ij}} = \varepsilon_{ik}\varepsilon_{kj}\varepsilon_{ij}\tau_{\alpha_{ik}}\tau_{\alpha_{kj}}, \tag{2.4.8}$$

$$D_2^{(k)} \tau_0 \cdot \tau_{\alpha_{ij}} = \varepsilon_{ij}\varepsilon_{kj}\varepsilon_{ik} D_1^{(k)} \tau_{\alpha_{ik}} \cdot \tau_{\alpha_{kj}}. \tag{2.4.9}$$

(Note that (2.4.6) is a special case of (2.4.9) where $k = j$.)

$n \geq 4$. There is an algebraic equation of energy 2 for each ordered quadruple of distinct indices i, j, k, ℓ:

$$\varepsilon_{ij}\varepsilon_{k\ell}\tau_0\tau_{\alpha_{ik}+\alpha_{j\ell}} + \varepsilon_{i\ell}\varepsilon_{jk}\tau_{\alpha_{ik}}\tau_{\alpha_{j\ell}} + \varepsilon_{ik}\varepsilon_{j\ell}\tau_{\alpha_{i\ell}}\tau_{\alpha_{jk}} = 0. \tag{2.4.10}$$

Equations (2.4.4–10), together with an algebraic equation of energy 3 for each ordered sixtuple of distinct indices similar to (2.4.10), form a complete list of non-trivial Hirota bilinear equations of energy ≤ 3 of the n-component KP hierarchy.

§3. The algebra of formal pseudo-differential operators and the n-component KP hierarchy as a dynamical system.

3.0. The KP hierarchy and its n-component generalizations admit several formulations. The one given in the previous section obtained by the field theoretical approach is the τ-function formulation given by Date, Jimbo, Kashiwara and Miwa [DJKM1]. Another well-known formulation, introduced by Sato [S], is given in the language of formal pseudo-differential operators. We will show that this formulation follows from the τ-function formulation given by equation (2.3.3).

3.1. We shall work over the algebra $\mathcal{A}$ of formal power series over $\mathbb{C}$ in indeterminates $x = (x_k^{(j)})$, where $k = 1, 2, \ldots$ and $j = 1, \ldots, n$. The indeterminates $x_1^{(1)}, \ldots, x_1^{(n)}$ will be viewed as variables and $x_k^{(j)}$ with $k \geq 2$ as parameters. Let

$$\partial = \frac{\partial}{\partial x_1^{(1)}} + \ldots + \frac{\partial}{\partial x_1^{(n)}}.$$

A formal $n \times n$ matrix pseudo-differential operator is an expression of the form

$$P(x, \partial) = \sum_{j \leq N} P_j(x)\partial^j, \tag{3.1.1}$$

where P_j are $n \times n$ matrices over $\mathcal{A}$. The largest N such that $P_N \neq 0$ is called the *order* of $P(x, \partial)$ (write *ord* $P(x, \partial) = N$). Let Ψ denote the vector space over $\mathbb{C}$ of all expressions (3.1.1). We have a linear isomorphism $s : \Psi \to \mathrm{Mat}_n(\mathcal{A}((z)))$ given by $s(P(x, \partial)) = P(x, z)$. The matrix series $P(x, z)$ in indeterminates x and z is called the *symbol* of $P(x, \partial)$.

Now we may define a product $\circ$ on Ψ making it an associative algebra:

$$s(P \circ Q) = \sum_{n=0}^{\infty} \frac{1}{n!} \frac{\partial^n s(P)}{\partial z^n} \partial^n s(Q).$$

We shall often drop the multiplication sign $\circ$ when no ambiguity may arise. Letting $\Psi(m) = \{P \in \Psi | ord\, \Psi \leq m\}$, we get a $\mathbb{Z}$-filtration of the algebra Ψ:

$$\cdots \Psi(m+1) \supset \Psi(m) \supset \Psi(m-1) \supset \cdots \tag{3.1.2}$$

One defines the differential part of $P(x,\partial)$ by $P_+(x,\partial) = \sum_{j=0}^{N} P_j(x)\partial^j$, and let $P_- = P - P_+$. We have the corresponding vector space decomposition:

$$\Psi = \Psi_- \oplus \Psi_+. \tag{3.1.3}$$

One defines a linear map $* : \Psi \to \Psi$ by the following formula:

$$(\sum_j P_j \partial^j)^* = \sum_j (-\partial)^j \circ {}^t P_j. \tag{3.1.4}$$

Here and further ${}^tP$ stands for the transpose of the matrix P. Note that $*$ is an anti-involution of the algebra Ψ. In terms of symbols the anti-involution $*$ can be written in the following closed form:

$$P^*(x,z) = (\exp \partial \frac{\partial}{\partial z})\, {}^tP(x,-z). \tag{3.1.5}$$

It is clear that the anti-involution $*$ preserves the filtration (3.1.2) and the decomposition (3.1.3).

3.2. Introduce the following notation

$$z \cdot x^{(j)} = \sum_{k=1}^{\infty} x_k^{(j)} z^k, \quad e^{z \cdot x} = diag(e^{z \cdot x^{(1)}}, \ldots, e^{z \cdot x^{(n)}}).$$

The algebra Ψ acts on the space U_+ (resp. U_-) of formal oscillating matrix functions of the form

$$\sum_{j \leq N} P_j z^j e^{z \cdot x} \quad (\text{resp. } \sum_{j \leq N} P_j z^j e^{-z \cdot x}), \text{ where } P_j \in \mathrm{Mat}_n(\mathcal{A}),$$

in the obvious way:

$$P(x)\partial^j e^{\pm z \cdot x} = P(x)(\pm z)^j e^{\pm z \cdot x}.$$

We can now prove the following fundamental lemma.

Lemma 3.2. *If $P, Q \in \Psi$ are such that*

$$Res_{z=0}(P(x,\partial)e^{z \cdot x})\, {}^t(Q(x',\partial')e^{-z \cdot x'})dz = 0, \tag{3.2.1}$$

then $(P \circ Q^*)_- = 0$.

Proof. Equation (3.2.1) is equivalent to

$$Res_{z=0} P(x,z) e^{z(x-x')}\, {}^tQ(x',-z)dz = 0. \tag{3.2.2}$$

The (i,m)-th entry of the matrix equation (3.2.2) is

$$Res_{z=0} \sum_{i=1}^{n} P_{ij}(x,z) Q_{mj}(x',-z) e^{z \cdot (x^{(j)} - x'^{(j)})} dz = 0.$$

Letting $y_k^{(j)} = x_k^{(j)} - x_k'^{(j)}$, this equation can be rewritten by applying Taylor's formula to Q:

$$\text{(3.2.3)} \qquad Res_{z=0}\sum_{j=1}^{n} P_{ij}(x,z)\exp\sum_{\ell=1}^{n}\sum_{k=1}^{\infty} y_k^{(\ell)}(\delta_{\ell j}z^k - \frac{\partial}{\partial x_k^{(\ell)}})Q_{mj}(x,-z)dz = 0.$$

Letting $y_k^{(\ell)} = 0$ for $k > 1$ and $y_1^{(\ell)} = y$ for all ℓ, we obtain from (3.2.3):

$$\text{(3.2.4)} \qquad Res_{z=0}P(x,z)\sum_{k\geq 0}\frac{(-1)^k}{k!}\partial^k({}^tQ)(x,-z)y^k e^{yz}dz = 0.$$

Notice that $y^k e^{yz} = (e^{yz})^{(k)}$. Here and further we write $\varphi^{(k)}$ for $\frac{\partial^k \varphi}{\partial z^k}$. Using integration by parts with respect to z, (3.2.4) becomes:

$$\text{(3.2.5)} \qquad Res_{z=0}\sum_{k\geq 0}\frac{1}{k!}(P(x,z)\partial^k({}^tQ)(x,-z))^{(k)}e^{yz}dz = 0.$$

Using Leibnitz formula, the left-hand side of (3.2.5) is equal to

$$\begin{aligned}
&Res_{z=0}\sum_{k\geq 0}\sum_{\ell=0}^{k}\frac{(-1)^{k-\ell}}{\ell!(k-\ell)!}P^{(\ell)}(x,z)\left(\partial^k({}^tQ)\right)^{(k-\ell)}(x,-z)e^{yz}dz \\
&= Res_{z=0}\sum_{\ell\geq 0}\frac{1}{\ell!}P^{(\ell)}(x,z)\partial^\ell\left(\sum_{k=0}^{\infty}\frac{(-1)^k}{k!}\partial^k({}^tQ^{(k)})(x,-z)\right)e^{yz}dz \\
&= Res_{z=0}\sum_{\ell\geq 0}\frac{1}{\ell!}P^{(\ell)}(x,z)\left(\frac{\partial^\ell Q^*(x,z)}{\partial x^\ell}\right)e^{yz}dz = Res_{z=0}(P\circ Q^*)(x,z)e^{yz}dz.
\end{aligned}$$

So we obtain that

$$\text{(3.2.6)} \qquad Res_{z=0}(P\circ Q^*)(x,z)e^{yz}dz = 0.$$

Now write $(P\circ Q^*)(x,z) = \sum_j A_j(x)z^j$ and $e^{yz} = \sum_{\ell=0}^{\infty}\frac{(zy)^\ell}{\ell!}$. Then from (3.2.6) we deduce:

$$0 = Res_{z=0}\sum_j\sum_{\ell=0}^{\infty}A_j(x)\frac{y^\ell}{\ell!}z^{\ell+j}dz = \sum_{\ell=0}^{\infty}A_{-\ell-1}(x)\frac{y^\ell}{\ell!}.$$

Hence $A_j(x) = 0$ for $j < 0$, i.e. $(P\circ Q^*)_- = 0$. □

3.3. We proceed now to rewrite the formulation (2.3.3) of the n-component KP hierarchy in terms of formal pseudo-differential operators.

Let $1 \leq a, b \leq n$ and recall formula (2.3.3) where α is replaced by $\alpha + \delta_a$ and β by $\beta - \delta_b$:

$$\text{(3.3.1)} \qquad \begin{aligned}
&\mathrm{Res}_{z=0}(dz\sum_{j=1}^{n}\varepsilon(\delta_j, \alpha+\delta_a-\beta+\delta_b)z^{(\delta_j|\alpha+\delta_a-\beta+\delta_b-2\delta_j)} \\
&\times\exp(\sum_{k=1}^{\infty}(x_k^{(j)'} - x_k^{(j)''})z^k)\exp(-\sum_{k=1}^{\infty}(\frac{\partial}{\partial x_k^{(j)'}} - \frac{\partial}{\partial x_k^{(j)''}})\frac{z^{-k}}{k}) \\
&\tau_{\alpha+\alpha_{aj}}(x')\tau_{\beta-\alpha_{bj}}(x'')) = 0 \quad (\alpha,\beta\in M).
\end{aligned}$$

For each $\alpha \in$ supp τ we define the (matrix valued) functions

$$\text{(3.3.2)} \qquad V^{\pm}(\alpha,x,z) = (V_{ij}^{\pm}(\alpha,x,z))_{i,j=1}^{n}$$

as follows:

$$(3.3.3)\qquad \begin{aligned} V_{ij}^{\pm}(\alpha,x,z) &\stackrel{\text{def}}{=} \varepsilon(\delta_j,\alpha+\delta_i)z^{(\delta_j|\pm\alpha+\alpha_{ij})} \\ &\times \exp(\pm\sum_{k=1}^{\infty} x_k^{(j)} z^k)\exp(\mp\sum_{k=1}^{\infty}\frac{\partial}{\partial x_k^{(j)}}\frac{z^{-k}}{k})\tau_{\alpha\pm\alpha_{ij}}(x)/\tau_\alpha(x). \end{aligned}$$

It is easy to see that equation (3.3.1) is equivalent to the following bilinear identity:

$$(3.3.4)\qquad Res_{z=0}V^+(\alpha,x,z)\ {}^tV^-(\beta,x',z)dz = 0 \text{ for all } \alpha,\beta\in M.$$

Define $n\times n$ matrices $W^{\pm(m)}(\alpha,x)$ by the following generating series (cf. (3.3.3)):

$$(3.3.5)\qquad \sum_{m=0}^{\infty} W_{ij}^{\pm(m)}(\alpha,x)(\pm z)^{-m} = \varepsilon_{ji} z^{\delta_{ij}-1}(\exp\mp\sum_{k=1}^{\infty}\frac{\partial}{\partial x_k^{(j)}}\frac{z^{-k}}{k})\tau_{\alpha\pm\alpha_{ij}}(x))/\tau_\alpha(x).$$

Note that

$$(3.3.6)\qquad W^{\pm(0)}(\alpha,x) = I_n,$$

$$(3.3.7)\qquad W_{ij}^{\pm(1)}(\alpha,x) = \begin{cases} \varepsilon_{ji}\tau_{\alpha\pm\alpha_{ij}}/\tau_\alpha & \text{if } i\neq j \\ -\tau_\alpha^{-1}\frac{\partial\tau_\alpha}{\partial x_1^{(i)}} & \text{if } i=j, \end{cases}$$

$$(3.3.8)\qquad W_{ij}^{\pm(2)}(\alpha,x) = \begin{cases} \mp\varepsilon_{ji}\frac{\partial\tau_{\alpha\pm\alpha_{ij}}}{\partial x_1^{(j)}}/\tau_\alpha & \text{if } i\neq j, \\ (\mp\frac{1}{2}\frac{\partial\tau_\alpha}{\partial x_2^{(i)}} + \frac{1}{2}\frac{\partial^2\tau_\alpha}{\partial x_1^{(i)2}})/\tau_\alpha & \text{if } i=j. \end{cases}$$

We see from (3.3.3) that $V^{\pm}(\alpha,x,z)$ can be written in the following form:

$$(3.3.9)\qquad V^{\pm}(\alpha,x,z) = (\sum_{m=0}^{\infty} W^{\pm(m)}(\alpha,x)R^{\pm}(\alpha,\pm z)(\pm z)^{-m})e^{\pm z\cdot x},$$

where

$$(3.3.10)\qquad R^{\pm}(\alpha,z) = \sum_{i=1}^{n}\varepsilon(\delta_i,\alpha)E_{ii}(\pm z)^{\pm(\delta_i|\alpha)}.$$

Here and further E_{ij} stands for the $n\times n$ matrix whose (i,j) entry is 1 and all other entries are zero. Now it is clear that $V^{\pm}(\alpha,x,z)$ can be written in terms of formal pseudo-differential operators

$$(3.3.11)\qquad P^{\pm}(\alpha)\equiv P^{\pm}(\alpha,x,\partial) = I_n + \sum_{m=1}^{\infty} W^{\pm(m)}(\alpha,x)\partial^{-m} \text{ and } R^{\pm}(\alpha) = R^{\pm}(\alpha,\partial)$$

as follows:

$$(3.3.12)\qquad V^{\pm}(\alpha,x,z) = P^{\pm}(\alpha)R^{\pm}(\alpha)e^{\pm z\cdot x}.$$

Since obviously

$$(3.3.13)\qquad R^-(\alpha,\partial)^{-1} = R^+(\alpha,\partial)^*,$$

using Lemma 3.2 we deduce from the bilinear identity (3.3.4):

$$(3.3.14)\qquad (P^+(\alpha)R^+(\alpha-\beta)P^-(\beta)^*)_- = 0 \text{ for any } \alpha,\beta\in \text{supp}\ \tau.$$

Furthermore, (3.3.14) for $\alpha = \beta$ is equivalent to

$$P^-(\alpha) = (P^+(\alpha)^*)^{-1}, \tag{3.3.15}$$

since $R^{\pm}(0) = I_n$ and $P^{\pm}(\alpha) \in I_n + \Psi_-$. Equations (3.3.14 and 15) imply

$$(P^+(\alpha)R^+(\alpha-\beta)P^+(\beta)^{-1})_- = 0 \text{ for all } \alpha, \beta \in \operatorname{supp} \tau. \tag{3.3.16}$$

In the rest of this paper we sometimes write $P(\alpha)$ instead of $P^+(\alpha)$.

Proposition 3.3. *Given $\beta \in$ supp τ, all the pseudo-differential operators $P(\alpha)$, $\alpha \in$ supp τ, are completely determined by $P(\beta)$ from equations (3.3.16).*

Proof. We have for $i \neq j:\ R(\alpha_{ij}) = A\partial + B + C\partial^{-1}$, where

$$A = \varepsilon_{ij}E_{ii},\ B = \sum_{\substack{k=1\\ k\neq i,j}}^{n} \varepsilon_{ik}\varepsilon_{jk}E_{kk},\ C = \varepsilon_{ji}E_{jj}. \tag{3.3.17}$$

For $P = I_n + \sum_{j=1}^{\infty} W^{(j)}\partial^{-j}$ we have

$$P^{-1} = I_n - W^{(1)}\partial^{-1} + (W^{(1)2} - W^{(2)})\partial^{-2} + \cdots. \tag{3.3.18}$$

Let $\alpha, \beta \in M$ be such that $\alpha - \beta = \alpha_{ij}$. It follows from (3.3.18) and (3.3.16) that $P(\alpha)R(\alpha-\beta)P(\beta)^{-1} = (P(\alpha)R(\alpha-\beta)P(\beta)^{-1})_+ = A\partial + B + W^{(1)}(\alpha)A - AW^{(1)}(\beta)$, or equivalently:

$$P(\alpha)(A\partial + B + C\partial^{-1}) = (A\partial + B + W^{(1)}(\alpha)A - AW^{(1)}(\beta))P(\beta).$$

Equating coefficients of ∂^{-m}, $m \geq 1$, we obtain:

$$\begin{aligned} &W^{(m+1)}(\alpha)A + W^{(m)}(\alpha)B + W^{(m-1)}(\alpha)C = \\ &= A(\partial W^{(m)}(\beta)W^{(m+1)}(\beta) - W^{(1)}(\beta)W^{(m)}(\beta)) + BW^{(m)}(\beta) + W^{(1)}(\alpha)AW^{(m)}(\beta). \end{aligned}$$

Substituting expressions (3.3.17) for A, B and C, we obtain an explicit form of matching conditions ($m \geq 1$):

$$\begin{aligned} &\varepsilon_{ij}W^{(m+1)}(\alpha)E_{ii} + \sum_{k\neq i,j} \varepsilon_{ik}\varepsilon_{jk}W^{(m)}(\alpha)E_{kk} + \varepsilon_{ji}W^{(m-1)}(\alpha)E_{jj} \\ &= \varepsilon_{ij}E_{ii}(\partial W^{(m)}(\beta) + W^{(m+1)}(\beta) - W^{(1)}(\beta)W^{(m)}(\beta)) \\ &+ \sum_{k\neq i,j} \varepsilon_{ik}\varepsilon_{jk}E_{kk}W^{(m)}(\beta) + \varepsilon_{ij}W^{(1)}(\alpha)E_{ii}W^{(m)}(\beta). \end{aligned} \tag{3.3.19}$$

It follows from (3.3.19) that $W^{(m+1)}(\alpha)$ for $m \geq 1$ can be expressed in terms of the $W^{(s)}(\beta)$ with $s \leq m+1$ and $W^{(1)}(\alpha)$. Looking at the (k,ℓ)-entry of (3.3.19) for $k, \ell \neq i, j$, we see that $W^{(1)}(\alpha)$ can be expressed in terms of $W^{(1)}(\beta)$ and $W^{(1)}_{ki}(\alpha)$, where $k \neq i, j$. The (k,j)-entry of (3.3.19) for $m = 1$ gives: $W^{(1)}_{ki}(\alpha)W^{(1)}_{ij}(\beta) = \varepsilon_{ik}\varepsilon_{jk}W^{(1)}_{kj}(\beta)$, and since the (j,j)-entry of this equation is $W^{(1)}_{ji}(\alpha)W^{(1)}_{ij}(\beta) = -1$, we see that $W^{(1)}_{ij}(\beta)$ is invertible, hence $W^{(1)}_{ki}(\alpha)$ is expressed in terms of $W^{(1)}(\beta)$.

Due to Proposition 2.3 for any $\alpha, \beta \in$ supp τ there exist a sequence $\gamma_1, \ldots, \gamma_k$ such that $\alpha = \gamma_1$, $\beta = \gamma_k$ and $\gamma_i - \gamma_{i+1} \in \Delta$ for all $i = 1, \ldots, k-1$. The proposition now follows. □

Remark 3.3. The functions $P^+(\alpha, x, z)$ ($\alpha \in M$) determine the τ-function $\sum_\alpha \tau_\alpha(x)e^\alpha$ up to a constant factor. Namely, we may recover $\tau_\alpha(x)$ from functions $P^+_{jj}(\alpha, x, z)$ as follows. We have from (3.3.5):

$$\log P^+_{jj}(\alpha,x,z) = \log \tau_\alpha(x^{(p)}_\ell - \frac{\delta_{jp}}{\ell z^\ell}) - \log \tau_\alpha(x^{(p)}_\ell).$$

Applying to both sides the operator $\frac{\partial}{\partial z} - \sum_{k\geq 1} z^{-k-1}\frac{\partial}{\partial x^{(j)}_k}$ (that kills the first summand on the right), we obtain:

$$(\frac{\partial}{\partial z} - \sum_{k\geq 1} z^{-k-1}\frac{\partial}{\partial x^{(j)}_k})\log P^+_{jj}(\alpha,x,z) = \sum_{k\geq 1} z^{-k-1}\frac{\partial}{\partial x^{(j)}_k}\log \tau_\alpha(x).$$

Hence

$$\frac{\partial}{\partial x^{(j)}_k}\log \tau_\alpha(x) = Res_{z=0}\ dz\ z^k(\frac{\partial}{\partial z} - \sum_{k\geq 1} z^{-k-1}\frac{\partial}{\partial x^{(j)}_k})\log P^+_{jj}(\alpha,x,z). \tag{3.3.20}$$

This determines $\tau_\alpha(x)$ up to a constant factor. It follows from (3.3.7) and Proposition 2.3 that these constant factors are the same for all α.

3.4. Introduce the following formal pseudo-differential operators $L(\alpha)$, $C^{(j)}(\alpha)$, $L^{(j)}(\alpha)$ and differential operators $B_m(\alpha)$ and $B^{(j)}_m(\alpha)$:

$$\begin{aligned} L(\alpha) &\equiv L(\alpha,x,\partial) = P^+(\alpha)\circ\partial\circ P^+(\alpha)^{-1}, \\ C^{(j)}(\alpha) &\equiv C^{(j)}(\alpha,x,\partial) = P^+(\alpha)E_{jj}P^+(\alpha)^{-1}, \\ L^{(j)}(\alpha) &\equiv C^{(j)}(\alpha)L(\alpha) = P^+(\alpha)E_{jj}\circ\partial\circ P^+(\alpha)^{-1}, \\ B_m(\alpha) &\equiv (L(\alpha)^m)_+ = (P^+(\alpha)\circ\partial^m\circ P^+(\alpha)^{-1})_+, \\ B^{(j)}_m(\alpha) &\equiv (L^{(j)}(\alpha)^m)_+ = (P^+(\alpha)E_{jj}\circ\partial^m\circ P^+(\alpha)^{-1})_+. \end{aligned} \tag{3.4.1}$$

Using Lemma 3.2 we can now derive the Sato equations from equation (3.3.4):

Lemma 3.4. *Each formal pseudo-differential operator $P = P^+(\alpha)$ satisfies the Sato equations:*

$$\frac{\partial P}{\partial x^{(j)}_k} = -(PE_{jj}\circ\partial^k\circ P^{-1})_-\circ P. \tag{3.4.2}$$

Proof. Notice first that

$$\begin{aligned} &(\frac{\partial}{\partial x^{(j)}_k} - B^{(j)}_k(\alpha))V^+(\alpha,x,z) = (\frac{\partial}{\partial x^{(j)}_k} - B^{(j)}_k(\alpha))P^+(\alpha)R^+(\alpha)e^{z\cdot x} \\ &= (\frac{\partial P^+(\alpha)}{\partial x^{(j)}_k}R^+(\alpha) + P^+(\alpha)R^+(\alpha)E_{jj}\partial^k - B^{(j)}_k(\alpha)P^+(\alpha)R^+(\alpha))e^{z\cdot x} \\ &= (\frac{\partial P^+(\alpha)}{\partial x^{(j)}_k} + P^+(\alpha)E_{jj}\partial^k - B^{(j)}_k(\alpha)P^+(\alpha))R^+(\alpha)e^{z\cdot x} \\ &= (\frac{\partial P^+(\alpha)}{\partial x^{(j)}_k} + L^{(j)}(\alpha)^k P^+(\alpha) - B^{(j)}_k(\alpha)P^+(\alpha))R^+(\alpha)e^{z\cdot x} \\ &= (\frac{\partial P^+(\alpha)}{\partial x^{(j)}_k} + (L^{(j)}(\alpha)^k)_- P^+(\alpha))R^+(\alpha)e^{z\cdot x} \end{aligned}$$

Next apply $\frac{\partial}{\partial x_k^{(j)}} - B_k^{(j)}(\alpha)$ to the equation (3.3.4) for $\alpha = \beta$ to obtain:

$$Res_{z=0}\, dz\, \left(\frac{\partial P^+(\alpha)}{\partial x_k^{(j)}} + (L^{(j)}(\alpha)^k)_-\right)(P^+(\alpha)R^+(\alpha)e^{z\cdot x})\ {}^t(P^-(\alpha)R^-(\alpha)e^{-z\cdot x'}) = 0.$$

Now apply Lemma 3.2 and (3.3.15) to obtain:

$$\left(\left(\frac{\partial P^+(\alpha)}{\partial x_k^{(j)}} + (L^{(j)}(\alpha)^k)_- P^+(\alpha)\right)P^+(\alpha)^{-1}\right)_- = 0$$

which proves the lemma. □

Proposition 3.4. *Consider the formal oscillating functions $V^+(\alpha, x, z)$ and $V^-(\alpha, x, z)$, $\alpha \in M$, of the form (3.3.12), where $R^\pm(\alpha, z)$ are given by (3.3.10) and $P^\pm(\alpha, x, \partial) \in I_n + \Psi_-$. Then the bilinear identity (3.3.4) for all $\alpha, \beta \in$ supp τ is equivalent to the Sato equation (3.4.2) for each $P = P^+(\alpha)$ and the matching condition (3.3.14) for all $\alpha, \beta \in$ supp τ.*

Proof. We have proved already that the bilinear identity (3.3.4) implies (3.4.2) and (3.3.14). To prove the converse, denote by $A(\alpha, \beta, x, x')$ the left-hand side of (3.3.4). The same argument as in the proof of Lemma 3.4 shows that:

$$\left(\frac{\partial}{\partial x_k^{(j)}} - B_k^{(j)}(\alpha)\right) A(\alpha, \beta, x, x') = 0, \tag{3.4.3}$$

$$A(\alpha, \beta, x, x') = 0,.\ \text{ if } x_k^{(i)} = x_k'^{(i)} \text{ for } k \geq 2, \tag{3.4.4}$$

where $B_k^{(j)}(\alpha)$ is defined by (3.4.1).

Denote by $A_1(\alpha, \beta)$ the expression for $A(\alpha, \beta, x, x')$ in which we set $x_k^{(j)} = x_k'^{(j)} = 0$ if $k \geq 2$ and $x_1^{(1)} = \ldots = x_1^{(n)} = x_1$, $x_1'^{(1)} = \ldots = x_1'^{(n)} = x_1'$. Expanding $A(\alpha, \beta, x, x')$ in a power series in $x_k^{(i)} - x_k'^{(i)}$ for $k \geq 2$ and $x_1^{(i)} - x_1^{(j)}$, $x_1'^{(i)} - x_1'^{(j)}$, we see from (3.4.3) and (3.4.4) that it remains to prove

$$A_1(\alpha, \beta) = 0. \tag{3.4.5}$$

But the same argument as in the proof of Lemma 3.2 shows that

$$A_1(\alpha, \beta) = \text{Res}_{z=0} W^+(\alpha, x_1, \partial) R^+(\alpha - \beta, \partial) W^-(\beta, x_1, \partial)^*\ e^{yz} dz,$$

where $y = x_1 - x_1'$. Hence, as at the end of the proof of Lemma 3.2, (3.4.5) follows from (3.3.14). □

3.5. Fix $\alpha \in M$; we have introduced above a collection of formal pseudo-differential operators $L \equiv L(\alpha)$, $C^{(i)} \equiv C^{(i)}(\alpha)$ of the form:

$$\begin{aligned} L &= I_n \partial + \sum_{j=1}^{\infty} U^{(j)}(x) \partial^{-j}, \\ C^{(i)} &= E_{ii} + \sum_{j=1}^{\infty} C^{(i,j)}(x) \partial^{-j},\ i = 1, 2, \cdots, n, \end{aligned} \tag{3.5.1}$$

subject to the conditions

$$(3.5.2)\qquad \sum_{i=1}^{n} C^{(i)} = I_n,\ C^{(i)}L = LC^{(i)},\ C^{(i)}C^{(j)} = \delta_{ij}C^{(i)}.$$

They satisfy the following set of equations for some $P \in I_n + \Psi_-$:

$$(3.5.3)\qquad \begin{cases} LP = P\partial \\ C^{(i)}P = PE_{ii} \\ \dfrac{\partial P}{\partial x_k^{(i)}} = -(L^{(i)k})_- P, \text{ where } L^{(i)} = C^{(i)}L. \end{cases}$$

Notice that the first equation of (3.5.3) follows from the last one, since $L = I_n\partial + \sum_i (L^{(i)})_-$.

Proposition 3.5. *The system of equations (3.5.3) has a solution $P \in I_n + \Psi_-$ if and only if we can find a formal oscillating function of the form*

$$(3.5.4)\qquad W(x,z) = (I_n + \sum_{j=1}^{\infty} W^{(j)}(x)z^{-j})e^{z\cdot x}$$

that satisfies the linear equations

$$(3.5.5)\qquad LW = zW,\ C^{(i)}W = WE_{ii},\ \frac{\partial W}{\partial x_k^{(i)}} = B_k^{(i)}W.$$

Proof (3.5.3) $\Rightarrow$ (3.5.5): Put $W = Pe^{z\cdot x}$. Then we have:

$$\begin{aligned} LW &= LPe^{z\cdot x} = P\partial e^{z\cdot x} = zPe^{z\cdot x} = zW; \\ C^{(i)}W &= C^{(i)}Pe^{z\cdot x} = PE_{ii}e^{z\cdot x} = Pe^{z\cdot x}E_{ii} = WE_{ii}; \\ \frac{\partial W}{\partial x_k^{(i)}} &= \frac{\partial P}{\partial x_k^{(i)}} + P\frac{\partial e^{z\cdot x}}{\partial x_k^{(i)}} = -(L^{(i)k})_- Pe^{z\cdot x} + z^k PE_{ii}e^{z\cdot x} \\ &= -(L^{(i)k})_- W + PE_{ii}\partial^k e^{z\cdot x} = -(L^{(i)k})_- W + C^{(i)}P\partial^k e^{z\cdot x} \\ &= -(L^{(i)k})_- W + C^{(i)}L^k Pe^{z\cdot x} = -(L^{(i)k})_- W + L^{(i)k}W = B_k^{(i)}W. \end{aligned}$$

(3.5.5) $\Rightarrow$ (3.5.3): Define $P \in \Psi$ by $W = Pe^{z\cdot x}$. If $LW = zW$, then $LPe^{z\cdot x} = zPe^{z\cdot x} = P\partial e^{z\cdot x}$, hence $LP = P\partial$.

If $C^{(i)}W = WE_{ii}$, then $C^{(i)}Pe^{z\cdot x} = Pe^{z\cdot x}E_{ii} = PE_{ii}e^{z\cdot x}$, hence $C^{(i)}P = PE_{ii}$.

Finally, the last equation of (3.5.5) gives: $\frac{\partial}{\partial x_k^{(i)}}(Pe^{z\cdot x}) = -(L^{(i)k})_- Pe^{z\cdot x} + L^{(i)k}Pe^{z\cdot x}$. Since we have already proved the first two equations of (3.5.3), we derive (as above): $L^{(i)k}Pe^{z\cdot x} = z^k Pe^{z\cdot x} = P\frac{\partial e^{z\cdot x}}{\partial x_k^{(i)}}$, hence: $\frac{\partial P}{\partial x_k^{(i)}}e^{z\cdot x} = -(L^{(i)k})_- Pe^{z\cdot x}$, which proves that P satisfies the Sato equations. □

Remarks 3.5. (a) It is easy to see that the collection of formal pseudo-differential operators $\{L, C^{(1)}, \ldots, C^{(n)}\}$ of the form (3.5.1) and satisfying (3.5.2) can be simultaneously conjugated to the trivial collection $\{\partial, E_{11}, \ldots, E_{nn}\}$ by some $P \in I_n + \Psi_-$. It follows that the solution of the form (3.5.4) to the linear problem (3.5.5) is unique up to multiplication on the right by a diagonal matrix of the form

$$(3.5.6)\qquad D(z) = \exp - \sum_{j=1}^{\infty} a_j z^{-j}/j,$$

where the a_j are diagonal matrices over $\mathbb{C}$ (indeed, this is the case for the trivial collection). The space of all solutions of (3.5.5) in formal oscillating functions is obtained from one of the form (3.5.4) by multiplying on the right by a diagonal matrix over $\mathbb{C}((z))$. For that reason the (matrix valued) functions

$$W^+(\alpha, x, z) = P^+(\alpha)e^{x\cdot z}, \ \alpha \in \text{supp}\ \tau, \tag{3.5.7}$$

are called the *wave functions* for τ. The formal pseudo-differential operator $P^+(\alpha)$ is called the wave operator. The functions $W^-(\alpha, x, z) = P^-(\alpha)e^{-x\cdot z}$ are called the adjoint wave functions and the operators $P^-(\alpha)$ (which are expressed via $P^+(\alpha)$ by (3.3.15)) are called the adjoint wave operators. Note that $V^+(\alpha, x, z)$ are solutions of (3.5.5) as well since they are obtained by multiplying $W^+(\alpha, x, z)$ on the right by $R^+(\alpha, z)$. (b) Multiplying the wave function $W^+(\alpha, x, z)$ on the right by $D(z)$ given by (3.5.6) corresponds to multiplying the corresponding τ-function by $\exp tr \sum_{k=1}^{\infty} a_k x_k$, where $x_k = \text{diag}\,(x_k^{(1)}, \dots, x_k^{(n)})$.

(c) The collection $\{L, C^{(1)}, \dots, C^{(n)}\}$ determines uniquely $P \in I_n + \Psi_-$ up to the multiplication of P on the right by a formal pseudo-differential operator with constant coefficients from $I_n + \Psi_-$.

3.6. In this section we shall rewrite the compatibility conditions of the system (3.5.3) (or equivalent compatibility conditions of the system (3.5.5)) in the form of Lax equations and Zakharov-Shabat equations.

Lemma 3.6. *If for every $\alpha \in M$ the formal pseudo-differential operators $L \equiv L(\alpha)$ and $C^{(j)} \equiv C^{(j)}(\alpha)$ of the form (3.5.1) satisfy conditions (3.5.2) and if the equations (3.5.3) have a solution $P \equiv P(\alpha) \in I_n + \Psi_-$, then the differential operators $B_k^{(j)} \equiv B_k^{(j)}(\alpha) = (L^{(j)}(\alpha)^k)_+$ satisfy one of the following equivalent conditions:*

$$\begin{cases} \dfrac{\partial L}{\partial x_k^{(j)}} = [B_k^{(j)}, L] \\ \dfrac{\partial C^{(i)}}{\partial x_k^{(j)}} = [B_k^{(j)}, C^{(i)}] \end{cases} \tag{3.6.1}$$

$$\frac{\partial L^{(i)}}{\partial x_k^{(j)}} = [B_k^{(j)}, L^{(i)}] \tag{3.6.2}$$

$$\frac{\partial B_\ell^{(i)}}{\partial x_k^{(j)}} - \frac{\partial B_k^{(j)}}{\partial x_\ell^{(i)}} = [B_k^{(j)}, B_\ell^{(i)}]. \tag{3.6.3}$$

Proof (cf. [Sh]). To derive the first equation of (3.6.1) we differentiate the equation $LP = P\partial$ by $x_k^{(j)}$:

$$\frac{\partial L}{\partial x_k^{(j)}} P + L \frac{\partial P}{\partial x_k^{(j)}} = \frac{\partial P}{\partial x_k^{(j)}} \partial,$$

and substitute Sato's equation (see (3.5.3)). Then one obtains:

$$\frac{\partial L}{\partial x_k^{(j)}} P = (B_k^{(j)} L - L B_k^{(j)}) P$$

from which we derive the desired result. The second equation of (3.6.1) is proven analogously: differentiate $C^{(i)}P = PE_{ii}$, substitute Sato's equation and use the fact that $[L^{(j)k}, C^{(i)}] = 0$.

Next we prove the equivalence of (3.6.1), (3.6.2) and (3.6.3). The implication (3.6.1) $\Rightarrow$ (3.6.2) is trivial. To prove the implication (3.6.2) $\Rightarrow$ (3.6.1) note that $L = \sum_{j=1}^n L^{(j)}$ implies that the first equation of (3.6.1) follows immediately. As for the second one, we have:

$$\frac{\partial C^{(i)}}{\partial x_k^{(j)}} = \left(\frac{\partial L^{(i)}}{\partial x_k^{(j)}} - C^{(i)} \frac{\partial L}{\partial x_k^{(j)}}\right) L^{-1}$$

$$= ([B_k^{(j)}, L^{(i)}] - C^{(i)}[B_k^{(j)}, L])L^{-1}$$
$$= ([B_k^{(j)}, C^{(i)}]L)L^{-1} = [B_k^{(j)}, C^{(i)}].$$

Next, we prove the implication (3.6.2) $\Rightarrow$ (3.6.3). Since both $\frac{\partial}{\partial x_k^{(j)}}$ and $ad\ B_k^{(j)}$ are derivations, (3.6.2) implies:

$$\frac{\partial L^{(i)\ell}}{\partial x_k^{(j)}} = [B_k^{(j)}, L^{(i)\ell}].$$

Hence:

$$\left(\frac{\partial B_\ell^{(i)}}{\partial x_k^{(j)}} - \frac{\partial B_k^{(j)}}{\partial x_\ell^{(i)}} - [B_k^{(j)}, B_\ell^{(i)}]\right) + \left(\frac{\partial (L^{(i)\ell})_-}{\partial x_k^{(j)}} - \frac{\partial (L^{(j)k})_-}{\partial x_\ell^{(i)}} + [(L^{(j)k})_-, (L^{(i)\ell})_-]\right)$$
$$= [B_k^{(j)}, L^{(i)\ell}] - [B_\ell^{(i)}, L^{(j)k}] - [B_k^{(j)}, B_\ell^{(i)}] + [(L^{(j)k})_-, (L^{(i)\ell})_-]$$
$$= [L^{(j)k}, L^{(i)\ell}] = 0.$$

Since $\Psi_- \cap \Psi_+ = \{0\}$, both terms on the left-hand side are zero proving (3.6.3).

Finally, we prove the implication (3.6.3) $\Rightarrow$ (3.6.2). We rewrite (3.6.3):

$$\frac{\partial L^{(i)\ell}}{\partial x_k^{(j)}} - [B_k^j, L^{(i)\ell}] = \frac{\partial (L^{(i)\ell})_-}{\partial x_k^{(j)}} + \frac{\partial B_k^{(j)}}{\partial x_\ell^{(i)}} - [B_k^{(j)}, (L^{(i)\ell})_-]$$

This right-hand side has order $k-1$, hence

(3.6.4) $$\frac{\partial L^{(i)\ell}}{\partial x_k^{(j)}} - [B_k^{(j)}, L^{(i)\ell}] \in \Psi(k-1) \text{ for every } \ell > 0.$$

Now suppose that $\frac{\partial L^{(i)}}{\partial x_k^{(j)}} - [B_k^{(j)}, L^{(i)}] \neq 0$. Then:

$$\lim_{\ell\to\infty} ord(\frac{\partial L^{(i)\ell}}{\partial x_k^{(i)}} - [B_k^{(j)}, L^{(i)\ell}]) = \infty$$

which contradicts (3.6.4). □

Equations (3.6.1) and (3.6.2) are called *Lax type* equations. Equations (3.6.3) are called the *Zakharov-Shabat type* equations. The latter are the compatibility conditions for the linear problem (3.5.5). Indeed, since $\frac{\partial}{\partial x_k^{(j)}}\frac{\partial}{\partial x_\ell^{(i)}} W = \frac{\partial}{\partial x_\ell^{(i)}}\frac{\partial}{\partial x_k^{(j)}} W$, one finds

$$0 = \frac{\partial}{\partial x_k^{(j)}}(B_\ell^{(i)}W) - \frac{\partial}{\partial x_\ell^{(i)}}(B_k^{(j)}W) = (\frac{\partial B_\ell^{(i)}}{\partial x_k^{(j)}} - \frac{\partial B_k^{(j)}}{\partial x_\ell^{(i)}} - [B_k^{(j)}, B_\ell^{(i)}])W.$$

Notice that as a byproduct of the proof of Proposition 3.6, we obtain complementary Zakharov-Shabat equations:

(3.6.5) $$\frac{\partial (L^{(i)\ell})_-}{\partial x_k^{(j)}} - \frac{\partial (L^{(j)k})_-}{\partial x_\ell^{(i)}} = [(L^{(i)\ell})_-, (L^{(j)k})_-].$$

Proposition 3.6. *Sato equations (3.4.2) on $P \in I_n + \Psi_-$ imply equations (3.6.3) on differential operators $B_k^{(i)} = (L^{(i)k})_+$.*

Proof is the same as that of the corresponding part of Lemma 3.6. □

Remark 3.6. The above results may be summarized as follows. The n-component KP hierarchy (2.3.7) of Hirota bilinear equations on the τ-function is equivalent to the bilinear equation (3.3.4) on the wave function, which is related to the τ-function by formula (3.3.3) and Remark 3.3. The bilinear equation (3.3.4) for each $\alpha = \beta$ implies the Sato equation (3.4.2) on the formal pseudo-differential operator $P \equiv P(\alpha)$. Moreover, equation (3.4.2) on $P(\alpha)$ for each α together with the matching conditions (3.3.14) are equivalent to the bilinear identity (3.3.4). Also, the Sato equation (or rather (3.5.3)) is a compatibility condition for the linear problem (3.5.5) for the wave function. The Sato equation in turn implies the system of Lax type equations (3.6.2) (or equivalent systems (3.6.1) or (3.6.3), which is the most familiar form of the compatibility condition) on formal pseudo-differential operators $L^{(i)}$ (resp. L and $C^{(i)}$ satisfying constraints (3.5.2)). The latter formal pseudo-differential operators are expressed via the wave function by formulas (3.4.1), (3.3.9–12).

3.7. In this section we write down explicitly some of the Sato equations (3.4.2) on the matrix elements $W_{ij}^{(s)}$ of the coefficients $W^{(s)}(x)$ of the pseudo-differential operator

$$P = I_n + \sum_{m=1}^{\infty} W^{(m)}(x)\partial^{-m}.$$

We shall write W_{ij} for $W_{ij}^{(1)}$ to simplify notation. We have for $i \neq k$:

$$\frac{\partial W_{ij}}{\partial x_1^{(k)}} = W_{ik}W_{kj} - \delta_{jk}W_{ij}^{(2)}, \tag{3.7.1}$$

$$\frac{\partial W_{ij}^{(2)}}{\partial x_1^{(k)}} = W_{ik}W_{kj}^{(2)} - \delta_{jk}W_{ij}^{(3)}. \tag{3.7.2}$$

Next, calculating $\frac{\partial W_{ij}}{\partial x_2^{(k)}}$ from (3.4.2) and substituting (3.7.1) and (3.7.2) in these equations, we obtain:

$$\frac{\partial W_{ij}}{\partial x_2^{(k)}} = W_{ik}\frac{\partial W_{kj}}{\partial x_1^{(k)}} - \frac{\partial W_{ik}}{\partial x_1^{(k)}}W_{kj} \text{ if } k \neq i \text{ and } k \neq j, \tag{3.7.3}$$

$$\frac{\partial W_{ij}}{\partial x_2^{(j)}} = 2\frac{\partial W_{jj}}{\partial x_1^{(j)}}W_{ij} - \frac{\partial^2 W_{ij}}{\partial x_1^{(j)2}} \text{ if } i \neq j, \tag{3.7.4}$$

$$\frac{\partial W_{ij}}{\partial x_2^{(i)}} = -2\frac{\partial W_{ii}}{\partial x_1^{(i)}}W_{ij} + \frac{\partial^2 W_{ij}}{\partial x_1^{(i)2}} \text{ if } i \neq j, \tag{3.7.5}$$

$$\frac{\partial W_{ii}}{\partial x_2^{(i)}} = \frac{\partial^2 W_{ii}}{\partial x_1^{(i)2}} + 2\sum_{p \neq i} W_{ip}\frac{\partial W_{pi}}{\partial x_1^{(i)}} - 2W_{ii}\partial W_{ii} + 2\partial W_{ii}^{(2)}. \tag{3.7.6}$$

Remark 3.7. Substituting expressions for the $W_{ij} = W_{ij}^{(1)}(\alpha = 0, x)$ given by (3.3.7), the above equations turn into the Hirota bilinear equations found in §2.4 as follows:

(3.7.1) for $i = j$ $\Rightarrow$ (2.4.4)
(3.7.1) for $i \neq j$ $\Rightarrow$ (2.4.8)
(3.7.5) $\Rightarrow$ (2.4.5),
(3.7.4) $\Rightarrow$ (2.4.6),
(3.7.3) for $i = j$ $\Rightarrow$ (2.4.7) (with j replaced by k),
(3.7.3) for $i \neq j$ $\Rightarrow$ (2.4.9).

3.8. In this section we write down explicitly some of the Lax equations (3.6.1) of the n-component KP hierarchy and auxiliary conditions (3.5.2) for the formal pseudo-differential operators

$$(3.8.1)\qquad L = I_n\partial + \sum_{j=1}^{\infty} U^{(j)}(x)\partial^{-j} \text{ and } C^{(i)} = E_{ii} + \sum_{j=1}^{\infty} C^{(i,j)}(x)\partial^{-j}\ (i = 1,\dots,n).$$

For the convenience of the reader, recall that x stands for all indeterminates $x_i^{(k)}$, where $i = 1, 2, \dots$ and $k = 1, \dots, n$, that the auxiliary conditions are

$$(3.8.2)\qquad \sum_{i=1}^{n} C^{(i)} = I_n,\ C^{(i)}C^{(j)} = \delta_{ij}C^{(i)},\ C^{(i)}L = LC^{(i)},$$

and that the Lax equations of the n-component KP hierarchy are

$$(3.8.3\text{a})_i\qquad \frac{\partial L}{\partial x_i^{(k)}} = [B_i^{(k)}, L],$$

$$(3.8.3\text{b})_i\qquad \frac{\partial C^{(\ell)}}{\partial x_i^{(k)}} = [B_i^{(k)}, C^{(\ell)}],$$

where $B_i^{(k)} = (C^{(k)}L^i)_+$. For example, we have:

$$(3.8.4)\qquad B_1^{(k)} = E_{kk}\partial + C^{(k,1)},\ B_2^{(k)} = E_{kk}\partial^2 + C^{(k,1)}\partial + 2E_{kk}U^{(1)} + C^{(k,2)}.$$

Denote by $C_{ij}^{(k,\ell)}$ and $U_{ij}^{(k)}$ the (i,j)-th entries of the $n \times n$ matrices $C^{(k,\ell)}$ and $U^{(k)}$ respectively. Then the ∂^{-1} term of the second equation (3.8.2) gives:

$$(3.8.5)\qquad C_{ij}^{(k,1)} = 0 \text{ if } i \neq k \text{ and } j \neq k, \text{ or } i = j = k,$$

$$(3.8.6)\qquad C_{kj}^{(k,1)} = -C_{kj}^{(j,1)}.$$

Hence the matrices $C^{(j,1)}$ are expressed in terms of the functions

$$A_{ij} := C_{ij}^{(j,1)} \quad (\text{note that } A_{ii} = 0).$$

The ∂^{-2} term of the second equation (3.8.2) allowes one to express most of the $C_{ij}^{(k,2)}$ in terms of the A_{ij}:

$$(3.8.7)\qquad C_{ij}^{(k,2)} = -A_{ik}A_{kj} \text{ if } i \neq k \text{ and } j \neq k,$$

$$(3.8.8)\qquad C_{k,k}^{(k,2)} = \sum_{p=1}^{n} A_{kp}A_{pk}.$$

Furthermore, the ∂^{-1} term of the Lax equation $(3.8.3\text{b})_1$ gives:

$$(3.8.9)\qquad \frac{\partial A_{ij}}{\partial x_1^{(k)}} = A_{ik}A_{kj} \text{ for distinct } i, j, k,$$

$$(3.8.10)\qquad C_{ij}^{(j,2)} = -\frac{\partial A_{ij}}{\partial x_1^{(j)}} \text{ for } i \neq j,$$

$$(3.8.11)\qquad C_{ij}^{(i,2)} = \sum_{\substack{p=1\\ p\neq i}}^{n} \frac{\partial A_{ij}}{\partial x_1^{(p)}} \text{ for } i \neq j.$$

The ∂^{-2} term of that equation gives for $i \neq j$ (recall that $\partial = \frac{\partial}{\partial x_1^{(1)}} + \ldots + \frac{\partial}{\partial x_1^{(n)}}$.):

$$(3.8.12) \qquad C_{ij}^{(i,3)} = -\frac{\partial C_{ij}^{(i,2)}}{\partial x_1^{(j)}} + A_{ij}C_{jj}^{(i,2)} - \sum_{p=1}^{n}(A_{ip}\partial A_{pj} + C_{ip}^{(i,2)}A_{pj}),$$

$$(3.8.13) \qquad C_{ij}^{(j,3)} = -\sum_{\substack{p=1\\p\neq i}}^{n} \frac{\partial C_{ij}^{(j,2)}}{\partial x_1^{(p)}} - A_{ij}C_{ii}^{(j,2)} + \sum_{p=1}^{n} A_{ip}C_{pj}^{(j,2)}.$$

Substituting (3.8.7, 8 and 11) (resp. (3.8.7, 8 and 10)) in (3.8.12) (resp. in (3.8.13)) we obtain for $i \neq j$:

$$(3.8.14) \qquad C_{ij}^{(i,3)} = -(\partial - \frac{\partial}{\partial x_1^{(i)}})^2 A_{ij} - 2\sum_{\substack{p=1\\p\neq i}}^{n} A_{ip}A_{pi}A_{ij},$$

$$(3.8.15) \qquad C_{ij}^{(j,3)} = \frac{\partial^2 A_{ij}}{\partial x_1^{(j)2}} + 2\sum_{\substack{p=1\\p\neq j}}^{n} A_{ij}A_{jp}A_{pj}.$$

Furthermore, the ∂^0 and ∂^{-1} terms of the Lax equation $(3.8.3a)_1$ give respectively for $i \neq j$:

$$(3.8.16) \qquad U_{ij}^{(1)} = -\partial A_{ij},$$

$$(3.8.17) \qquad \frac{\partial U_{ii}^{(1)}}{\partial x_1^{(j)}} = -\partial(A_{ij}A_{ji}).$$

Finally, the ∂^{-1} term of the Lax equation $(3.8.3b)_2$ gives

$$(3.8.18) \qquad \frac{\partial A_{ij}}{\partial x_2^{(j)}} = -2A_{ij}U_{jj}^{(1)} - C_{ij}^{(j,3)} \text{ for } i \neq j,$$

$$(3.8.19) \qquad \frac{\partial A_{ij}}{\partial x_2^{(i)}} = \partial^2 A_{ij} - 2\partial C_{ij}^{(i,2)} - C_{ij}^{(i,3)} + 2U_{ii}^{(1)}A_{ij} \text{ for } i \neq j,$$

$$(3.8.20) \qquad \frac{\partial A_{ij}}{\partial x_2^{(k)}} = A_{ik}\frac{\partial A_{kj}}{\partial x_1^{(k)}} - A_{kj}\frac{\partial A_{ik}}{\partial x_1^{(k)}} \text{ for } i \neq k \text{ and } j \neq k.$$

3.9. Finally, we write down explicitly expressions for $U^{(1)}$ and $C^{(i,1)}$ in terms of τ-functions. Recall that

$$P = I_n + \sum_{j=1}^{\infty} W^{(j)}(x)\partial^{-j},$$

$$L = P\partial P^{-1} = I_n\partial + \sum_{j=1}^{\infty} U^{(j)}\partial^{-j},$$

$$C^{(i)} = PE_{ii}P^{-1} = E_{ii} + \sum_{j=1}^{\infty} C^{(i,j)}\partial^{-j}.$$

Using (3.3.18) we have:

$$(3.9.1) \qquad U^{(1)} = -\partial W^{(1)},$$

(3.9.2)
$$U^{(2)} = W^{(1)}\partial W^{(1)} - \partial W^{(2)},$$

(3.9.3)
$$C^{(i,1)} = [W^{(1)}, E_{ii}],$$

(3.9.4)
$$C^{(i,2)} = [W^{(2)}, E_{ii}] + [E_{ii}, W^{(1)}]W^{(1)}.$$

Using (3.3.7) we obtain from (3.9.1) and (3.9.3) respectively:

(3.9.5)
$$U_{ij}^{(1)} = \begin{cases} -\varepsilon_{ji}\partial(\tau_{\alpha+\alpha_{ij}}/\tau_\alpha) & \text{if } i \neq j, \\ \partial(\frac{\partial \tau_\alpha}{\partial x_1^{(i)}}/\tau_\alpha) & \text{if } i = j. \end{cases}$$

(3.9.6)
$$A_{ij} \equiv C_{ij}^{(j,1)} = \varepsilon_{ji}\tau_{\alpha+\alpha_{ij}}/\tau_\alpha.$$

(Recall that by (3.8.5 and 6) all the matrices $C^{(k,1)}$ can be expressed via the functions A_{ij}.) Using (3.3.7 and 8) and (3.9.2 and 4) one also may write down the matrices $U^{(2)}$ and $C^{(i,2)}$ in terms of τ-functions, but they are somewhat more complicated and we will not need them anyway.

§4. The n-wave interaction equations, the generalized Toda chain and the generalized Davey-Stewartson equations as subsystems of the n-component KP.

4.0. In this section we show that some well-known soliton equations, as well as their natural generalizations, are the simplest equations of the various formulations of the n-component KP hierarchy. To simplify notation, let

(4.0.1)
$$t_i = x_2^{(i)},\ x_i = x_1^{(i)}, \text{ so that } \partial = \sum_{i=1}^{n} \frac{\partial}{\partial x_i}.$$

4.1. Let $n \geq 3$. Then the n-component KP in the form of Sato equation contains the system (3.7.1) of $n(n-1)(n-2)$ equations on $n^2 - n$ functions W_{ij} ($i \neq j$) in the indeterminates x_i (all other indeterminates being parameters):

(4.1.1)
$$\frac{\partial W_{ij}}{\partial x_k} = W_{ik}W_{kj} \text{ for distinct } i, j, k.$$

The τ-function is given by the formula (3.3.7) for a fixed $\alpha \in M$:

(4.1.2)
$$W_{ij} = \varepsilon_{ji}\ \tau_{\alpha+\alpha_{ij}}/\tau_\alpha.$$

Substituting this in (4.1.1) gives the Hirota bilinear equation (2.4.8):

(4.1.3)
$$D_1^{(k)}\tau_\alpha \cdot \tau_{\alpha+\alpha_{ij}} = \varepsilon_{ik}\varepsilon_{kj}\varepsilon_{ij}\tau_{\alpha+\alpha_{ik}}\tau_{\alpha+\alpha_{kj}}$$

Note that due to (3.9.3), $W_{ij} = A_{ij}$ if $i \neq j$, hence (4.1.1) is satisfied by the A_{ij} as well.

One usually adds to (4.1.1) the equations

(4.1.4)
$$\partial W_{ij} = 0,\ i \neq j.$$

We shall explain the group theoretical meaning of this constraint in §6.

Let now $a = \text{diag}(a_1, \dots, a_n)$, $b = \text{diag}(b_1, \dots, b_n)$ be arbitrary diagonal matrices over $\mathbb{C}$. We reduce the system (4.1.1) to the plane [D]:

(4.1.5)
$$x_k = a_k x + b_k t.$$

A direct calculation shows that (4.1.1) reduces then to the following equation on the matrix $W = (W_{ij})$ (note that its diagonal entries don't occur):

$$[a, \frac{\partial W}{\partial t}] - [b, \frac{\partial W}{\partial x}] = [[a, W], [b, W]] + b\partial W a - a\partial W b. \tag{4.1.6}$$

Hence, imposing the constraint (4.1.4), we obtain the famous $1+1$ n-wave system (cf. [D], [NMPZ]):

$$[a, \frac{\partial W}{\partial t}] - [b, \frac{\partial W}{\partial x}] = [[a, W], [b, W]]. \tag{4.1.7}$$

Let now

$$x_k = a_k x + b_k t - y. \tag{4.1.8}$$

Then equation (4.1.6) gives

$$[a, \frac{\partial W}{\partial t}] - [b, \frac{\partial W}{\partial x}] - a\frac{\partial W}{\partial y}b + b\frac{\partial W}{\partial y}a = [[a, W], [b, W]]. \tag{4.1.9}$$

If we let

$$Q_{ij} = -(a_i - a_j)W_{ij}.$$

equation (4.1.9) turns into the following system, which is called in [AC, (5.4.30a,c)] the $2+1$ n-wave interaction equations ($i \neq j$):

$$\frac{\partial Q_{ij}}{\partial t} = a_{ij}\frac{\partial Q_{ij}}{\partial x} + b_{ij}\frac{\partial Q_{ij}}{\partial y} + \sum_k (a_{ik} - a_{kj})Q_{ik}Q_{kj}, \tag{4.1.10}$$

where

$$a_{ij} = (b_i - b_j)/(a_i - a_j), \ b_{ij} = b_i - a_i a_{ij}. \tag{4.1.11}$$

On the other hand, letting (we assume that $a_1 > \ldots > a_n$):

$$w_{ij} = W_{ij}/(a_i - a_j)^{1/2}, \tag{4.1.12}$$

the equation (4.1.6) gives for $i \neq j$:

$$\frac{\partial w_{ij}}{\partial t} - a_{ij}\frac{\partial w_{ij}}{\partial x} - b_{ij}\frac{\partial w_{ij}}{\partial y} = \sum_k \varepsilon_{ijk} w_{ik} w_{kj}, \tag{4.1.13}$$

where

$$\varepsilon_{ijk} = \frac{a_i b_k + a_k b_j + a_j b_i - a_k b_i - a_j b_k - a_i b_j}{((a_i - a_k)(a_k - a_j)(a_i - a_j))^{1/2}}. \tag{4.1.14}$$

Imposing the constraint $\overline{w}_{ij} = -w_{ji}$, we obtain from (4.1.13) the following Hamiltonian system (considered in [NMPZ, pp 175, 242] for $n = 3$ and called there the $2+1$ 3-wave system) ($i < j$):

$$\frac{\partial w_{ij}}{\partial t} - a_{ij}\frac{\partial w_{ij}}{\partial x} - b_{ij}\frac{\partial w_{ij}}{\partial y} = \frac{\partial H}{\partial \overline{w}_{ij}}, \tag{4.1.15}$$

where

$$H = \sum_{\substack{i,k,j \\ i<k<j}} \varepsilon_{ijk}(w_{ik} w_{kj} \overline{w}_{ij} + \overline{w}_{ik}\overline{w}_{kj} w_{ij}). \tag{4.1.16}$$

Finally, let $n = 3$ and let $u_1 = iw_{13}$, $u_2 = i\overline{w}_{13}$, $u_3 = iw_{12}$, $a_1 = -a_{23}$, $b_1 = -b_{23}$, $a_2 = -\overline{a}_{13}$, $b_2 = -\overline{b}_{13}$, $a_3 = -a_{12}$, $b_3 = -b_{13}$. Then, after imposing the constraint $\varepsilon_{132} = 1$, equations (4.1.15) turn into the well-known $2+1$ 3-wave interaction equations (see [AC, (5.4.27)]):

$$\frac{\partial u_j}{\partial t} + a_j \frac{\partial u_j}{\partial x} + b_j \frac{\partial u_j}{\partial y} = i\overline{u}_k \overline{u}_\ell, \tag{4.1.17}$$

where (j,k,ℓ) is an arbitrary cyclic permutation of $1,2,3$.

4.2. Let $n \geq 2$. Then the n-component KP in the form of Sato equations contains the following subsystem of the system of equations (3.7.1) for arbitrary $\alpha \in M$ on the functions $W_{ij}(\alpha)$ in the indeterminates x_i (all other indeterminates being parameters):

$$\frac{\partial W_{ii}(\alpha)}{\partial x_j} = W_{ij}(\alpha) W_{ji}(\alpha) \text{ if } i \neq j. \tag{4.2.1}$$

The τ-function is given by (3.3.7) ($\alpha \in M$):

$$W_{ij}(\alpha) = \begin{cases} \varepsilon_{ji} \tau_{\alpha+\alpha_{ij}}/\tau_\alpha & \text{if } i \neq j \\ -\frac{\partial}{\partial x_i} \log \tau_\alpha & \text{if } i = j. \end{cases} \tag{4.2.2}$$

Substituting this in (4.2.1) gives the Hirota bilinear equations (2.4.4):

$$D_i D_j \tau_\alpha \cdot \tau_\alpha = 2\tau_{\alpha+\alpha_{ij}} \tau_{\alpha-\alpha_{ij}}. \tag{4.2.3}$$

In order to rewrite (4.2.1) in a more familiar form, let for $i \neq j$:

$$U_{ij}(\alpha) = \log \varepsilon_{ji} W_{ij}(\alpha) = \log(\tau_{\alpha+\alpha_{ij}}/\tau_\alpha). \tag{4.2.4}$$

Note that $\log(\tau_{\alpha+\alpha_{ij}}/\tau_\alpha) = -\log(\tau_{(\alpha+\alpha_{ij})-\alpha_{ij}}/\tau_{\alpha+\alpha_{ij}}$. Hence from (4.2.2) we obtain

$$U_{ij}(\alpha) = -U_{ji}(\alpha + \alpha_{ij}) \text{ if } i \neq j. \tag{4.2.5}$$

Furthermore, we have:

$$\begin{aligned} \frac{\partial^2}{\partial x_i \partial x_j} U_{ij}(\alpha) &= \frac{\partial^2}{\partial x_i \partial x_j} \log \tau_{\alpha+\alpha_{ij}} - \frac{\partial^2}{\partial x_i \partial x_j} \log \tau_\alpha \\ &= \frac{\partial W_{ii}(\alpha)}{\partial x_j} - \frac{\partial W_{ii}(\alpha+\alpha_{ij})}{\partial x_j} = W_{ij}(\alpha)W_{ji}(\alpha) - W_{ij}(\alpha+\alpha_{ij})W_{ji}(\alpha+\alpha_{ij}) \\ &= -\frac{\tau_{\alpha+\alpha_{ij}}}{\tau_\alpha}\frac{\tau_{\alpha-\alpha_{ij}}}{\tau_\alpha} + \frac{\tau_{\alpha+2\alpha_{ij}}}{\tau_{\alpha+\alpha_{ij}}}\frac{\tau_\alpha}{\tau_{\alpha+\alpha_{ij}}} = e^{U_{ij}(\alpha+\alpha_{ij})-U_{ij}(\alpha)} - e^{U_{ij}(\alpha)-U_{ij}(\alpha-\alpha_{ij})}. \end{aligned}$$

Thus the functions $U_{ij}(\alpha)$ ($i \neq j$) satisfy the following generalized Toda chain (with constraint (4.2.5)):

$$\frac{\partial^2 U_{ij}(\alpha)}{\partial x_i \partial x_j} = e^{U_{ij}(\alpha+\alpha_{ij})-U_{ij}(\alpha)} - e^{U_{ij}(\alpha)-U_{ij}(\alpha-\alpha_{ij})}. \tag{4.2.6}$$

Note also that (4.1.1) for distinct i,j and k becomes:

$$\frac{\partial U_{ij}(\alpha)}{\partial x_k} = \varepsilon_{ik}\varepsilon_{kj}\varepsilon_{ji} e^{U_{ij}(\alpha)+U_{ik}(\alpha)+U_{kj}(\alpha)}. \tag{4.2.7}$$

One should be careful about the boundary conditions. Let $S =$ supp τ; recall that by

Proposition 2.4, S is a convex polyhedron with vertices in M and edges parallel to roots. It follows that (4.2.6) should be understood as follows:

(i) if $\alpha \notin S$, then $U_{ij}(\alpha) = 0$ and (4.2.6) is trivial,

(ii) if $\alpha \in S$, but $\alpha + \alpha_{ij} \notin S$, then (4.2.6) is trivial,

(iii) if $\alpha \in S$, but $\alpha - \alpha_{ij} \notin S$, then the second term on the right-hand side of (4.2.6) is removed,

(iv) if $\alpha \in S$, $\alpha + \alpha_{ij} \in S$, but $\alpha + 2\alpha_{ij} \notin S$, then the first term on the right-hand side of (4.2.6) is removed.

Let now $n = 2$, and let $u_n = U_{12}(n\alpha_{12})$. Then we get the usual Toda chain:

$$\frac{\partial^2 u_n}{\partial x_1 \partial x_2} = e^{u_{n+1}-u_n} - e^{u_n - u_{n-1}}, \ n \in \mathbb{Z}. \tag{4.2.8}$$

It is a part of the Toda lattice hierarchy discussed in [UT].

4.3. Let $n \geq 2$. Then the n-component KP in the form of Sato equations contains the system of equations (3.7.4), (3.7.5), (3.7.3) and (3.7.1) for $j \neq k$ on n^2 functions W_{ij} in the indeterminates x_k and t_k $(k = 1, \ldots, n)$ (all other indeterminates being parameters):

$$\frac{\partial W_{ij}}{\partial t_j} = -\frac{\partial^2 W_{ij}}{\partial x_j^2} + 2\frac{\partial W_{jj}}{\partial x_j} W_{ij} \text{ if } i \neq j, \tag{4.3.1}$$

$$\frac{\partial W_{ij}}{\partial t_i} = \frac{\partial^2 W_{ij}}{\partial x_i^2} - 2\frac{\partial W_{ii}}{\partial x_i} W_{ij} \text{ if } i \neq j, \tag{4.3.2}$$

$$\frac{\partial W_{ij}}{\partial t_k} = W_{ik}\frac{\partial W_{kj}}{\partial x_k} - \frac{\partial W_{ik}}{\partial x_k} W_{kj} \text{ if } i \neq k \text{ and } j \neq k, \tag{4.3.3}$$

$$\frac{\partial W_{ij}}{\partial x_k} = W_{ik} W_{kj} \text{ if } i \neq k \text{ and } j \neq k. \tag{4.3.4}$$

This is a system of $n^3 - n$ evolution equations (4.3.1–3) and $n(n-1)^2$ constraints (4.3.4) which we call the generalized Davey-Stewartson system.

Note that the τ-functions of this system are given by (3.3.7), where we may take $\alpha = 0$. The corresponding to (4.3.1)–(4.3.4) Hirota bilinear equations are (2.4.6); (2.4.5); (2.4.7) if $i = j$ and (2.4.9) if $i \neq j$; (2.4.4) if $i = j$ and (2.4.8) if $i \neq j$, respectively.

Now, note that letting

$$\varphi_{ij} = \frac{1}{2}\left(\frac{\partial W_{ii}}{\partial x_i} + \frac{\partial W_{jj}}{\partial x_j} + \frac{\partial W_{ii}}{\partial x_j} + \frac{\partial W_{jj}}{\partial x_i}\right) (= \varphi_{ji}),$$

and subtracting (4.3.2) from (4.3.1) we obtain using (4.3.4):

$$\frac{\partial W_{ij}}{\partial t_j} - \frac{\partial W_{ij}}{\partial t_i} = -\left(\frac{\partial^2}{\partial x_i^2} + \frac{\partial^2}{\partial x_j^2}\right) W_{ij} + 2W_{ij}(\varphi_{ij} - W_{ij}W_{ji}). \tag{4.3.5}$$

Also, from (4.3.4) we obtain

$$\frac{\partial^2 \varphi_{ij}}{\partial x_i \partial x_j} = \frac{1}{2}\left(\frac{\partial}{\partial x_i} + \frac{\partial}{\partial x_j}\right)^2 (W_{ij}W_{ji}). \tag{4.3.6}$$

Let now $n = 2$; to simplify notation, let

$$q = W_{12}, \ r = W_{21}, \ \varphi = \varphi_{12} = \varphi_{21}.$$

Then, making the change of indeterminates

(4.3.7) $$s=-2i(t_1+t_2),\ t=-2i(t_1-t_2),\ x=x_1+x_2,\ y=x_1-x_2,$$

equations (4.3.5 and 6) turn into the decoupled Davey-Stewartson system:

(4.3.8) $$\begin{cases} i\frac{\partial q}{\partial t}=-\frac{1}{2}(\frac{\partial^2 q}{\partial x^2}+\frac{\partial^2 q}{\partial y^2})+q(\varphi-qr)\\ i\frac{\partial r}{\partial t}=\frac{1}{2}(\frac{\partial^2 r}{\partial x^2}+\frac{\partial^2 r}{\partial y^2})-r(\varphi-qr)\\ \frac{\partial^2\varphi}{\partial x^2}-\frac{\partial^2\varphi}{\partial y^2}=2\frac{\partial^2(qr)}{\partial x^2}.\end{cases}$$

Due to (3.3.7), the corresponding τ-functions are given by the following formulas, where we let $\tau_n=\tau_{n\alpha_{12}}$:

(4.3.9) $$q=-\tau_1/\tau_0,\ r=\tau_{-1}/\tau_0,\ \varphi=-\frac{\partial^2}{\partial x^2}\log\tau_0,$$

the Hirota bilinear equations being (cf. [HH]):

(4.3.10) $$\begin{aligned}&(iD_t+\frac{1}{2}D_x^2+\frac{1}{2}D_y^2)\tau_1\cdot\tau_0=0\\ &(-iD_t+\frac{1}{2}D_x^2+\frac{1}{2}D_y^2)\tau_{-1}\cdot\tau_0=0\\ &(D_x^2-D_y^2)\tau_0\cdot\tau_0=2\tau_1\tau_{-1}.\end{aligned}$$

Finally, imposing the constraint

(4.3.11) $$r=\kappa\bar{q},\ \text{where } \kappa=\pm1,$$

we obtain the classical Davey-Stewartson system

(4.3.12) $$\begin{cases} i\frac{\partial q}{\partial t}+\frac{1}{2}(\frac{\partial^2 q}{\partial x^2}+\frac{\partial^2 q}{\partial y^2})=(\varphi-\kappa|q|^2)q\\ \frac{\partial^2\varphi}{\partial x^2}-\frac{\partial^2\varphi}{\partial y^2}=2\kappa\frac{\partial^2|q|^2}{\partial x^2}.\end{cases}$$

Remark 4.3. It is interesting to compare the above results with that obtained via the Lax equations. To simplify notation, let $U_i=U_{ii}^{(1)}$. Substituting (3.8.15) (resp. (3.8.14)) in (3.8.18) (resp. (3.8.19)), we obtain for $i\neq j$:

(4.3.13) $$\frac{\partial A_{ij}}{\partial t_j}=-\frac{\partial^2 A_{ij}}{\partial x_j^2}-2A_{ij}U_j-2\sum_{k\neq j}A_{ij}A_{jk}A_{kj}$$

(4.3.14) $$\frac{\partial A_{ij}}{\partial t_i}=\frac{\partial^2 A_{ij}}{\partial x_i^2}+2A_{ij}U_i+2\sum_{k\neq i}A_{ij}A_{ik}A_{ki}$$

These equation together with (3.8.9, 17 and 20) give a slightly different version of the generalized DS system (recall that $A_{ij}=W_{ij}$ if $i\neq j$ and $U_i=-\partial W_{ii}$). For $n=2$ we get again the classical DS system after the change of indeterminates (4.3.7) if we let $\varphi=-\frac{1}{2}(U_1+U_2)$.

4.4. Finally, we explain what happens in the well-known case $n=1$. In this case $C^{(1)}=1$ and auxiliary conditions (3.8.2) are trivial. Lax equation (3.8.3b) is trivial as well, and Lax equation (3.8.3a) becomes

(4.4.1) $$\frac{\partial L}{\partial x_i}=[B_i,L],\ i=1,2,\dots,$$

where $L = \partial + \sum_{j=1}^{\infty} u_j(x)\partial^{-j}$, $\partial = \frac{\partial}{\partial x_1}$ and $B_i = (L^i)_+$. Thus, the KP hierarchy is a system of partial differential equations (4.4.1) on unknown functions $u_1, u_2, \dots$ in indeterminates $x_1, x_2, \dots$. By Lemma 3.6, (4.4.1) is equivalent to the following system of Zakharov-Shabat equations:

$$(4.4.2)_{k,\ell} \qquad \frac{\partial B_\ell}{\partial x_k} - \frac{\partial B_k}{\partial x_\ell} = [B_k, B_\ell].$$

By (3.8.4) we have:

$$(4.4.3) \qquad B_1 = \partial, \; B_2 = \partial^2 + 2u_1.$$

Furthermore, we have:

$$(4.4.4) \qquad B_3 = \partial^3 + 3u_1\partial + 3u_2 + 3\frac{\partial u_1}{\partial x_1}$$

Thus we see that equations $(4.4.2)_{k,1}$ are all trivial, the first non-trivial equation of (4.4.2) being

$$\frac{\partial B_2}{\partial x_3} - \frac{\partial B_3}{\partial x_2} = [B_2, B_3].$$

Substituting in it (4.4.3 and 4), the coefficients of ∂^0 and ∂^1 give respectively:

$$(4.4.5) \qquad 2\frac{\partial u_1}{\partial x_3} - 2\frac{\partial^2 u_1}{\partial x_1 \partial x_2} - 6u_1\frac{\partial u_1}{\partial x_1} = 3\frac{\partial u_2}{\partial x_2} - 3\frac{\partial^2 u_2}{\partial x_1^2}$$

$$(4.4.6) \qquad 6\frac{\partial u_2}{\partial x_1} = 3\frac{\partial u_1}{\partial x_2} - \frac{\partial^2 u_1}{\partial x_1^2}.$$

Differentiating (4.4.5) by x_1 and substituting $\frac{\partial u_2}{\partial x_1}$ from (4.4.6) gives a PDE on $u = 2u_1$, where we let $x_1 = x$, $x_2 = y, x_3 = t$:

$$(4.4.7) \qquad \frac{3}{4}\frac{\partial^2 u}{\partial y^2} = \frac{\partial}{\partial x}\left(\frac{\partial u}{\partial t} - \frac{3}{2}u\frac{\partial u}{\partial x} - \frac{1}{4}\frac{\partial^3 u}{\partial x^3}\right).$$

This is the classical KP equation. Due to (3.9.5), the connection between u and the τ-function is given by the famous formula

$$(4.4.8) \qquad u = 2\frac{\partial^2}{\partial x^2}\log\tau.$$

Substituting u in (4.4.7) gives the Hirota bilinear equation (2.4.3).

§5. Soliton and dromion solutions.

5.1. We turn now to the construction of solutions of the n-component KP hierarchy. As in [DJKM3] we make use of the vertex operators (2.1.14). When transported via the n-component boson-fermion correspondence σ from F to $B = \mathbb{C}[x] \otimes \mathbb{C}[L]$, they take the following form:

$$(5.1.1) \qquad \psi^{\pm(i)}(z) = Q_i^{\pm 1} z^{\pm\alpha_0^{(i)}} (\exp \pm \sum_{k=1}^{\infty} z^k x_k^{(i)})(\exp \mp \sum_{k=1}^{\infty} \frac{z^{-k}}{k}\frac{\partial}{\partial x_k^{(i)}}).$$

Note that for $z, w \in \mathbb{C}^\times$ such that $|w| < |z|$ we have ($\lambda, \mu = +$ or $-$):

$$\psi^{\lambda(i)}(z)\psi^{\mu(j)}(w) = (z-w)^{\delta_{ij}\lambda\mu} Q_i^{\lambda 1} Q_j^{\mu 1} z^{\alpha_0^{(i)}} w^{\alpha_0^{(j)}}$$

(5.1.2)

$$\times \exp \sum_{k=1}^{\infty} (\lambda z^k x_k^{(i)} + \mu w^k x_k^{(j)}) \exp - \sum_{k=1}^{\infty} (\lambda \frac{z^{-k}}{k} \frac{\partial}{\partial x_k^{(i)}} + \mu \frac{w^{-k}}{k} \frac{\partial}{\partial x_k^{(j)}}).$$

We let for $0 < |w| < |z|$:
(5.1.3)

$$\Gamma_{ij}(z,w) \stackrel{\text{def}}{=} \psi^{+(i)}(z)\psi^{-(j)}(w) = (z-w)^{-\delta_{ij}} Q_i Q_j^{-1} z^{\alpha_0^{(i)}} w^{-\alpha_0^{(j)}} \exp \sum_{k=1}^{\infty} (z^k x_k^{(i)} - w^k x_k^{(j)})$$

$$\times \exp - \sum_{k=1}^{\infty} (\frac{z^{-k}}{k} \frac{\partial}{\partial x_k^{(i)}} - \frac{w^{-k}}{k} \frac{\partial}{\partial x_k^{(j)}}).$$

Using (5.1.2), we obtain for $|z_1| > |z_2| > \ldots > |z_{2N-1}| > |z_{2N}| > 0$:

$$\Gamma_{i_1 i_2}(z_1,z_2)\ldots\Gamma_{i_{2N-1}i_{2N}}(z_{2N-1},z_{2N}) = \prod_{1\le k<\ell\le 2N} (z_k - z_\ell)^{(-1)^{k+\ell}\delta_{i_k i_\ell}}$$

(5.1.4)

$$\times Q_{i_1} Q_{i_2}^{-1} \ldots Q_{i_{2N-1}} Q_{i_{2N}}^{-1} \prod_{m=1}^{2N} z_m^{(-1)^{m-1}\alpha_0^{(i_m)}} \exp(-\sum_{m=1}^{2N}\sum_{k=1}^{\infty} (-1)^m z_m^k x_k^{(m)})$$

$$\times \exp(\sum_{m=1}^{2N}\sum_{k=1}^{\infty} (-1)^m \frac{z_m^{-k}}{k} \frac{\partial}{\partial x_k^{(m)}}).$$

We may analytically extend the right-hand side of (5.1.4) to the domain $\{z_i \neq 0,\ z_i \neq z_j \text{ if } i \neq j,\ i,j = 1,\ldots,2N\}$. Then we deduce from (5.1.4) for $N=2$ that in this domain we have:

(5.1.5)
$$\Gamma_{i_1 i_2}(z_1,z_2)\Gamma_{i_3,i_4}(z_3,z_4) = \Gamma_{i_3 i_4}(z_3,z_4)\Gamma_{i_1 i_2}(z_1,z_2),$$

(5.1.6)
$$\Gamma_{ij}(z_1,z_2)^2 \equiv \lim_{\substack{z_3\to z_1\\ z_4\to z_2}} \Gamma_{ij}(z_1,z_2)\Gamma_{ij}(z_3,z_4) = 0.$$

Remark 5.1. Let $A = (a_{ij})$ be a $n \times n$ matrix over $\mathbb{C}$ and let z_i, w_i $(i = 1,\ldots,n)$ be non-zero complex numbers such that $z_i \neq w_j$. Due to (1.2.3) the sum

(5.1.7)
$$\Gamma_A(z,w) = \sum_{i,j=1}^{n} a_{ij}\Gamma_{ij}(z_i,w_j)$$

lies in a completion of $r(g\ell_\infty)$.

By (5.1.5–6) we obtain:

(5.1.8)
$$\exp \Gamma_A(z,w) = \prod_{i,j=1}^{n} (1 + a_{ij}\Gamma_{ij}(z_i,w_j)).$$

Lemma 5.1. *(a) If τ is a solution of the n-component KP hierarchy (2.3.7) of Hirota bilinear equations, then* $(\exp\Gamma_A(z,w))\tau$ *is a solution as well for any complex $n \times n$ matrix A and any $z = (z_1,\ldots z_n)$, $w = (w_1,\ldots,w_n) \in \mathbb{C}^{\times n}$ such that $z_i \neq w_j$.*

(b) For any collection of complex $n \times n$-matrices $A_1,\ldots,A_N$ and any collection $z^{(1)},\ldots,z^{(N)},w^{(1)},\ldots,w^{(N)} \in \mathbb{C}^{\times n}$ with all coordinates distinct, the function

$$\exp \Gamma_{A_1}(z^{(1)}, w^{(1)}) \dots \exp \Gamma_{A_N}(z^{(N)}, w^{(N)}) \cdot 1 \tag{5.1.9}$$

is a solution of the n-component KP hierarchy (2.3.7).

Proof. (a) follows from Proposition 1.3 and Remark 5.1. (b) follows from (a) since the function $1 = \sigma|0\rangle$ satisfies (1.3.1). □

We call (5.1.9) the *N-solitary τ-function* (of the n-component KP hierarchy).

In order to write down (5.1.9) in a more explicit form, introduce the lexicographic ordering on the set S of all triples $s = (p, i, j)$, where $p \in \{1, \dots, N\}$, $i, j \in \{1, \dots, n\}$ (i.e. $s_1 < s_2$ if $p_1 < p_2$, or $p_1 = p_2$ and $i_1 < i_2$ or $p_1 = p_2$, $i_1 = i_2$ and $j_1 < j_2$). Given N $n \times n$ complex matrices $A_p = (a_{ij}^{(p)})$, we let $a_s = a_{ij}^{(p)}$ for $s = (p, i, j) \in S$; given in addition two sets of non-zero complex numbers z_s and w_s, all distinct, parametrized by $s \in S$, introduce the following constants

$$\begin{aligned} c(s_1, \dots, s_r) &= \prod_{k=1}^{r} a_{s_k} \prod_{\ell=k+1}^{r} \varepsilon_{i_k i_\ell} \varepsilon_{i_k j_\ell} \varepsilon_{j_k i_\ell} \varepsilon_{j_k j_\ell} \\ &\times \prod_{1 \le k < \ell \le r} \frac{(z_{s_k} - z_{s_\ell})^{\delta_{i_k i_\ell}} (w_{s_k} - w_{s_\ell})^{\delta_{j_k j_\ell}}}{(z_{s_k} - w_{s_\ell})^{\delta_{i_k j_\ell}} (w_{s_k} - z_{s_\ell})^{\delta_{j_k i_\ell}}}. \end{aligned} \tag{5.1.10}$$

Then the N-solitary solution (5.1.9) can be written as follows

$$\begin{aligned} &1 + \sum_{r=1}^{Nn^2} \sum_{(1,1,1) \le s_1 < \dots < s_r \le (N,n,n)} c(s_1, \dots, s_r) \\ &\times \left(\exp \sum_{k=1}^{r} \sum_{m=1}^{\infty} (z_{s_k}^m x_m^{(i_k)} - w_{s_k}^m x_m^{(j_k)})\right) e^{\sum_{k=1}^{r} \alpha_{i_k j_k}}. \end{aligned} \tag{5.1.11}$$

5.2. Let $n = 1$. Then the index set S is naturally identified with the set $\{1, \dots, N\}$, the two sets of complex numbers we denote by z_{2j-1} and z_{2j}, $j = 1, \dots, N$, and we let $A_p = (z_{2p-1} - z_{2p})^{-1} a_p$, where a_p are some constants. Then (5.1.11) becomes the well-known formula (see [DJKM3]) for the τ-function of the N-soliton solution:

$$\begin{aligned} \tau^{(N)} &= 1 + \sum_{r=1}^{N} \sum_{1 \le j_1 < \dots < j_r \le N} \prod_{k=1}^{r} a_{j_k} \prod_{1 \le k < \ell \le 2r} (z_{j_k} - z_{j_\ell})^{(-1)^{k+\ell}} \\ &\times \exp \sum_{k=1}^{r} \sum_{m=1}^{\infty} (z_{j_{2k-1}}^m - z_{j_{2k}}^m) x_m. \end{aligned} \tag{5.2.1}$$

Letting $x_1 = x$, $x_2 = y$, $x_3 = t$ and all other indeterminates constants $x_4 = c_4, \dots$, we obtain, due to (4.4.8), the soliton solution of the classical KP equation (4.4.7):

$$u(t, x, y) = 2 \frac{\partial^2}{\partial x^2} \log \tau^{(N)}(x, y, t, c_4, c_5, \dots). \tag{5.2.2}$$

In particular, the τ-function of the 1-soliton solution is

$$\tau^{(1)}(x, y, t) = 1 + \frac{a}{z_1 - z_2} \exp((z_1 - z_2)x + (z_1^2 - z_2^2)y + (z_1^3 - z_2^3)t + \text{ const.}) \tag{5.2.3}$$

and we get the corresponding 1-soliton solution of the classical KP equation (4.4.7):

$$u(x, y, t) = \frac{(z_1 - z_2)^2}{2} \cosh^{-2}\left(\frac{1}{2}((z_1 - z_2)x + (z_1^2 - z_2^2)y + (z_1^3 - z_2^3)t) + \text{ const.}\right). \tag{5.2.4}$$

5.3. Let $n = 2$. Then any $\tau \in \mathbb{C}[x] \otimes \mathbb{C}[M]$ can be written in the form

$$\tau = \sum_{\ell \in \mathbb{Z}} \tau_\ell e^{\ell \alpha_{12}}, \text{ where } \tau_\ell \equiv \tau_{\ell \alpha_{12}}.$$

For a N-solitary solution $\tau^{(N)}$ given by (5.1.11) we then have

$$\tau_\ell^{(N)} = \delta_{\ell,0} + \sum_{r=1}^{4N} \sum_{(s_1,\dots,s_r)} c(s_1,\dots,s_r) \exp \sum_{k=1}^{r} \sum_{m=1}^{\infty} (z_{s_k}^m x_m^{(i_k)} - w_{s_k}^m x_m^{(j_k)}), \tag{5.3.1}$$

where $(s_1,\dots,s_r)$ run over the subset $(5.3.2)_2$ of S^r, where

$$\begin{cases} (1,1,1) \le s_1 < s_2 \dots < s_r \le (N,n,n) \\ \#\{(i_k,j_k) | i_k > j_k\} - \#\{(i_k,j_k) | i_k < j_k\} = \ell. \end{cases} \tag{$(5.3.2)_n$}$$

Letting (cf. (4.3.9)):

$$\begin{aligned} q &= -\frac{\tau_1(x,y,t,c,c_{3,\dots}^{(1)})}{\tau_0(x,y,t,c,c_3^{(1)},\dots)}, \quad r = \frac{\tau_{-1}(x,y,t,c,c_3^{(1)},\dots)}{\tau_0(x,y,t,c,c_3^{(1)},\dots)}, \\ \varphi &= -\frac{\partial^2}{\partial x^2}(\log \tau_0(x,y,t,c,c_3^{(1)},\dots)), \end{aligned} \tag{5.3.3}$$

where $x = x_1^{(1)} + x_2^{(1)}$, $y = x_1^{(1)} - x_2^{(1)}$, $t = -2i(x_2^{(1)} - x_2^{(2)})$, $c = -2i(x_2^{(1)} + x_2^{(2)})$ and all other indeterminates $x_k^{(j)}$ are arbitrary constants $c_k^{(j)}$, we obtain a N-solitary solution of the decoupled Davey-Stewartson system (4.3.8).

We turn now to the classical Davey-Stewartson system (4.3.12) for $\kappa = -1$. The constraint (4.3.11) gives

$$\tau_1/\tau_0 = \overline{\tau_{-1}/\tau_0}.$$

One way of satisfying this constraint is to let

$$\begin{aligned} a_{ij}^{(p)} &= (-1)^{i+j} \overline{a}_{ji}^{(p)}, \quad z_{(p,i,j)} = -\overline{w}_{(p,j,i)}, \\ c_2 &= 0, \quad c_k^{(j)} \in i^{k+1}\mathbb{R}. \end{aligned} \tag{5.3.4}$$

We shall concentrate now on the case $N = 1$. It will be convenient to use the following notation:

$$\begin{aligned} &x_1 = x_1^{(1)}, \quad x_2 = x_1^{(2)}, \\ &z_{ij} = z_{(1,i,j)}, \quad a_i = a_{(1,i,i)} \in \mathbb{R}_+ \ (1 \le i,j \le 2), \quad a_3 = a_{(1,1,2)} \in \mathbb{C}, \\ &C(z,w) = \frac{z-w}{z+\overline{w}}, \quad D(z,w) = \frac{|a_3|^2}{(z+\overline{z})(w+\overline{w})}, \\ &A_j(z) = (z+\overline{z})(x_j - (-1)^j it\frac{z-\overline{z}}{4}) + \sum_{k=3}^{\infty}(z^k - (-\overline{z})^k)c_k^{(j)} \ (j = 1,2), \\ &A_3(z,w) = zx_1 + \overline{w}x_2 + it(\frac{z^2}{4} + \frac{\overline{w}^2}{4}) + \sum_{k=3}^{\infty}(z^k c_k^{(1)} - (-\overline{w})^k c_k^{(2)}). \end{aligned}$$

Then $q = -\tau_1/\tau_0$ and $\varphi = -\frac{1}{2}(\frac{\partial}{\partial x_1} + \frac{\partial}{\partial x_2})^2 \log \tau_0$ is a solution of (4.3.12), where

$$\tau_1 = a_3 e^{A_3(z_{12},z_{21})}(1 + a_1 C(z_{12},z_{11}) e^{A_1(z_{11})})(1 + a_2 C(\overline{z}_{21},\overline{z}_{22}) e^{A_2(z_{22})}), \tag{5.3.5a}$$

$$
\begin{aligned}
\tau_0 &= (1+a_1 e^{A_1(z_{11})})(1+a_2 e^{A_2(z_{22})}) \\
&+ D(z_{12},z_{21}) e^{A_3(z_{12},z_{21})+\overline{A_3(z_{12},z_{21})}}(1+a_1|C(z_{12},z_{21})|^2 e^{A_1(z_{11})}) \\
&\times (1+a_2|C(z_{21},z_{22})|^2 e^{A_2(z_{22})}).
\end{aligned} \tag{5.3.5b}
$$

Consider now two special cases of (5.3.5a,b):

$(D)\quad z_1 \equiv z_{11} = z_{12}$ and $z_2 \equiv z_{22} = z_{21}$,

$(S)\quad a_i = 0 \quad (i=1,2)$,

and let $T = D$ or S. Then (5.3.5a and b) reduce to

$$
\tau_1 = a_3 e^{A_3(z_1,z_2)} \text{ in both cases,} \tag{5.3.6a}
$$

$$
\tau_0^{(T)} = (1+\delta_{TD} a_1 e^{A_1(z_1)})(1+\delta_{TD} a_2 e^{A_2(z_2)}) + D(z_1,z_2) e^{A_1(z_1)+A_2(z_2)}, \tag{5.3.6b}
$$

so that $q^{(T)} = -\tau_1/\tau_0^{(T)}$, $\varphi^{(T)} = -\frac{1}{2}(\frac{\partial}{\partial x_1}+\frac{\partial}{\partial x_2})^2 \log \tau_0^{(T)}$ is a solution of (4.3.12).

In order to rewrite $q^{(T)}$ in a more familiar form, let ($j = 1,2$ and $a_j(z_j+\overline{z}_j) > 0$):

$$
\begin{aligned}
& p_j^{(T)} = (a_j(z_j+\overline{z}_j))^{-1/2} \text{ if } T = D \text{ and } = 1 \text{ if } T = S, \\
& \mu_1 = \mu_{1R} + i\mu_{1I} = \tfrac{1}{2}\overline{z}_1,\ \mu_2 = \mu_{2R} + i\mu_{2I} = \tfrac{1}{2} z_2, \\
& m_j^{(T)} = 2\sqrt{2}\mu_{jR} p_j^{(T)} e^{-\sum_{k=3}^{\infty}(-1)^j((-1)^j 2\mu_j)^k c_k^{(j)}}, \\
& \xi_j = 2x_j + 2\mu_{jI} t,\ \tilde{\xi}_j = \frac{1}{\mu_{jR}} \log \frac{|m_j^{(T)}|}{\sqrt{2\mu_{jR}}}, \\
& \rho^{(T)} = -a_3 p_1^{(T)} p_2^{(T)}.
\end{aligned}
$$

Then we obtain the following expression for $q^{(T)}$:

$$
\frac{4\rho^{(T)}(\mu_{1R}\mu_{2R})^{1/2} \exp\{-(\mu_{1R}(\xi_1-\tilde{\xi}_1)+\mu_{2R}(\xi_2-\tilde{\xi}_2))+i(-(\mu_{1I}\xi_1+\mu_{2I}\xi_2)+(|\mu_1|^2+|\mu_2|^2)t+\arg m_1 m_2)\}}{((\delta_{TD}+\exp(-2\mu_{1R}(\xi_1-\tilde{\xi}_1))(\delta_{TD}+\exp(-2\mu_{2R}(\xi_2-\tilde{\xi}_2))+|\rho^{(T)}|^2)}
$$

The function $q^{(D)}$ is precisely the $(1,1)$-dromion solution of the Davey-Stewartson equations (4.3.12) with $\kappa = -1$ found in [FS] (provided that $\mu_{jR} \in \mathbb{R}_+$). On the other hand, if we let $\mu_{1I} = \mu_{2I} = 0$, then $q^{(T)}$ reduces to the 2-dimensional breather solution found in [BLMP]. Finally, $q^{(S)}$ is a 1-soliton solution.

Recall that the dromion solutions of the DS equation were originally discovered in [BLMP] and [FS] (see also [HH]). The dromion solutions of the DS equation were first studied from the point of view of the spinor formalism by [HMM].

5.4. Similarly, we obtain the following solutions of the 2-dimensional Toda chain (4.2.8):

$$
u_\ell = \begin{cases} \log(\tau_{\ell+1}^{(N)}/\tau_\ell^{(N)}) & \text{if } -N \le \ell \le N-1 \\ 0 & \text{otherwise} \end{cases} \tag{5.4.1}
$$

where the τ-functions $\tau_\ell^{(N)}$ are obtained from (5.3.1) by letting all indeterminates $x_m^{(j)}$ with $m > 1$ to be arbitrary constants:

$$\tau_\ell^{(N)} = \delta_{\ell,0} + \sum_{r=1}^{4N} \sum_{(s_1,\dots,s_r)} c(s_1,\dots,s_r) c_{s_1} \dots c_{s_r} \exp \sum_{k=1}^{r} (x_{i_k} z_{s_k} - x_{j_k} w_{s_k}), \tag{5.4.2}$$

where $(s_1,\dots,s_r)$ runs over $(5.3.2)_2$ and c_s $(s \in S)$ are arbitrary constants.

5.5. Let now $n \geq 3$. Then we obtain solutions of the $2+1$ n-wave system (4.1.9) as follows. For $1 \leq i,j \leq n$ let

$$\begin{aligned} \tau_{ij}^{(N)} = \delta_{ij} + \sum_{r=1}^{Nn^2} \sum_{(s_1,\dots,s_r)} c_{s_1} \dots c_{s_r} \\ \times \exp \sum_{k=1}^{r} (a_{i_k} x + b_{i_k} t - y) z_{s_k} - (a_{j_k} x + b_{j_k} t - y) w_{s_k}, \end{aligned} \tag{5.5.1}$$

where $(s_1,\dots,s_r)$ runs over $(5.3.2)_n$ and c_s $(s \in S)$ are arbitrary constants. Then $W_{ij} = \varepsilon_{ji} \tau_{ij}/\tau_0$ $(i \neq j)$ is a solution of (4.1.9), and $Q_{ij} = \varepsilon_{ij}(a_i - a_j)\tau_{ij}/\tau_0$ $(i \neq j)$ is a solution of (4.1.10).

§6. m-reductions of the n-component KP hierarchy.

6.1. Fix a positive integer m and let $\omega = \exp \frac{2\pi i}{m}$. Introduce the following mn^2 fields $(1 \leq i,j \leq n,\ 1 \leq k \leq m)$ [TV]:

$$\alpha^{(ijk)}(z) \equiv \sum_{p \in \mathbb{Z}} \alpha_p^{(ijk)} z^{-p-1} =: \psi^{+(i)}(z) \psi^{-(j)}(\omega^k z) :, \tag{6.1.1}$$

where the normal ordering is defined by (2.1.6). Note that

$$\alpha^{(ijm)}(z) = \alpha^{(ij)}(z), \tag{6.1.2}$$

where $\alpha^{(ij)}(z)$ are the bosonic fields, defined by (2.1.5), which generate the affine algebra $g\ell_n(\mathbb{C})^\wedge$ with central charge 1 (see (2.1.7)). It is easy to check that for arbitrary m, the fields $\alpha^{(ijk)}(z)$ generate the affine algebra $g\ell_{mn}(\mathbb{C})^\wedge$ with central charge 1. More precisely, all the operators $\alpha_p^{(ijk)}$ $(1 \leq i,j \leq n,\ 1 \leq k \leq m,\ p \in \mathbb{Z})$ together with 1 form a basis of $g\ell_{mn}(\mathbb{C})^\wedge$ in its representation in F with central charge 1, the charge decomposition being the decomposition into irreducibles. Hence, using (2.1.14), (2.2.5 and 8), we obtain the vertex operator realization of this representation of $g\ell_{mn}(\mathbb{C})^\wedge$ in the vector space B (see [TV] for details).

Now, restricted to the subalgebra $s\ell_{mn}(\mathbb{C})^\wedge$, the representation in $F^{(0)}$ is not irreducible any more, since $s\ell_{mn}(\mathbb{C})^\wedge$ commutes with all the operators

$$\beta_k^{(n)} \overset{\text{def}}{=} \frac{1}{n} \sum_{j=1}^{n} \alpha_k^{(j,j,m)}, \quad k \in m\mathbb{Z}. \tag{6.1.3}$$

In order to describe the irreducible part of the representation of $s\ell_{mn}(\mathbb{C})^\wedge$ in $B^{(0)}$ containing the vacuum 1, we choose the complementary generators of the oscillator algebra $\mathfrak{a}$ contained in $s\ell_{mn}(\mathbb{C})^\wedge$ $(k \in \mathbb{Z})$:

(6.1.4)

$$\beta_k^{(j)} = \begin{cases} \alpha_k^{(jjm)} & \text{if } k \notin m\mathbb{Z}, \\ \frac{1}{j(j+1)}(\alpha_k^{(11m)} + \dots + \alpha_k^{(jjm)} - j\alpha_k^{(j+1j+1m)}) & \text{if } k \in m\mathbb{Z} \text{ and } 1 \leq j < n, \end{cases}$$

so that the operators (6.1.3 and 4) also satisfy relations (2.1.9). Then the operators $1, \alpha_p^{(ijk)}$

for $i \neq j$, together with operators (6.1.4) form a basis of $s\ell_{mn}(\mathbb{C})^\wedge$. Hence, introducing the new indeterminates

$$(6.1.5)\qquad y_k^{(j)} = \begin{cases} x_k^{(j)} & \text{if } k \notin m\mathbb{N}, \\ \frac{1}{j(j+1)}(x_k^{(1)} + \ldots + x_k^{(j)} - jx_k^{(j+1)}) & \text{if } k \in m\mathbb{N} \text{ and } j < n, \\ \frac{1}{n}(x_k^{(1)} + \ldots + x_k^{(n)}) & \text{if } k \in m\mathbb{N} \text{ and } j = n, \end{cases}$$

we have: $\mathbb{C}[x] = \mathbb{C}[y]$ and

$$(6.1.6)\qquad \sigma(\beta_k^{(j)}) = \frac{\partial}{\partial y_k^{(j)}} \text{ and } \sigma(\beta_{-k}^{(j)}) = k y_k^{(j)} \text{ if } k > 0.$$

Now it is clear that the irreducible with respect to $s\ell_{mn}(\mathbb{C})^\wedge$ subspace of $B^{(0)}$ containing the vacuum 1 is the vector space

$$(6.1.7)\qquad B_{[m]}^{(0)} = \mathbb{C}[y_k^{(j)} | 1 \leq j < n,\ k \in \mathbb{N}, \text{ or } j = n,\ k \in \mathbb{N}\backslash m\mathbb{Z}] \otimes \mathbb{C}[M].$$

The vertex operator realization of $s\ell_{mn}(\mathbb{C})^\wedge$ in the vector space $B_{[m]}^{(0)}$ is then obtained by expressing the fields $\alpha^{(ijk)}(z)$ for $i \neq j$ in terms of vertex operators (2.1.14), which are expressed via the operators (6.1.4), the operators $Q_i Q_j^{-1}$ and $\alpha_0^{(i)} - \alpha_0^{(j)}$ $(1 \leq i < j \leq n)$ (see [TV] for details).

The n-component KP hierarchy of Hirota bilinear equations on $\tau \in B^{(0)} = \mathbb{C}[y] \otimes \mathbb{C}[M]$ when restricted to $\tau \in B_{[m]}^{(0)}$ is called the *m-th reduced* KP *hierarchy*. It is obtained from the n-component KP hierarchy by making the change of variables (6.1.5) and putting zero all terms containing partial derivates by $y_m^{(n)}, y_{2m}^{(n)}, y_{3m}^{(n)}, \ldots$.

It is clear from the definitions and results of §3 that the condition on the n-component KP hierarchy to be m-th reduced implies the following equivalent conditions (cf. [DJKM3]):

$$(6.1.8)\qquad L(\alpha)^m \text{ is a differential operator,}$$

$$(6.1.9)\qquad \sum_{j=1}^{n} \frac{\partial W(\alpha)}{\partial x_m^{(j)}} = z^m W(\alpha),$$

$$(6.1.10)\qquad \sum_{j=1}^{n} \frac{\partial \tau}{\partial x_m^{(j)}} = \lambda\tau, \quad \text{for some } \lambda \in \mathbb{C}.$$

It follows from (6.1.8) that these conditions automatically imply that all of them hold if m is replaced by any multiple of m.

The totality of solutions of the m-th reduced KP hierarchy is given by the following

Proposition 6.1. *Let $\mathcal{O}_{[m]}$ be the orbit of 1 under the (projective) representation of the loop group $SL_{mn}(\mathbb{C}[t, t^{-1}])$ corresponding to the representation of $s\ell_{mn}(\mathbb{C})^\wedge$ in $B_{[m]}^{(0)}$. Then*

$$\mathcal{O}_{[m]} = \sigma(\mathcal{O}) \cap B_{[m]}^{(0)}.$$

In other words, the τ-functions of the m-th reduced KP hierarchy are precisely the τ-functions of the KP hierarchy in the variables $y_k^{(j)}$, which are independent of the variables $y_{m\ell}^{(n)}$, $\ell \in \mathbb{N}$.

Proof is the same as of a similar statement in [KP2]. □

Remark 6.1. The above representation of $s\ell_{mn}(\mathbb{C})^\wedge$ in $B^{(0)}_{[m]}$ is a vertex operator construction of the basic representation corresponding to the element of the Weyl group S_{mn} of $s\ell_{mn}(\mathbb{C})$ consisting of n cycles of length m (see [KP1] and [TV]). In particular, for $n=1$ this is the principal realization [KKLW], and for $m=1$ this is the homogeneous realization [FK]. The m-th reduced 1-component KP was studied in a great detail in [DJKM2] (see also [KP2]).

6.2. Let $n=1$. Then the 2-reduced KP hierarchy becomes the celebrated KdV hierarchy on the differential operator $S \equiv (L^2)_+ = \partial^2 + u$, where $u = 2u_1$:

$$\frac{\partial}{\partial x_{2n+1}} S^{\frac{1}{2}} = [(S^{n+\frac{1}{2}})_+, S^{\frac{1}{2}}],\ n=1,2,\ldots, \tag{6.2.1}$$

the first equation of the hierarchy being the classical Korteweg-deVries equation

$$4\frac{\partial u}{\partial t} = \frac{\partial^3 u}{\partial x^3} + 6u\frac{\partial u}{\partial x}. \tag{6.2.2}$$

Of course, the 3-reduced KP is the Boussinesq hierarchy, and the general m-reduced KP are the Gelfand-Dickey hierarchies.

6.3. Let $n=2$. The equations of the 1-reduced 2-component KP are independent of x, hence equation (4.3.8) becomes independent of x and φ becomes 0 (see (4.3.9)). Thus, equation (4.3.8) turns into the decoupled non-linear Schrödinger system (called also the AKNS system):

$$\begin{aligned} i\frac{\partial q}{\partial t} &= -\frac{1}{2}\frac{\partial^2 q}{\partial y^2} - q^2 r \\ i\frac{\partial r}{\partial t} &= \frac{1}{2}\frac{\partial^2 r}{\partial y^2} + q r^2. \end{aligned} \tag{6.3.1}$$

Thus (6.3.1) is a part of the 1-reduced 2-component KP. For that reason the 1-reduced 2-component KP is sometimes called the non-linear Schrödinger hierarchy. Of course, under the constraint (4.3.11), we get the non-linear Schrödinger equation

$$i\frac{\partial q}{\partial t} = -\frac{1}{2}\frac{\partial^2 q}{\partial y^2} - \kappa|q|^2 q. \tag{6.3.2}$$

Similarly, under the same reduction the 2-dimensional Toda chain (4.2.8) turns into the 1-dimensional Toda chain

$$\frac{\partial^2 u_n}{\partial x^2} = e^{u_n - u_{n-1}} - e^{u_{n+1}-u_n}\ (\text{here } x = 2y_1^{(1)}). \tag{6.3.3}$$

Thus, the 1-dimensional Toda chain is a part of the non-linear Schrödinger hierarchy. It was studied from the representation theoretical point of view in [TB].

6.4. Let $n \geq 3$. Since the constraint (4.1.4) is contained among the constraints of the 1-reduced n-component KP hierarchy, we see that the $1+1$ n-wave system (4.1.7) is a part of the 1-reduced n-component KP hierarchy. Note also that the 1-reduction of the n-component KP reduces the $2+1$ n-wave interaction system (4.1.10) into the $1+1$ system (4.1.7).

6.5. Since the non-linear Schrödinger system (6.3.1) is a part of the 1-reduced 2-component KP hierarchy, the 1-reduced n-component KP hierarchy will be called the n-component NLS. Let us give here its formulation since it is especially simple.

Given a $n\times n$ matrix $C(z) = \sum_j C_j z^j$, we let

$$C(z)_- = \sum_{j<0} C_j z^j,\ C(z)_+ = \sum_{j\geq 0} C_j z^j.$$

Also, given a diagonal complex matrix $a = \text{diag}\,(a_1, \dots, a_n)$ we let

$$x_k^a = \sum_{j=1}^{n} a_k x_k^{(j)}, \quad \frac{\partial}{\partial x_k^a} = \sum_{j=1}^{n} a_k \frac{\partial}{\partial x_k^{(j)}}.$$

Let $\mathfrak{h}$ denote the set of all traceless diagonal matrices over $\mathbb{C}$.

The n-component NLS hierarchy is the following system on matrix valued functions

$$P(\alpha) \equiv P(\alpha, x, z) = 1 + \sum_{j>0} W^{(j)}(\alpha, x) z^{-j}, \ \alpha \in M,$$

where $x = \{x_k^{(a)} | a \in \mathfrak{h}, \ k = 1, 2, \dots\}$:

$$\frac{\partial P(\alpha)}{\partial x_k^{(a)}} = -(P(\alpha) a P(\alpha)^{-1} z^k)_- P(\alpha) \tag{6.5.1}$$

with additional matching conditions

$$(P(\alpha) R(\alpha - \beta, z) P(\beta)^{-1})_- = 0, \ \alpha, \beta \in M, \tag{6.5.2}$$

where $R(\gamma, z) \equiv R^+(\gamma, z)$ is defined by (3.3.10).

This formulation implies the Lax form formulation if we consider $C^{(a)}(x,z) = P(\alpha) a P(\alpha)^{-1}$ for each $a \in \mathfrak{h}$ and fixed α. Consider a family of commuting matrix valued functions of the form

$$C^{(a)} \equiv C^{(a)}(x, z) = a + \sum_{j>0} C_j^{(a)}(x) z^{-j},$$

depending linearly on $a \in \mathfrak{h}$, and let $B_k^{(a)} = (C^{(a)} z^k)_+$. Then the Lax form of the n-component NLS is

$$\frac{\partial C^{(a)}}{\partial x_k^{(b)}} = [B_k^{(b)}, C^{(a)}], \ a, b \in \mathfrak{h}, \ k = 1, 2, \dots. \tag{6.5.3}$$

The equivalent zero curvature form of the n-component NLS is

$$\frac{\partial B_\ell^{(a)}}{\partial x_k^{(b)}} - \frac{\partial B_k^{(b)}}{\partial x_\ell^{(a)}} = [B_k^{(b)}, B_\ell^{(a)}], \ a, b \in \mathfrak{h}, \ k, \ell = 1, 2, \dots. \tag{6.5.4}$$

Since for the 1-reduced n-component KP one has: $L = \partial$, i.e. all $U^{(j)} = 0$, we see from Remark 4.3 that the n-component NLS in the form (6.5.3) contains the following system of equations on functions $A_{ij} \equiv (C_1^{E_{jj}})_{ij}\ (i \neq j)$:

$$\begin{aligned}
\frac{\partial A_{ij}}{\partial t_j} &= -\frac{\partial^2 A_{ij}}{\partial x_j^2} - 2 \sum_{k \neq j} A_{ij} A_{jk} A_{kj}, \\
\frac{\partial A_{ij}}{\partial t_i} &= \frac{\partial^2 A_{ij}}{\partial x_i^2} + 2 \sum_{k \neq i} A_{ij} A_{ik} A_{ki}, \\
\frac{\partial A_{ij}}{\partial t_k} &= A_{ik} \frac{\partial A_{kj}}{\partial x_k} - A_{kj} \frac{\partial A_{ik}}{\partial x_k} \qquad \text{if } i \neq k, \ j \neq k, \\
\frac{\partial A_{ij}}{\partial x_k} &= A_{ik} A_{kj} \qquad \text{if } i \neq k, \ j \neq k, \\
\sum_k \frac{\partial A_{ij}}{\partial x_k} &= \sum_k \frac{\partial A_{ij}}{\partial t_k} = 0.
\end{aligned} \tag{6.5.5}$$

This reduces to (6.3.1) if $n = 2$.

Remark 6.5. Equations (6.5.1), (6.5.3) and (6.5.4) still make sense if we consider an arbitrary algebraic group G and a reductive commutative subalgebra $\mathfrak{h}$ of its Lie algebra $\mathfrak{g}$. The functions $P(\alpha)$ take values in $G(\mathcal{A}((z)))$ and the functions $C^{(a)}$ take values in $\mathfrak{g}(\mathcal{A}((z)))$. If G is a simply laced simple Lie group, the element $R(\gamma, z) \in G(\mathbb{C}[z, z^{-1}])$ in matching conditions (6.5.2) can be generalized as follows. Let $\mathfrak{h}$ be a Cartan subalgebra of $\mathfrak{g}$, normalize the Killing form on $\mathfrak{g}$ by the condition that $(\alpha|\alpha) = 2$ for any root α, and identify $\mathfrak{h}$ with $\mathfrak{h}^*$ using this form. Let M (resp. L) $\subset \mathfrak{h}^* = \mathfrak{h}$ be the root (resp. weight) lattice and let $\varepsilon(\alpha,\beta) : M \times M \to \{\pm 1\}$ be a bimultiplicative function such that $\varepsilon(\alpha,\alpha) = (-1)^{\frac{1}{2}(\alpha|\alpha)}$, $\alpha \in M$. Define $R(\alpha, z) \in H(\mathbb{C}[z, z^{-1}])$ for each α as follows:

$$R(\alpha, z) = c_\alpha z^\alpha, \tag{6.5.6}$$

where in any finite-dimensional representation V of G, $c_\alpha \in H$ and $z^\alpha \in H$ for $z \in \mathbb{C}^\times$ are defined by

$$c_\alpha(v) = \varepsilon(\beta, \alpha)v, \;\; z^\alpha(v) = z^{(\alpha|\beta)}v \text{ if } v \in V_\beta. \tag{6.5.7}$$

Note that this GNLS hierarchy is closely related to the Bruhat decomposition in the loop group $G(\mathbb{C}((z)))$.

6.6. It is clear that we get the τ-function of the m-th reduced n-component KP hierarchy if we let in (5.1.9)

$$w_s = \omega_s z_s, \;\; s \in S, \tag{6.6.1}$$

where ω_s are arbitrary m-th roots of 1. The totality of τ-functions is (a completion of) the orbit of $1 \in B^{(0)}$ under the group $SL_{mn}(\mathbb{C}[t, t^{-1}])$.

References

[AC] M.J. Ablowitz and P.A. Clarkson, *Solitons, nonlinear evolution equations and inverse scattering*, London Math. Soc. Lecture Note Series 149, Cambridge University Press, 1991.

[BLMP] M. Boiti, J. Leon, L. Martina and F. Pempinelli, *Scattering of localized solitons in the plane*, Phys. Lett. **132A** (1988), 432–439.

[D] B.A. Dubrovin, *Completely integrable Hamiltonian systems related to matrix operators, and abelian manifolds*, Funct. Anal. Appl. 11:4 (1977), 28–41.

[DJKM1] E. Date, M. Jimbo, M. Kashiwara and T. Miwa, *Operator approach to the Kadomtsev-Petviashvili equation. Transformation groups for soliton equations. III*, J. Phys. Soc. Japan **50** (1981), 3806–3812.

[DJKM2] E. Date, M. Jimbo, M. Kashiwara and T. Miwa, *Transformation groups for soliton equations. Euclidean Lie algebras and reduction of the KP hierarchy*, Publ. Res. Inst. Math. Sci. **18** (1982), 1077–1110.

[DJKM3] E. Date, M. Jimbo, M. Kashiwara and T. Miwa, *Transformation groups for soliton equations*, in: Nonlinear integrable systems—classical theory and quantum theory eds M. Jimbo and T. Miwa, World Scientific, 1983, 39–120.

[Di] L.A. Dickey, *On Segal-Wilson's definition of the τ-function and hierarchies of AKNS-D and mcKP*, preprint 1991.

[FK] I.B. Frenkel and V.G. Kac, *Basic representations of affine Lie algebras and dual resonance models*, Invent. Math. **62** (1980), 23–66.

[FS] A.S. Fokas and P.M. Santini, *Dromions and boundary value problem for the Davey-Stewartson 1 equation*, Physica **D44** (1990), 99–130.

[H] R. Hirota, *Direct method in soliton theory,*, in Solitons, eds. R.K. Bullough and P.J. Caudrey, Springer Verlag, 1980.

[HH] J. Hietarinta, R. Hirota, *Multidromion solutions of the Davey-Stewartson equation*, Phys. Lett A **145** (1990), 237–244.

[HMM] R. Hernandes Heredero, L. Martinez Alonso and E. Medina Reus, *Fusion and fission of dromions in the Davey-Stewartson equation*, Phys. Lett. A **152** (1991), 37–41.

[JM] M. Jimbo and T. Miwa, *Solitons and infinite dimensional Lie algebras*, Publ. Res. Inst. Math. Sci. **19** (1983), 943–1001.

[K] V.G. Kac, *Infinite dimensional Lie algebras*, Progress in Math., vol. 44, Brikhäuser, Boston, 1983; 2nd ed., Cambridge Univ. Press, 1985; 3d ed., Cambridge Univ. Press, 1990.

[KKLW] V.G. Kac, D. A. Kazhdan, J. Lepowsky, and R.L. Wilson, *Realization of the basic representations of the Euclidean Lie algebras*, Adv. in Math. **42** (1981), 83–112.

[KP1] V.G. Kac and D.H. Peterson, *112 constructions of the basic representation of the loop group of E_8*, in Proc. of the Symposium "Anomalies, Geometry, Topology", Argonne, eds. W.A. Bardeen, A.R. White, World Scientific, 1985, pp. 276–298.

[KP2] V.G. Kac and D.H. Peterson, *Lectures on the infinite wedge representation and the MKP hierarchy*, Sem. Math. Sup., vol. 102, Presses Univ. Montreal, Montreal, 1986, pp. 141–184.

[KR] V.G. Kac and A.K. Raina, *Bombay lectures on highest weight representations of infinite-dimen-*
sional Lie algebras, Advanced Ser. in Math. Phys., vol. 2, World Scientific, 1987.

[KW] V.G. Kac and M. Wakimoto, *Exceptional hierarchies of soliton equations*, Proc. Symp. Pure Math. **49** (1989), 191–237.

[NMPZ] S. Novikov, S. Manakov, L. Pitaevskii, V. Zakharov, *Theory of solitons*, Consultants Bureau, 1984.

[PK] D.H. Peterson and V.G. Kac, *Infinite flag varieties and conjugacy theorems*, Proc. Nat. Acad. Sci. U.S.A. **80** (1983), 1778–1782.

[S] M. Sato, *Soliton equations as dynamical systems on infinite dimensional Grassmann manifolds*, Res. Inst. Math. Sci. Kokyuroku **439** (1981), 30–46.

[SW] G. Segal and G. Wilson, *Loop groups and equations of KdV type*, Inst. Hautes Etudes Sci. Publ. Math. **63** (1985), 1–64.

[Sh] T. Shiota, *Characterization of Jacobian varieties in terms of soliton equations*, Invent. Math. **83** (1986), 333–382.

[Sk] T.H.R. Skyrme, *Kinks and the Dirac equation*, J. Math. Phys. **12** (1971), 1735–1743.

[TB] A.P.E. ten Kroode and M.J. Bergvelt, *The homogeneous realization of the basic representation of $A_1^{(1)}$ and the Toda lattice*, Lett. Math. Phys. **12** (1986), 139–147.

[TV] F. ten Kroode and J. van de Leur, *Bosonic and fermionic realizations of the affine algebra $\hat{gl}_n$*, Comm. Math. Phys. **137** (1991), 67–107.

[UT] K. Ueno and K. Takasaki, *Toda lattice hierarchy*, Adv. stud. Pure Math., vol. 4, North-Holland, 1984, 1–95.

[W1] G. Wilson, *On two construction of conservation laws for Lax equations*, The Quarterly Journal of Math. **32** (1981), 491–512.

[W2] G. Wilson, *The τ-function of the AKNS equations*, preprint, 1992.

Compatible Brackets in Hamiltonian Mechanics

*H.P. McKean**

Courant Institute of Mathematical Sciences, New York, USA

1. Raising and Lowering

The notion of "raising and lowering" was not noticed classically: in fact, the first instance was discovered by A. Lenard in connection with KdV, as will be described below. I explain the idea at once. Let d (= the number of degrees of freedom) be fixed (=1,2,3 *etc.*) and let $C^\infty(R^{2d})$ be a Lie algebra under the *bracket* $[H_1, H_2] = \nabla H_1 J \nabla H_2$. $\nabla = \partial/\partial x$ is the gradient. $J : R^{2d} \to GL(2d, R)$ is smooth and also skew so that $[H_1, H_2] = -[H_2, H_1]$, and Jacobi's identity is satisfied:

$$\big[[H_1, H_2], H_3\big] + \big[[H_2, H_3]\,,\, H_1\big] + \big[[H_3, H_1],\, H_2\big] = 0\,, \tag{1}$$

The latter can be expressed as a system of identities of the form

$$(J\nabla)_1 J_{23} + (J\nabla)_2 J_{31} + (J\nabla)_3 J_{12} = 0 \tag{2}$$

or

$$\partial J^{-1}_{12}/\partial x_3 + \partial J^{-1}_{23}/\partial x_1 + \partial J^{-1}_{31}/\partial x_2 = 0\,, \tag{3}$$

one such for each triple of indices 123, 134, *etc.*, 3) being the same as to say that J^{-1} is closed, by which I mean

$$\sum_{i<j} J^{-1}_{ij}\, dx_i \wedge dx_j \text{ is closed.} \tag{4}$$

I write $[H_1, H_2]_J$ to distinguish this bracket from a second bracket $[H_1, H_2]_K = \nabla H_1 K \nabla H_2$ and I look for Hamiltonians $H_0 \in C^\infty(R^{2d})$ such that the corresponding K-field $X : f \in C^\infty(R^{2d}) \to [f, H_0]_K$ can also be expressed as a J-field, *i.e.*, $Xf = [f, H_1]_J$ with a new Hamiltonian H_1: to spell it out, $J^{-1}K\nabla H_0 = \nabla H_1$. I sum up this state of affairs by saying that H_0 *raises (lifts) to* H_1 or that H_1 *lowers (drops) to* H_0: in symbols, $H_0 \uparrow H_1$ or $H_1 \downarrow H_0$. Now suppose not only that $H_0 \uparrow H_1$ but also that $H_1 \uparrow H_2 \uparrow H_3 \uparrow$ *etc.*, *i.e.*, that raising is unobstructed. Then, for $i < j$,

$$[H_i, H_j]_J = [H_i, H_{j-1}]_K = -[H_{j-1}, H_i]_K = -[H_{j-1}, H_{i+1}]_J = [H_{i+1}, H_{j-1}]_J\,,$$

* The work presented here was performed at the Courant Institute of Mathematical Sciences with the support of the National Science Foundation under Grant MCS-7607039, which is gratefully acknowledged.

** Lenard's rule was first published in Gardner-Greene-Kruskal-Miura [1974].

and repetitions either raise i, resp. lower j, to the same level *or else* reverse them, according as i and j have or have not the same parity: in the first case, $[H_i, H_j] = [H_k, H_k] = 0$ with $k = (1/2)(i+j)$; in the second case, $[H_i, H_j] = [H_j, H_i] = -[H_i, H_j] = 0$, the upshot being that H_0, H_1, H_2, *etc.* commute, so that if ∇H_j $(0 \le j \le d)$ are independent, then the H_0-flow $\dot{x} = J\nabla H_0$ is integrable.

2. Lenard on KdV

This was Lenard's observation for KdV in which case $x \in R^{2d}$ is replaced by* $x \in C_\downarrow^\infty(R)$, $\nabla = \partial/\partial x$ is the functional gradient, $J = D : x \to x'$, and $K = xD + Dx - (1/2)D^3$. The J-bracket $\int \nabla H_1 D \nabla H_2$ looks queer but is not: it is just the push-forward of the classical bracket $\int [\frac{\partial H_1}{\partial x}\frac{\partial H_2}{\partial y} - \frac{\partial H_1}{\partial y}\frac{\partial H_2}{\partial x}]$ for $C_\downarrow^\infty \times C_\downarrow^\infty$ to the (1/2)-dimensional submanifold: $y =$ the (indefinite) integral of x. Why $\int \nabla H_1 K \nabla H_2$ is also a bracket is simple enough: it is just the vanishing, after a cyclic permutation of $h_1, h_2, h_3 \in C_\downarrow^\infty(R)$ and a sum, of the cubic form**

$$
\begin{aligned}
\int h_1 K\nabla \int h_2 K h_3 &= \int h_1 K \nabla \int [x(h_2h_3' - h_2'h_3) - \frac{1}{2}h_2h_3''']\\
&= \int h_1 K[h_2h_3' - h_2'h_3]\\
&= \int h_1(xD + Dx - \frac{1}{2}D^3)(h_2h_3' - h_2'h_3)\\
&= 2\int xh_1(h_2h_3'' - h_2''h_3) + \int x'h_1(h_2h_3' - h_2'h_3)\\
&\quad - \frac{1}{2}\int h_1''(h_2h_3'' - h_2''h_3)\,,
\end{aligned}
$$

which is plain. The discovery of Lenard is that if $H_0 = \int x$, then

$$K\nabla H_0 = K1 = Dx = J\nabla\frac{1}{2}\int x^2\,,$$

i.e., $H_0 \uparrow H_1 = \int \frac{1}{2}x^2$, and likewise

$$K\nabla H_1 = Kx = D(\frac{3}{2}x^2 - \frac{1}{2}x'') = J\nabla \int [\frac{1}{2}x^3 + \frac{1}{4}(x')^2]\,,$$

i.e., $H_1 \uparrow H_2 = \int [\frac{1}{2}x^3 + \frac{1}{4}(x')^2]$, *etc.*; in fact, repetitions produce the whole series of standard KdV invariants. But what is happening? What is it about

* $C^\infty(R)$ is the class of smooth rapidly vanishing functions on the line.

** H_1, H_2, H_3 are 3 smooth functions from $C_\downarrow^\infty$ to R.

J, K, and $H_0 = \int x$ that permits H_0 to be raised without obstruction? The answer is the *compatibility* of J and K (and a lucky choice of H_0). The idea is presented in the classical format. I demand (of J and K) that

$$H_0 \in C^\infty(R^{2d}) \textit{ can be raised only if it can be lowered and vice versa:} \quad (5)$$

in that case, $H_0 \uparrow H_1$ implies $H_1 \uparrow H_2$ in view of $H_1 \downarrow H_0$, and similarly for $H_2 \uparrow H_3 \uparrow H_4$ *etc.* What does this mean? Only that the conditions
a+) $J^{-1}K\nabla H_0$ is a gradient (of $H_1 = H_0$ raised).
a−) $K^{-1}J\nabla H_0$ is a gradient (of $H_{-1} = H_0$ lowered)
are the same, *i.e.*, that either both or neither of
b+) $\left[(\partial/\partial x_i)(J^{-1}K\nabla H_0)_j : 0 \le i,j \le 2d\right]$
b−) $\left[(\partial/\partial x_i)(K^{-1}J\nabla H_0)_j : 0 \le i,j \le 2d\right]$
is symmetric. This takes place if

$$K\left[\text{skew part of } b-)\right]K + J\left[\text{skew part of } b+)\right]J \quad (6)$$

vanishes for every H_0, which can be reduced, by use of (2) for J and for K, to the condition that $J + K$ also satisfies (2): in short, that J, K, *and* $J + K$ *produce honest brackets.* This is the *compatibility* cited before; it guarantees that *raising, resp. lowering, once begun, continues unobstructed.* I omit the computation which is tiresome but straightforward.

Amplification 1. (5) and (6) would be the same if the gradients of raisable functions $H_0 \in C^\infty(R^{2d})$ would span R^{2d} at each point. But, as will be seen below, all such functions commute, and it is a general fact that more than d commuting functions cannot be independent: otherwise, you would have more than d independent flows $\dot{x} = J\nabla H(x)$ preserving the less than d-dimensional manifold determined by fixing the function values $H = h$.

Amplification 2. J^{-1} and K^{-1} are closed and compatible (meaning that J and K are such) if and only if $L = J - cK$ satisfies (2) for 3 values of c, the former being of degree 2 in the latter. Now

$$L^{-1} = (1 - cJ^{-1}K)^{-1}J^{-1} = J^{-1} + cJ^{-1}KJ^{-1} + c^2J^{-1}KJ^{-1}KJ^{-1} + etc.$$

so the matter can be restated thus: J^{-1} and K^{-1} are closed and compatible if and only if $(J^{-1}K)^nJ^{-1}$ is closed for $n \ge 0$, or even for $n = 0, 1$, and 2; this will be helpful in art. 10. Note also that $(KJ^{-1})^nK$ is compatible with J for $n = 1, 2, 3$, *etc.* as soon as K is such.

Amplification 3. J and K being compatible, the class of functions which can be raised ($\uparrow$) is one and the same as the class of functions which can be lowered ($\downarrow$) I note that $\uparrow=\downarrow$ is a subalgebra of $C^\infty(R^{2d})$ under either bracket: indeed, if H_1 and H_2 lift, then* $[H_1, H_2]_J \uparrow [H_1^+, H_2]_J + [H_1, H_2^+]_J$, etc. I omit the easy proof.

---

* H^+ is the lift of H.

4. Back to KdV.

The compatibility of $J = D$ and $K = xD + Dx - (1/2)D^3$ comes down to the vanishing, after a cyclic permutation of $h_1, h_2, h_3 \in C^\infty_\downarrow(R)$ and a sum, of the cubic form

$$\int h_1 K\nabla \int h_2 J h_3 + \int h_1 J\nabla \int h_2 K h_3 = \int h_1(h_2 h_3'' - h_2'' h_3) ,$$

so that explains Lenard; it is still a matter of luck that raising $H_0 = \int x$ produces *all* the integrals of motion. The recognition of this special case led to the discovery of compatible brackets for all the fashionable integrable systems: Toda, sine-Gordon, cubic Schrödinger, *etc.*, and also to investigations in the classical format; see, esp., Dorfman-Gelfand [1979], Gelfand-Zakharevich [1989 & 1991], Olver [1990], and the literature cited there. F. Magri [1978] was the first in the field and, in my opinion, has understood the matter best. The rest of this talk is an account of his work with some small additions of my own.

5. Spec $J^{-1}K$ Lifts

The spectrum of $J^{-1}K$ plays a central role. The fact is that

$$D \equiv \ell g \det(\lambda I - J^{-1}K) \uparrow \lambda D + \text{sp}\, J^{-1}K ,$$

with the implication that the power sums $\pi_n = \text{sp}(J^{-1}K)^n$ $(n \geq 0)$ lift according to the rule $\pi_1 \uparrow \pi_2/2 \uparrow \pi_3/3$ *etc.*; in particular, spec $J^{-1}K$ commutes in the sense that $[\pi_i, \pi_j] = 0$ $(i < j)$.
PROOF. The first statement is immediate from the second. Now take λ so large that $\lambda J - K \equiv L$ is invertible. Then

$$\begin{aligned}(L\nabla\, \ell g \det L)_i &= L_{ij}\ \text{sp}\ L^{-1}\partial L/\partial x_j \text{ with summation on } 1 \leq j \leq d \\ &= L_{ij} L^{-1}_{ab}\ \partial L_{ba}/\partial x_j \text{ with summation on } a, b \\ &\overset{*}{=} -L^{-1}_{ab}\left[L_{bj}\,\partial L_{ai}/\partial x_j + L_{aj}\,\partial L_{ib}/\partial x_j\right] \\ &= 2\,\partial L_{ij}/\partial x_j .\end{aligned}$$

The lifting of the π's follows by developing the left-hand side in descending powers λ^n and interpreting the fact that $n \leq -1$ contributes nothing.

6. Spec Commutes with $\uparrow$ & $\downarrow$

It is to be proved that spec $J^{-1}K$ commutes with any function H_0 which can be lifted.

Item 1 $H \in \uparrow$ & $\downarrow$ if and only if $[J, H]_K - [K, H]_J + J\nabla^2 H K - K\nabla^2 H J = 0$.

Proof $[(\partial/\partial x_i)(J^{-1}K\nabla H)_j\colon\ i, j \leq d]$ must be symmetric. The identity expresses this fact.

---

* L satisfies (2).

Item 2 sp $J^{-1}[K,H]_J J^{-1}K =$ sp $J^{-1}[K,H]_K$.

Proof (2) is used, once for J and once for K.

Item 3 The identity of item 1 is pre/post-multiplied by $J^{-1}/J^{-1}K$ and the trace is taken:

$$\begin{aligned}&\mathrm{sp}(J^{-1}[J,H]_K J^{-1}K - J^{-1}[K,H]_J J^{-1}K + \nabla^2 HKJ^{-1}K - J^{-1}K\nabla^2 HK)\\ &= \mathrm{sp}(-[J^{-1},H]_K K - J^{-1}[K,H]_K) = -\mathrm{sp}[J^{-1}K,H]_K = -[\pi_1,H]_K\end{aligned}$$

i.e., $[\pi_1,H]_K =^* [\pi_1,H^+]_J = 0$ and so also $[\pi_1,H]_J = 0$ since $(\uparrow)^+ = \uparrow$.

Item 4 The rest is easy: $0 = [\pi_1,H]_K = [\pi_1^+,H]_J = [\pi_2/2,H]_J$ and so also $0 = [\pi_2/2,H^+]_J = [(\pi_2/2)^+,H]_J = [\pi_3/3,H]_J$, *etc.*

7. Ampleness.

$H_0 \in \uparrow\downarrow$ is said to be *ample* if the lifts $H_0 \uparrow H_1 \uparrow \ldots \uparrow H_{d-1}$ have independent gradients.** I assume spec $J^{-1}K$ is ample in the sense that π_1 is such, *i.e.*, the gradients $\nabla\pi_i$ ($i \le d$) are independent. Fixing the values of the π's now defines a smooth d-dimensional "spectral" submanifold $M \subset R^{2d}$ to which $\nabla\pi_i$ ($i \le d$) is normal and $J\nabla\pi_j$ ($j \le d$) is tangent, by the commutativity of the π's. This has important consequences which occupy the next 2 articles.

8. Spec is Double.

Let n_i ($i \le d$) be a unit perpendicular base of the normal space of M at a fixed point. Then, with $G = [(Jn_i, Jn_j) : 1 \le i,j \le d]$, $n_i' = \sum G_{ij}^{-1/2} Jn_j$ is a unit perpendicular base of the tangent space, and

$$D = \det(\lambda I - J^{-1}K) = \det \begin{bmatrix} \lambda\delta_{ij} - (n_i, J^{-1}Kn_j) & -(n_i, J^{-1}Kn_j') \\ -(n_i', J^{-1}Kn_j) & \lambda\delta_{ij} - (n_i', J^{-1}Kn_j') \end{bmatrix}$$

with $1 \le i,j \le d$ in each block; moreover, the lower left-hand block vanishes, $J^{-1}Kn_j$ being normal and n_i' tangent to M, so that D is the product of the 2 diagonal blocks. But with the abbreviation $(a,b) = [(a_i,b_j) : 1 \le i,j \le d]$, the

* H^+ is the lift of H.

** No further lifts can be independent by the general principle noted in amplif. 3.1.

second factor is

$$\begin{aligned}\det[\lambda I - (n', J^{-1}Kn')] &= \det[\lambda I - G^{-1/2}(KJ^{-1}Jn, Jn)G^{-1/2}] \\ &=^* \det[\lambda I - (JKn, n)(n, J^{-2}n)] \\ &= \det[\lambda I - (J^{-1}Kn, n)] \\ &= \det[\lambda I - (n, J^{-1}Kn)] :\end{aligned}$$

in short, the two factors are the same and spec $J^{-1}K$ is double.

NOTE. The double nature of spec $J^{-1}K$ accords with the fact that no more than d of the commuting functions $\pi_1, \pi_2/2$, *etc.* can be functionally independent; see amplif. 3.1. From now on λ_i ($i \le d$) is a list of spec $J^{-1}K$ in the normal space.

9. Structure of $\uparrow\downarrow$.

The amplitude of spec $J^{-1}K$ and the fact that it commutes with $H \in \uparrow\downarrow$ implies that H is invariant under the flows $\dot{x} = [x, \pi_n/n]$ ($n \le d$), and as these are transitive on the spectral manifold in the small, so H is a function h of spec $J^{-1}K$ alone. Now move x a bit so as to make the spectrum of precise multiplicity 2, as you may do in view of its ampleness. Then the individual eigenvalues commute and $\lambda_i \uparrow \lambda_i^2/2$ ($i \le d$)**, so that $H = h(\lambda_1, \dots, \lambda_d) \in \uparrow$ implies $J^{-1}K\nabla H = (\partial h/\partial\lambda)(\partial\lambda^2/2/\partial x) =^{***} (\partial h^+/\partial\lambda)(\partial\lambda/\partial x)$, whence $\partial h^+/\partial\lambda_i = \lambda_i \partial h/\partial\lambda_i$ and $[\lambda_i \partial^2 h/\partial\lambda_i\partial\lambda_j : 1 \le i,j \le d]$ is symmetric. But now the simplicity of spec $J^{-1}K$ implies $\partial^2 h/\partial\lambda_i\partial\lambda_j = 0$ ($i \ne j$), *i.e.*, h is a 'split" function $h = h_1(\lambda_1) + h_2(\lambda_2) + etc.$: in fact, $H \in \uparrow$ if and only if it has this split form. It is a bonus that $\uparrow = \downarrow$ *commutes*. The discussion assumes that spec is *real*, which need not be so, but the general picture is not much changed: it is only that to each complex pair $\lambda, \lambda^* \in$ spec is associated one summand $h_0(\lambda)$ which is harmonic is in the variables real λ, imag λ. $\uparrow\downarrow$ still commutes.

10. Magri's Form of K.

Let J and K be compatible and spec $J^{-1}K$ ample. Then $H_i = \pi_i/i$ ($i \le d$) commute and can be completed by supplementary functions G_j ($j \le d$), canonically conjugate relative to J^*. Then

a) $\qquad [H_i, H_j]_K = [H_i, H_j^+]_J = 0\ ,$

b) $\qquad [H_i, G_j]_K = [H_i^+, G_j]_J = 1 \quad \text{if } j = i+1 \le d$

---

* $G^{-1} = -(n, J^{-2}n)$.

** Use $\pi_1, \dots, \pi_d$ and the fact that $[\lambda_i^{j-1} : 1 \le i \le j \le d]$ is nonsingular.

*** $H^+ = h^+(\lambda_1, etc.)$ is the lift of H.

* This is Darboux's theorem; see Arnold [1978]. G cannot be raised as it is not a split function of spec $J^{-1}K$.

$$= 0 \text{ if } j \neq i+1 \leq d$$
$$= \frac{\partial H_{d+1}}{\partial H_j} \text{ if } i = d ,$$

and
c) $[G_i, G_j]_K = c_{ij}(H_1, \dots, H_d)$ is constant on M in view of 5):

$$\begin{aligned}[][G_i,G_j]_K,H_k]_J &= [[G_i,G_j]_K,H_k^-]_K \\ &=^{**} \ -[[G_j,H_k^-]_K,G_i]_K \\ &\quad -[[H_k^-,G_i]_K,G_j]_K \\ &= -[[G_j,H_k]_J,G_i]_K \\ &\quad -[[H_k,G_i]_J,G_j]_K \\ &=^{***} \ 0 .\end{aligned}$$

a), b), c) determine K since ∇H_i $(i \leq d)$ and ∇G_j $(j \leq d)$ span R^{2d}. The formula of Magri [198]:

$$\begin{aligned} K = &\sum_{n=1}^{d} \left[J\nabla H_n^+ \otimes J\nabla G_n - J\nabla G_n \otimes J\nabla H_n^+\right] \\ &+ \sum_{1\leq i,j\leq d} c_{ij}\left[J\nabla H_i \otimes J\nabla H_j - J\nabla H_j \otimes J\nabla H_i\right]\end{aligned}$$

is checked in this style. The fact is more neatly expressed by saying that $J^{-1}KJ^{-1}$ produces the closed form

$$\sum_{n=1}^{d} dH_n^+ \wedge dG_n + \sum_{1\leq i<j\leq d} c_{ij}\, dH_i \wedge dH_j \ .$$

Amplification 1 $[c_{ij} : 1 \leq i < j \leq d]$ is in the nature of a gauge: it has no effect on raising/lowering since $J\nabla H_i \otimes J\nabla H_j$ kills ∇H_0 for any $H_0 \in \uparrow\downarrow$.

Amplification 2 $c_- = \sum c_{ij}\, dH_i \wedge dH_j$ is closed since $J^{-1}KJ^{-1}$ is such. It has additional features, too: for example,

$$\begin{aligned}[][G_i,G_j]_K,G_k]_K &= [c_{ij},G_k]_K \\ &= \sum (\partial c_{ij}/\partial H_\ell)\,[H_\ell,G_k]_K \\ &= \sum (\partial c_{ij}/\partial H_\ell)\,[H_\ell^+,G_k]_J \\ &= \frac{\partial c_{ij}}{\partial H_d}\frac{\partial H_{d+1}}{\partial H_1} \quad \text{if } k=1 \\ &= \frac{\partial c_{ij}}{\partial H_{k-1}} + \frac{\partial c_{ij}}{\partial H_d}\frac{\partial H_{d+1}}{\partial H_k} \quad \text{if } k \geq 2 ,\end{aligned}$$

** $H_k^- = H_{k-1}$ with $H_0 = \ell g \det J^{-1}K$.

*** $[H,G]_J = 0$ or 1; *esp.*, it is constant!

and this must vanish after cyclic permutation of ijk and a sum. What does this mean? It is just the condition that the form

$$c_+ = \sum_{i<j} c_{ij}(dH_i^+ \wedge dH_j + dH_i \wedge dH_j^+) \quad \text{be closed,}$$

as you will easily check.

Amplification 3 The fact elicited in amplif. 2 completes the specification of K: indeed, if H_i $(1 \le d)$ is *any* family of d commuting functions, if the supplementary family G_j $(j \le d)$ is canonically paired, if H_j $(j > d)$ stands in the same relation to H_i $(1 \le i \le d)$ as π_j/j $(j > d)$ does to π_i/i $(1 \le i \le d)$, and if $[c_{ij}(H_1, \dots, H_d)$: $1 \le i < j \le d]$ is such that both forms c_- and c_+ of amplif. 2 are closed, then, with $H_n^+ = H_{n+1}$ $(1 \le n \le d)$, *the skew form K produced by Magri's rule is compatible with J; moreover, $H_1 \uparrow H_2 \uparrow \dots \uparrow H_d$.*

Proof $J^{-1}KJ^{-1}$is closed and so is $J^{-1}KJ^{-1}KJ^{-1}$,the associated form being $\sum dH_n^{+2} \wedge G_n + c_+$. The computation employs the following remark: π_j/j is a polynomial p_j of $\pi_1, \pi_2/2, \dots, \pi_d/d$ if $j > d$ and H_j is the same polynomial of $H_1, \dots H_d$ for such j, by definition, so from $J^{-1}K\nabla H_i = H_{i+1}$, $(i \le d)$ you find, for $j > d$, that

$$J^{-1}K\nabla H_j = \sum_i (\partial p_j/\partial H_i)\nabla H_{i+1} = \nabla p_{j+1}(H_1, \dots, H_d) = \nabla H_{j+1} .$$

The rest comes from amplif. 3.2: J^{-1}, $J^{-1}KJ^{-1}$, and $J^{-1}KJ^{-1}KJ^{-1}$ are closed, so K is compatible with J.

Amplification 4 The discussion of amplif. 3 is completed by checking that *the functions H_i $(i \le d)$ must fall in with the power sums π_i/i $(i \le d)$ for* spec $J^{-1}K$, up to unimportant additive constants. H_1 is ample so $J^{-1}K$ acts upon the space $N =$ span$[\nabla H_i : i \le d]$ normal to the "energy" surface M obtained by fixing the values $H_i = h_i$ $(i \le d)$. The action $J^{-1}K : \nabla H_i \to \nabla H_{i+1}$ is expressed by the matrix displayed at the left. Now the discussion of art. 8 applies: spec $J^{-1}K$ is double and can be computed from the action

$$\begin{bmatrix} 01000 \\ 00100 \\ 00010 \\ 00001 \\ \partial H_{d+1}/\partial H_i\,(i \le d) \end{bmatrix}$$

on N. The rest follows from the choice of $H_{d+1} = p_{d+1}(H_1, \dots, H_d)$ in amplif. 3: in fact $\partial H_{d+1}/\partial H_d = H_1 =$ sp $J^{-1}K = \pi_1$; now raise.

Amplification 5 To clarify the significance of the gauge $[c_{ij} : i, j \le d]$, I note that the closure of the forms $c_\pm$ of amplif. 2 is the same as to say that

$$c_{ij} \text{ is of the form } \frac{\partial^2 E}{\partial H_i^- \partial H_j} - \frac{\partial^2 E}{\partial H_j^- \partial H_i} \quad \text{for } 1 \le i < j \le d ,$$

in which $\partial/\partial H_i^-$ $(i \le d)$ means that you take partials relative to $H_0 =$

$\ell g \det J^{-1}K, H_1, H_2, \ldots, H_{d-1}$ instead of $H_1, \ldots, H_d$. It is a nice exercise in calculus to prove that $\partial/\partial H_i^- = 0$ or the old $\partial/\partial H_{i-1}$ according as $i = 1$ or $i \geq 2$ *plus* $(\partial H_{d+1}/\partial H_i)\,\partial/\partial H_d$, *i.e.*, it is nothing but the field $[\ \ , G_i]_K$ figuring in amplif. 2.

Amplification 6 The preceding amplif. 5 leads to the pretty fact that the gauge can be removed by a canonical map $H_i' = H_i$, $G_i' = G_i + F_i(H_1, \ldots, H_d)$ $(i \leq d)$. The adjective *canonical* means that $(F_1, \ldots, F_d)$ is the gradient of some E (relative to $H_1, H_2, etc.$). Now compute

$$\begin{aligned} c_{ij}' = [G_i', G_j']_K &= c_{ij} + [F_i, G_j]_K + [G_i, F_j]_K \\ &= c_{ij} + \frac{\partial F_i}{\partial H_j^-} - \frac{\partial F_j}{\partial H_i^-} \\ &= c_{ij} + \frac{\partial^2 E}{\partial H_j^- \partial H_i} - \frac{\partial^2 E}{\partial H_i^- \partial H_j} \end{aligned}$$

and compare amplif. 5.

Amplification 7 The discussion makes plain that if $\dot{x} = J\nabla H(x)$ is an integrable flow on R^{2d} with d independent commuting integrals of motion $H = H_1, H_2, \ldots, H_d$, then Magri's formula provides a compatible K with the pleasing feature that $H_1 = \pi_1 \uparrow H_2 = \pi_2/2 \uparrow$ *etc.* You can even make spec $J^{-1}K$ real and simple (in the small) by adding inessential constants to $H_1, \ldots, H_d$ and moving x a little. Then $\lambda_i \uparrow \lambda_i^2/2 \uparrow \lambda_i^3/3\, etc.$ individually, for $i \leq d$, and K can be recovered from the form

$$\sum_{i<j} (J^{-1}KJ^{-1})_{ij}\, dx_i \wedge dx_j = \sum_{n=1}^{d} d(\lambda_n^2/2) \wedge d\mu_n$$

in which λ_i $(i \leq d)$ and μ_j $(j \leq d)$ are canonically paired. The gauge c_{ij} now takes the form $(\lambda_i - \lambda_j)\,\partial^2 E/\partial\lambda_i\partial\lambda_j$ $(1 \leq i < j \leq d)$, and the removal of the gauge (by canonical transformation, as in amplif. 6) has the pleasant effect that, besides $J^{-1}K\nabla\lambda_i = \lambda_i\nabla\lambda_i$, you have $J^{-1}K\nabla\mu_i = \lambda_i\nabla\mu_i$ as well, confirming the double character of spec $J^{-1}K$.*

Moral. The class of integrable Hamiltonians is (locally) coextensive with the class of Hamiltonians which are ample relative to a pair off compatible symplectic forms J^{-1} and K^{-1}, and a little more: the latter may be adjusted so that spec $J^{-1}K$ is real and (precisely) double. This is the geometrical content of Magri's splendid work.

11. Back to KdV.

Let $x \in C_\downarrow^\infty(R)$ be fixed and let $L = -D^2 + x$ have transmission coefficient s_{11} and (left-hand) reflection coefficient s_{21}. In the absence of bound states

* This does *not* mean that the supplementary μ_i $(i \leq d)$ lift: $\nabla\lambda \otimes \nabla\mu$ is not symmetric so $\lambda\nabla\mu$ is not a gradient.

$\ell g\,|s_{11}(k)| : k \geq 0$ is a complete list of commuting integrals of motion and with* $J = 2D$ and $K = xD + Dx - (1/2)D^3$, you have $K\nabla \ell g\,|s_{11}| = k^2 J \nabla \ell g |s_{11}|$ so that $\ell g|s_{11}| \uparrow k^2 \ell g|s_{11}|$. The supplementary variables$(1/\pi)$ph $s_{21}(k') : k' \geq 0$ commute and satisfy

$$[\ell g|s_{11}(k)|\ ,\ (1/\pi)\ \text{ph}\ s_{21}(k')]\ d(k')^2 = \ \text{the unit mass at } k' = k\ ,$$

i.e., the latter are canonically paired to the former.** K can now be expressed in Magri's form:

$$K = -\int_0^\infty k^2 (J\nabla \frac{1}{\pi}\ \text{ph}\ s_{21}) \otimes (J\nabla \ell g|s_{11}|) - \text{ the reverse}]\ dk^2\ ,$$

but that is too cumbersome to be of any use.

12. Grassmann Picture.

The amplitude of spec $J^{-1}K$ means that $x \in R^{2d}$ lies on a d-dimensional spectral manifold M with normal space $N = \text{span}[\nabla \pi_i : i \leq d]$, permitting a natural map of $x \to N$ from R^{2d} into the Grassmannian $G_{1/2}$ of $d = (1/2) \times (2d)$-dimensional subspaces of R^{2d}. I believe that this map is 1:1 in the small and that $D = \ell g \det(\lambda I - J^{-1}K)$ is Sato's tau function [1981 & 1983], more or less.

13. Complex Structure.

It is an intriguing fact that, for every *explicitly integrable* flow, a natural complex structure makes its appearance. This must have something deep to do with the overdetermined character of the system $[H_i, H_j] = 0$ $(1 \leq i < j \leq d)$: in fact, $d(d-1)/2$ relations are imposed upon d functions for an over-determination of, *e.g.*, 35 if $d = 10$ and 4850 if $d = 100$. Anyhow, it suggests searching for projective curves, divisors, Jacobi varieties, theta functions and all that, *esp.* in connection with the tau function cited above.

14. Lax Pairs.

It is natural to ask if the integrable flow $\dot{x} = J\nabla H_0$ is equivalent to*** $L^\bullet = [P, L] \equiv PL - LP$ in which $x \to L$ is a 1:1 map to $d \times d$ or $2d \times 2d$ symmetric matrices, and P is of like size and skew. The conventional wisdom requires L to have the 'same" spectrum as $J^{-1}K$ and $\nabla\lambda$ to be the "squares" of the associated eigenvectors of L, but I believe this is naive:* $\ell g|s_{11}(k)| : k \geq 0$

---

* $[H_1, H_2]_J = \int [\nabla H_1 D \nabla H_2 - \nabla H_2 D \nabla H_1]$, *etc.*

** Buslaev-Faddeev [1986] provide details.

*** I adopt Gelfand's notation: PL = Peter Lax.

* McKean [1986] clarifies this.

is not the spectrum of anything; it is, instead (the logarithm of) a spectral density.** Likewise, for periodic KDV, $\lambda J - K = 2\lambda D - xD - Dx + (1/2)D^3$ has always the "supernumerary" null function $e = \partial\Delta/\partial x$ ** with multiplier $m = +1$,*** but the only (isolated and robust) *double spectrum* of $\lambda J - K$ is the roots of $\Delta = 0$, i.e., with the multiplier $m = -1$. They form a complete list of integrals of motion, but have, on the face of it, nothing to do with the conventional (band or periodic/antiperiodic) spectrum of $L = -D^2 + x$. In short, the role of Lax pairs still needs to be clarified.

References

Arnold, V.: Mathematical Methods of Classical Mechanics, Springer Verlag, New York/ Heidelberg/Berlin, 1978.

Dorfman, I. Ya., and I. M. Gelfand: Hamiltonian operators and algebraic structures related to them, *Func. Anal. Appl. 13* (1979) 248–262.

Gelfand, I. M., and I. S. Zakharevich: Spectral theory of a pencil of skew-symmetric differential operators of the third order on S^1, *Func. Anal. Appl.* **23** (1989) 85–93.

————: Webs, Veronese curves and biHamiltonian systems. Preprint, 1991.

Gardner, C.S., J.M. Greene, M.D. Kruskal, and R. Miura: Korteweg-deVries equations and generalizations VI. Methods for exact solutions. *CPAM*, **27**, (1974), 97-133.

Magri, F.: A simple model of the integrable Hamiltonian equation, *J. Math. Phys. 19* (1978) 1156–1162.

————: Private communication, 1989.

McKean, H. P.: Geometry of KdV (1), *Revista Mat. Iberoamericana* **2** (1986) 235–261.

Oliver, P.: Canonical forms and integrability of biHamiltonian systems. *Physics Letters A*, **148**, (1990) 177-187.

Sato, M.: Soliton equations as dynamical systems on an ∞-dimensional Grassmann manifold, *RIMS Kokyuroku 439* (1981) 30–46.

Sato, M., and Y. Sato.: Soliton equations as dynamical systems on an ∞-dimensional Grassmann manifold, *Lecture Notes Numer. Appl. Math. 5* (1983) 259–272.

** Δ is Hill's discriminant.

*** $e(x+1) = me(x)$.

Symmetries – Test of Integrability

A.B. Shabat and A.V. Mikhailov

L.D. Landau Institute for Theoretical Physics, Moscow, Russia, and Institute of Scientific Interchange, Turin, Italy

1 Introduction

At the beginning of the history of solitons it was commonly accepted that any PDE with higher order conservation law or symmetries is integrable. All classical equations like KDV, NLS, Sine-Gordon and the Toda chain were discovered as equations with extra conservation laws (see [1], [2], [3], [4]). In the recently published book "What is integrability?" [5] one may find a few more points of view for this basic question but we still prefer the original one.

The idea of using these notions to develop a rigorous definition belongs to many authors [6], [7], [8], [9], [10]. In these papers one can find the first "exhaustive" lists of integrable equations (i.e. equations satisfying the definition of integrability).

It's rather curious that the consideration of an extra symmetry (a conservation law) of fixed but smallest order makes the problem essentially more complicated. Moreover it may lead to missing some interesting equations (cf. [6] and [9]). The question "what is more basic: conservation laws or symmetries?" proved not to be so important, the answer just depends on the sort of equations under consideration.

In our article with V.V.Sokolov [11] there was an attempt to present a mathematical theory of higher order symmetries and conservation laws for PDE's with two independent variables. Here we bypass this theory and only overview known and new results.

2 Basic Definitions

For a system of PDE's

$$\mathbf{F}(\mathbf{x},\mathbf{u},\mathbf{u_x},\,\dots)=0,\qquad \mathbf{x}=(x^1,\,\dots,\,x^d),\ \mathbf{u}=(u^1,\,\dots,\,u^N) \tag{2.1}$$

the notion of symmetry can be introduced with the help of the infinitesimal transformation

$$\mathbf{u}'(\mathbf{x}') = \mathbf{u}+\tau\mathbf{U}(\mathbf{x},\mathbf{u},\mathbf{u_x},\,\dots), \tag{2.2}$$
$$\mathbf{x}' = \mathbf{x}+\tau\mathbf{X}(\mathbf{x},\mathbf{u},\mathbf{u_x},\,\dots), \tag{2.3}$$

where τ is a small parameter. The transformation (2.2), (2.3) can be put in the form

$$\mathbf{u}'(\mathbf{x}) = \mathbf{u} + \tau\mathbf{G}(\mathbf{x}, \mathbf{u}, \mathbf{u_x}, \dots), \tag{2.4}$$

$$\mathbf{x}' = \mathbf{x} \tag{2.5}$$

with the same accuracy in τ, where $\mathbf{G} = \mathbf{U} - \mathbf{u_x} \cdot \mathbf{X}$. The function $\mathbf{G}(\mathbf{x}, \mathbf{u}, \mathbf{u_x}, \dots)$ defines a symmetry[1] if

$$\mathbf{F}(\mathbf{x}, \mathbf{u}', \mathbf{u_x}', \dots) = O(\tau^2). \tag{2.6}$$

It follows from (2.6) that

$$\mathbf{F}_*\mathbf{G} = 0, \tag{2.7}$$

where $\mathbf{F}_*$ is Frechét derivative, i.e.

$$\mathbf{F}_*(\mathbf{G}) = \frac{d}{d\varepsilon}\mathbf{F}(\mathbf{x}, \mathbf{u} + \varepsilon\mathbf{G}, \mathbf{u_x} + \varepsilon\mathbf{D} \cdot \mathbf{G}, \dots)|_{\varepsilon=0}, \tag{2.8}$$

and $\mathbf{D}$ is a total derivative

$$\mathbf{D}(\mathbf{G}) = \mathbf{G_x} + \mathbf{G_u} \cdot \mathbf{u_x} + \mathbf{G_{u_x}} \cdot \mathbf{u_{xx}} \dots \tag{2.9}$$

We shall use equation (2.7) as a definition of a symmetry. In equation (2.7) all variables $\mathbf{x}, \mathbf{u}, \mathbf{u_x}, \dots$ are treated as independent, modulo equation (2.1) and its differential consequences. This definition contains the classical concept of Lie symmetries, as well as the modern notion of higher symmetries.

A fairly general definition of a local conservation law is the following: A vector function $\mathbf{j}(\mathbf{x}, \mathbf{u}, \mathbf{u_x}, ...) = (j^1, \dots, j^d)$ is called a local conservation law if

$$\mathbf{div}(\mathbf{j}) = \sum_{k=1}^{d} \mathrm{D}_k j^k = 0, \tag{2.10}$$

modulo of equation (2.1) and its differential consequences. A conservation law (2.10) is called trivial if the corresponding differential form $j^1dx^1 + \dots + j^d dx^d$ is exact one[12].

Symmetries (conservation laws) form a linear space, since a linear combination with any constant coefficients is a symmetry (conservation law). Moreover symmetries form a Lie algebra. The commutator of two symmetries is defined as follows

$$[G, H] = G_*(H) - H_*(G).$$

These general concepts go back to classical works of S.Lie and E.Noether (see [12]).

In the case of evolution systems of equations

$$\mathbf{u}_t = \mathbf{F}(t, \mathbf{x}, \mathbf{u}, \mathbf{u_x}, \mathbf{u_{xx}}, \dots). \tag{2.11}$$

the above definitions of symmetries and conservation laws can be set in a more

[1] In the book of P.Olver this function is called the characteristic of a symmetry [12]

constructive way. Without loss of generality, symmetries and conservation laws may be assumed to depend on $t, x, \mathbf{u}$ and finite number of space derivatives only, since all time derivatives can be eliminated due to (2.11). With any symmetry $\mathbf{G} = \mathbf{G}(t, \mathbf{x}, \mathbf{u}, \mathbf{u_x}, \ldots)$ one can associate evolution equation

$$\mathbf{u}_\tau = \mathbf{G}(t, \mathbf{x}, \mathbf{u}, \mathbf{u_x}, \ldots). \tag{2.12}$$

Equation (2.7) which defines a symmetry of (2.11) in more "symmetric" form

$$\frac{\partial \mathbf{G}}{\partial t} + \mathbf{G}_*(\mathbf{F}) = \mathbf{F}_*(\mathbf{G}). \tag{2.13}$$

The order of a symmetry is defined as the order of differential operator $\mathbf{G}_*$ defined in (2.8) A symmetry whose order is higher then the order of the equation (2.11) is called a higher symmetry. Equation (2.13) expresses the commutativity of two differentiations (vector fields) $[D_t, D_\tau] = 0$,

$$\begin{aligned} \mathrm{D}_t &= \frac{\partial}{\partial t} + \mathbf{F}\frac{\partial}{\partial \mathbf{u}} + \mathbf{D}(\mathbf{F})\frac{\partial}{\partial \mathbf{u_x}} + \cdots, \\ \mathrm{D}_t &= \frac{\partial}{\partial t} + \mathbf{F}\frac{\partial}{\partial \mathbf{u}} + \mathbf{D}(\mathbf{F})\frac{\partial}{\partial \mathbf{u_x}} + \cdots, \end{aligned}$$

related with (2.11) and (2.12) reactively. These differentiations commute with $\mathbf{D}$ (2.9).

Later on we shall consider only the case of one space and one time variables. Here and below $\mathbf{u}_k$ denotes k^{th} x-derivative of $\mathbf{u}$. The operator of total differentiation with respect to x in this notations have the following form

$$\mathrm{D} = \frac{\partial}{\partial x} + \mathbf{u}_1\frac{\partial}{\partial \mathbf{u}} + \mathbf{u}_2\frac{\partial}{\partial \mathbf{u}_1} + \cdots$$

In the case of evolution system of equations (2.11) of one space variable conservation laws can be written in the form

$$\mathrm{D}_t(\rho) = \mathrm{D}(\sigma) \tag{2.14}$$

The function $\rho = \rho(t, x, \mathbf{u}, \mathbf{u}_1, \ldots)$ is called *a density* of the conservation law (2.14), both functions ρ and σ assumed to be functions of $t, x, \mathbf{u}$ and finite number of derivatives $\mathbf{u}_1, \mathbf{u}_2, \ldots$. A density which itself is a total derivative $\rho = \mathrm{D}(h)$ corresponds to a trivial conservation law (in this case $\sigma = \mathrm{D}_t(h)$).

3 Scalar evolution equation and canonical densities

A general mathematical theory of higher symmetries has mainly been developed in the case of scalar evolution equations of one spatial variable

$$u_t = F(x, t, u, u_1, u_2, \ldots, u_n), \quad n \geq 2. \tag{3.15}$$

Despite the fact that a generic equation (3.15) does not have any symmetries, equations which come from applications usually do possess symmetries. How many symmetries of a given order equation (3.15) can have in principle?

There is the theorem of B.Magadeev [13] which states that equation (3.15) has an infinite dimension linear space of symmetries of order less or equal a fixed constant only in the case if (3.15) can be turn into a linear equation via a point or contact transformation.

Thus, if we are interested with essentially nonlinear equations of the form (3.15) we have to consider equations which possess a finite number of symmetries of a fixed order.

Presence of classical (i.e. point or contact) symmetries helps to find particular solutions but does not guarantee integrability, even if the group is very reach. Indeed, the following equation [14]

$$u_t = (u_2)^{-1/3}$$

has seven parameter group of classical symmetries, but it is not integrable in any known sense. However it seems to be true that any equation (3.15) of even order n can be reduced to a linear one if it possess at least one higher symmetry. That general conjecture had been proved in some cases by S.Svinolupov (see ([15]), ([16])). Moreover, it seems to be true that the scalar equation (3.15) is integrable (i.e. linearizable or integrable by the inverse transform method) if it has at least one higher symmetry.

The definition of symmetries given above is, as a matter of fact, a definition of *local symmetries*. There also exist the so called *nonlocal symmetries* which occur when the function G (2.12) depends on quadratures as well. Nonlocal symmetries are typical for the case of explicitly time dependent symmetries, and for the multidimensional case. We shall return to discussion of nonlocal symmetries later in this section.

From now on, we shall assume that equation

$$u_t = F(x, u, u_1, u_2, \dots, u_n), \quad n \geq 2. \tag{3.16}$$

and its symmetries do not depend on t explicitly. This assumption simplifies the theory drastically. A subspace of symmetries of order less or equal k shall be denoted by S_k. The dimension of factor space S_k/S_{k-1} is a natural measure for the number of symmetries of order k. It can be easily proved that there exist at most one higher local symmetry in any order. As the matter of fact it is very often happen that there are lacunas in the sequence of symmetries. For example the KDV equation

$$u_t = u_3 + uu_1 \tag{3.17}$$

has no symmetries of even order, meanwhile the following equation

$$u_t = u_5 + 5uu_3 + 5u_1u_2 + 5u^2u_1$$

does not possess symmetries of orders which are multiples of 2 or 3. The structure of Lie algebra of higher symmetries has been studied by V.V.Sokolov ([14]).

For applications one of the most interesting result of the theory is a set of necessary conditions for existence of higher symmetries which can be formulated in terms of *canonical conservation laws.* For instance, if equation (3.16) has a symmetry of order higher than n then the function

$$\rho = \left(\frac{\partial F}{\partial u_n}\right)^{-\frac{1}{n}} \tag{3.18}$$

is a density of a conservation law [7]. If the equation

$$u_t = u_n + f(x, t, u, u_1, u_2, \ldots, u_m), \ \partial f/\partial u_m \neq 0, \quad m < n \tag{3.19}$$

has a symmetry of order higher than $n+m$ then the function

$$\rho = \frac{\partial f}{\partial u_m} \tag{3.20}$$

must be a density of a conservation law. According to the definition of conservation laws the necessary conditions (3.18), (3.20) mean that there have to exist a function σ which depends on x, u and finite number of derivatives $u_1, u_2, \ldots$, such that $\mathrm{D}_t(\rho) = \mathrm{D}(\sigma)$ (2.14). How one can use these conditions?

Example 1. Consider the KDV type equations

$$u_t = u_3 + f(x, u, u_1). \tag{3.21}$$

If there exists a local symmetry of (3.21) of order higher than five, then in virtue of (3.20)

$$h \stackrel{\text{def}}{=} \mathrm{D}_t(f_{u_1}) = u_4 f_{u_1u_1} + u_3 f_{u_1u} + u_2 f_{u_1u_1} f_{u_1} + u_1 f_{u_1u_1} f_u + f_{u_1u_1} f_x + f_{u_1u} f \tag{3.22}$$

must be a total derivative. That imposes constraints on f. To find these constraints we shall transform (3.22) using the idea of integration by parts. For example

$$u_4 f_{u_1u_1} = -u_3 \mathrm{D}(f_{u_1u_1}) + \mathrm{D}(u_3 f_{u_1u_1}) \cong -u_3 \mathrm{D}(f_{u_1u_1})$$

(the notation $p \cong q$ for functions p, q means that $p - q \in \mathrm{Im}(\mathrm{D})$). We drop out all total derivatives since we are interested with obstacles for h to belong to $\mathrm{Im}(\mathrm{D})$. Thus (3.22) can be rewritten as

$$\begin{aligned} h \ \cong\ & \frac{1}{2} u_2^3 f_{u_1u_1u_1u_1} + \frac{3}{2} u_2^2 (u_1 f_{u_1u_1u_1u} + f_{u_1u_1u_1x}) + u_2 (u_1^2 f_{u_1u_1uu} + 2u_1 f_{u_1u_1ux}) \\ & - u_1 f_{u_1uu} + f_{u_1u_1xx} - f_{u_1ux} + f_{u_1u_1} f_{u_1} + u_1 f_{u_1u_1} f_u + f_{u_1u_1} f_x + f_{u_1u} f. \end{aligned} \tag{3.23}$$

It is obvious that (3.23) can be a total derivative only if

$$f_{u_1u_1u_1u_1} = 0 \quad \text{and} \quad u_1 f_{u_1u_1u_1u} + f_{u_1u_1u_1x} = 0.$$

In other words the function f should be a polynomial in u_1 of order 3, whose leading coefficient is a constant

$$f(x,u,u_1) = \alpha u_1^3 + a(x,u)u_1^2 + b(x,u)u_1 + c(x,u), \quad \alpha \in \mathbb{C}\ .$$

The further analysis of the condition is rather straightforward. Nevertheless we stop here and suggest to the interested reader to complete it and find all restrictions on function f which follows from the condition $h \in \mathrm{Im}(\mathrm{D})$.

Example 2. If the equation

$$u_t = a(u)u_n + f(u, u_1, u_2, \ldots, u_m), \quad m < n$$

has a higher symmetry, then, according to (3.18), the function $\rho = 1/\sqrt[n]{a(x,u)}$ defines a conservation law $\mathrm{D}_t(\rho) = \mathrm{D}(\sigma)$. The change of variables

$$dt' = dt, \quad dx' = \rho dx + \sigma dt, \quad u(t,x) = u'(t',x')$$

gives the equation

$$u'_{t'} = u'_n + f'(u', \ldots, u'_{n-1})$$

where $f' = f - \rho^{-1}\sigma u_1$ and all variables $u, u_1, \ldots$ are expressed through $u', u'_1 \ldots$ via invertible transformation:

$$u' = u, \quad u'_1 = \rho^{-1}u_1, \quad \ldots, \quad u'_k = (\rho^{-1}\mathrm{D})^k u, \quad \ldots.$$

Densities (3.18) and (3.20) are just two examples of so called *canonical densities* in terms of which we formulate necessary conditions of existence of higher symmetries. The canonical densities can be computed for any evolution equation (3.16) regardless whether this equation is integrable not. Only a few fist canonical densities have a reasonably short form (see Appendix). Despite the fact that the algorithm is rather involved, in the scalar case even a PC is capable to perform this job. The second (also very algorithmic) problem is to verify the fact that for a certain equation these densities indeed defines conservation laws (i.e. the time derivative of a density is a total space derivative).

The following equations have been studied in great detail:

1. The general second order equation [15]

$$u_t = F(x, u, u_1, u_2) \tag{3.24}$$

2. Third order equations of the form, [11]

$$u_t = A(x, u, u_1, u_2)u_3 + B(x, u, u_1, u_2). \tag{3.25}$$

3. Fourth order equations of form [16]

$$u_t = u_4 + F(x, u, u_1, u_2, u_3). \tag{3.26}$$

4. Fifth order equations of the form [16], [17], [11]

$$u_t = u_5 + F(x, u, u_1, u_2, u_3, u_4). \tag{3.27}$$

Of course the conditions discussed above are only necessary ones, but in all listed cases it turns out that the first few conditions are already sufficient for having higher symmetry. Thus canonical conservation laws give us a very fine test and even a criteria for having higher symmetries. Canonical densities for equations (3.24), (3.25) and (3.27) which can be used for testing integrability as well as for classification, are given in the Appendix.

The concept of canonical densities unifies the two basic notions of higher symmetries and higher conservation laws and seems to be more abstract and basic than local symmetries and conservation laws. Higher conservation laws impose additional constraints. But we delay discussion of that question to the next section. In practice, the few first conditions are very crucial; if a given equation satisfies them in a nontrivial way (i.e. the corresponding canonical conservation laws are nontrivial), it means that this equation is very close to an integrable one (or integrable itself) and higher canonical densities will give higher conservation laws. If an equation satisfies these conditions but the corresponding conservation laws are all trivial or of zero order, then it often means that the equation can be linearized (via a composition of invertible and Cole-Hopf type transformations)[11].

To illustrate the last statement we formulate the following theorem of S.Svinolupov [15]:

For a second order equation (3.24) the following statements are equivalent:

Equation (3.24) has conservation laws (may be trivial) with densities (7.63), (7.64), (7.65).

Equation (3.24) has at least one higher symmetry.

Equation (3.24) has infinity many higher symmetries.

Via invertible transformation equation (3.24) can be reduced to one of the following:

$$u_t = u_2 + q(x)u, \tag{3.28}$$

$$u_t = u_2 + 2uu_1 + q(x), \tag{3.29}$$

$$u_t = \mathrm{D}(u^{-2}u_1 + \alpha xu + \beta u), \tag{3.30}$$

$$u_t = \mathrm{D}(u^{-2}u_1 - 2x), \tag{3.31}$$

where α, β are constants and $q(x)$ is a function of x.

Equations (3.29), (3.30), (3.31) can be linearized by a transformation of Cole-Hopf type [15], [11]. But far not all linearizable equations can be put in the above form via invertible transformation.

Example 3. Equation [18]

$$u_t = u^2u_2 + x^2u_1 - 3xu \tag{3.32}$$

does not have higher local symmetries, since the first canonical density $\rho = 1/u$ does not provide a local conservation law:

$$(1/u)_t = \mathrm{D}(-u_1 + x^2/u) + x/u \notin \mathrm{Im}(\mathrm{D}),$$

but can be linearized via the following transformation

$$x = -2\mathrm{D}^2(\log w), \qquad u = 2\mathrm{D}^3(\log w).$$

All higher symmetries of equation (3.32) are nonlocal! More precisely, higher symmetries depend on x, u derivatives u_k and $\mathrm{D}^{-1}(x/u)$. Despite this, one can construct infinitely many equations of this type; there is a uniform way to make appropriate (nonlocal) extension of variables, which allows one to plug all of them in the frame of Svinolupov's theorem [11].

Nevertheless a straightforward approach fails because of nonlocalities. There exist more sophisticated examples, when the "order" of nonlocality grouping up with the order of symmetry. The well known KDV equation (3.17) besides of higher local symmetries has a sequence of nonlocal ones, which can be obtained from $f = 3tu_t + xu_x + 2u$ (the Gallilean boost) by application of the recursion operator $\Lambda = D^2 + 4uD + 2u_1 D^{-1}$ and its powers [19]. These time dependent symmetries and the close notion of a master symmetry are well known in the theory of integrable equations (see for example [19]). The problem of nonlocality is of the main unsolved problems in the symmetry approach. The future investigation in multidimension will go through the development of the concept of nonlocality.

4 Systems of two equations of second order

In this section we consider systems of two evolution equations of second order

$$\mathbf{u}_t = A(\mathbf{u}) \cdot \mathbf{u}_2 + \Phi(\mathbf{u}, \mathbf{u}_1), \quad \det(A(\mathbf{u})) \neq 0, \quad \mathbf{u} = (u^1, u^2). \tag{4.33}$$

A symmetry $\mathbf{G}$ of (4.33) is called *a nondegenerate* one if

$$\det\left(\frac{\partial \mathbf{G}}{\partial \mathbf{u}_m}\right) \neq 0, \qquad m = \mathrm{ord}(\mathbf{G}).$$

Compare to the scalar case, for system of equations the structure of higher local symmetries becomes much more sophisticated and it is natural to impose extra conditions or restrictions on the rhs of the system. Even in the case of simple triangular systems

$$\begin{aligned} u_t &= u_2 + f(x, u, v, u_1, v_1), \\ v_t &= \lambda v_2, \qquad \lambda \in \mathbb{C}, \end{aligned}$$

properties of higher symmetries are not known well. Extra requirements of existence of higher conservation laws and a nontriangular coupling, which seems to be quite natural for applicability of the system, simplify the problem a lot and allow to solve it with the same completeness as in the scalar case.

Conservation laws form a linear space, but it is natural to identify two conservation laws whose difference is a trivial conservation law ($\rho_1 - \rho_2 \in \mathrm{Im}(\mathrm{D})$), $\rho_1 \cong \rho_2$. The definition of the order of a conservation law should be invariant with respect to adding of total derivatives. By *the order of a conservation law with density* ρ we shall mean the order of the differential operator

$$R = \left(\frac{\delta \rho}{\delta u}\right)_*, \tag{4.34}$$

where the variational derivative is defined as

$$\frac{\delta}{\delta u} = \sum_{k=0}(-1)^k \mathrm{D}^k \frac{\partial}{\partial u_k}. \tag{4.35}$$

. A conservation law is called nondegenerate if the leading coefficient of the operator $R = r_m \mathrm{D}^m + r_{m-1}\mathrm{D}^{m-1} + \cdots$, $m = \mathrm{ord}\rho$ (4.34) is a nonsingular matrix ($\det r_m \neq 0$).

Existence of nondegenerated higher conservation laws implies existence of canonical conservation laws. But conservation laws impose more constraints on the rhs of equations than symmetries. For instance, the trace of the matrix $A(\mathbf{u})$ must be equal to zero

$$\mathrm{trace}(A(\mathbf{u})) = 0, \tag{4.36}$$

otherwise equation (4.33) does not have a conservation law of order higher than one [20]. Thus, eigenvalues of the matrix $A(\mathbf{u})$ are $\lambda(\mathbf{u}), -\lambda(\mathbf{u})$. Like the scalar case, it can be shown that $\rho = \lambda^{-1/2}$ has to be a conserved density. Via a composition of a space coordinate transformation defined by (compare *Example 2*)

$$dx' = \rho_{-1}dx - \sigma_{-1}dt, \qquad \mathbf{u}' = \mathbf{u}$$

and a point transformation one can reduce equation (4.33) into the form [20]

$$\begin{aligned} u_t &= u_2 + f(u, v, u_1, v_1), \\ -v_t &= v_2 + g(u, v, u_1, v_1). \end{aligned} \tag{4.37}$$

There is a regular way to obtain constraints like (4.36) [21], [11]. Higher symmetries imply canonical conservation laws, and higher conservation laws imply the existence of *canonical potentials.* For instance, equations (4.37) will have a conservation law of order higher than one only in the case that the function

$$\omega_0 = -(\partial f/\partial u_1 + \partial g/\partial v_1)/2$$

is a total derivative ($\omega_0 \in \mathrm{Im}(\mathrm{D})$). An existence of a conservation law of order $N > 1$ for equations (4.37) implies that functions ω_k, $k = 0, \ldots N-2$ are total derivatives $\omega_k = \mathrm{D}(\phi_k)$. There is an algorithmic way to compute this sequence of canonical potentials [21], [11]. The canonical densities and potentials for (4.37) are given in Appendix.

Sure, conditions $\omega_k, \mathrm{D}_t(\rho_k) \in \mathrm{Im}(\mathrm{D})$ are only necessary for the existence of higher symmetry and conservation laws, but a tiresome analysis has shown that if they are satisfied for $k = 0, 1, 2, 3, 4$ then the system (4.37) is indeed integrable (i.e. possesses infinitely many higher symmetries and conservation laws)[21], [11]! So we have more than a test, we have necessary and sufficient conditions of integrability (i.e. of having local higher conservation laws) of nontriangular equations (4.37).

A first result that follows from these conditions is a general statement about functions f, g in rhs of (4.37). Namely, these functions have to be polynomial in first derivatives u_1, v_1, or more precisely, they must be of the form:

$$\begin{aligned} f &= a u_1^2 v_1 + b_u u_1^2 + c_v u_1 v_1 + r u_1 + h v_1^2 + p v_1 + s, \\ g &= -a v_1^2 u_1 + b_v v_1^2 + c_u u_1 v_1 - r v_1 + k u_1^2 + q u_1 + n, \end{aligned} \tag{4.38}$$

where $a, b, c, r, h, k, p, q, s, n$ are functions of u, v only. Thus, in many cases, it is sufficient just to look at the equation to be sure that it is not integrable. Like the scalar case, it is often enough to check only a few first simple conditions. For instance, if $a \neq 0$ it is sufficient to check conditions with $k = 0, 1$; the conditions with $k = 4$ one needs to verify only in a very special cases.

Example 4. For system of equations

$$u_t = u_2 + f(u, v), \quad -v_t = v_2 + g(u, v), \tag{4.39}$$

$\rho_1 = \omega_1 = 0$ and the first nontrivial canonical density and potential are (cf. (7.66)– (7.68)):

$$\rho_2 = f_u + g_v, \quad \omega_2 = f_u - g_v.$$

It follows from $\omega_2 \in \mathrm{Im(D)}$ that $\partial f / \partial u = \partial g / \partial v$ that $f = z_v$, $g = z_u$. Now the canonical densities (7.69) – (7.73) can be put in the form

$$\rho_1 = 2 z_{uv}, \quad \rho_2 = 2(z_{uuv} u_1 - z_{vvu} v_1, \quad \rho_3 = \sigma_3 + z_{uv}^2 - 4 z_{uu} z_{vv}.$$

If $z_{uu} = 0$ (or $z_{vv} = 0$) the system of equations (4.39) becomes triangular. Assuming $z_{uu} z_{vv} \neq 0$ one can verify that conditions $\mathrm{D}_t(\rho_k) \in \mathrm{Im(D)}$, $k = 1, 2, 3$ are equivalent to the following constraints on z:

$$z_{vuuu} = z_{uvvv} = z_{uuuu} = zvvvv = 0, \quad z_v (z_{uu} z_{vv})_u = z_u (z_{uu} z_{vv})_v.$$

If $z_{uuvv} \neq 0$ it leads to the Nonlinear Schrödinger equation

$$u_t = u_2 + u^2 v, \quad -v_t = v_2 + v^2 u, \tag{4.40}$$

otherwise $(z_{uuvv} = 0)$ to the Boussinesq equation

$$u_t = u_2 + (u + v)^2, \quad -v_t = v_2 + (v + u)^2. \tag{4.41}$$

Now we are going to discuss briefly equations (4.33), which do not possess higher conservation laws but still have higher symmetries. Among of these equations there are a vector generalization of the Burgers equation

$$\mathbf{u}_t = \mathbf{u}_2 + \mathbf{F}(\mathbf{u}, \mathbf{u}_1), \quad \mathbf{u} = (u^1, \ldots, u^N) \tag{4.42}$$

and equations with distinct eigenvalues of the matrix $A(\mathbf{u})$. The problem of description and classification of integrable equations (4.42) has been solved[2] in [22]. Among of integrable systems of two equations (4.42) there is only one nontriangular case

$$\begin{aligned} u_t &= u_2 - 2vu_1 - u^3 - uv^2, \\ v_t &= v_2 - 2uu_1 - 4vv_1. \end{aligned}$$

The above equation can be linearized via a generalized Cole-Hopf transformation

$$u = U_1/\sqrt{V - U^2}, \quad v = (2UU_1 - V_1)/(V - U^2),$$

which turns it in two uncoupled heat equations.

An example of the integrable equation (4.33) with distinct eigenvalues of $A(\mathbf{u})$ has been found in [23]:

$$\begin{aligned} u_t &= \lambda_1 u_2 + 2\lambda_1(u+v)u_1 + (\lambda_1 - \lambda_2)uv_1 + (\lambda_1 - \lambda_2)(u+v)uv, \\ v_t &= \lambda_2 v_2 + 2\lambda_2(u+v)v_1 + (\lambda_2 - \lambda_1)vu_1 + (\lambda_2 - \lambda_1)(u+v)uv. \end{aligned}$$

The corresponding Cole-Hopf type transformation

$$u = U_1/(U+V), \qquad v = V_1/(U+V)$$

turns it to $U_t = \lambda_1 U_2$, $V_t = \lambda_2 V_2$.

In the most general form, the problem of classification of all equations (4.33) with higher symmetries and without higher conservation laws has not been solved. It seems very unlikely to find any interesting result on this way, therefore nobody wants to waist a time, in spite of the fact all difficulties are of a technical character.

5 Chain equations

We consider here infinite-dimensional dynamical systems

$$\mathrm{D}_x(\mathbf{u}_n) = F(\mathbf{u}_{n+1}, \mathbf{u}_n, \mathbf{u}_{n-1}), n = o, \pm 1, \pm 2, ... \tag{5.43}$$

which are invariant under shift $n \mapsto n+1$. In that case the role of higher symmetry play an additional dynamical system

$$\mathrm{D}_t \mathbf{u}_n = f(\mathbf{u}_{n+m}, ..., \mathbf{u}_{n-m}) \tag{5.44}$$

satisfying the commutation relation $\mathrm{D}_x\mathrm{D}_t = \mathrm{D}_t\mathrm{D}_x$ of corresponding to (5.43), (5.44) differentiations. On first glance the classification problems of PDE's and of dynamical systems (5.43) appeared unrelated. The first important hint for understanding how they are actually related comes from the remark that by some conditions one can express shifts in terms of derivatives: use (5.43) to put these expressions in (5.44) and then write down the last one in a differential

[2] With one extra but natural assumption on the form of higher symmetries.

form. For example the pair

$$\mathrm{D}_x(u_n) = u_{n+1}, \mathrm{D}_t u_n = u_{n+2}$$

corresponds evidently to the PDE $u_\tau = u_{xx}$.

In the papers [24],[25] one can find the lists of dynamical systems

$$\mathrm{D}_x(u_n) = F(u_{n+1}, u_n, u_{n-1}) \tag{5.45}$$

and

$$\mathrm{D}_x^2(u_n) = F(\mathrm{D}_x(u_n), u_{n+1}, u_n, u_{n-1}) \tag{5.46}$$

which can be considered as generalization of well known discrete KDV

$$\mathrm{D}_x(u_n) = u_n(u_{n+1} - u_{n-1})$$

and Toda's chain

$$\mathrm{D}_x^2(u_n) = \exp(u_{n+1} - u_n) - \exp(u_n - u_{n-1})$$

correspondingly. The listed chains posses surely higher symmetries and in both cases (5.46), (5.45) after introducing

$$u = u_0, \qquad v = u_1$$

one can express all dynamical variables u_n in terms of derivatives of fixed two u and v.The coupled PDE's systems for u and v obtained from symmetries of dynamical systems (5.46), (5.45) were considered in a paper ([26]). It was discovered that the two classification problems connected with (4.37) and with (5.46), (5.45) are in some sense are equivalent. Roughly speaking, the exhaustive list of PDE's of Schrödinger type equation can be obtained from lists of equations (5.46), (5.45) coupled with their symmetries.

That strange coincidence needs some explanation. Let us compare the definitions on which the two classification problems are based. As in the case of Schrödinger type equations, for equations (5.46), (5.45) we chose the following analog of the notion of local conservation law (see Section 1). We shall call a function of the dynamical variables $p(\mathbf{u}_0, \mathbf{u}_{\pm 1}, \ldots)$ a density of local conservation law for (5.43) if there exist a function q depending on finite number of dynamical variables such that

$$\mathrm{D}_x(p) = q_1 - q$$

where q_1 is obtained from q by shift $\mathbf{u}_k \mapsto \mathbf{u}_{k+1}$. As a general rule the function p is also a conserved density for symmetry (5.44) i.e.

$$\mathrm{D}_t p = \hat{q}_1 - \hat{q}.$$

Cross differentiation of above relations leads us to local conservation law with density q for PDE which corresponds to the pair (5.43), (5.44) (see([26])).In this way conserved densities for Schrödinger type equations may be obtained from "conservation laws" of infinite-dimensional dynamical systems (5.46), (5.45).

The chains (5.46), (5.45) with a lot of conserved densities have great interest in their own right; also after imposing periodic conditions they become integrable finite dimensional systems. These integrable (by Liouville's theorem) periodic chains can be used for an investigation of finite-gap solutions of the corresponding PDE's ([26]). The discussion above explains the link between infinite dimensional dynamical systems and PDE's and allows one to understand in more details the nature of integrability of PDE's.

The invariance under a shift in the chains under consideration, corresponds to invariance of the related PDE with respect to some discrete group of transformations which one used to call a group of auto-Bäcklund transformations. Many ambiguities in the theory of Bäcklund transformations disappear if one can formulate properties of these in the terms of properties of corresponding infinite dimensional dynamical system (5.44). The problem of building up infinite dimensional dynamical models starting with some description of auto-Bäcklund transformations was considered in ([27]), ([28]), ([29]).A well known Bäcklund transformation for KdV equation leads one to following infinite system of ODE's

$$\mathrm{D}_x(u_n + u_{n+1}) = u_n^2 - u_{n+1}^2 + \lambda_n - \lambda_{n+1} \tag{5.47}$$

A role of the symmetry (5.44) for equation (5.47) is played by the MKdV equation

$$\mathrm{D}_t u_n = \mathrm{D}_x^3 u_n + 6(\lambda_n + u_n^2)\mathrm{D}_x u_n \tag{5.48}$$

See the above cited paper [26] in which it is proven also that all finite-gap solutions of KdV equation can be obtained from the periodic case of (5.47), (5.48). Analogous to (5.47) chain representations for groups of auto-Bäcklund transformations were obtained for another KdV-type equations in the paper [30].

In the problem of building up an exhaustive list of integrable chains of the form (5.46), (5.45) one have to overcome same difficulties that were described in the previous section. First of all in accordance with our general scheme an appropriate necessary (for existence of conservation laws) conditions must be formulated. In the case of (5.46) these conditions can be written down in following way

$$\mathrm{D}_x(\rho^k) = \sigma_1^k - \sigma^k, \ k = 1, 2, 3 \tag{5.49}$$

$$\omega^k = \phi_1^k - \phi^k, \ k = 1, 2 \tag{5.50}$$

The formulae for "canonical densities and potentials" in terms of right hand side of (5.46) are given below

$$\rho^1 = \log\left(\frac{\partial F}{\partial u_{n+1}}\right), \quad \rho^2 = 2\sigma^1 + \frac{\partial F}{\partial(\mathrm{D}_x u_n)},$$

$$\rho^3 = 2\sigma^2 - \frac{1}{2}\mathrm{D}_x\left(\frac{\partial F}{\partial(\mathrm{D}_x u_n)}\right) + \frac{1}{4}\left[\left(\frac{\partial F}{\partial(\mathrm{D}_x u_n)}\right)^2 + (\rho^2)^2\right] + \frac{\partial F}{\partial u_n},$$

$$\omega^1 = \log\left(\frac{\partial F}{\partial u_{n+1}} / \frac{\partial F}{\partial u_{n-1}}\right), \quad \omega^2 = D_x(\phi^1) + \frac{\partial F}{\partial (D_x u_n)}.$$

The second and harder part of the work is an investigation of the overdetermined system for the function F obtained from (5.49), (5.50). The main point is that the indicated above necessary conditions had been chosen in such way that at the end they also proved to be sufficient conditions. Analogous conditions for the case (5.45) can be found in ([30]).

As in the case of systems of two equations (see previous section) the general problem of classification of all equations with higher symmetries seems to be difficult. In particular one unsolved problem is to prove that a chain of the form

$$D_x(u_n) = F(u_n, u_{n-1})$$

can be reduced to the linear one iff it posses higher symmetry.

6 Vector equations with higher symmetries

It is absolutely impossible to describe all integrable systems of equations in the whole generality. Fortunately in applications monstrous equations appear very seldom, if any. Therefore it is quite natural to fix an algebraic form of equations and link the problem of listing equations with higher symmetries and conservation laws with an algebraic one.

Consider vector generalizations of classical integrable equations

$\mathbf{u}_t = \mathbf{u}_{xxx} + \langle \mathbf{u}, \mathbf{u}_x \rangle,$ – vector Korteweg de Vries equation

$$\begin{aligned} i\mathbf{u}_t &= \mathbf{u}_{xx} + \langle \mathbf{u}, \mathbf{v}, \mathbf{u} \rangle \\ -i\mathbf{v}_t &= \mathbf{v}_{xx} + \langle \mathbf{v}, \mathbf{u}, \mathbf{v} \rangle^{\dagger} \end{aligned},$$ – vector Nonlinear Schrödinger equation

$\mathbf{u}_{xy} = \hat{A} \exp \mathbf{u},$ – vector Liouville equation

where $\mathbf{u}, \mathbf{v}$ are vectors belonging to vector spaces $V, \tilde{V}$ respectively, $\hat{A}$ is a constant matrix and $\langle ... \rangle$, $\langle ... \rangle^{\dagger}$ denotes bilinear or tri-linear mappings of appropriate vector spaces or, in other words, a kind of products. What constraints on these mappings do appear if the system of equations possess higher symmetries and conservation laws? These problems have been solved in [31]), [32], [33] and the result can be formulated in pure algebraic way.

A binary multiplication is defined by the products of basis vectors $\mathbf{e}_i$, $i = 1, ..., N$

$$\langle \mathbf{e}_j, \mathbf{e}_k \rangle = a^i_{jk} \mathbf{e}_i, \tag{6.51}$$

where a usual summation rule is assumed. The tri-linear mapping related to (6) can be defined as

$$\langle \mathbf{e}_j, \tilde{\mathbf{e}}_k, \mathbf{e}_m \rangle = a^i_{jkm} \mathbf{e}_i, \quad \langle \tilde{\mathbf{e}}_j, \mathbf{e}_k, \tilde{\mathbf{e}}_m \rangle^{\dagger} = \tilde{a}^i_{jkm} \tilde{\mathbf{e}}_i, \tag{6.52}$$

vectors $\mathbf{e}_k, \tilde{\mathbf{e}}_k$ are basis vectors of spaces V and $\tilde{V}$ respectively. Obvious symmetry of the mappings

$$\langle \mathbf{e}_j, \tilde{\mathbf{e}}_k, \mathbf{e}_m \rangle = \langle \mathbf{e}_m, \tilde{\mathbf{e}}_k, \mathbf{e}_j \rangle, \quad \langle \tilde{\mathbf{e}}_j, \mathbf{e}_k, \tilde{\mathbf{e}}_m \rangle^\dagger = \langle \tilde{\mathbf{e}}_m, \mathbf{e}_k, \tilde{\mathbf{e}}_j \rangle^\dagger \tag{6.53}$$

is assumed without loss of generality.

For irreducible systems of vector KDV equations, i.e. equations that can not be put in a triangular form via a special choice of the basis vectors,

$$u^i_t = u^i_{xxx} + a^i_{jk} u^j u^k_x \tag{6.54}$$

it can be proved that the matrix coefficient $K^i_r u^r_n$ of a higher symmetry

$$\mathrm{D}_t u^i = K^i_r u^r_n + f^i(\mathbf{u}, \mathbf{u}_1, ..., \mathbf{u}_{n-1}), \quad n \geq 1 \tag{6.55}$$

satisfy the following conditions

$$\text{i) } \det(K^i_r) \neq 0, \quad \text{ii) all eigenvalues of } (K^i_r) \text{ are coincide}$$

(a similar statement can be proved for vector nonlinear Schrödinger equation as well). Moreover, for general system (6.54) (not necessary irreducible)

$$K^i_r a^r_{jk} = a^i_{jr} K^r_k.$$

A criteria of integrability of (6.54) can be formulated in a nice algebraic form. Namely, the vector KDV has higher symmetries if and only if the product (6.51) satisfies the following conditions:

$$\langle \mathbf{uv} \rangle = \langle \mathbf{vu} \rangle, \quad \langle \langle \mathbf{u}^2 \mathbf{v} \rangle \mathbf{u} \rangle = \langle \mathbf{u}^2 \langle \mathbf{vu} \rangle \rangle. \tag{6.56}$$

A linear space with the product (6.56) is called a Jordan algebra [34], [35]. An example of a Jordan algebra is the usual matrix algebra if the product is defined in the following way

$$\langle \mathbf{uv} \rangle = \mathbf{u} \cdot \mathbf{v} + \mathbf{v} \cdot \mathbf{u}.$$

Irreducible equations correspond to simple Jordan algebras (i.e. algebras without any nontrivial ideals).

The vector generalization of the Schrödinger equation

$$i\mathrm{D}_t u^i = u^i_{xx} + a^i_{jkm} u^j v^k u^m, \quad -i\mathrm{D}_t v^i = v^i_{xx} + \tilde{a}^i_{jkm} v^j u^k v^m \tag{6.57}$$

has higher symmetries if and only if the tri-linear mapping (6.52) satisfies the following identities

$$\begin{aligned} \langle \mathbf{u}\tilde{\mathbf{v}} \langle \mathbf{v}\tilde{\mathbf{u}}\mathbf{w} \rangle \rangle + \langle \mathbf{v} \langle \tilde{\mathbf{v}}\mathbf{u}\tilde{\mathbf{u}} \rangle^\dagger \mathbf{w} \rangle &= \langle \mathbf{v}\tilde{\mathbf{u}} \langle \mathbf{u}\tilde{\mathbf{v}}\mathbf{w} \rangle \rangle + \langle \mathbf{w}\tilde{\mathbf{u}} \langle \mathbf{u}\tilde{\mathbf{v}}\mathbf{v} \rangle \rangle \\ \langle \tilde{\mathbf{u}}\mathbf{v} \langle \tilde{\mathbf{v}}\mathbf{u}\tilde{\mathbf{w}} \rangle^\dagger \rangle^\dagger + \langle \tilde{\mathbf{v}} \langle \mathbf{v}\tilde{\mathbf{u}}\mathbf{u} \rangle \tilde{\mathbf{w}} \rangle^\dagger &= \langle \tilde{\mathbf{v}}\mathbf{u} \langle \tilde{\mathbf{u}}\mathbf{v}\tilde{\mathbf{w}} \rangle^\dagger \rangle^\dagger + \langle \tilde{\mathbf{w}}\mathbf{u} \langle \tilde{\mathbf{u}}\mathbf{v}\tilde{\mathbf{v}} \rangle^\dagger \rangle^\dagger. \end{aligned} \tag{6.58}$$

A pair of vector spaces with product satisfying (6.58) algebraists call as Jordan's pair [36]. An example of Jordan pair is provided by a graded Lie algebra

$$G = G_1 + G_0 + G_{-1}, \quad [G_i, G_j] \subset G_{i+j}.$$

In this case $V = G_1$, $\tilde{V} = G_{-1}$ and

$$\langle \mathbf{u}\tilde{\mathbf{v}}\mathbf{w}\rangle = [[\mathbf{u}, \tilde{\mathbf{v}}], \mathbf{w}], \quad \langle \tilde{\mathbf{u}}\mathbf{v}\tilde{\mathbf{w}}\rangle^{\dagger} = [[\tilde{\mathbf{u}}, \mathbf{v}], \tilde{\mathbf{w}}].$$

Irreducible systems correspond to the simple (without ideals) Jordan pairs.

This algebraic formulation allows one to use the results of the well developed and reach theory of Jordan algebras [34], [35], [36].

An open problem is to find an algebraic description of lower order inhomogeneous terms which can be added to equation (6.54) and (6.57).

A vector generalization of the Liouville equation can be written in following way

$$D_x D_y u^i = a_{i1} e^{u_1} + \dots + a_{ir} e^{u_r}, \quad \det(a_{ij}) \neq 0 \tag{6.59}$$

By simple shifts $u^i \mapsto u^i + c^i$, c^i is a const, one can make

$$a_{ii} = 2 \quad \text{or} \quad 0. \tag{6.60}$$

As in previous sections before we shall use the notations

$$\mathbf{u}_n = D_x^n \mathbf{u}, \quad u_n^i = D_x^n u^i.$$

The existence of "first integrals" is an important characteristic feature of the scalar Liouville equation $D_x D_y u = 2e^u$ [9]. Namely, $w = u_2 - (u_1)^2$ is a first integral because

$$D_y(w) = D_y D_x^2 u - 2u_1 D_x D_y u = 0.$$

One can straightforwardly check that functions

$$f(x, u, u_1, u_2, u_3, \dots) = (D_x + u_1) W(x, w, D_x(w), D_x^2(w), \dots) \tag{6.61}$$

define higher symmetries of the Liouville equation where W is an arbitrary function.

The main result of paper ([33]) states that for vector irreducible systems (6.59), (6.60)

$$D_x D_y \mathbf{u} = \mathbf{F}(\mathbf{u})$$

one has a maximal set of solutions $w = w(\mathbf{u}_1, \mathbf{u}_2, \dots)$ of equation

$$D_y(w) = (\mathbf{F}\frac{\partial}{\partial \mathbf{u}_1} + D_x(\mathbf{F})\frac{\partial}{\partial \mathbf{u}_2} + \dots)(w) = 0 \tag{6.62}$$

only in the case if matrix (a_{ij}) is a Cartan matrix of a simple Lie algebra. Moreover this simple Lie algebra and the Lie algebra of first order PDE's (6.62) are connected in a canonical way: the latter is the nilpotent subalgebra of the former in the Weyl decomposition.

Integrability in quadratures of vector Liouville equation (6.59) related to the Cartan matrices has been discovered earlier in [37]. For particular solutions depending on $x+y$ only equation (6.59) becomes a finite dimensional dynamical system, its integrability (Painlevé analysis) has been discussed in [38]

7 Conclusion

An existence of higher order local symmetries can be considered as the rigorous definition of integrability of PDE's with two independent variables. The mathematical theory developed on that solid base leads to necessary and sufficient conditions of integrability which proved to be strikingly efficient in some cases. For example the integrability conditions gave us a possibility to describe explicitly the whole class of integrable systems of the form (4.37). Despite the fact that the approach based on the notion of L-A pair proved to be very successful in "fishing" of integrable equations, it seems to be inappropriate for the description of all equations of a given order. Classification of integrable chains discussed in Section 4 provides another successful example of the application of the symmetry approach. The direct and explicit link between infinite dimensional dynamical systems and PDE's allows one to understand deeper the nature of integrability of PDE's.

On the other hand the more flexible and more general notion of L-A pairs, though not very well defined yet, works almost without exception. As a matter of fact the vector generalizations of KDV and NLS equations were described originally in the terms of L-A pairs in the papers [39], [40]. In cases like these, the symmetry approach gives an independent and more thorough verification of previously obtained lists of integrable equations[3]. Also, what is more important, it gives immediate answer to the question whether a given equation has higher conservation laws (symmetries) or not.

Appendix

i) Canonical densities for equations of the form (3.24) are:

$$\rho_{-1} = F_2^{-1/2}, \tag{7.63}$$

$$\rho_0 = F_1/F_2 - \sigma_{-1}F_2^{-1/2}, \tag{7.64}$$

$$\rho_1 = F_2^{-1/2}\left(\frac{1}{2}F_0 + \frac{1}{4}\sigma_0 + \frac{1}{8}\sigma_{-1}^2 - \frac{1}{32}F_2^{-1}(2F_1 - (\mathrm{D}(F_2))^2)\right), \tag{7.65}$$

where

$$F_2 = \frac{\partial F}{\partial u_2}, \quad F_1 = \frac{\partial F}{\partial u_1}, \quad F_0 = \frac{\partial F}{\partial u}$$

and σ_{-1}, σ_0 can be recursively found from the conditions $\mathrm{D}_t(\rho_k) = \mathrm{D}(\sigma_k)$.

ii) Canonical densities for equations of the form (3.25):

$$\rho_{-1} = F_3^{-1/3},$$

$$\rho_0 = F_2/F_3,$$

[3] A correspondence between simple Lie algebras and L-A pairs used in [39], [40] proved to be noncanonical and a lot of systems connected with simple algebras in fact are triangular.

$$
\begin{aligned}
\rho_1 &= \rho_{-1}\sigma_{-1} - \rho_{-1}^2 F_1 + F_2\sigma_{-1}\mathrm{D}(\sigma_{-1}) + \sigma_{-1}^5 F_2^5/3 \\
&\quad \sigma_{-1}^3(\mathrm{D}(\sigma_1))^2 + \mathrm{D}(\sigma_{-1}^2 F_2 + 2\sigma_{-1}^{-2}\mathrm{D}(\sigma_{-1})), \\
\rho_2 &= \rho_{-1}\sigma_0/3 + 2\sigma_{-1}^7 F_2^3/27 + F_2^2\sigma_{-1}^3\mathrm{D}(\sigma_1)/3 \\
&\quad -\sigma_{-1}^4 F_1 F_2/3 - F_2(\mathrm{D}(\sigma_{-1}))^2/\sigma_{-1} \\
&\quad +F_0\sigma_{-1} - F_1\mathrm{D}(\sigma_{-1}) - F_2\mathrm{D}^2(\sigma_{-1})/3, \\
\rho_3 &= \rho_{-1}\sigma_1 - \rho_1\sigma_{-1},
\end{aligned}
$$

where $F_0 = \partial F/\partial u,\ F_k = \partial F/\partial u_k$.

iii) The first six canonical densities for equations of the form (3.27) are:

$$
\begin{aligned}
\rho_0 &= F_4, \\
\rho_1 &= 2F_4^2 - 5F_3, \\
\rho_2 &= 4F_4^3 - 15F_3F_4 + 25F_2, \\
\rho_3 &= (\mathrm{D}(f_4))^2 - F_3\mathrm{D}(F_4) + 7F_4^4/25 - 5F_1 \\
&\quad -7F_3F_4^2/5 + 2F_4F_2 + F_3^2, \\
\rho_4 &= 4F_4(\mathrm{D}(f_4))^2/5 - \mathrm{D}(F_3)\mathrm{D}(F_4) + 3\mathrm{D}(F_3)F_4^2/10 \\
&\quad +44F_4^5/625 - 11F_3F_4^3/25 + 3F_4^2F_2/5 + \mathrm{D}(F_4)F_2 \\
&\quad -F_1F_4 - F_3F_2 + 5F_0 + \sigma_0, \\
\rho_5 &= \sigma_1.
\end{aligned}
$$

Here as above we use notation $F_0 = \partial F/\partial u,\ F_k = \partial F/\partial u_k$.

iv) For system of equations (4.37) it is convenient to present these canonical potentials (ω_k) together with densities of canonical conservation laws (ρ_k):

$$\omega_0 = -(\partial f/\partial u_1 + \partial g/\partial v_1)/2, \tag{7.66}$$

$$\rho_0 = (\partial f/\partial u_1 - \partial g/\partial v_1)/2, \tag{7.67}$$

$$\omega_1 = \mathrm{D}_t(\phi_0) - \phi_0\rho_0 - (\partial f/\partial v_1)(\partial g/\partial u_1) + \partial f/\partial u - \partial g/\partial v, \tag{7.68}$$

$$
\begin{aligned}
\rho_1 &= \sigma_0 - [(\partial f/\partial u_1)^2 + (\partial g/\partial v_1)^2]/4 - (\partial f/\partial v_1)(\partial g/\partial u_1) \\
&\quad +\partial f/\partial u + \partial g/\partial v,
\end{aligned} \tag{7.69}
$$

$$
\begin{aligned}
\omega_2 &= \mathrm{D}_t(\phi_1) + 2\omega_0(\partial f/\partial v_1)(\partial g/\partial u_1) \\
&\quad -2[(\partial g/\partial u)(\partial f/\partial v_1) + (\partial f/\partial v)(\partial g/\partial u_1)],
\end{aligned} \tag{7.70}
$$

$$\rho_2 = \sigma_1, \tag{7.71}$$

$$
\begin{aligned}
\omega_3 &= \mathrm{D}_t(\phi_2) + \rho_1\omega_1 - \rho_0(\omega_2 - \mathrm{D}_t(\phi_1)) + \mathrm{D}_t(\partial f/\partial u + \partial g/\partial v) \\
&\quad +\omega_0[(\partial g/\partial u_1)\mathrm{D}(\partial f/\partial v_1) - (\partial f/\partial v_1)\mathrm{D}(\partial g/\partial u_1)] \\
&\quad +(\partial g/\partial v)\mathrm{D}(\partial g/\partial v_1) - (\partial f/\partial u)\mathrm{D}(\partial f/\partial u_1) + \mathrm{D}(\omega_0)\mathrm{D}(\rho_0) \\
&\quad -2(\partial g/\partial u)\mathrm{D}(\partial f/\partial v_1) + 2(\partial f/\partial v)\mathrm{D}(\partial g/\partial u_1),
\end{aligned} \tag{7.72}
$$

$$
\begin{aligned}
\rho_3 &= \sigma_2 + (\rho_1^2 + \omega_1^2) - \omega_0(\omega_2 - \mathrm{D}_t(\phi_1)) + \mathrm{D}_t(\partial f/\partial u - \partial g/\partial v) \\
&\quad +\mathrm{D}_t(\partial g/\partial u_1)(\partial f/\partial v_1) - \mathrm{D}_t(\partial f/\partial v_1)(\partial g/\partial u_1) - 4(\partial f/\partial v)(\partial g/\partial u)
\end{aligned}
$$

$$
\begin{aligned}
&-[(\partial f/\partial v_1)(\partial g/\partial v_1)]^2 + 2(\partial f/\partial v_1)(\partial g/\partial v_1)(\partial f/\partial u + \partial g/\partial v)\\
&-2\mathrm{D}(\partial f/\partial v_1)\mathrm{D}(\partial g/\partial v_1) + [(\mathrm{D}\omega_0)^2 + (\mathrm{D}\rho_0)^2]/2\\
&+\rho_0[(\partial g/\partial u_1)\mathrm{D}(\partial f/\partial v_1) - (\partial f/\partial v_1)\mathrm{D}(\partial g/\partial u_1)]\\
&+2(\partial g/\partial u)\mathrm{D}(\partial f/\partial v_1) + 2(\partial f/\partial v)\mathrm{D}(\partial g/\partial u_1)\\
&-(\partial g/\partial v)\mathrm{D}(\partial g/\partial v_1) - (\partial f/\partial u)\mathrm{D}(\partial f/\partial u_1), \qquad (7.73)
\end{aligned}
$$

where functions ϕ_k, σ_k can be found from $\mathrm{D}(\phi_k) = \omega_k$, $\mathrm{D}(\sigma_k) = \mathrm{D}_t(\rho_k)$.

References

[1] Miura R. M., Gardner C. S., Kruskal M. D.: Korteweg – de Vries equation and generalizations. II. Existence of conservation laws and constants of motion. *J. Math. Phys.*, 9(8):1204–1209, 1968.

[2] Zakharov V. E., Shabat A. B.: Exact theory of two dimensional self focusing and one dimensional automodulation of waves in nonlinear medias. *ZhETF (in Russian)*, 61(1):118–134, 1971.

[3] Lamb G. L.: Higher conservation laws in ultrashort optical pulse propagation. *Phys. Lett A*, 32:251–252, 1970.

[4] Toda M.: *Theory of nonlinear lattices.* Volume 20 of *Solid-State Sci.*, Springer-Verlag, Berlin, Heidelberg, 1981.

[5] Zakharov V.E., editor : *What is Integrability?* Springer-Verlag, Berlin, Heidelberg, New-York, London, Paris, Tokio, Hong Kong, Barselona, 1990.

[6] Kulish P.P. : Factorization of classical and quantum S – matrix and conservation laws. *Teor. Mat. Fiz.*, 26(2):198–205, 1976.

[7] Ibragimov N.Kh., Shabat A.B.: On infinite dimension Lie Bäclund Algebras. *Funk. Anal. i ego Pril. (in Russian)*, 14(4):79–80, 1980.

[8] Fokas A. S.: A symmetry approach to exactly solvable evolution equations. *J. Math. Phys.*, 21:1318–1325, 1980.

[9] Zhiber A.V. and Shabat A.B.: Nonlinear Klein-Gordon equations with nontrivial group. *Soviet Phys. Dokl.*, 24(5):1104–1107, 1979.

[10] Abellanas L., Galindo A.J.: Conserved Densities for Nonlinear Evolution Equations I. Even order Case. *J. Math. Phys.*, 20(6):1239–1243, 1979.

[11] Mikhailov A.V., Shabat A.B., Sokolov V.V.: The symmetry approach to classification of integrable equations. In Zakharov V.E., editor, *What is Integrability?*, pages 115–184, Springer-Verlag, Berlin, Heidelberg, New-York, London, Paris, Tokio, Hong Kong, Barselona, 1990.

[12] Olver P.J.: *Application of Lie Groups to Differential Equations.* Springer-Verlag, New-York, Berlin, Heidelberg, Tokio, 1986.

[13] Magadeev B.A.: *On Symmetries of Evolution Equations.* PhD thesis, Bashkir. State Univ., Ufa, November 1990.

[14] Sokolov V.V.: On symmetries of evolutionary equations. *Usp. Mat. Nauk. (in Russian)*, 43(5):133–163, 1988.

[15] Svinolupov S.I.: Evolutionary equations of second order with symmetries. *Usp. Mat. Nauk. (in Russian)*, 40(5):263–264, 1985.

[16] Svinolupov S.I.: On the analogues of the Burgers equation of any Order. *Teor. Mat. Fiz. (in Russian)*, 65(2):303–307, 1985.

[17] Drinfeld V.G., Svinolupov S.I., Sokolov V.V.: Classification of Evolution Equations of fifth Order with Infinite sequense of Conservation Laws. *Dokl. Akad. Nauk Ukr.SSR (in Russian)*, (10):8–10, 1985.

[18] Svinolupov S.I.: Private communication.

[19] Fokas A. S.: Symmetries. *Stud. Appl. Math.*, 77, 253, 1987.

[20] Mikhailov A.V., Shabat A.B.: Integrability Conditions for Sytems of Equations of the Form $\mathbf{u}_t = A(\mathbf{u})\mathbf{u}_{xx} + \mathbf{F}(\mathbf{u}, \mathbf{u}_x)$.I. *Teor. Mat. Fiz. (in Russian)*, 62(2):163–185, 1985.

[21] Mikhailov A.V., Shabat A.B., Yamilov R.I.: The symmetry approach to classification of nonlinear equations. complete lists of integrable systems. *Usp. Mat. Nauk. (in Russian)*, 42(4):3–53, 1987.

[22] Svinolupov S.I.: On the analogues of the Burgers equation. *Phys. Lett A.*, 135(1):32–36, 1989.

[23] Borovik A.E., Popkov V.Yu. Robuk V.N.: . *Dcl. Acad. Nauk. SSSR (in Russian)*, 1989.

[24] Yamilov R.I.: About Classification of Discrete Evolution Equations. *Usp. Mat. Nauk (in Russian)*, 38(6):155–156, 1983.

[25] Yamilov R.I.: Generalized Toda Chains and Conservation Laws. *Preprint Bashkir Branch Acad. Nauk SSSR, Ufa.*, 1–21, 1989.

[26] Shabat A.B., Yamilov R.I.: Symmetries of nonlinear chains. *Algebra and Analisis, (in Russian)*, 2(2):183–208, 1990; see also Lattice representations of integrable systems, *Phys. Lett. A*, 138:271–275, 1988.

[27] Levi D.: Nonlinear differential difference equations as Bäcklund transformations. *J. Phys. A: Math. Gen.*, 14:1083–1098, 1981.

[28] Weiss J.: Periodic fixed points of Bäcklund transformations and the Korteweg-de Vries equation. *J. Math. Phys.*, 27:2647–2656, 1986.

[29] Konopelchenko B.G.: The non-abelian 1+1-dimensional Toda lattice as the periodic fixed point of the Laplace transform for the 2+1-dimensional integrable system. *Physics Letters A*, to appear,1991.

[30] Yamilov R.I.: Invertible substitutions generated by the Bäcklund transformations. *Teor. Mat. Fiz. (in Russian)*, 85(3):368–375, 1990.

[31] Svinolupov S.I.: Jordan algebras and generalized Korteweg de Vries equations. *Teor.Mat.Phys.*, 1991.

[32] Svinolupov S.I.: Generalized Schrödinger equations and Jordan Pairs. *Commun. Math. Phys.*, 1991. to appear.

[33] Shabat A.B. and Yamilov R.I.: Exponential systems of type 1 and Cartan matrices. *Preprint Bashkir Branch Acad. Nauk SSSR, Ufa.*, 1–22, 1981.

[34] Jordan P., von Neumann J., Wigner E.: *Ann. of Math.*, 36:29–64, 1934.

[35] Jacobson N.: *Structure and representations of Jordan algebras.* Providence, 1968.

[36] Bachturin Yu.A., Slin'ko A.M., Shestakov I.P.: Nonassociative Rings. In *Sovr. Probl. Matem.*, pages 3–72, VINITI, 1981. v.18.

[37] Leznov A.N.: About Complete Integrability of a Nonlinear System of Partial Differential Equations on a Plain. *Teor. Mat. Fiz. (in Russian)*, 42(3):343–349, 1980.

[38] Adler M., van Moerbeke P.: . *Commun. Math. Phys.*, 83, 1982.

[39] Fordy A.P., Kulish P.P.: Nonlinear Schrodinger Equations and Simple Lie Algebras. *Commun. Math. Phys.*, 89:427–443, 1983.

[40] Athorne C., Fordy A. P.: . *J. Phys. A*, 20:1377–1386, 1987.

Conservation and Scattering in Nonlinear Wave Systems

V. Zakharov[1], *A. Balk*[1], *and E. Schulman*[2]
[1]L.D. Landau Institute for Theoretical Physics, Russian Academy of Sciences, Moscow, Russia
[2]P.P. Shirshov Institute for Oceanology, Russian Academy of Sciences, Moscow, Russia

1 Introduction

Perturbational approaches have a long history of application to dynamical systems. A special role in this subject belongs to Hamiltonian systems with a small parameter ϵ, which are integrable when $\epsilon = 0$. Expanding in powers of ϵ is a common tool in celestial mechanics. When one tries to find the solution as a power series in ϵ, one encounters the classical problem of resonances. The same happens when one attempts to find additional invariants of motion. Attempts to overcome these difficulties stimulated the development of the theory of canonical transformations, simplifying the system by mapping it to some "normal form". These and related issues were elaborated in "New Methods of Celestial Mechanics" by A.Poincare [1] and in "Dynamic Systems" by D.D.Birkhoff [2].

The classical theory of Poincare and Birkhoff was developed for a system with a finite number of degrees of freedom, described by ordinary differential equations. This article contains a survey of results of pertubational approach to Hamiltonian systems with infinite number of degrees of freedom obtained by our group since 1980.

We speak actually about weakly nonlinear models of wave propagation in spacially uniform media. More rigorously, we consider systems with Hamiltonians having nonzero quadratic part, which have the following form in the normal coordinates:

$$H = H_0 + H_{int}; \; H_0 = \sum_{\alpha=0}^{N} \omega_k^{(\alpha)} a_k^{(\alpha)} a_k^{*(\alpha)} dk. \tag{1.1}$$

Here $a_k^{(\alpha)}$ are complex canonical amplitudes of the α-th linear mode (usually simply expressed through Fourier components of physical fields) and $k = (k_1, \cdots, k_d)$ is the wave vector, d is the dimension of the medium and $\omega_k^{(\alpha)}$ is the dispersion law of the α - th mode, H_{int} is the Hamiltonian of interaction (about Hamiltonian formalim for nonlinear waves see [3,4,5]). Equations of motion for the canonical amplitudes are of the form

$$\dot{a}_k^{(\alpha)} = -i\frac{\delta H}{\delta a_k^{*(\alpha)}} = -i\omega_k^{(\alpha)} a_k^{(\alpha)} - i\frac{\delta H_{int}}{\delta a_k^{*(\alpha)}}. \tag{1.2}$$

We shall call Hamiltonian PDEs of this type "Hamiltonian Wave Systems" (HWS). The approach reviewed in this paper is based on considering H_{int} in (1.2) as pertubation so that the characteristic amplitude of waves or "nonlinearity level" is analogous to the small parameter in the qualitative theory of O.D.E. The goal of this survey is to illuminate the main ideas of pertubrational methods developed by the authors. These ideas could be divided into four groups:

1. Invariants of resonance wave intractions and conservation laws of the kinetic equations for waves.

2. Coinciding of singular manifolds and equations solvable by the Inverse scattering method.

3. Classical scattering matrix and integrals of motion.

4. Canonical transformations and normal form for Hamiltonian Wave Systems.

Here we shall concentate on the key ideas; we refer the interested reader to the original papers for details. We will not strive here for generality and will mainly consider 2-dimensional systems of waves of only one type, whose H_{int} contains nonzero cubic part. We will point out what differences arise in other situations.

Thus we take

$$
\begin{aligned}
H_0 &= \int \omega_k a_k a_k^* dk, \quad d = \dim k = 2, && (1.3)\\
H_{int} &= H^{(3)} + H^{(4)} + \cdots, && (1.4)\\
H^{(3)} &= \frac{1}{3!}\sum_{ss_1s_2} \int V_{012} a a_1 a_2 \delta(sk + s_1k_1 + s_2k_2) dk dk_1 dk_2, &&\\
H^{(4)} &= \frac{1}{4!}\sum_{s_1s_2s_3s_4} \int W_{1234} a_1 a_2 a_3 a_4 \delta(s_1k_1 + s_2k_2 + s_3k_3 + s_4k_4) dk_1 dk_2 dk_3 dk_4 &&
\end{aligned}
$$

where

$$
\begin{aligned}
a_j &= a_{k_j}^{s_j} = \begin{cases} a_{k_j}, & s_j = 1, \\ a_{k_j}^*, & s_j = -1, \end{cases}\\
V_{012} &= V_{kk_1k_2}^{ss_1s_2}, \; W_{1234} = W_{k_1k_2k_3k_4}^{s_1s_2s_3s_4},
\end{aligned}
$$

δ - functions arise in (1.4) due to homogeneity of the media under consideration; the dots signify the terms of higher order. The dynamic equation of this wave system is

$$
i\dot{a}_k = \frac{\delta H}{\delta a_k^*}. \qquad (1.5)
$$

2 Integrals of Motion for Nonlinear Wave Systems

It is easy to check that the equation (1.5) conserves energy H and momentum

$$P = \int k|a_k|^2 dk \tag{2.1}$$

Can it conserve any additional quantity? Following the classical results of A.Poincare and G.Birkhoff for finite-dimensional systems we will search for motion invariants I in the form of functional power series (similarly to natural physical ones, H and P):

$$I = \int \phi_k |a_k|^2 dk + \frac{1}{3!}\sum_{ss_1s_2} \int F_{012} a_1 a_2 a_3 \delta(sk + s_1k_1 + s_2k_2) dk dk_1 dk_2 +$$
$$+\frac{1}{4!}\int G_{1234}\delta(s_1k_1 + s_2k_2 + s_3k_3 + s_4k_4) a_1 a_2 a_3 a_4 dk_1 dk_2 dk_3 dk_4 + \ldots \tag{2.2}$$

We require

$$\frac{dI}{dt} = 0$$

with time evolution of a_k governed by (1.5). Differentiating (2.2) in time by virtue of (1.5) and collecting terms of one and the same order we obtain a set of conditions on the coefficient functions $F_{012}, G_{1234}, \ldots$ of the power series (2.2). If these conditions are satisfied in all orders the quantity (2.2) is an exact integral of motion. If the series (2.2) diverges but all these conditions are satisfied we will call it the formal exact integral of motion. If

$$\frac{dI}{dt} = O(a^{m+3}), a \to 0,$$

the quantity I will be called the approximate integral of motion of the order m ($m = 0, 1, 2, \ldots$). Let's write down the first terms of the time derivative of the quantity I which we will use later.

$$\begin{aligned}
\frac{dI}{dt} &= \frac{1}{3!}\sum_{ss_1s_2}\int [F_{012}(s\omega + s_1\omega_1 + s_2\omega_2) - V_{012}(s\phi + s_1\phi_1 + s_2\phi_2)] \times \\
&\times \delta(sk + s_1k_1 + s_2k_2) a a_1 a_2 dk dk_1 dk_2 + \qquad (2.3) \\
&+ \frac{1}{4!}\sum_{s_1s_2s_3s_4}\int [G_{1234}(s_1\omega_1 + s_2\omega_2 + s_3\omega_3 + s_4\omega_4) \\
&-W_{1234}(s_1\psi_1 + s_2\psi_2 + s_3\psi_3 + s_4\psi_4) + \\
&+\sum_{s_7}(F_{712}V_{-734} - F_{-734}V_{712})\, s_7 + \\
&+\sum_{s_5}(F_{514}V_{-523} - F_{-523}V_{514})\, s_5 +
\end{aligned}$$

$$+\sum_{s_6}(F_{613}V_{-624} - F_{-624}V_{613})\, s_6] \times$$

$$\times\delta(s_1k_1 + s_2k_2 + s_3k_3 + s_4k_4)a_1a_2a_3a_4dk_1dk_2dk_3dk_4 + O(a^5) \quad (2.4)$$

where the wave vectors k_7, k_5, k_6 are defined by the equalities:

$$\begin{aligned} s_7k_7 &= -s_1k_1 - s_2k_2 = s_3k_3 + s_4k_4 \\ s_5k_5 &= -s_1k_1 - s_4k_4 = s_2k_2 + s_3k_3 \\ s_6k_7 &= -s_1k_1 - s_3k_3 = s_2k_2 + s_4k_4 \end{aligned} \quad (2.5)$$

and the following notation is used: $V_{-012} = V_{kk_1k_2}^{-ss_1s_2}$.

3 First Order Approximate Integrals of Motion. Conservation Laws of the Kinetic Equations for Waves.

From (2.4) we see that the quantity I is the first order approximated integral of motion:

$$\frac{dI}{dt} = O(|a|^4), a \to 0, \quad (3.1)$$

if

$$F_{012} = \frac{s\phi + s_1\phi_1 + s_2\phi_2}{s\omega + s_1\omega_1 + s_2\omega_2} V_{012}. \quad (3.2)$$

The function F_{012} has a pole singularity: it is infinite on the manifold Γ_3 defined by the conditions

$$sk + s_1k_1 + s_2k_2 = 0, \;\; s\omega + s_1\omega_1 + s_2\omega_2 = 0. \quad (3.3)$$

The existence of the invariant I depends on the properties of the field a_k. It is the Fourier transform of some fuction $A(r)$ (describing the wave system in real space). Suppose that $A(r)$ tends to zero at $|r| \to \infty$ so fast that a_k is a continuous function. Then one can replace in (3.1)

$$\frac{1}{s\omega + s_1\omega_1 + s_2\omega_2} \text{ by } \mathcal{P}\frac{1}{s\omega + s_1\omega_1 + s_2\omega_2}$$

and understand integrals in (2.2) in the sense of the principal value. In this case the invariant (2.2) can be constructed for any function $\phi(k)$ (which determines the quadratic term in (2.2). But it is much more interesting from the physical point of view to assume $A(r)$ to be a periodic or a quasiperiodic function or a statistically homogeneous random field. In both cases $a(k)$ is a distribution, and the integral (2.2) becomes infinite if the coefficient functions $F_{012}, G_{1234}, ...$ have pole singularities. Th e integral exists only if

$F_{012}, G_{1234}, \ldots$ are limited functions. It leads us to the following alternative [6,8,9]:

Alternative. *On the resonance manifold Γ_3 at least one of the two following conditions should be fulfilled*

$$\text{A) } V_{012} = 0 \tag{3.4}$$

or

$$\text{B) } s\phi + s_1\phi_1 + s_2\phi_2 = 0. \tag{3.5}$$

Let the condition A) be satisfied. Then the approximate invariant I exists for an arbitrary choice of the function $\phi(k)$. In this case three wave interaction is "illusive" - cubic terms in the Hamiltonian (1.4) can be excluded by a proper cano nical transformation. The existence of an additional approximate invariant is a trivial fact in this case.

In the opposite case B) we have functional equation on function of two variables $\phi(k) = \phi(p, q)$. Its resolvability imposes strong restrictions on the form of dispersion law $\omega(k)$.

By definition a dispersion law is degenerate if the equation (3.4) has a nontrivial solution $\phi = A\omega + (B, k)$, (A, B are constant).

In a typical physical situation $\omega_k \geq 0$ at all k. Then the conditions (3.3),(3.5) have the form

$$k = k_1 + k_2, \; \omega = \omega_1 + \omega_2 \tag{3.6}$$

and

$$\phi = \phi_1 + \phi_2. \tag{3.7}$$

The resonant conditions (3.5) are the basis for three wave interactions in nonlinear media.

In more rare cases ω_k can change sign in k-space. Then the conditions

$$\omega + \omega_1 + \omega_2 = 0, \; k + k_1 + k_2 = 0 \tag{3.8}$$

have to be taken into consideration as well. These conditions describe a so-called "explosive instability" of the ground state $a \equiv 0$.

Let us consider an ensemble of weakly-nonlinear waves with random phases. Such an ensemble represents a "wave turbulence" that has to be described statistically by introducing a correlation function

$$< a_k a^*_{k'} >= n_k \delta_{k-k'}. \tag{3.9}$$

The evolution of n_k is determined by the kinetic equation for waves[10, 11,12]

$$\frac{\partial n_k}{\partial t} = \int (R_{012} - R_{102} - R_{210}) dk_1 dk_2 \tag{3.10}$$

where

$$R_{012} = 2n \left| V^{-1,1,1}_{k,k_1,k_2} \right|^2 \delta(k - k_1 - k_2)\delta(\omega_k - \omega_{k_1} - \omega_{k_2})(n_1 n_2 - n n_1 - n n_2).$$

The kinetic equation has "natural" invariants energy $\mathcal{E}$ and momentum $\mathcal{P}$

$$\left.\begin{array}{ll} \mathcal{E} = \int \omega_{\|}\backslash_{\|}\Gamma\|, & \frac{d\mathcal{E}}{\sqcap} = 0, \\ \mathcal{P} = \int \|\backslash_{\|}\Gamma\|, & \frac{d\mathcal{P}}{\sqcap} = 0. \end{array}\right\} \tag{3.11}$$

If the condition (3.7) is satisfied on the resonant manifold (3.6), the kinetic equation (3.10) has an additional invariant

$$I = \int \phi(k) n_k dk. \tag{3.12}$$

It is important to stress that this invariant is exact. So the approximate invariant

$$\begin{aligned} I &= \int \phi_k a_k a_k^* dk + \int \frac{\psi_k - \psi_{k_1} - \psi_{k_2}}{\omega_k - \omega_{k_1} - \omega_{k_2}} \delta(k - k_1 - k_2) \\ &\times \left(V^{-1,1,1}_{k,k_1,k_2} a_k^* a_{k_1} a_{k_2} + V^{1,-1,-1}_{k,k_1,k_2} a_k a_{k_1}^* a_{k_2}^* \right) dk_1 dk_2 \end{aligned} \tag{3.13}$$

of the dynamical system (1.5) generates the exact invariant (3.12) of the kinetic equation (3.10). This fact explains why approximate integrals of the nonlinear wave system are so important for physical applications.

If the condition (3.4) is satisfied, then $R_{012} \equiv 0$, and a three-wave kinetic equation does not exist.

4 Degenerate Dispersion Laws and Web Theory

The first and the most important example of degenerate dispersion law is given by the following construction [13]. Let the dispersion law $\omega = \omega(k) = \omega(p, q)$ be a function that can be parametrized in the form

$$\begin{aligned} p &= \alpha(\xi) - \alpha(\eta), \\ q &= \beta(\xi) - \beta(\eta), \\ \omega &= \gamma(\xi) - \gamma(\eta) \end{aligned} \tag{4.1}$$

where α, β, γ are real functions of one variable. Then $\omega(k)$ is a degenerate dispersion law.

To see it let us parametrize:

$$\begin{aligned} p_1 &= \alpha(\xi_1) - \alpha(\eta_1), \qquad & p_2 &= \alpha(\xi_2) - \alpha(\eta_2), \\ q_1 &= \beta(\xi_1) - \beta(\eta_1), & q_2 &= \beta(\xi_2) - \beta(\eta_2), \\ \omega_1 &= \gamma(\xi_1) - \gamma(\eta_1), & \omega_2 &= \gamma(\xi_2) - \gamma(\eta_2) \end{aligned}$$

The conditions (3.5) define a three-dimensional manifold Γ_3 in the six dimensional space $(\xi, \eta, \xi_1, \eta_1, \xi_2, \eta_2)$. This manifold has the component, defined by the conditions

$$\xi = \xi_1, \eta = \eta_2, \eta_1 = \xi_2. \tag{4.2}$$

Let the function $\phi = \phi(p, q)$ be parametrized as follows

$$\begin{aligned} p &= \alpha(\xi) - \alpha(\eta), \\ q &= \beta(\xi) - \beta(\eta), \\ \phi &= f(\xi) - f(\eta), \end{aligned} \tag{4.3}$$

where f is an arbitrary function. It is clear that on the component (4.2) of the manifold Γ_3 the function ϕ satisfies the condition (3.7).

Examples.

1) The parametrization

$$p = \xi - \eta, q = \xi^2 - \eta^2, \omega = 4\xi^3 - 4\eta^3, \tag{4.4}$$

defines the dispersion law

$$\omega = p^3 + \frac{3q^2}{p}$$

of the Kadomtsev-Petviashvili equation

$$(u_t + 6uu_x + u_{xxx})_x = 3u_{yy}. \tag{4.5}$$

2) The parametrization

$$\begin{aligned} p &= \sin\xi - \sin\eta, \\ q &= \cos\xi - \cos\eta, \\ \omega &= \xi - \eta, \end{aligned} \tag{4.6}$$

define the dispersion

$$\omega = \pm 2 \arcsin \frac{\sqrt{p^2 + q^2}}{2} \tag{4.7}$$

of the waves in the Toda chain.

The second example shows that in the general case the parametrization (4.1) gives a many-valued function $\omega(p, q)$. Only in some particular situations, like (4.4), it is single-valued function; then it is an odd function: $\omega(-k) = -\omega(k)$. It does not necessarily mean that the system under consideration contains explosive instability. It can mean that the values of the normal variable a_k in the points k and $-k$ are connected, e.g. $a_k = a^*_{-k}$. In this case one can consider that the normal variable a_k is defined only on half plane $0 < p < \infty, -\infty < q < \infty$. This happens for the Kadomtsev-Petviashvili equation if u is a real function.

The parametrization (4.1) is invariant with respect to permutation

$$\xi \to \eta, p \to p, q \to q, \omega \to -\omega.$$

It means that in general the parametrization (4.1) define some number of odd dispersion laws $\omega(-k) = -\omega(k)$ and some number of pairs $\omega = \pm\omega(k)$.

One can say that a dispersion law is degenerate to degree N if there exist N linearly independent (in aggregate with $\omega(k), k$) functions $\phi_n(k)(n = 1, ..., N)$ satisfying the condition (3.4) on the resonanse manifold (3.2). Dispersion laws belonging to the class (4.1) are infinitely degenerate, because $f(\xi)$ in (4.3) is arbitrary.

Another example of infinitely degenerate dispersion law $\omega(k)$ could be obtained from (4.1) by the following limiting transition. Let us replace ξ, p, q, ω in (4.1) by $\eta + \varepsilon, \varepsilon p, \varepsilon q, \varepsilon\omega$ and pass to the limit $\varepsilon \to 0$. Then we get the following dispersion law

$$\omega = pg(\frac{q}{p}), \tag{4.8}$$

where $g(\eta)$ is a real function of one variable. So, any homogeneous function of the first degree is a degenerate dispersion law. In this case vectors k, k_1, k_2 in the points of the resonance manifold (3.2) are parallel.

There exist dispersion laws of a finite degree of degeneracy. The dispersion law

$$\omega = \frac{p}{1 + p^2 + q^2} \tag{4.9}$$

is degenerate to degree $N = 2$ [14] and the invariants have the form

$$\begin{aligned} \phi^1 &= \arctan\frac{q + \sqrt{3}p}{k^2} - \arctan\frac{q - \sqrt{3}p}{k^2}, \\ \phi^2 &= \log\frac{(q + \sqrt{3}p)^2 + k^4}{(q - \sqrt{3}p)^2 + k^4}. \end{aligned}$$

This example is remarkable as far as (4.8) is the dispersion law for Rossby waves in atmosphere and for drift waves in magnetized plasma. In both these cases the phase space of the dynamical system is reduced ($a_k = a^*_{-k}$) and only an odd function ϕ^1 defines a non-zero invariant. So the effective degree of degeneracy is $N = 1$.

The concept of degenerate dispersion laws can be naturally extended for systems containing waves of several types. In this situation 3-wave resonance manifold has the form (3.2), where $\omega_j = \omega^j(k_j)$ is the dispersion law of waves of j -th type.

A system of dispersion laws $\omega^j(k)(j = 1, 2, 3)$ is called degenerate if there exists such set of functions

$$\phi^j(k) \neq A\omega^j(k) + (B, k) + C_j \; (j = 1, 2, 3; A, B, C_j \text{ are constants}),$$

that the equality (3.4) (where $\phi_j = \phi^j(k_j)$) holds on the resonance manifold (3.2).

The single type wave system (4.1) with infinite number of extra invariants (4.3) may be naturally generalized for the case of several type wave systems. Let the resonance manifold (3.2) be parametrized in the form [9]

$$\begin{aligned} sp &= \alpha_1(x) - \alpha_2(y), s_1p_1 = \alpha_2(y) - \alpha_3(z), s_2p_2 = \alpha_3(z) - \alpha_1(x), \\ sq &= \beta_1(x) - \beta_2(y), s_1q_1 = \beta_2(y) - \beta_3(z), s_2q_2 = \beta_3(z) - \beta_1(x), \\ s\omega &= \gamma_1(x) - \gamma_2(y), s_1\omega_1 = \gamma_2(y) - \gamma_3(z), s_2\omega_2 = \gamma_3(z) - \gamma_1(x). \end{aligned} \tag{4.10}$$

Then three arbitrary function f_1, f_2, f_3 of one variable define an invariant (3.4), where

$$s\phi = f_1(x) - f_2(y), s_1\phi_1 = f_2(y) - f_3(z), s_2\phi_2 = f_3(z) - f_1(x).$$

The parametrization (4.10) takes place e.g. for the system of three wave packets.

Some limiting cases of parametrization (4.10) also yield wave systems with infinite number of invariants, similar to the system (4.7).

There exist as well essentially different wave systems with infinite number of invariants [15]. Let us consider a system of waves of two types with the same quadratic dispersion law

$$\Omega(k) = \omega(k) = k^2. \tag{4.11}$$

Let the only possible 3-wave resonance interactions consist of a decay of a wave of the first type into two waves of the second type (and inverse processes), so that the resonance manifold has the form

$$k = k_1 + k_2, \ \Omega(k) = \omega(k_1) + \omega(k_2)$$

It turns out [15] that an arbitrary function $\phi(\xi)$, possessing such property $\phi(\xi) = -\phi(-\frac{1}{\xi})$, gives an invariant of this system: the equality

$$0 = \phi(\frac{p_1}{q_1}) + \phi(\frac{p_2}{q_2})$$

holds on the resonance manifold.

The above example naturally arises in practice when one considers a single type wave system with quadratic dispersion law $\omega(k) = k^2$ and supposes that the energy spectrum of this system is concentrated in two regions, so that all possible resonance interactions are decays of some wave from the first region into two waves from the second region (and inverse processes).

The paper [16] gives a complete description of two-dimensional wave systems with infinite number of invariants. It is based on the circumstance, disclosed in [15], that the problem of degenerate dispersion laws and wave system invariants is closely connected with the classical differential geometry, namely with the theory of webs.

The web is defined as follows [17]:

The collection of three families of curves in some 3-dimensional domain is called a 3-web (of curves in 3-dimensional space) if for any point of this domain there is exactly one curve from each family passing through this point. Two such 3-webs are consid ered to be congruent (or the same) if there is a one-to-one transformation mapping each family of one web onto the corresponding family of the other web.

Three dimensional resonance manifold (3.2) in some coordinates x, y, z is given by the vector-functions $k_j(x, y, z)(j = 0, 1, 2)$ that satisfy equations (3.2) identically. The following three families of curves

$$k_j(x, y, z) = C_j \ (j = 0, 1, 2) \tag{4.12}$$

(C_j is an arbitrary constant) form a 3-web, which is called the web associated with the resonance manifold Γ_3 [15]. The associated web contains all the necessary information about resonance manifold. All resonance manifolds, parametriz ed in the form (4.9), correspond to one and the same web; it is the simplest so-called coordinate web that consists of three families of straight lines parallel to some of the 3 coordinative axis:

$$\begin{aligned} j = 0 &: x = \text{const}, y = \text{const}; \\ j = 1 &: y = \text{const}, z = \text{const}; \\ j = 2 &: z = \text{const}, x = \text{const}. \end{aligned} \tag{4.13}$$

We get essentially different associated webs in the case (4.9) of Rossby waves and in the case (4.11) of the waves with quadratic dispersion.

Do there exist systems with dispersion (4.9) or (4.11) that possess exact integral of motion (not only approximate one)? A negative answer to this question will be obtained in the next section. It turns out that only systems whose associated web is coordinate can possess exact integrals of motion.

The presence of extra invariants (3.4) essentially depends on the dimension d of the medium. If $d = 1$ any 3-wave resonance manifold (3.2) possesses infinite number of invariants. If $d \geq 2$ only very peculiar resonance manifolds Γ_3 admit extr a invariants. That was the reason to call the collection of dispersion laws $\omega^j(k)(j = 1, 2, 3)$, for which the resonance manifold Γ_3 possesses extra invariants (3.4), degenerate. Such dispersion laws form a set of zero measure among all dispersion laws. However they arise often enough in physical systems.

For $d \geq 3$ nontrivial examples of degenerate dispersion laws are not known at all.

The concept of degenerate dispersion laws can be extended naturally to higher nonlinear processes involving m waves ($m \geq 3$):

$$s_1k_1 + \ldots + s_mk_m = 0, \; s_1\omega_1 + \ldots + s_m\omega_m = 0 \tag{4.14}$$

($s_j = \pm 1, \omega_j = \omega^j(k_j)$). The set of functions $\phi^j(k)(j = 1, \ldots, m)$ is called an invariant of the process (4.14) if the equality

$$s_1\phi_1 + \ldots + s_m\phi_m = 0 \; (\phi_j = \phi^j(k_j)) \tag{4.15}$$

holds on the manifold (4.14).

The web geometry is applied to the general situation with arbitrary m and d (see [15]). The paper [18] completely describes all systems (4.14) with $m = 4, d = 1$, that possess extra invariants. Any such system proves to have either infinite num ber of invariants or only one invariant.

Single type wave systems with interaction "2 to 2"

$$k_1 + k_2 = k_3 + k_4, \; \omega(k_1) + \omega(k_2) = \omega(k_3) + \omega(k_4) \tag{4.16}$$

provide an example of systems with infinite number of invariants [9]. Indeed, the resonance manifold (4.16) in typical situation consists of two components

$$k_1 = k_3, k_2 = k_4 \text{ and } k_1 = k_4, k_2 = k_3$$

and, obviously, for an arbitrary function $\phi(k)$ the equality

$$\phi(k_1) + \phi(k_2) = \phi(k_3) + \phi(k_4)$$

holds on the resonance manifold.

The following resonance manifold

$$k_1 + k_2 = k_3 + k_4, \; \sqrt{a^2 + k_1^2} + \sqrt{a^2 + k_2^2} = \sqrt{b^2 + k_3^2} + \sqrt{b^2 + k_4^2} \tag{4.17}$$

($a \neq b$) admits only one extra invariant

$$\arcsin\frac{k_1}{a} + \arcsin\frac{k_2}{a} = \arcsin\frac{k_3}{b} + \arcsin\frac{k_4}{b} \tag{4.18}$$

(see [15]). This example is interesting for the theory of electromagnetic wave propagation in optical fibers, anisotropic in its cross-section; herewith a and b are different prolongation constants for the two polarizations of light. Among other ap plications of the example (4.17-18) are interactions of relativistic particles (then a is a rest mass) and transverse oscillations in plasma.

The paper [19] considers the concept of degenerate dispersion laws on a class of analytic functions; it studies invariants with only pole singularities.

5 Second Order Approximate Integrals. Coinciding of Singular Manifolds.

Let the quantity (2.2) be a second order approximated integral of motion:

$$\frac{dI}{dt} = O(|a|^5), a \to 0, \tag{5.1}$$

and the second term in (2.4) be zero. Found from this condition G_{1234} can be transformed to the expression

$$G_{1234} = \frac{s_1\phi_1 + s_2\phi_2 + s_3\phi_3 + s_4\phi_4}{s_1\omega_1 + s_2\omega_2 + s_3\omega_3 + s_4\omega_4} R_{1234} + \check{G}_{1234} \tag{5.2}$$

where

$$\begin{aligned} R_{1234} &= W_{1234} \qquad\qquad (5.3)\\ &- \frac{1}{2}\sum_{s_7} s_7 V_{712} V_{-734} \left(\frac{1}{s_7\omega_7 + s_1\omega_1 + s_2\omega_2} + \frac{1}{s_7\omega_7 - s_3\omega_3 - s_4\omega_4}\right)\\ &- \frac{1}{2}\sum_{s_5} s_5 V_{514} V_{-523} \left(\frac{1}{s_5\omega_5 + s_1\omega_1 + s_4\omega_4} + \frac{1}{s_5\omega_5 - s_2\omega_2 - s_3\omega_3}\right)\\ &- \frac{1}{2}\sum_{s_6} s_6 V_{612} V_{-634} \left(\frac{1}{s_6\omega_6 + s_1\omega_1 + s_3\omega_3} + \frac{1}{s_6\omega_6 - s_2\omega_2 - s_4\omega_4}\right),\\ \check{G}_{1234} &= \frac{1}{2}\sum_{s_7} s_7 \left(\frac{V_{712}F_{-734}}{s_7\omega_7 + s_1\omega_1 + s_2\omega_2} + \frac{V_{-734}F_{712}}{s_7\omega_7 - s_3\omega_3 - s_4\omega_4}\right)\\ &+ acl2\sum_{s_5} s_5 \left(\frac{V_{514}F_{-523}}{s_5\omega_5 + s_1\omega_1 + s_4\omega_4} + \frac{V_{-523}F_{514}}{s_5\omega_5 - s_2\omega_2 - s_3\omega_3}\right)\\ &+ \frac{1}{2}\sum_{s_6} s_6 \left(\frac{V_{613}F_{-624}}{s_6\omega_6 + s_1\omega_1 + s_3\omega_3} + \frac{V_{-624}F_{613}}{s_6\omega_6 - s_2\omega_2 - s_4\omega_4}\right) \qquad (5.4) \end{aligned}$$

For the existence of the integral (2.2), the function G_{1234} should be bounded. Hence the product

$$(s_1\phi_1 + s_2\phi_2 + s_3\phi_3 + s_4\phi_4) R_{1234}$$

must equal zero on the four-wave resonance manifold Γ_4

$$\begin{aligned} &s_1k_1 + s_2k_2 + s_3k_3 + s_4k_4 = 0,\\ &s_1\omega_1 + s_2\omega_2 + s_3\omega_3 + s_4\omega_4 = 0 \quad (\omega_j = \omega(k_j)) \end{aligned} \tag{5.5}$$

(of dimension $N = 5$). As far as degenerate dispersion laws for the processes (5.5) don't exist for $d \geq 2$, we get

$$R_{1234} = 0 \text{ on } \Gamma_4. \tag{5.6}$$

The condition (5.6) determines matrix element W_{1234} on the resonance manifold (5.5)

$$W_{1234} = \sum_{s_7} \frac{s_7 V_{712} V_{-734}}{s_7\omega_7 + s_1\omega_1 + s_2\omega_2}$$
$$+ \sum_{s_5} \frac{s_5 V_{514} V_{-523}}{s_5\omega_5 + s_1\omega_1 + s_4\omega_4} + \sum_{s_7} \frac{s_6 V_{613} V_{-624}}{s_6\omega_6 + s_1\omega_1 + s_3\omega_3} \quad (5.7)$$

Outside manifold (5.5) the function W_{1234} can be changed arbitrarily with the aid of canonical transformations

$$\tilde{a}_0 = a_0 + \sum_{s_1 s_2 s_3} \int S_{0123} a_1 a_2 a_3 \delta(-s_0 k_0 + s_1 k_1 + s_2 k_2 + s_3 k_3) dk_1 dk_2 dk_3 + \ldots$$

(the dots signify the terms of higher order). These transformations don't change W_{1234} on Γ_4.

The expression (5.7) has singularities on the following three manifolds:

$$\left. \begin{array}{ll} s_7 k_7 + s_1 k_1 + s_2 k_2 = 0, & s_7\omega_7 + s_1\omega_1 + s_2\omega_2 = 0, \\ s_7 k_7 - s_3 k_3 - s_4 k_4 = 0, & s_7\omega_7 - s_3\omega_3 - s_4\omega_4 = 0; \end{array} \right\} \quad (5.8)$$

$$\left. \begin{array}{ll} s_5 k_5 + s_1 k_1 + s_4 k_4 = 0, & s_5\omega_5 + s_1\omega_1 + s_4\omega_4 = 0, \\ s_5 k_5 - s_2 k_2 - s_3 k_3 = 0, & s_5\omega_5 - s_2\omega_2 - s_3\omega_3 = 0; \end{array} \right\} \quad (5.9)$$

$$\left. \begin{array}{ll} s_6 k_6 + s_1 k_1 + s_3 k_3 = 0, & s_6\omega_6 + s_1\omega_1 + s_3\omega_3 = 0, \\ s_6 k_6 - s_2 k_2 - s_4 k_4 = 0, & s_6\omega_6 - s_2\omega_2 - s_4\omega_4 = 0. \end{array} \right\} \quad (5.10)$$

All of them are submanifolds (of dimension 4 and codimension 1) of the major resonant manifold (5.5). The function $\check{G}_{1234}$ also has singularities on the manifold (5.8)-(5.10). These singularities must cancel each other in order for the invariant (2.2) to exist.

In the case of nontrivial three wave interaction ($V_{012} \neq 0$ on Γ_3) the only possibility for singularities to cancel each other is the pairwise coincidence of the manifolds (5.8)-(5.10). (In a typical situation each of these manifolds consi sts of two parts. E.g. one part of the manifold (5.8) coincides with some part of the manifold (5.9), and another part of (5.8) coincides with a part of the manifold (5.10)). Then the cancellation of the singularities occurs if the residues in coin ciding singularities are equal (by modulus) and have opposite signs.

The condition of coincidence of singular manifolds (5.8)-(5.10) and cancellation of the singularities allows to obtain very strong results [20] that will be formulated for the simplest case of single type wave system:

1) Singularities cancel each other simultaneously in both expressions for W_{1234} and for $\check{G}_{1234}$.

2) The dispersion law $\omega(p,q)$ can be parametrized in the form (4.1) or has the form (4.8).

3) The matrix element V_{012} is defined uniquely on the resonance manifold Γ_3 up to a constant factor and canonical transformation $a_k \to a_k \exp(i\psi_k)$,

fixing the initial phases of wave amplitudes. (Strictly speaking this is true only for that components of Γ_3, where the matrix element V_{012} differs from zero. On other components and outside Γ_3 it can be made arbitrary with the aid of cannonical transformation.)

Let us introduce:

$$d(\xi,\eta) = \begin{vmatrix} \alpha'(\xi) & \alpha'(\eta) \\ \beta'(\xi) & \beta'(\eta) \end{vmatrix} \tag{5.11}$$

$$D(\xi,\eta,x) = \begin{vmatrix} \alpha'(\xi) & \alpha'(\eta) & \alpha'(x) \\ \beta'(\xi) & \beta'(\eta) & \beta'(x) \\ \gamma'(\xi) & \gamma'(\eta) & \gamma'(x) \end{vmatrix}. \tag{5.12}$$

On the manifold Γ_3, parametrized in the form

$$sp = \alpha(\xi) - \alpha(\eta), s_1 p_1 = \alpha(\eta) - \alpha(x), s_2 p_2 = \alpha(x) - \alpha(\xi),$$
$$sq = \beta(\xi) - \eta(\eta), s_1 q_1 = \beta(\eta) - \beta(x), s_2 q_2 = \beta(x) - \beta(\xi),$$

V_{012} becomes a function of three variables $V_{012} = V(\xi,\eta,x)$. The condition of residues cancelling gives the following functional equation for the function $V(\xi,\eta,x)$

$$\frac{V(x,\xi,\eta)}{D(x,\xi,\eta)}\frac{V(y,\xi,\eta)}{D(y,\xi,\eta)} s_7 d(\xi,\eta) = \frac{V(\xi,x,y)}{D(\xi,x,y)}\frac{V(\eta,x,y)}{D(\eta,x,y)} s_5 d(x,y). \tag{5.13}$$

It was solved in [20]:

$$V(\xi,\eta,x) = \frac{D(\xi,\eta,x)}{\sqrt{|d(\xi,\eta)d(\eta,x)d(x,\xi)|}} \tag{5.14}$$

If the quantity I were exact integral of motion, the matrix elements of higher nonlinear processes would be also uniquely determined after choosing functions α, β, γ.

For Kadomtsev-Petviashvili equation the functions α, β, γ have the form (4.4) and according to formular (5.13):

$$V(z,x,y) = \sqrt{|(x-y)(y-z)(z-x)|} = \sqrt{pp_1p_2};$$

all matrix elements of higher nonlinear processes turn out to equal zero.

6 Exact Integrals and Scattering Matrix

Suppose that the system (1.3-5) has an exact integral of motion (2.2), conserved in all orders of a_k, a_k^*. What information about the behaviour of the system can be extracted from this, very nontrivial, fact? It could be easily formulated afte r introducing the classical scattering matrix.

Let us change the Hamiltonian of our system to the form

$$H = H_0 + e^{-\epsilon|t|}H_{int} \tag{6.1}$$

where ϵ is a positive parameter. At $t \to \pm\infty$ the system becomes linear, and

$$a(k,t) \to c^{\pm}(k)e^{\omega_k t} \tag{6.2}$$

The asymptotic states $c^{\pm}(k)$ are connected by a nonlinear operator $\hat{S}_\epsilon$

$$c^{+}(k) = \hat{S}_\epsilon[c^{-}(k)] \tag{6.3}$$

which can be expanded in a functional-power series

$$c^{s}(k) = c^{-s}(k) + \sum_{n=2}^{\infty} \sum_{s_1,\dots,s_n} \int S_{\epsilon;-0,1,\dots,n} c_1 \cdots c_n$$
$$\delta(-sk + s_1k_1 + \dots + s_nk_n)dk_1 \dots dk_n \tag{6.4}$$

All coefficients $S_{-0,1,\dots,n} = S_\epsilon^{-s,s_1,\dots,s_n}(k,k_1,\dots,k_n)$ can be found in an explicit form through the coefficients of the Hamiltonian H.

By definition, the scattering matrix $\hat{S}(k)$ is the limit (in the sense of formal power series)

$$\hat{S}(k) = \lim_{\epsilon\to 0} \hat{S}_\epsilon. \tag{6.5}$$

After the limiting transition the operator $\hat{S}$ becomes a formal functional power series. We care neither about its convergence nor about the existence of "real" asymptotic states $c^{\pm}(k)$ in our system at $\epsilon = 0$. It is only important that the operator $\hat{S}$ can be calculated explicitly to all orders.

The first order term is

$$S_{012} = 2\pi i\delta(sk + s_1k_1 + s_2k_2)\delta(s\omega + s_1\omega_1 + s_2\omega_2)V_{0,1,2}. \tag{6.6}$$

For the second order term we have

$$S_{1234} = S_{1234reg} + S_{1234sing} \tag{6.7}$$

where $S_{1234;reg}$ is the "regular" part of S_{1234}

$$S_{1234;reg} = \pi i\delta(s_1k_1 + s_2k_2 + s_3k_3 + s_4k_4)\delta(s_1\omega_1 + s_2\omega_2 + s_3\omega_3 + s_4\omega_4)R_{1234}, \tag{6.8}$$

concentrated on the "major" 4-wave resonance manifold Γ_4 (defined in(5.5); the expression for R see in (5.3)). $S_{1234sing}$ is the "singular" part concentrated on the 3-wave resonance manifolds (submanifolds of Γ_4; see (5.8-10))

$$S_{1234sing} = -2\pi^2 \sum_{s_7} \int dk_7$$

$$\{V_{712}V_{-734}\delta(s_7k_7+s_1k_1+s_2k_2)\delta(s_7\omega_7+s_1\omega_1+s_2\omega_2)\times$$
$$\delta(s_7k_7-s_3k_3-s_4k_4)\delta(s_7\omega_7-s_3\omega_3-s_4\omega_4)\times$$
$$+V_{714}V_{-723}\delta(s_7k_7+s_1k_1+s_4k_4)\delta(s_7\omega_7+s_1\omega_1+s_4\omega_4)\times$$
$$\delta(s_7k_7-s_2k_2-s_3k_3)\delta(s_7\omega_7-s_2\omega_2-s_3\omega_3)\times$$
$$+V_{713}V_{-724}\delta(s_7k_7+s_1k_1+s_3k_3)\delta(s_7\omega_7+s_1\omega_1+s_3\omega_3)\times$$
$$\delta(s_7k_7-s_2k_2-s_4k_4)\delta(s_7\omega_7-s_2\omega_2-s_4\omega_4)\}.$$

In this section we will assume that the dispersion law does not belong to the special class (4.8). In that special case the scattering matrix becomes infinite.

The element of the scattering matrix pertaining to the order m can be expanded in the following sum

$$S_{1,\dots,m}=S_{1,\dots,m;reg}+S^{(1)}_{1,\dots,m;sing}+\cdots+S^{(l)}_{1,\dots,m;sing}+\cdots+S^{(m-3)}_{1,\dots,m;sing} \quad (6.9)$$

where

$$S_{1,\dots,n;reg}=\pi\delta(s_1k_1+\dots+s_mk_m)\delta(s_1\omega_1+\dots+s_m\omega_m)R_{1,\dots,m} \quad (6.10)$$

is the most regular term, concentrated on the major resonance manifold Γ_m (defined by conditions (4.14)). Other terms in (6.9) are ordered according to degree of their singularity. They are concentrated on submanifolds of the major manifold Γ_m and determined by additional resonance conditions. The term $S^{(l)}_{1,\dots,m;sing}$ is concentrated on the submanifold of Γ_m of codimension l. The maximum codimension is $l=m-3$; the corresponding resonance manifold is determined by a system of three wave resonance conditions generalizing the conditions (5.8-10).

The existence of at least one additional integral of motion imposes very strong limitations on the scattering matrix $\hat{S}$. As it was shown in sect.5, $R_{1234}=0$ and

$$S_{1234;reg}=0. \quad (6.11)$$

In the papers [6,7,8,9] the following statements were proved.

Theorem 6.1. *If there exists at least one additional integral of motion with $\phi(k)\neq\alpha\omega(k)+(\beta,k)$, then all elements of scattering matrix S, except for the most singular are zero.*

So, in the expansion (6.11) only the last term $S^{(m-3)}_{1,\dots,m;sing}$ can differ from zero.

Vanishing or nonvanishing of the most singular term $S^{(m-3)}_{1,\dots,m;sing}$ is determined by the Alternative in sect. 3. In the case A: $V_{012}=0$ on Γ_3, and therefore $S^{(m-3)}_{1,\dots,m;sing}=0$. Thus we have the following theorem

Theorem 6.2. *If the nonlinear system (1.3-5) has an additional integral of motion, and the 3-wave interaction is trivial ($V_{012}=0$ on Γ_3; in particular, the dispersion law is of nondecay type) then the scattering matrix is trivial:*

the asymptotic states at $t \to \pm\infty$ *coincide*

$$c_k^+ = c_k^- . \tag{6.12}$$

If $V_{012} = 0$ on Γ_3 and the case B of the Alternative is realized, the scattering matrix $\hat{S}$ is nonzero. It is a sum of its the most singular elements $S_{1,\dots,m;sing}^{(m-3)}$. This sum could be explicitly computed.

Theorem 6.3. *If the nonlinear system (1.3-5) possess an additional integral of motion, and the 3-wave interaction is nontrivial, the asymptotic states are connected by the following nonlinear integral equation*

$$c_k^+ - c_k^- = -\frac{\pi}{2} \sum_{s_1 s_2} \int V_{012} \delta(sk + s_1 k_1 + s_2 k_2) \delta(s\omega + s_1 \omega_1 + s_2 \omega_2) \times \times (c_1 + c_{-1})(c_2 + c_{-2}) dk_1 dk_2 . \tag{6.13}$$

7 Integrability and Solvability

How does the existence of at least one additional integral of motion (besides energy, momentum, and, sometimes wave action) affect the integrability or solvability of the system?

Any wave system possesses additional integrals of motion if the wave field $A(r)$ tends to zero sufficiently fast while $r \to 0$, so that the complex amplitude a_k is a smooth function (the smoother the function a_k the more integrals of moti on the system possesses). So, in this section we will consider wave systems (1.3-5) with periodic boundary conditions.

In the periodic case all wave vectors belong to the latice

$$k = (p, q) = (\frac{2\pi m}{A}, \frac{2\pi n}{B}) \tag{7.1}$$

(A, B are space periods, m, n are integers) and all integrals over k-space are replaced by sums over the lattice. We consider such wave systems, that possess an additional motion integral for an arbitrary choice of positive A, B.

It turns out that the existence of one additional integral of motion implies the existence of an infinite set of motion integrals. However, the structure of this set essentially depends on the Alternative in sect. 3.

Suppose that the condition A of the Alternative is realized ($V_{012} = 0$ on Γ_3). Then the function (3.2) is bounded, and the first order approximate integral of motion exists for any function ϕ_k.

Further, let us consider the expression (5.2) for the coefficient G_{1234}, defining the term of fourth degree in some motion integral (2.2). The condition A) implies that $\tilde{G}_{1234} = 0$; and due to the existence of one additional integral of motion $R_{1234} = 0$ on Γ_4. Hence the coefficient G_{1234} is bounded, and the second order approximate integral of motion exists for any choice of the function ϕ_k.

It can be shown [8] that in the case A) there exists an exact integral of motion with an arbitrary function ϕ_k (defining the quadratic part of the integral). Thus the set of motion integrals depends on an arbitrary function of two variables $\phi(p, q)$. The system possesses an integral of motion for each degree of freedom; it has as many motion integrals as linearized system (1.3-5) with $H_{int} = 0$.

In the case B) (when $V_{012} \neq 0$ on Γ_3) the situation is completely different. In order for the quantity (2.2) to be the first order approximate integral of motion, the dispersion law should be degenerate, and only some particular func tions (satisfying the condition B) define the invariant. The number of such (linearly independent) functions ϕ_k can be infinite (as for KP-equation (4.5)) or finite as for the system of Rossby waves). However the quantity (2.2) can be a second ord er approximate integral of motion only if the dispersion law ω_k can be parametrized in the form (4.1) (or has the form (4.8)). Then there exists an infinite set of the functions ϕ_k, satisfying the condition B), and therefore there is a n infinite set of approximate integrals of motion of first and second orders. One can show that there exists an exact integral of motion with an arbitrary function ϕ_k, that can be parametrized in the form (4.3). Thus in the case B) the set of mo tion integrals depends on an arbitrary function of one variable $f(\xi)$. So the set of motion integrals in the case B) is essentially narrower than in the case A).

The system (1.3-5) with periodic boundary conditions in the case A) is integrable (in the classical Liouville sense). And one can find the action-angle variables [9]. Before the formulation of the rigorous statement about the normal form for the wave system in the case A) let us make the following remark.

If we take some wave vectors $k_1, \ldots, k_m$ from the lattice (7.1) obeying the relation

$$s_1 k_1 + \cdots + s_m k_m = 0 \tag{7.2}$$

they probably will not belong to the resonance surface

$$s_1 \omega_1 + \cdots + s_m \omega_m = 0. \tag{7.3}$$

Regardless of the values of the periods A, B the set $k_1, \ldots, k_m$ with large p_j, q_j $(j = 1, \ldots, m)$ may satisfy (7.2-3) with great accuracy, and we come to the problem of small denominators, when trying to construct the appropriate can onical transformation in the form of a power functional series. The condition A) of the Alternative guarantees that the coefficients in that series are finite in points of general position. However, there are special values of $k_1, \ldots, k_m$ that alway s belong to the resonance manifold (7.2-3); they correspond to billiard scattering with

$$m = 2l, s_1 = \ldots = s_l = 1, s_{l+1} = \ldots = s_{2l} = -1$$

and the set $(k_1, \dots, k_l)$ is a transposition of the set $(k_{l+1}, \dots, k_{2l})$. These billiard scattering processes are important for the construction of the normal form. The following theorem can be proved[21].

Theorem 7.1 *Let the system (1.3-5) with periodic boundary conditions posssess at least one additional integral of motion, and let the condition A) of the Alternative be realized. Then:*

1)If the interaction Hamiltonian (1.4) is cubic: $H_{int} = H^{(3)}$ *and* $\Omega(k) \to 0, V_{012} \to 0$ *while* $k \to 0$*, then there exists a canonical transformation*

$$\alpha_k^s = a_k^s + \sum_{m\geq 2} \sum_{s_1,k_1;\dots;s_m,k_m} Z_{-0,1,\dots,m}\delta(-s_0k_0+s_1k_1+\dots+s_mk_m)a_1\cdots a_m \quad (7.4)$$

mapping the system (1.3-5) to the normal form

$$is\dot{\alpha}_k^s = \Omega_k\alpha_k^s \quad (7.5)$$

with Hamiltonian

$$\mathcal{H} = \omega_k n_k + \sum_{m\geq 2} \sum_{k_1,\dots,k_m} h_{1,\dots,m} n_1 \cdots n_m; \quad (7.6)$$

$$n_k = |\alpha_k|^2; \Omega_k = \frac{\partial \mathcal{H}}{\partial \backslash_{||}}. \quad (7.7)$$

The quantities $Z_{-0,1,\dots,m}$ *can be obtained by recurrence or with the aid of diagram technique and are finite at any m, so that the zero denominators are absent in the canonical transformation (7.4).*

If the function ω_k *has singularity at* $k \to 0$*, then the results are the same when imposing additional constraints. (See the example of the KP-equation in the next section.)*

2)If the interaction Hamiltonian (1.4) is of the fourth order: $H_{int} = H^{(4)}$ *and there are finite limits of the functions:* $\omega(k)$ *while* $k \to 0$ *and* W_{1234} *while* $s_1k_1 - s_2k_2 \to 0$*, then there is a canonical transformation (7. 4), mapping this system to the form (7.5) and its Hamiltonian to the normal form (7.6).*

If W_{1234} *doesn't have a limit when* $s_1k_1 - s_2k_2 \to 0$*, then this singularity is to be analyzed separately (e.g. for the Davey-Stewartson equation this statement is valid without change).*

In the situation B (when $V_{012} \neq 0$ on Γ_3) the set of motion invariants is not enough for the integrability in the periodic case. So the system is not integrable, though it possesses an infinite set of motion integrals and it is solvable by Inverse Scattering Method (see sect. 9).

The theorem 7.1 remains true for any dimension $d \geq 2$.

8 Examples from Physics

Let us discuss some universal physical models in view of the above results.

Propagation of weakly nonlinear and weakly dispersive waves is described by the well known **Kadomtsev-Petviashvili equation** (KP-equation)

$$(u_t + 6uu_x + u_{xxx})_x = 3\alpha^2 u_{yy}. \tag{8.1}$$

Its dispersion law

$$\omega = p^3 + \frac{3\alpha^2 q^2}{p} \tag{8.2}$$

is of decay type for $\alpha^2 = 1$ (equation KP1) and of nondecay type for $\alpha^2 = -1$ (equation KP2).

In the form (8.1) KP equation is not quite defined, since (8.1) is equivalent to the equation

$$u_t + 6uu_x + u_{xxx} = 3\alpha^2 \int_0^x u_{yy} dx' + f(y,t) \tag{8.3}$$

with an arbitrary function $f(y,t)$, which should be specified.

Let us consider, at first, KP equation with periodic boundary conditions

$$u(x+A,y) = u(x,y+B) = u(x,y). \tag{8.4}$$

From (8.3-4) we have

$$\int_0^A u dx = \text{const}, \tag{8.5}$$

$$\int_0^A dx \{ \int_0 x u_{yy} dx' + f(y,t) \} = 0 \tag{8.6}$$

If the function f is defined from the condition (8.6):

$$f(y,t) = \frac{3\alpha^2}{A} \int_0^A x u_{yy} dx \tag{8.7}$$

then the condition (8.5) will be preserved under time evolution, governed by the equation (8.3). The additional term (8.7) was obtained in [22]. With regularization (8.3,5,7) the KP equation after the transform to the normal variables a_k

$$u(x,y,t) = \sum_{p\neq 0,q} a_k(t) \sqrt{|p|} e^{i(px+qy)} \; (p = \frac{2\pi}{A} m, q = \frac{2\pi}{B} n) \tag{8.8}$$

acquires the Hamiltonian form (1.3-5) with

$$H_{int} = \frac{1}{3} \sum_{p\neq 0,q,p_1\neq 0,q_1,p_2\neq 0,q_2} \sqrt{|pp_1p_2|} a a_1 a_2 \delta(k - k_1 - k_2) \tag{8.9}$$

(The amplitudes a_k with $p=0$ are excluded from this summation.) It is well known that the KP equation has an infinite set of motion integrals. However the structure of this set is essentially different for KP1 and KP2.

In the case of KP2 ($\alpha^2=-1$) the dispersion law has nondecay type, and therefore the condition A) of the Alternative (in sect.3) is realized. Hencc KP2 has an integral of motion (2.2) with an arbitrary function ϕ_k. The equation KP2 is integrable. Recently this conclusion was obtained on the basis of algebrogeometric approach [23]. We should stress that our results were obtained on the level of formal power series and that the convergence of these series has not been proved.

In the case of KP1 ($\alpha^2=1$) the dispersion law is of the decay type, and $V_{012}\neq 0$ on Γ_3. Hence, according to the Alternative the dispersion law (8.2) with ($\alpha^2=1$) should be degenerate. Indeed it can be parametrized in the form (4.4); the functions ϕ_k that can start the motion integrals (2.2) are parametrized in the form

$$p=\xi-\eta, q=\xi^2-\eta^2, \phi=f(\xi)-f(\eta) \tag{8.10}$$

where $f(\xi)$ is an arbitrary function of one variable.

Thus the equation KP1 is not integrable under periodic boundary conditions.

When studying the KP equation on the entire plane with boundary condition

$$u(r,t)\to 0 \text{ while } r\to\infty \tag{8.11}$$

one meets the well-known difficulties connected to the infinite speed of disturbance propagation. The same difficulties arise for the simple linear equation

$$u_{tx}=u_{yy} \tag{8.12}$$

and in general for any equation whose dispersion law is not analytic function.

The solutions of the KP equation cannot decrease rapidly (e.g. exponentially) when $r\to 0$. Even if the initial condition $u(r,0)$ is a function with finite support, the solution $u(r,t)$ for $t\geq 0$ will have tails decreasing like $1/r$.

In the case (8.11) the function $f(y,t)$ equal zero, and the KP equation implies infinite set of constraints; two first of them

$$\int_{-\infty}^{+\infty} u dx=0, \int_{-\infty}^{+\infty} dx \int_{-\infty}^{x} u_{yy} dx'=0 \tag{8.13}$$

are the same as for the linear equation (8.12) (compare with (8.5-6)); the next constraints are nonlinear.

All the solutions of KP, constructed by the inverse scattering transform method, automatically satisfy these constraints. (So that the set of solutions, satisfying all the constraints, is sufficiently large.) These constraints were obtained in [24] with the aid of inverse scattering transform method. They are preserved in time by the KP equation. One can say that they form an infinite set of motion integrals, which are all equal to zero. The KP equation possesses also an infinite set of "genuine" motion integrals (not necessarily equal zero), which can be obtained e.g. from the motion integals of periodic KP by limiting transition while $A, B \to +\infty$ (see also [25]).

One can expand these motion integrals in a power series (2.2). For KP1 the function ϕ_k in (2.2) has to be parametrized in the form (8.10) and therefore it is singular when $p \to 0$. Hence the coefficients $F_{012}, G_{1234}, \ldots$ in (2.2) are also singular in the case of the equation KP1. E.g. if $f(\xi) = 2\xi^4$ in (8.10) then

$$\phi_k = qp^2 + \frac{q^3}{p^2}, \quad F_{012} = \left(\frac{q}{p} + \frac{q_1}{p_1} + \frac{q_2}{p_2}\right)\sqrt{|pp_1p_2|} \tag{8.14}$$

Nevertheless the integrals are finite due to the infinite set of constraints. E.g. the constraints (8.13) guarantee that the integral (2.2) is finite in the case (8.14).

In the case of the equation KP2 there are motion integrals (2.2) with bounded coefficients $F_{012}, G_{1234}, \ldots$.

It is interesting and instructive to examine the **second equation in the hierachy of KP1 equations**, which are known to have infinite number of motion integrals. The Hamiltonian of this equation is the "first" motion integral for KP; it has dispersion law

$$\omega_k = qp^2 + \frac{q^3}{p^2} \tag{8.15}$$

and matrix element

$$V_{012} = \left(\frac{q}{p} + \frac{q_1}{p_1} + \frac{q_2}{p_2}\right)\sqrt{|pp_1p_2|}. \tag{8.16}$$

(see (8.14)).The dispersion law (8.15) has decay type; it is degenerate and can be parametrized in the form (4.1) with

$$\alpha = \xi, \beta = \xi^2, \gamma = 2\xi^4. \tag{8.17}$$

However the whole resonance manifold

$$p = p_1 + p_2, \quad q = q_1 + q_2, \tag{8.18}$$

$$qp^2 + \frac{q^3}{p^2} = q_1p_1^2 + \frac{q_1^3}{p_1^2} + q_2p_2^2 + \frac{q_2^3}{p_2^2} \tag{8.19}$$

cannot be parametrized in the form (4.10). The manifold (8.18-19) turns out to consist of two components; one of them coincides with the resonance manifold of KP1 and can be parametrized in the form (4.10); the other component is defined by the equations

$$p = p_1 + p_2,\ q = q_1 + q_2,\ qp^2 = q_1 p_1^2 + q_2 p_2^2 \tag{8.20}$$

(it comes from (8.18-19) when $q \ll p$). According to the results of sect. 6 the matrix element (8.16) should equal zero on the component (8.20). It really does! This turns out to be connected to the invariants of Rossby waves. In the limiting situation $|k| \ll 1, |p| \ll |q|$ the dispersion of Rossby waves and the invariants $\phi^{(1)}, \phi^{(2)}$ acquire the form

$$\omega = pq^2,\ \phi^{(1)} = \frac{p^3}{q^2},\ \phi^{(2)} = \frac{p}{q} \tag{8.21}$$

(see [14,26]); the function $\omega + \phi^{(1)} = pq^2 + p^3/q^2$ is an invariant as well. So, transposing p and q, we find that the equations

$$p + p_1 + p_2 = 0,\ q + q_1 + q_2 = 0,\ qp^2 + q_1 p_1^2 + q_2 p_2^2 = 0 \tag{8.22}$$

imply the following equations

$$(qp^2 + \frac{q^3}{p^2}) + (q_1 p_1^2 + \frac{q_1^3}{p_1^2}) + (q_2 p_2^2 + \frac{q_2^3}{p_2^2}) = 0, \tag{8.23}$$

$$\frac{q}{p} + \frac{q_1}{p_1} + \frac{q_2}{p_2} = 0. \tag{8.24}$$

Changing k for$-k$ in (8.22-24) we find that (8.20) implies: 1) the equality (8.19), so that manifold (8.20) really is a component of (8.18-19); 2) the equality (8.24), so that the matrix element in (8.16) equals zero on the component (8.20).

Consider also the Veselov-Novikov equation [27]

$$v_t = \partial^3 v + \bar\partial^3 v + \partial(uv) + \bar\partial(\bar u v),$$
$$\bar\partial u = -3\partial v, v = \bar v.$$

Here $\partial = \partial_z = \partial_x - i\partial_y, z = x + iy$. If u, v doesn't depend on y, this equation is reduced to the Kortveg-de-Vries equation. The Veselov-Novikov equation can be solved via the inverse scattering transform method[27] and poss esses an infinite number of motion integrals.

The dispersion law of this equation

$$\omega_k = 2(p^3 - 3q^2) \tag{8.25}$$

is nondegenerate and therefore the condition A) of the Alternative is realized. Hence the properties of the Veselov-Novikov equation are similar to the properties of the equation KP2: this equation is integrable in the periodic case, and can be transformed to the normal form (7.5-6), its scattering matrix is trivial.

The Davey-Stewartson equation

$$i\psi_t + (\partial_x^2 - \partial_y^2)\psi + u\psi = 0$$
$$\Delta u = (\partial_x^2 - \partial_y^2)|\psi|^2 \tag{8.26}$$

arises in the theory of two-dimensional long waves over a finite depth fluid [28]. Also, the problem of interaction of small-amplitude quasimonochromatic wave packets with acoustic waves leads in natural way to the equation (8.26).

The equation (8.26) is solvable by inverse scattering transform method and possesses an infinite number of motion integrals. Its 3-wave matrix element equals zero identically and the condition A of the Alternative is realized. Hence the Davey-Stewartson equa tion has an integral of motion (2.2) with an arbitrary function ϕ_k. This equation is integrable and reducible to the normal form (7.5-6); its scattering matrix is trivial.

Finally let us consider **the system of three resonantly interacting wave packets**

$$\begin{aligned} \dot{A}_0 + (v_0, \nabla)A_0 &= A_1 A_2, \\ \dot{A}_1 + (v_1, \nabla)A_1 &= A_0 A_2^*, \\ \dot{A}_2 + (v_2, \nabla)A_2 &= A_0 A_1^* \end{aligned} \tag{8.27}$$

The 3-wave resonance manifold for this system

$$k = k_1 + k_2, (v_0, k) = (v_1, k_1) + (v_2, k_2) \tag{8.28}$$

can be parametrized in the form (4.10) and admits an infinite number of invariants. The system of kinetic equations for the system (8.27) possesses an infinite number of conservation laws. The system (8.27) is known to be integrable for rapidly decreasing initial conditions. However, the system (8.27) is not integrable in the periodic case. It has no extra integral of motion (besides energy, momentum, Manly-Rough invariants). The condition of coincidence of singular manifolds (described in sect. 6) doesn't hold for the system of three wave packets[20]. So, this system doesn't have any second order approximate integral of motion.

9 Connection to Dressing Method

With the aid of dressing method [29] one can derive integrable, or at least solvable by inverse scattering transform method, nonlinear differential equations. It turns out that these equations possess Hamiltonian structure. What are the dispersion l aws and the matrix elements of the equations obtained by the dressing method?

The dispersion laws of such equations can be parametrized in the form (4.1), and the formular for their matrix elements is derived in [20].

Let us consider the dressing method for a single type wave system in the form of scalar $\bar{\partial}$-problem:

$$\frac{\partial \psi}{\partial \bar{\lambda}} = \int \psi(\mu) R(\mu, \lambda) d\mu \tag{9.1}$$

with prolonged derivatives

$$\left.\begin{aligned} D_1 &= \tfrac{\partial}{\partial x} + i\alpha(\lambda), \\ D_2 &= \tfrac{\partial}{\partial y} + i\beta(\lambda), \\ D_3 &= \tfrac{\partial}{\partial t} - i\gamma(\lambda); \end{aligned}\right\} \tag{9.2}$$

$$R(\mu, \lambda) = e^{i[\alpha(\mu)x + \beta(\mu)y - \gamma(\mu)t]} \mathcal{R}(\mu, \lambda) e^{-i[\alpha(\lambda)x + \beta(\lambda)y - \gamma(\lambda)t]}. \tag{9.3}$$

For definiteness we consider the function $\psi(\lambda)$ to be normed by 1 at $\lambda \to \infty$:

$$\psi(\lambda) = 1 + \chi(\lambda) = 1 + \frac{\Phi}{\lambda} + \frac{\Phi_1}{\lambda^2} + \cdots \tag{9.4}$$

($\chi(\lambda)$ is a continuous function over the whole complex plane). The dressing method leads us to some nonlinear equation for the function $\Phi = \Phi(x, y, t)$, which after Fourier transform

$$\Phi(x, y, t) = \frac{1}{2\pi} \int b_k(t) e^{i(px + qy)} dp dq \qquad (k = (p, q)) \tag{9.5}$$

can be written as follows

$$i\dot{b}_k^s = \omega b_k^s + \frac{1}{2\pi} \int U_{-0,1,2} \delta(-sk + s_1 k_1 + s_2 k_2) dk_1 dk_2 + \cdots \tag{9.6}$$

($s = \pm 1, b^1 = b, b_{-1} = b^*$, dots signify the terms of higher order). The equation (9.1) with condition (9.4) is equivalent to the following integral equation

$$\chi(\lambda) = \int \frac{1}{2\pi(\lambda - \mu)} [1 + \chi(\nu)] R(\mu, \lambda) d\mu d\nu \tag{9.7}$$

which one can solve (for sufficiently small R) by successive approximation method. In this way we get for the function Φ

$$\begin{aligned} \Phi(x, y, t) &= \int \frac{R(\nu, \mu)}{2\pi} d\mu d\nu \\ &+ \int \frac{R(\nu_1, \mu_1)}{2\pi} \frac{R(\nu_2, \mu_2)}{2\pi} \frac{1}{\nu_1 - \mu_2} d\mu_1 d\nu_1 d\mu_2 d\nu_2 + \cdots \end{aligned} \tag{9.8}$$

The first term in this series defines the solution Φ_0 of linearized equation (when we neglect nonlinear terms in (9.6))

$$\Phi_0 = \int \frac{\mathcal{R}(\nu, \mu)}{2\pi} e^{i\{[\alpha(\nu) - \alpha(\mu)]x + [\beta(\nu) - \beta(\mu)]y + [\gamma(\nu) - \gamma(\mu)]t\}} d\mu d\nu \tag{9.9}$$

(see [30]). Thus the dispersion law $\omega(p,q)$ of the equation (9.6) can be parametrized in the form of differences:

$$\begin{aligned} p &= \alpha(\nu) - \alpha(\mu), \\ q &= \beta(\nu) - \beta(\mu), \\ \omega &= \gamma(\nu) - \gamma(\mu) \end{aligned} \tag{9.10}$$

(compare with (4.1)).

The second term in the series (9.8) gives us the coefficient function U_{012} in (9.6)

$$U_{012} = \frac{s\omega + s_1\omega_1 + s_2\omega_2}{4\pi}\left(\frac{1}{\nu_1 - \mu_2} + \frac{1}{\nu_2 - \mu_1}\right) \tag{9.11}$$

where $\mu_1, \nu_1, \mu_2, \nu_2$ are defined from the following system of algebraic equations

$$\left.\begin{aligned} p_1 = \alpha(\nu_1) - \alpha(\mu_1), p_2 = \alpha(\nu_2) - \alpha(\mu_2), \\ q_1 = \beta(\nu_1) - \beta(\mu_1), q_2 = \beta(\nu_2) - \beta(\mu_2). \end{aligned}\right\} \tag{9.12}$$

On the resonance manifold (3.2) the function (9.11) acquires the form

$$U_{012} = \frac{1}{4\pi}\frac{D(\mu,\nu,\lambda)}{d(\mu,\nu)} \tag{9.13}$$

where μ, ν, λ are such numbers that

$$\begin{aligned} sp = \alpha(\mu) - \alpha(\nu), s_1p_1 = \alpha(\nu) - \alpha(\lambda), s_2p_2 = \alpha(\lambda) - \alpha(\mu), \\ sq = \beta(\mu) - \beta(\nu), s_1q_1 = \beta(\nu) - \beta(\lambda), s_2q_2 = \beta(\lambda) - \beta(\mu), \\ s\omega = \gamma(\mu) - \gamma(\nu), s_1\omega_1 = \gamma(\nu) - \gamma(\lambda), s_2\omega_2 = \gamma(\lambda) - \gamma(\mu) \end{aligned}$$

(the functions D and d are defined in (5.11-12)). After the following change of variables in the equation (9.6)

$$b_k = \frac{a_k}{\sqrt{d(\mu,\nu)}} \tag{9.14}$$

we get a similar equation for a_k with the function

$$V_{012} = \frac{1}{4\pi}\frac{D(\mu,\nu,\lambda)}{\sqrt{|d(\mu,\nu)d(\nu,\lambda)d(\lambda,\mu)|}} \tag{9.15}$$

instead of U_{012}. The function (9.15) possesses the following symmetries

$$V_{012} = V_{102} = V_{210} \tag{9.16}$$

and hence the equation (9.6) possesses Hamiltonian structure (up to the second order in wave amplitudes). The function (9.15) is the matrix element of 3-wave interaction.

The expression (9.15) coincides with the expression (5.14) which we obtained supposing that our nonlinear Hamiltonian equation (1.5) possesses a second order approximate integral of motion (see sect. 5). If the equation (1.5) had an exact extra integral of motion then the matrix elements of higher nonlinear processes would also be uniquely determined after a choice of functions α, β, γ. Each equation (1.5) with an extra integral of motion, for which the functions α, β, γ and a ll matrix elements are meromorphic functions, can be constructed by $\bar{\partial}$-problem with long derivatives (9.2).

It also follows from (9.9) that the support of the function $\mathcal{R}(\nu, \mu)$ should belong to the following 2-D manifold defined by the equations

$$\alpha(\nu) - \alpha(\mu) = \bar{\alpha}(\nu) - \bar{\alpha}(\mu), \beta(\nu) - \beta(\mu) = \bar{\beta}(\nu) - \bar{\beta}(\mu) \qquad (9.17)$$

in order that the solution Φ be finite, when $x^2 + y^2 \to \inf$ (see [30]). Thus the function $\mathcal{R}(\nu, \mu)$ actually depends on two *real* variables but not on two *complex* variables.

We know from section 5 that if there exists a second order approximate integral of motion in the case of nontrivial 3-wave interaction ($V_{012} \neq 0$ on Γ_3) then the dispersion law can be parametrized in the form (4.1) with real functions α, β:

$$\alpha(\lambda) = \bar{\alpha}(\lambda), \beta(\lambda) = \bar{\beta}(\lambda), \lambda = \bar{\lambda}. \qquad (9.18)$$

This condition is stronger than (9.17). In general case (9.10) (with complex functions α, β) the resonance manifold Γ_3 can consist of several components; some of them can be parametrized in the form (4.1) with real functions α, β, and on the other components the matrix element V_{012} turns into zero.

10 Conclusion

In the present paper we have aimed to show that an approach similar to the Poincare and Birkhoff analysis of the interability of dynamic systems, based on the study of the perturbation theory series, proves to be very effective to test integrability properties of n onlinear wave systems (see [6,7,8,9]). This approach leads to rather deep understanding of the difference between solvability in the inverse scattering sense and integrability in the Liouville sense and to classification of solvable models by their dispersion laws. The dispersion laws and resonance manifolds of the wave systems can be effectively analyzed with the aid of Web geometry (see [15,16,18]).

If you have some nonlinear Hamiltonian wave system (in dimension $d = 2$) you can analyze it in the following way.

First, one should investigate the resonance manifold of the system. If the dispersion law is degenerate the system possesses extra first order approximate integrals of motion regardless to the form of its matrix element. The corresponding kinetic equat ion for weak turbulence possesses extra (exact) integrals of motion. One should study the web associated with the resonance manifold to determine how many invariants the system has and what their form is.

Suppose that we know from some consideration that our system possesses at least one extra integral of motion. Then it has infinite set of motion integrals.

If in addition the matrix element V_{012} equals zero on the whole resonance manifold Γ_3 (the condition A of the alternative is realized (see sect. 3)) then the wave system is integrable in the classical Liouville sense: one can introduce action-an gle variables (see (7.5-6)); the scattering matrix of this system is trivial: the asymptotic states at $t \to \pm\infty$ coincide. This situation is realized e.g. for the equation KP2, Davey-Stewartson equation, Veselov-Novikov equation. 3-wave reso nance interaction is absent in such equations.

For the wave system with nontrivial 3-wave interaction to have extra integral of motion, the web associated with the resonance manifold should be of the simplest, coordinate, type and the matrix element V_{012} should have the form (5.14). In this case the set of motion integrals is much narrower and it is not enough for integrability, though this system is sovable by inverse scattering transform method.

A number of questions relating to the topic of this paper still remain open. For example, it is not known whether there exists a wave system that possesses an additional integral of motion with zero quadratic part.

References

[1] H. Poincare: *New Methods of Celestial Mechanics* (NASA, Springfield, Va., 1967) v.1.

[2] G. D. Birkhoff: *Dynamic Systems*, Colloquium publications, v.9 (American Mathematical Society, Providence, RI, 1966.

[3] V. E. Zakharov: *Hamiltonian Formalism in the Theory of Waves in Nonlinear Media with Dispersion*, Izvestya VUZov, Radiofizika **17** (1974) 431-453.

[4] V. E. Zakharov, S. L. Musher, A. M. Rubenchik: *Hamiltonian Approach to the Description of Nonlinear Plasma Phenomena*, Phys. Reports **129** (1985) 285-366.

[5] V. E. Zakharov, E. A. Kuznetsov: *Hamiltonian Formalism for Systems of hydrodynamic type*, Sov. Sci. Rev., SEc. C, Math. Phys. Rev. 4 (1984) 167-220.

[6] V. E. Zakharov, E. I. Schulman: Physica D1 (1980) 191-202.

[7] V. E. Zakharov: *Integrable Systems in Multidimensional spaces*, in: Lect. Notes Phys. **153** (Springer, Berlin, Heidelberg, 1983) p. 120-216.

[8] V. E. Zakharov, E. I. Schulman: Physica D29 (1988) 283-320.

[9] V. E. Zakharov, E. I. Schulman. *Integrability of Nonlinear Systems and Perturbation Theory*, in: *What is integrability?*, ed. by V. E. Zakharov (Springer-Verlag,1991) p.185-250.

[10] K. Hasselmann: *On the Nonlinear Energy Transfer in Gravity-Wave Spectrum*, J. Fluid Mech. **12** (1962) 481-500.

[11] B. B. Kadomtsev: *Plasma Turbulence* (Academic Press, New York 1965).

[12] V. E. Zakharov, V. S. L'vov: *Statistical Description of Nonlinear Wave Fields.* Izvestiya VUZov, Radiofizika **28** (1975) 1470.

[13] S. V. Manakov: Physica D3 (1981) 420.

[14] A. M. Balk: A new invariant for Rossby wave systems, Phys. lett. A155 (1991) 20-24.

[15] A. M. Balk, E. A. Ferapontov: *Invariants of Wave Systems and Web Geometry*, in: *Advances in Soviet Mathematics* (American Mathematical Society, 1992), in press.

[16] A. M. Balk, E. A. Ferapontov: *Two Dimensional Wave Systems with Infinite Number of Invariants*, Physica D, to be published.

[17] W. Blaschke: *Einführung in die Geometrie der Waben*, Birkhäuser-Verlag, 1955.

[18] A. M. Balk, E. A. Ferapontov: *Invariants of 1-dimensional wave systems*, Physica D, to be published.

[19] E. I. Schulman, D. P. Tskhakaja: Teor. Mat. Fiz, 1991.

[20] A. M. Balk, V. E. Zakharov: *Systems with an Extra Integral of Motion*, to be published.

[21] E. I. Schulman: Teor. Mat. Fiz. **76** (1988) 88-89.

[22] A. Reiman, M. Semionov-Tian Shanskii: Proc. of LOMI Scientific Seminars **133** (1984) 212-227.

[23] I. M. Krichever: *Spectral Theory of two-dimensional periodic operators*, Uspekhi Mat. Nauk (1989) 121-184.

[24] J. Lin, H. H. Chen: *Costraints and Conserved Quantities of the Kadomtsev-Petviashvili Equations*, Phys. Lett. **A89** (1982) 163-167.

[25] V. G. Bakurov: *On the Complete Integrability of the Kadomtsev-Petviashvili Equation*, Phys. Lett. **A160** (1991) 367-371.

[26] A. M. Balk, S. V. Nazarenko, V. E. Zakharov: *New invariant for drift Turbulence*, Phys. Lett. **A152** (1990) 276-280.

[27] A. P. Veselov, S. P. Novikov: Dokl. Akad. Nauk **279** (1984) 784-788.

[28] A. Daveay, K. Stewartson: Proc. Roy. Soc. Lond. **A338** (1974) 101-110.

[29] V. E. Zakharov, A. B. Shabat: *Integration of Nonlinear Evolution Equations of Mathematical Physics by the Inverse Scattering Transform Method*, Funct. Anal. Appl. **8** (1974) 43-53; **13** (1979) 13-22.

[30] L. V. Bogdanov, V. E. Zakharov: *The Decaying Solutions and Dispersion Laws in (2+1)-dimensional dressing method*, Algebra i Analiz **3** (1991) 49-56.

Part IV

Quantum and Statistical Mechanical Models

The Quantum Correlation Function as the τ Function of Classical Differential Equations

A.R. Its[1], *A.G. Izergin*[2], *V.E. Korepin*[3], *and N.A. Slavnov*[4]

[1]Department of Mathematics and Computer Science,
Clarkson Unversity, Potsdam, NY 13699-5818, USA
[2]St. Petersburg Branch (LOMI), Mathematical Institute,
Fontanka 27, 191011 St. Petersburg, Russia
[3]Institute for Theoretical Physics, SUNY at Stony Brook,
Stony Brook, NY 11794-3840, USA
[4]Mathematical Institute of Academy of Sciences, Moscow, Russia

We consider the quantum version of completely integrable differential equations. We show that quantum correlation functions can be described as τ functions of completely integrable classical differential equations. Our main example is the nonlinear Schrödinger equation.

1 Introduction

The Inverse Scattering Method helps to solve a lot of interesting nonlinear partial differential equations. We will be interested here in evolution differential equations. Many of these equations have Hamiltonian structure [1]. They can be quantized. After quantization these equations are still completely integrable. The Quantum Inverse Scattering Method is explained in detail in the book [2]. We shall concentrate on quantum correlation functions. The quantum correlation function can be described by another completely integrable differential equation (which is closely related to the original one that was quantized). This is the subject of our paper. There are a lot of similarities between different completely integrable differential equations. We shall pick the nonlinear Schrödinger equation as our example:

$$i\partial_t \Psi = -\partial_x^2 \Psi + 2c\Psi^\dagger \Psi^2. \tag{1.1}$$

The Hamiltonian H of the model is

$$H = \int dx \{\partial_x \Psi^\dagger \partial_x \Psi + c\Psi^\dagger \Psi^\dagger \Psi \Psi\} \tag{1.2}$$

After quantization Ψ becomes a quantum operator in a Fock space with the canonical Bose commutation relation

$$\left[\Psi(x), \Psi(y)^\dagger\right] = \delta(x-y). \tag{1.3}$$

The vacuum $\mid 0\rangle$ and dual vacuum $\langle 0 \mid$ are defined in the standard way

$$\Psi(x) \mid 0\rangle = 0$$

$$\langle 0 \mid \Psi^{\dagger}(y) = 0$$

The quantum version of the nonlinear Schrödinger equation is equivalent to the quantum mechanics of N particles with the Hamiltonian:

$$H_N = -\sum_{j=1}^{N} \frac{\partial^2}{\partial x_j^2} + 2c \sum_{1\leq j<k\leq N} \delta(x_j - x_k). \tag{1.4}$$

The temperature correlation function $\langle \Psi^{\dagger}(x_1,t_1)\Psi(x_2,t_2)\rangle_T$ is defined as

$$\langle \Psi^{\dagger}(x_1,t_1)\Psi(x_2,t_2)\rangle_T = \frac{tr\left(e^{(-\frac{H}{T})}\Psi^{\dagger}(x_1,t_1)\Psi(x_2,t_2)\right)}{tr\left(e^{-\frac{H}{T}}\right)} \tag{1.5}$$

Here tr means the trace in the whole Fock space. It was described by a completely integrable differential equation in [3,4]. To define the quantum correlation function at zero temperature one needs to construct the eigenfunction of the Hamiltonian $\mid \Omega\rangle$, which minimizes the energy in the sector with the fixed density. The zero temperature completely integrable differential equation is defined as follows:

$$\langle \Psi^{\dagger}(x_1,t_1)\Psi(x_2,t_2)\rangle = \frac{\langle \Omega \mid \Psi^{\dagger}(x_1,t_1)\Psi(x_2,t_2) \mid \Omega\rangle}{\langle \Omega \mid\mid \Omega\rangle}. \tag{1.6}$$

It was described by a completely integrable differential equation in [4,5]. In this paper we shall outline the main ideas behind the calculation of the quantum correlation function. First one should represent the quantum correlation function as the determinant of some integral operator (of Fredholm type). Actually all Fredholm integral operators which appear in this relation are very special; they are described in Sect. 2. In Sect. 3 it is explained how each of these Fredholm operators are related to a very special Riemann-Hilbert problem, that is used later for the evaluation of asymptotics of the quantum correlation function. In Sect. 2 it is also explained that to invert each of our integral operators we should consider some integral equation. We can differentiate this integral equation with respect to the parameters of the correlation function to obtain two linear differential equations for the solution of this integral equation. This is similar to [6,7]. This pair of linear differential equations can be considered as the Lax representation for some new completely integrable differential equation, which is associated with the quantum correlation function. In other words, we consider our Fredholm integral operator as the Gelfand-Levitan-Marchenko integral operator for this new completely integral differential equation. The whole case of the Fredholm integral operator which appears in relation to the completely integrable differential equation is described in Sect. 2.

2 Linear Integral Operators of the Fredholm Type

Every quantum correlation function can be represented as a determinant [see [15] [8-12]].

$$\det(1+\widehat{V}) \tag{2.1}$$

of some integral operator. Let us discuss the kernel $V(\lambda,\mu)$ of the integral operator $\widehat{V}$. Here λ and μ are integration variables. In the generic situation $\widehat{V}$ is an integral operator defined on the whole real axis $\{\lambda \epsilon R\}$ (an interval will be a special case, when our functions $e_j(\lambda), E_k(\mu)$ [defined below] are finite [equal to zero outside the interval]). All of these Fredholm integral operators belong to a special class: the kernel of each of them $V(\lambda,\mu)$ can be represented in the form

$$V(\lambda,\mu) = \frac{1}{\lambda-\mu}\sum_{j=1}^{N} e_j(\lambda)E_j(\mu)\,. \tag{2.2}$$

Sometimes we shall call such operators "integrable" integral operators. Here $e_j(\lambda)$ are N linearly independent functions (they are also supposed to be continuous and integrable, rapidly decaying at infinity). The functions $E_j(\lambda)$ have similar properties. In all examples related to correlation functions, the kernel (2.2) is nonsingular. This means that

$$\sum_{j=1}^{N} e_j(\lambda)E_j(\lambda) = 0\,. \tag{2.3}$$

But first we shall study a more general situation, where the condition (2.3) is not necessarily satisfied and the integral operator (2.2) is singular. Suppose we act with the integral operator on the function $\varphi(\mu)$; the result will be:

$$(\widehat{V}\varphi)(\lambda) = \int_{-\infty}^{\infty} V(\lambda,\mu)\varphi(\mu)d\mu\,.$$

Here we shall understand the integral as the principal value. We will assume that the operator $1+\widehat{V}$ is not degenerate.

It is interesting to mention that the product of two such operators has the same form. Let us take two different integrable operators, $\widehat{V}^{(1)}$ and $\widehat{V}^{(2)}$,

$$V^{(a)}(\lambda,\mu) = \frac{1}{\lambda-\mu}\sum_{j=1}^{N_a} e_j^{(a)}(\lambda)E_j^{(a)}(\mu)\,,\quad a=1,2. \tag{2.4}$$

Let us show that their product is also an "integrable" operator:

$$\begin{aligned} V^{(3)}(\lambda,\mu) &= \fint d\nu V^{(1)}(\lambda,\nu)V^{(2)}(\nu,\mu) \\ &= \frac{1}{\lambda-\mu}\left\{\sum_{j=1}^{N_1} e_j^{(1)}(\lambda)E_j^{(3)}(\mu) + \sum_{k=1}^{N_2} e_k^{(3)}(\lambda)E_k^{(2)}(\mu)\right\} \end{aligned} \tag{2.5}$$

with

$$\begin{aligned} E_j^{(3)}(\mu) &= \fint d\nu E_j^{(1)}(\nu)V^{(2)}(\nu,\mu)\,, \\ e_j^{(3)}(\lambda) &= \fint d\nu V^{(1)}(\lambda,\nu)e_j^{(2)}(\nu) \end{aligned} \tag{2.6}$$

which becomes obvious if one uses the relation

$$\frac{1}{(\lambda-\nu)(\nu-\mu)} = \frac{1}{\lambda-\mu}\left\{\frac{1}{\lambda-\nu}+\frac{1}{\nu-\mu}\right\} . \tag{2.7}$$

All integrals here are understood in the sense of the principal value. But if (2.3) is valid for each operator then it is also valid for the product.

This is a very important property. In particular, it results in the following universal construction for the resolvent operator $\widehat{R}$ (with kernel $R(\lambda,\mu)$). Let us define the resolvent as usual

$$(1+\widehat{V})(1-\widehat{R}) = 1; \quad (1+\widehat{V})\widehat{R} = \widehat{V} \tag{2.8}$$

$$R(\lambda,\mu) + ⨍ d\nu V(\lambda,\nu)R(\nu,\mu) = V(\lambda,\mu) . \tag{2.9}$$

To write down the resolvent more explicitly, one introduces functions $f_j^L(\lambda)$ and $f_j^R(\lambda)$ by the following integral equations:

$$f_j^L(\lambda) + ⨍ V(\lambda,\mu) f_j^L(\mu) d\mu = e_j(\lambda) \quad (j=1,\dots,N) \tag{2.10}$$

$$f_j^R(\lambda) + ⨍ f_j^R(\mu) V(\mu,\lambda) d\mu = E_j(\lambda) . \tag{2.11}$$

These relations can also be rewritten in the form

$$f_j^L(\lambda) = e_j(\lambda) - ⨍ R(\lambda,\mu) e_j(\mu) d\mu \tag{2.12}$$

$$f_j^R(\lambda) = E_j(\lambda) - ⨍ E_j(\mu) R(\mu,\lambda) d\mu . \tag{2.13}$$

Theorem The resolvent $\widehat{R}$ of the operator $\widehat{V}$ (2.2) is given in terms of the functions f_j^L, f_j^R by the following expression:

$$R(\lambda,\mu) = \frac{1}{\lambda-\mu} \sum_{j=1}^{N} f_j^L(\lambda) f_j^R(\mu) . \tag{2.14}$$

Proof: Equation (2.9) defining the resolvent kernel can be rewritten as

$$(\lambda-\mu)R(\lambda,\mu) + \sum_{j=1}^{N} e_j(\lambda) ⨍ E_j(\nu) R(\nu,\mu) d\nu +$$

$$+ ⨍ V(\lambda,\nu)[(\nu-\mu)R(\nu,\mu)] d\nu = \sum_{j=1}^{N} e_j(\lambda) E_j(\mu) \tag{2.15}$$

after multiplication of equation (2.9) by $(\lambda-\mu)$. Let us now move the second term in the l.h.s. to the r.h.s. and use (2.13); then we get

$$(\lambda-\mu)R(\lambda,\mu) \quad +$$

$$+ \quad ⨍ V(\lambda,\nu)[(\nu-\mu)R(\nu,\mu)] d\nu = \sum_{j=1}^{N} e_j(\lambda) f_j^R(\mu) . \tag{2.16}$$

Here we used (2.13). We can now act on this equation with the integral operator

$1 - \hat{R}$ from the left and we shall get

$$(\lambda - \mu)R(\lambda, \mu) = \sum_{j=1}^{N} f_j^L(\lambda) f_j^R(\mu) \tag{2.17}$$

by means of (2.8) and (2.12). In this way the theorem is proved. (All integrals are, as usual, to be understood in the principal value sense.)

If we consider the regular subgroup (2.3) then eq. (2.9) shows that the function $R(\lambda, \mu)$ has no singularities at $\lambda = \mu$. This means that the f's satisfy the identity

$$\sum_{j=1}^{N} f_j^L(\lambda) f_j^R(\lambda) = 0 \, . \tag{2.18}$$

Thus we have shown that the product of two operators of the form (2.2) have the same form, and that the inverse operator also has the same form. These properties define a group. In the next section we shall associate a Riemann problem with each of these operators.

3 Riemann-Hilbert Problem

Let us show that the integral operators introduced in the previous section are related in a natural way to a very special Riemann-Hilbert problem. First of all we shall consider the subgroup of nonsingular integral operators (see(2.3)). Only this subgroup is important for quantum correlation functions.

Let us define the following Riemann-Hilbert problem: One has to construct the matrix (of dimension N) function

$$\chi(\lambda) \tag{3.1}$$

It has to have the following properties:
1) It has to be an analytic function of λ in the upper ($\Im m\lambda > 0$) half-plane and in the lower one ($\Im m\lambda < 0$).
2) At large λ it must approach the unit matrix I

$$\chi(\lambda) \to I, \qquad \lambda \to \infty \tag{3.2}$$

3) On the real λ axis the matrix function $\chi(\lambda)$ can have a discontinuity. The limit from the lower half-plane $\chi_-(\lambda)$ and the limit from the upper half plane $\chi_+(\lambda)$ have to be related on the real axis by the following formula:

$$\chi_-(\lambda) = \chi_+(\lambda) G(\lambda) \tag{3.3}$$

4) The matrix (of dimension N) function $G(\lambda)$ is defined only on the real axis. Let us write down the formula for its matrix elements:

$$G_{jk}(\lambda) = \delta_{jk} - 2\pi i e_j(\lambda) E_k(\lambda) \tag{3.4}$$

This concludes the definition of our Riemann-Hilbert problem. Now let us show

how to relate it to the integral equation (2.10)

$$f_j^L(\lambda) \quad + \quad \int_{-\infty}^{\infty} V(\lambda,\mu) f_j^L(\mu) d\mu = e_j(\lambda) \tag{3.5}$$

$$V(\lambda,\mu) \quad = \quad \frac{1}{\lambda-\mu}\sum_{j=1}^{N} e_j(\lambda)E_j(\mu) \tag{3.6}$$

$$\sum_{j=1}^{N} e_j(\lambda)E_j(\lambda) = 0 \tag{3.7}$$

Let us prove that

$$f_j^L(\lambda) = \sum_{k=1}^{N} \chi_{jk}^{+}(\lambda)e_k(\lambda). \tag{3.8}$$

Proof. First we shall represent the Riemann-Hilbert problem as a singular linear integral equation (see [13,14]):

$$\chi_{+}(\lambda) = I + \frac{1}{2\pi i}\int_{-\infty}^{\infty} \frac{\chi_{+}(\mu)(1-G(\mu))}{\mu-\lambda-i0}\, d\mu. \tag{3.9}$$

Now let us substitute the explicit expression for $G(\lambda)$ and let us also write (3.9) in terms of matrix elements

$$\chi_{jk}^{+}(\lambda) = \delta_k^j + \int_{-\infty}^{\infty} \frac{\chi_{jl}^{+}(\mu)e_l(\mu)E_k(\mu)}{\mu-\lambda-i0}\, d\mu. \tag{3.10}$$

Let us now multiply this equation by $e_k(\lambda)$ and sum with respect to K. We shall get

$$\{\chi_{jk}^{+}(\lambda)e_k(\lambda)\} = e_j(\lambda) - \int \frac{\{\chi_{jl}^{+}(\mu)e_l(\mu)\}\sum_k e_k(\lambda)E_k(\mu)}{\lambda-\mu}\, d\mu. \tag{3.11}$$

Here we omit $(i0)$ due to (3.7). Now we can use definition (3.6). Finally it is evident that for $\sum_{k=1}^{N}\chi_{jk}^{+}(\lambda)e_k(\lambda)$ we get exactly the same equation as for $f_j^L(\lambda)$. But according to our definition the integral operator $(1+\widehat{V})$ is not degenerate. Thus (3.5) has a unique solution. This means that

$$f_j^L(\lambda) = \sum_{k=1}^{N} \chi_{jk}^{+}(\lambda)e_k(\lambda) \tag{3.12}$$

So we have proven (3.8).

In special cases (for particular correlation functions), the functions $e_j(\lambda)$ depend on the parameters of the quantum correlation functions (for example, $e(\lambda) = \exp\{ix\lambda + it\lambda^2\}$). In this case we can differentiate (3.5) with respect to these parameters (x,t) and get a pair of linear differential equations for $f_j(\lambda)$; see the next section. These linear differential equations can be considered as a Lax representation. We mean that the compatibility condition for these two equations will give us a completely integrable nonlinear partial differential equation which will derive correlation functions. From the general point of view this means that we consider (3.5) as a Gelfand-Levitan-Marchenko equation for some new completely integrable system.

4 Differential Equation for Space and Time Dependent Correlation Function

To illustrate the derivation of the differential equations for the quantum correlation function let us consider the simplest two-field correlation function at zero temperature (and $c = \infty$). In this case the correlation function was represented as the Fredholm minor in [15]. Let us summarize this result:

Introducing distance x and time t by

$$x \equiv \frac{1}{2}(x_1 - x_2), \quad t \equiv \frac{1}{2}(t_2 - t_1) \tag{4.1}$$

one rewrites the determinant formula from [15] as

$$\langle \Psi(x_2, t_2)\Psi^\dagger(x_1, t_1)\rangle = -\frac{1}{2\pi} e^{2iht} b_{++} \det(1 + \widehat{\mathcal{V}}) \,. \tag{4.2}$$

Let us explain the notation. The linear integral operator $\widehat{\mathcal{V}}$ acts on arbitrary functions φ on the integral $[-q, q]$ as follows:

$$(\widehat{\mathcal{V}}\varphi)(\lambda) = \int_{-q}^{q} \mathcal{V}(\lambda, \mu)\varphi(\mu) d\mu$$

where q is the Fermi momentum. The kernel $\mathcal{V}(\lambda, \mu)$ of $\widehat{\mathcal{V}}$ is given by

$$\mathcal{V}(\lambda, \mu) = \frac{e_+(\lambda)e_-(\mu) - e_-(\lambda)e_+(\mu)}{\lambda - \mu} \,, \tag{4.3}$$

the functions $e_\pm(\lambda)$ being given by

$$\begin{aligned} e_-(\lambda) &= \frac{1}{\pi} e^{it\lambda^2 + ix\lambda} \\ e_+(\lambda) &= e_-(\lambda)E(\lambda) \,. \end{aligned} \tag{4.4}$$

The function $E(\lambda) \equiv E(\lambda \mid x, t)$ (and also the function $G \equiv G(x, t)$, which is used later) are defined by

$$G \equiv G(x,t) \equiv \int_{-\infty}^{+\infty} d\mu e^{-2it\mu^2 - 2ix\mu} \tag{4.5}$$

$$E(\lambda) \equiv E(\lambda \mid x, t) \equiv \int_{-\infty}^{\infty} \frac{d\mu}{\mu - \lambda} e^{-2it\mu^2 - 2ix\mu} \,. \tag{4.6}$$

The operator $\widehat{\mathcal{V}}$ belongs to the class of integral operators described in Sect. 2. So it is natural to introduce functions $f_\pm(\lambda)$ (see (2.10)). As the operator $\widehat{\mathcal{V}}$ is symmetric,

$$\mathcal{V}(\lambda, \mu) = \mathcal{V}(\mu, \lambda) \,, \tag{4.7}$$

the functions f^L and f^R in (2.10) and (2.11) coincide up to sign. So one has

$$f_\pm(\lambda) + \int_{-q}^{q} \mathcal{V}(\lambda,\mu) f_\pm(\mu) d\mu = e_\pm(\lambda) . \tag{4.8}$$

The "potential" $b_{++} \equiv b_{++}(x,t)$ entering (4.2) is defined as

$$b_{++} = B_{++} - G \tag{4.9}$$

where B_{++} is one of the potentials B_{lm}

$$B_{lm} = \int_{-q}^{q} e_l(\mu) f_m(\mu) d\mu, \quad l,m = +,-. \tag{4.10}$$

So we have defined all the quantities entering the right-hand side of representation (4.2) for the correlator.

In what follows the potentials C_{lm}

$$C_{lm} = \int_{-q}^{q} \mu e_l(\mu) f_m(\mu) d\mu, \quad l,m = +,- \tag{4.11}$$

will also be used.

Our first aim is to obtain a linear differential equation (actually two of them) for the solution of the integral equations (4.8). They will play the role of a Lax representation. Below we follow the ideas of the papers [7,8]. This representation generates the nonlinear evolution partial differential equations for the potentials B. Before doing this , however, let us establish some important properties of the functions introduced. Our remark is that since the operator $\widehat{\mathcal{V}}$ is integrable, one easily constructs the kernel $\mathcal{R}(\lambda,\mu)$ of the resolvent operator $\widehat{\mathcal{R}}$

$$\begin{aligned} &(1+\widehat{\mathcal{V}})(1-\widehat{\mathcal{R}}) = 1, \quad (1+\widehat{\mathcal{V}})\widehat{\mathcal{R}} = \widehat{\mathcal{V}} , \\ &\mathcal{R}(\lambda,\mu) + \int_{-q}^{q} \mathcal{V}(\lambda,\nu)\mathcal{R}(\nu,\mu) d\nu = \mathcal{V}(\lambda,\mu) . \end{aligned} \tag{4.12}$$

It can be expressed in terms of the functions $f_\pm(\lambda)$ (4.8)

$$\mathcal{R}(\lambda,\mu) = \frac{f_+(\lambda)f_-(\mu) - f_-(\lambda)f_+(\mu)}{\lambda-\mu} , \tag{4.13}$$

$$\mathcal{R}(\lambda,\mu) = \mathcal{R}(\mu,\lambda) . \tag{4.14}$$

It is easily seen now that the following relation is valid:

$$B_{+-}(x,t) = B_{-+}(x,t) \tag{4.15}$$

so that only the potentials B_{++}, B_{--} and B_{+-} are independent. However, these potentials are complex and $B_{++} \neq B_{--}$.

Differentiating (4.8) with respect to x and t one can prove the following theorem.

Theorem The two-component function $\mathbf{F}(\lambda)$

$$\mathbf{F}(\lambda) = \begin{pmatrix} f_+(\lambda) \\ f_-(\lambda) \end{pmatrix} \tag{4.16}$$

(where the functions $f_\pm(\lambda)$ are defined by the integral equations (4.8)) satisfies, for any λ, the following equations:

$$(\partial_x + i\lambda\sigma_3 - 2iQ)\mathbf{F} = 0 \tag{4.17}$$

$$(\partial_t + i\lambda^2\sigma_3 - 2i\lambda Q - iV)\mathbf{F} = 0 \tag{4.18}$$

where the 2×2 matrices Q and V (λ-independent) are expressed in terms of the potentials B by

$$Q = \begin{pmatrix} 0 & b_{++} \\ B_{--} & 0 \end{pmatrix} \tag{4.19}$$

$$V = \begin{pmatrix} 2b_{++}B_{--} & i\partial_x b_{++} \\ -i\partial_x B_{--} & -2B_{--}b_{++} \end{pmatrix}, \quad (b_{++} = B_{++} - G)\,. \tag{4.20}$$

The matrix V can also be represented as

$$V = i\partial_x U\,, \quad U = \begin{pmatrix} -B_{+-} & b_{++} \\ -B_{--} & B_{+-} \end{pmatrix}. \tag{4.21}$$

The relations (4.17) and (4.18) can be regarded as a Lax representation for some classical nonlinear evolution partial differential equation for the potentials b_{++}, B_{++}. Actually this is exactly the Lax representation for the nonlinear Schrödinger equation.

The proof of this theorem is given in [4,2].

Relations (4.17), (4.18) are valid for any λ and can be regarded as a Lax representation for some nonlinear evolution partial differential equation for the potentials B_{++}, B_{--}. The compatibility condition for (4.17), (4.18),

$$[\partial_x + i\lambda\sigma_3 - 2iQ, \partial_t + i\lambda^2\sigma_3 - 2i\lambda Q - iV] = 0 \tag{4.22}$$

results in the following equations for the potentials:

Theorem The potentials b_{++}, B_{--} satisfy the following system of partial differential equations, which represent the nonlinear Schrödinger equation (however, it is to be remembered that the complex conjugation involution is absent in the real time case):

$$\begin{aligned} i\partial_t b_{++} &= -\frac{1}{2}\partial_x^2 b_{++} - 4b_{++}^2 B_{--} \\ i\partial_t B_{--} &= \frac{1}{2}\partial_x^2 B_{--} + 4B_{--}^2 b_{++}\,. \end{aligned} \tag{4.23}$$

Initial data for these equations can be extracted from the equal time correlator (see [5]).

The proof of this theorem is rather straightforward. One has merely to calculate the left-hand side of the compatibility condition (4.22).

So we have obtained the description of the potentials b_{++}, B_{--} entering the representation of the correlator. Our aim now is to relate the Fredholm determinant $\Delta = \det(1 + \widehat{V})$ to these potentials. It is convenient to introduce the function σ, which is the logarithm of the determinant and plays the role of τ-function for the integral system

$$\sigma = \ln\det(1 + \hat{\mathcal{V}} = \sigma(x,t)\,. \tag{4.24}$$

Derivatives of σ with respect to x and t can be expressed in terms of the potentials B, C:

Theorem The derivatives of the function $\sigma = \ln\det(1 + \hat{\mathcal{V}})$ are expressed in terms of the potentials B, C (4.10), (4.11) by the following formula:

$$\partial_x \sigma = -2iB_{+-}; \quad \partial_x^2 \sigma = 4b_{++}B_{--} \tag{4.25}$$

$$\partial_t \sigma = -2iGB_{--} - 2i(C_{+-} + C_{-+})\,. \tag{4.26}$$

All potentials entering the right-hand sides here can be expressed in terms of B_{++}, B_{--}, namely

$$\partial_x(C_{+-} + C_{-+}) = (B_{++} - 2G)\partial_x B_{--} - B_{--}\partial_x B_{++}$$

$$\partial_t B_{+-} = b_{++}\partial_x B_{--} - B_{--}\partial_x b_{++}\,. \tag{4.27}$$

At $t = 0$ function σ was completely described in [5] as the solution of the Painlevè differential equation. All the details of the above theorem are given in [4,2].

The Riemann-Hilbert problem can be used for evaluation of the asymptotics of the quantum correlation function. This asymptotics is explicitly evaluated in [16-20], [2].

Conclusion

We have explained here why the quantum correlation functions are τ functions of completely integrable classical differential equations. This is quite a remarkable development of the Quantum Inverse Scattering Method. We want to emphasize that these results are not restricted to the nonlinear Schrödinger equation. The determinant representation for Sine-Gordon was obtained in [21]. The differential equation for the autocorrelation of the Ising model in the critical magnetic field was obtained in [22]. Nonrelativistic fermions were considered in [23].

This work was supported by NSF grant PHY-9107261.

References

[1] L.D. Faddeev and L.A. Takhtajan, *Hamiltonian Methods in the Theory of Solitons*, Springer-Verlag, (1987).

[2] V.E. Korepin, A.G. Izergin, N.M. Bogolubov, "Quantum Inverse Scattering Method and Correlation Functions", Cambridge University Press, (1992).

[3] A.R. Its, A.G. Izergin and V.E. Korepin, Temperature correlators of the impenetrable Bose gas as an integrable system, *Comm. Math. Phys.* **129**, 205-222, (1990).

[4] A.R. Its, A.G. Izergin, V.E. Korepin and N.A. Slavnov, Differential equations for quantum correlation function, *Int. J. Mod. Phys.* **B4**, 1003-1037, (1990).

[5] A. Jimbo, T. Miwa, Y. Mori and M. Sato, Density matrix of an impenetrable Bose gas and the fifth Painlevè transcendent, *Physica* **1D**, 80-158, (1980).

[6] M.J. Ablowitz and A.S. Fokas, Linearization of a class of nonlinear evolution equations, *J. Math. Phys.* **27**, 2614-2619, (1964).

[7] A.S. Fokas and M.J. Ablowitz, *Phys. Rev. Lett.* **47**, 1096, (1981).

[8] V.E. Korepin and N.S. Slavnov, Correlation functions of fields in one-dimensional Bose gas, *Comm. Math. Phys.* **136**, 633-644, (1991).

[9] A. Lenard, One-dimensional impenetrable bosons in thermal equilibrium, *J. Math. Phys.* **7**, 1268-1272, (1966).

[10] A. Lenard, Momentum distribution in the ground state of the one-dimensional system of impenetrable bosons, *J. Math. Phys.* **5**, 930-943, (1964).

[11] V.E. Korepin, Dual field formulation of quantum integral models, *Comm. Math. Phys.* **113**, 177-190, (1987).

[12] V.E. Korepin, Generating functional of correlation functions for the nonlinear Schrödinger equation, *Funk. Analiz. i jego Prilozh* **23**, 15-23, (1989) (in russian).

[13] F.D. Gakhov, *Boundary problems*, Nauka, Moscow, 1977.

[14] I.Z. Gokhberg and M.P. Krein, Systems of integral equations on a half-axis with kernels depending on difference of variables, *Usp. Matem. Nauk* **13**, 3-72, (1958).

[15] V.E. Korepin and N.A. Slavnov, The time-dependent correlation function of an impenetrable Bose gas as a Fredholm minor, *Comm. Math. Phys.* **129**, 103-113, (1990).

[16] A.R. Its, A.G. Izergin and V.E. Korepin, Correlation radius for one-dimensional impenetrable bosons, *Phys. Lett.* **141A**, 121-125, (1989).

[17] A.R. Its, A.G. Izergin and V.E. Korepin, Long-distance asymptotics of temperature correlators for the impenetrable Bose gas, *Comm. Math. Phys.* **130**, 471-486, (1990).

[18] A.R. Its, A.G. Izergin and V.E. Korepin, Space correlation in the one-dimensional impenetrable Bose gas at finite temperature, Physica D **53**, 187-213, (1991).

[19] A.R. Its, A.G. Izergin, V.E. Korepin and G.G. Varzugin, Large time and distance asymptotics of the correlator of the impenetrable bosons at finite temperature, Physica D **54**, 351-395 (1992).

[20] A.R. Its, A.G. Izergin and V.E. Korepin, Large time and distance asymptotics of the temperature field correlator in the impenetrable Bose gas, *Nucl. Phys.* **B348**, 757-765, (191).

[21] H. Itoyama, H. Thacker and V.E. Korepin, Correlation functions of sine-Gordon at free fermion point as Fredholm determinant, preprint ITP-SB-90-92, SUNY Stony Brook, (1990).

[22] A.R. Its, A.G. Izergin, V.E. Korepin and N.Yu. Novokshenov, Temperature autocorrelations of the transverse Ising chain at the critical magnetic field, *Nucl. Phys.* **B340**, 752, (1990).

[23] A. Berkovich, Temperature and magnetic field dependent correlators of the exactly integrable $(1+1)$-dimensional gas of impenetrable fermions, *J. Phys.* **A24**, 1543-1556, (1991).

Lattice Models in Statistical Mechanics and Soliton Equations

B.M. McCoy

Institute for Theoretical Physics, State University of New York at Stony Brook, Stony Brook, NY 11794-3840, USA

The theory of soliton equations for the correlation functions of the Ising model, XY quantum spin chain, and the impenetrable Bose gas is reviewed. Possible generalizations to more general superintegrable and integrable models are discussed.

1 Introduction

The theory and applications of soliton equations have many independent starting points. The earliest is the famous observation by Russell [1] in 1838 of solitary waves in barge canals. This led, after half a century of controversy, to the introduction in 1895 by Korteweg and de Vries [2] of the nonlinear partial differential equation which now bears their names that provides an explanation of Russell's physical phenomena. For some 70 years the mathematical structure of this equation remained unexplored.

Almost contemporary with Korteweg and de Vries was the purely mathematical work of Painlevé [3] in 1902 on nonlinear ordinary differential equations. Painlevé's quest was the classification of all second order nonlinear ordinary differential equations whose branch points and essential singularities are at positions fixed by the differential equation independent of the boundary conditions. For 60 years the physical applications of these equations remained unknown.

At the time of their discovery there was no hint that the KdV nonlinear partial differential equation and the Painlevé nonlinear ordinary differential equations were related. However, starting in the '60's with the work of Gardner, Greene, Kruskal and Miura [4] for KdV and Myers [5] for Painlevé there has developed the truly remarkable theory of integrable soliton equations which incorporates both these partial and ordinary nonlinear differential equations into an immensely larger picture. The many and varigated approaches to these classical equations forms the subject of most of this book.

There also exists a third independent starting point which leads to integrable soliton equations, namely the free fermi representation of the two dimensional Ising model. This line of investigation was initiated in 1949 when Kaufmann [6] reduced the Ising model to a problem of free fermions on a lattice and when Kaufmann and Onsager [7] reduced the computation of the Ising two spin correlation to the computation of certain Toplitz determinants. The connection with soliton equations was first made in 1973 when it was discovered [8,9,10] that in the continuum limit near T_c the 2 spin correlation function could be computed in terms of a Painlevé III function. Since this initial discovery it has been shown that near T_c all spin correlations of the Ising model satisfy systems of holonomic nonlinear partial differential equations [11], that for all temperatures the correlations satisfy nonlinear partial difference equations [12,13] and that the diagonal 2 spin correlations satisfy a Painlevé VI equation in the temperature [14].

Moreover, the Ising model is not the only free fermi model. The closely related XY quantum spin chain is of this class and the 2 spin time dependent correlation functions satisfy Toda's equation [15,16] and more general sets of integrable partial difference-differential equations. Another free fermi model is the impenetrable Bose gas whose 2 point function was shown to satisfy a Painlevé V equation [17] in 1980. Thus by the early '80's the connection between correlation functions of free fermion solvable lattice models and soliton equations was richly developed. The details of these results will be presented in section 2 for the Ising model, section 3 for the XY chain and section 4 for the impenetrable Bose gas.

The importance of these achievements is that correlation functions are the building blocks of all quantum field theories and all condensed matter physics. Consequently, the discovery that the correlation functions of any nontrivial system can satisfy anything as simple as a nonlinear differential equation of finite order is a remarkably useful piece of insight and one is naturally lead to ask if there are other systems whose correlation functions can be simply characterized.

There are many lattice models in 2 dimensions which share with free fermi models the feature that their free energies, transfer matrix spectrum and (in some cases) their order parameters have been computed. These models are based either on the generalization of the operator algebra used by Onsager [18] to solve the Ising model in 1944 (in which case the models are called superintegrable [19]) or on the generalization of the star-triangle equation [20] also noted by Onsager to hold for the Ising model (in which case the models are called integrable [21]). The superintegrable class will be presented in section 5a and the integrable case in section 5b. These are the most natural models in which to search for further relations between soliton equations and correlation functions. We conclude in section 5c by reviewing the status of the search with particular emphasis on the linear differential equations satisfied by the correlation functions of the conformal field theories of Belavin, Polyakov and Zamolodchikov [22] and their relation to the nonlinear equations of soliton theory.

2 Free Fermions and the Ising Model

The study of the two dimensional Ising model by free fermi methods has a long and rich history originating with the work of Kaufmann [6] and Kaufmann and Onsager [7] in 1949 and extending through the discovery of the quadratic difference equations for all spin correlations on the lattice by McCoy, Perk and

Wu [12,13] in 1981. It is impossible to present this large body of work in any detail here; consequently, we content ourselves with the definitions and results and merely indicate a few of the basic tools. For any amount of detail the reader should consult the references. In addition, chapter 2 of the book by Itzykson and Drouffe [23] is recommended.

The Ising model is a classical statistical mechanical model of "spin" variables $\sigma_{j,k}$ located on the j, k (rows, columns) of a square lattice which take on the values $\sigma_{j,k} = \pm 1$. The most general classical interaction energy which involves only nearest neighbor spins is

$$\mathcal{E} = -\sum_{j,k}\{E^v(j,k)\sigma_{j,k}\sigma_{j+1,k} + E^h(j,k)\sigma_{j,k}\sigma_{j,k+1} + H\sigma_{j,k}\} \tag{2.1}$$

where the vertical (horizontal) interaction energies $E^v(j,k)(E^h(j,k))$ can be completely arbitrary.

The first object of interest to compute for any statistical system is the partition function

$$Z = \sum_{\{\sigma\}} e^{-\mathcal{E}/kT} \tag{2.2}$$

where $\sum_{\{\sigma\}}$ is the sum over all spin configurations, T is the temperature and k is Boltzmann's constant. More precisely, if the lattice has L_h columns and L_v rows we are interested in the free energy per site in the infinite lattice (thermodynamic) limit defined as

$$\mathcal{F} = -kT \lim_{\substack{L_v \to \infty \\ L_h \to \infty}} (L_v L_h)^{-1} \ln Z\,. \tag{2.3}$$

However, there in much physics beyond what is contained in the partition function and accordingly there is much interest in the n-spin correlation functions

$$\langle \sigma_{M_1,N_1} \cdots \sigma_{M_n,N_n} \rangle = \frac{1}{Z} \sum_{\{\sigma\}} \sigma_{M_1,N_1} \cdots \sigma_{M_n,N_n} e^{-\mathcal{E}/kT} \tag{2.4}$$

The simplest such spin correlation function is the one point function $\mathcal{M}(H)$, the magnetization per site. Of particular interest is the spontaneous magnetization $\mathcal{M}$ defined as

$$\mathcal{M} = \lim_{H \to 0+} \lim_{\substack{L_v \to \infty \\ L_h \to \infty}} \mathcal{M}(H) \tag{2.5}$$

If $\mathcal{M} > 0$ the system is said to be ordered and if $\mathcal{M} = 0$ the system is said to be disordered. Generally if $E^v(j,k)$ and $E^h(j,k)$ are positive $\mathcal{M}$ is a monotonic decreasing function of T. If there exists a value $T_c > 0$ of T such that $\mathcal{M} > 0$ for $T < T_c$ a phase transition is said to take place at T_c and T_c is referred to as the critical temperature. For the translationally invariant case

$$E^v(j,k) = E^v,\ E^h(j,k) = E^h \tag{2.6}$$

Yang [24] proved in 1952 that

$$\mathcal{M} = \{1 - [\sinh(2E^v/kT)\sinh(2E^h/kT)]^{-2}\}^{1/8} \tag{2.7}$$

and hence T_c is determined from

$$1 = \sinh(2E^v/kT_c)\sinh(2E^h/kT_c)\,. \tag{2.8}$$

The first tool needed to study the Ising model was developed by Kramers and Wannier [25] in 1941 when they introduced the concept of transfer matrix. We will here consider only $H = 0$ and define a transfer matrix $T^{(j)}_{\sigma,\sigma'}$ depending on the spins in the j row (called $(\sigma_1, \sigma_2, \cdots \sigma_{L_h})$ and the spins in the $j+1$ row (called $(\sigma'_1, \sigma'_2 \cdots \sigma'_{L_h})$) as

$$T^{(j)}_{\{\sigma\},\{\sigma'\}} = \prod_{k=1}^{L_h} W^h_j(\sigma_k, \sigma_{k+1}) W^v_j(\sigma_k, \sigma'_k) \tag{2.9}$$

where

$$W^v_j(\sigma_k, \sigma'_k) = e^{E^v(j,k)\sigma_k\sigma'_k/kT} \tag{2.10a}$$

$$W^h_j(\sigma_k, \sigma_{k+1}) = e^{E^h(j,k)\sigma_k\sigma_{k+1}/kT} \tag{2.10b}$$

Then the partition function Z can be written as

$$Z = TrT^{(1)}T^{(2)} \ldots T^{(L_v)} \tag{2.11}$$

so that for the translationally invariant case (2.6)

$$Z = TrT^{Lv} \tag{2.12}$$

It should be noted that there is no unique way to define a transfer matrix. For the Ising model is often useful to split the definition (2.9) into 2 pieces

$$T^{(j)}_1 = \prod_{k=1}^{L_h} W^v_j(\sigma_k, \sigma'_k)$$

and

$$T^{(j)}_2 = \delta_{\sigma_1,\sigma'_1} \cdots \delta_{\sigma_{L_h},\sigma'_{L_h}} \prod_{k=1}^{L_h} W^h_j(\sigma_k, \sigma_{k+1})\,. \tag{2.13}$$

The transfer matrix (2.9) is thus $T = T_1T_2$, but for some purposes the more symmetric transfer matrices $T' = T_2^{1/2}T_1T_2^{1/2}$ or $T'' = T_1^{1/2}T_2T_1^{1/2}$ are more useful.

The relation of T_1 and T_2 to fermions can be seen if we first write (2.13) in terms of operators. Define direct product matrices

$$\sigma^i_l = I \otimes \cdots \otimes I \times \sigma^i \otimes \cdots \times I \tag{2.14}$$

where $i = x, y, z$, the σ^i are the 2×2 Pauli spin matrices and in (2.14) the σ^i are at the site l. Then as shown in ref. 26 from (2.13)

$$T^{(j)}_1 = [\prod_{k=1}^{L_h} 2\sinh(2E^v(j,k)/kT)] \prod_{k=1}^{L_h} e^{K^{v*}(j,k)\sigma^z_k} \tag{2.15a}$$

and

$$T_2^{(j)} = \prod_{k=1}^{L_h} e^{K^h(j,k)\sigma_k^x\sigma_{k+1}^x} \tag{2.15b}$$

where

$$\tanh K^{v*}(j,k) = e^{-2E^v(j,k)/kT} . \tag{2.15c}$$

and

$$K^h(j,k) = E^h(j,k)/kT .$$

It is at this point that the connection is made with fermions by introducing operators γ_k which satisfy [16]

$$\gamma_k\gamma_l + \gamma_l\gamma_k = \delta_{kl} . \tag{2.16}$$

Then the spin operators can be written as

$$\sigma_j^x = \frac{1}{\sqrt{2}}\prod_{k=1}^{j-1}(i2\gamma_{2k-1}\gamma_{2k})\gamma_{2j-1} , \tag{2.17a}$$

$$\sigma_j^y = \frac{1}{\sqrt{2}}\prod_{k=1}^{j-1}(i2\gamma_{2j-1}\gamma_{2k})\gamma_{2j} , \tag{2.18b}$$

and

$$\sigma_j^z = -i\gamma_{2j-1}\gamma_{2j} . \tag{2.18c}$$

The transfer matrix now becomes a quadratic form in the γ_k and the eigenvalues of T' can be evaluated [6] by free fermi methods such as direct diagonalization [26] or Grassman integrals [13,16]. For the translationally invariant case all of these methods may be explicitly carried out and the eigenvalues of T' are found to be of the form

$$T' = (\sinh 2K^v)^{L_h/2}\prod_{l=1}^{L_h}\exp \pm\frac{1}{2}\epsilon_{q_l} \tag{2.19}$$

where ϵ_q is the positive solution of

$$\cosh\epsilon_q = \cosh 2K^h\cosh 2K^{v*} - \sinh 2K^h\sinh 2K^{v*}\cos q \tag{2.20}$$

the L_h distinct q_l satisfy either

$$e^{iqL_h} = 1 \tag{2.21a}$$

or

$$e^{iqL_h} = -1 , \tag{2.21b}$$

and all the $\pm$ signs are chosen independently. There is a correlation between the $\pm$ sign and the use of (2.21a) or (2.21b). For L_h even the relation is that if the number of $+$ signs is even then q satisfies (2.21b) and if an odd number of $+$ signs is chosen then (2.21a) must be used.

The exact spectrum (2.19) on the finite lattice with independent choices of $\pm\epsilon_q$ for each q_l is identical with that of free fermions in a finite box.

The great power of the fermi formalism lies in the fact that it may be extended to correlation functions. This is accomplished by first writing spin correlation functions in terms of the transfer matrix T. For example for a 2 spin corrrelation

$$\langle\sigma_{M_1,N_1}\sigma_{M_2,N_2}\rangle = \frac{1}{Z}Tr[(\prod_{j=1}^{M_1-1} T^{(j)})\sigma^x_{N_1}$$

$$(\prod_{j=M_1}^{M_2-1} T^{(j)})\sigma^x_{N_2}(\prod_{j=M_2}^{L_h} T^{(j)})] . \tag{2.22}$$

This can now be rewritten in terms of the γ_k and the traces evaluated as Grassman integrals using the Wick theorem or the theorem on compound Pfaffians [13,16]. We do not attempt to give derivations here. However, to give the most symmetric form of the resulting partial difference equations we need to introduce the concept of the disorder operater $\mu_{M,N}$. More specifically we define the expectation value of two disorder operators, at (M_1, N_1) and (M_2, N_2) by considering an arbitrary path connecting the center of the square at $(M_1+1/2, N_1+1/2)$ and $(M_2+1/2, N_2+1/2)$. Replace each bond $E^v(j,k)$ and $E^h(j,k)$ intersected by the path by $-E^v(j,k)$ and $-E^h(j,k)$. Call the partition function of the resulting lattice Z'. The expectation value $\langle\mu_{M_1,N_1}\mu_{M_2,N_2}\rangle$ is now defined as

$$\langle\mu_{M_1,N_1}\mu_{M_2,N_2}\rangle = Z'/Z . \tag{2.23}$$

This definition is path independent. We can now state the result of [12]. We use the notation

$$\sigma[M_1, N_1; \cdots; M_n, N_n] \equiv \langle\sigma_{M_1,N_1}\sigma_{M_2,N_2}\cdots\sigma_{M_n,N_n}\rangle \tag{2.24}$$

and

$$\mu_{ij}[M_1, N_1; \cdots; M_n, N_n] \equiv \langle\sigma_{M_1,N_1}\cdots\mu_{M_i,N_i}\cdots\mu_{M_j,N_j}\cdots\sigma_{M_n,N_n}\rangle , \tag{2.25}$$

where in (2.25) the two disorder variables are associated with sites (M_i, N_i) and (M_j, N_j), the arguments of $\sigma[\cdots]$ and $\mu_{ij}[\cdots]$ will not be explicitly given except when needed, and μ_{ij} is defined to be antisymmetric in i and j.

We introduce the notation $S_1(M,N) = \sinh[2E^h(M,N)/kT]$ and $S_2(M,N) = \sinh[2E^v(M,N)/kT]$ and obtain the following results:

$$S_1(M_i,N_i)S_2(M_j,N_j)\{\sigma[\cdots]\sigma[M_i,N_i+1;M_j+1,N_j]-\sigma[M_i,N_i+1]\sigma[M_j+1,N_j]\}$$

$$= \mu_{ij}[\cdots]\mu_{ij}[M_i-1,N_i;M_j,N_j-1]-\mu_{ij}[M_i-1,N_i]\mu_{ij}[M_j,N_j-1] \tag{2.26a}$$

for $i \neq j$;

$$S_1(M_i,N_i)S_1(M_j,N_j)\{\sigma[\cdots]\sigma[M_i,N_i+1;M_j,N_j+1]-\sigma[M_i,N_i+1]\sigma[M_j,N_j+1]\}$$

$$= \mu_{ij}[M_i-1,N_i]\mu_{ij}[M_j-1,N_j]-\mu_{ij}[\cdots]\mu_{ij}[M_i-1,N_i;M_j-1,N_j] \tag{2.26b}$$

for $i \neq j$ and $(M_i, N_i) \neq (M_j, N_j)$;

$$S_2(M_i,N_i)S_2(M_j,N_j)\{\sigma[\cdots]\sigma[M_i{+}1,N_i;M_j{+}1,N_j]-\sigma[M_i{+}1,N_i]\sigma[M_j{+}1,N_j]\}$$

$$= \mu_{ij}[M_i,N_i-1]\mu_{ij}[M_j,N_j-1]-\mu_{ij}[\cdots]\mu_{ij}[M_i,N_i-1;M_j,N_j-1] \quad (2.26\tilde{b})$$

for $i \neq j$ and $(M_i, N_i) \neq (M_j, N_j)$;

$$S_2(M_l,N_l)\{\sigma[\cdots]\mu_{ij}[M_l+1,N_l]-\sigma[M_l+1,N_l]\mu_{ij}[\cdots]\}$$

$$= \mu_{lj}[\cdots]\mu_{lj}[M_l,N_l-1]-\mu_{li}[M_l,N_l-1]\mu_{lj}[\cdots] \quad (2.26c)$$

for $l \neq i, l \neq j$, and $i \neq j$; and

$$S_1(M_l,N_l)\{\sigma[\cdots]\mu_{ij}[M_l,N_l+1]-\sigma[M_l,N_l+1]\mu_{ij}[\cdots]\}$$

$$= -\mu_{li}[\cdots]\mu_{lj}[M_l-1,N_l]+\mu_{li}[M_l-1,N_l]\mu_{lj}[\cdots] \quad (2.26\tilde{c})$$

for $l \neq i, l \neq j$, and $i \neq j$.

In addition, because of the arbitrariness of the placement of the string in the definition of the disorder variable a relative minus sign is appended to some μ_{ij} wherever lattice sites on both sides of the string are involved.

These five equations cannot by themselves completely determine the correlation because no reference to T_c is yet included. The needed extra equation(s) will not be the same for all arbitrary lattices. We here specialize our attention to the translationally invariant case (2.6). In this case it is notationally convenient to absorb a factor of $(S_1S_2)^{1/2}$ into μ_{ij} and to define $\gamma_1 = 2z_2(1-z_1^2), \gamma_2 = 2z_1(1-z_2^2)$, and $z_i = \tanh(E_i/kT)$, and rewrite Eqs. (2.26), respectively, as

$$\sigma[\cdots]\sigma[M_i,N_i+1;M_j+1,N_j]-\sigma[M_i,N_i+1]\sigma[M_j+1,N_j]$$

$$= \mu_{ij}[\cdots]\mu_{ij}[M_i-1,N_i;M_j,N_j-1]-\mu_{ij}[M_i-1,N_i]\mu_{ij}[M_j,N_j-1], \quad (2.27a)$$

$$\sigma[\cdots]\sigma[M_i,N_i+1;M_j,N_j+1]-\sigma[M_i,N_i+1]\sigma[M_j,N_j+1]$$

$$= (\gamma_1/\gamma_2)\{\mu_{ij}[M_i-1,N_i]\mu_{ij}[M_j-1,N_j]-\mu_{ij}[\cdots]\mu_{ij}[M_i-1,N_i;M_j-1,N_j]\}\,, \quad (2.27b)$$

$$\sigma[\cdots]\sigma[M_i+1,N_i;M_j+1,N_j]-\sigma[M_i+1,N_i]\sigma[M_j+1,N_j]$$

$$= (\gamma_2/\gamma_1)\{\mu_{ij}[M_i,N_i-1]\mu_{ij}[M_j,N_j-1]-\mu_{ij}[\cdots]\mu_{ij}[M_i,N_i-1;M_j,N_j-1]\}\,, \quad (2.27\tilde{b})$$

$$\sigma[\cdots]\mu_{ij}[M_l+1,N_l]-\sigma[M_l+1,N_l]\mu_{ij}[\cdots]$$

$$= (\gamma_2/\gamma_1)^{1/2}\{\mu_{li}[\cdots]\mu_{lj}[M_l,N_l-1]-\mu_{li}[M_l,N_l-1]\mu_{lj}[\cdots]\}\,, \quad (2.27c)$$

and

$$\sigma[\cdots]\mu_{lj}[M_l,N_l+1]-\sigma[M_l,N_l+1]\mu_{lj}[\cdots]$$

$$= -(\gamma_1/\gamma_2)^{1/2}\{\mu_{li}[\cdots]\mu_{lj}[M_l-1,N_l]-\mu_{li}[M_l-1,N_l]\mu_{lj}[\cdots]\}\,. \quad (2.27\tilde{c})$$

For this translationally invariant case we obtain the extra equation

$$
\begin{aligned}
&(\gamma_2/\gamma_1)^{1/2}\{\sigma[M_i,N_i+1]\mu_{ij}[M_i,N_i-1]+\sigma[M_j,N_j+1]\mu_{ij}[M_j,N_j-1]\}\\
&\quad+(\gamma_1/\gamma_2)^{1/2}\{\sigma[M_i+1,N_i]\mu_{ij}[M_i-1,N_i]+\sigma[M_j+1,N_j]\mu_{ij}[M_j-1,N_j]\}\\
&\quad+\sum_{l\neq i,j}\{\mu_{li}[M_l,N_l-1]\mu_{lj}[M_l-1,N_l]-\mu_{li}[M_l-1,N_l]\mu_{lj}[M_l,N_l-1]\}\\
&\quad=2a(\gamma_1\gamma_2)^{-1/2}\sigma[\cdots]\mu_{ij}[\cdots]\,,
\end{aligned}
\tag{2.28}
$$

where $a=(1+z_1^2)(1+z_2^2)$.

There are several properties of these equations which should be noted:
(i) Because of the arbitrariness inherent in specifying the relation of the order to disorder variables the equations remain valid if for any disorder variable the replacements $M_i\to M_i-1$ or $N_i\to N_i-1$, or both, are made. Thus, for example, from Eq. (2.26c) we obtain

$$
\begin{aligned}
&S_2(M_l,N_1)\{\sigma[\cdots]\mu_{ij}[M_l+1,N_l;M_l-1,N_i]-\sigma[M_l+1,N_l]\mu_{ij}[M_i-1,N_i]\}\\
&\quad=\mu_{li}[M_i-1,N_i]\mu_{lj}[M_l,N_l-1]-\mu_{li}[M_l,N_l-1;M_i-1,N_i]\mu_{lj}[\cdots]\,.
\end{aligned}
\tag{2.29}
$$

(ii) There is also a symmetry if we interchange order and disorder variables. For example, if in Eq. (2.26b) we make the replacements $\sigma[\cdots]\to\mu_{jk}[\cdots]$, $\mu ij[\cdots]\to\mu_{ik}[M_j+1,N_j+1]$, and $S_1(M_j,N_j)\to S_2^{-1}(M_j,N_j+1)$, we obtain

$$
\begin{aligned}
&S_1(M_i,N_i)\{\mu_{jk}[\cdots]\mu_{jk}[M_i,N_i+1;M_j,N_j-1]-\mu_{jk}[M_i,N_i+1]\mu_{jk}[M_j,N_j-1]\}\\
&=S_2(M_j,N_j)\{\mu_{ik}[\cdots]\mu_{ik}[M_i-1,N_i;M_j+1,N_j]-\mu_{ik}[M_i-1,N_i]\mu_{ik}[M_j+1,N_j]\}\,.
\end{aligned}
\tag{2.30}
$$

Equations (2.29) and (2.30) are consequences of Eqs. (2.26).
(iii) Correlations involving four or more disorder variables can be expressed as Pfaffians of correlations with only two disorder variables, e.g.,

$$
\sigma\mu_{ijkl}=\mu_{ij}\mu_{kl}-\mu_{ik}\mu_{jl}+\mu_{il}\mu_{jk}\,. \tag{2.31}
$$

Finally we take the scaling limit $T\to T_c$. Define

$$
y=[2(a-\gamma_1-\gamma_2)\gamma_1^{-1}]^{1/2}M,\ x=[2(a-\gamma_1-\gamma_2)\gamma_2^{-1}]^{1/2}N,\ \bar\sigma=\mathcal{M}^{-n}\sigma,\ \bar\mu_{ij}=\mathcal{M}^{-n}\mu_{ij}\,, \tag{2.32}
$$

where $\mathcal{M}$ is given by (2.7).

Then in the limit $a-\gamma_1-\gamma_2\to 0$ (where $\mathcal{M}\to 0$) and $M,N\to\infty$, with $x,y,\bar\sigma$, and $\bar\mu_{ij}$ fixed, we find that Eqs. (2.27) become, respectively,

$$
\bar\sigma\frac{\partial^2\bar\sigma}{\partial x_i\partial y_j}-\frac{\partial\bar\sigma}{\partial x_i}\frac{\partial\bar\sigma}{\partial y_j}=\bar\mu_{ij}\frac{\partial^2\bar\mu_{ij}}{\partial x_j\partial y_i}-\frac{\partial\bar\mu_{ij}}{\partial x_j}\frac{\partial\bar\mu_{ij}}{\partial y_i}\,, \tag{2.33a}
$$

$$
\bar\sigma\frac{\partial^2\bar\sigma}{\partial x_i\partial x_j}-\frac{\partial\bar\sigma}{\partial x_i}\frac{\partial\bar\sigma}{\partial x_j}=\frac{\partial\bar\mu_{ij}}{\partial y_i}\frac{\partial\bar\mu_{ij}}{\partial y_j}-\bar\mu_{ij}\frac{\partial^2\bar\mu_{ij}}{\partial y_i\partial y_j}\,, \tag{2.33b}
$$

$$
\bar\sigma\frac{\partial^2\bar\sigma}{\partial y_i\partial y_j}-\frac{\partial\bar\sigma}{\partial y_i}\frac{\partial\bar\sigma}{\partial y_j}=\frac{\partial\bar\mu_{ij}}{\partial x_i}\frac{\partial\bar\mu_{ij}}{\partial x_j}-\bar\mu_{ij}\frac{\partial^2\bar\mu_{ij}}{\partial x_i\partial x_j}\,, \tag{2.33$\tilde{b}$}
$$

$$
\bar\sigma\frac{\partial\bar\mu_{ij}}{\sigma y_l}-\frac{\partial\bar\sigma}{\partial y_l}\bar\mu_{ij}=-\mu_{li}\frac{\partial\bar\mu_{lj}}{\partial x_l}+\frac{\partial\bar\mu_{lj}}{\partial x_l}\bar\mu_{lj}\,, \tag{2.34}
$$

and

$$\bar{\sigma}\frac{\partial \bar{\mu}_{ij}}{\partial x_l} - \frac{\partial \bar{\sigma}}{\partial x_l}\bar{\mu}_{ij} = \bar{\mu}_{li}\frac{\partial \bar{\mu}_{lj}}{\partial y_l} - \frac{\partial \bar{\mu}_{li}}{\partial y_l}\bar{\mu}_{lj} \, . \tag{2.34a}$$

Furthermore, Eq. (2.28) simplifies in this limit through the use of the above equations to

$$\frac{1}{2}\bar{\sigma}\left[\frac{\partial^2 \bar{\mu}_{ij}}{\partial x_i^2} + \frac{\partial^2 \bar{\mu}_{ij}}{\partial y_i^2} + \frac{\partial^2 \bar{\mu}_{ij}}{\partial x_j^2} + \frac{\partial^2 \bar{\mu}_{ij}}{\partial y_j^2}\right] + \frac{1}{2}\bar{\mu}_{ij}\left[\frac{\partial^2 \bar{\sigma}}{\partial x_i^2} + \frac{\partial^2 \bar{\sigma}}{\partial y_i^2} + \frac{\partial^2 \bar{\sigma}}{\partial x_j^2} + \frac{\partial^2 \bar{\sigma}}{\partial y_j^2}\right]$$

$$-\frac{\partial \bar{\sigma}}{\partial x_i}\frac{\partial \bar{\mu}_{ij}}{\partial x_i} - \frac{\partial \bar{\sigma}}{\partial y_i}\frac{\partial \bar{\mu}_{ij}}{\partial y_i} - \frac{\partial \bar{\sigma}}{\partial x_j}\frac{\partial \bar{\mu}_{ij}}{\partial x_j} - \frac{\partial \bar{\mu}}{\partial y_j}\frac{\partial \bar{\mu}_{ij}}{\partial y_j} = \bar{\sigma}\bar{\mu}_{ij} \, . \tag{2.34b}$$

It can be shown that Eqs. (2.34) are equivalent to those of Sato, Miwa, and Jimbo [11].

These above results for Ising correlations are the most general known . However, they are perhaps not the most familiar form of soliton and Painlevé equations. It is thus enlightening to present the results when specialized to the 2 point function. The reduction from lattice equation to continuum limit is done in [27] and the original relation of the scaled Ising 2 point function to PIII is found of [8-10]

$$G_{\pm}(r) = \frac{1}{2}[1 \mp \eta(\frac{r}{2})]\eta^{-1/2}(r/2)$$
$$\exp \int_{r/2}^{\infty} d\theta \frac{1}{4}\theta\eta^{-2}[(1-\eta^2)^2 - (\eta')^2] \tag{2.35}$$

where η satisfies the special case of PIII

$$\eta'' = \frac{1}{\eta}(\eta')^2 - \frac{1}{\theta}\eta' + \eta^3 - \eta^{-1} \, . \tag{2.36}$$

There is one other remarkable result for the Ising 2 point function that must be mentioned derived in 1980 by Jimbo and Miwa [14]. Namely if we define

$$t = \left(\sinh\frac{2E^v}{kT}\sinh\frac{2E^h}{kT}\right)^2 \quad \text{for} \quad T < T_c \tag{2.37a}$$

and

$$\left(\sinh\frac{2E^v}{kT}\sinh\frac{2E^h}{kT}\right)^{-2} \quad \text{for} \quad T > T_c \tag{2.37b}$$

and set

$$\sigma_+(t) = t(t-1)\frac{d}{dt}\ln\langle\sigma_{0,0}\sigma_{N,N}\rangle - \frac{1}{4} \quad \text{for} \quad T < T_c \tag{2.38a}$$

and

$$\sigma_-(t) = t(t-1)\frac{d}{dt}\ln\langle\sigma_{0,0}\sigma_{N,N}\rangle - \frac{1}{4}t \quad \text{for} T > T_c \tag{2.38b}$$

then σ satisfies for either $\pm$

$$[t(t-1)\frac{d^2\sigma}{dt^2}]^2$$

$$= N^2[(t-1)\frac{d\sigma}{dt} - \sigma]^2$$
$$-4\frac{d\sigma}{dt}[(t-1)\frac{d\sigma}{dt} - \sigma - \frac{1}{4}][t\frac{d\sigma}{dt} - \sigma] \quad (2.39)$$

which is the equation for the τ function of a PVI.

In this form if we scale $t \to 1$ we find that in addition to (2.35) the scaled 2 point function also obeys

$$\zeta = r\frac{d \ln G_\pm}{dr} \quad (2.40)$$

with

$$(r\zeta'')^2 = 4(r\zeta' - \zeta)^2 - 4(\zeta')^2(r\zeta' - \zeta) + (\zeta')^2 \quad (2.41)$$

This is the τ function equation for PV. For a derivation of this PV form from the PIII form see ref. [28].

It must be realized, of course, that in addition to deriving these differential equations we still need boundary conditions to complete the specification of the problem. Consider for example the PV formulation for $G_\pm$ of (2.41). There are two obvious (or convenient) places to specify the boundary conditions, either at $r = 0$ or $r = \infty$. However, once one set of boundary conditions is specified the other must be expressed in terms of it and cannot be given independently. Thus for a complete understanding of the correlation functions we need to solve a connection problem.

Consider first the short distance behavior of (2.41) $r \to 0$. In general (2.41) has one two parameter family of solutions with the behavior $G(r) \sim Br^{\sigma-1/4}$. However, for the Ising model it is independently known that $\sigma = 0$. In this case we find that (2.41) has two independent one parameter families of solutions

$$G_\pm = \text{const}\ r^{-1/4}\{1 \pm \frac{1}{2}[\ln r + C_\pm] + \frac{1}{16}r^2 + \cdots\} \quad (2.42)$$

The 2 families are distinguished by the $\pm$ sign in (2.42) and we have so picked the signs to correspond to $G_\pm$ defined by (2.35).

As $r \to \infty$ (long distance) we find that the behavior of $G_\pm$ depends on $\pm$ in a much more complicated fashion. For $T < T_c$ we know by the definition of spontaneous magnetization and correlation length that

$$G_- \to 1 - \text{exponentially small terms.} \quad (2.43)$$

In this case ζ is exponentially small and hence (2.41) linearizes. There is only a one parameter family of solutions of the linearized equation which in fact is exponentially small and we find

$$G_- = 1 - \lambda_-[r^2(K_1(r)^2 - K_0(r)^2)$$
$$-rK_0(t)K_1(r) + \frac{1}{2}K_0(r)^2] + \cdots \quad (2.44)$$

where λ_- is the one boundary condition and $K_n(r)$ is the modified Bessel function of order n.

For $T > T_c$, however, G_+ does not approach a nonzero constant as $r \to \infty$ but rather vanishes exponentially. Now ζ is not exponentially small but rather is only algebraic. Hence no linearization of (2.41) is possible. Nevertheless it is expected from general arguments of rotational invariance and isolated one particle states that (just as in quantum field theory) $G_+(t)$ should have the form

$$G_+(r) \sim \lambda_+ K_0(r) \,. \tag{2.45}$$

Indeed if we try

$$\zeta = rK_0'(r)/K_0(r) \tag{2.46}$$

in (2.41) we find that this an exact solution. This exact solution for G_+ does not agree with (2.42) as $r \to 0$. Nevertheless

$$G_+(t) = \lambda_+ K_0(r)(1 + \quad \text{exponentially small terms}) \tag{2.47}$$

will be a solution of (2.41) which has (2.45) as its $r \to \infty$ behavior. On the other hand it would be quite surprising if all values of C_+ in (2.42) would lead to solutions of the form (2.42). Thus to complete our understanding of the correlation G_+ we would like to compute the $r \to 0$ behavior (2.47) in terms the λ_+ of the $r \to \infty$ expansion. This involves a connection problem and was solved (in the PIII formulation) in ref. 29. In particular we find

$$\sigma = (2/\pi) \quad \arcsin \quad (\pi\lambda_+) \tag{2.48}$$

so that only $\lambda_+ = \pi^{-1}$ corresponds to the family (2.42). We also find that when $\lambda_+ = \pi^{-1}$ that C_+ is uniquely determined to be

$$C_+ = \ln\frac{1}{8} + \gamma \tag{2.49}$$

where γ is Euler's constant.

3 The XY spin chain

The second free fermi model we discuss is the XY quantum spin chain whose Hamiltonian is

$$\mathcal{H}_{xy} = -\frac{1}{2}\sum_{l=1}^{L_h}\{\frac{1}{2}(1+\gamma)\sigma_l^x\sigma_{l+1}^x + \frac{1}{2}(1-\gamma)\sigma_l^y\sigma_{l+1}^y + h\sigma_l^z\} \,. \tag{3.1}$$

The close relation of this quantum spin chain to the Ising model was already seen in the original papers of Onsager and Kaufmann [6,7,18]. A particularly transparent relation is the commutation relation [30]

$$[T'', H_{xy}] = 0 \,, \tag{3.2}$$

with T'' defined as in section 2, which holds if

$$\cosh 2K^{v*} = \gamma^{-1} \tag{3.3a}$$

and

$$\tanh 2K^h = (1-\gamma^2)^{1/2}/h\,. \tag{3.3b}$$

Thus T'' and H_{xy} have a common set of eigenvectors and as an example we may derive [30]

$$\begin{aligned}\langle\sigma_{M,N}\sigma_{M,N'}\rangle &= \cosh^2 K^{v*}\langle\sigma^x_N\sigma^x_{N'}\rangle_{xy}\\ &\quad - \sinh^2 K^{v*}\langle\sigma^y_N\sigma^y_{N'}\rangle_{xy}\,.\end{aligned} \tag{3.4}$$

For the spin chain we are interested in the correlation functions depending on space, time, and temperature and here we specifically concentrate on

$$\begin{aligned}X_n(t) &= \langle e^{-i\mathcal{H}t}\sigma^x_0 e^{i\mathcal{H}t}\sigma^x_n\rangle\\ &= \frac{Tr(e^{-i\mathcal{H}t}\sigma^x_0 e^{i\mathcal{H}t}\sigma^x_n e^{-\mathcal{H}/kT})}{Tr e^{-\mathcal{H}/kT}}\end{aligned} \tag{3.5}$$

It should be remarked that the temperature of the spin chain in (3.5) has nothing to do with the temperature of the two dimensional statistical model. The temperature dependence of (3.5) is a new effect not seen in the Ising correlations of the previous section.

The correlations (3.5) satisfy systems of differential difference equations independent of T [15]. The simplest occurs for $h=1$ where X_n obeys the Toda equation

$$\frac{d^2\ln X_n}{dt^2} = \frac{X_{n-1}X_{n+1}}{X_n^2} - 1 \tag{3.6}$$

with the initial condition [31]

$$\frac{dX_n(t)}{dt}\,\Big|_{t=0} = -i\delta_{n,0}2\pi^{-1}\int_0^{\pi/2} d\theta \sin\theta \tanh\left(\frac{\sin\theta}{kT}\right) \tag{3.7}$$

and the equal time correlations $X_n(0)$ are given by $n\times n$ Toplitz determinants [31,32]. Here the physical question of greatest interest is the behavior of $X_n(t)$ as $t\to\infty$.

There are 3 distinct cases of interest.

1) $T=\infty$
Here all correlations are zero except for $n=0$ and one can find directly [33] or from (3.6) that

$$X_n(t) = \delta_{n,0}e^{-t^2/2}\,. \tag{3.8}$$

2) $0<T<\infty$
Further partial differential equations involving T have been derived for this case [34] and a recent report [35] is that as $t\to\infty$

$$\ln X_0(t) = t\frac{1}{2\pi}\int_{-1}^{1} dz \ln|\tanh(z/kT)| + 0(1)\,. \tag{3.9}$$

3) $T=0$
Here the equal time correlation has a very simple exact form

$$X_n(0)=\left(\frac{2}{\pi}\right)^{|n|}\prod_{l=1}^{|n|-1}\left[1-\frac{1}{4l^2}\right]^{l-|n|} \tag{3.10}$$

and further

$$X_n(0)=-i\delta_{n,0}2\lambda \quad \text{with} \quad \lambda=1/\pi\,. \tag{3.11}$$

In this case it was shown that [36] $X_n(t)$ is the τ function of a PV equation. Explicitly

$$X_n(t)=X_n(0)e^{-t^2/2}\exp\int_0^t dt'\frac{\sigma_n(t')}{t'} \tag{3.12}$$

where

$$\left(t\frac{d^2\sigma_n}{dt^2}\right)^2=16\left(t\frac{d\sigma_n}{dt}-\sigma_n-n^2\right)$$
$$\left[t\frac{d\sigma_n}{dt}-\sigma_n-\frac{1}{4}\left(\frac{d\sigma_n}{dt}\right)^2\right]\,. \tag{3.13}$$

As $t\to 0$ the correlation function $X_n(t)$ must be analytic. This restricts (3.13) to a one parameter family of solutions

$$\sigma_n(t)=\sum_{l=1}^{\infty}a_l(n)t^{2l}+it^{2n+1}\sum_{l=0}^{\infty}b_l(n)t^{2l} \tag{3.14}$$

where $a_l(n)$ for $l\le n$ are independent of λ. In particular for $n\ge 1$

$$a_1(n)=\frac{4n^2}{4n^2-1},\, b_0(n)=-2\lambda\prod_{l=1}^{n}\left(\frac{2l}{(2l-1)^2(2l+1)}\right)\,. \tag{3.15}$$

Let us use (3.13) to study the $t\to\infty$ behavior of the system. There are 3 regions to be considered:

$$I \qquad 0\le n/t<1, \qquad \text{time like}\,; \tag{3.16a}$$
$$II \qquad t=n+n^{1/3}(z/2) \qquad \text{with } z=0(1), \qquad \text{light cone}\,; \tag{3.16b}$$
$$III \qquad 1<n/t, \qquad \text{space like}\,. \tag{3.16c}$$

The asymptotic behavior of $\sigma_n(t)$ [and hence $X_n(t)$] is drastically different in these 3 regions.

I. Time like region

A local analysis of (3.13) shows that there is a two parameter expansion

$$\sigma_n(t)=t^2-4ab(t^2-n^2)^{1/2}$$
$$+t(t^2-n^2)^{-1/4}\sum_{l=0}^{\infty}[t(t^2-n^2)^{-3/4}]^l\sum_{m=0}^{l}(n/t)^{2m}F_{l,m}(s) \tag{3.17}$$

where

$$s = \sqrt{t^2 - n^2} - n \ \arccos{(n/t)} + ab \ln t^2 (t^2 - n^2)^{-3/2} \tag{3.18}$$

with, for example

$$F_{0,0} = a e^{is} + b e^{-2is} \tag{3.19}$$

It still remains to express the two parameters a and b of the $t \to \infty$ expansion in terms of the parameter λ of the $t \to 0$ expansion. This connection problem has been solved in ref. 37 where we found (for $\lambda \leq \pi^{-1}$)

$$ab = -(2\pi)^{-1} \ln \pi\lambda \tag{3.20}$$

$$\left\{ \begin{array}{c} a(\lambda) \\ b(\lambda) \end{array} \right\} = \frac{1}{2}\pi^{-1/2}\Gamma(1 \pm 2iab) e^{\pm \pi i/4} e^{\mp 6iab \ln 2} e^{\pi ab}(1 \mp \pi\lambda) \, . \tag{3.21}$$

Thus in the physical case $\lambda \to \pi^{-1}$

$$a = 0, \qquad b = \pi^{-1/2} e^{-\pi i/4} \, . \tag{3.22}$$

II. Light cone

If we degenerate the PV τ equation (3.13) using (3.16b) and define

$$f(z) = \lim_{\substack{t \to \infty \\ n \to \infty}} n^{-2/3} \{\sigma_n[n + n^{1/3}(z/2)] - t^2\} \tag{3.23}$$

we find that

$$f(z) = -1/4z^2 - 2g(-z) \tag{3.24}$$

where

$$g(x) = \frac{1}{2}\left\{ \left(\frac{dy}{dx}\right)^2 - (y^2 + \frac{x}{2})^2 \right\} \tag{3.25}$$

and y obeys the PII equation

$$\frac{d^2 y}{dx^2} = 2y^3 + xy + a \qquad \text{with } a = 1/2 \tag{3.26}$$

This light cone connection formula is to be compared with the PII connection formula for the usual (soliton free) long time behavior of the $mKdV$ equation [38] which has $a = 0$.

III. Space like

Here the asymptotic expansion of (3.13) for which $\sigma_n - t^2$ is bounded at infinity has only one free parameter. Explicitly

$$\sigma_n(t) = \sigma_n^r(t) + i\sigma_n^i(t) \tag{3.27}$$

where

$$\sigma_n^r = t^2 + \frac{1}{4}\,t^2(n^2-t^2)^{-1}$$
$$+\frac{1}{16}\,t^6(n^2-t^2)^{-4}(1+7n^2t^{-2}+n^4t^{-4})+\cdots \tag{3.28}$$

and

$$\sigma_n^i = -\frac{1}{2}\pi^{1/2}\lambda e^{-2x}t(n^2-t^2)^{1/4}+\cdots \tag{3.29}$$

with

$$x = -\sqrt{n^2-t^2}+n \quad \arccos\,(n/t)\,. \tag{3.30}$$

In writing (3.29) we have chosen the normalization factors that the solution of the related connection problem is rendered trivial.

4 The impenetrable Bose gas

The third free fermion model we discuss is the impenetrable Bose gas in one dimension. The study of the problem was initiated by Girardeau [39] in 1960 who found that the spectrum has the same free fermi form as the XY model. The study of the correlation functions was initiated by Schultz [40] and Lenard [41] and much further progress was made by Vaidya and Tracy [42] who reduced the study of the correlation functions to that of the XY model is the limit $\gamma \rightarrow 0, h \rightarrow 1$. The complete connection with Painlevé functions was made by Jimbo, Miwa, Mori and Sato in ref. 17. It is their results we now summarize.

The impenetrable Bose gas is the $c \rightarrow \infty$ limit of the N body Hamiltonian

$$\mathcal{H} = -\frac{1}{2}\sum_{l=1}^{N}\frac{\partial^2}{\partial x_l^2}+c\sum_{l<j}\delta(x_l-x_j) \tag{4.1}$$

on the interval $0 < x < L$ with periodic boundary conditions. The objects of physical interest are the elements of the n- particle reduced density matrix

$$\rho_{n,N,L}(x_1\cdots x_n;x_1',\cdots x_n')$$
$$=\frac{N!}{(N-n))!}\int_0^L\cdots\int_0^L dy_{n+1}\cdots dy_n$$
$$\psi_{N,L}^*(x_1\cdots x_n,y_{n+1}\cdots y_N)\psi_{N,L}(x_1',\cdots x_n',y_{n+1}\cdots y_N) \tag{4.2}$$

where $\psi_{N,L}$ is the ground state solution of Schroedinger's equation normalized so that

$$\rho_{1,N,L}(x;x) = N/L = \rho_0 \tag{4.3}$$

is the density of particles. The important result of ref. 17 is that all the n particle reduced density matrices satisfy sets of holonomic equations. In particular the one particle density matrix satisfies

$$\rho(x) = \rho_0\exp\int_0^x dx'\left(\frac{x'}{4y(1-y)^2}\left(\left(\frac{dy}{dx'}\right)^2+4y^2\right)-\frac{(1+y)^2}{4x'y}\right) \tag{4.4}$$

where y is the solution of the PV equation.

$$\frac{d^2y}{dx^2} = \left(\frac{1}{2y}+\frac{1}{y-1}\right)\left(\frac{dy}{dx}\right)^2 - \frac{1}{x}\frac{dy}{dx} + \frac{(y-1)^2}{x^2}(\alpha y + \beta/y) + \frac{\gamma y}{x} + \frac{\delta y(y+1)}{y-1} \tag{4.5}$$

with $\alpha = 1/2, \beta = -1/2, \gamma = -2i$, and $\delta = 2$.
The boundary condition is

$$y \to -1 - \frac{2i}{3}x + \left(\frac{2}{9}+\frac{2i}{\pi}\right)x^2 + 0(x^3) \text{ as } x \to 0\,.$$

Setting

$$\sigma_1(x) = x\frac{d}{dx}\ln\rho(x) \tag{4.6}$$

so that

$$\rho(x) = \rho_0 \exp\int_0^x dx'\sigma_1(x')/x' \tag{4.7}$$

they show that σ_1 satisfies

$$\left(x\frac{d^2\sigma_n}{dx^2}\right)^2 = -4\left(x\frac{d\sigma}{dx} - n^2 - \sigma_n\right)\left(x\frac{d\sigma_n}{dx} + \left(\frac{d\sigma_n}{dx}\right)^2 - \sigma_n\right) \tag{4.8}$$

with $n = 1$ and the boundary condition

$$\sigma_1(x) = -\frac{1}{3}x^2 + \frac{1}{3\pi}x^3 + 0(x^4) \text{ as } x \to 0\,. \tag{4.9}$$

Equation (4.8) for the impentrable Bose gas is identical with equation (3.14) for the XY model if in (3.13) we set $n = 1$,

$$t = -i\frac{1}{2}x \tag{4.10}$$

with a boundary condition

$$\sigma(x) = \frac{4}{3}t^2 + \frac{4\lambda}{3}it^3 + 0(t^4) \tag{4.11}$$

and

$$\lambda = 2/\pi\,. \tag{4.12}$$

As far as the small x or t behavior of $\sigma(x)$ is concerned the relation between the Bose gas problem and the XY problem is smooth and analytic in both x and t and λ and one problem can be considered on the analytic continuation of the other. However, for the large x and large t behavior there is no possibility of analytically continuing one problem to the other and a completely separate analysis is required.

This analysis is given in ref. 17 directly for the Bose gass with $\lambda = 2/\pi$ and in a more general context for arbitrary λ in ref. 43. In this latter paper it is demonstrated that as $x \to \infty$ the form of the asymptotic expansion is different if $\lambda < 1/\pi$ or $\lambda > 1/\pi$ and the connection problem is solved completely in both cases. When $\lambda = 2/\pi$ this reduces to the final result of ref. 17 that

$$\begin{aligned}\rho_0^{-1}\rho(x) &= \rho_\infty x^{-1/2}\Big\{1 + \frac{1}{2^3x^2}\left(\cos 2x - \frac{1}{4}\right) \\ &+ \frac{3}{2^4x^3}\sin 2x + \frac{3}{2^8x^4}\left(\frac{11}{2^3} - 31\cos 2x\right) - \frac{3\cdot 151}{2^9x^5}\sin 2x \\ &+ \frac{3^2}{2^{14}\cdot x^6}\left(3\cdot 1579\cos 2x - \frac{17\cdot 19}{2^2}\right) + \frac{3^3\cdot 5\cdot 7\cdot 311}{2^{15}\cdot x^7}\sin 2x \\ &+ \frac{1}{2^{13}\cdot x^8}\left(c_8 - \frac{3^2\cdot 2064719}{2^6}\cos 3x + \frac{3^2}{2^2}\cos 4x\right) + \cdots\Big\} \qquad (4.13)\end{aligned}$$

where c_8 is a constant yet to be determined and $\rho_\infty = \pi e^{1/2}2^{-1/3}A^{-6}$ (A= Glaisher's constant).

There is one further remark that should be made before we leave the subject of free fermi models. It is shown in ref. 17 that the Painlevé equation (4.8) with $n = 0$ and a boundary condition of

$$\sigma_0(x) = -\lambda x - \lambda^2x^2 + 0(x^3) \text{ as } x \to 0 \qquad (4.14)$$

with $\lambda = 1/\pi$ plays a deep role in the theory of level spacings of random matrices [44]. Moreover the derivatives of σ as a function of λ at $\lambda = 1/\pi$ also play an important role. The crossing point $\lambda = 1/\pi$ that separates the regions of λ studied in ref. 43 is of great interest. Indeed, it has recently been conjectured by Odlyzko [45] that the zeroes of the Riemann zeta function have the same distribution as this random matrix problem. Further discussion is contained in ref. 46. If this conjecture proves true it is clearly of the greatest importance.

5 Generalizations

We have now completed our survey of the known results for the connection between lattice models and soliton equations and thus, in the narrow sense, have discharged the mandate of the title of this article. But as explained in the introduction there are several avenues of potential gneralization of these results that seem to have much promise. In this concluding section we will discuss some of these potential generalizations in detail. More specifically we discuss the 3 separate topics of a) superintegrable models, b) integrable models, and 3) conformal field theory.

a) Superintegrable models

As discussed in the introduction, the free fermion construction was not the original approach used by Onsager to solve the Ising model in 1944 [18]. Instead he computed the partition function of the Ising model and more generally the entire eigenvalue spectrum of the row transfer matrix T or T' by embedding the problem in a larger algebraic structure. To see this algebra it is perhaps most convenient to consider a second transfer matrix along the diagonal of the lattice defined as

$$T_D = \prod_{j=1}^{M} W^v(\sigma_j, \sigma_j')W^h(\sigma_j, \sigma_{j+1}') \,. \tag{5.1}$$

When $E^v \to \infty$ and $E^h \to 0$ this transfer matrix reduces to the identity and to first order in an appropriate anisotropy variable u (measuring E^h/E^v) T_D may be expanded as

$$T_D = 1 + u \text{ const } + u\mathcal{H} \tag{5.2}$$

where

$$\mathcal{H} = A_0 + k'A \,. \tag{5.3}$$

with

$$A_0 = \sum_{g=1}^{M} \sigma_j^z \sigma_{j+1}^{z\dagger} \tag{5.4a}$$

and

$$A_1 = \sum_{j=1}^{M} \sigma_j^x \,. \tag{5.4b}$$

There operators have the remarkable property that

$$[A_0, [A_0, [A_0, A_1]]] = \text{ const } [A_0, A_1] \tag{5.5a}$$

and

$$[A_1, [A_1, [A_1, A_0]]] = \text{ const } [A_1, A_0] \tag{5.5b}$$

(where the constant is 16) and this is sufficient to recursively define a set of operators A_k and G_k such that

$$[A_k, A_m] = 4G_{k-m} \tag{5.6a}$$

$$[G_m, A_k] = 2(A_{k+m} - A_{k-m}) \tag{5.6b}$$

$$[G_m, G_k] = 0 \tag{5.6c}$$

This algebra is now referred to as Onsager's algebra. Onsager further showed that for the Ising model a periodicity relation holds of

$$A_{n+M} = -CA_n = -A_nC \tag{5.7}$$

when C is the operator that reverses the sign of all spins. From this algebra alone Onsager computed the spectrum (2.19).

For 41 years this algebraic solution was regarded as a curiosity because no other representations were found. However in 1985 von Gehlen and Rittenberg [47] made the remarkable discovery that if for arbitrary N

$$A_0 = \sum_{l=1}^{M} \sum_{n=1}^{N-1} \frac{e^{i(2l-N)\pi/2N}}{\sin(\pi n/N)} (Z_l Z_{l+1}^{\dagger})^n \tag{5.8a}$$

$$A_1 = \sum_{l=1}^{M} \sum_{n=1}^{N-1} \frac{e^{i(2l-N)\pi/2N}}{\sin(\pi n/N)} X_l^n \tag{5.8b}$$

where X_l and Z_l are of the direct product form (2.1) with σ^x replaced by

$$X_{j,k} = \delta_{j,k+1} \qquad (\text{modN}) \tag{5.9a}$$

and σ^z replaced by

$$Z_{j,h} = \delta_{j,h} w^{j-1} \tag{5.9b}$$

with $\omega = e^{2\pi i/N}$ then the relation (5.5) holds with the constant $= 4N^2$ and thus (with a rescaling) (5.6) will hold. Subsequently it was proven by several authors [48,49] that this algebra alone is sufficient to guarantee that the eigenvalues of $\mathcal{H}$ decompose into a collection of sets, each of the form

$$E = \alpha + \beta k' + 2N \sum_{j=1}^{M} m_j \sqrt{1 + k'^{\,2} + k' a_j} \tag{5.10}$$

where m_j takes on the values

$$-S_j \leq m_j \leq S_j \tag{5.11}$$

for S_j some integer or $\frac{1}{2}$ integer, a_j are constants independent of k', and α and β are constants. For the specific case of (5.8) $S_j = \frac{1}{2}$ and the equation determining the a_j for each set have been reduced to a Bethe's equation [50-52]. The set structure of the $N = 3$ case has been studied in great detail in ref. 50.

The spectrum (5.10) is surely the simplest generalization of the free fermi spectrum that it is possible to obtain. It is because of this remarkable similarity to a free fermi spectrum that these superintegrable models seem to be the most natural place to look for an extension of the soliton equations obtained in the previous 3 sections.

(b) Integrable models

As of the date of this writing the only known superintegrable models are (5.8). However there is a much larger collection of models which may be studied by the third tool discovered by Onsager to hold for the Ising model, the star triangle equation.

The star triangle equation is a local condition on Boltzmann weights which allows us to prove that the diagonal transfer matrix (5.1) has the remarkable property that if the interaction energies in 2 successive diagonals are called E^v, E^h and $\bar{E}^v, \bar{E}^h$ and if

$$\sinh \frac{2E^v}{kT} \sinh \frac{2E^h}{kT} = \sinh \frac{2\bar{E}^v}{kT} \sinh \frac{2\bar{E}^h}{kT} \tag{5.12}$$

then parameterizing E^v, E^h by some appropriate variable u we have the commutation relation.

$$[T_D(u), T_D(\bar{u})] = 0 \; . \tag{5.13}$$

Thus in conjunction with (5.2) we see that

$$[T_D(u), \mathcal{H}] = 0 \tag{5.14}$$

and hence the Hamiltonian $\mathcal{H}$ possesses an infinite number of constants of the motion which commute with themselves. This is obviously connected with the concept of Lax pair which plays such a prominant role for classical soliton equations. Because of these infinite constants of the motion systems which obey the commutation relation (5.13) are called integrable.

Unlike the case of superintegrable models there are an enormous number of models now known which satisfy (5.13). The search for these models has given birth to the entire new field of quantum groups. These are the second class of models where it is natural to look for soliton equations for correlation functions.

It should be noted that the language strongly suggests every superintegrable model should be integrable. It is in fact the case that the superintegrable model (5.8) are in fact integrable but there is no general proof of this following from the mere definition of superintegrability alone.

c) Conformal field theory

Having now defined the two classes of models we feel are most likely to generalize the soliton equations for correlation functions we should conclude by presenting some evidence that such equations do in fact exist. In the strictest sense we cannot do this but, with a slight amount of indulgence, a most useful plausibility argument can be given.

All 3 of the cases, free fermi, superintegrable, and integrable possess infinte parameter symmetry groups and it is surely the case that for the free fermion case one can say that it is this infinite parameter group which controls the correlations. The problem would thus seem to be in getting mathematical control over the infinite group.

In 1984 a remarkable advance in the study of such infinite parameter groups was made by Belavin, Polyakov and Zamolodchikov [22] who initiated the study of continuum quantum field theories in 2 dimensions invariant under the infinite parameter conformal group. It is surely not a coincidence that the classification of these conformal field theories bears an intense relationship to the classification of integrable lattice models. Conformal field theories are by their very construction massless and so they should correspond to the continuum limit of lattice models at T_c. All the equations found for correlation functions in conformal field theory are linear equations in distinction to the nonlinear equation of soliton theory. However, at least for the Ising model 2 point function the nonlinear PV equation [28] presented in section 2 has been shown to reduce to the linear equation for second level degenerate operators of conformal field theory when T is set equal to T_c. It is certain (though surprisingly the derivation has never been published) that all the n point function equations (2.27) and (2.28) will also linearize when the scale constant a in (2.28) is set equal to zero. In addition there are large classes of nonlinear soliton equations presented by Miwa and Jimbo [53] which so far have not found any physical applications (except for the scattering from a broken corner, which was shown by Myers [54] to lead to a matrix generalization of PIII). It seems most likely that the equations of ref. 53 will also linearize in some appropriate conformal limit. If such a linearization gives the linear equation for an nth order degenerate field theory it would start to give very suggestive evidence to support the grand *conjecture* that the correlation function of all integrable lattice models satisfy classical soliton equations.

Acknowledgments

The author wishes to thank Prof. V. Korepin for bringing the unpublished work of P. Deift to his attention. This work was supported in part by the National Science Foundation grant DMR-9106648.

References

1. J.S. Russell, Report of the committee on waves, Report of the 7th meeting of British Association for the Advancement of Science, Liverpool (1838) pp. 417-498.
2. D.J. Korteweg and G. de Vries, Philos. Mag. Ser. 5, 39 (1895) 422.
3. P. Painlevé, *Acta Math.* **25** (1902) 1.
4. C.S. Gardner, J.M. Greene, M.D. Kruskal and R.M. Miura, *Phys. Rev. Lett.* **19** (1967) 1095.
5. J. Myers, Ph.D. Thesis Harvard Univ. 1962 (unpublished); *J. Math. Phys.* **6** (1965) 1839.
6. B. Kaufman, *Phys. Rev.* **76** (1949) 1232.
7. B. Kaufman and L. Onsager, *Phys. Rev.* **76** (1949) 1244.
8. E. Barouch, B.M. McCoy and T.T. Wu, *Phys. Rev. Lett.* **31** (1973) 1409.
9. C.A. Tracy and B.M. McCoy, *Phys. Rev. Lett.* **31** (1973) 1500.
10. T.T. Wu, B.M. McCoy, C.A. Tracy and E. Barouch, *Phys. Rev.* **B13** (1976) 316.
11. M. Sato, T. Miwa and M. Jimbo, *Proc. Japan Acad.* **53A** (1977) 6; *Publ. Res. Int. Math. Sci.* **14** (1978) 223; **15** (1979) 201, 577, 871.
12. B.M. McCoy, J.H.H. Perk and T.T. Wu, *Phys. Rev. Lett.* **46** (1981) 757.
13. J.H.H. Perk, in Proc. II International Symposium in Selected Topics in Statistical Mechanics, Dubna (1981) 165.

14. M. Jimbo and T. Miwa, *Proc. Japan Acad.* **56A** (1980) 405; **57A** (1981) 347.

15. J.H.H. Perk, *Phys. Lett.* **79A** (1980)1,3.

16. J.H.H. Perk, H.W. Capel, G.R.W. Quispel and F.W. Nijhoff, *Physica* **123A** (1984) 1.

17. M. Jimbo, T. Miwa, Y. Mori and M. Sato, *Physica* **1D** (1980) 80.

18. L. Onsager, *Phys. Rev.* **65** (1944) 117.

19. G. Albertini, B.M. McCoy, J.H.H. Perk and S. Tang, *Nucl. Phys.* **B314** (1989) 741.

20. G.H. Wannier, *Rev. Mod. Phys.* **17** (1945) 50.

21. R.J. Baxter, Exactly Solved Models in Statistical Mechanics, Academic Press (1982).

22. A.A. Belavin, A.M. Polyakov and A.B. Zamolodchikov, *J. Stat. Phys.* **34** (1984) 763; *Nucl. Phys.* **B241** (1984) 333.

23. C. Itzykson and J-M. Drouffe, "Statistical field theory", Cambridge Monographs on Mathematical Physics (1989).

24. C.N. Yang, *Phys. Rev.* **85** (1952) 808.

25. H.A. Kramers and G.H. Wannier, *Phys. Rev.* **60** (1941) 252.

26. T.D. Schultz, D.C. Mattis and E.H. Lieb, *Rev. Mod. Phys.* **36** (1964) 856.

27. B.M. McCoy and T.T. Wu, *Phys. Rev. Lett.* **45** (1980) 675.

28. B.M. McCoy and J.H.H. Perk, *Nucl. Phys.* **B285** [FS19] (1987) 279.

29. B.M. McCoy, C.A. Tracy and T.T. Wu, *J. Math. Phys.* **18** (1977) 1058.

30. M. Suzuki, *Phys. Lett.* **A34** (1971) 94.

31. E. Lieb, T. Schultz and D. Mattis, *Ann. Phys.* **16** (1961) 407; S. Katsura, *Phys. Rev.* **127** (1962) 1508; **129** (1963) 2835.

32. E. Barouch and B.M. McCoy, *Phys. Rev.* **A3** (1971) 786.

33. H.W. Capel and J.H.H. Perk, *Physica* **87A** (1977) 211.

34. A.R. Its, A.G. Izergin, V.E. Korepin and V.Yu Novokshenov, *Nucl. Phys.* **B340** (1990) 752.

35. P. Deift (private communication for V.E. Korepin).

36. B.M. McCoy, J.H.H. Perk and R.E. Shrock, *Nucl. Phys.* **B220** [FS8] (1983) 35, 269.

37. B.M. McCoy and S. Tang, *Physica* **19D** (1986) 42.

38. M.J. Ablowitz and H. Segur, *Stud. Appl. Math.* **57** (1977) 13 and "Solitons and the inverse scattering transform" SIAM, Philadelphia (1981) ch. 1; H. Segur and M.J. Ablowitz, *Physica* **3D** (1981) 165.

39. M. Girardeau, *J. Math. Phys.* **1** (1960) 516.

40. T.D. Schultz, *J. Math. Phys.* **4** (1963) 666.

41. A. Lenard, *J. Math. Phys.* **5** (1964) 930; **7** (1966) 1268.

42. H.G. Viadya and C.A. Tracy, *Phys. Lett.* **A68** (1978) 378; *Phys. Rev. Lett.* **42** (1979) 3.

43. B.M. McCoy and S. Tang, *Physica* **20D** (1986) 187.

44. F.J. Dyson, *Comm. Math. Phys.* **47** (1976) 171.

45. A. Odlyzko, "The 10^{20} -th zero of the Rieman zeta function and 175 million of its neighbors", preprint.

46. E.L. Basor and C.A. Tracy, preprint RIMS to appear in Mathematical Science (in Japanese).

47. G. von Gehlen and V. Rittenberg, *Nucl. Phys.* **B257** [FS14] (1985) 351.

48. B. Davies, *J. Phys. A* **23** (1990) 2245; preprint SMS-004-91 Australian National University (1991).

49. S.S. Roan preprint MPI 91/70 Max Planck Institute (1991).

50. G. Albertini, B.M. McCoy, J.H.H. Perk and S. Tang, *Nucl. Phys.* **314** (1989) 741.

51. G. Albertini, B.M. McCoy and J.H.H. Perk, Advanced Studies in Pure Mathematics **19**, Kinokuniya - Academic Press, (1989) 1.

52. V.O. Tarasov, *Phys. Lett.* **A147** (1990) 487.

53. M. Jimbo, T. Miwa and K. Ueno, **Physica** 2D (1981) 306; M. Jimbo and T. Miwa, *Physica* **2D** (1981) 407; **4D** (1981) 26.

54. J. Myers, *Physica* **11D** (1984) 51

Elementary Introduction to Quantum Groups

L.A. Takhtajan

Program in Applied Mathematics, University of Colorado, Boulder, CO 80309-0526, USA

1 Classical Integrable Systems and Poisson-Lie Groups

In this elementary survey we introduce and study basic properties of quantum groups – new mathematical objects emerged from the theory of quantum integrable systems. Their classical counterpart – Poisson-Lie groups – play a fundamental role in the theory of classical integrable systems. In the first two sections we briefly recall the basic facts from classical theory in such a way that the concept of quantum groups will appear quite naturally in the last two sections. In numerous remarks will indicate further references to this broad and rapidly developing subject.

1.1 Poisson Manifolds

Let X be a C^∞ manifold – *a classical phase space* – and $A = C^\infty(X)$ be the (commutative) algebra of smooth functions on X – *algebra of classical observables.*

Definition 1 X is called a *Poisson manifold* if there exists a Poisson bracket mapping

$$\{\ ,\ \} : A \otimes A \mapsto A$$

(i.e. a bilinear map from $A \times A$ into A) with the following properties:

i) skew-symmetry

$$\{\phi, \psi\} = -\{\psi, \phi\}$$

ii) Leibniz rule

$$\{\phi\psi, \chi\} = \phi\{\psi, \chi\} + \psi\{\phi, \chi\}$$

iii) Jacobi identity

$$\{\phi, \{\psi, \chi\}\} + \{\chi, \{\phi, \psi\}\} + \{\psi, \{\chi, \phi\}\} = 0$$

for all $\phi, \psi, \chi \in A$.

In other words A has an additional structure of *infinite dimensional Lie algebra* with the property that a Lie bracket (given by $\{\ ,\ \}$) is a derivation with respect to the usual pointwise multiplication in A. Typically (e.g. in classical mechanics) A is considered to be a $\mathbf{R}$-algebra, i.e. an algebra with unit over real numbers. From an algebraic point of view A could be complexified, i.e.

considered as a C-algebra (algebra with unit over complex numbers) as well (algebras over reals are specified by certain involutions). We adopt this point of view and, until otherwise specified explicitly, A will always be a C-algebra.

Example 1 $X = \mathbf{R}^{2n}$ with coordinates $x_1, \ldots, x_n, y_1, \ldots, y_n$ and canonical Poisson bracket

$$\{\phi, \psi\} = \sum_{i=1}^{n} \left(\frac{\partial \phi}{\partial x_i}\frac{\partial \psi}{\partial y_i} - \frac{\partial \phi}{\partial y_i}\frac{\partial \psi}{\partial x_i}\right)$$

is a Poisson manifold.

Example 2

$$X = \mathbf{S}^2 = \{(x_1, x_2, x_3) \in \mathbf{R}^3 \,|\{\phi, \psi\} = \sum_{i,j,k=1}^{3} \epsilon_{ijk} x_i \frac{\partial \phi}{\partial x_j}\frac{\partial \psi}{\partial x_k},$$

where ϵ_{ijk} is a totally anti-symmetric tensor with $\epsilon_{123} = 1$, is a Poisson manifold.

In terms of the local coordinates $x = (x_1, \ldots, x_n)$ on the Poisson manifold X Poisson bracket $\{\ ,\ \}$ takes the form

$$\{\phi, \psi\} = \sum_{i,j=1}^{n} J_{ij}(x) \frac{\partial \phi}{\partial x_i}\frac{\partial \psi}{\partial x_j},$$

where 2-tensor $J_{ij}(x)$ is called a *Poisson tensor*. It is skew-symmetric and satisfies a relation which follows from the Jacobi identity. If this tensor is non-degenerate, i.e. if $\det(J_{ij}(x)) \neq 0$, then one can define the following 2-form on X

$$\omega = \sum_{i,j=1}^{n} J^{ij}(x)\, dx_i \wedge dx_j$$

where $J^{ij} = (J^{-1})^{ij}$, i.e. $\sum_{k=1}^{n} J^{ik} J_{kj} = \delta_{ij}$. It it easy to show that Jacobi identity is equivalent to the condition $d\omega = 0$, i.e. ω is a closed 2-form.

A non-degenerate closed 2-form ω on X is called a *symplectic form* and X is called a *symplectic manifold.* Symplectic manifolds are Poisson manifolds with a non-degenerate Poisson tensor.

Remarks Excellent introduction to the mathematical methods of classical mechanics and to the theory of symplectic and Poisson manifolds can be found in [1], [2].

1.2 Hopf Algebras

Let $X = G$ be a Lie group. Due to the multiplication law $G \times G \mapsto G$, i.e. $(g_1, g_2) \mapsto g_1 g_2$ for all $g_1, g_2 \in G$, C-algebra $A = C^{\infty}(G)$ carries an additional structure. It is called a *coalgebra* structure and is represented by the mapping

$$\Delta : A \mapsto A \otimes A$$

defined by

$$\Delta(f)(g_1, g_2) = f(g_1 g_2)$$

for all $f \in A$, where we have used that $A \otimes A = C^\infty(G) \otimes C^\infty(G) \cong C^\infty(G \times G)$. Mapping Δ is called a *coproduct* (or *comultiplication*) and satisfies the following properties.

1) *Coassociativity:*

$$(\mathrm{id} \otimes \Delta) \circ \Delta = (\Delta \otimes \mathrm{id}) \circ \Delta,$$

where both sides map A into $A \otimes A \otimes A$ and $\mathrm{id} : A \mapsto A$ is the identity mapping. This property immediately follows from the associativity law of group multiplication: $g_1(g_2 g_3) = (g_1 g_2) g_3$ for all $g_1, g_2, g_3 \in G$.

2) Comultiplication is a C-algebra homomorphism, i.e.

$$\Delta(f_1 f_2) = \Delta(f_1)\Delta(f_2)$$

for all $f_1, f_2 \in A$.

In addition, the existence of the unit element $e \in G$ and of the inverse g^{-1} for all $g \in G$ leads to mappings $\epsilon : A \mapsto \mathrm{C}$ called a *counit* and $S : A \mapsto A$ – an *antipode* defined by

$$\epsilon(f) = f(e), \quad S(f)(g) = f(g^{-1})$$

for all $f \in A$. The counit and antipode satisfy the following properties.

3)

$$(\mathrm{id} \otimes \epsilon) \circ \Delta = (\epsilon \otimes \mathrm{id}) \circ \Delta = \mathrm{id}.$$

This property is equivalent to the definition of the unit: $ge = eg = g$ for all $g \in G$. Namely,

$$(\mathrm{id} \otimes \epsilon)(\Delta(f)) = f$$

because if $\Delta(f) = \sum_i f_i' \otimes f_i''$, or

$$\Delta(f)(g_1, g_2) = f(g_1 g_2) = \sum_i f_i'(g_1) f_i''(g_2),$$

(strickly speaking, due to the definition of the tensor product of linear topological spaces, $\Delta(f)$ is approximated by such sums) so that application of $\mathrm{id} \otimes \epsilon$ results in $\sum_i f_i'(g_1) f_i''(e) = f(g_1)$. The second equation above is proved analogously.

4)

$$m((\mathrm{id} \otimes S) \circ \Delta) = m((S \otimes \mathrm{id}) \circ \Delta) = \epsilon 1,$$

where in our case $m : A \otimes A \mapsto A$ is a usual pointwise multiplication in A (i.e. $m(f_1, f_2) = f_1 f_2$ for all $f_1, f_2 \in A$) and 1 is a unit in A. This property follows from the definition of the inverse: $gg^{-1} = g^{-1}g = e$ for all $g \in G$ and is proved as follows. Write $\Delta(f) = \sum_i f_i' \otimes f_i''$ so that

$$(\mathrm{id} \otimes S)(\Delta(f))(g_1, g_2) = \sum_i f_i'(g_1) f_i''(g_2^{-1}),$$

therefore, since $m(F)(g) = F(g, g)$ for all $F(g_1, g_2)$ in $A \otimes A$, we get

$$m(\mathrm{id}\otimes S)(\Delta(f))(g)=\sum_i f_i'(g)f_i''(g^{-1})=f(gg^{-1})=f(e)=\epsilon(f)1.$$

The second equation above is proved analogously.

In addition, we have the following properties.

5) S is C-algebra and coalgebra anti-homomorphism, i.e.

$$S(f_1f_2)=S(f_2)S(f_1),\quad \Delta\circ S=\sigma\circ(S\otimes S)\circ\Delta.$$

Here $\sigma: A\otimes A\mapsto A\otimes A$ is a flip homomorphism (permutation):

$$\sigma(f_1\otimes f_2)=f_2\otimes f_1$$

for all $f_1, f_2\in A$. The latter equation immediately follows from the property $(g_1g_2)^{-1}=g_2^{-1}g_1^{-1}$ for all $g_1, g_2\in G$.

Properties 1) - 5) define what is called a *Hopf algebra.* It consists of a C-algebra A with a product m, unit 1 and a coproduct Δ, counit ϵ, antipode S satisfying all the axioms 1) - 5) above.

Example 3 Algebra $A=C^\infty(G)$ with a product, coproduct, counit and antipode as above is a commutative Hopf algebra.

Hopf algebra is called *cocommutative* if

$$\sigma\circ\Delta=\Delta.$$

Algebra $A=C^\infty(G)$ is cocommutative if and only if Lie group G is abelian. This example actually exhausts all commutative Hopf algebras. Our second example realizes (in a certain sense) all cocommutative Hopf algebras and is a dualized version of the previous one.

Example 4 Let g be a Lie algebra and Ug be its universal enveloping algebra, i.e. a C-algebra of "non-commutative" polynomials (with elements of g as variables) satisfying relations $xy-yx=[x,y]$ for all $x,y\in g$, where multiplication in the left hand side is that in Ug and $[\ ,\ ]$ is a Lie bracket in g. Set

$$\Delta(x)=x\otimes 1+1\otimes x,\ S(x)=-x,\ \epsilon(x)=0,\ \epsilon(1)=1,\ x\in g$$

and extend these operations to Ug using properties 1) - 5). Then Ug becomes a cocommutative Hopf algebra (it is obvious that $\sigma\circ\Delta=\Delta$) which is commutative if and only if g is abelian.

Remarks For an introduction to Hopf algebras see, e.g. [3].

1.3 Poisson-Lie Groups

Here we will consider the case when a Poisson manifold X has an additional group structure, i.e. when $X=G$ is a Lie group.

Definition 2 A Lie group G is called a *Poisson-Lie group* if it is a Poisson manifold such that the Poisson bracket $\{\ ,\ \}$ and the coproduct Δ in $A=C^\infty(G)$

satisfy *Poisson-Lie property*

$$\Delta(\{f_1, f_2\}) = \{\Delta(f_1), \Delta(f_2)\}_2,$$

where $\{\ ,\ \}_2$ is a Poisson bracket on $G \times G$ induced by $\{\ ,\ \}$.

(Recall that $\{f_1 \otimes f_3, f_2 \otimes f_4\}_2 = \{f_1, f_2\} \otimes f_3 f_4 + f_1 f_2 \otimes \{f_3, f_4\}$ for all $f_1, f_2, f_3, f_4 \in A$).

Let G be a Poisson-Lie group and g be its Lie algebra with a basis x_i, $i = 1, \dots, n$. Denote by ∂_i corresponding basis of left-invariant vector fields on G:

$$(\partial_i f)(g) = \frac{df}{dt}(ge^{tx_i})|_{t=0},$$

We can represent a Poisson bracket $\{\ ,\ \}$ in the left-invariant frame:

$$\{\phi, \psi\} = \sum_{i,j=1}^{n} \eta^{ij}(g)\partial_i \phi \partial_j \psi$$

and introduce the mapping $\eta : G \mapsto \wedge^2 g$ by

$$\eta(g) = \sum_{i,j=1}^{n} \eta^{ij}(g) x_i \wedge x_j.$$

Such representation of $\{\ ,\ \}$ is always possible since vector fields ∂_i, $i = 1, \dots, n$, trivialize the tangent bundle TG of a Lie group G.

Theorem 1 The Poisson-Lie property for the bracket

$$\{\phi, \psi\} = \sum_{i,j=1}^{n} \eta^{ij}(g)\partial_i \phi \partial_j \psi$$

is equivalent to the condition that η is a 1-cocycle for the G-module $\wedge^2 g$ with adjoint G-action, i.e.

$$\eta(g_1 g_2) = \eta(g_2) + \mathrm{Ad}^{-1} g_2 \cdot \eta(g_1)$$

for all $g_1, g_2 \in G$.

Proof Simply rewrite the Poisson-Lie property in the form

$$\{\phi, \psi\}(g_1 g_2) = \{L_{g_1}(\phi), L_{g_1}(\psi)\}(g_2) + \{R_{g_2}(\phi), R_{g_2}(\psi)\}(g_1),$$

where L_g and R_g are respectively left and right translations: $(L_g f)(g') = f(gg')$, $(R_g f)(g') = f(g'g)$ and use the formulas

$$L_g \partial_x = \partial_x L_g, \quad R_g \partial_x = \partial_{\mathrm{Ad} g \cdot x} R_g, \quad x \in g.$$

If G is a simple Lie group, then corresponding cohomology group $H^1(G, \wedge^2 g)$ is trivial ($= \{0\}$) so that any 1-cocycle is a coboundary. In other words, there exists an element $r \in \wedge^2 g$ such that

$$\eta(g) = -r + \mathrm{Ad}^{-1} g \cdot r, \; g \in G.$$

This gives the following expression for the Poisson bracket $\{\ ,\ \}$:

$$\{\phi, \psi\} = \sum_{i,j=1}^{n} r^{ij} (\partial'_i \phi \partial'_j \psi - \partial_i \phi \partial_j \psi),$$

where $r = \sum_{i,j=1}^{n} r^{ij} x_i \otimes x_j$ and ∂'_i are corresponding right-invariant vector fields, i.e.

$$\partial'_i f(g) = \frac{df}{dt}(e^{tx_i} g)|_{t=0}, \; i = 1, \ldots, n.$$

To get this formula for $\{\ ,\ \}$ from the formula in the left-invariant frame one should use the relation

$$(\partial'_x f)(g) = (\partial_{\mathrm{Ad}^{-1} g \cdot x} f)(g).$$

It converts right-invariant vector fields (as first-order differential operators on G) at a point $g \in G$ into the left-invariant ones.

From now on we assume that Poisson structure on a Lie group G is always of this form (there is no loss of generality in the case of simple Lie groups).

Now we examine the case when the defined above bracket $\{\ ,\ \}$ satisfies Jacobi identity, i.e. is a Poisson bracket.

Theorem 2 The bracket

$$\{\phi, \psi\} = \sum_{i,j=1}^{n} r^{ij} (\partial'_i \phi \partial'_j \psi - \partial_i \phi \partial_j \psi)$$

satisfies Jacobi identity, i.e. provides a Lie group G with a structure of a Poisson-Lie group, if and only if the following 3-tensor

$$< r, r >= [r_{12}, r_{13}] + [r_{12}, r_{23}] + [r_{13}, r_{23}] \in \Lambda^3 g$$

is invariant with respect to adjoint g-action:

$$[< r, r >, \; x \otimes 1 \otimes 1 + 1 \otimes x \otimes 1 + 1 \otimes 1 \otimes x] = 0$$

for all $x \in g$. Here

$$r_{12} = \sum_{i,j=1}^{n} r^{ij} x_i \otimes x_j \otimes 1, \quad r_{13} = \sum_{i,j=1}^{n} r^{ij} x_i \otimes 1 \otimes x_j,$$

$$r_{23} = \sum_{i,j=1}^{n} r^{ij} 1 \otimes x_i \otimes x_j \in Ug \otimes Ug \otimes Ug.$$

Proof Consider the left hand side of Jacobi identity $\{\phi, \{\psi, \chi\}\} + \{\chi, \{\phi, \psi\}\} + \{\psi, \{\chi, \phi\}\}$ and develop it using the definition of the bracket $\{\ ,\ \}$. Use the fact that left- and right-invariant vector fields commute and commutation relations

$$[\partial_i, \partial_j] = \sum_{k=1}^{n} c_{ij}^k \partial_k, \quad [\partial'_i, \partial'_j] = -\sum_{k=1}^{n} c_{ij}^k \partial'_k,$$

where (pay attention to the negative sign in the second formula!) c_{ij}^k are the structure constants of a Lie algebra g, i.e. $[x_i, x_j] = \sum_{k=1}^n c_{ij}^k x_k$. A straightforward calculation shows that the left hand side of the Jacobi identity can be written as

$$\sum_{i,j,k=1}^{n} < r, r >^{ijk} (\partial_i' \phi \partial_j' \psi \partial_k' \chi - \partial_i \phi \partial_j \psi \partial_k \chi),$$

where $< r, r >= \sum_{i,j,k=1}^n < r, r >^{ijk} x_i \wedge x_j \wedge x_k$. Finally, the formula expressing right-invariant vector fields in terms of the left-invariant ones shows that Jacobi identity is equivalent to $< r, r >$ being invariant under the adjoint G-action. This proves the theorem.

Notations introduced above are extremely convenient and we will use them throughout this paper.

Remarks Poisson-Lie groups were introduced by V.Drinfeld; see his ICM-86 – address [4] and references there. Our elementary presentation follows lecture courses [5, 6]. Basic facts from the Lie groups cohomology (they will be used in the next section as well) can be found, e.g., in [2].

2 Classical Yang-Baxter Equation and Integrable Systems

2.1 Classical Yang-Baxter Equation

Let G be a simple Lie group. In this case we will formulate explicitly necessary and sufficient conditions for the element $r \in \wedge^2 g$ to define a Poisson-Lie structure. Let g be a Lie algebra of a Lie group G, $(\, , \,)$ be its Cartan-Killing form, i.e. a non-degenerate invariant (under the adjoint g-action) bilinear form on g, and x_μ be an orthonormal basis with respect to this form.

Denote by $c \in S^2 g \subset Ug$ the quadratic Casimir element:

$$c = \sum_\mu x_\mu^2,$$

which has the fundamental property

$$[c, x] = 0, \ x \in g.$$

Since Δ is Ug-homomorphism we also get $[\Delta(c), \Delta(x)] = 0$. Using the explicit form of Δ (see Example 4) we conclude that the element

$$t = \frac{1}{2}(\Delta(c) - c \otimes 1 - 1 \otimes c) \in S^2 g$$

satisfies the important relation

$$[t, x \otimes 1 + 1 \otimes x] = 0, \ x \in g.$$

Next we use a nontrivial fact from the Lie algebra cohomology: for a simple Lie algebra g the invariant subspace in the vector space $\wedge^3 g$ (with respect to the adjoint g-action) is one-dimensional and is generated by the element $[t_{13}, t_{23}]$. (We are using notations introduced at the end of Section 1). The property of t makes it obvious that $[t_{13}, t_{23}]$ is invariant, i.e. it commutes with $x \otimes 1 \otimes 1 + 1 \otimes x \otimes 1 + 1 \otimes 1 \otimes x$ for all $x \in g$ (although both t_{13} and t_{23} are invariant only their commutator belongs to $\wedge^3 g$); it is less obvious that it is the only (up to a scalar) element in $\wedge^3 g$ with this property!

Now using Theorem 2 and this fact we can state that the bracket

$$\{\phi, \psi\} = \sum_{i,j=1}^{n} r^{ij} (\partial_i' \phi \partial_j' \psi - \partial_i \phi \partial_j \psi)$$

satisfies Jacobi identity if and only if the element $r \in \wedge^2 g$ – classical r-matrix – satisfies the so-called *modified classical Yang-Baxter equation* (mCYBE)

$$< r, r >= \alpha \, [t_{13}, t_{23}]$$

for some constant α.

One can distinguish the two major cases.

Case A This is when $\alpha = 0$ so that r-matrix satisfies what is called "usual" *classical Yang-Baxter equation* (CYBE)

$$< r, r >= 0$$

together with the *classical unitarity condition* $r \in \wedge^2 g$, i.e. $r^{ij} = -r^{ji}$, or $r_{12} = -r_{21}$.

Using CYBE it is straightforward to show that

i) maximal subspace $g_0 \subset g$ where r is non-degenerate (i.e. a restriction r on $\wedge^2 g_0$ is a non-degenerate 2-tensor) is a subalgebra of the Lie algebra g so that one can consider (restricting r, if necessary) only this case;

ii) in terms of the inverse tensor r_{ij}, i.e. $\sum_{k=1}^{n} r^{ik} r_{kj} = \delta_{ij}$, CYBE can be written as a 2-cocycle condition for the mapping $\omega : \wedge^2 g \mapsto \mathbf{C}$,

$$\omega(x, y) = \sum_{i,j=1}^{n} r_{ij} u^i v^j,$$

where $x = \sum_{i=1}^{n} u^i x_i, \ y = \sum_{i=1}^{n} v^i x_i \in g$. In other words the following equation

$$\omega(x, [y, z]) + \omega(z, [x, y]) + \omega(y, [z, x]) = 0$$

(for all $x, y, z \in g$) is equivalent to the CYBE for a non-degenerate r-matrix.

Since for the simple Lie algebra $H^2(g, \mathbf{C})$ is trivial (another fact from Lie algebra cohomology), all 2-cocycles are coboundaries so that their bilinear forms are degenerate. Therefore r-matrices satisfying CYBE and classical unitarity condition can not "reside" in a simple Lie algebra: they are necessarily degenerate and reside in a non simple subalgebras of the Lie algebra g. Since in here we primarily deal with simple Lie groups and Lie algebras this case is not so interesting to us.

Case B Here $\alpha \neq 0$, so without loss of generality we can assume that $\alpha = -1/4$ (say). Setting $\tilde{r} = r + \frac{1}{2}t$ and using the basic property of t

$$[t_{12}, A_{13} + A_{23}] = 0,$$

for all $A \in Ug \otimes Ug$, we get

$$\begin{aligned} <\tilde{r}, \tilde{r}> &= <r, r> + \frac{1}{4}[t_{12}, t_{13} + t_{23}] + \frac{1}{4}[t_{13}, t_{23}] \\ &+ \frac{1}{2}([r_{12}, t_{13} + t_{23}] + [t_{12}, r_{13} + r_{23}] + [r_{13}, t_{23}] + [t_{13}, r_{23}]) \\ &= \frac{1}{2}([r_{12} + r_{13}, t_{23}] + [t_{12}, r_{13} + r_{23}] + [r_{12} + r_{32}, t_{13}]) = 0, \end{aligned}$$

so that $\tilde{r}$ satisfies the usual CYBE. However this "tradeoff" of mCYBE for CYBE changes the classical unitarity condition: instead of $r_{12} + r_{21} = 0$ $\tilde{r}$ satisfies

$$\tilde{r}_{12} + \tilde{r}_{21} = t_{12}.$$

Nevertheless due to the basic property of t replacement $r \mapsto \tilde{r}$ does not affect our Poisson brackets at all. Therefore dealing with Poisson-Lie groups we can use without any preference either unitary r-matrices satisfying mCYBE or non-unitary r-matrices satisfying usual CYBE and the condition above.

Remarks Classical r-matrices originated from the theory of classical integrable models; for a comprehensive presentation of this theory (with many examples) see [7]. Our presentation follows [5, 6].

2.2 Classical r-Matrix as an Operator

Let g be a simple Lie algebra; using Cartan-Killing form we can identify a linear space g with its dual g^*, i.e. $g \cong g^*$ so that element $r \in \wedge^2 g$ gives rize to a skew-symmetric operator $\hat{r}$ acting in g, $\hat{r} \in g^* \otimes g = \mathrm{End}(g)$. Then it can be shown that mCYBE for the element r is equivalent to the following quadratic equation for the operator $\hat{r}$:

$$[\hat{r}x, \hat{r}y] - \hat{r}([\hat{r}x, y] + [x, \hat{r}y]) = -[x, y]$$

for all $x, y \in g$. This equation is a sufficient condition for the skew-symmetric bracket

$$[x, y]_r = \frac{1}{2}([\hat{r}x, y] + [x, \hat{r}y])$$

to satisfy the Jacobi identity. Therefore $[\ ,\ ]_r$ defines a new Lie bracket on a Lie algebra g. This means that, actually, g has two Lie algebra structures with brackets [,] and [, $]_r$. Such objects were called a *Lie bialgebras* by Drinfeld; they are infinitesimal versions of Poisson-Lie groups.

Now let g be a simple Lie algebra, n_+, n_-, h be its nilpotent and Cartan subalgebras respectively and $P_\pm$ be corresponding projectors on the nilpotent subalgebras $n_\pm$ in the Cartan decomposition of g:

$$g = n_+ \oplus h \oplus n_-.$$

Then we have the following result.

Theorem 3 The r-matrix $\hat{r} = P_+ - P_-$ satisfies mCYBE and provides a simple Lie group G (a Lie algebra g) with a Poisson-Lie structure (with a bialgebra structure).

Proof We need to verify the equation for $\hat{r}$ which is equivalent to mCYBE. Let

$$x = x_+ + x_0 + x_-, \ y = y_+ + y_0 + y_-$$

be Cartan decompositions of elements $x, y \in g$. Since $P_\pm(x) = x_\pm, P_\pm(y) = y_\pm$, we have

$$[\hat{r}x, \hat{r}y] = [x_+ - x_-, y_+ - y_-]$$

and

$$\begin{aligned} & \hat{r}([\hat{r}x, y] + [x, \hat{r}y]) \\ = & (P_+ - P_-)([x_+ - x_-, y_+ + y_0 + y_-] + [x_+ + x_0 + x_-, y_+ - y_-]) \\ = & (P_+ - P_-)(2[x_+, y_+] + [x_+, y_0] - [x_-, y_0] - 2[x_-, y_-] + [x_0, y_+] - [x_0, y_-]) \\ = & 2[x_+, y_+] + [x_+, y_0] + [x_-, y_0] + 2[x_-, y_-] + [x_0, y_+] + [x_0, y_-], \end{aligned}$$

where we used that $[h, n_\pm] \subset n_\pm$. Therefore

$$\begin{aligned} [\hat{r}x, \hat{r}y] - \hat{r}([\hat{r}x, y] + [x, \hat{r}y]) & = -([x_+, y_+] + [x_+, y_0] + [x_+, y_-] + [x_-, y_+] \\ & \quad + [x_-, y_0] + [x_-, y_-] + [x_0, y_+] + [x_0, y_-]) \\ & = -[x, y], \end{aligned}$$

which completes the proof.

Let

$$g = h \oplus \sum_\alpha \oplus g_\alpha$$

be the root space decomposition of a simple Lie algebra g (sum over all roots) and choose $x_\alpha \in g_\alpha$ such that $(x_\alpha, x_{-\alpha}) = 1$, where $\alpha \in \Delta_+$ (all positive roots). Then operator $\hat{r} = P_+ - P_-$ gives rise to the element r

$$r = \sum_{\alpha \in \Delta_+} (x_{-\alpha} \otimes x_\alpha - x_\alpha \otimes x_{-\alpha}) \in \wedge^2 g,$$

which is called *canonical r-matrix* for a Lie group G.

Remarks The "operator approach" to classical r-matrices was developed by M.A.Semenov-Tian-Shansky [8, 9]; see also [7], [4] and [5, 6].

2.3 Connection with Integrable Models

It is well-known that for integrable models on a lattice the most important object is the so-called local Lax L-operator L_n which is a $k \times k$-matrix depending on the dynamic variables of the model (classical fields on a lattice) at the lattice site $n = 1, \ldots, N$. The main property of integrable models, which really distinguishes them from the general ones, is that they are Hamiltonian with Poisson brackets of a very special form. Namely, a Poisson structure for integrable models is defined by the following Poisson brackets between elements of L_n:

$$\{L_n \otimes, L_m\} = [r, L_n \otimes L_n]\delta_{mn}.$$

Here $\{L_n \otimes, L_m\}$ stands for the $k^2 \times k^2$-matrix consisting of all possible Poisson brackets $\{l(n)_{ab}, l(m)_{cd}\}$ and organized into a matrix in the same way as all possible products $l(n)_{ab} l(m)_{cd}$, $a, b, c, d = 1, \ldots, k$, are organized into the tensor product $L_n \otimes L_m$ of the $k \times k$-matrices. On the right hand side r is $k^2 \times k^2$-constant matrix (i.e. it depends on, but does not contain dynamical variables of the model); Kronecker's δ stands for the fact that dynamical variables at different lattice sites have vanishing Poisson brackets – this is the *ultrolocality* property of integrable models. In order that these Poisson brackets satisfy Jacobi identity it is sufficient (but by no means necessary) that matrix r satisfies CYBE! It turns out that for ultra-local integrable models corresponding matrices r do satisfy CYBE.

Another remarkable property of these Poisson brackets is that the product of L's

$$T_N = L_N \cdots L_1$$

– a so-called *monodromy matrix* – satisfies the same Poisson brackets

$$\{T_N \otimes, T_N\} = [r, T_N \otimes T_N].$$

What looks miraculous, but in fact is rather a simple observation, is that this property is nothing but the Poisson-Lie property! Indeed, realize a Poisson-Lie group G with a bracket $\{\ ,\ \}$ (see Section 1.3) as a subgroup of $GL(k)$, i.e. as a matrix group and denote by $\rho : g \mapsto Mat_k(\mathbb{C})$ corresponding representation of its Lie algebra g by $k \times k$-matrices. Let t_{ab} be coordinate functions on G, i.e. $t_{ab}(g) = g_{ab}$, $g \in G$, $a, b = 1, \ldots, k$, and $T = (t_{ab})$ be the corresponding matrix-function on G with these functions as elements. Using the formulas

$$\partial_x t_{ab} = (Tx)_{ab}, \ \partial'_x t_{ab} = (xT)_{ab},$$

which follow immediately from the definition of left- and right-invariant vector fields, it is possible to evaluate Poisson bracket of coordinate functions explicitly. We obtain

$$\{t_{ab}, t_{cd}\} = [\tilde{r}, T \otimes T]_{ac,bd},$$

where

$$\tilde{r} = (\rho \otimes \rho)(r) = \sum_{i,j=1}^{n} r^{ij} \rho(x_i) \otimes \rho(x_j).$$

Using our notations we can rewrite this formula in the form

$$\{T \otimes, T\} = [\tilde{r}, T \otimes T],$$

which coincides with the corresponding expression for the Poisson brackets for integrable models!

Thus we have seen that classical integrable models on a lattice produce Poisson brackets which define Poisson-Lie groups. Conversely, we can treat L-operators for such models as matrices of coordinate functions on Poisson-Lie groups. It is also clear that the Poisson-Lie property can be formulated as follows. Consider two copies of coordinate matrix-functions T' and T'' with pairwise vanishing Poisson brackets and with individual Poisson brackets given by the formula above. Then their matrix product $T = T'T''$ satisfies the same Poisson brackets! It is follows directly from Definition 2 specialized to coordinate functions since $\Delta(t_{ab}) = \sum_{c=1}^{k} t_{ac} \otimes t_{cb}$.

Example 5 Let $G = SL(2)$. The r-matrix from Theorem 3 (recall that $\hat{r} = P_+ - P_-$) in the standard root basis for $g = sl(2)$

$$x_+ = \begin{pmatrix} 0 & 1 \\ 0 & 0 \end{pmatrix}, \quad x_- = \begin{pmatrix} 0 & 0 \\ 1 & 0 \end{pmatrix}$$

is given by the following 4×4-matrix:

$$r = \begin{pmatrix} 0 & 0 & 0 & 0 \\ 0 & 0 & 1 & 0 \\ 0 & -1 & 0 & 0 \\ 0 & 0 & 0 & 0 \end{pmatrix}.$$

Introducing the matrix of coordinate functions T for $SL(2)$:

$$T = \begin{pmatrix} a & b \\ c & d \end{pmatrix}, \quad \det T = ad - bc = 1,$$

(do not confuse coordinate functions with matrix indices above!) we can write down all Poisson brackets of coordinate functions:

$$\{a, b\} = ab, \ \{b, c\} = 0, \ \{b, d\} = bd, \ \ \{a, c\} = ac, \ \{c, d\} = cd, \ \{a, d\} = 2bc.$$

One can easily describe symplective leaves of this Poisson structure on $G = SL(2)$. Essentially they are parametrized by the values of the ratio b/c, which has vanishing Poisson brackets with all functions on G. In particular, symplectic leaf $b = c$ contains a simply connected component $a, b, c, d > 0$ which can be parametrized by canonical coordinates P, Q with Poisson bracket $\{P, Q\} = 1$ in the following way:

$$a = e^P\sqrt{1+e^{2Q}},\ b = c = e^Q,\ d = e^{-P}\sqrt{1+e^{2Q}}.$$

Corresponding matrix L_n takes the form

$$L_n = \begin{pmatrix} e^{P_n}\sqrt{1+e^{2Q_n}} & e^{Q_n} \\ e^{Q_n} & e^{-P_n}\sqrt{1+e^{2Q_n}} \end{pmatrix}$$

and is the L-operator for the Liouville model on a lattice. This model is a discrete integrable version of the famous Liouville equation

$$\frac{\partial^2\phi}{\partial t^2} - \frac{\partial^2\phi}{\partial x^2} + e^\phi = 0,$$

which has interesting properties.

Remarks For applications of classical r-matrices to integrable models see [7]; Liouville model on a lattice was introduced in [10].

3 Quantum Integrable Models and Quantum Yang-Baxter Equation

3.1 Quantization of the Liouville Model on a Lattice, Commutation Relations and Quantum Yang-Baxter Equation

Recall that the paradigm of quantum mechanics consists of the *correspondence principle.* It states that the commutative algebra of classical observables on the phase space should be "deformed" ("quantized") into the non-commutative algebra of quantum observables (operators in a Hilbert space of states) in such a way that a classical Poisson bracket $\{\ ,\ \}$ is replaced (in a leading order in Planck's constant $\hbar$) by $[\ ,\ ]/i\hbar$, where $[\ ,\ ]$ stands for operator's commutator. Applied to the simplest case of the phase space $\mathbf{R}^2$ with coordinates P, Q and canonical Poisson bracket $\{P, Q\} = 1$ the correspondence principle produces famous Heisenberg commutation relations $[\hat{P}, \hat{Q}] = i\hbar$ for the quantum mechanical momentum and position operators.

Using this approach one can say that quantum groups are nothing but quantization (deformation in algebraic language) of commutative Hopf algebras of classical observables on very special phase spaces – Poisson-Lie groups. In addition to the correspondence principle it is required that the deformed non-commutative algebra should preserve the property of the Hopf algebra with the same coproduct (and counit) as in the commutative case $\hbar = 0$.

We will define these algebras later. Before doing it we will consider an interesting example – quantization of the Liouville model described in Example 5.

Example 6 Replace classical canonical variables P, Q by quantum operators $\hat{P}, \hat{Q}$ satisfying Heisenberg commutation relations and assume that there exists a quantum version of the classical L-operator L_n introduced in Example 5. Using certain "aesthetical" arguments it is rather natural to write down the quantum L-operator $\hat{L}_n$ in the following form (supressing index n)

$$\hat{L} = \begin{pmatrix} \sqrt{1+e^{2\hat{Q}+i\hbar}}e^{\hat{P}} & e^{\hat{Q}} \\ e^{\hat{Q}} & e^{-\hat{P}}\sqrt{1+e^{2\hat{Q}+i\hbar}} \end{pmatrix} = \begin{pmatrix} A & B \\ C & D \end{pmatrix}.$$

Heisenberg commutation relations imply

$$e^{\hat{P}}e^{\hat{Q}} = e^{i\hbar}e^{\hat{Q}}e^{\hat{P}}$$

– famous commutation relations of Hermann Weyl – and using them it is easy to prove that operators A, B, C, D satisfy the following algebra of commutation relations

$$AD = 1 + qBC, \; DA = 1 + q^{-1}BC$$

and

$$BC = CB, AB = qBA, \; AC = qCA, \; BD = qDB, \; CD = qDC,$$

where $q = e^{i\hbar}$. In particular, first two equations imply that

$$AD - DA = (q - q^{-1})BC$$

and

$$AD - qBC = 1.$$

Why are these relations so remarkable? First of all, it is easy to see that in *semi-classical limit* $\hbar \to 0$ i.e. $A \mapsto a, \; B \mapsto b, \; C \mapsto c, \; D \mapsto d$ such that $[\;,\;] \mapsto i\hbar\{\;,\;\}$ variables a, b, c, d become commutative and enjoy the same Poisson brackets as coordinate functions on the Poisson-Lie group $SL(2)$ (see Example 5). To verify it, expand q and $[\;,\;]$ in powers of $\hbar$ as $\hbar \to 0$ using the correspondence principle (say $0 = AB - qBA = [A, B] - i\hbar BA + O(\hbar^2)$, etc.). The terms containing no $\hbar$ give rise to commutativity of a, b, c, d (i.e. in the leading order $ab = ba$, etc.) while the terms linear in $\hbar$ yield $SL(2)$ Poisson brackets. The latter equation above results in the unimodularity condition $ad - bc = 1$.

However, we know that Poisson brackets for coordinate functions on a Poisson-Lie group can be written in elegant compact form as a single matrix equation

$$\{T \otimes, T\} = [r, T \otimes T].$$

What about corresponding quantum commutation relations, can we rewrite them as a single equation? It turns out that the answer is affirmative and the resulting equation yields a new structures of *quantum group* and *Quantum Yang-Baxter Equation* (QYBE).

Namely, let

$$\hat{L}_1 = \hat{L} \otimes I, \; \hat{L}_2 = I \otimes \hat{L}$$

be tensor products of the 2×2-matrix L (with non-commutative entries A, B, C, D) with the 2×2-unit matrix I in two opposite orders. If elements of L were commutative then we would get $\hat{L}_1\hat{L}_2 = \hat{L}_2\hat{L}_1$, where the product is understood as a matrix product of 4×4-matrices (this equation is necessary and sufficient for elements of L to be commutative). In our case the elements of L are non-commutative – they satisfy commutation relations displayed above. It turns out that the second and the third sets of these relations can be written as a single matrix equation

$$R_q\hat{L}_1\hat{L}_2 = \hat{L}_2\hat{L}_1R_q,$$

where R_q – called *quantum R-matrix* – is the following 4×4-matrix:

$$R_q = \begin{pmatrix} q & 0 & 0 & 0 \\ 0 & 1 & 0 & 0 \\ 0 & q - q^{-1} & 1 & 0 \\ 0 & 0 & 0 & q \end{pmatrix}.$$

Again, what is so special about this relation (which is a straightforward but necessary exercise to do!) and the matrix R_q ?

First, we have at the semi-classical limit

$$R_q = I - h\tilde{r} + O(h^2), \quad h = i\hbar,$$

where

$$\tilde{r} = r - P = -\begin{pmatrix} 1 & 0 & 0 & 0 \\ 0 & 0 & 0 & 0 \\ 0 & 2 & 0 & 0 \\ 0 & 0 & 0 & 1 \end{pmatrix}$$

where r is canonical r-matrix for the Poisson-Lie group $SL(2)$ (see Example 5) and P is a permutation matrix in $\mathbf{C}^4 = \mathbf{C}^2 \otimes \mathbf{C}^2$:

$$P = \begin{pmatrix} 1 & 0 & 0 & 0 \\ 0 & 0 & 1 & 0 \\ 0 & 1 & 0 & 0 \\ 0 & 0 & 0 & 1 \end{pmatrix}$$

In terms of the canonical element t (see Section 2.1) it has the following expression:

$$P = t + \frac{1}{2}I,$$

where I is 4×4-unit matrix. Therefore in the semi-classical limit quantum R-matrix R_q goes into r-matrix for the Poisson-Lie group $SL(2)$ which is not skew-symmetric (due to a shift by a matrix P) but satisfies CYBE (see discussion in case **B** of Section 2.1). In addition, commutation relations with matrix R_q go into Poisson brackets with classical r-matrix.

Indeed, as $h \to 0$, we get

$$\hat{L}_1\hat{L}_2 - \hat{L}_2\hat{L}_1 = h(\tilde{r}\hat{L}_1\hat{L}_2 - \hat{L}_2\hat{L}_1\tilde{r}) + O(h^2).$$

The matrix on the left hand side consists of all possible commutators of elements of $\hat{L}$ organized in the same way as the tensor product of 2×2-matrices, so that in the semi-classical limit it goes into $h\{L \otimes_, L\}$. Since $\hat{L} \to L$ with commutative entries we have $L_1 L_2 = L_2 L_1 = L \otimes L$ so that the right hand side is $h[\tilde{r}, L \otimes L] = h[r, L \otimes L]$, which completes the proof.

Second, quantum R-matrix itself satisfies an interesting equation. To understand its origin, consider the following problem. Assume that $\hat{L}$ satisfies commutation relations above and introduce matrices $\hat{L}_1 = \hat{L} \otimes I \otimes I$, $\hat{L}_2 = I \otimes \hat{L} \otimes I$, $\hat{L}_3 = I \otimes \otimes I \otimes \hat{L}$, where I is 2×2-unit matrix. Now consider the triple product $\hat{L}_1 \hat{L}_2 \hat{L}_3$ and try to interchange $\hat{L}$'s using commutation relations with quantum R-matrix in order to obtain their product in the opposite order: $\hat{L}_3 \hat{L}_2 \hat{L}_1$. It is clear that there are two different ways of doing it. The first way is realized as follows:

$$\hat{L}_1 \hat{L}_2 \hat{L}_3 = R_{12}^{-1} \hat{L}_2 \hat{L}_1 \hat{L}_3 R_{12} = R_{12}^{-1} R_{13}^{-1} \hat{L}_2 \hat{L}_3 \hat{L}_1 R_{13} R_{12} =$$

$$R_{12}^{-1} R_{13}^{-1} R_{23}^{-1} \hat{L}_3 \hat{L}_2 \hat{L}_1 R_{23} R_{13} R_{12},$$

where we have used that R_{12} (say) commutes with $\hat{L}_3$, etc. (because they act in different spaces).

The second way is realized as the following:

$$\hat{L}_1 \hat{L}_2 \hat{L}_3 = R_{23}^{-1} \hat{L}_1 \hat{L}_3 \hat{L}_2 R_{23} = R_{23}^{-1} R_{13}^{-1} \hat{L}_3 \hat{L}_1 \hat{L}_2 R_{13} R_{23} =$$

$$R_{23}^{-1} R_{13}^{-1} R_{12}^{-1} \hat{L}_3 \hat{L}_2 \hat{L}_1 R_{12} R_{13} R_{23}.$$

Certainly, these two apparently different ways yield the same result which means that commutation relations with matrix R_q (a set of quadratic relations for A, B, C, D) imply additional cubic relations on elements of $\hat{L}$. It is clear that aesthetically the most appealing case is when these cubic relations are satisfied identically (i.e. we prefer not to impose addition cubic relations on the elements of $\hat{L}$). The simplest way to achieve it is to impose the following equation for the matrix R_q:

$$R_{12} R_{13} R_{23} = R_{23} R_{13} R_{12}.$$

Then it is obvious that all cubic relations (and further relations of higher order) are satisfied identically (by no means it is a necessary condition; it is only the simplest sufficient condition).

Equation displayed above is nothing but the famous *quantum Yang-Baxter Equation* (QYBE)! It is simple to check (do it!) that our matrix R_q satisfy QYBE.

What is more, CYBE can be considered as a semi-classical limit of QYBE. Indeed, substitute the expansion $R_q = I - h\tilde{r} + O(h^2)$ into the QYBE we observe that linear terms in h cancel out identically and quadratic terms in h can be presented as two separate groups – quadratic expressions in $\tilde{r}$ and unspecified terms $O(h^2)$. The latter terms drop out automatically and the resulting equation in the order h^2 is reduced to CYBE!

To summarize:
1) quantum L-operator for the Liouville model on a lattice satisfies commutation relations with quantum R-matrix:

$$R_q\hat{L}_1\hat{L}_2 = \hat{L}_2\hat{L}_1R_q;$$

2) quantum R-matrix R_q satisfies QYBE and has the semi-classical expansion

$$R_q = I - h\tilde{r} + O(h^2),$$

where $\tilde{r}$ satisfies CYBE.

In addition, if we introduce the matrix $\hat{R}_q = PR_q$,

$$\hat{R}_q = \begin{pmatrix} q & 0 & 0 & 0 \\ 0 & q-q^{-1} & 1 & 0 \\ 0 & 1 & 0 & 0 \\ 0 & 0 & 0 & q \end{pmatrix},$$

then it satisfies the following quadratic equation:

$$\hat{R}_q^2 = I + (q-q^{-1})\hat{R}_q,$$

which is called the *Hecke condition.* Note when $q = 1$ (i.e. $h = 0$) matrix $\hat{R}_q$ turns into the permutation matrix P.

Remarks QYBE (as well as its name) was introduced in the theory of quantum integrable systems in [11]. A comprehensive account of classical and quantum integrable models can be found in lectures [12]. More about QYBE and its role in different areas of mathematics and physics can be found in [13].

3.2 Connection with E.Artin Braid Group

It is instructive to rewrite QYBE in terms of the matrix $\hat{R}_q$ introduced above. Using the basic properties of the permutation operator P

$$P(A_1 \otimes A_2) = (A_2 \otimes A_1)P, \quad P^2 = I,$$

(for all $n \times n$ matrices A_1, A_2) we can rewrite QYBE in the following form (supressing q for a while)

$$(\hat{R}\otimes I)(I\otimes\hat{R})(\hat{R}\otimes I) = (I\otimes\hat{R})(\hat{R}\otimes I)(I\otimes\hat{R})$$

Indeed, multiply QYBE from the left successively by P_{23}, P_{12} and P_{23} and use the basic property of P.

What is truly remarkable is that this equation is nothing but a defining relation for the generators of E.Artin braid group!

Definition 3 Braid group B_N of N-braids is a group generated by $N-1$ generators $b_1, \ldots, b_{N-1}$ satisfying relations

$$b_i b_j = b_j b_i$$

for $i, j = 1, \ldots, N-1,\ |i-j| > 1$ and

$$b_i b_{i+1} b_i = b_{i+1} b_i b_{i+1},$$

for $i = 1, \ldots, N-2$.

Geometrically, B_N can be realized as a group consisting of braidings of N strings – N-braids – in a three-dimensional space. Generators b_i describe braidings between i-th and $i+1$-th strings; the first set of relations imply that braiding of two separated pairs of strings can be performed in any order. The latter set characterizes braiding of three consecutive strings – in this case the order is important but the two operations indicated above produce the same result. It is very illuminating to visualize this picture and to make all relations in B_N topologically clear.

Now we can say that QYBE for matrix $\hat{R}$ represents the defining relation for the braid group B_3. Namely, denoting $\hat{R} \otimes I,\ I \otimes \hat{R}$ by $\tilde{b}_1,\ \tilde{b}_2$ respectively, we can rewrite QYBE as

$$\tilde{b}_1 \tilde{b}_2 \tilde{b}_1 = \tilde{b}_2 \tilde{b}_1 \tilde{b}_2$$

which is a relation in B_3!

Strictly speaking this formula means that there is a representation ρ of the braid group B_3 in the linear space $\mathbf{C}^n \otimes \mathbf{C}^n \otimes \mathbf{C}^n$ defined by

$$\rho(b_1) = \tilde{b}_1,\ \rho(b_2) = \tilde{b}_2.$$

This definition is consistent (i.e. ρ is a representation of B_3) because of QYBE.

In general, any $n^2 \times n^2$ solution of the QYBE gives rize to a representation ρ of the braid group B_N in the linear space $\mathbf{C}^n \otimes \mathbf{C}^n \cdots \otimes \mathbf{C}^n$ (N-fold tensor product) defined by

$$\rho(b_i) = I \otimes I \cdots \otimes \hat{R} \otimes \cdots \otimes I,$$

where matrix $\hat{R}$ occupies i-th and $i+1$-th factors of the tensor product, $i = 1, \ldots, N-1$. It is easy to see that the first set of relations for B_N is satisfied by the construction of ρ whereas the validity of the second set follows from QYBE for R-matrix $\hat{R}$!

Since braid groups are related to knots, links, etc., this simple (but very important!) observation explains why QYBE played such remarkable role in recent breakthrough in the "production" of knots and links invariants. Not all the solutions of QYBE are equally interesting; in particular, if the matrix $\hat{R}$ satsfies the condition $\hat{R}^2 = I$, i.e. $R_{12}R_{21} = I$ – called *unitarity condition*, then representation ρ descends to the representation of the symmetric group $Symm(N)$ (which is defined by imposing additional relations $b_1^2 = \cdots = b_{n-1}^2 = 1$ in B_N) and is not so relevant for braid groups. The first non-trivial example of such representation ρ is provided by R-matrix R_q from the previous section. It gives rise to the famous Jones' polynomial for knots!

Remarks More about braids, knots, links and QYBE can be found in [14].

4 Quantum Matrix Algebras, Quantum Groups and Quantum Vector Spaces

4.1 Quantum Matrix Algebras

Let R be a non-degenerate $n^2 \times n^2$-matrix satisfying QYBE

$$R_{12}R_{13}R_{23} = R_{23}R_{13}R_{12}.$$

Denote by $A = \mathbb{C} < t_{ij} >$ the C-algebra with free generators $t_{ij},\ i,j = 1,\ldots,n$. In other words, A is a free associative algebra with unit over complex numbers with generators t_{ij} – algebra of non-commutative polynomials in n^2 variables. Denote by T the $n \times n$ matrix with elements t_{ij} and set $T_1 = T \otimes I,\ T_2 = I \otimes T$, where I is $n \times n$-unit matrix and $\otimes$ stands for the tensor product of matrices.

Definition 4 A quantum matrix algebra A_R associated with matrix R is a quotient algebra of A by a two-sided ideal generated by all elements of $RT_1T_2 - T_2T_1R$, i.e.

$$A_R = A/(RT_1T_2 - T_2T_1R).$$

In other words, algebra A_R is defined by the generators t_{ij} and relations

$$RT_1T_2 = T_2T_1R.$$

When R is a unit matrix (which obviously satisfies QYBE) this relation implies that variables t_{ij} are commutative (see the previous section) and, therefore, T is a matrix with commutative elements. In this case, we get a usual matrix algebra (strictly speaking, $A_I = \mathbb{C}[t_{ij}]$ – usual polynomials in n^2 variables, or, which is the same, A_I is an algebra of polynomial functions on a matrix algebra $Mat_n(\mathbb{C})$ with elements t_{ij} as coordinate functions). This is why we call A_R a quantum matrix algebra.

Now recall the general discussion of Hopf algebras in the first section. It implies that A_I has an additional coalgebra structure with the coproduct Δ induced by the usual product of matrices. For coordinate functions t_{ij} we have explicitly (c.f. Section 2.3)

$$\Delta(t_{ij}) = \sum_{k=1}^{n} t_{ik} \otimes t_{kj},$$

where now $\otimes$ stands for the tensor product $A_I \otimes A_I$! These formulas can be organized into a single matrix formula using the convenient notation $\dot{\otimes}$ which is organized as a matrix product in which products of elements are replaced by tensor products. This is the quantum analog of the notation $\otimes$,, introduced in Section 2.3. Using it, we can rewrite the coproduct formula as

$$\Delta(T) = T\dot{\otimes}T.$$

In addition to the coproduct Δ algebra A_I has a counit ϵ induced by the unit element in $Mat_n(\mathbb{C})$ – a $n \times n$-unit matrix. On elements t_{ij} it is given by the formula $\epsilon(t_{ij}) = \delta_{ij}$, or

$$\epsilon(T) = I.$$

These formulas for the coproduct Δ and counit ϵ, defined on generators, can be extended to the whole algebra A_I using the homomorphism property.

Now what about algebra A_R for general R? Remarkably, it still carries the coalgebra structure while its C-algebra structure is no longer commutative! What is more, this coalgebra structure on generators is defined by the same formula as in the commutative case. We have

Theorem 4 A C-algebra A_R is a coalgebra with the coproduct and counit given by

$$\Delta(T) = T\dot{\otimes}T, \quad \epsilon(T) = I$$

and extended to the whole algebra A_R using the homomorphism property.

Proof Since A_R is a quotient algebra of a free C-algebra A by two-sided ideal J generated by all elements of $RT_1T_2 - T_2T_1R$ it is sufficient to check that $\Delta(J) \subset J \otimes 1 + 1 \otimes J$. This is a straightforward exercise using the definition of Δ and the fact that it is a C-algebra homomorphism. A more illuminating (but less formal) proof goes as follows (c.f. with discussion in Section 2.3). Let T', T'' be two pairwise commuting sets of generators t'_{ij}, t''_{ij} both satisfying commutation relations with R-matrix. Since $\Delta(T) = T'T''$ it is sufficient to check that $T'T''$ satisfies the same commutation relations. Indeed,

$$\begin{aligned} RT_1T_2 = RT_1'T_1''T_2'T'' &= RT_1'T_2''T_1''T_2'' = T_2'T_1'RT_1''T_2'' \\ &= T_2'T_1'T_2''T_1''R = T_2'T_2''T_1'T_1''R = T_2T_1R, \end{aligned}$$

where we have used commutation relations for T' and T'' and the fact that T_1'' commutes with T_2' and T_1' – with T_2''. Remaining properties of the counit ϵ are trivial to check. This completes the proof.

Now let g be a simple Lie algebra. We know from Theorem 3 that there exists canonical r-matrix which provides a Lie group G with a Poisson-Lie structure (for the case $g = sl(2)$ this r-matrix was displayed in Example 5). A non-trivial fact is that all such r-matrices can be "quantized", i.e. for every simple Lie algebra g there exists a one-parameter family of quantum R-matrices R_q which satisfy QYBE and have the property

$$R_q = I - h\tilde{r} + O(h^2), \quad h = \log q \to 0,$$

where $\tilde{r}$ is a non-unitary classical r-matrix (obtained from a canonical matrix r by adding a multiple of t and of a unit matrix such that $\tilde{r}$ satisfies CYBE instead of mCYBE). For the case $g = sl(2)$ such R-matrix was presented in Example 6.

Using this approach we can we can associate with the matrix R_q a quantum matrix algebra $A_q = A_{R_q}$ for a simple Lie group G. It has a coalgebra structure (see Theorem 5), but is not a Hopf algebra, i.e. it has no antipode. Imposing additional relations on A_q, i.e. considering its quotient algebra by a certain two-sided ideal one can turn this quotient algebra into a Hopf algebra, i.e. to define an antipode. (This procedure is similar to the definition of $GL(n)$ as the set of all invertible matrices in $Mat_n(\mathbb{C})$ or of $SL(n)$ as the set of all matrices with unit determinant). The resulting Hopf algebra is called the quantum group G_q which corresponds to a Poisson-Lie group G with canonical r-matrix. (Note that we are deforming not a group G itself but an algebra of functions on it; therefore G_q should be treated as a non-commutative Hopf algebra of "quantum functions" so that a quantum group itself should be a "non-commutative spectrum" of this algebra). Here I will illustrate these ideas considering the simplest example of the Poisson-Lie group $SL(2)$ and its quantum group $SL_q(2)$.

Remarks General concept of quantum matrix algebras was introduced in [15].

4.2 Quantum Special Linear Group, Quantum Plane and Quantum Exterior Algebra

Let R_q be the R-matrix from the previous lecture. Corresponding quantum matrix algebra A_q is generated by the elements A, B, C, D satisfying following the explicit relations:

$$\begin{aligned} AB &= qBA, \;\; AC = qCA, \;\; BD = qDB, \;\; CD = qDC \\ AD - DA &= (q - q^{-1})BC, \;\; BC = CB. \end{aligned}$$

Coproduct Δ and counit ϵ are given by the following formulas:

$$\Delta(A) = A \otimes A + B \otimes C, \;\; \Delta(B) = A \otimes B + B \otimes D,$$
$$\Delta(C) = C \otimes A + D \otimes C, \;\; \Delta(D) = C \otimes B + D \otimes D;$$
$$\epsilon(1) = \epsilon(A) = \epsilon(D) = 1, \;\; \epsilon(B) = \epsilon(C) = 0.$$

It is clear that the element

$$\det_q T = AD - qBC$$

commutes with A, B, C, D and, therefore, with all elements in A_q. In other words it belongs to a center of the algebra A_q. The element $\det_q T$ is called *quantum determinant*; this terminology is justified by the following definition and the theorem.

Definition 5 A quotient algebra $SL_q(2) = A_q/(\det_q T - 1)$ is called a *quantum group* corresponding to the Poisson-Lie group $SL(2)$.

In other words, in $SL_q(2)$ we have an additional relation $\det_q T = 1$, i.e. quantum determinant is 1 as expected.

Theorem 5 The quantum group $SL_q(2)$ is a Hopf algebra with the antipode S given explicitly by

$$S(T) = \begin{pmatrix} D & -q^{-1}B \\ -qC & A \end{pmatrix}.$$

and extended to $SL_q(2)$ using the anti-homomorphism property.

Proof We need to check axiom 4) of the antipode (see Section 1.2). Since the coproduct Δ is induced by a matrix multiplication we can rewrite this property for the element T as

$$S(T) \cdot T = T \cdot S(T) = I,$$

where the dot stands for the usual matrix product (2×2-matrices with non-commutative elements). Using defining relations of the algebra A_q and relation $\det_q T = AD - qBC = 1$ the latter equations can be checked immediately.

Definition 6 C-algebra $\mathbf{C}_q^2$ with generators x, y and relation $xy = qyx$ is called a *quantum plane.*

We know that group $SL(2)$ acts on a two-dimensional complex plane $\mathbf{C}^2$; what about quantum group $SL_q(2)$ and quantum plane $\mathbf{C}_q^2$?

Theorem 6 There is a *coaction* of a quantum group $SL_q(2)$ on a quantum plane $\mathbf{C}_q^2$ given by the C-algebra homomorphism

$$\delta : \mathbf{C}_q^2 \mapsto SL_q(2) \otimes \mathbf{C}_q^2$$

defined by the formula

$$\delta(X) = T\dot{\otimes}X,$$

where $X = \begin{pmatrix} x \\ y \end{pmatrix}$, or

$$\delta(x) = A \otimes x + B \otimes y,$$
$$\delta(y) = C \otimes x + D \otimes y.$$

Proof Set $x' = \delta(x)$, $y' = \delta(y)$. One needs to verify that $x'y' = qy'x'$. It is a very simple calculation using commutation relations defining quantum group $SL_q(2)$ and quantum plane $\mathbf{C}_q^2$ (note that while commuting x' and y' one should use – this is a definition of the tensor product of algebras – that all A, B, C, D commute with x, y).

We also have

Definition 7 A C-algebra $\bigwedge \mathbf{C}_q^2$ with generators ξ, η and relations

$$\xi^2 = \eta^2 = 0, \quad \xi\eta = -q^{-1}\eta\xi$$

is called *quantum exteriour algebra of a quantum plane.*

Theorem 7 Formulas

$$\delta(\Xi) = T\dot{\otimes}\Xi,$$

where $\Xi = \begin{pmatrix} \xi \\ \eta \end{pmatrix}$ define a coaction $\delta : \bigwedge \mathbf{C}_q^2 \mapsto SL_q(2) \otimes \bigwedge \mathbf{C}_q^2$ of the quantum group $SL_q(2)$ on the quantum exteriour algebra.

Proof Analogous to the previous one.

As a corollary we obtain

$$\begin{aligned} \delta(\xi\eta) &= (A \otimes \xi + B \otimes \eta)(C \otimes \xi + D \otimes \eta) \\ &= AD \otimes \xi\eta + BC \otimes \eta\xi \\ &= \det_q T \otimes \xi\eta \end{aligned}$$

which shows that the quantum determinant emerges in the same natural way as the classical determinant does.

As another application of quantum plane consider the following

Example 7 "Quantum" binomial theorem. Let x, y be the generators of the quantum plane $\mathbf{C}_q^2$. We have

$$(x+y)^2 = x^2+(1+q)yx+y^2,\ (x+y)^3 = x^3+(1+q+q^2)y^2x+(1+q+q^2)yx^2+y^3$$

and it is easy to prove by induction (using relation $xy = qyx$) that

$$(x+y)^n = \sum_{k=0}^{n} q^{\frac{k(n-k)}{2}} \begin{bmatrix} n \\ k \end{bmatrix}_q y^k x^{n-k},$$

where

$$\begin{bmatrix} n \\ k \end{bmatrix}_q = \frac{[n]_q!}{[k]_q![n-k]_q!}$$

are "quantum" binomial coefficients and

$$[n]_q! = [n]_q[n-1]_q \cdots [1]_q, \quad [n]_q = \frac{q^{n/2} - q^{-n/2}}{q^{1/2} - q^{-1/2}}.$$

However, this formula had been known long before quantum groups appeared (in fact, it was already known to Gauss!) and it is better to call it a q-binomial theorem.

Remarks Our presentation follows [15] and [5, 6, 16]. Quantum vector spaces were also considered in [17]. There is vast literature on q-special functions (which were studied long before quantum groups); see, e.g., [18]. Recently they reappear as the matrix elements of the representations of quantum groups (in the same manner as classical special functions appear in representation theory of $SU(2)$, say); see, e.g. [19].

4.3 Real Forms of $SL_q(2)$

So far we have defined quantum group $SL_q(2)$ which is a deformation of the algebra of (polynomial) functions on the Lie group $SL(2, \mathbf{C})$. What about quantization of its compact form $SU(2)$ and real form $SL(2, \mathbf{R}) \simeq SU(1,1)$? They are defined using additional structure on a Hopf algebra $SL_q(2)$ – $*$ - anti-involution. (Recall that a $*$ - anti-involution of the Hopf algebra A is the mapping $* : A \mapsto A$ which is $\mathbf{C}$-algebra anti-homomorphism, coalgebra homomorphism and satisfies the properties $(a^*)^* = a$, $S(S(a^*)^*) = a$ for all $a \in A$).

We have two cases.

Case **A**

In this case $q \in \mathbf{R}$ and R-matrix R_q is real-valued. Defining

$$T^* = S(T)^t$$

or

$$T^* = \sigma S(T)^t \sigma, \quad \sigma = \begin{pmatrix} 1 & 0 \\ 0 & -1 \end{pmatrix},$$

where ^{t} denotes the transposition, we get quantum groups $SU_q(2)$ and $SU_q(1,1)$ respectively.

Explicitly $*$ - anti-involution for the quantum unitary group $SU_q(2)$ is given by

$$A^* = D, \quad B^* = -qC;$$

corresponding group $SU_q(1,1)$ is defined by

$$A^* = D, \quad B^* = qC.$$

(Check that this definition of $*$ is consistent with defining relations in $SL_q(2)$; condition $q \in \mathbf{R}$ is important).

Case **B**

Here $|q| = 1$ and we can define the $*$ - anti-involution by

$$T^* = T$$

which gives rise to the quantum group $SL_q(2, \mathbf{R})$. We have explicitly

$$A^* = A, \ B^* = B, \ C^* = C, \ D^* = D.$$

Again it is easy to prove that this definition of $*$ is consistent with relations in $SL_q(2)$ (condition $|q| = 1$ is important).

Note that classical isomorphism $SL(2, \mathbf{R}) \simeq SU(1,1)$ in the quantum case is no longer valid since corresponding quantum groups are now defined for different values of q! This shows that the quantum case is more "rigid" than the classical one – certain isomorphisms between low dimensional classical Lie groups break down at the quantum level.

Remarks We follow the approach in [15]. Quantum group $SU_q(2)$ was also introduced by Woronowicz in [20] using the ideas of non-commutative geometry and C*-algebras.

4.4 Quantized Universal Enveloping Algebra

So far we dealt with quantum groups as deformations of algebras of functions on Lie groups. What about corresponding Lie algebras, can one define their quantum version? Certainly it is not natural to deform a simple Lie algebra itself (it has no nontrivial deformations) but it is possible to deform its universal enveloping algebra as a Hopf algebra. Namely, starting from Ug (see Example 2) it is possible to define what is called *quantized universal enveloping algebra* $U_q g$ of a simple Lie algebra g which is no longer cocommutative. Instead of giving a general definition for arbitrary simple Lie group g I will present here a simple construction for the case $g = sl(2)$.

Definiton 8 A C-algebra $U_q sl(2)$ with generators H, X_+, X_- and relations

$$[H, X_\pm] = \pm 2X_\pm, \quad [X_+, X_-] = \frac{\sinh(hH)}{\sinh(h)}$$

is called *quantized universal enveloping algebra* of a Lie algebra $sl(2)$.

Here, as usual, $q = e^h$ and $\sinh(hH)/\sinh(h)$ should be considered as formal power series expansion in h.

Why is this particular C-algebra so remarkable?

First, in the semi-classical limit $h \to 0$ its defining relations go into usual $sl(2)$-commutators; in other words it is a deformation of $Usl(2)$. Second, algebra $U_q sl(2)$ (as well as $Usl(2)$) is a Hopf algebra.

Theorem 8 Algebra $U_q sl(2)$ is a Hopf algebra with a coproduct Δ, counit ϵ and antipode S defined by the following formulas

$$\begin{aligned}
\Delta(H) &= H \otimes 1 + 1 \otimes H, \quad \Delta(X_\pm) = X_\pm \otimes e^{-hH/2} + e^{hH/2} \otimes X_\pm, \\
\epsilon(H) &= \epsilon(X_\pm) = 0, \\
S(H) &= -H, \quad S(X_\pm) = -e^{-hH/2} X_\pm e^{hH/2},
\end{aligned}$$

and extended to $U_q sl(2)$ using the homomorphism property.

Proof It is sufficient to check that these formulas are consistent with the defining relations of $U_q sl(2)$. This is left as an exercise.

In a semi-classical limit Hopf algebra structure of $U_q sl(2)$ goes into that of $Usl(2)$ (see Example 4).

Finally, let us mention that there is an interesting duality between quantum groups and quantized universal enveloping algebras which explains special commutation relations for the latter. However, it is better to stop at this point; a lot more can be found in the references below!

Remarks Algebra $U_q sl(2)$ was introduced in [21]; in [22] it was proved that it is a Hopf algebra. This example serves as a starting point for Drinveld and Jimbo [4], [23] in their general definition. The fundamental idea that Hopf algebra's language is a language of quantum groups belongs to Drinfeld, who presented the following definition "quantum groups are non commutative and non cocommutative Hopf algebras". According to it, $U_q g$ is also a quantum group (in a broder sense); however sometimes it is convenient to distinguish between quantum groups (in a narrow sense) and quantized universal enveloping algebras. The mentioned above duality between them was introduced in [15]. Interesting applications of quantum groups to conformal field theory, together with many references, can be found in conference proceedings [24]. For those looking for mathematically oriented papers, the references below provide some of them: [25, 26], [27, 29], [21], [30], [31, 32].

References

[1] V.Arnold. *Mathematical Methods of Classical Mechanics.* Springer-Verlag, 1978.

[2] B.Dubrovin, A.Fomenko and S.Novikov. *Modern Geometry – Methods and Applications.* Springer-Verlag, 1984.

[3] E.Abe. *Hopf Algebras.* Cambridge Univ. Press, 1980.

[4] V.Drinfeld. *Quantum Groups.* Proc. ICM-86, **1**, p.798-820, Academic Press, 1986.

[5] L.Takhtajan. *Introduction to Quantum Groups*, Springer Lecture Notes in Physics, **370**, 3-28, 1990.

[6] L.Takhtajan. *Lectures on Quantum Groups*, Nankai Institute Series on Mathematical Physics, World Scientific, 1990.

[7] L.Faddeev, L.Takhtajan. *Hamiltonian Methods in the Theory of Solitons.* Springer-Verlag, 1987.

[8] M.A.Semenov-Tian-Shansky. *What is a Classical R-Matrix?* Funct. Analysis i ego Pril., **17**, No.3, p.17-33, 1983 (in Russian).

[9] M.A.Semenov-Tian-Shansky. *Dressing Transformations and Poisson Group Actions.* Publ. RIMS, **21**, No.6, 1237-1260, 1985.

[10] L.Faddeev, L.Takhtajan. *Liouville Model on the Lattice*, Springer Lecture Notes in Physics, **246**, p.166-179, 1986.

[11] L.Takhtajan, L.Faddeev. *Quantum Inverse Transform Method and XYZ-Heisenberg Model*, Uspechi Mat. Nauk, **34**, No.5, p.13-63, 1979 (in Russian).

[12] L.Faddeev. *Integrable Models in* $1+1$ *Dimensional Quantum Field Theory.* Les Houches Lectures 1982, Elsevier, 1984.

[13] M.Jimbo, Ed., *Yang-Baxter Equation in Integrable Models.* World Scientific, 1989.

[14] C.Yang, M. Ge, Eds., *Braid Group, Knot Theory and Statistical Mechanics.* World Scientific, 1989.

[15] N.Reshetikhin, L.Takhtajan and L.Faddeev. *Quantization of Lie Groups and Lie Algebras.* Leningrad Math. Journal, **1**, No.1, p.178-206, 1990.

[16] L.Takhtajan. *Quantum Groups and Integrable Models.* Adv. Studies in Pure Math. **19**, p.435-457, 1989.

[17] Yu.Manin. *Quantum Groups and Non-Commutative Geometry.* preprint CRM-1561, Montreal, 1988.

[18] G.Gasper, M.Rahman. *Basic Hypergeometric Series.* Cambridge Univ. Press, 1990.

[19] G.Lusztig. *Quantum Deformations of Certain Simple Modules over Enveloping Algebras.* Adv. Math., **70**, p.237-249, 1988.

[20] S.Woronowicz. *Twisted $SU(2)$-Group. An Example of Noncommutative Differential Calculus.* Publ. RIMS, **23**, p.117-181, 1987.

[21] P.Kulish, N.Reshetikhin. *Quantum Linear Problem for the Sine-Gordon Equation and Higher Representations.* Proc. Steklov Math. Inst. at Leningrad, **101**, p.101-110, 1981 (in Russian).

[22] E.Sklyanin. *On a Certain Algebra Generated by Quadratic Relations.* Uspechi Mat. Nauk, **40**, No.2, p.214, 1985 (in Russian).

[23] M.Jimbo. *A Q-Analog of $U(gl(N+1))$, Hecke Algebras, and Yang-Baxter Equation.* Lett. Math. Phys., **11**, p.247-252, 1986.

[24] H.Doebner, J.Henning, Eds. *Quantum Groups*, Springer Lecture Notes in Physics, **370**, 1990

[25] V.Drinfeld. *On Almost Cocommutative Hopf Algebras.* Leningrad Math. Journal, **1**, 321-342, 1990.

[26] V.Drinfeld. *Quasi-Hopf Algebras*, Leningrad Math. Journal, **1**, 1419-1457, 1990.

[27] D.Kazhdan, G.Lusztig. *Affine Lie Algebras and Quantum Groups.* Duke Math. Journal, **62**,21-29, 1991.

[28] D.Kazhdan, N.Reshetikhin. *Balanced Categories and Invariants of 3-Manifolds.* Preprint 1991.

[29] A.N.Kirillov, N.Yu.Reshetikhin. *Q-Weyl Group and a Multiplicative Formula for Universal R-Matrices.* Commun. Math. Phys., **134**, 421-431, 1991.

[30] N.Reshetikhin. *Quasitriangular Hopf Algebras and Invariants of Links.* Leningrad Math. Journal, **1**, 1989.

[31] M.Rosso. *Finite Dimensional Representations of the Quantum Analogue of the Enveloping Algebra of a complex Simple Lie Algebra.* Commun. Math. Phys., **117**, 581-593, 1988.

[32] M.Rosso. *An Analogue of P.B.W. Theorem and the Universal R-Matrix for $U_h sl(N+1)$.* Commun. Math. Phys., **124**, 307-318, 1989.

Knot Theory and Integrable Systems

M. Wadati

Department of Physics, Faculty of Sciences, University of Tokyo, Hongo 7-3-1, Bunkyo-ku, Tokyo 113, Japan

This review article concerns the construction of link polynomials, topological invariants for knots and links, from exactly solvable(integrable) models. Through a general theory which is based on fundamental properties of the models, representations of the braid group and the Markov traces on the representations are made. In addition, the equivalence of algebraic and graphical formulations is proved. Various examples including Alexander, Jones, Kauffman and new link polynomials are explicitly shown. To sum up, the soliton theory contains the essence of the knot theory.

1. Introduction

In 1965, Zabusky and Kruskal [1] introduced a new concept, the ***soliton***, in the study of nonlinear dynamics. Soliton is a term for a nonlinear wave with particle properties. Since its discovery, the soliton theory has been an "inexhaustible spring" of new physics and mathematics. What will be described in this article may be one of the best examples.

By the inverse scattering method which is the extension of the Fourier transformation to nonlinear problems, the soliton system is proved to be a completely integrable system. A remarkable property of the system is the existence of an infinite number of conserved quantities. The inverse scattering method is also applicable to quantum systems. This generalization is called the quantum inverse scattering method. The quantum inverse scattering method encompasses the conventional methods in theoretical physics, the Bethe ansatz method and the transfer matrix method. Then, there emerges a unified viewpoint on exactly solvable models in (1+1)-dimensional field theory and in 2-dimensional statistical mechanics. To each model we can associate a family of commuting transfer matrices which are generators of an infinite number of conserved quantities. A sufficient condition for the commutability of transfer matrices and then for the solvability of models is the *Yang-Baxter relation* [2,3].

The Yang-Baxter relation is a key to the recent developments in mathematical physics. Among them, a novel connection between physics and mathematics has been found. Namely, the theory of exactly solvable models provides us with a new approach to the knot theory. This development will be explained in an elementary way.

The organization is as follows: In section 2, we briefly summarize the knot theory. Basic concepts such as knot, link, braid, braid group, closed braid, Markov trace and link polynomial are introduced. In section 3, we consider solvable models in 2-dimensional statistical mechanics and define the Yang-Baxter relation for vertex and IRF models. We show that the representations of the braid group are obtained from the Yang-Baxter operators. In section 4, we present explicit forms of the Markov traces. A sufficient condition for the existence of the Markov trace is given. Then, we have a general theory to construct link polynomials from exactly solvable models. In addition, the equivalence of algebraic and graphical formulations is proved. Section 5 is devoted to applications of the theory. We present solvable models which lead to Alexander, Jones, Kauffman and new link polynomials. Two-variable extensions are discussed. The last section contains concluding remarks.

2. Knot theory

We shall introduce some basic knowledge on knot theory. The knot theory is a branch of topology, which deals with 1-dimensional objects (curves) in 3-dimensional space. Let us call a 1-dimensional object a string. In general, a string has an orientation. A *knot* is a closed self-avoiding string in 3-dimensional space. An assembly of knots is called a *link* (Fig.1). In other words, link is a general terminology for knot and link.

Classification of knots and links is one of the most fundamental problems in topology. Given two knots/links, we want to judge whether they are different or not. Two knots/links are said to be *topologically equivalent* or *ambient isotopic* when they are transformed into each other by continuous deformations without tearing string(s). It is not difficult to convince ourselves that a trefoil knot (a) and knot (c) in Fig.1 are equivalent.

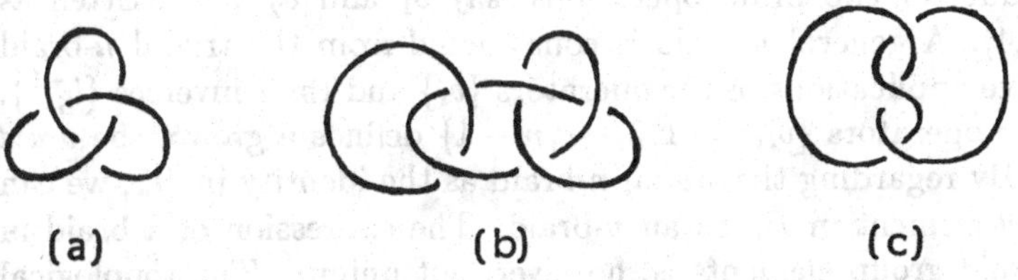

Fig.1 (a) Trefoil knot, (b) a link consisting of trivial and trefoil knots, (c) this knot is equivalent to the trefoil knot (a).

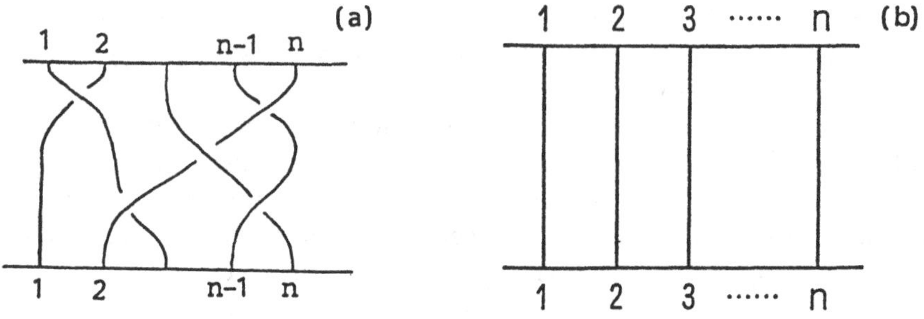

Fig.2 (a) A general n-braid, (b) trivial n-braid.

In physics, there are many string-like objects, for instance, polymers, vortex filaments, dislocations, magnetic fluxes and particle trajectories. Recently, string theory in particle physics has attracted much attention from physicists. It is interesting to recall that a similar (but, of course different) idea was proposed in the last century. Lord Kelvin tried to explain the periodic law assuming that atoms are knots in the ether field. This idea was not successful, but motivated P.G.Tait to make a first list of knots.

To classify knots and links in a systematic way, it is necessary to use a topological invariant, that is, a quantity which does not change under continuous deformations of strings. The Gauss linking number is a well-known example. When the invariant is expressed in the form of a polynomial in some variable, it is called a *link polynomial.*

We introduce now braid and braid group. Prepare two horizontal bars on which n base points are marked. A *braid* is formed when n points on the upper bar are connected to n points on the lower bar by n strings (Fig.2(a)). A trivial n-braid is a configuration where no intersection between the strings is present (Fig.2(b)). The operation of making an intersection, where the i-th string (a string from point i on the upper bar) passes over the $(i+1)$-th string (a string from point $i+1$ on the upper bar), is denoted by b_i. And the inverse operation is denoted by b_i^{-1} (Fig.3).

A product of two braid operations, say b_1 and b_2^{-1}, is written as $b_1 b_2^{-1}$ (Fig.4). A general n-braid is constructed from the trivial n-braid by successive applications of the operators $\{b_i\}$ and their inverses $\{b_i^{-1}\}$. A set of the operators $\{b_i, i = 1, 2, \cdots, n-1\}$ defines a group, the *braid group* B_n. By regarding the trivial n-braid as the identity in B_n, we can identify any element in B_n as an n-braid. The expression of a braid in terms of braid group elements is, however, not unique. The topological equivalence between different expressions of a braid is guaranteed by the following relations (Fig.5):

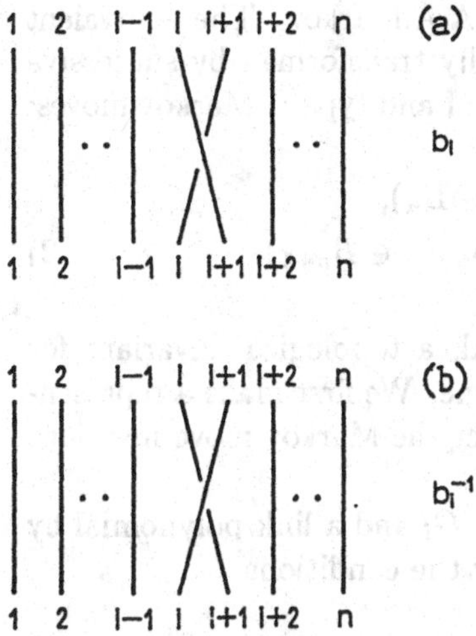

Fig.3 (a) Braid operation b_i, (b) inverse operation b_i^{-1}.

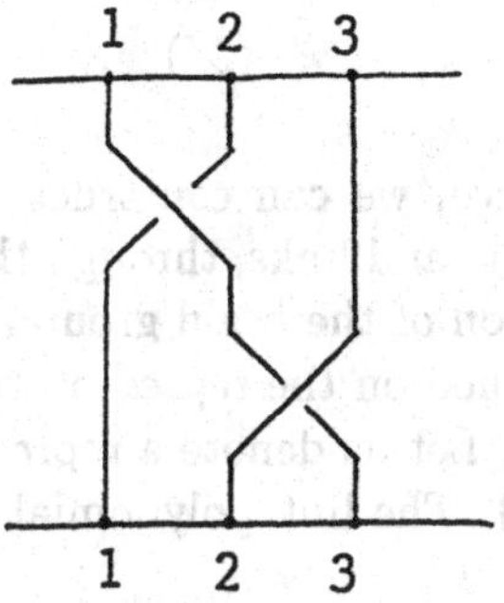

Fig.4 A product $b_i b_j$, for instance, $b_1 b_2^{-1}$.

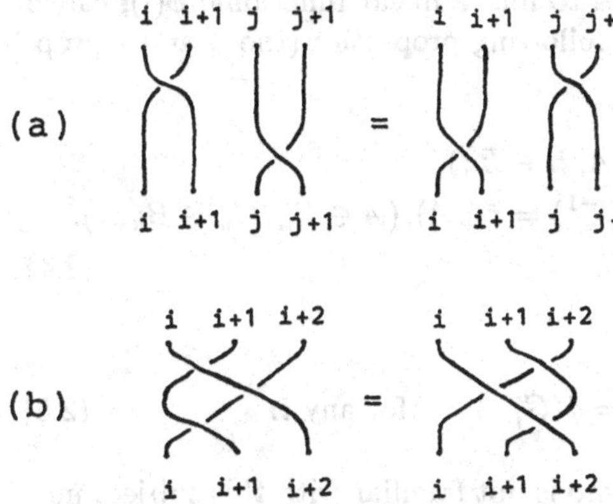

Fig.5 Braid group. The defining relation (2.1) is illustrated.

$$b_i b_j = b_j b_i, \quad | i - j | \geq 2, \qquad b_i b_{i+1} b_i = b_{i+1} b_i b_{i+1}. \tag{2.1a, b}$$

This is the defining relation of the braid group given by E.Artin. Under the relation (2.1), each topologically equivalent class of braids is identified with an element in B_n. Therefore, any n-braid is expressed as a word on B_n.

Given a braid, one may form a link by tying opposite ends. Conversely, any (oriented) link is represented by a *closed braid.* This theorem due to J.Alexander gives the braid group a fundamental role in knot theory. However, the representation of a link as a closed braid is highly non-unique. This difficulty was solved by A.A.Markov. The equivalent braids expressing the same link are mutually transformed by successive applications of two types of operations, type I and type II Markov moves:

$$\begin{aligned} &(I) \quad AB \to BA \qquad (A, B \in B_n), \\ &(II) \quad A \to Ab_n^{\pm 1} \qquad (A \in B_n, b_n \in B_{n+1}). \end{aligned} \tag{2.2}$$

Hence, we can construct a link polynomial, a topological invariant for knots and links, through the following scheme. We first make a representation of the braid group and then construct the Markov move invariant defined on the representation.

Let us denote a representation of b_i by G_i and a link polynomial by $\alpha(\cdot)$. The link polynomial $\alpha(\cdot)$ must satisfy the conditions

$$\begin{aligned} &(I) \quad \alpha(AB) = \alpha(BA), \qquad (A, B \in B_n), \\ &(II) \quad \alpha(AG_n) = \alpha(AG_n^{-1}) = \alpha(A), \qquad (A \in B_n, G_n \in B_{n+1}). \end{aligned} \tag{2.3}$$

To construct $\alpha(\cdot)$, it is convenient to find a linear functional $\phi(\cdot)$, called the *Markov trace*, which has the following properties (the Markov properties):

$$\begin{aligned} &(I) \quad \phi(AB) = \phi(BA), \qquad (A, B \in B_n), \\ &(II) \quad \phi(AG_n) = \tau\phi(A), \phi(AG_n^{-1}) = \bar{\tau}\phi(A), (A \in B_n, G_n \in B_{n+1}). \end{aligned} \tag{2.4}$$

Here the factors τ and $\bar{\tau}$ are given by

$$\tau = \phi(G_i), \quad \bar{\tau} = \phi(G_i^{-1}) \qquad \text{for any } i. \tag{2.5}$$

(For the time being, the reader who is not familiar with this subject may imagine the Markov trace as the usual trace of matrix).

In terms of the Markov trace $\phi(\cdot)$, the link polynomial $\alpha(\cdot)$ is expressed as

$$\alpha(A) = (\tau\bar{\tau})^{-(n-1)/2}(\bar{\tau}/\tau)^{e(A)/2}\phi(A), \qquad A \in B_n, \tag{2.6}$$

where $e(A)$ is the exponent sum of the $\{b_i\}$ operators appearing in the braid A; e.g., for $A = b_1^3 b_2^{-1} b_3^2$, $e(A) = 3 - 1 + 2 = 4$. It is not difficult to prove that $\alpha(\cdot)$ defined by (2.6) satisfies the condition (2.3).

In 1928, the link polynomial was first introduced by Alexander (Alexander polynomial [4]). V.F.R.Jones [5] presented a new link polynomial (Jones polynomial) in 1985. This discovery had a great impact on the mathematics community: After a lapse of nearly 60 years, a more powerful link polynomial (for instance, mirror images are detectable) was found from a seemingly different field, the operator algebra. Subsequently, the Jones polynomial was extended into a two-variable polynomial [6] (HOMFLY polynomial after six researchers' names). In 1986, L.H.Kauffman [7] introduced another new link polynomial (Kauffman polynomial). Systematic research into the construction of link polynomials from exactly solvable models started in the same year [8].

3. Exactly solvable models and braids

We consider statistical mechanical models defined on a 2-dimensional square lattice [9]. Depending on the way of assigning state (spin, charge, color etc.) variables, there are two types of models, vertex models and IRF (Interaction Round a Face) models. The most familiar solvable model in statistical mechanics, the Ising model, can be described as both types. The models are exactly solvable when the Boltzmann weights (statistical weights) satisfy the Yang-Baxter relation [9,10].

In vertex models, state variables are located on the edges. The Boltzmann weight $w(i,j,k,l;u)$ is defined for a configuration $\{i,j,k,l\}$ round a vertex (Fig.6(a)). The parameter u in the weight is called the spectral parameter (this terminology is reminiscent of the inverse scattering method), which controls the coupling constants in the model. The Yang-Baxter relation for the vertex model is given by (Fig.7) :

$$\sum_{lmn} w(m,n,q,r;u)w(l,k,p,n;u+v)w(i,j,l,m;v)$$
$$=\sum_{lmn} w(l,m,p,q;v)w(i,n,l,r;u+v)w(j,k,m,n;u). \qquad (3.1)$$

It is known that S-matrices of integrable systems are equivalent to the Boltzmann weights of solvable vertex models. Therefore, for notational convenience, we shall often write

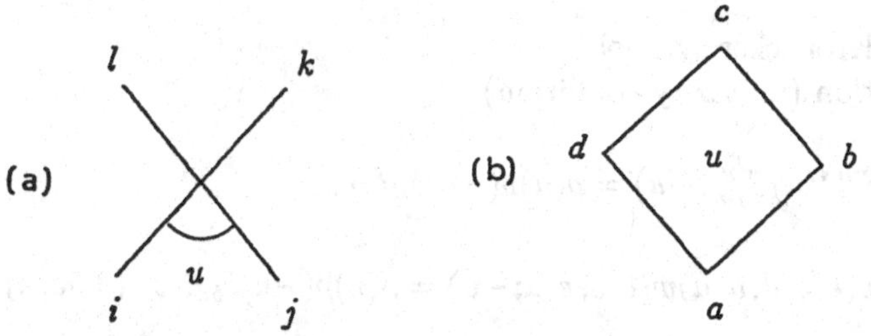

Fig.6 (a) Boltzmann weight $w(i,j,k,l;u) = S_{jl}^{ik}(u)$ for vertex model, (b) Boltzmann weight $w(a,b,c,d;u)$ for IRF model.

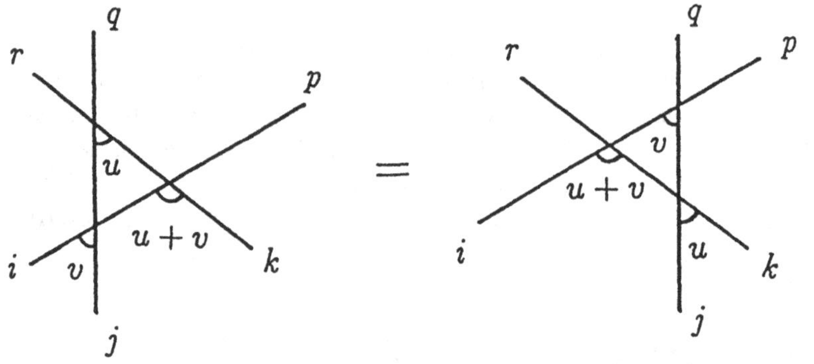

Fig.7 Yang-Baxter relation for vertex model.

$$w(i,j,k,l;u) = S_{jl}^{ik}(u). \tag{3.2}$$

For S-matrices, S_{jl}^{ik}(u) means the amplitude of the scattering process: $i \to k, j \to l$, and u is the rapidity difference of the scattering particles.

In IRF models, state variables are located on the lattice sites. The Boltzmann weight $w(a,b,c,d;u)$ is defined for a configuration $\{a,b,c,d\}$ round a face (unit square) (Fig.6(b)). For IRF models the Yang-Baxter relation is written as

$$\begin{aligned}&\sum_c w(b,d,c,a;u)w(d,e,f,c;u+v)w(c,f,g,a;v)\\ &=\sum_c w(d,e,c,b;v)w(b,c,g,a;u+v)w(c,e,f,g;u).\end{aligned} \tag{3.3}$$

The Boltzmann weights of solvable models satisfy the following relations in addition to the Yang-Baxter relation:

1) Standard initial condition

$$S_{jl}^{ik}(u=0) = \delta_{il}\delta_{jk}, \quad w(a,b,c,d;u=0) = \delta_{ac}. \tag{3.4a, b}$$

where δ_{ij} is the Kronecker symbol.
2) Inversion relation (unitarity condition)

$$\begin{aligned}&\sum_{mp} S_{pl}^{mk}(u)S_{jm}^{ip}(-u) = \rho(u)\rho(-u)\delta_{il}\delta_{jk},\\ &\sum_e w(e,c,d,a;u)w(b,c,e,a;-u) = \rho(u)\rho(-u)\delta_{bd},\end{aligned} \tag{3.5a, b}$$

where $\rho(u)$ is a normalization function.

3) Second inversion relation (second unitarity condition)

$$\sum_{mp} S_{pl}^{im}(\lambda-u)S_{mj}^{kp}(\lambda+u)r(p)r(m)/r(k)r(j)$$
$$=\rho(u)\rho(-u)\delta_{ij}\delta_{kl},$$
$$\sum_{e} w(c,e,a,b;\lambda-u)w(a,e,c,d;\lambda+u)\psi(e)\psi(b)/\psi(a)\psi(c)$$
$$=\rho(u)\rho(-u)\delta_{bd}. \qquad (3.6a,b)$$

We call λ a crossing parameter, $r(i)$ and $\psi(a)$ crossing multipliers. The term *crossing* originates from scattering theory. Replacement of u by $\lambda-u$ implies a 90 degree rotation of the lattice, which corresponds to a relation between scattering channel and crossing channel in the scattering theory.

4) Crossing symmetry

$$S_{jl}^{ik}(u)=S_{k\bar{i}}^{j\bar{l}}(\lambda-u)\cdot[r(i)r(l)/r(j)r(k)]^{1/2},$$
$$w(a,b,c,d;u)=w(b,c,d,a;\lambda-u)\cdot[\psi(a)\psi(c)/\psi(b)\psi(d)]^{1/2}. \qquad (3.7a,b)$$

Here, we have introduced the notation $\bar{k}=-k$ for the "antiparticle" of k. The crossing multiplier satisfies $r(\bar{k})=1/r(k)$.

5) Reflection symmetry

$$S_{jl}^{ik}(u)=S_{kj}^{li}(u), \qquad w(a,b,c,d;u)=w(c,b,a,d;u). \qquad (3.8a,b)$$

6) Charge (or spin) conservation condition

$$S_{jl}^{ik}(u)=0 \qquad \text{unless} \quad i+j=k+l. \qquad (3.9)$$

The Yang-Baxter relation indicates invariance of the graph under a displacement of one of three lines over an intersection of the other two lines (see Fig.7). Interestingly, this diagram looks similar to the graphical illustration Fig.5(b) of the braid group. To develop this observation, we define the Yang-Baxter operator for vertex models by

$$X_i(u)=\sum_{klmp} S_{lp}^{km}(u)I^{(1)}\otimes\cdots\otimes e_{pk}^{(i)}\otimes e_{ml}^{(i+1)}\otimes I^{(i+2)}\otimes\cdots\otimes I^{(n)}, \quad (3.10)$$

where $\otimes$ means a tensor product, $I^{(i)}$ an identity matrix, e_{pk} a matrix whose elements are given by $(e_{pk})_{ab}=\delta_{pa}\delta_{kb}$. For IRF models we define the Yang-Baxter operator by

$$[X_i(u)]^{p_0p_1\cdots p_n}_{l_0l_1\cdots l_n} = \prod_{j=0}^{i-1} \delta(p_j,l_j) w(l_j,l_{j+1},p_j,l_{j-1};u) \prod_{j=i+1}^{n} \delta(p_j,l_j)$$

$$\text{for} \quad p_{j+1} \sim p_j, l_{j+1} \sim l_j,$$

$$=0 \qquad \text{otherwise.} \qquad (3.11)$$

Here, $a \sim b$ means that state a is admissible to state b, and $\delta(p,l)$ is the Kronecker delta. The Yang-Baxter operators $\{X_j(u)\}$ satisfy the following relation (*Yang-Baxter algebra*):

$$X_i(u)X_j(v) = X_j(v)X_i(u), \qquad |i-j| \geq 2,$$
$$X_i(u)X_{i+1}(u+v)X_i(v) = X_{i+1}(v)X_i(u+v)X_{i+1}(u). \quad (3.12a,b)$$

The relation (3.12a) is obvious from the definition, and the relation (3.12b) is nothing but the Yang-Baxter relation in operator form.

We notice that, if $u = u + v = v$, (3.12) reduces to (2.1). This means that by eliminating the spectral parameters we have the braid representation G_i from the Yang-Baxter operator. Thus, we get a formula for the representation of the braid group:

$$G_j = \lim_{u\to\infty} X_j(u)/\rho(u), \quad G_j^{-1} = \lim_{u\to\infty} X_j(-u)/\rho(-u), \quad I = X_j(0). \qquad (3.13)$$

Solutions of the Yang-Baxter relation are given in terms of (i)elliptic functions, (ii)hyperbolic/trigonometric functions, and (iii)rational functions. The case (ii) (respectively,(iii)) corresponds to the critical situation of the case (i) (respectively,(ii)). The limit $u \to \infty$ requires that, to have a meaningful representation, the Boltzmann weights be parametrized by hyperbolic or trigonometric functions [8]. Then, in statistical mechanics, it implies that the model is at the criticality.

4. Construction of link polynomials

4.1 Markov trace

In order to obtain link polynomials we shall make the Markov trace on the braid group representation. For vertex models, the Markov trace has the following form,

$$\phi(A) = \mathrm{Tr}(H^{(n)}A)/\mathrm{Tr}(H^{(n)}), \qquad A \in B_n \qquad (4.1)$$

where $H^{(n)} = h^{(1)} \otimes h^{(2)} \otimes \cdots \otimes h^{(n)}$, and $h^{(i)}$ is a diagonal matrix whose elements are

$$h_{pq} = r^2(p)\delta_{pq}. \qquad (4.2)$$

The trace $\phi(\cdot)$ defined in (4.1) satisfies the Markov property (2.4) under the condition

$$\lim_{u\to\infty}\sum_{l} S_{lk}^{kl}(\pm u)/\rho(\pm u)\cdot r^2(l) = \chi(\pm) \qquad \text{(independent of } k\text{)}. \quad (4.3)$$

For IRF models we introduce a constrained trace $\hat{\text{Tr}}(A)$ by

$$\hat{\text{Tr}}(A) = \sum_{l_0 l_1 \cdots l_n} A_{l_0 l_1 \cdots l_n}^{l_0 l_1 \cdots l_n}\cdot[\psi(l_n)/\psi(l_0)], \quad (l_0 : \text{fixed}) \qquad (4.4)$$

where the summation is over admissible indices $\{l_i : l_{i+1} \sim l_i\}$. The Markov trace $\phi(\cdot)$ is written as

$$\phi(A) = \hat{\text{Tr}}(A)/\hat{\text{Tr}}(I(n)), \qquad A \in B_n, \qquad (4.5)$$

where $I(n)$ is the identity operator for n strings. The trace $\phi(\cdot)$ defined in (4.5) satisfies the Markov property (2.4) under the condition

$$\begin{aligned}&\lim_{u\to\infty}\sum_{b\sim a} w(a,b,a,c;\pm u)/\rho(\pm u)\cdot\psi(b)/\psi(a)\\ =&\chi(\lambda) \qquad \text{(independent of } k\text{)}.\end{aligned} \qquad (4.6)$$

We find from (4.1) and (4.4) that the Markov trace is the "weighted" trace of matrix with the crossing multipliers.

Conditions (4.3) and (4.6) are extended into the relations which hold for finite spectral spectral parameter u:

$$\sum_{l} S_{lk}^{kl}(u)r^2(l) = H(u;\lambda)\rho(u) \quad \text{(independent of } k\text{)},$$

$$\sum_{b\sim a} w(a,b,a,c;u)\psi(b)/\psi(a) = H(u;\lambda)\rho(u) \quad \text{(independent of } a, c\text{)}.$$

$$(4.7a, b)$$

We call (4.7) the extended Markov property and $H(u;\lambda)(= H(u))$ the characteristic function [11,12]. To summarize, the extended Markov property is sufficient for the existence of the Markov trace. This completes the algebraic construction of link polynomials by the use of exactly solvable models. It should be added that Witten's formula [13] for the link invariants in S^3 is essentially the same as the constrained trace (4.4).

4.2 Graphical calculation

The crossing symmetry is significant in algebraic and graphical aspects of the knot theory. For solvable models with the crossing symmetry (recall (3.7)), the Yang-Baxter operator becomes the Temperley-Lieb operator at the point $u = \lambda$ [11,14]. In fact, by setting

$$E_i = X_i(\lambda), \tag{4.8}$$

we find that the operators $\{E_i\}$ satisfy the following relations (*Temperley-Lieb algebra*):

$$\begin{aligned} &E_i E_{i\pm1} E_i = E_i, \quad E_i^2 = q^{1/2} E_i, \\ &E_i E_j = E_j E_i, \quad | i - j | \geq 2. \end{aligned} \tag{4.9}$$

Here the quantity $q^{1/2}$ is related to the crossing multiplier as

$$\begin{aligned} q^{1/2} &= \sum_j r^2(j), \qquad \text{for vertex model} \\ &= \sum_{b\sim a} \psi(b)/\psi(a), \quad \text{for IRF model.} \end{aligned} \tag{4.10}$$

Let us consider the graphical meaning of the Temperley-Lieb operator. Hereafter in this section, we discuss only vertex models, because all the arguments go parallel for IRF models. From the crossing symmetry and the standard initial condition we have

$$\begin{aligned} S_{jl}^{ik}(\lambda) &= [r(i)r(l)/r(j)r(k)]^{1/2} S_{ki}^{jl}(0), \\ &= r(i)\delta(i,\bar{j}) \cdot r(l)\delta(l,\bar{k}). \end{aligned} \tag{4.11}$$

The elements $r(i)\delta(i,\bar{j})$ and $r(l)\delta(l,\bar{k})$ are regarded as the weights for the pair-annihilation diagram and the pair-creation diagram, respectively (Fig.8(a),(b)). Then, the Yang-Baxter operator at $u = \lambda$ is depicted as

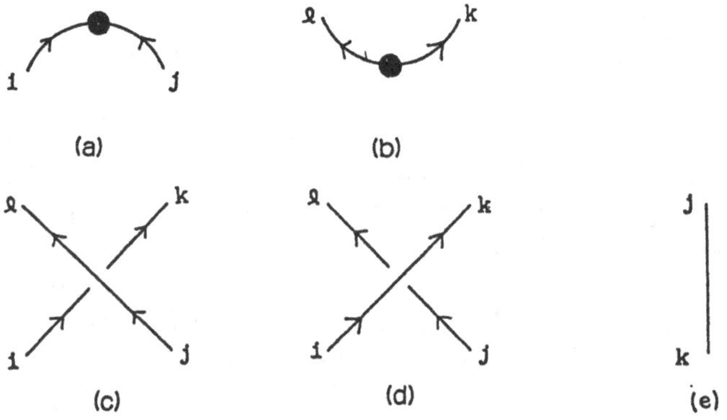

Fig.8 Elements of link diagrams for vertex models.
(a)annihilation diagram $r(i)\delta(i,\bar{j})$, (b)creation diagram $r(l)\delta(l,\bar{k})$,
(c)braid diagram $\sigma_{lk,ij}^{(+)}$, (d)inverse braid diagram $\sigma_{lk,ij}^{(-)}$,
(e)line diagram $\delta(j,k)$. Dots denote the crossing multipliers.

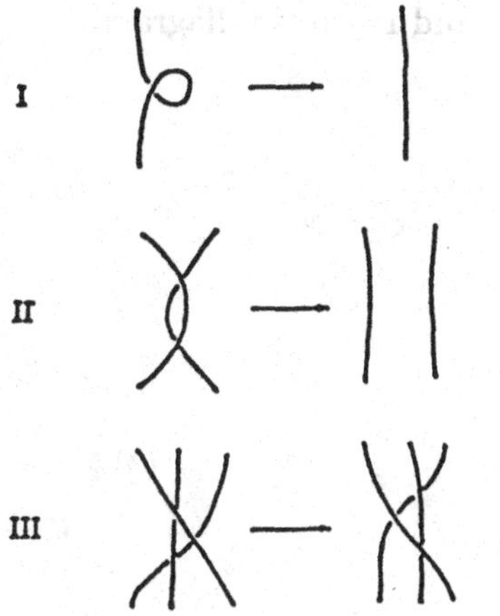

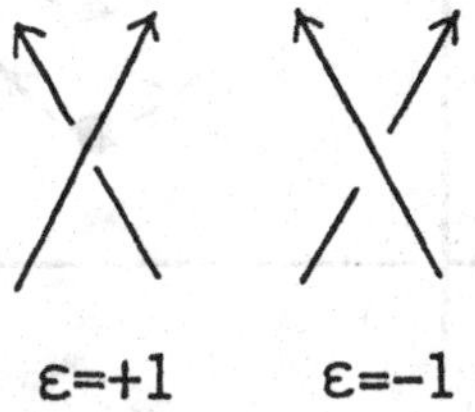

Fig.9 Reidemeister moves I, II and III.   Fig.10 Sign $\varepsilon(C)$ of crossing C.

the monoid diagram [7]. This identification is consistent with the S-matrix theory that the energy at $u = \lambda$ is related to the pair-creation energy.

We can formulate link polynomials directly on link diagrams. To explain this, we add some terminology of knot theory. The link diagram $\hat{L}$ is a 2-dimensional projection of a link L. Link diagrams expressing ambient isotropic links are transformed into each other by a finite sequence of Reidemeister moves (Fig.9). Crossings in the link diagram are classified into two types. A sign $\varepsilon(C)$ at crossing C is defined as in Fig.10. The writhe $w(\hat{L})$ of a link diagram $\hat{L}$ is the sum of signs for all crossings in the link diagram,

$$w(\hat{L}) = \sum_{C} \varepsilon(C). \tag{4.12}$$

The writhe is invariant under the Reidemeister moves II and III, but is changed by the Reidemeister move I. Link diagrams are said to be *regular isotopic* if they are transformed into each other by a finite sequence of the Reidemeister moves II and III. Thus, the writhe is a regular isotopy invariant. The link diagram is decomposed into the following elements (elementary diagrams): annihilation, creation, braid, inverse braid and line diagrams (Fig.8). To them, we assign weights $r(i)\delta(i,\bar{j})$, $r(l)\delta(l,\bar{k})$, $\sigma^{(+)}_{lk,ij}$, $\sigma^{(-)}_{lk,ij}$ and $\delta(j,k)$, where the weights $\sigma^{(\pm)}_{lk,ij}$ are defined by

$$\sigma^{(\pm)}_{lk,ij} = \lim_{u\to\infty} S^{ik}_{jl}(\pm u)/[\rho(\lambda \mp u)\rho(\pm u)]^{1/2}. \tag{4.13}$$

Taking the charge conservation condition into account, we take a summation over all possible state variables for the link diagram $\hat{L}$. We

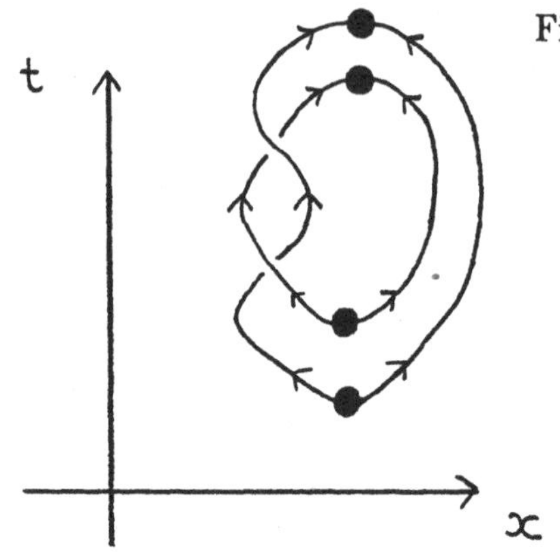

Fig.11 Knot and Feynman diagram.

denote this sum by $\mathrm{Tr}(\hat{L})$. Notice that the Reidemeister moves II and III are nothing but the unitarity condition and the Yang-Baxter relation. Then, $\mathrm{Tr}(\hat{L})$ is a regular isotopy invariant. We multiply $\mathrm{Tr}(\hat{L})$ by a factor $c^{-w(\hat{L})}$, where c is defined by (the normalization of the braid operator is the same as (4.13))

$$G_i E_i = E_i G_i = c E_i, \tag{4.14}$$

or, in terms of the characteristic function,

$$c = \lim_{u \to \infty} H(\lambda - u)[\rho(\lambda - u)\rho(u)]^{1/2}. \tag{4.15}$$

The factor $c^{-w(\hat{L})}$ is also a regular isotopy invariant. Then, we find that a link polynomial is expressed as

$$\alpha(L) = c^{-w(\hat{L})}\mathrm{Tr}(\hat{L})/\mathrm{Tr}(\hat{K}_0), \tag{4.16}$$

where $\hat{K}_0$ is the trivial knot diagram (a loop).

We have shown that the link polynomials constructed from solvable models with the crossing symmetry are also graphically calculable. The graphical formulation offers an alternative method to evaluate the Markov trace and is equivalent to the algebraic formulation based on the Markov trace.

It is intriguing to interpret link diagrams as the Feynman diagrams for the charged particles and link polynomials as the scattering amplitudes. We consider a plane (space-time plane), the time direction being upwards. At the lowest point in the diagram there occurs a pair-creation and at the highest point a pair-annihilation (Fig.11). Then, we arrive at a "physicist's view" of the knot theory; strings in the knot theory are trajectories of charged particles in a completely integrable system.

5. Examples

In the previous sections a general theory to construct link polynomials by the use of exactly solvable models has been given. To various solvable models, we shall apply formulae; (3.13) for the braid representation, (4.1) or (4.5) for the Markov trace, and then (2.6) for link polynomial.

5.1 N-state vertex model

The 6-vertex model is a solvable model [9] where state variables assume two values; this is considered to be a critical limit of the 8-vertex model. The 6-vertex model satisfies the charge conservation condition while the 8-vertex model does not. The N-state vertex model which we shall consider is a straightforward generalization of the 6-vertex model. For $N = 3$, it is the 19-vertex model found by Zamolodchikov and Fateev. For general N, we have the recursion formula [15] for the Boltzmann weights and the algebraic construction [16] for the Yang-Baxter operator.

It is instructive to start with the $N = 2$ case (the 6-vertex model). The non-zero Boltzmann weights are

$$
\begin{aligned}
&S^{1/2\ 1/2}_{1/2\ 1/2}(u) = S^{-1/2\ -1/2}_{-1/2\ -1/2}(u) = \sinh(\lambda - u)/\sinh\lambda, \\
&S^{1/2\ 1/2}_{-1/2\ -1/2}(u) = S^{-1/2\ -1/2}_{1/2\ 1/2}(u) = \sinh u/\sinh\lambda, \\
&S^{1/2\ -1/2}_{-1\ 2\ 1\ 2}(u) = S^{-1/2\ 1/2}_{1\ 2\ -1\ 2}(u) = 1.
\end{aligned} \tag{5.1}
$$

We introduce a transformation (*symmetry breaking transformation*);

$$
S^{ik}_{jl}(u) \to \tilde{S}^{ik}_{jl}(u) = \mathrm{e}^{1/2\cdot(j+k-i-l)u} S^{ik}_{jl}(u). \tag{5.2}
$$

The transformed weights $\tilde{S}^{ik}_{jl}(u)$ also satisfy the Yang-Baxter relation when the model has the charge conservation property. Applying formula (3.13) to the asymmetrized weights $\tilde{S}^{ik}_{jl}(u)$ with $\rho(u) = \sinh(\lambda - u)/\sinh\lambda$, we get

$$
\begin{aligned}
G_i =& e_{11} \otimes e_{11} + e_{22} \otimes e_{22} - t^{1/2} e_{21} \otimes e_{12} \\
&- t^{1/2} e_{12} \otimes e_{21} + (1-t) e_{22} \otimes e_{11}, \quad t = \mathrm{e}^{2\lambda}.
\end{aligned} \tag{5.3}
$$

The braid operator G_i in (5.3) obeys the quadratic relation (*Hecke algebra*).

Using the N-state vertex model (asymmetrized by the symmetry breaking transformation), we get the braid operators $\{G_i\}$ which are characterized by an N-th order relation [8,10]:

$$
(G_i - C_1)(G_i - C_2) \cdots (G_i - C_N) = 0, \tag{5.4}
$$

where for $j = 1, 2, \cdots, N$

$$C_j = (-1)^{N+j} t^{N(N-1)/2 - j(j-1)/2}, \quad t = e^{2\lambda}. \tag{5.5}$$

We call a relation for G_i such as (5.4) a *reduction relation* of the braid operator. The crossing multiplier for the asymmetrized N-state vertex model is

$$r(k) = e^{-\lambda k} = t^{-k/2}, \quad k = -s, -s+1, \cdots, s, \tag{5.6}$$

where $s = (N-1)/2$ is the "spin". The extended Markov property (4.7) is satisfied with the characteristic function given by

$$H(u; \lambda) = \sinh(N\lambda - u)/\sinh(\lambda - u). \tag{5.7}$$

The factors τ and $\bar{\tau}$ are

$$\begin{aligned} \tau &= 1/(1 + t + \cdots + t^{N-1}), \\ \bar{\tau} &= t^{N-1}/(1 + t + \cdots + t^{N-1}). \end{aligned} \tag{5.8}$$

It is remarkable that there exists an infinite sequence of independent link polynomials corresponding to the N-state vertex models ($N = 2, 3, 4, \cdots$). The $N = 2$ case corresponds to the Jones polynomial. It is believed that the link polynomial for a larger N is more powerful. In fact, an example of two different links, which cannot be classified by the Jones polynomial but can be by the $N = 3$ link polynomial, has been studied in [17].

5.2 *ABCD* IRF models

The IRF model corresponding to affine Lie algebra $A^{(1)}_{m-1}$ ($B^{(1)}_m$, $C^{(1)}_m$, $D^{(1)}_m$) is called the $A^{(1)}_{m-1}$ ($B^{(1)}_m$, $C^{(1)}_m$, $D^{(1)}_m$) model [18]. The crossing parameter λ and sign σ are defined as

$$\begin{aligned} &\lambda = m\omega/2, \quad \sigma = 1 \quad \text{for} \quad A^{(1)}_{m-1} \\ &\lambda = (2m-1)\omega/2, \quad \sigma = 1 \quad \text{for} \quad B^{(1)}_m \\ &\lambda = (m+1)\omega, \quad \sigma = -1 \quad \text{for} \quad C^{(1)}_m \\ &\lambda = (m-1)\omega, \quad \sigma = 1 \quad \text{for} \quad D^{(1)}_m, \end{aligned} \tag{5.9}$$

where ω is a parameter.

It is shown [12] that the reduction relations are

$$\begin{aligned} &(G_i - 1)(G_i + \gamma^2) = 0 \quad \text{for} \quad A^{(1)}_m, \\ &(G_i - 1)(G_i - \beta)(G_i + \gamma^2) = 0 \quad \text{for} \quad B^{(1)}_m, C^{(1)}_m \text{ and } D^{(1)}_m, \end{aligned} \tag{5.10}$$

with $\gamma = e^{-i\omega}$ and $\beta = \sigma e^{-[2\lambda+\omega(1+\sigma)]}$. The extended Markov property is proved and the characteristic functions are calculated as [12]

$$\begin{aligned} H(u) &= \frac{\sin(m\omega - u)}{\sin(\omega - u)} \quad \text{for} \quad A^{(1)}_{m-1}, \\ H(u) &= \frac{\sigma \sin(2\lambda - u)\sin(\sigma\omega + \lambda - u)}{\sin(\lambda - u)\sin(\omega - u)} \quad \text{for} \quad B^{(1)}_m, C^{(1)}_m \text{ and } D^{(1)}_m. \end{aligned} \tag{5.11}$$

The explicit forms of the crossing multipliers are given in [12]. Using the reduction relation and the Markov traces, we obtain the generalized skein relations;

$$\begin{aligned} \alpha(L_+) =& (1-t)t^{(m-1)/2}\alpha(L_0) + t^m \alpha(L_-) \quad \text{for} \quad A^{(1)}_{m-1}, \\ \alpha(L_{2+}) =& (1 - t + \beta)e^{-i(2\lambda+\omega(\sigma-1))}\alpha(L_+) \\ &+ (t + t\beta - \beta)e^{-2i(2\lambda+\omega(\sigma-1))}\alpha(L_0) \\ &- t\beta e^{-3i(2\lambda+\omega(\sigma-1))}\alpha(L_-) \\ &\text{for} \quad B^{(1)}_m, C^{(1)}_m \text{ and } D^{(1)}_m, \end{aligned} \tag{5.12}$$

where $t = e^{-2i\omega}$. In (5.12), by L_+, L_0 and L_- we have denoted links which have the same configuration except b_i, b_i^0 and b_i^{-1} at an intersection. L_{2+}, L_+, L_0 and L_- should be understood similarly.

Link polynomials thus obtained are one-variable invariants for each fixed m. We may regard m as a continuous parameter which is independent of t. Then, the link polynomial constructed from $A^{(1)}_{m-1}$ model is the two-variable extension of the Jones polynomial (HOMFLY polynomial). The link polynomials constructed from $B^{(1)}_m$, $C^{(1)}_m$, $D^{(1)}_m$ models correspond to the Kauffman polynomial. Those polynomials were also made from the vertex models by V.Turaev [19]. Two-variable extensions of the link polynomials constructed from A type composite (fusion) models are given in [16,10]. This series of polynomials contains the HOMFLY polynomial as the simplest case.

5.3 $gl(M \mid N)$ vertex model

We consider a family of solvable vertex models with graded symmetry $gl(M \mid N)$. This symmetry corresponds to a system with M-kinds of bosons and N-kinds of fermions. We prepare a set of signs $\{\epsilon_i\}$;

$$\epsilon_i = 1 \quad \text{or} \quad -1 \quad \text{for} \quad i = 1, 2, \cdots, M+N. \tag{5.13}$$

The number of positive (negative) signs is M (N). For any set of signs $\{\epsilon_i\}$ we have a solution of the Yang-Baxter relation. Non-zero Boltzmann weights are

$$
\begin{aligned}
w(a,a,a,a;u) =& \sinh(\eta - \epsilon_a u)/\sinh\eta, \\
w(a,b,b,a;u) =& \exp(-u) \quad \text{for} \quad a < b \\
=& \exp u \quad \text{for} \quad a > b, \\
w(a,b,a,b;u) =& \sinh u/\sinh\eta \quad \text{for} \quad a \neq b,
\end{aligned} \tag{5.14}
$$

where η is a parameter. The Markov trace is given by (4.1) with the diagonal matrix h [20] :

$$
\begin{aligned}
h(j) =& \epsilon_j \exp(\eta \sum_{k=1}^{j-1} (2\epsilon_k + \epsilon_j - M + N)), \\
& \text{for} \quad j = 1, 2, \cdots, M + N.
\end{aligned} \tag{5.15}
$$

Note that in the limit $\eta \to 0$, the trace with the matrix h reduces to the supertrace $strA = \sum_i \epsilon_i A_{ii}$. The link polynomial satisfies the skein relation:

$$
\alpha(L_+) = t^{p/2}(1 - t)\alpha(L_0) + t^{p+1}\alpha(L_-), \tag{5.16}
$$

where

$$
p = M - N - 1. \tag{5.17}
$$

It is important to notice that as far as p is common we have the same link polynomial [20]. The link polynomial for $p = -1$ (super symmetric case) is the Alexander polynomial, while the case $p = 1$ is the Jones polynomial. Two comments are in order. First, for $M = N$ ($p = -1$) case the Markov trace has to be modified since the "naive" trace vanishes identically due to the cancellation of boson parts and fermion parts (see [21] for a prescription). Second, we can construct the composite (fusion) models from the $gl(M \mid N)$ model.

6. Concluding remarks

We have shown that various link polynomials are systematically constructed from exactly solvable (integrable) models. The theory of solvable models offers all the information which is necessary for the knot theory:

1) The Yang-Baxter operator $X_i(u)$ reduces to the braid group representation G_i when the spectral parameter u is sent to infinity.
2) A variable t of link polynomial $\alpha(\cdot)$ is given by the crossing parameter λ, and the Markov trace $\phi(\cdot)$ is written in terms of the crossing multipliers.
3) For models with the crossing symmetry, the Yang-Baxter operator at $u = \lambda$ is the monoid operator E_i (Temperley-Lieb operator in physics).

The relations among the operators $\{G_i, E_i\}$ (***braid-monoid algebra***) gives a basis for the graphical formulation.

It is now established that there exists a list of braid group representations and link polynomials. These results should be useful in many areas of physics and mathematics. For instance, the Jones polynomial and a series of new link polynomials have been used for classification of 3-manifolds [22].

In the 1970s, there were important extensions and successful applications of the soliton theory. The 1980s were rather extraordinary in the following sense. The Fields Medals for 1990 were awarded to S.Mori, V.F.R.Jones, E.Witten and V.G.Drinfeld. The achievements of the latter three researchers are more or less related to mathematical structures of integrable systems. From the viewpoint of physics, what were clarified are properties of the critical models. Since the theory of solvable models describes more general situations (off-critical, higher dimensional, etc.), it is natural to expect further developments. In conclusion, the soliton theory will continue to be exciting in the next decade.

Ackowledgements

The author thanks Y.Akutsu and T.Deguchi for fruitful collaborations. He also thanks C.N.Yang, L.H.Kauffman and K.Murasugi for continuous encouragements.

References

[1] N.J.Zabusky and M.D.Kruskal, Phys.Rev.Lett.15(1965)240.
[2] C.N.Yang, Phys.Rev.Lett.19(1967)1312.
[3] R.J.Baxter, Ann.of Phys.70(1972)323.
[4] J.W.Alexander, Trans.Amer.Math.Soc.30(1928)275.
[5] V.F.R.Jones, Bull.Amer.Math.Soc.12(1985)103.
[6] P.Freyd,D.Yetter,J.Hoste,W.B.R.Lickorish,K.Millett and A.Ocneanu, Bull.Amer.Math.Soc.12(1985)239. J.H.Przytycki and K.P.Traczyk, Kobe J.Math.4(1987)115.
[7] L.H.Kauffman, On Knots (Princeton University Press,1987).
[8] Y.Akutsu and M.Wadati, J.Phys.Soc.Jpn.56(1987)839,3039: Commun.Math.Phys.117(1988)243.
[9] R.J.Baxter, Exactly Solved Models in Statistical Mechanics (Academic Press,1982).
[10] M.Wadati,T.Deguchi and Y.Akutsu, Phys.Reports 180(1989)427.
[11] Y.Akutsu,T.Deguchi and M.Wadati, J.Phys.Soc.Jpn.57(1988)1173.

[12] T.Deguchi,M.Wadati and Y.Akutsu, J.Phys.Soc.Jpn.57(1988)2921.
[13] E.Witten, Commun.Math.Phys.121(1989)351.
[14] T.Deguchi,M.Wadati and Y.Akutsu, J.Phys.Soc.Jpn.57(1988)1905.
[15] K.Sogo,Y.Akutsu and T.Abe, Prog.Theor.Phys.70(1983)730,739.
[16] T.Deguchi,Y.Akutsu and M.Wadati, J.Phys.Soc.Jpn.57(1988)757.
[17] Y.Akutsu,T.Deguchi and M.Wadati, J.Phys.Soc.Jpn.56(1987)3464.
[18] M.Jimbo,T.Miwa and M.Okado, Commun.Math.Phys.116(1988)353.
[19] V.G.Turaev, Invent.Math.92(1988)527.
[20] T.Deguchi and Y.Akutsu, J.Phys.A:Math.Gen.23(1990)1681.
[21] Y.Akutsu,T.Deguchi and T.Ohtsuki, Journal of Knot Theory and its Ramifications 1(1992)161.
[22] N.Yu.Reshetikhin and V.G.Turaev, Invent.Math.103(1991)547.
T.Kohno, Topology 31(1992)203.
V.G.Turaev and O.Y.Viro, LOMI preprint ,1990.

Part V

Near-Integrable Models and Computational Aspects

Solitons and Computation

M.J. Ablowitz[1] *and B.M. Herbst*[2]
[1]Program In Applied Mathematics, University of Colorado at Boulder, Boulder Colorado 80309-0526, USA
[2]Department of Applied Mathematics, University of the Orange Free State, Bloemfontien 9300

The theory of solitons, the Inverse Scattering Transform, and related methods have had a profound effect upon many fields. In this article, the application of soliton theory and integrability to certain classes of discrete problems is discussed. It has been found that different discretizations of, for example, the nonlinear Schrödinger equations (NLS), can produce dramatically different results. The phenomena of numerically induced chaos and numerical homoclinic instabilities are illustrated via the NLS equation.

It turns out that solitons also arise in totally discrete systems, i.e. cellular automata. Two different cellular automata rules are presented. One of them is the well known Parity Rule Filter Automata (PRFA), which is irreversible in time. The second is a novel reversible rule which possesses all particles contained in the PRFA plus many more. The new rule allows for particle production.

1. Introduction

During the past 25 years an important class of nonlinear equations have been found to be integrable by means of the Inverse Scattering Transform (IST). (For a review see, for example, [1,2].) These equations have numerous features in common, one being the existence of solitons. Solitons are stable localized waves which interact elastically with each other. Applications are diverse, ranging from classical physics such as fluid dynamics and plasma physics to modern physics: particle physics, quantum field theory and relativity. There are significant applications in pure mathematics as well, e.g. analysis, knot theory, algebraic geometry and group theory. Recently there have been valuable connections with certain computational problems. Some of these will be discussed in this short review, in which we address two issues, the first being numerically induced chaos and the second solitons in cellular automata.

The phenomena of numerically induced chaos, described in the first part of this article, involves numerical calculations of various equations which are solvable by the IST method (and consequently are integrable). We have found that, corresponding to various classes of initial data, specific "standard" simulations of these equations appear to possess chaotic solutions. If we call h a parameter which measures the mesh size of the simulation, then for h above a certain critical size h_*, the simulation behaves chaotically. For $h \ll h_*$ the calculations show convergence to solutions obtained analytically. We compare different numerical schemes and show that one of them, derived via IST Theory, has superlative characteristics. Although we have concentrated on simulations of nonlinear partial differential equations in one space and one time dimension, we believe that these ideas will apply to higher dimensional problems as well. An important aspect of this work is that a natural class of Hamiltonian perturbations to integrable nonlinear Hamiltonian PDEs is obtained and studied. We try to understand PDE phenomena using concepts from ODEs, with suitable generalizations. The study of homoclinic orbits and homoclinic chaos in PDEs (see [3]) is central.

The second portion of this article describes novel, entirely discrete nonlinear evolution systems, which possess a class of particles which behave like solitons: i.e. they interact elastically with each other. Here we shall describe two such systems. The first one, referred to as the Parity Filter Rule Automata (PFRA) was originally proposed by Park, Steiglitz and Thurston [4] and subsequently was shown to satisfy an alternative but equivalent rule (Fast Rule) [5] with which one can demonstrate the soliton property analytically. However careful analysis of this rule [6] shows that it possesses some dissipation; there is an energy-like quantity E^t, which satisfies an entropy condition: $E^{t+1} \leq E^t$. Hence this rule is generally irreversible. Recently, we [7] have developed a simple modification of the above rules which are stable, multi-state and *reversible*. The new system (rule) admits all of the previous soliton and particle modes as special solutions, possesses an even broader array of stable particles, and allows interesting particle interaction behavior such as particle production.

2. Numerical Chaos and the Nonlinear Schrödinger Equation

A paradigm nonlinear evolution equation which arises in a wide variety of physical applications is the Nonlinear Schrödinger Equation (NLS)

$$i\, u_t + u_{xx} + 2u^2u^* = 0 \tag{2.1}$$

where u^* is the complex conjugate of u. In applications u represents a complex envelope of a rapidly varying wave train in weakly nonlinear media. The numerical simulations were carried out on what we shall refer to as the "standard" scheme

$$i\, u_{n_t} + (u_{n+1} + u_{n-1} - 2u_n)/h^2 + 2u_n^2u_n^* = 0 \tag{2.2}$$

and an "integrable" scheme

$$i\, u_{n_t} + (u_{n+1} + u_{n-1} - 2u_n)/h^2 + uu_n^*(u_{n+1} + u_{n-1}) = 0. \tag{2.3}$$

The standard and integrable schemes are interesting in their own right. The standard scheme (2.2) has been considered as models of nonlinear dimers [8], models of nonlinear self-trapping phenomena [9] and models of biological systems [10]. The integrable scheme was derived in 1976 [11] (see also [1]) as one of the most natural and simple yet solvable nonlinear differential-difference equations. It is associated with a 2×2 discrete spectral problem (see also section 3 of this paper) in the same way as the Toda lattice is associated with a second order scalar discrete spectral problem. We consider the equations on a finite interval with periodic boundary conditions, $u(x,t) = u(x + L, t)$, subject to given initial data $u(x,0) = f(x)$. The discretization in (2.2 - 2.3) is formed in the usual way: $u_n(t) = u(nh,t)$ where $h = L/N$ and periodicity: $u_{n+N}(t) = u_n(t)$, $n = 0,1,2,\dots,N$.

Both schemes are Hamiltonian. For (2.2) the Hamiltonian is

$$H = -i \sum_{j=0}^{N-1} \{|u_{j+1} - u_j|^2/h^2 - |u_j|^4\} \tag{2.4}$$

with standard Poisson brackets. The L^2 norm, $\sum_{j=0}^{N-1} |u_j|^2$, is conserved as well. In (2.3) the Hamiltonian is given [12] by

$$H = -i \sum_{j=0}^{N-1} \{u_j^*(u_{j+1} + u_{j-1}) - 2 \log (1 + h^2 u_j u_j^*)\}. \tag{2.5}$$

Together with the nonstandard Poisson brackets $\{u_n, u_m^*\} = h^{-1}(1 + h^2 u_n u_n^*)\delta_{m,n}$ and $\{u_n, u_m\} = \{u_n^*, u_m^*\} = 0$, the scheme (2.3) is a Hamiltonian system. When $N = \infty$, (2.3) has an infinity of constants of the motion [11]; the periodic case has N constants (the solution of the periodic problem has been considered in [13]), one of the simplest being $\sum_{j=0}^{N-1} u_j^*(u_{j+1} + u_{j-1})$.

In our calculations [14-15] we have found significant differences between the numerical simulations of the NLS equation via (2.2) vs. (2.3). These differences are substantially more than one might expect merely by analyzing the truncation errors of the two schemes- which are both second order accurate. We find that (2.3) preserves the main qualitative features of the true analytical solution whereas (2.2) develops instabilities and chaos for intermediate values of the mesh size. Eventually (2.2) does converge, as one would expect, for h small enough. Although in this review only finite difference simulations are discussed, we have found similar behavior in Fourier Spectral schemes as well [16].

Some typical calculations are now described. All numerical calculations were performed using the Runge-Kutta-Merson routine in the NAG (Numerical Algorithms Group) soft-

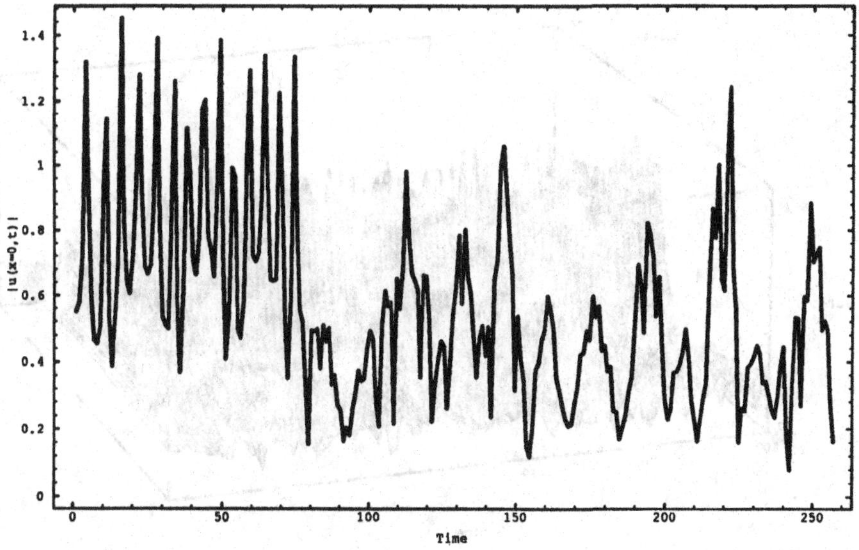

FIGURE 1. Standard scheme, $N = 32$

ware library with sufficiently high accuracy specified in the computations to ensure that the results were not consequences of the time integration. In figures 1 and 2 we used the following initial values; $u(x,0) = a(1+\epsilon\cos\mu x)$, with $a = 1/2$, $\epsilon = 0.1$, $\mu = 2\pi/L$, $L = 2\sqrt{2}\pi$. In figures 3-6 we used the parameters $a = 1/2, \mu = \mu_m = 2\pi/L_m, L_m = 2\sqrt{2}m\pi$ with $m = 2$ in figures 3-4, and $m = 3$ in figures 5-6. Figures 1-6 refer to calculations of equations (2.2) and (2.3) respectively. In figures 1 and 2 we plot the amplitude of the solution at $x = 0$ vs. time. In figures 3–6 we plot the amplitudes of the function u (in all space) versus time.

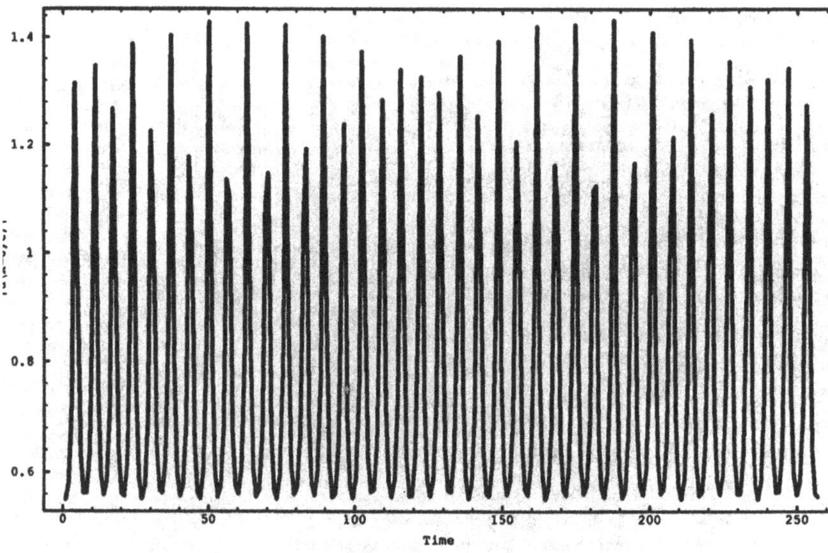

FIGURE 2. Integrable scheme, $N = 32$

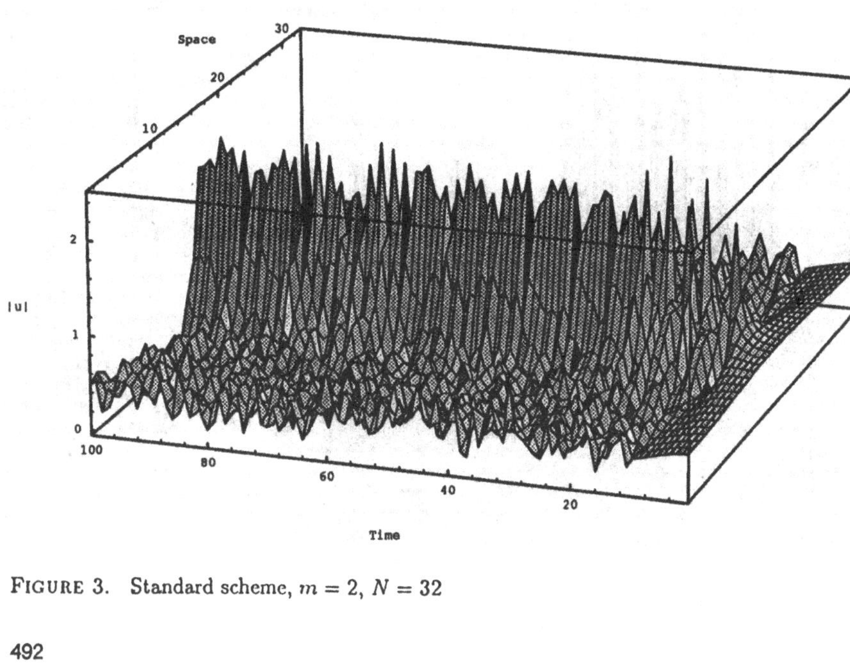

FIGURE 3. Standard scheme, $m = 2$, $N = 32$

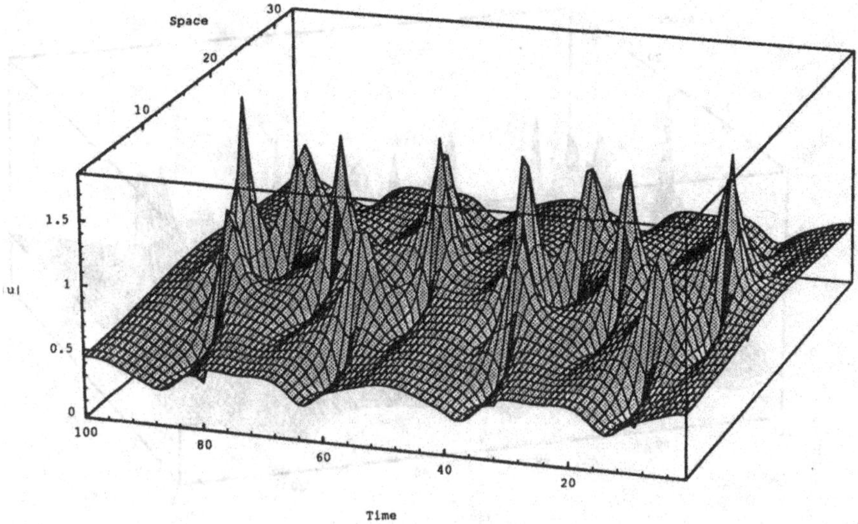

FIGURE 4. Integrable scheme, $m = 2$, $N = 32$, $a = 1/2$, $t = 0.1$,
$u(x,0) = a(1 + \epsilon \cos \mu_m x), \quad \mu_m = 2\pi/L_m, \quad L_m = 2\sqrt{2}m\pi$

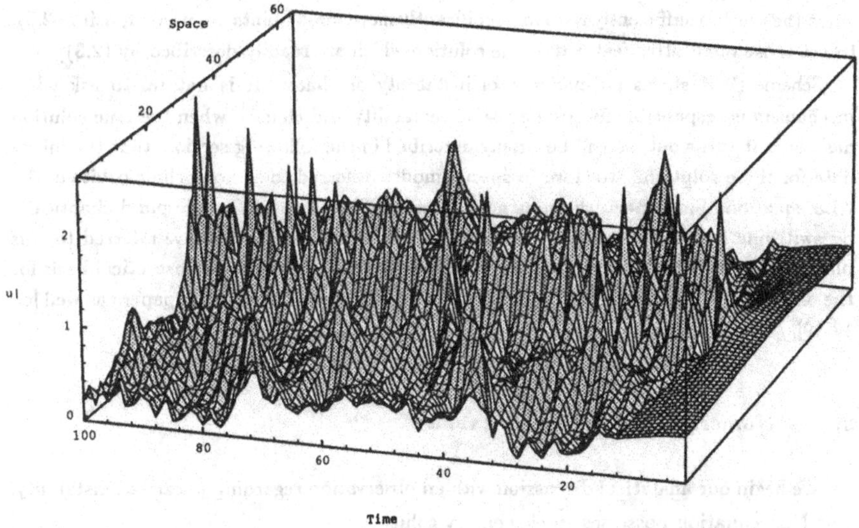

FIGURE 5. Standard scheme, $m = 3$, $N = 64$

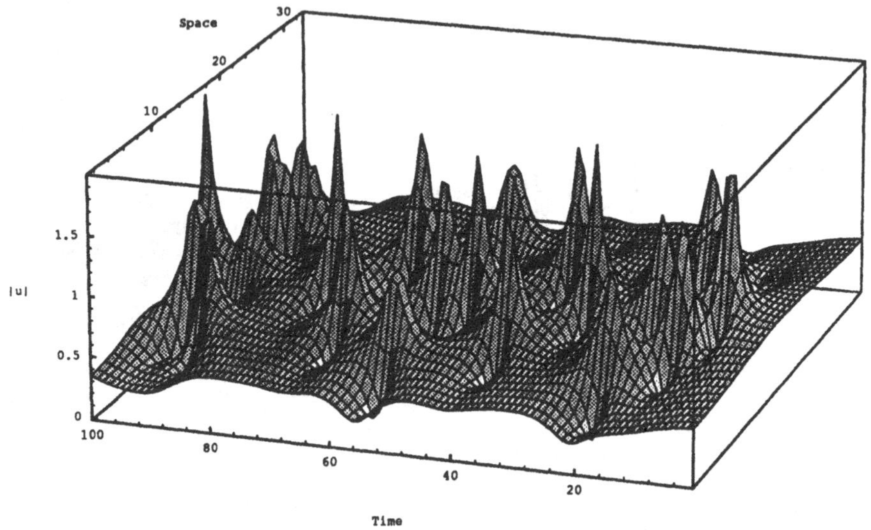

FIGURE 6. Integrable scheme, $m = 3$, $N = 32$

Even though the solutions calculated via the standard scheme (2.2) eventually converge when the mesh is sufficiently refined, significantly more mesh points are needed, using (2.2), to get those qualitative features of the solution which are readily described by (2.3).

Scheme (2.3) shows no evidence of instability or chaos. It is natural to ask what mechanism is responsible for giving rise to instability and chaos – when the true solution has none. It turns out, as will be briefly described in the following section, that the initial data for these solutions are close to special modes, referred to as homoclinic orbits of the NLS equation. Small perturbations near these homoclinic orbits can respond chaotically by switching irregularly between the "two sides" of the orbit. We have referred to this phenomenon as Numerical Homoclinic Instability (NHI). It provides a theoretical basis for the observations presented here and numerical results presented in other papers as well [cf. 17-19].

3. Numerical Homoclinic Instability

We begin our analytical discussion with an observation regarding linearized instability. The NLS equation possesses an elementary solution

$$u = u_0(t) = ae^{2i|a|^2t+\gamma}, \tag{3.1}$$

for constant a, γ, about which small perturbations can respond with unstable growth. Letting $u(x,t) = u_0(t)(1 + \epsilon(x,t))$, where for $|\epsilon(x,t)| \ll 1$, the function ϵ satisfies the following linear equation

$$i\epsilon_t + \epsilon_{xx} + 2|a|^2(\epsilon + \epsilon^*) = 0. \tag{3.2}$$

The modal representation of (3.2),

$$\epsilon(x,t) = \hat{\epsilon}_n(t)\exp(i\mu_n x) + \hat{\epsilon}_{-n}(t)\exp(-i\mu_n x) \tag{3.3}$$

with $\mu_n = 2\pi n/L$, has the time dependence $\hat{\epsilon}_{\pm n}(t) = \hat{\epsilon}_{\pm n}(0)\exp(\pm\Omega_n t)$ where

$$\Omega_n = \mu_n\sqrt{4|a|^2 - \mu_n^2}\,. \tag{3.4}$$

Consequently any mode having $4|a|^2 - \mu_n^2 > 0$ will be unstable. The maximum number of unstable modes M is the largest integer satisfying $0 < M \le |a|L/\pi$. After relating $\hat{\epsilon}_n(0)$ and $\hat{\epsilon}_{-n}(0)$ the solution of the nth mode of the linear problem ($h > 0$) is given by

$$\begin{aligned}\epsilon(x,t) =&\alpha_+(\mu_n^2 + i\Omega_n)\cos(\mu_n x + \phi_+)e^{\Omega_n t}\\ &+\alpha_-(\mu_n^2 - i\Omega_n)\cos(\mu_n x + \phi_-)e^{-\Omega_n t}\end{aligned} \tag{3.5}$$

where $\alpha_\pm$ and $\phi_\pm$ are arbitrary constants.

It is natural to ask how this linear instability saturates due to the nonlinearity. In this regard we consider the initial condition $u(x,0) = a(1+\epsilon_0\cos\mu x)$, with $\mu = 2\pi/L$, $a = 1/2$, $L = 2\sqrt{2}\pi$ ($m = 1$) and $\epsilon_0 = |\epsilon|e^{i\phi_0}$. We take three cases; (a) $\phi_0 = \pi/4$ (see figure 7a) which corresponds to the "purely unstable" case (where the coefficient of $e^{-\Omega_n t}$ vanishes, i.e. $\alpha_- = 0$); (b) $\phi_0 = \pi/4 - 0.1$ (see figure 7b); and (c) $\phi_0 = \pi/4 + 0.1$ (see figure 7c). We consider cases (b) and (c) as "inside" and "outside" the homoclinic orbit: $\phi_0 = \pi/4$. The case $\phi_0 = \pi/4$ acts as a "separatrix" between the two behaviors. Note that the frequency of maxima (at $x = 0$) in case (c) is roughly half that of case (b), much the same way as it is for a cubic nonlinear oscillator.

It turns out that the analytical description of homoclinic orbits, or separatrices for NLS, may be expressed in terms of elementary functions. In [15] we have shown that the homoclinic orbits with the "fixed point" solution (3.1) can be expressed via the formula,

$$u(x,t) = \left(ae^{2i|a|^2t}\right)\left(\frac{g(x,t)}{f(x,t)}\right) \tag{3.6a}$$

where $g(x,t)$, $f(x,t)$ are polynomials in the exponentials $e^{\pm i\mu_j x}$ and $e^{\Omega_j t}$, and as $|t| \to \infty$, $|g/f| \to 1$. The simplest such orbit (we call this the $N = 1$ case, i.e., the first homoclinic orbit) is given by the formula:

$$u_1(x,t) = ae^{2i|a|^2t}\left(\frac{1 + 2\cos\mu x\; e^{\Omega t + 2i\phi + \gamma} + A_{12}e^{2(\Omega t + 2i\phi + \gamma)}}{1 + 2\cos\mu x\; e^{\Omega t + \gamma} + A_{12}e^{2(\Omega t + \phi)}}\right) \tag{3.6b}$$

where $A_{12} = \sec^2\phi$, $\Omega = \mu\sqrt{4|a|^2 - \mu^2}$, $\mu = 2\pi/L = 2|a|\sin\phi$ (i.e. given a and L this determines ϕ). We note the identification of Ω with the unstable linearized growth rate Ω_1 in (3.4). As time gets large ($|t| \to \infty$):

$$u_1(x,t) \sim ae^{2i|a|^2t}\begin{cases} 1 & t \to -\infty \\ e^{4i\phi} & t \to +\infty \end{cases}$$

We see that asymptotically the amplitude of u_1 is unity, hence our reference to homoclinic orbit. We also note that for $|\epsilon_0| = ce^{\gamma} \ll 1$ with $a = 1/2$, $L = 2\pi\sqrt{2}$,

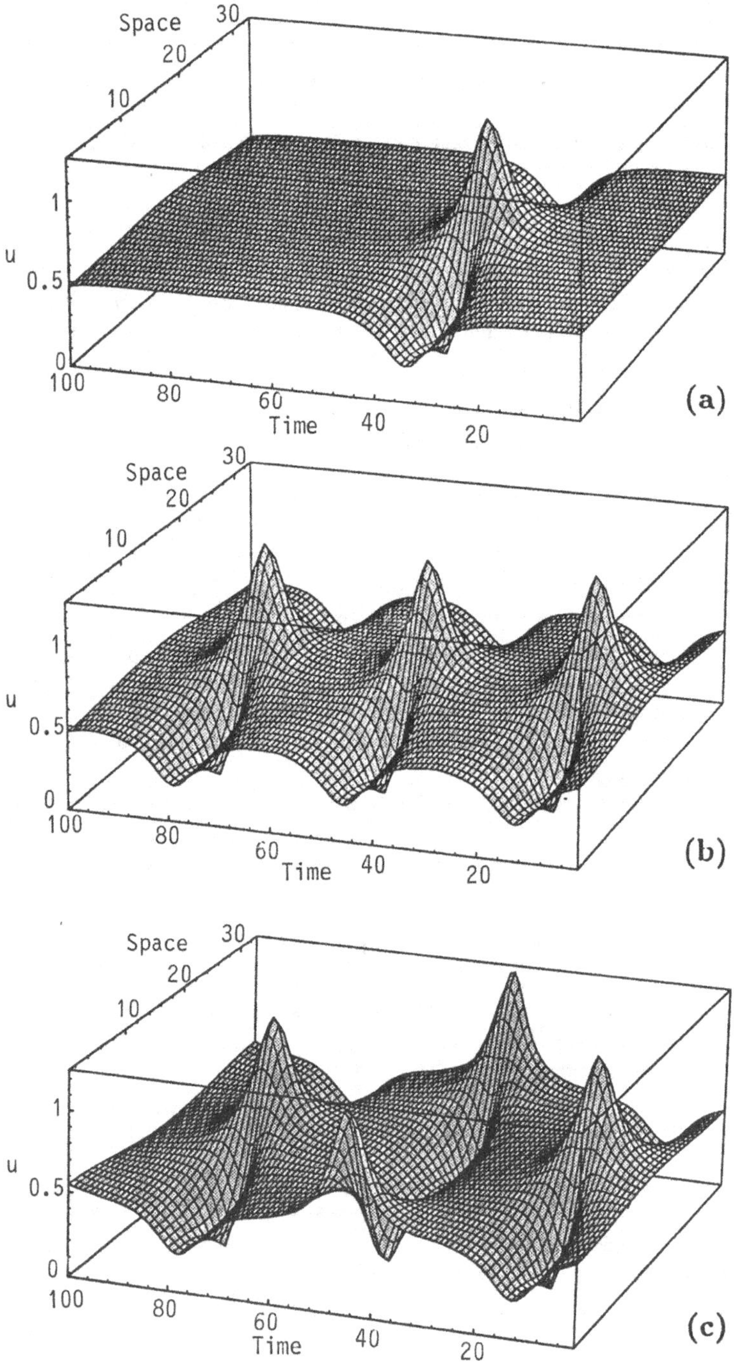

FIGURE 7. (a) homoclinic, (b) 'inside', $\phi_0 = \pi/4 - 0.1$, (c), 'outside', $\phi_0 = \pi/4 + 0.1$

$u_1(x,0) \sim (1+|\epsilon_0|e^{i\pi/4}\cos\mu x)/2$, which agrees with the numerical result described above (see figure 7a).

Formulae governing the higher homoclinic orbits are somewhat cumbersome, hence we do not give them here [cf. 15]. However they are obtained from known multi-soliton solutions by a suitable rotation of coordinates and identification of parameters. We note that if there were N unstable modes in the linearized problem, then there would be an associated "N" homoclinic orbit with appropriate growth rates. Alternative methods of finding these solutions are available, e.g. Backlund transformations [20,21]. In figure 8, an $N=2$ homoclinic orbit is given. We also note that this mode is more complicated than the one with $N=1$ (figure 7a). Indeed for higher N the respective homoclinic orbits become more and more complicated - and indeed they are much harder to resolve numerically. The difficulty of numerical evaluation of homoclinic modes and nearby perturbations is even better understood by appealing to the related linearization of the NLS equation; i.e. the scattering theory.

The NLS equation is solved via the associated linear scattering problem

$$\begin{aligned} v_{1x} + i\zeta v_1 &= u(x,t)v_2 \\ v_{2x} + i\zeta v_2 &= -u^*(x,t)v_1 \end{aligned} \tag{3.7}$$

where ζ is a scattering parameter (eigenvalue) which is *invariant in time*, i.e., $\partial\zeta/\partial t = 0$. The periodic boundary value problem $u(x,t) = u(x+L,t)$ means that we are investigating a Floquet type linear problem in (3.7). The relevant spectral function to analyze is the Bloch function $\psi(x,\zeta)$, satisfying [cf. 22-24] the equation $\psi(x+L,\zeta) = \lambda\psi(x,\zeta)$. The spectra $\lambda = \lambda(\zeta)$ satisfy $|\lambda| = 1$ and in particular we concentrate on the periodic/antiperiodic eigenvalues $\lambda = \pm 1$. It is standard to fix a basis, e.g.

$$\phi(x_0,\zeta) = \begin{pmatrix} 1 \\ 0 \end{pmatrix} \qquad \text{and} \qquad \overline{\phi}(x_0,\zeta) = \begin{pmatrix} 0 \\ 1 \end{pmatrix}$$

and introduce the monodromy data

$$\phi(x+L,\zeta) = \alpha(\zeta)\overline{\phi}(x,\zeta) + \beta(\zeta)\phi(x,\zeta), \tag{3.8}$$

whereby $\lambda(\zeta)$ satisfies $\lambda^2 - 2\alpha_R(\zeta)\lambda + 1 = 0$, where α_R is the real part of α, and $\alpha_R(\zeta) = \pm 1$ yields periodic/antiperiodic eigenvalues. The homoclinic modes are associated with eigenvalues with multiplicity two. These modes can be viewed as small perturbations from the *fixed point solution* $u(x,t) = u_0(t) = ae^{2i|a|^2 t}$. Corresponding to $u_0(t)$, we can solve (3.7) for the basis ϕ, $\overline{\phi}$; the data $\alpha_R(\zeta)$, and analytically study the perturbations [cf. 25].

We find that $\alpha_R(\zeta) = \cos\sqrt{\zeta^2 + |a|^2}\, L$, and the periodic/antiperiodic eigenvalues satisfy

$$\zeta_n^2 = (n\pi/L)^2 - |a|^2. \tag{3.9}$$

Each of these points, except $n=0$, are double points (eigenvalues), i.e. $\partial\alpha_R/\partial\zeta|_{\zeta_n} = 0$. Given L, there are a finite number of complex double points (pure imaginary) and an infinite number of real double points. The number and location of complex double points correspond exactly to the number of unstable linear modes.

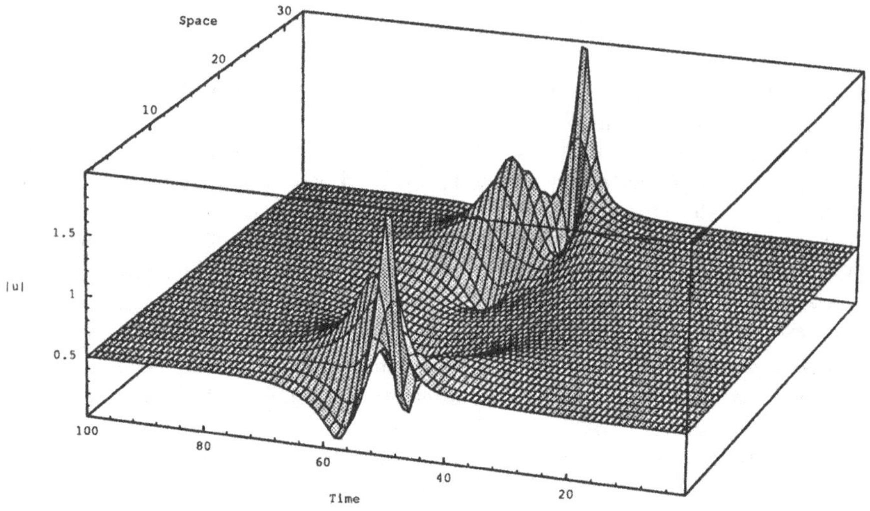

FIGURE 8. homoclinic orbit, $N = 2$

When we perturb off the fixed point by $u(x,0) = a + \epsilon_0 e^{i\phi_0} \cos \mu_n x$, where $\mu_n > (2\pi n/L)$, the eigenvalues perturb as do the eigenfunctions ($\epsilon = \epsilon_0 e^{i\phi_0}$)

$$\zeta_n = \zeta_n^{(0)} + \epsilon \zeta_n^{(1)} + \cdots, \qquad v_n = v_n^{(0)} + \epsilon v_n^{(1)} + \cdots.$$

We do not go through the analysis here, but merely note the important results; for details see [25]. Given $\phi_0 \neq \pi/4$, each double eigenvalue splits under perturbation. The *real* eigenvalues split in *one direction*: they move with $\zeta_n^{(1)}$ allowed to be purely imaginary. The complex eigenvalues split in *two* directions; i.e. $\zeta^{(1)}$ can be either real or pure imaginary. The splitting of the complex eigenvalues for $\phi_0 \neq \pi/4$ correspond to being *inside* ($\phi_0 < \pi/4$) and *outside* ($\phi_0 > \pi/4$) the homoclinic orbit (see figures 7b,c). For $\phi_0 = \pi/4$, the eigenvalue remains double – in agreement with what we expect from the exact solution.

This analysis qualitatively explains the numerical observations made earlier. Namely homoclinic orbits correspond to multiple eigenvalues and, as expected, they are difficult to resolve numerically. Generic small perturbations split the complex eigenvalues out to either side of the homoclinic orbit. When we attempt to solve the NLS equation by the standard scheme and we have *two* or more unstable modes in the initial conditions, as is the case of the numerical results described in section 2, then the perturbation, due to the standard scheme being a perturbation of the NLS equation, influences the modes to "interact" with each other. We observe chaotic switching between the various "sides" of the homoclinic orbits. When we use the standard scheme and perturb an $N = 1$ homoclinic orbit in the initial data, then we find only a weak phase instability [15]. In order to drive the chaos we need at least two unstable modes in the initial data (i.e. two complex eigenvalues).

As mentioned earlier, the integrable scheme (2.3) does not allow the instabilities or chaos observed in the standard scheme. Spectral theory also provides a basis by which we can understand this phenomenon. In [11] it is shown that the linearization of (2.2) is associated with the following scattering problem,

$$v_{1,n+1} = z v_{1,n} + u_n(t) v_{2,n}$$
(3.10)
$$v_{2,n+1} = \frac{1}{z} v_{2,n} - u_n^*(t) v_{1,n}$$

where the eigenvalue z satisfies $z = \exp(-i\zeta h)$, ζ being the eigenvalue of the continuous problem (3.7). The eigenvalue z is isospectral: $\partial z/\partial t = 0$. Each eigenvalue ζ of the continuous problem has a counterpart in the discrete problem - real ζ corresponds to values z on the unit circle and imaginary ζ ($\mathrm{Im}\,\zeta > 0, \mathrm{Im}\,\zeta < 0$) corresponds to real z ($|z| < 1, |z| > 1$). The solution of the discrete integrable scheme (2.3) ***preserves*** double eigenvalues and consequently provides a uniformly small perturbation to the continuous NLS equation. Preservation of the multiple eigenvalues does not allow the homoclinic switching prevalent in the standard scheme.

In recent work McLaughlin and Schober [26] have also investigated the phenomenon of numerically induced chaos for the NLS equation. They have numerically followed the time evolution of the periodic/antiperiodic eigenvalues of NLS which evolve due to the perturbation caused by the standard scheme. They have found that the complex eigenvalues begin to switch chaotically between real and pure imaginary directions, i.e. between the inside and outside regions of the homoclinic orbits. They also investigate the existence of the chaos via Mel'nikov type arguments. Important in this consideration is that the motion of the complex eigenvalues and their associated nonlinear structures comprise the dominant part of solution. In a sense the system behaves like a finite-dimensional nonlinear oscillator. In section 4 of this paper we discuss a Mel'nikov type analysis as it applies to ODE reductions of NLS and related difference approximations.

Finally we remark that there are other interesting nonlinear wave equations which provide similar and even more exotic behavior than the NLS. For example we have studied the complex modified KdV (CMKdV) equation

(3.11)
$$u_t + 6|u|^2 u_x + u_{xxx} = 0$$

along with *three* discretizations [27];
(i) "standard"

(3.12)
$$u_{nt} + \Delta_n^3 u_n + 6|u_n|^2 (u_{n+1} - u_{n-1})/2h = 0$$

(ii) "standard with Hamiltonian structure"

(3.13)
$$u_{nt} + \Delta_n^3 u_n + 3u_n^* (u_{n+1}^2 - u_{n-1}^2)/2h = 0$$

(iii) "integrable"

$$u_{nt} + (1 + h^2|u_n|^2)\Delta_n^3 u_n + u_n^*(u_{n+1}^2 - u_{n-1}^2)/2h + \left(|u_{n+1}|^2 u_{n+2} - |u_{n-1}|^2 u_{n-2}\right)/2h$$
(3.14)
$$- u_n \left(u_{n+1}^* u_{n-1} - u_{n+1} u_{n-1}^*\right)/2h = 0$$

where $\Delta_n^3 u_n = (u_{n+2} - 2u_{n+1} + 2u_{n-1} - u_{n-2})/2h^3$. The difference between (i) and (ii) is that we have found a Hamiltonian and associated Poisson brackets in case (ii). We have developed formulae governing the homoclinic orbits of (3.11) and have associated these orbits with the linearly unstable modes of CMKdV. Discretizations (3.13) and (3.14) behave in a similar way to how (2.2) and (2.3) behave for NLS. Namely, (3.13) introduces a chaotic response to initial data which are nearby homoclinic orbits, whereas (3.14) provides a "nice" solution. On the other hand (3.12) develops singularities in finite time! We are currently investigating why singularities are obtained in (3.12).

4. ODE Reductions, Maps, and Mel'nikov Functions

It is natural to consider ordinary differential, and their corresponding ordinary difference, reductions of the NLS equation and its discretizations. In particular we let $u(x,t) = e^{it}v(x)$ in the NLS equation (2.1) and similarly the reduction $u_n(t) = e^{it}v_n$ in (2.2-2.3). The NLS equation reduces to the Duffing oscillator (for convenience we take u and v_n to be real)

$$v_{xx} - v + 2v^3 = 0 \tag{4.1}$$

and the difference analogues satisfy (a) the "standard" scheme:

$$v_{n+1} + v_{n-1} - 2v_n - h^2(v_n - 2v_n^3) = 0 \tag{4.2}$$

and (b) the "integrable" scheme:

$$v_{n+1} + v_{n-1} - 2v_n - h^2\left(v_n - v_n^2(v_{n+1} + v_{n-1})\right) = 0 \tag{4.3}$$

We note that (4.3) is sometimes referred to as the "Macmillan Map" [28]; here it is obtained naturally from the integrable version of the discrete NLS equation. Equations (4.2) and (4.3) are readily converted to 2-D maps: (a') "standard":

$$\begin{aligned} u_{n+1} &= v_n \\ v_{n+1} &= -u_n + (2+h^2)v_n - 2h^2v_n^3 \end{aligned} \tag{4.4}$$

and (b') "integrable":

$$\begin{aligned} u_{n+1} &= v_n \\ v_{n+1} &= -u_n + \frac{(2+h^2)v_n}{(1+h^2v_n^2)}. \end{aligned} \tag{4.5}$$

We see that the origin, $u_n = v_n = 0$ is a hyperbolic fixed point and the maps are area preserving (i.e. the Jacobian, $J(u_n, v_n)$, is unity). Given initial values u_0, v_0 we can calculate iterates u_n and v_n corresponding to given values of h. In the standard case we find that chaotic-like trajectories are obtained for sufficiently large values of h, e.g. $h > 0.31$. As h decreases to approximately 0.26, the chaos disappears, and then for $h < 0.26$ a smooth trajectory, which is virtually indistinguishable from that of the integrable case, is obtained.

Both schemes are second order accurate approximations to the Duffing oscillator. As such, we may view the standard discretization as a small perturbation of the integrable case, i.e.

$$\begin{aligned} u_{n+1} &= v_n \\ v_{n+1} &= -u_n + \frac{(2+h^2)v_n}{1+h^2v_n^2} + \frac{2\epsilon h^2v_n^3(\frac{1}{2}-v_n^2)}{1+h^2v_n^2} \end{aligned} \tag{4.6}$$

where $\epsilon = h^2$. In order to apply well-known perturbation methods to (4.6) it is convenient to treat ϵ and h as independent small parameters. For the unperturbed problem, i.e. the integrable case (formally obtained by taking $\epsilon = 0$) we proceed most easily if we have a constant of motion and an explicit representation for the homoclinic orbit. Indeed the NLS equation and its integrable difference version provide the framework by which both pieces of information are obtained. The first comes from the conservation laws of the differential difference equation (2.3). They are of the form $\partial/\partial t\ T_n + \Delta_n H_n = 0$; in the ordinary difference reduction they reduce to $\Delta_n H_n = 0$ where $\Delta_n H_n = H_{n+1} - H_n$.

Moreover, the homoclinic orbit is nothing but a discrete NLS soliton reduced to the time-independent case. In [29] we derive these formulae, here we only quote the result. The constant of motion of (4.5) or (4.3) is given by

$$H(u_n, v_n) = \frac{1}{2}\left(u_n^2 v_n^2 + \frac{1}{h^2}(u_n^2 + v_n^2 - 2\mu u_n v_n)\right), \tag{4.7}$$

where $\mu = 1 + h^2/2$, and the homoclinic orbit by

$$u_n(\xi) = \sinh\omega\,\mathrm{sech}\,(\tilde{\omega}n + \xi) \tag{4.8}$$

where $\sinh\omega = \sinh\tilde{\omega}/h$, and $\cosh\tilde{\omega} = 1 + h^2/2$; ξ is a phase parameter. For $h \to 0, \tilde{\omega} \sim h$.

Given perturbations of maps of the form

$$\mathbf{x}_{n+1} = F(\mathbf{x}_n) + \epsilon G(\mathbf{x}_n, \epsilon) \tag{4.9}$$

where $F, G : \mathbf{R}^2 \to \mathbf{R}^2$, and for $\epsilon = 0$ there is a homoclinic orbit, one computes the splitting distance between the stable and unstable orbits. To leading order in ϵ, this is frequently referred to as Mel'nikov analysis [cf. 30]. Assume we have a constant of motion for the unperturbed problem: $H(F(\mathbf{x}_n) = \mathbf{x}_{n+1}) = H(\mathbf{x}_n)$. Then the splitting distance between the stable and unstable orbit is given by the function

$$\epsilon M(\xi;\epsilon) = \sum_{n=-\infty}^{\infty} \Delta_n H = \sum_{n=-\infty}^{\infty} (H(\mathbf{x}_{n+1}) - H(\mathbf{x}_n)).$$

The Mel'nikov function measures the change of the constant of motion, due to the perturbation, over the homoclinic orbit. We note the following;

$$\begin{aligned} H(\mathbf{x}_{n+1}) &= H(F(\mathbf{x}_n) + \epsilon G(\mathbf{x}_n)) \\ &= H(F(\mathbf{x}_n)) + \epsilon\nabla H(F(\mathbf{x}_n)) \cdot G(\mathbf{x}_n) + \cdots, \\ &= H(\mathbf{x}_n) + \epsilon\nabla H(\mathbf{x}_{n+1}) \cdot G(\mathbf{x}_n) + \cdots. \end{aligned}$$

Consequently,

$$M(\xi;\epsilon) = \sum_{n=-\infty}^{\infty} \nabla H(\mathbf{x}_{n+1}(\xi)) \cdot G(\mathbf{x}_n(\xi);\epsilon) \tag{4.10a}$$

or

$$M(\xi;\epsilon) = \sum_{n=-\infty}^{\infty} G(\mathbf{x}_n(\xi);\epsilon) \wedge \hat{\mathbf{v}}_{n+1}(\xi) \tag{4.10b}$$

where $\mathbf{u} \wedge \mathbf{v} = u^{(1)}v^{(2)} - u^{(2)}v^{(1)}$.

For our problem,

$$(4.11)\qquad G(\xi,h)=\begin{pmatrix}0\\ 2h^2v_n^3\left(\frac{1/2-v_n^2}{1+h^2v_n^2}\right)\end{pmatrix}$$

$$\hat{\mathbf{v}}_{n+1}=\begin{pmatrix}\frac{\partial H}{\partial v}\\ -\frac{\partial H}{\partial u}\end{pmatrix}=\begin{pmatrix}u^2v+\frac{1}{h^2}(v-\mu u)\\ -uv^2-\frac{1}{h^2}(u-\mu v)\end{pmatrix}_{n+1}$$

and consequently the Mel'nikov function is given by

(4.12)

$$M(\xi,h)=-2h^2\sum_{n=-\infty}^{\infty}u_n^3\left(\frac{1/2-u_n^2}{1+h^2u_n^2}\right)\left(u_n^2v_n+\frac{1}{h^2}(v_n-\mu u_n)\right)=h\sum_{n=-\infty}^{\infty}m(\tilde{\omega}n+\xi).$$

The latter relation defines $m(\xi)$.

We may establish the following properties of $M(\xi,h)$;

(a) $\lim_{h\to 0}M(\xi,h)=0$,

(b) $M(\xi,h)=M(\xi+\tilde{\omega},h)$, i.e. it is periodic in ξ with period $\tilde{\omega}$,

(c) $M(0;h)=0,\ M'(0;h)\neq 0$.

The fact that the Mel'nikov vanishes (with simple zeroes) and is periodic, is sufficient to establish that the underlying map is chaotic – i.e. the intersections of the stable and unstable orbits form a "horseshoe" map. However, as $h\to 0$ we observe from numerical experiments that the chaos disappears rapidly. Indeed as we show below, the Mel'nikov function approaches zero *exponentially fast*, and hence the unperturbed homoclinic structure is restored at an exponential rate.

We next estimate the rate at which $M(\xi,h)$ vanishes as $h\to 0$. For this purpose, we write $M(\xi,h)$ as a Fourier series,

$$(4.13a)\qquad M(\xi,h)=\sum_{j=-\infty}^{\infty}\hat{M}_je^{2\pi ij\xi/\tilde{\omega}}$$

where

$$(4.13b)\qquad \hat{M}_j=\frac{h}{\tilde{\omega}}\int_0^{\tilde{\omega}}\left(\sum_{n=-\infty}^{\infty}m(\tilde{\omega}n+\xi)\right)e^{-2\pi ij\xi/\tilde{\omega}}\,d\xi$$

By interchanging the sum and integral, and using the periodicity property of $m(\xi)$ we have that

$$(4.14)\qquad \hat{M}_j=\frac{h}{\tilde{\omega}}\int_{-\infty}^{\infty}m(\xi)e^{2\pi ij\xi/\tilde{\omega}}\,d\xi.$$

We know that $\hat{M}_0=0$, since we have an area-preserving map. Hence it is sufficient to estimate (4.14) as $h\to 0$ for $j\neq 0$. The function $m(\xi)$ is analytic, decaying rapidly as $\xi\to\infty$. The major contribution to the integral comes from the nearest singularity, which in this case is a pole at $\xi=i\pi/2$. We have therefore that

$$(4.15)\qquad \hat{M}_j\sim \text{const}\ \exp(-\pi^2j/\tilde{\omega})$$

and consequently the leading contribution to $M(\xi,h)$ arises from the $j=1$ Fourier coefficient,

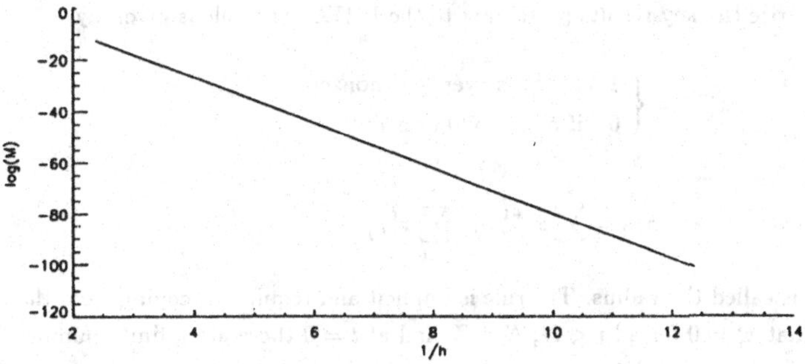

FIGURE 9. Mel'nikov calculation, 50 digits

$$M(\xi, h) \sim \text{const } \exp(-\pi^2/h)e^{2\pi i\xi/h} \tag{4.16}$$

(recall that $\tilde{\omega} \sim h$). Numerical calculations confirm this estimate. We performed the numerical calculation of (4.12) to 50 digits on Mathematica (see figure 9).

What we see from this example is that whereas the order of accuracy of the original standard scheme is second-order, nevertheless the *qualitative* features of the true solution are restored at an exponentially fast rate. The analysis, while done for this specific problem, can be generalized to a wide class of weakly integrable area-preserving maps. Moreover, Nekhoroshev-type estimates [31,32] show that the action variables for weakly perturbed integrable Hamiltonian (or area-preserving) systems where perturbations are Hamiltonian (or area-preserving) change only over exponentially long time intervals. Consequently both Mel'nikov type analysis and Nekhóroshev estimates provide considerable insight as to why "Symplectic Integrators" [33-35] have been so successful. Symplectic schemes successfully predict the actions or amplitudes of the underlying motion far better than the order of accuracy of the scheme suggest. Higher order accurate symplectic integrators do not appear to more accurately predict "amplitudes". However, the angle variables can be expected to improve according to the order of accuracy of the underlying numerical scheme.

5. Solitons and Cellular Automata

Generally speaking, cellular automata (CA) are dynamical systems in discrete space and time, whose field variables take on finitely many values, e.g. 0 and 1, or values in a finite field. Many workers believe that CA, with may of their remarkable properties, may play a fundamental role in describing nature (cf. [36] for a review). Important applications include fluid dynamics, models of chemical reactions, biological modeling, etc. In [4] a remarkable rule referred to as the Parity Rule Filter Automata (PRFA) was proposed. The PRFA exhibits a wide range of particle-like structures, including an important subclass of particles with soliton-like behavior. The PRFA and associated properties have been studied by a number of authors (cf [5,6,37-39]). The PRFA is, in general, irreversible in time, and as such we expect that there may be yet a more fundamental CA rule (or class of rules) which share many of the desirable features but is time reversible. We construct such a time-reversible rule in [7] and briefly discuss them in this section.

First we summarize the key results pertaining to the PRFA. The rule is given by

$$x_i^{t+1} = \begin{cases} 1 & \text{if } S_{i,2}^t \text{ is even and nonzero} \\ 0 & \text{if } S_{i,2}^t \text{ is odd or zero} \end{cases} \tag{5.1}$$

where

$$S_{i,2}^t = \sum_{j=1}^{r} x_{i-j}^{t+1} + \sum_{j=0}^{r} x_{i+j}^t.$$

The integer $r \geq 1$ is called the radius. The rule is implicit and requires sweeping from the left. We assume that $x_i^t = 0$ for all $i < N$, $N \in \mathbb{Z}$, and at $t = 0$ there are a finite number of nonzero entries in the initial data.

An equivalent formulation of the PRFA, more suitable for analysis, called the Fast Rule Theorem (FRT) was given in [5] and consists of the following,

$$x_{i-r}^{t+1} = \begin{cases} x_i^t & \text{if } i \notin B(t) \\ 1 - x_i^t & \text{if } i \in B(t), \end{cases} \tag{5.2}$$

where $B(t) \subset \mathbb{Z}$ is given by the following inductive procedure.
(i) Sweeping from the left, place the index (i.e. the spatial coordinate) of the first nonzero site of $\{x_i^t\}$ in $B(t)$;
(ii) If $i \in B(t)$, then place the index $(i + r + 1)$ in $B(t)$, if any $x_{i+j}^t \neq 0$ for $j = 1, 2, \ldots, r$;
(iii) If $i \in B(t)$ and $x_{i+j}^t = 0$ for $j = 1, 2, \ldots, r$ then place the index of the next nonzero x_i^t in $B(t)$.
(iv) Continue by repeating steps (ii) and (iii) until all the nonzero sites of x_i^t are exhausted.

The following properties can be established by use of the FRT (e.g. see [5-6], [38-39]).

I. Stability. For every t the set $B(t)$ is finite and therefore at every time step there are finitely many 1's.

II. Time-Irreversibility. As a dynamical system the PRFA contains dissipation, i.e. is nonconservative, and is therefore time-irreversible. The simplest initial data, a single 1 in an infinity of zeros, evolves to the zero configuration in one time step. This state is called a prenull. More generally, if the configuration C^t contains prenulls, i.e. there exists (at least one) $i \in B(t)$ such that $x_i^t = 1$ and $x_{i+j}^t = 0$, $j = 1, 2, \ldots, r$, then the one step evolution $C^t \to C^{t+1}$ is irreversible. Hence the state C^t cannot be uniquely determined from C^{t+1} by "going backwards".

We call C_r the (reversible) subset of set C of all initial data consisting of those states in which there appears no prenull in evolution forward and backward in time, and C_i (irreversible) its complement in C. The restriction of the PRFA to C_r is time-reversible. The irreversibility is responsible for the complications discussed in part IV(b).

III. Particle Content. The time-evolution of PRFA exhibits many different coherent structures. Each evolves in a time-reversible way; i.e. there are integers p, called the period, and d, the displacement, such that for all $i \in \mathbb{Z}$, $x_{i-d}^{t+p} = x_i^t$. Such modes are called particles. The time-evolution of a particle is translation to the left by d units after p successive time steps. its velocity is defined as $v = d/p$, where $0 \leq v \leq r - 1$. When $p = 1$ the particle moves like a point-like object; a localized wave retaining its "shape". Particles with $p > 1$ have additional internal structure (inner degrees of freedom) which exhibits itself during a period. There exist many particles with various p and d, and a systematic way of constructing them via linear difference equations is given in [39]. Among all particles, one can distinguish basic particles which behave like solitons [38]. They correspond to initial data of length $r+1$ units; the zero configuration and prenull are excluded.

IV. Interaction and Scattering of Particles. The large time asymptotic picture is described by different coherent structures which depend on the interaction between particles. There are two major cases.

IV(a). Solitonic (time-reversible and elastic) scattering. Here the scattering of basic particles is similar to the scattering in $1+1$ integrable solitonic dynamical systems, i.e. an initial state of spatially separated basic particles $A_1, A_2, \ldots, A_n$ ordered by their velocities $v_1 < v_2 < \ldots < v_n$ evolves through a complicated interaction into an asymptotic state describing the free motion of the same particles $A_n, \ldots, A_2, A_1$ arranged in the opposite order.

IV(b). Non-Solitonic (time-irreversible) scattering. This phenomenon is due to the existence of time-irreversible initial data. In the process of evolution, states containing prenulls are produced; they die in the next time step. As mentioned in the introduction, the prenull state is responsible for the non-conservation of the energy-type functional introduced in [6]. There are three typical examples of the non-solitonic scattering:

Gluing. Particles A and B interact and form another single particle C (e.g. see figure 10a);

Inelastic. Scattering of two particles A and B results in two different particles A' and B', neither of which is A or B.

Irreversible Solitonic. Scattering of two particles A and B results in the same particles B and A, but this process, due to the appearance of prenulls, is not time-reversible.

We also note that the parity rule (5.1) is equivalent to the following difference relation;

$$(5.3)\qquad \sum_{j=0}^{r} x_{i-j}^{t+1} \equiv \sum_{j=0}^{r} x_{i+j}^{t} + \delta_2(x_i^t)\prod_{j=1}^{r}\delta_2(x_{i-j}^{t+1})\delta_2(x_{i+j}^{t}) - 1$$

Here equality $\equiv$ is understood mod 2, i.e. $(1+1) \mod 2 = 1+1 \equiv 0$, and $\delta_q(x)$ is a "delta function mod q"; i.e. $\delta_2(x) = 1$ if $x = 0 \mod 2$ and 0 otherwise. Rewriting (5.3) in the form $x_i^{t+1} \equiv S_{i,2}^t + \delta_2(x_i^t)\prod_{j=1}^{r}\delta_2(x_{i-j}^{t+1})\delta_2(x_{i+j}^t) - 1$, we can interpret $S_{i,2}^t$ as a linear term and the rest as a highly nonlinear perturbation.

Formulation (5.2) admits a natural multi-state generalization, x_i^t, consisting of q elements $0, 1, \ldots, q-1$, and define its time evolution by

$$(5.4)\qquad \sum_{j=0}^{r} x_{i-j}^{t+1} \equiv \sum_{j=0}^{r} x_{i+j}^{t} + \delta_q(x_i^t)\prod_{j=1}^{r}\delta_q(x_{i-j}^{t+1})\delta_q(x_{i+j}^{t}) - 1$$

where now equality $\equiv$ is understood mod q. Alternatively, (5.4) is equivalent to the following q-state parity rule

$$(5.5)\qquad x_i^{t+1} = \begin{cases} (S_{i,q}^t - 1) \mod q & \text{if } S_{i,q}^t \neq 0 \\ 0 & \text{if } S_{i,q}^t = 0 \end{cases}$$

where now $S_{i,q}^t = (q-1)\sum_{j=1}^{r} x_{i-j}^{t+1} + \sum_{j=0}^{r} x_{i+j}^t$, and (5.5) is equivalent to the following q-state Fast Rule Theorem

(5.6) $$x_{i-r}^{t+1} = \begin{cases} x_i^t & \text{if } i \notin B(t) \\ (x_i^t - 1) \mod q & \text{if } i \in B(t), \end{cases}$$

where the set $B(t)$ was defined earlier by (i)-(iii).

This multi-state CA has all of the properties I-IV of the PRFA (with yet a wider particle content); but as with the PRFA it is time-irreversible; i.e. a prenull vanishes at the next time step.

An inspection of equation (5.3) shows that the reason for dissipation and time-irreversibility is due to the presence of the factor $\delta_2(x_i^t)$ in the product, which clearly violates the symmetry $x_{i+j}^t \leftrightarrow x_{i-j}^{t+1}$. Thus we define a new two-state CA by the following difference relation

(5.7) $$\sum_{j=0}^{r} x_{i-j}^{t+1} \equiv \sum_{j=0}^{r} x_{i+j}^{t} + \prod_{j=1}^{r} \delta_2\left(x_{i-j}^{t+1}\right) \delta_2\left(x_{i+j}^{t}\right) - 1$$

with equality understood to be mod 2, which has the symmetry $x_{i+j}^t \leftrightarrow x_{i-j}^{t+1}$, and therefore is time-reversible. Backward evolution in time, from C^{t+1} to C^t is given by the same equation (5.7) which we solve by sweeping from the *right*. Equivalently, the time-evolution in this new CA can be described by the reversible parity rule

(5.8) $$x_i^{t+1} = \begin{cases} 1 & \text{if } S_{i,2}^t \text{ is even and non-zero or } x_i^t = S_{i,2}^t = 1 \\ 0 & \text{otherwise,} \end{cases}$$

or alternatively by the Reversible Fast rule theorem (RFRT), given by (5.2) where the set $B(t)$ is replaced by the set $\tilde{B}(t)$ defined as follows: steps (i), (ii), (iv) in the definition of $B(t)$ remain unchanged, whereas step (iii) is replaced by the following:

(iii') If $i \in \tilde{B}(t)$ and $x_{i+j}^t = 0$, $j = 1, 2, \ldots, r$, then either $i + r \in \tilde{B}(t)$ if $x_i^t = 1$ or, if $x_i^t = 0$, the index of the next non-zero x_i^t in C^t belongs to $\tilde{B}(t)$.

The RFRT shows that the new CA is stable. It follows from (5.7,5.8) that restriction of the new CA to the subset C_r of initial data coincides with the PRFA, hence it possesses all the desirable properties I-IV(a)! Moreover, due to time-reversibility, it "cures" the bad property IV(b) of the PRFA. Namely, in this new CA the prenulls do not vanish but instead are now elementary basic particles that are stationary in time. They represent the simplest of all particles, which we call prebasic particles. With these particles, the reversible rule possesses an infinite collection of particles, with highly non-trivial interaction properties.

For instance, even in the simplest case $r = 1$ (which is trivial for the PRFA), appropriate initial data produce particles with zero velocity and arbitrarily large periods exhibiting complicated internal motion.

In addition to the solitonic interactions, the new CA exhibits highly non-trivial particle interactions. We note the change from the "gluing" situation that occurs in the irreversible rule to particle production in the reversible rule. In figure 10 we take the same initial conditions; namely AOB: $100111, z(11), 1011$ (where $z(11)$ denotes 11 zeroes), with radius $r = 3$. This represents two particles, $A : v = 1, p = 6$, and $B : v = 5/3, p = 3$, which eventually collide. The irreversible rule produces a final "glued" state of a single particle of velocity 1, period 6, whereas the reversible rule results in *three* particles; a large one with $v = 1$, $p = 18$ moving away from two stationary (and well-separated) prebasic particles. This example illustrates that scattering of particles A and B results in a new particle C

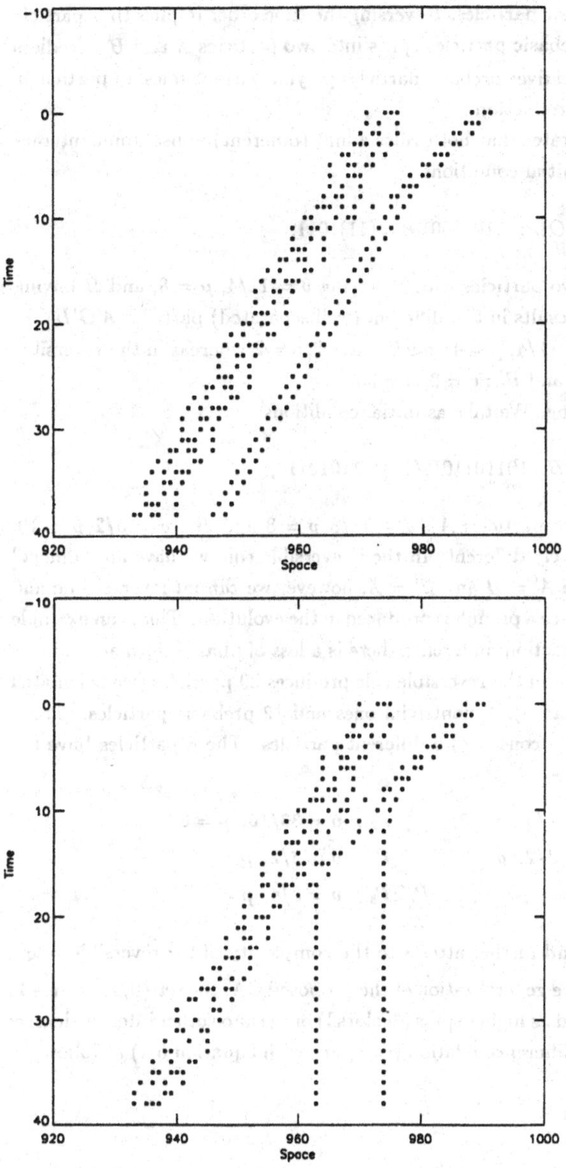

FIGURE 10. **(a)** Particle interaction defined by equations (1) and (2), bound state irreversible; **(b)** Reversible Particle interaction defined by equations (7) and (8) and RFRT, particle production.

and a number of emitted prebasic particles. Reversing the time order implies that particle C bombarded by a beam of prebasic particles splits into two particles A and B. Needless to say, this picture is very suggestive: prebasic particles playing various roles; in particular being responsible for particle production.

The next example demonstrates that both rules admit (different) nonsolitonic interactions. We take, for both, the initial conditions

$$AOB: \; 11000001011z(11)1011,$$

with $r = 5$. This represents two particles with A having $v = 11/4$, $p = 8$, and B having $v = 3$, $p = 3$. The interaction results in two different (well separated) particles $A'O'B'$. In the irreversible rule A' has $v = 11/4$, $p = 4$ and $B' : v = 2, p = 2$ whereas in the reversible rule A' has $v = 54/21$, $p = 42$, and $B' : v = 2$, $p = 30$.

The final example is startling. We take as initial conditions

$$AOB: 101101101z(14)1111010111,$$

with $r = 4$; this represents two particles $A : v = 17/8, p = 8$ and $B : v = 5/2, p = 10$. The interaction picture is entirely different. In the irreversible rule we have an "almost" solitonic picture $A'O'B'$ where $A' = B$ and $B' = A$; however we *cannot* reverse time and recover the initial conditions since a prenull is produced in the evolution. This is an example of an irreversible solitonic interaction; in a sense there is a loss of phase information, i.e. a "phase shock". On the other hand the reversible rule produces 20 particles (we calculated this evolution to 23 000 time steps): 8 nontrivial ones and 12 prebasic particles. The 8 well-separated particles actually consist of 6 different particles. The 8 particles have the following characteristics:

$$\begin{aligned} P_1 &: \; v = 7/3, \; p = 3, & P_2 &: \; v = 33/16, \; p = 32 \\ P_3 &: \; v = 3/2, \; p = 28, & P_4 &: \; v = 3/2, \; p = 2 \\ P_5\&P_6 &: \; v = 1, \; p = 6, & P_7\&P_8 &: \; v = 3/4, \; p = 4 \end{aligned}$$

This behavior is unexpected and further attests to the complexity of the reversible rule.

Finally, we give a multi-state generalization of the proposed CA; the set $\{0, 1, \ldots, q-1\}$ of values x_i^t might be considered as higher spins ("colors") or perhaps other internal degrees of freedom. We define a new difference relation (compare with equation 5.4) as follows:

$$\begin{aligned} \sum_{j=0}^{r} x_{i-j}^{t+1} &\equiv \sum_{j=0}^{r} x_{i+j}^{t} \quad + \\ &\left[\delta_q\left(x_i^t\right)\delta_q\left(x_i^{t+1}\right) - \delta_q\left(x_i^t - 1\right)\delta_q\left(x_i^{t+1} - (q-1)\right)\right] \prod_{j=1}^{r} \delta_q\left(x_{i-j}^{t+1}\right)\delta_q\left(x_{i+j}^{t}\right) - 1. \end{aligned} \tag{5.9}$$

with equality understood to be mod q. The corresponding q-state RFRT takes the form

$$x_{i-r}^{t+1} = \begin{cases} x_i^t & \text{if } i \notin \tilde{B}(t), \\ (x_i^t - 1) \mod q & \text{if } i \in \tilde{B}(t), \end{cases} \tag{5.10}$$

where the set $\tilde{B}(t)$ was defined earlier by (i)-(iii') and (iv). Qualitatively, it has the same properties as the two-state CA, with even richer particle content and interactions due to the "color".

The FRT for the backwards time-evolution is given by

$$x^t_{i+r} = \begin{cases} x_i^{t+1} & \text{if } i \notin \widehat{B}(t) \\ (x_i^{t+1}+1) \mod q & \text{if } i \in \widehat{B}(t), \end{cases} \tag{5.11}$$

where the set $\widehat{B}(t)$ is defined in the same manner as $\tilde{B}(t)$ but with the sweeping going from right to left.

This novel class of proposed time-reversible multi-state CA's exhibits intriguing regularity and a large variety of coherent structures. Given the vast array of coherent particle-like solitons, we believe that this new CA, and models like this one, may find valuable applications in many areas of science, e.g. nonlinear physics, information and computation theory, etc.

Acknowledgements

This work (MJA) was partially supported by the NSF, Grants No. DMS-9024528, and the Air Force Office of Scientific Research, Grant No. AFOSR-90-0039.

It is our pleasure to acknowledge the important contributions of our colleagues J. Keiser and L. Takhtajan, who collaborated with us on portions of the research described in this review article.

References

[1] M. J. Ablowitz and H. Segur, *Solitons and the Inverse Scattering Transform*, SIAM (1981).

[2] Novikov S. P., Manakov S. V., Pitaevskii L. P. and Zakharov V. E., *Theory of Solitons. The Inverse Scattering Method*, Plenum, New York, 1984.

[3] N. Ercolani, M. G. Forest and D. W. McLaughlin, *Geometry of the Modulational Instability part III: Homoclinic Orbits for the Periodic Sine-Gordon Equation*, Physica D 43 (1990), 349–384, Also see the article by D.W.McLaughlin in this volume, and the references therein.

[4] J. Park, K. Steiglitz, and W. Thurston, Physica D 19 (1986), 423.

[5] T. S. Papatheodorou, M. J. Ablowitz, and Y. G. Saridakis, Stud. Appl. Math 79 (1988), 173.

[6] C. H. Goldberg, Complex Systems 2 (1988), 91.

[7] M. J. Ablowitz, J. M. Keiser, L. A. Takhtajan, *A class of multi-state time-reversible cellular automata with rich particle content (PAM* report #74)*, Phys. Rev. A (1991) (to appear Nov. 15 1991).

[8] V. M. Kenkre and D. K. Campbell, Phys. Rev. B 34 (1986), 4959;V. M. Kendre and G. P. Tsironis, Phys. Rev. B 35 (1987), 1473.

[9] J. C. Eilbeck, P. S. Lomdahl, and A. C. Scott, Physica D 16 (1985), 318.

[10] A. S. Daveydov, J. Theor. Biol. 38 (1973), 559; Usp. Fiz. Nauk 138 (1982), 603; Sov. Phys. Usp. 25 (1982).

[11] M. J. Ablowitz and J. F. Ladik, Stud. Appl. Math. 55 (1976), 213.

[12] P. P. Kulish, *Quantum difference nonlinear Schrödinger equation*, Letters in Mathematical Physics 5 (1981), 191–197.

[13] N. N. Bogolyubov and A. K. Prikarpat'skii, Sov. Phys. Dokl. 27 (1982), 113.

[14] B. M. Herbst and M. J. Ablowitz, *Numerically induced chaos in the nonlinear Schrödinger equation*, Phys. Rev. Lett. 62 (1989), 2065–2068.

[15] M. J. Ablowitz and B. M. Herbst, SIAM J.Appl. Math. 50 (1990), 339–351.

[16] B. M. Herbst and M. J. Ablowitz, *On Numerical Chaos in the Nonlinear Schrödinger Equation*, Integrable Systems and Applications, M. Balabane, P. Lochak and C. Sulem (editors), Lecture Notes in Physics #342, Springer-Verlag, Berlin, 1989.

[17] E. A. Caponi, P. G. Saffman and H. C. Yuen, *Instability and confined chaos in a nonlinear dispersive scheme*, Phys. Fluids 25 (1982), 2159–2166.

[18] J. A. C. Weideman, *Computation of instability and recurrence phenomena in the nonlinear Schrödinger equation*, Ph.D.Thesis, University of the Orange Free State, Bloemfontein (1986).

[19] Mei-Mei Shen and D. R. Nicolson, *Stochasticity in numerical solutions of the nonlinear Schrödinger equation*, Phys.Fluids 30 (1987), 3150–3154.

[20] N. N. Akhmediev, V. M. Eleonskii, and N. E. Kulagin, *Generation of periodic trains of picosecond pulses in an optical fiber; exact solutions*, Sov.Phys.JETP 62 (1985), 894–899.

[21] N. Ercolani, D. W. McLaughlin and M. G. Forest, *Notes on the Mel'nikov integral for models of the driven pendulum chain* (1989), (Preprint, 1989).

[22] V. Kotlyarov and A. R. Its, D.Akad.Nauk. UKRSR A 11 (1976).

[23] Y. C. Ma and M. J. Ablowitz, *The periodic cubic Schrödinger equation*, Stud. Appl.Math 65 (1981), 113–158.

[24] M. G. Forest and J. E. Lee, *Geometry and Modulation Theory for the periodic nonlinear Schrödinger Equation*, IMA Volumes in Math & its applications 2 (1986), 35–70.

[25] M. J. Ablowitz and B. M. Herbst, *On Homoclinic Boundaries in the Nonlinear Schrödinger Equation*, Hamiltonian Systems, Tranformation Groups and Spectral Transform Methods, J. Harnad and J. E. Marsden (editors), CRM, Montreal, 1990.

[26a] D. W. McLaughlin and C. M. Schober, *Chaotic and homoclinic behavior for numerical discretizations of the nonlinear Schrödinger equation* (1990), (Preprint, 1990).

[26b] C. M. Schober, *Numerical and Analytical aspects of the discrete nonlinear Schrödinger equation*, Ph.D.thesis, U. of Arizona (1991).

[27] B. M. Herbst, M. J. Ablowitz, E. Ryan, *Numerical Homoclinic Instabilities and the complex modified Korteweg de Vries equations*, Comp.Phys.Commun. 65 (1991), 137–142.

[28] E. M. McMillan, *A Problem in the Stability of Periodic Systems*, Topics in Modern Physics, Colorado University Press, Boulder, Colorado, 1971, pp. 219–244.

[29a] B. M. Herbst and M. J. Ablowitz, *Mel'nikov Analysis and Numerically Induced Chaos*, PAM* report #19 (1990); Proc.Conf.Chaos in Australia (Feb.1990) (to appear).

[29b] B. M. Herbst and M. J. Ablowitz, *Numerical chaos, symplectic integrators and exponentially small splitting distance*, PAM* report #78 (1991).

[30] R. W. Easton, *Computing the Dependence on a Parameter of a Family of Unstable Manifolds: Generalized Mel'nikov Formulas*, Nonlinear Analysis, Theory, Methods and Appl. 8 (1984), 1.

[31] N. N. Nekhorovoshev, Russ.Math.Surveys 32 (1977), 1–65.

[32] G. Bennetin, L. Calgani, A. Giorgilli, *A proof of Nekhoroshev's Theorem for the Stability times in nearly integrable Hamiltonian systems*, Celestial Mech. 37 (1985), 1–25.

[33] F. Kang, *Difference schemes for Hamiltonian formalism and symplectic geometry*, J. Comput. Math. 4 (1986), 279–289.

[34] P. J. Channel and C. Scovel, *Symplectic integration of hamiltonian systems*, nonlinearity 3 (1990), 231–259.

[35] J. M. Sanz-Serna, *Runge-Kutta schemes for Hamiltonian systems*, BIT 28 (1988), 877–883.

[36] S. Wolfram, *Theory and Applications of Cellular Automata*, World Scientific (1986).

[37] K. Steiglitz, I. Kamal, and A. Watson, IEEE Trans. Comp. 37 (1988).

[38] A. S. Fokas, E. Papadopoulou, Y. G. Saridakis, M. J. Ablowitz, Stud.Appl.Math. 81 (1989), 153.

[39] J. M. Keiser and M. J. Ablowitz, *On Particles and Interaction Properties of the Parity Rule Filter Automata*, PAM* report #68 (1990).

*Program in Applied Mathematics Report, University of Colorado at Boulder

Symplectic Aspects of Some Eigenvalue Algorithms

P. Deift[1], *L.-C. Li*[2], *and C. Tomei*[3]
[1]Courant Institute, New York University, New York, NY 10012, USA
[2]Department of Mathematics,
Pennsylvania State University, PA 16802, USA
[3]Universidade Católica do Rio de Janeiro, Brazil

Introduction

It turns out that many well known algorithms in numerical analysis can be reformulated in a useful and intrinsic way, by making use of basic ideas in symplectic geometry. Actually, from a modern perspective, it is surprising to see how many examples of important techniques in the theory of dynamical systems with special symmetries were known to numerical analysts. The methods of symplectic geometry enable us to extend our understanding of the algorithms, by providing a conceptual setup in which to interpret a number of algebraic procedures. This text is an attempt to illustrate the above claims.

In the first section, we describe the key ingredient of the numerical algorithms we consider, the QR step, together with some of its basic properties. Next we show how the smallest natural phase space for the QR step, the set of Jacobi matrices, can be given a symplectic structure through which some special Hamiltonians (the Toda hierarchy) generate completely integrable flows, one of which has the additional property of interpolating the QR iteration. Well known phenomena for numerical analysts are given rather unusual interpretations and, in the process, the set of tridiagonal matrices will be thoroughly scrutinized by special parametrizations (inverse variables / action-angle variables). The constructions therein will serve as a motivation for the more complicated extensions in the subsequent sections. In sections 2 and 3, we show how most of the properties of the QR iteration and its continuous counterpart over Jacobi matrices carry through for much larger phase spaces, first (essentially arbitrary) symmetric matrices, and then (essentially), arbitrary matrices: each step presents new technical difficulties requiring more sophisticated (but still familiar) constructions in symplectic geometry. In the fourth section, we describe some remarkable stability properties of an algorithm suggested recently by Demmel and Kahan to compute the singular value decomposition of a matrix. First we again provide a Hamiltonian continuous interpolation for the iteration defined by the algorithm. The stability properties then follow from the perturbation theory of symplectic matrices.

The text is not complete in any sense. There are essentially no proofs, which can be found in the quoted references, and we did not intend to exhaust the description of the interplay between both areas. Also, at many points of the paper, we use known constructions in symplectic geometry, without describing them in detail.

Somehow, our understanding of the algorithms in the text seems imbalanced: there is a substantial amount of theoretical knowledge that is still not operational — an outstanding problem is how to convert this knowledge into better algorithms or better implementations of known algorithms.

Finally we would like to mention the work of Batterson and his collaborators ([BS], [B1], [BDa], [B2]) on the convergence properties of the shifted QR algorithm. Although these authors do not use the symplectic structure associated with the QR algorithm, the papers contain beautiful applications of the methods of dynamical systems to problems arising in numerical analysis.

Section 1

The basic ingredient in many algorithms currently used to compute eigenvalues of a matrix M is a "flip step" as follows. Factor M as a product AB, where A and B have specific properties and define a new matrix N to be the product BA. So, for example, the QR step applied to a real, invertible matrix M consists of factoring M as a product of an orthogonal matrix Q and an upper triangular matrix R with positive diagonal entries and then setting $N = RQ$. Different factorizations may be used: A and B can be the factors in the LU decomposition (in this case, A is lower triangular and B is upper triangular with diagonal elements equal to one), or as the factors in the Cholesky decomposition of a positive matrix M (and then A would be a lower triangular matrix with positive diagonal and B would be its transpose). Other factorizations are useful for special classes of matrices (see for example [WE1], [B-G,B,M]). In this paper, however, we will concentrate on the first flip step described above, the QR step.

It is a well known fact of linear algebra that the matrices $M = AB$ and $N = BA$ have the same spectrum (for invertible A, write $N = A^{-1}MA$ and now extend by taking limits). In numerical analysis one considers flip steps with an additional property which makes them useful in the computation of eigenvalues: the iteration of the step starting from a matrix M gives rise to a sequence of matrices convergent to a diagonal matrix D. Clearly, M and D have the same eigenvalues, and we are led to a simple (and rather primitive) algorithm to approximate the spectrum of M. A situation in which convergence to a diagonal matrix is guaranteed is the case when M is a real symmetric matrix with simple spectrum and one takes the QR step.

We list some properties of the QR step which are well known to numerical analysts ([Par1], [GvL]).

QR1. It is well defined for real invertible matrices M: the columns of the orthogonal matrix Q are obtained by applying the Gram-Schmidt procedure to the columns of M.

QR2. It preserves symmetry and the spectrum of the starting matrix M: as seen above, the QR step applies an orthogonal conjugation to M.

QR3. It preserves the band structure of M: so, for example, if M is symmetric tridiagonal or upper Hessenberg (i.e., elements below the first lower subdiagonal are zero), then N is too. (This fact is not immediate: it is one of the properties of the QR algorithm which is evident after we provide the appropriate symplectic structure.)

There is another natural procedure which is frequently used to compute approximate eigenvalues and eigenvectors of a matrix M. Start with a generic vector u and compute a sequence of normalized vectors $v_n = (M^n v)/\|M^n v\|$. For a symmetric matrix M, the sequence $\{v_n\}$ converges to an eigenvector u of M associated to an eigenvalue of largest absolute value (simply use the spectral theorem to compute $M^n v$ for a v with nonzero component along the direction u). Similarly, for a generic choice of orthonormal basis, the iteration of a step consisting of applying M to the basis and re-orthonormalizing gives rise to a sequence of bases converging to an orthonormal basis of eigenvectors of M, with associated eigenvalues appearing in decreasing order of absolute value. This procedure and the QR iteration are closely related by the statement below, which generalizes QR1.

QR4. The n-th iteration M_n of the QR step starting with M is equal to $Q_n^T M Q_n$, where Q_n is defined in the QR factorization $M^n = Q_n R_n$.

The convergence properties of the iteration now should be clear.

QR5. If M is a symmetric matrix whose eigenvalues have distinct moduli, then the sequence of QR iterates $\{M_n\}$ converges to a diagonal matrix.

The hypothesis ensuring convergence of the QR iteration can be relaxed (see, e.g., [Par2]), but we will not be concerned with this issue in this paper.

Tridiagonal matrices are especially important: numerical analysts know very effective algorithms ([Par1]) to convert an arbitrary symmetric matrix into a tridiagonal matrix with the same spectrum. Rather surprisingly, this can be achieved by conjugating the original $n \times n$ matrix by $n-2$ simple hyperplane reflections (Householder transformations) — the whole procedure only requires occasional computations of square roots. Instead of appling the QR iteration to a symmetric matrix whose eigenvalues are to be determined, one first tridiagonalizes it to obtain a matrix to which the QR iteration is applied at a substantially lower cost. We will begin our description of the symplectic structures related to the QR step by studying its action on symmetric, tridiagonal matrices.

Historically, the symplectic approach started from a very different source, Flaschka's proof of the complete integrability of the Toda lattice ([Fla]: also [Man]). The Toda lattice describes the motion of n particles on the line induced by the Hamiltonian

$$H(x,y) = \frac{1}{2}\sum_{k=1}^{n} y_k^2 + \sum_{k=1}^{n-1} \exp(x_k - x_{k+1}) ,$$

where $x, y \in \mathbf{R}^n$ are the position and velocity vectors of the n particles. Flaschka changed variables,

$$\begin{aligned} a_k &= -y_k/2 , \qquad k = 1,\ldots,n , \\ b_k &= (\exp[(x_k - x_{k+1})/2])/2 , \quad k = 1,\ldots,n-1 , \end{aligned}$$

and recast the equations of motion for x and y in Lax pair form,

$$(*) \qquad \dot{L} \equiv \frac{dL}{dt} = [B,L] \equiv BL - LB , \quad L(0) = L_0 ,$$

where L and B are the tridiagonal matrices below,

$$L = \begin{pmatrix} a_1 & b_1 & & & & \\ b_1 & a_2 & b_2 & & O & \\ & b_2 & \cdot & \cdot & & \\ & & \cdot & \cdot & & \\ & O & & & & b_{n-1} \\ & & & & b_{n-1} & a_n \end{pmatrix},$$

$$B = \begin{pmatrix} 0 & b_1 & & & & \\ -b_1 & 0 & b_2 & & 0 & \\ & -b_2 & \cdot & \cdot & & \\ & & \cdot & \cdot & & \\ & O & & & & b_{n-1} \\ & & & & -b_{n-1} & 0 \end{pmatrix}.$$

As Lax had already pointed out in his remarkable paper on the conserved quantities for the Korteweg-deVries equation ([Lax]), an operator changing in time according to $(*)$ undergoes a conjugation by an orthogonal matrix, thus preserving its initial symmetry and spectrum. The eigenvalues of the initial matrix then provide n conserved quantities which can be shown to be in involution with respect to the usual symplectic structure $\sum_{i=1}^{n} dx_i \wedge dy_i$ in (x, y) space. Later, the Toda lattice was explicitly linearized by Moser ([Mos]) by making use of a parametrization of the set of Jacobi matrices (i.e., real, symmetric tridiagonal matrices L with $b_i > 0$). This parametrization, analogous to the inverse variables employed to study the Schrödinger operator in an interval with Dirichlet boundary conditions ([PT]), is described as follows. First notice that a Jacobi matrix cannot have an eigenvectors with first coordinate equal to zero. In particular, eigenvectors can be normalized in a unique way so that their first coordinate is a strictly positive number. Also it follows that the spectrum of a Jacobi matrix is simple. Moser showed that the map

$$\text{Jacobi matrix} \mapsto \begin{pmatrix} \text{ordered spectrum, first} \\ \text{coordinates of normalized eigenvectors} \end{pmatrix}$$

is a diffeomorphism between the set of Jacobi matrices and the product of the set of n-uples with increasing coordiantes with the positive octant of the unit sphere in $\mathbf{R}^n$. Moreover, he computed the inverse map explicitly (by use of continued fractions, following an idea of Stieltjes) and proved that, along an orbit of $(*)$, the matrix $L(t)$ becomes diagonal at $\pm\infty$ with diagonal entries ordered decreasingly (increasingly) at $+\infty$ $(-\infty)$.

In the late seventies, Adler and Kostant ([Adl], [Kos]) showed that the Toda lattice in (a, b) variables is a very special differential equation, obtained through a natural (and very general) construction. Let G be a Lie group with Lie algebra g, having g^* as its dual vector space. Then the adjoint action Ad of G on g induces a coadjoint action Ad^* of G on g^* with the following remarkable property: each orbit in g^* has a natural symplectic structure, induced by the so-called Lie-Poisson structure. To obtain the Toda lattice, start with the group of real lower triangular matrices and identify g^* with the set of symmetric matrices by the nondegenerate pairing $\langle A, B\rangle = \text{tr } A^T B$. The coadjoint action is then given by

$$\begin{array}{rccc} Ad^*: & G \times g^* & \longrightarrow & g^* \\ & (h, S) & \mapsto & (h^{-T} S h^T)_- + (h^{-T} S h^T)_0 + ((h^{-T} S h^T)_-)^T , \end{array}$$

where M_-, M_0 and M_+ are the strictly lower, diagonal and strictly upper parts of the matrix M. Moreover, the set of Jacobi matrices with fixed trace is a coadjoint orbit, its natural symplectic structure coincides with the structure obtained by pushing forward the usual structure in (x, y) variables, and the Toda lattice is induced by the Hamiltonian

$$H = \frac{1}{2}\left(\sum_{i=1}^{n} a_i^2 + 2\sum_{i=1}^{n-1} b_i^2\right) = \frac{1}{2}\ \mathrm{tr}\ L^2 ,$$

which is of course, up to a factor, the original Hamiltonian expressed in (a, b) variables. The theory of Lie groups could then be used to provide a better understanding of the integrability of the equations. In particular, Adler, Kostant and Symes ([Kos], [Sym1], [Adl]) used Lie algebraic techniques to show that the flows commuting with the Toda lattice are described by the equations

$$(**) \qquad \dot{L} = [\pi_k(f(L)), L] , \qquad L(0) = L_0 ,$$

where $\pi_k(M) = M_+ - (M_+)^T$ and f is analytic on the spectrum of L. Reyman and Semenov-Tian-Shansky [Rey-STS1], Olshanetsky and Perelomov [OP], Symes [Sym1] and (implicitly) Kostant [Kos1] also (independently) obtained a remarkable factorization formula to solve $(**)$: indeed,

$$L(t) = Q^T(t)\, L_0\, Q(t) ,$$

where $Q(t)$ is obtained from the QR factorization of the matrix $\exp(tf(L_0)) = Q(t)\, R(t)$.

Both of the preceding subjects seem unrelated: there is an algorithm to compute eigenvalues of Jacobi matrices and there is an interesting differential equation for which the Jacobi matrices provide a natural phase space. Symes ([Sym2]) made the first connection between the subjects (but see the beginning of the next section for additional historical remarks) by proving that the QR iterates $M_0 = e^{L_0}, M_1, M_2, \ldots$ of the exponential of an initial condition L_0, are equal to the exponentials of the values of the orbit $L(t)$ of the Toda lattice starting at L_0 at integer times, i.e., $M_k = e^{L(t=k)}$. Soon after, Deift, Nanda and Tomei ([DNT]) showed with different techniques from those employed by Symes that actually the QR iteration itself is the evaluation at integer times of the orbit of the differential equation $(**)$ for the choice $f(x) = \log x$.

There are elementary proofs of all the statements above which make no use of Lie group theory or inverse variables whatsoever, and we sketch them now. First notice that the conjugation

$$L(t) = Q^T(t)\, L_0\, Q(t)$$

is the (unique) solution of the differential equation

$$\dot{L} = [B, L] , \qquad L(0) = L_0 ,$$

where $B = -Q^T\dot{Q}$ is skew-symmetric, since $Q(t)$ is orthogonal. It suffices to show that the recipe above for $Q(t)$ gives rise to the B defined in (**). This is a simple differentiation: since

$$\exp(t\, f(L_0)) = Q(t)\, R(t) ,$$

we must have

$$Q^T f(L_0)\, Q = Q^T\dot{Q} + \dot{R}R^{-1} ,$$

where the LHS is $f(L(t))$ and the RHS is the sum of a (unique) skew-symmetric and an upper triangular matrix. The skew-orthogonal component is exactly the operator appearing in (**). From the explicit solution of (**), one obtains that the flows commute for different functions f, and for $f(x) = \log x$, the time one map associated to (*) is the QR step.

The inverse algorithm also can be performed in a simple way. The matrix K of eigenvectors of the Jacobi matrix L can be computed from its first row v and the spectrum of L, $\Lambda = \operatorname{diag}(\lambda_1, \ldots, \lambda_n)$, as follows. Let $\widetilde{K}$ be the matrix with rows equal to $v, v\Lambda, v\Lambda^2, \ldots, v\Lambda^{n-1}$. Now, Gram-Schmidt on the rows of $\widetilde{K}$ obtains K, up to choice of sign for each row, chosen to make the off-diagonal entries of $L = K\Lambda K^T$ positive. This construction appeared in ([DNT]) and Parlett (personal communication) pointed out that this is a well known procedure used to tridiagonlize a symmetric matrix ([GvL]) — in this case, rather ironically, the matrix which is being tridiagonalized is the diagonal matrix Λ.

For a symplectic manifold M of dimension $2n$, in which there are n commuting independent globally defined flows induced by Hamiltonians $H_1, \ldots, H_n$, the celebrated Liouville-Arnold theorem ([Arn]) states that the connected components of the level sets of the moment map

$$\begin{array}{rccc} \phi: & M & \longrightarrow & \mathbf{R}^n \\ & L & \mapsto & (H_1(L), \ldots, H_n(L)) \end{array}$$

are products of lines and circles. In the case of the Toda hierarchy (**) acting on Jacobi matrices, one can take

$$H_j(L) = \frac{1}{j} \operatorname{tr} L^j , \qquad j = 2, \ldots, n ,$$

and from the inverse variables, it is clear that the Liouville-Arnold invariant set is diffeomorphic to $\mathbf{R}^{n-1}$. Also, by converting the differential equations in (a, b) variables to inverse variables, one can show that the flows act transitively on the invariant sets. More precisely, it can be shown that the values of ϕ parametrize (bijectively) the family of Liouville-Arnold invariant sets, and each invariant set is parameterized (bijectively) by the remaining inverse variables. It should be no surprise that the inverse variables can be used to describe the so-called action-angle variables ([Arn]) of the equations (**) on Jacobi matrices. We will present the action-angle variables of (**) in larger phase spaces in the following sections.

From the symplectic setup , some facts are easier to explain: the QR step preserves the tridiagonal form since the (globally defined) flows (**) have to stay within the symplectic orbit of Jacobi matrices and the QR step is interpolated by the flow associated to $f(x) = \log x$. Also, the fact that the flows (**) do indeed commute with each other is a consequence of a general result, the Kostant-Symes theorem . The solution by factorization follows by a general procedure ([Kos2], [Sym1], [Rey-STS2], [STS]) for solving a large class of differential equations on coadjoint orbits.

For the special case of Jacobi matrices, the phase space of the flows induced by (**) is well understood. The Liouville-Arnold invariant set $\mathcal{J}_\Lambda$ of Jacobi matrices with simple spectrum Λ (homeomorphic to $\mathbf{R}^{n-1}$, by the argument using inverse variables given before) sits inside the vector space of tridiagonal matrices so that its closure is homeomorphic to the convex span of the $n!$ points in $\mathbf{R}^n$ obtained by permuting the n elements of Λ ([Tom]). The map was explicitly constructed by Bloch, Flaschka and Ratiu ([BFR]) as follows. For $L \in \mathcal{J}_\Lambda$, diagonalize

$$L = K \operatorname{diag}(\lambda_1 \ldots, \lambda_n) K^{-1} ,$$

where $\lambda_1 > \cdots > \lambda_n$ and K is unitary with positive entries in the first row, and now set $\widetilde{L} = K^{-1} \operatorname{diag}(\lambda_1, \ldots, \lambda_n) K$. The desired map takes L to $(\widetilde{L})_0$, the diagonal part of $\widetilde{L}$. It can be shown to be a diffeomorphism extending bijectively to the closure of the domain taking diagonal matrices to the $n!$ extreme points and matrices with zero off-diagonal elements to the boundary of the convex set. In the 3×3 case, for example, with $\Lambda = \{2, 4, 8\}$, the closure of $\mathcal{J}_\Lambda$ is homeomorphic to the hexagon in Figure 1 (see [DNT]). The vertices and edges of the hexagon are invariant under the flows (**); orbits of the usual Toda flow are indicated in the picture. For 4×4 matrices with $\Lambda = (1, 2, 3, 4)$, the closure of $\mathcal{J}_\Lambda$ is isomorphic to the polyhedron in Figure 2 (see [DNT]), where some orbits of the Toda flow are drawn. Notice the durations near equilibria of initially close orbits. Not surprisingly, orbits passing too close to an equilibrium spend a long time to run a small stretch, and, roughly, two diagonal entries flip their positions. Table I (see [DNT]) shows 19 iterations of the QR step, in which this phenomenon is evident. The numerical literature refers to this behavior as 'root confusion'.

Real life algorithms make use of the possible closeless to an equilibrium, or more generally, to the boundary of $\mathcal{J}_\Lambda$, by a process named deflation ([Par1]), and exaggerated root confusion is rarely seen in practice. There are many rather simple but very effective tricks which convert the QR iteration into an extremely efficient procedure to compute approximate eigenvalues of matrices, but we will give no details here (see [Par1], [GvL]).

The full manifold M_n of $n \times n$ real, symmetric, tridiagonal matrices with fixed simple spectrum (and no restrictions on the off-diagonal entries) has been studied in detail ([Dav], [Fri], [Tom]). Its universal covering is $\mathbf{R}^{n-1}$ and the sum s of the partial traces, the sum s of the partial traces, $s = \sum_{j=1}^{n} (n - j) a_j$, is a height function for the Toda flow: it is strictly increasing along orbits, with the only exceptions at equilibria. Actually, s is a perfect Morse function for M_n: the k-the Betti number of M_n is the number of equilibria of the Toda flow

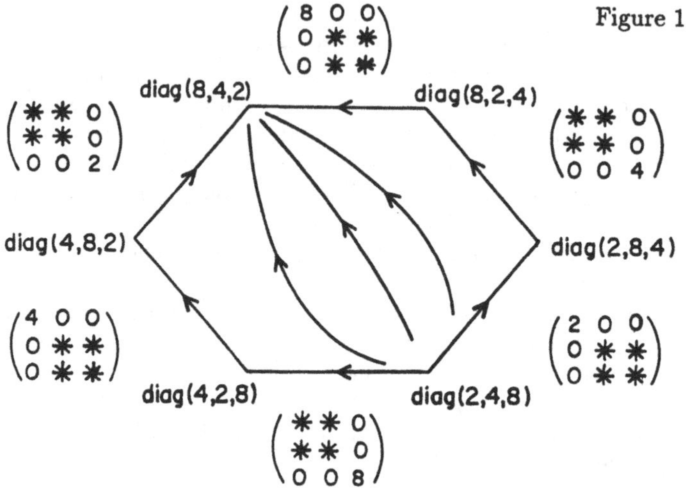

Figure 1

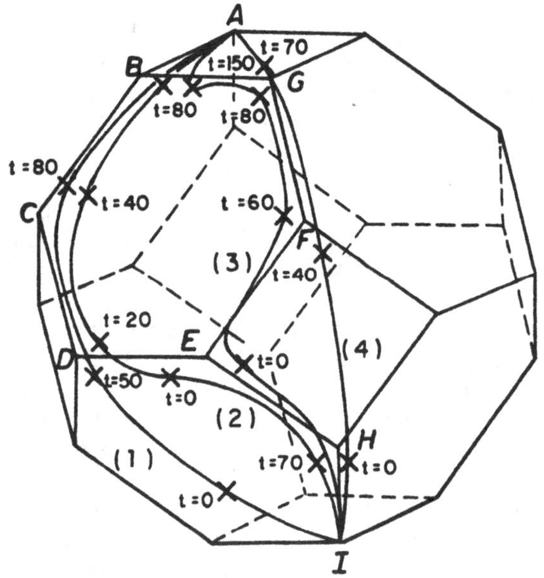

Figure 2

at which the Hessian of s has signature k, and is equal to the so-called Euler number $\left\langle {n \atop k+1} \right\rangle$ (see [Knu] for notation). The Euler characteristic of M_n, the alternate sum of the Betti numbers, is $(-1)^{n+1}B_{n+1}\,2^{n+1}(2^{n+1}-1)/(n+1)$, where B_n is the n-th Bernoulli number. The full cohomology ring of M_n has been computed by Davis and Fried. As a curiosity, M_3 is a bitorus.

It is very simple to see that the function s increases along a Toda orbit: from the evolution in (a,b) variables, we immediately have the positivity of the derivative. The result extends to orbits starting with arbitrary symmetric

TABLE I

Time	a_1	a_2	a_3	b_1	b_2
0	2.0000E+00	7.6000E+00	4.4000E+00	4.3346E−10	1.2000E+00
1.0000E+00	2.0000E+00	7.9999E+00	4.0001E+00	2.4206E−07	2.4420E−02
2.0000E+00	2.0000E+00	3.0000E+00	4.0000E+00	5.7653E−05	4.4728E−04
3.0000E+00	2.0003E+00	7.9997E+00	4.0000E+00	3.9394E−02	8.1925E−06
4.0000E+00	7.2516E+00	2.7484E+00	4.0000E+00	1.9825E+00	4.2484E−07
5.0000E+00	8.0000E+00	2.0000E+00	4.0000E+00	5.6146E−03	2.9369E−00
6.0000E+00	8.0000E+00	2.0000E+00	4.0000E+00	1.3917E−05	2.1701E−05
7.0000E+00	8.0000E+00	2.0000E+00	4.0000E+00	3.4497E−00	1.6035E−03
8.0000E+00	8.0000E+00	2.0000E+00	4.0000E+00	8.5510E−11	1.1848E−03
9.0000E+00	8.0000E+00	2.0000E+00	4.0000E+00	2.1196E−13	8.7515E−01
1.0000E+01	8.0000E+00	2.0021E+00	3.9979E+00	5.2567E−16	6.4621E−02
1.1000E+01	8.0000E+00	2.1081E+00	3.8919E+00	1.3390E−18	4.5216E−01
1.2000E+01	8.0000E+00	3.5144E+00	2.4856E+00	6.5512E−21	8.5756E−01
1.3000E+01	8.0000E+00	3.9883E+00	2.0117E+00	1.0172E−22	1.5238E−01
1.4000E+01	8.0000E+00	3.9998E+00	2.0002E+00	1.9125E−24	2.0741E−02
1.5000E+01	8.0000E+00	4.0000E+00	2.0000E+00	3.5026E−26	2.8073E−03
1.6000E+01	8.0000E+00	4.0000E+00	2.0000E+00	6.4153E−28	3.7993E−04
1.7000E+01	8.0000E+00	4.0000E+00	2.0000E+00	1.1250E−29	5.1417E−05
1.8000E+01	8.0000E+00	4.0000E+00	2.0000E+00	2.1521E−31	6.9586E−06
1.8970E+01	8.0000E+00	4.0000E+00	2.0000E+00	4.4442E−33	9.9998E−07

matrices. A nice application of this fact is a short proof of a well known theorem in perturbation theory, the Wielandt-Hoffman theorem ([Parl]). Let A and B be symmetric matrices with eigenvalues $\{a_i\}$ and $\{b_i\}$. Then there is an ordering of the eigenvalues for which $\sum(a_i - b_i)^2 \leq \operatorname{tr}(A-B)^2$. To see this, assume without loss that A is a positive diagonal matrix with decreasing diagonal entries, and write the above inequality in the equivalent form $\sum a_i b_i \geq \operatorname{tr} AB$. Now take B to be an initial condition for the Toda flow, and one obtains the result at ∞, by the monotonicity of tr $AB(t)$! Lagarias obtained similar monotonicity properties for a large class of functionals under the Toda family and suggested a conjecture which was later proved by Deift, Rivera, Tomei and Watkins ([Lag], [DRTW]). Bloch, Flaschka and Ratiu showed that the Toda flow actually is a gradient system for an appropriate metric on the set of symmetric matrices with fixed spectrum — again, this provides an abstract framework for some related monotonicity properties.

The abundance of structure associated with the tridiagional Toda flow is quite overwhelming. Bloch, Flaschka and Ratiu ([BFR]) showed that by flipping matrices $L(t)$ along orbits of (**)

$$L(t) = K \wedge K^{-1} \mapsto \widetilde{L}(t) = K^{-1}(t) \wedge K(t)$$

one obtains analytic continuations (in time) of the very simple orbits induced by diagonal unitary conjugations on the manifold O_Λ of self-adjoint matrices with spectrum equal to Λ: the analytic continuation is to be taken with respect to the complex structure in O_Λ arising from the fact that, as an adjoint orbit, it is a Kahler manifold ([Bes]).

Section 2.

As seen in the previous section, Jacobi matrices provide a natural phase space for the QR iteration as well as for the Toda flows, but there is nothing to prevent us from considering other phase spaces. For example, it is clear from the properties of the QR step described above that a QR step on an arbitrary real invertible symmetric matrix preserves its symmetry together with its specctrum. Also, the equations (**) make perfectly good sense for matrices which are not necessarily tridiagonal. In this section, we introduce an open dense set of symmetric matrices for which all the structure associated with the integrability of the QR iteration (and of the flows) is still present. In particular, we obtain action-angle variables for both the QR step and the Toda flow, and show that indeed the QR step is the evaluation at time 1 of an integrable flow arising from (**).

¿From information on the QR step well known to numerical analysts, it is rather easy to find an interpolating flow as follows. As in property QR4, in which one takes powers of the QR step by conjugating by the orthogonal part of the QR factorization of a power of the initial conditon, one can take a square root of the QR step, for example (i.e., a diffeomorphism on positive matrices which applied twice gives the QR step), by conjugating by the orthogonal part in the QR factorization of the positive square root of the initial condition. One can obtain in a similar fashion nth roots of the QR step and the interpolating flow now is obtained by taking the appropriate limit. In more detail, define

$$M_{1/n} = Q_{1/n}^T \, M_0 \, Q_{1/n} \, ,$$

where $Q_{1/n}$ is the orthogonal matrix in the QR decomposition $(M_0)^{1/n} = Q_{1/n} R_{1/n}$. One can see, as in the proof of property QR4, that indeed $M_0 \mapsto M_{1/n}$ is an n-th root of the QR step. The interpolating vector field at M_0, if it exists, has to be

$$\lim_{n \to \infty} (M_{1/n} - M_0) \, / \, (1/n) \, .$$

To compute the limit, notice that

$$(M_0)^{1/n} = I + (\log M_0)/n + O(1/n^2) \, ,$$

and that the QR decomposition of $I + \epsilon X$ is

$$I + \epsilon X = \left(I + \epsilon \pi_a X + O(\epsilon^2)\right) \left(I + \epsilon \pi_u X + O(\epsilon^2)\right) \, ,$$

where $\pi_a X = X_- - X_-^T$ and $\pi_u = \mathrm{id} - \pi_a$. So we must have

$$M_0^{1/n} = (I + \frac{\pi_a \log M_0}{n} + O(\frac{1}{n^2}))\,(I + \frac{\pi_u \log M_0}{n} + O(\frac{1}{n^2}))$$

and

$$\begin{aligned} \lim_{n\to\infty} n(M_{1/n} - M_0) &= \left[\, -\pi_a \log M_0, M_0\right] \\ &= \left[\pi_k \log M_0,\, M_0\right] \,, \end{aligned}$$

in the notation of the previous section.

It is very surprising that Rutishauser in 1954 ([Rut]), by taking a different limit for the LU iteration (with shifts, to be correct), stumbled over the equation

$$\frac{dA}{dt} = [-A_-, A] = [A_+ + A_0, A] \,,$$

which is one equation in the family analogous to (**) for the LU step. Remarkably, he showed that the equation could be solved by a factorization procedure (predating the solution by factorization of the Toda flow by 25 years!), but did not realize that the LU step itself admits an interpolation given by a differential equation in the hierarchy

$$\frac{dA}{dt} = [(f(A))_-, A] \,,$$

suggested by Watkins ([Wat]). For a discussion of Rutishauser's work the reader should refer to [WE2], in which the authors show that Rutishauser had also obtained the differential equations for the Toda flow in (a, b) variables by taking a continuous limit of his QD algorithm.

We now describe phase space for the Toda flow on symmetric matrices (see [Adl], [Kos]). Let L be the identity component of the group of lower triangular, invertible real matrices, with Lie algebra ℓ and interpret the set of symmetric matrices to be its dual ℓ^* through the nondegenerate pairing $\langle A, B\rangle = \mathrm{tr}\ A^T B$. As already noted in Section 1, the coadjoint action of L on ℓ^* is given by

$$\mathrm{Ad}^*_{h^{-1}}(S) = \pi^*_\ell((h^{-1})^T\, S\, h^T) \,,\, h \in L \,,\, S \in \ell^* \,,$$

where

$$\pi^*_\ell(M) = M_0 + M_- + (M_-)^T \,.$$

The orbits of the coadjoint action carry natural symplectic structures, being induced by the Lie-Poisson structure

$$\{F, G\}_{\ell^*}(S) = \ \mathrm{tr}(S[\pi_\ell \nabla F(S) \,,\, \pi_\ell \nabla G(S)]) \,,$$

where $\pi_\ell(M) = M_0 + M_- + (M_+)^T$ and $(\nabla F)_{ij}(M) = \frac{\partial F}{\partial M_{ij}}$. The Jacobi matrices with fixed trace are an example of a rather small dimensional orbit. We now describe the generic orbits, which are also the orbits of largest dimension — details can be found in [DLNT]. First, some additional notation. Let $(M)_{kp}$ be the $(n-k) \times (n-p)$ matrix obtained by deleting the first k rows and the last p columns of M. If $k = p$, set $(M)_{kk} = (M)_k$. For $0 \le k \le [n/2]$, define

$$P_k(M,\lambda) \equiv \det(M-\lambda I)_k \equiv \sum_{r=0}^{n-2k} E_{rk}(M)\,\lambda^{n-2k-r} .$$

It turns out that the signs of the values $E_{0k}(M)$ are invariant under the coadjoint action on M, so, in particular, the functions

$$I_{rk}(M) = E_{rk}(M) \,/\, E_{0k}(M) \;,\; 0 \le k \le [(n-1)/2] \;,\; 1 \le r \le n-2k \;,$$

are well defined throughout a coadjoint orbit O_M if they happen to make sense at a point M, i.e. if $E_{0k}(M) \neq 0$.

Proposition. *The generic orbit O_S through $S \in \ell^*$ is the set*

$$\{\widetilde{S} = \widetilde{S}^T\!: \textit{sign}\; E_{0k}(\widetilde{S}) = \textit{sign}\; E_{0k}(S) \;,\; 1 \le k \le [n/2] \;,$$
$$I_{1k}(\widetilde{S}) = I_{1k}(S) \;, \qquad 0 \le k \le [(n-1)/2]\} \;,$$

and has dimension $2[n^2/4]$.

Notice that the I_{rk}'s are symmetric functions of the (generalized) eigenvalues λ_{rk} satisfying

$$(M-\lambda)_k v_{rk} = \lambda_{rk} v_{rk} \;, \quad 0 \neq v_{rk} \in \mathbf{C}^{n-k}$$

The generic orbit is described by some discrete and some continuous invariants, in a way similar to the Jacobi orbit (positive off diagonal elements, fixed trace). The I_{rk}'s are essentially the action variables for the Toda flow.

Theorem. *The Toda flow ((**) in the previous section with $f(x) = x$) is induced by the Hamiltonian $\frac{1}{2}$ tr $S^2 = \frac{1}{2}((I_{10}(S))^2 - I_{20}(S),)$. Moreover, on a generic orbit O_{S_0}, the Hamiltonians I_{rk}, $0 \le k \le [n/2]+1$, $2 \le r \le n-2k$, provide $[n^2/4]$ commuting vector fields which are independent on an open subset of O_{S_0}. The flow induced by I_{rk} with initial condition S_0 is given by*

$$S(t) = Z(t)^T\, S_0\, Z(t) = \pi_\ell^*\big((X(t))^{-T}\, S_0 (X(t))^T\big)$$

where $\exp(t\,\nabla I_{rk}(S_0)) = Z(t)\,X(t)$ *is the QL factorization of the LHS (Z is orthogonal and X is lower triangular with positive diagonal).*

An immediate consequence of the solution by factorization is that the flow induced by (**) with $f(x) = \log x$ interpolates the QL iteration. It then follows that the QL iteration on real symmetric matrices is the integer time evaluation of a completely integrable Hamiltonian flow commuting with the Toda flow.

Remark. We leave the derivation of the elementary relationship between the QL and the QR algorithms to the reader.

Both expressions for $S(t)$ have an abstract interpretation in terms of Ad*-actions, compatible with the general factorization procedure described in [STS]. One can also verify that

$$Z(t) = \begin{pmatrix} I & 0 & 0 \\ 0 & [Z(t)]_k & 0 \\ 0 & 0 & I \end{pmatrix} ,$$

where the $(n-2k) \times (n-2k)$ matrix $[Z(t)]_k$ is obtained from the QL decomposition

$$\exp(t[\nabla I_{rk}(S_0)]_k) = [Z(t)]_k \, [X(t)]_k$$

of the $(n-2k)\times(n-2k)$ matrix obtained by exponentiating the central $(n-2k)\times(n-2k)$ block of $t \nabla I_{rk}(S_0)$. The I_{r0} flows are familiar. They appear already in the low dimensional Jacobi orbit. The new flows are more complicated — the above formulae imply their global existence and indicate that they do not converge necessarily to diagonal matrices.

What are the Liouville-Arnold invariant sets for this integrable system? This can be answered after we provide the angle variables in O_{S_0}. Call a matrix M *generic* if it belongs to a generic orbit; call M *simple* if all the generalized eigenvalues λ_{rk} are distinct amongst each other. We need a technical lemma to construct the angle variables.

Lemma. *For M simple, the solution of*

$$(M - \lambda_{rk})_{k-1} \, u_{rk} = e_1$$

is of the form $u_{rk} = \binom{w_{rk}}{0}$. Moreover,

$$(M - \lambda_{rk})_k \, w_{rk} = 0$$

and, if M is real and symmetric, the last entry f_{rk} of w_{rk} is not zero. □

By Cramer's rule,

$$f_{rk}(M) = \left(e_{n-k} \, , \, ((M - \lambda_{rk})_{k-1})^{-1} \, e_1\right),$$

where e_1 and e_{n-k} are vectors in the canonical basis of $\mathbf{C}^{n-k+1}$, and the inner product $(\cdot,\cdot)$ is real. Now, set $f_{r0}(M) = \det(B(M) - \lambda_{r0}(M))$, where $B(M)$ is the $(n-1)\times(n-1)$ matrix obtained by deleting the last row and last column of M. For a simple symmetric M, define, as allowed by the lemma above,

$$\mu_{rk}(M) = f_{rk}(M) \,/\, f_{1k}(M) \, ,$$
$$0 \le k \le [n/2] - 1 \, , \quad 2 \le r \le n - 2k \, .$$

Theorem. *At a simple, generic S,*

$$\{\log \mu_{rk} \, , \, \log \mu_{r'k'}\}_{\ell^\bullet}(S) = 0 \, ,$$
$$\{\lambda_{rk} \, , \, \lambda_{r'k'}\}_{\ell^\bullet}(S) = 0 \, ,$$
$$\{\log \mu_{rk} \, , \, \lambda_{r'k'}\}_{\ell^\bullet}(S) = \delta_{kk'}(\delta_{rr'} - \delta_{1r'})(1 + \delta_{0k})$$
$$0 \le k, k' \le [n/2] - 1 \, ; \, 2 \le r \le n - 2k \, , \, 2 \le r' \le n - 2k' \, . \; \square$$

Notice that the action variables λ_{1k} are superfluous, as the sum I_{1k} of the generalized eigenvalues $\sum_r \lambda_{rk}$ is an orbit invariant. Also, the putative angle variable $\log \mu_{rk}$ is not necessarily a real number and we have to take care in selecting the real and imaginary parts of $\log \mu_{rk}$ (as well as of λ_{rk}!) in order to obtain a set of real action-angle variables (for details, see [DLNT]).

There are also angle variables for the actions I_{rk} ([DLNT]). The idea of taking the last coordinates of eigenvectors to construct angle variables is clearly suggested by Moser's inverse variables for the tridiagonal case. There is an effective inverse algorithm for symmetric matrices too:

Theorem. *Let*

$$\alpha_{rk}\ ,\ \beta_{r'k'}\ ,\quad \begin{array}{l} 0 \le k \le [(n-1)/2]\ , \\ 0 \le k' \le [n/2]-1\ , \\ 1 \le r \le n-2k\ , \\ 1 \le r' \le n-2k'\ , \end{array}$$

be numbers satisfying the properties below.

(i) The α_{rk}'s are distinct and the β_{rk}'s are nonzero

(ii) The α_{r0}'s and β_{r0}'s are real, $\alpha_{10} < \alpha_{20} < \cdots < \alpha_{n0}$ and the β_{10}'s alternate in sign.

(iii) For $1 \le k \le [n/2]-1$, the pairs $(\alpha_{rk}, \beta_{rk})$ come in complex conjugate pairs: if α_{jk} is not real for some j, then there is a unique j' such that $\alpha_{jk} = \overline{\alpha_{j'k}}$ and $\beta_{jk} = \overline{\beta_{j'k}}$. Moreover, if α_{jk} is real, then so is β_{jk}.

Then there are exacty 2^{n-1} real regular symmetric matrices S for which

$$\lambda_{rk}(S) = \alpha_{rk}\ ,\quad \mu_{rk}(S) = \beta_{rk}/\beta_{1k}\ .$$

These matrices are obtained from the set $\{\alpha_{rk}, \beta_{r'k'}\}$ by solving a sequence of linear systems with a quadratic constraint at each step. □

In other words, the inverse procedure is completely explicit, as in the inverse algorithm for Jacobi matrices.

Theorem. *Let $S \in \ell^*$ be generic and simple. Then the connected component of the level set $\cap_{r,k} I_{rk}^{-1}(I_{rk}(S))$ is diffeomorphic to*

$$\mathbf{R}^{[n^2/4]-\Sigma c_k} \times (S^1)^{\Sigma c_k}\ ,$$

where c_k is the number of conjugate pairs in $\{\lambda_{rk}\}$.

The diffeomorphism can be given explicitly ([DLNT]). Notice that in different regions of the same orbit O_S, one can find invariant sets of different topologies. Separatrices consist of matrices for which the action-angle variables above are not defined, and there the numbers c_k may change. A detailed analysis using the inverse variables shows that there are exactly $2^{[(n-1)/2]}$ disjoint copies of the connected level set, all having the same image under the I_{rk}'s.

Summing up, we have described an explicit set of coordinates in an open dense set of ℓ^*, the symmetric matrices, which are constructed from (generalized) eigenvalues and (generalized) eigenvectors. The open question is this: how can these variables be used to provide better algorithms to compute the eigenvalues of general symmetric matrices?

Section 3.

We again enlarge the phase space by considering the differential equation

$$(***) \qquad \dot{M} = [\pi(f(M))\,,\, M]\;,\; M(0) = M_0\;,$$

where now $\pi(M) = ((M^T)_+)^T - (M^T)_+$ and M is a real matrix, with no required symmetry. Clearly for a symmetric initial condition, we obtain the differential equation studied in the previous section. We begin by presenting the symplectic manifold on which $(***)$ will be shown to be completely integrable. Consider the diffeomorphism

$$m\colon O^+(n,\mathbf{R}) \times L^+(n,\mathbf{R}) \to G_{QL} \subset GL^+(n,\mathbf{R})\,,$$
$$(O,L) \mapsto O^{-1}L$$

where G^+ denotes the identity component of a group G. Induce in the set G_{QL} a group structure through m, and let g_{QL} be its Lie algebra. As a set, g_{QL} is $M(n,\mathbf{R})$, the set of all real $n \times n$ matrices. The Lie bracket, however, is not the usual one:

$$[M,N]_{g_{QL}} = [\pi_\ell M, \pi_\ell N] - [\pi_k M, \pi_k N]\,,$$

where $\pi_\ell X = X_- + X_0 + X_+^T$ and $\pi_k X = X_+ - X_+^T$ as before. Identify the dual g_{QL}^* with g_{QL} itself through the pairing $\langle A,B\rangle = \operatorname{tr} A^T B$. Standard computations give the action and coadjoint actions of G_{QL}: for $g = m(g_0^{-1}, g_L) = g_0 g_L$, $\mathrm{Ad}_g(M) = (g_0)^{-1}(\pi_k M)g_0 + g_L(\pi_\ell M)(g_L)^{-1}$ and $\mathrm{Ad}^*_{g^{-1}}(M) = \pi_{\ell\perp}((g_0)^{-1}Mg_0) + \pi_{k\perp}((g_L)^{-T}M(g_L)^T)$, where $\pi_{\ell\perp}X = X_+ - X_-^T$ and $\pi_{k\perp}X = X_- + X_0 + X_-^T$. The fact that G_{QL} is a product is reflected in the equivalent formula

$$\mathrm{Ad}^*_{g^{-1}}(M) = \pi_{\ell\perp}((g_0)^{-1}(\pi_{\ell\perp}M)g_0) + \pi_{k\perp}((g_L)^{-T}(\pi_k M)(g_L)^T)\,.$$

The description of the generic coadjoint orbit follows by considering each term separately, but we only give the final result (details for this section can be found in [DLT]). Define, for $M \in g_{QL}^*$,

$$J(M,h,z) \equiv \det((1-h)M + hM^T - z) \equiv \det M(h,z)$$
$$\equiv \sum_{r=0}^{n}\sum_{k=0}^{[r/2]} J_{rk}(M)\,(h(1-h))^k\, z^{n-r}\,,$$

and let $A(M) = (M - M^T)/2$ be the skew-symmetric part of M.

Proposition. *Let $M \in g_{QL}^*$ satisfy*

(G1) $E_{0k}(M) \neq 0$, $k = 1, \cdots, [n/2]$ ($E_{0k}(M)$ was defined in the previous section).

(G2) The spectrum of $A(M)$ is simple.

Then the coadjoint orbit O_M through M has dimension $n^2 - n$ and is given by

$$O_M = \{N \in M(n,\mathbf{R}): sgn\ E_{0k}(N) = sgn\ E_{0k}(M)\ ,\ 1 \le k \le [n/2]\ ,$$
$$I_{1k}(N) = I_{1k}(M)\ ,\ 1 \le k \le [(n-1)/2]\ ,$$
$$J_{10}(N) = J_{10}(M)\ ,\ J_{2k,k}(N) = J_{2k,k}(M)\ ,\ 1 \le k \le [n/2]\ ,$$

and, if n is even, $Pfaffian(A(N)) = Pfaffian(A(M))\}$. □

The orbits described in the proposition are generic and of maximal dimension. A set of interesting nongeneric orbits is given by the upper Hessenberg matrices — they are matrices with entries below the lower subdiagonal set equal to zero. As in the case of tridiagonal symmetric matrices, the Hessenberg matrices are invariant under the QR step, a well known fact in numerical analysis.

The Lie-Poisson bracket for g^*_{QL} is

$$\{F,H\}_{g^*_{QL}}(M) = \mathrm{tr}(M^T[\nabla F(M), \nabla H(M)]_{g_{QL}})$$

and it induces a symplectic structure on the coadjoint orbits.

Proposition. *The Toda flow*

$$\dot{M} = -[(\pi_\ell M^T)^T, M] = -[(\pi_k M^T), M]$$

is generated by the Hamiltonian $H(M) = \frac{1}{2}\ tr\ M^2$.

It is clear from these formulae that if the initial condition is symmetric, or (strict) upper triangular, then so is the solution. For general initial conditions, however, there is nontrivial coupling.

We now prove the integrability of (***). In the symmetric case, the λ_{rk}'s for $k \neq 0$ were the additional conserved quantities in involution. This is still true for the nonsymmetric orbit, but we need more conserved quantities. Similarly, the angle variables $\log \mu_{rk}$ for $k \neq 0$ are still angle variables for the λ_{rk}'s in the nonsymmetric case. There are some technicalities in the definition of the variables μ_{rk}. A simple matrix (i.e., a matrix satsisfying (G3) below) has nonzero f_{rk}'s, where, as in the symmetric case,

$$f_{rk}(M) = (e_{n-k}\ ,\ ((M - \lambda_{rk})_{k-1})^{-1} e_1)\ ,$$
$$1 \le k \le [n/2] - 1\ ,\quad 1 \le r \le n - 2k\ ,$$

and again we can define the μ_{rk}'s exactly as before.

It turns out that the Toda flow not only preserves the spectrum of $M(t)$ (and of $(M(t))^T$), but preserves the spectrum of any linear combination $a M(t) + b(M(t))^T$! This was the motivation for the definition of the polynomial $J(M,h,z)$ above, whose coefficients J_{rk} are new integrals for the flow. Moreover, inspired by the inverse theory of the periodic tridiagonal symmetric problem ([Van-Moer]), one should look for angle variables in the Riemann surface corresponding to the spectral curve $J(M(t),h,z) = 0$, which is invariant under the Toda flow. We make two additional genericity hypotheses on $M \in g^*$:
(G3) The $\lambda_{rk}(M)$'s are all distinct.
(G4) Zero is a regular value of $J(M,\cdot,\cdot)$.
For the (nonempty) open set of matrices in O_M satisfying (G1)–(G4) we will construct (essentially) the action-angle variables for the Toda flow.

Lemma. *Let M_0 satisfy* (G1)–(G4).

(a) *to a Riemann surface $C(M_0) \subset \mathbf{CP}^2$, of genus $g = (n-1)(n-2)/2$.*

(b) *The 1-forms*

$$\omega_{rk}(M_0) = \frac{(h(1-h))^k z^{n-r}}{J_z(h,z)} dh \ ,$$

for $0 \le k \le [(n-3)/2]$, $2k+3 \le r \le n$, are holomorphic in $C(M_0)$, and, together with the forms $\omega^e_{rk} = (1-2h)\omega_{rk}$, $0 \le k \le [(n-4)/2]$, $2k+4 \le r \le n$, provide a basis for the space of holomorphic 1-forms on $C(M_0)$.

(c) *For $p \in C(M_0)$, dim ker $M(h,z) = 1$ and the map*

$$\begin{aligned} C(M_0) &\longrightarrow \mathbf{CP}^{n-1} \\ p &\mapsto \text{kernel of } M(p) \end{aligned}$$

is holomorphic.

(d) *The points in $C(M_0)$ for which the last coordinate of the eigenvector map vanishes form a divisor $D(M_0)$ of degree $n(n-1)/2$.*

Integrable systems with periodic conditions such as the periodic Toda flow on $n \times n$ tridiagonal periodic matrices and KdV for periodic potentials, are known to be linearized by the Abel-Jacobi map in the following sense. The flow is first shown to preserve a Riemann surface C ('a spectral curve') and then an effective divisor D of degree g on C is displayed with the property that, as D changes according to the flow, its image under the Abel-Jacobi map

$$\begin{aligned} \text{Jac}: C^g &\longrightarrow \mathbf{C}^g/(\text{periodic lattice}) \\ D &\mapsto \text{Jac}(D) \end{aligned}$$

moves along straight lines on the Jacobi variety. In those examples, the spectral curve, together with the divisor, can be taken to be data for an inverse problem, and they end up being the building blocks for the action-angle variables of the original flow. In our case, however, there are too few holomorphic 1-forms and we need some additional variables, which are obtained by considering the set of meromorphic 1-forms

$$\omega^m_j = \frac{z^j}{hJ_z(h,z)} dh \ , \qquad 0 \le j \le n-2 \, .$$

The ω^m_j's are regular at infinity and have poles only at $(h = 0,\ z = \lambda_{i0})$, where $\lambda_{i0} \in$ spec M_0. We now define the extended Jacobian of $C(M_0)$, which is the key object in the description of the missing angle variables. Notice that the coadjoint invariance of the coefficients $J_{2k,k}$ is equivalent to the coadjoint invariance of the spectrum of $A(M)$, the skew-symmetric part of M. Take $\eta \in$ spec $A(M)$, and define $P_\infty \equiv [(1,-2\eta,0)] \in C(M_0)$ (in homogeneous coordinates), $D_\infty \equiv \frac{1}{2}n(n-1)\,P_\infty$. For an open set U of divisors of degree $g+n-1$, set

$$\widetilde{\text{Jac}}: U \longrightarrow \mathbf{C}^{g+n-1}/\Lambda$$

$$D \mapsto (\phi_{rk} = \int_{D_\infty}^{D} \omega_{rk} \ , \int_{D_\infty}^{D} \omega^e_{rk} \ , \nu_j = \int_{D_\infty}^{D} \omega^m_j)^T \ ,$$

where the lattice Λ takes care of the multivaluedness as follows. Fix a basis of $H_1(C(M_0), \mathbf{Z})$ consisting of $2g$ cycles and add to it small simple curves surrounding the n points $[(0, \lambda_{i0}, 1)]$ ($\lambda_{i0} \in$ spec M_0). These points are the poles of the forms ω_j^m, and are distinct for matrices satisfying (G3). Let L be the $(g+n-1) \times (2g+n)$ matrix whose entries are the values of the integrals of ω_{rk}, ω_{rk}^e and ω_j^m over the chosen cycles (different rows (columns) correspond to different forms (cycles)). The lattice Λ is the set of integral linear combinations of the columns of L.

Proposition. *The map $M \mapsto \widetilde{Jac}(D(M))$ is well defined and smooth for matrices M satisfying (G1)–(G4). In particular, for such matrices, $D(M)$ does not intersect the poles of ω_j^m and the lattice Λ generates a discrete additive subgroup of $\mathbf{C}^{g+n-1}$. Matrices satisfsying (G1)–(G4) form an open dense set of full measure in $M(n, \mathbf{R})$ and in each generic orbit.*

Theorem. *For an appropriate ordering of the variables, the table of Poisson brackets is given by the matrix below.*

	λ_{rk}	J_{rk}	J_{r0}	$\log \mu_{rk}$	$\frac{1}{2}\phi_{rk}$	ν_j
λ_{rk}	0	0	0	I	0	0
J_{rk}	0	0	0	0	I	0
J_{r0}	0	0	0	0	0	I
$\log \mu_{rk}$	$-I$	0	0	0	0	0
$\frac{1}{2}\phi_{rk}$	0	$-I$	0	0	*	*
ν_j	0	0	$-I$	0	*	*

The range of indices in each row is as follows. In the first row, $1 \leq k \leq [n/2]-1$, $2 \leq r \leq n-2k$; in the second, $1 \leq r \leq n$, $1 \leq k \leq [r/2]$, $r \neq 2k$; then, $2 \leq r \leq n$; $1 \leq k \leq [n/2]-1$, $2 \leq r \leq n-2k$; $0 \leq k \leq [(n-3)/2]$, $2k+3 \leq r \leq n$; $0 \leq j \leq n-2$.

The generic symmetric orbit can be thought of as a limit of generic orbits — the Riemann surface $C(M_0)$ degenerates into n disjoint complex lines and the angle variables associated with the Toda integrals tr S^k are obtained as a limit of the ν_j's.

On our way to the coordinatization of the Liouville-Arnold invariant set $\mathcal{J}_{M_0} \subset O_{M_0}$ defined as the intersection of the M_0-level surfaces of the action variables and coadjoint invariants λ_{rk} and J_{rk}, we first observe that all matrices in $\mathcal{J}_{M_0}$ satisfy (G1)–(G4), and so the action-angle variables are well defined in a thickening of $\mathcal{J}_{M_0}$. By considering real and imaginary parts of the action-angle variables displayed in the table of Poisson brackets, one obtains a set of variables which parametrizes bijectively the set $\mathcal{J}_{M_0}^c$, the connected component of $\mathcal{J}_{M_0}$ containing M_0. This is proved by solving explicitly (in terms of theta functions) an inverse problem containing all the ingredients already present in the symmetric case and all those present in the standard (tridiagonal) periodic inverse problems, together with an additional feature: one has to invert the extended Jacobian map (taking values on a noncompact set!) in the proper way. The details (and there are many) can again be found in [DLT]. Analysis of the extended period matrix then provides the correct counting of lines and circles in $\mathcal{J}_{M_0}^c$.

Theorem. *$\mathcal{J}^c_{M_0}$ is diffeomorphic to $\mathbf{R}^\ell \times (S^1)^c$, where $c = [(n-1)^2/4] + n_c$, $\ell = n(n-1)/2 - c$ and n_c is the number of complex conjugate pairs in the set $\{\lambda_{rk}, 0 \le k \le [n/2]-1, 1 \le r \le n-2k\}$.* $\square$

There are solutions for the λ_{rk}-flows given by factorization with formulae analogous to those described for the same Hamiltonians in the symmetric case. The J_{rk} flows are harder to solve and require a factorization over a loop group. This procedure is best described by making use of another interpretation of the bracket on G_{QL} through the so-called R-matrix theory ([STS]). Let g be a Lie algebra with bracket $[\cdot,\cdot]$. A linear operator R: $g \to g$ is a (classical) R-matrix if the bilinear operation

$$[X,Y]_R = \frac{1}{2}([RX,Y] + [X,RY])$$

is also a Lie bracket, i.e., $[\cdot,\cdot]_R$ satisfies the Jacobi identity. Denote by g_R the algebra g equipped with the bracket $[\cdot,\cdot]_R$. There are now two Lie-Poisson brackets on $C^\infty(g^*)$:

$$\{F_1,F_2\}(\alpha) = \alpha([X_1,X_2])$$

and

$$\{F_1,F_2\}_R(\alpha) = \alpha([X_1,X_2]_R)\ ,$$

for $\alpha \in g^*$, $X_i = dF_i(\alpha) \in g$. It is easy to see that $[\cdot,\cdot]_{g_{QL}}$ above is the bracket induced by $R = \pi_k - \pi_\ell$. Let $g_\pm \equiv (R \pm I)g$. In our case, $g_+ = \pi_k g$, $g_- = \pi_\ell g$ and we have a direct sum decomposition of g into Lie subalgebras,

$$g = g_+ \oplus g_-\ ,$$

which yields a group factorization

$$G = G_+ \cdot G_-\ .$$

A sufficient condition that $[\cdot,\cdot]_R$ is a Lie bracket, is that R satisfies the so-called modified Yang-Baxter equation (see [STS]). For such R the subspaces $g_\pm$ are always subalgebras as in the case $R = \pi_k - \pi_\ell$ above, and we have the following result.

Lemma. (Solution by factorization; R-matrix version). *Let F: $g^* \to \mathbf{C}$ be ad^*-invariant with respect to the basic bracket on g:*

$$\alpha([dF(\alpha)\ ,X]) = 0\ , \quad \textit{for all} \quad \alpha \in g^*\ , X \in g\ .$$

Then the equation of motion on g^* generated by F under the new bracket,

$$\dot\alpha = \{F,\alpha\}_R = \frac{1}{2}(ad^* R(dF(\alpha)))(\alpha)\ ,$$

is solved by

$$\alpha(t) = (Ad^*_{g+(t)})(\alpha(0)) = (Ad^*_{g^*_-(t)})(\alpha(0))$$

where g_+ and g_- are given by the factorization

$$\exp(t\,dF(\alpha(0))) = g_+(t)\,g_-(t)\ , \qquad g_\pm \in G_\pm$$

and the Ad^* *action is that of the group* G.

The solution by factorization of the λ_{rk}-flows, for example, can be rederived as an application of the above lemma. We will now use it to provide the expression for the J_{rk}-flows. As in the more familiar case of the periodic tridiagonal Toda flow, the equation of interest will be considered as one particular case in a parameterized set of equations, which will be solved simultaneously by factorization on a substantially larger (dual) algebra. Let Δ denote the ring of rational functions with poles only at 0,1 and ∞, and let $g\ell_n(\Delta)$ be the ring of $n \times n$ matrix functions $\{X(\cdot)\}$ with entries in Δ, with the natural Lie bracket

$$[X,Y](h) \equiv [X(h),Y(h)] = X(h)Y(h) - Y(h)X(h)\ , \quad h \in \mathbf{C}\setminus\{0,1\}\ .$$

Let Δ_- be the subring of elements vanishing at ∞, and consider the direct sum decomposition into subalgebras

$$g\ell_n(\Delta) = \tilde{\ell} \oplus \tilde{k}\ ,$$

where

$$\tilde{\ell} = \{\sum_{j=0}^{N} A_j h^j \in g\ell(\mathbf{C}[h]) \colon A_0^T \text{ is lower triangular}\}\ ,$$

$$\tilde{k} = \{A(h) \in g\ell(\Delta_- \oplus \mathbf{C}) \colon A(\infty) \text{ is skew-adjoint}\}\ .$$

Let $\pi_{\tilde{\ell}}$ and $\pi_{\tilde{k}}$ be the associated projections and define the R-matrix

$$\widetilde{R} = \pi_{\tilde{k}} - \pi_{\tilde{\ell}}\ .$$

One can check that $\widetilde{R}$ (satisfies the modified Yang-Baxter equation) and indeed gives rise to a Lie bracket with $\mathrm{Ran}(\widetilde{R}+I) = \mathrm{Ran}\ \pi_{\tilde{k}}$, $\mathrm{Ran}(\widetilde{R}-I) = \mathrm{Ran}\ \pi_{\tilde{\ell}}$. The algebra $g\ell_n(\Delta)$ carries a nondegenerate, ad-invariant pairing

$$\langle X,Y\rangle = \oint_{|h|=2} \mathrm{tr}(X(h)\,Y(h)\,)\frac{dh}{2\pi i h}\ ,$$

and we identify (a subset of) $g\ell_n^*(\Delta)$ with $g\ell_n(\Delta)$ through the pairing.

Theorem. *The* J_{rk}*-flow in* $g\ell_n(\mathbf{R})$

$$\dot{M} = [M,\ (\pi_\ell \nabla J_{rk}(M)\,)^T]\ , \qquad M(0) = M_0\ ,$$

lifts to the flow

$$\dot{X} = [X,\ \pi_{\tilde{\ell}}(h(1-h)\,)^{-k}(1-2h)\,\nabla E_r^T(X(h)\,)\,]\ ,$$

$$X(0) = M_h(0) = M_0 + h(M_0^T - M_0)\ ,$$

on $g\ell_n(\Delta)$*, where the* E_r*'s are the usual symmetric functions of the spectrum,*

$$det(A - z) = \sum_{r=0}^{n} E_r(A) z^{n-r} .$$

The lifted flow is induced by the R-matrix bracket through the Hamiltonian

$$F(Y) = \oint_{|h|=2} (h(1-h))^{-k} (2h-1) E_r(Y(h)) \frac{dh}{2\pi i h} , \quad 0 \leq k \leq [r/2] ,$$

and can be solved by factorization:

$$\begin{aligned} M_h(t) &= (Ad^*_{g_+(t)}) M_h(0) = g_+^{-1} M_h(0) g_+ , \\ &= (Ad^*_{g_-(t)}) M_h(0) = g_-^{-1} M_h(0) g_- , \end{aligned}$$

where

$$\exp(t\, dF(M_h(0))) = g_+(h,t)\, g_-(h,t) ,$$

with $g_\pm$ are in the subgroups of G corresponding to the subalgebras $\tilde{\ell}$ and $\tilde{k}$. In particular, the evaluation at $h = 0$ provides the solution of the J_{rk}-flow on $g\ell_n(\mathbf{R})$.

The proof of the theorem is in [DL], where the factorization is explicitly presented in terms of theta functions over the spectral curve.

Again, the QR step for general matrices admits an interpolation by a completely integrable Hamiltonian flow induced by a differential equation of the form $(***)$. Furthermore, there are differential equations analogous to $(***)$ which are completely integrable, have an equally explicit set of action-angle variables and provide interpolations for LU and Cholesky-type algorithms on nonsymmetric matrices. We again refer the reader to [DLT].

Section 4.

In this section, we make use of our knowledge of the symplectic structure underlying a particular numerical algorithm to obtain a detailed understanding of its remarkable and rather unexpected stability properties.

The singular value decomposition (SVD) of a real $n \times n$ matrix A is the factorization $A = U \Sigma V^T$, where U and V are orthogonal matrices, $\Sigma = \text{diag}\{\sigma_1, \cdots, \sigma_n\}$ and $\sigma_1 \geq \sigma_2 \geq \cdots \geq \sigma_n \geq 0$. The σ_i's are the singular values of A, the columns v_j of V the right singular vectors of A, and the columns u_i of U the left singular vectors of A. We will consider the SVD of the bidiagonal matrix B,

$$B = \begin{pmatrix} b_1 & a_1 & & & O \\ & \ddots & \ddots & & \\ & & \ddots & \ddots & \\ & & & & a_{n-1} \\ O & & & & b_n \end{pmatrix} ,$$

where we assume without loss that the a_i and b_i are positive. Such matrices arise as the final stage in the computation of the SVD of a general matrix A ([GK], [GvL]), as well as in computing the eigendecomposition of a symmetric positive definite tridiagonal matrix T ([BD]).

How accurately can the SVD of a general matrix A be computed? By regular perturbation theory,

$$|\sigma_i - \sigma_i'| \le \eta \|A\| ,$$

for singular values of matrices A and $A + \delta A$ respectively, with $\|\delta A\|/\|A\| \le \eta$. This bound is useful for large singular values σ_j, $\sigma_j \sim \sigma_1 = \|A\|$, but for $\sigma_j \ll \sigma_1$, the bound may be far from realistic. For a bidiagonal matrix B, introduce the *relative error* for δB,

$$\eta_r \equiv (2n-1) \max_{i,j} |\log(\frac{B_{ij} + \delta B_{ij}}{B_{ij}})| .$$

A perturbation argument now shows that

$$e^{-\eta_r} \le \frac{\sigma_i'}{\sigma_i} \le e^{\eta_r} ,$$

for the singular values σ_i, σ_i' of B and $B + \delta B$. When $\eta_r \ll 1$, this implies

$$|\sigma_i' - \sigma_i| \le \eta_r \sigma_i .$$

Thus small relative perturbations in the entries of B only cause small relative perturbations in the σ_i, independent of their magnitudes. The new bound is always at least about as strong as the previous one, and is substantially better for estimating the effect of small relative perturbations on small singular values. Using η_r in place of η, there are similar improvements in the error bounds for singular vectors (the reader is referred to [DDLT] for additional information on this section).

Given the much greater accuracy to which singular values and vectors of bidiagonal matrices are determined by the data, it is desirable to have an algorithm which computes these quantities to their inherent accuracy. The desired algorithm, a modification of the basic algorithm in [GK], was introduced in [DK] and shown in [DK] and [DDLT] to be optimally accurate in the following sense: if the original bidiagonal matrix B contains a small relative error, then the errors introduced by the algorithm are essentially of the same size.

The main step in the algorithm is very similar to the QR step. Let B_0 be our initial bidiagonal matrix. Given B_i, compute the QR decompositions $B_i B_i^T = Q_1 R_1$ and $B_i^T B_i = Q_2 R_2$. Then let $B_{i+1} = Q_1^T B_i Q_2$. Then B_i is always bidiagonal and converges as $i \to \infty$ to a diagonal matrix with the (common) singular values of $\{B_i\}$ on the diagonal.

We think of the SVD step as a mapping from $(\mathbf{R}^+)^{2n-1}$ (the entries of B_i) to $(\mathbf{R}^+)^{2n-1}$ (the entries of B_{i+1}). To understand how errors propagate, it is natural to look at the Jacobian of this map. However, since we are interested in the propagation of relative errors, we follow [DK] and look at the Jacobian of the map F taking the logarithms of the entries of B_i to the logarithm of the entries of B_{i+1}. Let $M(j,i)$, for $j > i$ be the Jacobian of $F^{(j-i)}$, the application of $(j-i)$ SVD-steps, starting from B_i. By the chain rule, $M(j,i) = M(j,j-1)\cdots M(i+1,i)$.

The following four facts were observed by J. Demmel and W. Kahan during initial numerical experiments.

Fact 1: The eigenvalues of $M(j,i)$ appear in reciprocal pairs. In other words, if λ is an eigenvalue, so is $1/\lambda$.

Fact 2: For large enough i, the eigenvalues of $M(i+1,i)$ are simple and lie on the unit circle.

Fact 3: As $i \to \infty$, $M(i+1,i)$ converges to the constant matrix

$$M_\infty = \begin{pmatrix} I_{n-1} & \Gamma_n \\ 0 & I_n \end{pmatrix}, \quad \text{where } \Gamma_n = \begin{pmatrix} -2 & 2 & & & \\ & -2 & 2 & & \\ & & \cdot & \cdot & \\ & O & & \cdot & \cdot \\ & & & -2 & 2 \end{pmatrix}.$$

Fact 4: $\|M(j,i)\|$ grows linearly in $(j-i)$. More precisely, Demmel observed numerically that, for a large class of problems,

$$\|M(j,i)\|_\infty \leq 5.06n\,(j-i)\,.$$

This is the essential property of roundoff error propagation which makes it possible to prove that the algorithm computes singular vectors to the desired accuracy. The linear growth is consistent with the asymptotic behavior of $M(i+1,i)$.

Rather surprisingly, the relevant symplectic structure for this algorithm is not related in a simple way to the one used in the study of the QR step on tridiagonal matrices (but see [LP]).

Theorem. *The set B_Δ of bidiagonal matrices with positive entries b_p, a_p and fixed determinant $\Delta = \prod_{p=1}^{n} b_p$ is a $(2n-2)$-dimensional symplectic leaf for the Sklyanin bracket*

$$\{\phi,\psi\}_{SKL}(A) \equiv (R(D\phi(A))\,,\,D\psi(A)) - (R(D'\phi(A)\,,\,D'\psi(A))\,,$$

where $(A,B) = \operatorname{tr} AB$, $R(A) = A_+ - A_-$, $D'\phi(A) = (\nabla\phi(A))^T A$ and $D\phi(A) = A(\nabla\phi(A))^T$. Moreover, B_Δ has a (global) Darboux chart given by the variables

$$x_i \equiv \log a_i\,, \qquad 1 \leq i \leq n-1\,,$$

$$y_i = \log \prod_{j=1}^{i} b_j\,, \quad 1 \leq i \leq n-1\,,$$

$$\{x_i,x_j\}_R = 0\,,\ \{y_i,y_j\}_R = 0\,,\ \{x_i,y_j\}_R = \delta_{ij}\,.$$

□

Theorem. *The equation*

$$\frac{d}{dt}\phi(A(t)) = \{\phi, H_{SVD}\}_{SKL}(A(t)) \ , \ A(t=0) = A_0 \ ,$$

generated by the Hamiltonian

$$H_{SVD}(A) = -\frac{1}{4}\ tr\ (\log A^T A)^2 \ ,$$

is equivalent to

$$\frac{dA}{dt} = [A,\ \pi_0(\log A^T A)] \ , \quad A(t=0) = A_0 \ ,$$
$$\textit{where} \ \ \pi_0(A) = A_- - A_-^T \ .$$

Moreover, the time one map of the flow induced by this equation is the SVD step. In other words, $A(k) = A_k$, $k = 0, 1, 2, \ldots$, *where* $A_0, A_1, A_2, \ldots$ *are the iterates of the SVD step.*

The equation which interpolates the SVD step is due to Chu ([Chu]), and the bracket is a special case of a construction of Sklyanin ([STS]).

From the theorem, we must have that the SVD step in (x, y) variables is a symplectic map, and hence its Jacobian is a symplectic matrix. In particular, its spectrum is closed under taking inverses and from simple additional algebra we obtain Fact 1. Fact 4 follows by studying the asymptotic behavior of the Jacobian, once the asymptotics $b_i(t) = \sigma_i + o(1)$, $a_i(t) \sim a_i^\infty (\sigma_{i+1}/\sigma_i)^{2t}$, as $t \to \infty$ (where $a_i^\infty > 0$) are established: the upshot is that

$$\|M(t,0)\|_\infty \le (8n-4)t + O(1) \ , \qquad \text{as} \quad t \to \infty \ ,$$

in good agreement with the bound obtained experimentally. Facts 2 and 3 follow from detailed estimates for the eigenvalues of the Jacobian $M(i+1, i)$ and a perturbation argument for symplectic matrices essentially due to Krein. Actually, more can be proved: the eigenvalues can enter the unit circle only at the points $\pm$ 1. If one could rule out the possibility of entering at $+1$, this would imply that the spectrum of the Jacobian lies in the unit circle at a very early stage of the iteration, which provides an additional theoretical argument for the remarkable stability properties of the algorithm.

Acknowledgements.

The research of the first author was supported in part by NSF Grant # DMS–9001857. The third author acknowledges the support of CNPq, Brazil.

References

[Adl] M. Adler, On a trace functional for formal pseudo-differential operators and the symplectic structure of the Korteweg-deVries type equations, Inventiones Math. 50 (1979) 219–248.

[Arn] V. I. Arnold, Mathematical methods of classical mechanics, GTM 60, Springer-Verlag, New York, 1978.
[B1] S. Batterson, Convergence of the shifted QR algorithm in 3×3 normal matrices, Numer. Math. 58 (1990) 341–352.
[B2] S. Batterson, Convergence of the Francis shifted QR algorithm on normal matrices, Emory University preprint, 1991.
[BDa] S. Batterson and D. Day, Linear convergence in the shifted QR algorithm, Emory University preprint, 1991.
[BD] J. Barlow, J. Demmel, Computing accurate eigensystems of scaled diagonally dominant matrices, SIAM J. Num. Anal. 27 (1990) 762–791.
[Bes] A. Besse, Einstein manifolds, Ergebrisse der Mathematik undeŕ ihrer Grenzgebiete, 3, Folge, Springer-Verlag, 1987.
[BFR] A. M. Bloch, H. Flaschka, T. Ratiu, A convexity theorem for isospectral manifolds of Jacobi matrices in a compact Lie algebra, Duke Math. J. 61(1) (1990) 41–65.
[B-G,B,M] A. Bunse-Gerstner, R. Byers and V. Mehrmann, A chart of numerical methods for structured eigenvalue problems, preprint.
[BS] S. Batterson and J. Smillie, The dynamics of Rayleigh quotient iteration, SIAM J. Num. Anal. 26 (1989) 624–636.
[Chu] M. Chu, A differential equation approach to the singular value decomposition of bidiagonal matrices, Lin. Alg. Appl. 80 (1986) 71–80.
[Dav] M. W. Davis, Some aspherical manifolds, Duke Math. J. 55 (1987) 105–139.
[DDLT] P. A. Deift, J. Demmel, L. C. Li, C. Tomei, The bidiagonal singular value decomposition and Hamiltonian mechanics, Argonne National Laboratory, Math. and Comp. Sci. Div., TM 133 (1989) 1–66. To appear in SIAM J. Num. Anal.
[DL] P. A. Deift, L. C. Li, Generalized affine Lie algebras and the solution of a class of flows associated with the QR eigenvalue algorithm, Comm. Pure Appl. Math. 42 (1989) 963–991.
[DLNT] P. A. Deift, L. C. Li, T. Nanda, C. Tomei, The Toda flow on a generic orbit is integrable, Comm. Pure Appl. Math. 39 (1986) 183–232.
[DLT] P. A. Deift, L. C. Li, C. Tomei, Matrix factorizations and integrable systems, Comm. Pure Appl. Math. 42 (1989) 443–521.
[DNT] P. A. Deift, T. Nanda, C. Tomei, Differential equations for the symmetric eigenvalue problem, SIAM J. Num. Anal. 20 (1983) 1–22.
[DRTW] P. A. Deift, S. Rivera, C. Tomei, D. Watkins, A monotonicity property for Toda-type flows, preprint, SIAM J. Matrix Anal. Appl. 12 (1991) 463–468.
[DK] J. Demmel, W. Kahan, Accurate singular values of bidiagonal matrices, SIAM J. Sci. Stat. Comput. 11 (1990) 873–912.
[Fla] H. Flaschka, The Toda lattice, I, Phys. Rev. B 9 (1974) 1924–1925.
[Fried] D. Fried, The cohomology of an isospectral flow, Proc. Amer. Math. Soc. 98(2) (1986) 363–368.
[GK] E. Golub, W. Kahan, Calculating the singular values and pseudo-inverse of a matrix, SIAM J. Num. Anal. (Series B), 2(2) (1965) 205–224.
[GvL] E. Golub, C. Van Loan, Matrix Computations, Johns Hopkins University Press, Baltimore, MD, 1983.
[Kir] A. A. Kirillov, Elements of the Theory of Representations, Grundlehren 220, Springer-Verlag, New York, 1976.
[Knu] D. E. Knuth, Sorting and searching, the art of computer programming, Vol. 3, Addison-Wesley, New York, 1973.

[Kos1] B. Kostant, Quantization and representation theory, in Proc. Research Symposium on representations of Lie groups, Oxford, 1977, London Math. Soc. Lecture Notes Series 34 (1979).

[Kos2] B. Kostant, The solution to a generalized Toda lattice and representation theory, Advances in Math. 34 (1979) 139–338.

[Lag] J. C. Lagarias, Monotonicity properties of the generalized Toda flow and QR flow, SIAM J. Matrix Anal. Appl. 12 (1991) 449–462.

[Lax] P. D. Lax, Integrals of nonlinear equations of evolution and solitary waves, Comm. Pure Appl. Math. 21 (1968) 467–490.

[LP] L. C. Li and S. Parmentier, Nonlinear Poisson structures and r-matrices, Comm. Math. Phys. 125 (1989) 545–563.

[Man] S. V. Manakov, Complete integrability and stochastization of discrete dynamical systems, Sov. Phys. JETP 40 (1974) 269–274.

[Mos] J. Moser, Finitely many mass points on the line under the influence of an exponential potential — an integrable system, Lecture Notes in Physics 38, J. Moser, ed., Springer-Verlag, New York, 1975, 467–497.

[OP] M. A. Olshanetsky and A. M. Perelomov, Explicit solutions of classical generalized Toda models, Inv. Math. 54 (1979) 261–269.

[Par1] B. Parlett, The symmetric eigenvalue problem, Prentice-Hall, Englewood Cliffs, NJ, 1980.

[Par2] B. Parlett, Global convergence of the basic QR algorithm on Hessenberg matrices, Math. of Comp. 22(104) (1968) 803–817.

[PT] J. Pöschel and E. Trubowitz, Inverse Spectral Theory, Academic Press, New York, 1987.

[Rey-STS1] A. G. Reyman and M. A. Semenov-Tian-Shansky, Reduction of Hamiltonian systems, affine Lie algebras and Lax equations, Inv. Math. 54 (1979) 81–100.

[Rey-STS2] A. G. Reyman and M. A. Semenov-Tian-Shansky, Reduction of Hamiltonian systems, affine Lie algebras and Lax equations, II, Inv. Math. 63 (1981) 423–432.

[Rut] H. Rutishauser, Solution of eigenvalue problems with the LR-transformation, National Bureau of Standards Applied Math. Series 49 (1958) 47–81.

[STS] M. A. Semenov-Tyan-Shansky, What is a classical R-matrix? Func. Anal. Appl. (1984) 259–272.

[Sym1] W. W. Symes, Hamiltonian group actions and integrable systems, Physica 1D (1980) 339–374.

[Sym2] W. W. Symes, The QR algorithm and scattering for the finite nonperiodic Toda lattice, Physica 4D (1982) 275–280.

[Tom] C. Tomei, The topology of isospectral manifolds of tridiagonal matrices, Duke Math. J. 51 (1984) 981–996.

[Van-Moer] P. van Moerbeke, The spectrum of Jacobi matrices, Inv. Math. 37 (1976) 45–81.

[Wat] D. S. Watkins, Isospectral flows, SIAM Review, 26 (1984) 379–391.

[WE1] D. S. Watkins and L. Elsner, Selfsimilar Flows, Lin. Alg. Appl. 110 (1988) 213–242.

[WE2] D. S. Watkins and L. Elsner, On Rutishauser's approach to self-similar flows, SIAM J. Matrix Anal. Appl. 11(2) (1990) 301–311.

Whiskered Tori for NLS Equations

*D.W. McLaughlin**

Department of Mathematics, Princeton University, Princeton, New Jersey 08540, USA

Spectral theory is used to display instabilities and hyperbolic structure for certain periodic soliton equations, as well as to generate representations of whiskered tori for these equations. The NLS equation is discussed as a primary model, with references to other equations which possess a similar hyperbolic structure. This chapter in the theory of integrable soliton equations is described in the terminology of dynamical systems theory, in anticipation of its future use in the study of near integrable perturbations.

1 Introduction

Long wavelength instabilities are known to be related to the formation of spatially localized coherent structures, as well as to chaotic behavior of such structures as they respond to each other and to external forces. Such chaotic behavior of coherent structures occurs throughout physics – for example, in fluid mechanics, plasma physics, laser optics, quantum chemistry – and on many spatial scales – from atomic and molecular, through microscopic and atmospheric, to astrophysical. Realistic physical systems are usually modeled with complicated equations which make any mathematical study of chaotic behavior with them extremely difficult. Simplified problems, such as one dimensional nonlinear wave equations, provide model pde's whose solutions display some of the chaotic behavior of realistic physical systems, yet for which detailed mathematical studies are currently feasible. These features and advantages of model wave equations are particularly valid for near-integrable soliton equations.

During the past decade, numerical experiments have carefully documented the existence and properties of spatially coherent, yet temporarily chaotic, behavior in near-integrable pde's. Simultaneously, theoretical developments in soliton mathematics have identified and represented "hyperbolic structure" in the integrable pde, which has then been used to help design, organize, interpret, and understand chaotic behavior in near-integrable numerical experiments.

*Funded in part by AFOSR-90-0161 and NSF DMS 8922717 A01.

In particular, during the 1980's the periodic spectral transform was used to identify linearized instabilities in integrable soliton systems. More importantly, this transform was used to construct representations of global objects such as homoclinic orbits and whiskered tori [8] [10]. By 1986, this spectral theory had been used to measure directly from numerical data the presence of hyperbolic structures in perturbed time series[4]. By the end of the decade, studies were beginning which used these integrable representations in geometric perturbation descriptions of near integrable systems [18] [25] [5]. Such analytical studies of near-integrable pde's are a subject for the 1990's.

In this article we will describe explicit representations of "whiskered tori" for periodic soliton equations. In addition, we will indicate potential uses for these representations in the investigation of perturbations of soliton equations. While the representations themselves are constructed with periodic spectral theory, we will attempt to describe them in the language of dynamical systems theory in the hope that the methods of that theory will be useful when studying perturbations.

Soliton pde's are infinite dimensional Hamilitonian systems with an infinite number of commuting constants of the motion. The common level sets of these constants of motion are generically infinite dimensional tori of maximal dimension (one half the dimension of the phase space in a very precise sense). Typically, these constants of the motion have linearly independent gradients; however, at certain *critical tori* these gradients become linearly dependent. These critical tori have dimension which is lower than maximal, but otherwise arbitrary. They can be either stable or unstable. (In the unstable case, the constants of motion have a saddle structure in a neighborhood of the critical torus.) An unstable critical torus has unstable manifold of dimension N + M, where M denotes the (possibly infinite) dimension of the torus itself and N will be shown to be finite. A phase point on this unstable (= stable) manifold approaches the critical torus as $t \longrightarrow \pm\infty$. Such a phase point is said to lie on an orbit which is *homoclinic to the critical torus.* These unstable manifolds, with their boundary tori, are the "whiskered tori" of the title.

These whiskered tori are sources of sensitivity which can produce chaotic reponses when an integrable equation is perturbed. In [26] we summarize numerical experiments on damped driven perturbations of integrable systems which possess chaotic behavior which, on the one hand, display an irregular jumping of the coherent excitation between distinct spatial locations [Figure 1], and, on the other hand, can be correlated numerically to the presence and continual crossings of critical level sets in the nearby integrable system. The presence of "whiskered tori" in an integrable pde has also been shown to generate instabilities in conservative difference schemes [2] [1] [30] [27]. In finite dimensional Hamilitonian systems, these whiskers provide paths for "Arnold diffusion" [3]. Recently, the presence of whiskered

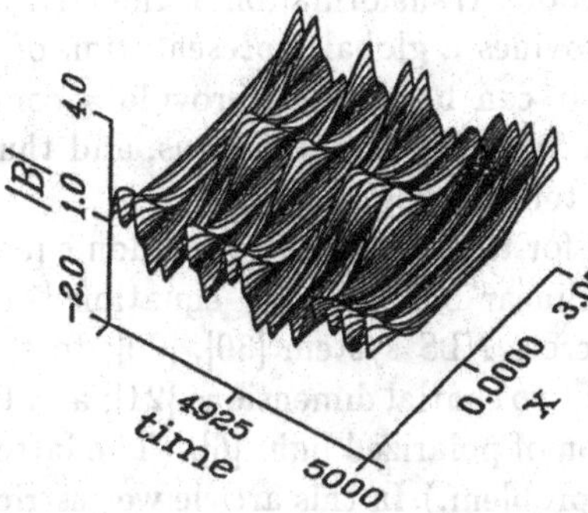

Figure 1. A solution surface for one sample damped-driven NLS experiment: At these parameter values, the solution is temporally chaotic, yet spatially coherent. Note the jumping of the localized coherent excitation between center and edge locations.

tori has been correlated with striking observations of enhanced "diffusion" in conservative perturbations of the sine-Gordon equation [14].

Limitations of space prohibit us from describing in more detail those numerical experiments which correlate the presence of hyperbolic structure in integrable pde's with chaotic response under near integrable perturbations. Such experiments are discussed, some in considerable detail, in reference [26]. In the present article we will focus upon analytical representations of the whiskered tori for the periodic soliton equations, attempting to emphasize a dynamical systems setting for the integrable pde and to indicate its possible use under perturbations. While most of this material may also be found in the long article [26], we hope that a concise description of whiskered tori for soliton equations, together with an emphasis upon a dynamical systems description *for the pde* will be useful.

It is rare to have explicit representations of whiskered tori in natural settings. Periodic soliton equations provide a large family of natural examples for finite discrete systems of arbitrary dimension, as well as for pde's in both one and two spatial dimensions. For these periodic soliton equations, inverse spectral theory permits very explicit representations of their whiskered tori. (In fact, the representations of the whiskers are somewhat more explicit that the representations of the tori themselves.) The procedure for constructing these representation is as follows: *First*, one identifies and represents the critical tori. If they are finite dimensional, this may be done in terms of theta functions of several complex variables. If the critical tori are infinite dimensional, less is known – although, in principle, they could be represented by infinite genus theta functions. *Second*, given a critical torus, one investigates its linearized stability with a "squared-eigenfunction" basis. This procedure provides a represention of the tangent spaces for the stable and unstable manifolds at the critical torus. *Third*, one then constructs the global whisker by a Bäcklund or Darboux transformation of the critical torus. This Bäcklund transformation provides a global representation of a homoclinic orbit to the critical torus. It can be used to provide a complete representation of the unstable manifold of the critical torus, and thus a global representation of the whiskered torus.

We first carried out this construction for the periodic sine-Gordon equation [10] and later for the periodic nonlinear Schroedinger equation [24]. Recently, it has been extended to a discrete NLS system [30], [20]; to the well posed Davey-Stewartson equation in two spatial dimensions [21]; and to a coupled NLS system for the propagation of polarized light [6]. (The latter is integrated with a third order spectral problem.) In this article we restrict our description to the simplist case of nonlinear Schroedinger equation under even, periodic boundary conditions.

The construction is carried out with inverse spectral theory. Some background spectral material is summarized in Section 2. The key terms are Floquet discriminent; "action" information such as periodic points, critical and double points; "angle" information such as Dirichlet eigenvalues; counting lemmas and the control of the integer "N". In Section 3, we use the Floquet discriminant as a Morse function to identify critical tori; and then we use a Bäcklund transformation to construct their unstable manifolds. In Section 4, we explicitly set up the construction for a simple example of a whiskered circle. In Section 5, we extend this example to a whiskered torus of arbitrary dimension. Our description will be in the terminology of dynamical systems theory. In that final Section 5, we will briefly mention potential uses of this material for perturbations.

2 Spectral Preliminaries

Consider the NLS equation

$$2\,i\,q_\tau = q_{yy} + \frac{1}{2}\,qq^*q, \tag{2.1}$$

under even, periodic boundary conditions

$$\begin{aligned} q(y+\ell,\tau) &= q(y,\tau), \\ q(-y,\tau) &= q(y,\tau), \end{aligned}$$

as a Hamiltonian system on the phase space $\mathcal{F} = H^1$ of even square integrable functions with a square integrable derivative:

$$\begin{aligned} -i\,q_\tau &= \frac{\delta H}{\delta\, q^*}, \\ H &\equiv \frac{1}{2}\int_0^\ell [q_y q_y^* - \frac{1}{4}(qq^*)^2]dy. \end{aligned} \tag{2.2}$$

This is an integrable Hamiltonian system which can be integrated with the aid of a linear operator $\hat{L}$ for which NLS is an isospectral flow:

$$\hat{L} \equiv -i\sigma_3 \frac{d}{dy} + \frac{1}{2}\begin{pmatrix} o & q \\ -q^* & 0 \end{pmatrix}. \tag{2.3}$$

Here σ_3 denotes the third Pauli matrix $\equiv$ diag(1,-1). Viewing $\hat{L}$ as a differential operator in $L^2(\mathbb{R})$ (with dense domain H^1), we consider the eigenvalue problem

$$\hat{L}\,\vec{\psi} = \zeta\,\vec{\psi}. \tag{2.4}$$

The spectrum of the differential operator$\hat{L}$, $\sigma(\hat{L})$, is defined as the closure of the set of complex ζ for which there exists an eigenfunction for equation(2.4) which is bounded for all $y \in \mathbb{R}$. Since the coefficients $q(y)$ and $q^*(y)$ of $\hat{L}$ are periodic functions of y, this spectrum $\sigma(\hat{L})$ can be studied with Floquet theory.

2.1 The Floquet Discriminant and the Spectrum

Floquet theory begins from the fundamental matrix $M = M(y;\zeta;q)$, defined by the initial value problem

$$(\hat{L} - \zeta)\ M = 0, \tag{2.5}$$

$$M|_{y=0} = \begin{pmatrix} 1 & 0 \\ 0 & 1 \end{pmatrix}.$$

Through M one defines the "Floquet discriminant" $\Delta(\zeta;q)$:

$$\Delta: \boldsymbol{C} \times \mathcal{F} \to \boldsymbol{C} \tag{2.6}$$

$$\Delta(\zeta;q) = tr\ M(\ell;\zeta;q).$$

This scalar functional $\Delta = \Delta(\zeta;q)$ is central to the theory; in fact, the complete integration of NLS is accomplished through an interplay between the ζ and the q dependence of the Floquet discriminant Δ. This function $\Delta(\zeta;q)$ is entire in both $\zeta \in \boldsymbol{C}$ and (u,v), $u + iv \equiv q \in \mathcal{F}$. On the one hand, for every value of ζ, $\Delta(\zeta,q)$ is a constant of the motion for the NLS flow since it Poisson commutes with the NLS Hamiltonian; thus, Δ generates an infinite set of NLS invariants (one for each ζ). On the other hand, for each q, the spectrum $\sigma(\hat{L})$ is characterized in terms of the ζ dependence of the Floquet discriminant by

$$\sigma(\hat{L}) = \left\{\zeta \in \boldsymbol{C} \,\middle|\, \Delta(\zeta) \text{ is real and } -2 \leq \Delta \leq 2\right\}. \tag{2.7}$$

From this characterization (2.7), one sees that the spectrum is continuous, and consists in closed curves in the complex ζ plane which themselves are subsets of the curves of real Δ. These curves of real Δ can be used to

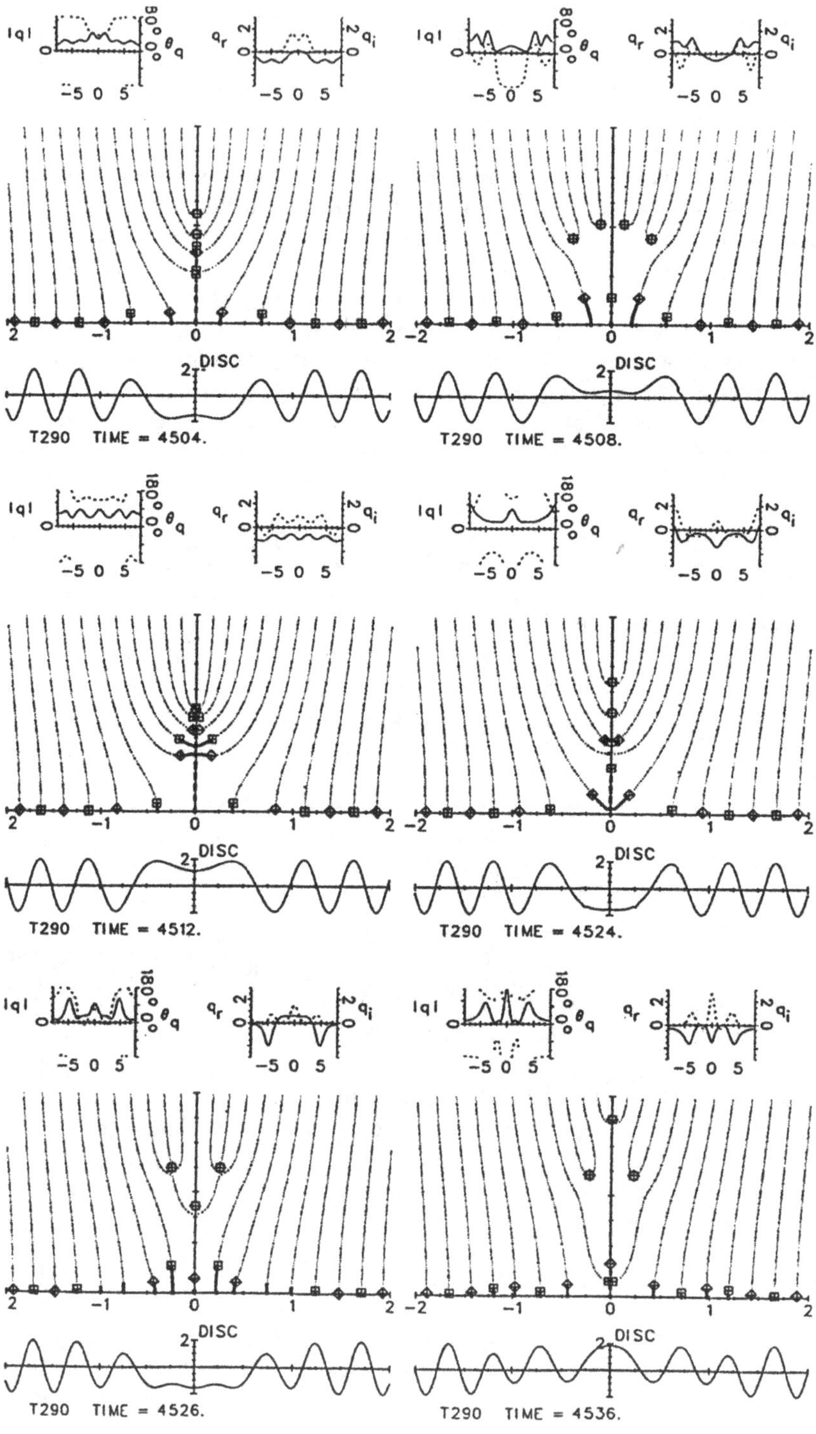
T290 TIME = 4504.
T290 TIME = 4508.
T290 TIME = 4512.
T290 TIME = 4524.
T290 TIME = 4526.
T290 TIME = 4536.
DISC
|q|
θq
qr
qi

organize the spectrum of $\hat{L}$. Curves of real Δ, for many different coefficients $q(y)$, are shown in [26]; several samples are shown in Figure 2.

The operator $\hat{L}$ is not self adjoint; however, it does inherit certain "reality constraints" from the (q, q^*) symmetry of its coefficient matrix:

$$\begin{aligned} &(i) && \mathbb{R} \subset \sigma(\hat{L}), \\ &(ii) && \zeta \in \sigma(\hat{L}) \Rightarrow \zeta^* \in \sigma(\hat{L}), \\ &(iii) && \zeta \in \sigma(\hat{L}) \Rightarrow -\zeta \in \sigma(\hat{L}) \quad (\text{ even } q). \end{aligned} \tag{2.8}$$

Analyticity of Δ as a function of ζ shows that any critical point ζ^c,

$$\left.\frac{d\,\Delta}{d\,\zeta}\right|_{\zeta=\zeta^c} = 0, \tag{2.9}$$

for which

$$-2 < \Delta(\zeta^c) < 2,$$

is a point of bifurcation of the spectrum of $\hat{L}$. Furthermore, perturbation expansions near $q = 0$ show that there are a countable number of such critical points on the real axis which approach $j\pi/l$ as $j \to \infty$, and at which a short "spine" of spectrum bifurcates from the real axis into the complex ζ plane. In addition to these short spines of complex spectrum which are connected to the real axis, there can exist (for larger q) curves of complex spectrum which are not connected (through spectrum) to the real axis.

To anticipate the presence of such complex spectrum which is not connected to the real axis, consider the well known relationship between solitons and bound state eigenvalues (in the spectrum of $\hat{L}$ when the coefficient $q(y) \to 0$ as $y \to \infty$). Recall that, in this case of focusing NLS, these discrete eigenvalues are points in the complex plane. In textbooks on quantum mechanics, the transition from whole line scattering theory to periodic band-gap (Floquet) theory is frequently made by constructing a periodic potential through a periodic translation and superposition of single wells. Under this construction, bound state eigenvalues spread into small bands of continuous spectrum. Here one begins with a single soliton $q(y)$ and, by periodic translation and superposition, constructs a periodic soliton wave-train. Under this construction, the complex discrete eigenvalue associated with the soliton spreads into a short complex band of spectrum which is disconnected from the real axis.

Figure 2. The spectrum $\sigma[\hat{L}(q)]$ for six(6) different spatial profiles. (Each of these profiles is taken from a single chaotic time series for a damped-driven NLS experiment, at six different times.) Plotted for each of the six cases are (i) the amplitude(solid), phase(dotted), real(solid) and imaginary(dotted) parts of the amplitude $q(y)$; (ii) curves of real Δ in the complex ζ plane; (iii) Δ versus the real part of ζ.

2.2 Counting Lemmas

Since the operator $\hat{L}$ is not self adjoint, its spectrum is not constrained to the real axis. Indeed, as described in the preceding paragraph, periodic solitons provide examples of short bands of complex spectrum which are disconnected from the real axis. However, the possible locations in the complex plane for spectrum are not as free as one might initially think. With "counting lemmas" one can show that all complex critical points must reside in a disc D in the complex ζ plane [see Figure 3], centered at the origin, with radius $r = (2N+1)\pi/2l$. Moreover, this integer N is controlled be the first three NLS invariants (the L^2 norm, the linear momentum, and the energy); or equivalently, by the H^1 norm of q. Outside of the disc D, the situation is tame: all spectrum resides on the short spines of spectrum which bifurcate from real critical points, themselves located near $j\pi/\ell$; inside the disc D, one finds exactly $2N+1$ bands of spectrum. *All eccentric, nonselfadjoint behavior occurs inside the disc and is due to a finite number* ($\leq 2N+1$) *of complex bands of spectrum.*

The important curves of real Δ bifurcate at the critical points ζ_j^c, which are the zeros of Δ'. Controlling these, together with the periodic and antiperiodic eigenvalues which are the zeros of $\Delta^2 \mp 2$,(and thus the endpoints of the bands of spectrum,) will enable one to control the spectrum *even in this nonselfadjoint case.* This control is accomplished through "counting lemmas" which should be compared to the Gerschgorin theorems of linear algebra [31]. Here, these counting lemmas procede by locating zeros of analytic functions by the use of Roche's theorem from complex analysis. For example, for Δ' we have [21], [28] :

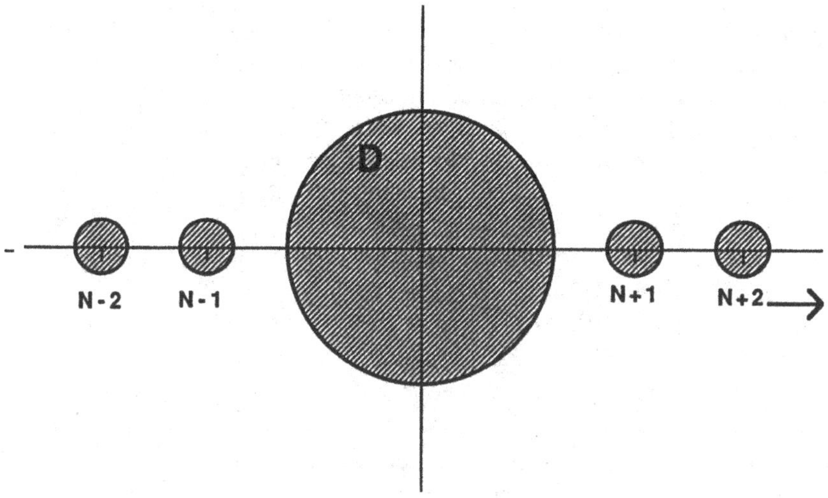

Figure 3. The discs of the "counting lemmas" in the complex ζ plane. (The horizontal scale is in units of π/l.)

Lemma 1 *For $q \in H^1$, set $N = N(\|q\|_{H^1}) \in Z^+$ by*

$$N(\|q\|_{H^1}) = \left[\ell\|q\|_2^2 \cosh\left(\sqrt{\ell}\|q\|_2/2\right) + 6\ell\|q\|_{H^1} \sinh\left(\sqrt{\ell}\|q\|_2/2\right)\right],$$

where $[x] \equiv$ first integer greater than x. Consider

$$\Delta'(\zeta;q) = \frac{d}{d\zeta}\Delta(\zeta;q).$$

Then

- *(i) $\Delta'(\zeta;q)$ has exactly $2N+1$ zeros (counted according to multiplicity) in the interior of the disc $\{\zeta \in C: |\zeta| < (2N+1)\frac{\pi}{2\ell}\}$;*
- *(ii) $\forall k \in Z, |k| > N, \Delta'(\zeta,q)$ has exactly one zero in each disc $\{\zeta \in C: |\zeta - k\frac{\pi}{\ell}| < \frac{\pi}{4\ell}\}$.*
- *(iii) $\Delta'(\zeta;q)$ has no other zeros.*
- *(iv) For $|\zeta| > (2N+1)\frac{\pi}{2\ell}$, the zeros of Δ', $\{\zeta_j^c, |j| > N\}$, are all real, simple, and satisfy the asymptotics*

$$\zeta_j^c = \frac{j\pi}{\ell} + o(1) \quad \textit{as } |j| \to \infty.$$

Similiar counting lemmas exist for the periodic and antiperiodic eigenvalues. From these counting lemmas, one obtains very good control of the spectrum outside of the disc and some control of the size of the disc. Within the disc, little is known about possible spectral configurations; however, the problem classifying all possible configurations has been reduced to a finite counting problem – possibly a very difficult one.

Since Δ is an NLS constant of the motion, the spectral information generated by Δ provides "action" information for the NLS Hamiltonian system. On the other hand, "angle" information is generated by Dirichlet spectrum. The Dirichlet eigenvalues are defined as the zeros of $\tilde{M}_{21}$,

$$\tilde{M}_{21}(\ell;\zeta;q)\Big|_{\zeta=\mu_j(q)} = 0, \tag{2.10}$$

where $\tilde{M}$ is a unitary transformation of the fundamental matrix M,

$$\tilde{M}(y;\zeta;q) = U\, M(y;\zeta;q)\, U^\dagger \tag{2.11}$$
$$U = \frac{1}{\sqrt{2}}\begin{pmatrix} 1 & -1 \\ i & i \end{pmatrix}.$$

The Dirichlet eigenvalues μ_j are not NLS invariants; rather, they execute oscillatory motion under the NLS flow. As indicated by the following "trace formulas", Dirichlet eigenvalues provide the additional "angle" information, which together with the periodic and antiperiodic eigenvalues, provide a representation of $q(y)$:

$$\begin{aligned} q(y) - q^*(y) &= 2\sum_{k\in Z}\Big(\zeta_{2k} + \zeta_{2k-1} - 2\mu_k(y)\Big) \\ q(y) + q^*(y) &= -2i\sum_{k\in Z}\Big(\zeta_{2k} + \zeta_{2k-1} - 2\nu_k(y)\Big). \end{aligned} \tag{2.12}$$

Here ν_j are the Dirichlet eigenvalues for iq,

$$\tilde{M}_{21}(\ell;\zeta;iq)\Big|_{\zeta=\nu_j} = 0, \tag{2.13}$$

and $(\mu_j(y), \nu_j(y))$ denote eigenvalues for the translated coefficient $q^{(y)}(y') = q(y' + y)$. ζ_k denote the periodic points. (Representation (2.12) assumes that certain "reality constraints", which are not well understood, [22]; [15], [7], are satisfied.)

A counting lemma for μ_j, similar to lemma 2.1 for the critical points ζ_j^c, shows that the Dirichlet eigenvalues can be placed in a natural one-to-one correspondence with the critical points. In particular, for each j greater than the "N" of the counting lemma, there exists exactly one μ_j near the jth (real) critical point ζ_j; with exactly 2N+1 $\mu's$ inside the disc of radius $(2N+1)\pi/2l$. Moreover, symmetries of M show that at each real *double point* ζ^d (a zero of Δ' at which $\Delta^2 = 4$), the associated Dirichlet eigenvalue must be locked to that double point. On the other hand, at a *complex* double point (which is necessarily within the disc D), the associated Dirichlet eigenvalue may either be *locked* or it may be *free* to move. It turns out that on a whiskered torus, these Dirichlet eigenvalues are locked at the complex double points when the phase point q lies on the torus itself, and are free to move in C when the phase point lies on the whisker. Thus, the whisker is coordinatized by those free Dirichlet eigenvalues which are associated to complex double points and which approach these double points as $t \to \pm\infty$.

In summary, counting lemmas provide good control of the infinite number of action and angle variables outside of the disc, but little control over the finite number within the disc. The crucial point is that this finite number $2N + 1$ is controlled by only the H^1 norm of q, (or equivalently, through Sobolev inequalities, by the invariants of energy and L^2 norm). Within this disc, certain complex Dirichlet eigenvalues are associated with complex *double* points. These move about within the disc, provide coordinates for the whiskers of the whiskered torus, and approach the complex double points as $t \to \pm\infty$.

3 Whiskered Tori

Integrable theory of periodic NLS focuses attention upon level sets of the constants of the motion. These level sets are also the isospectral sets for the operator $\hat{L}$; thus, they can be defined in terms of the Floquet discriminant:

$$\mathcal{M}(q) \equiv \left\{ r \in \mathcal{F} \middle| \Delta(\zeta; r) = \Delta(\zeta; q) \, \forall \zeta \in C \right\} \tag{3.1}$$

For generic $q \in \mathcal{F}$, the isospectral level set $\mathcal{M}(q)$ is an (infinite) product of circles. At issue here is how these tori change with the (values of) the spectrum, and how they stratify (or fill out) the phase space $\mathcal{F}$. Clearly such a stratification by tori will be organized by neighborhoods of those "critical tori" for which one or more of the circles in the infinite product has pinched off.

3.1 Critical Tori

To identify these critical tori, one begins from a sequence of functionals $\boldsymbol{F}_j$:

$$\boldsymbol{F}_j \colon \mathcal{F} \to C \quad \text{by} \quad \boldsymbol{F}_j(q) = \Delta(\zeta_j^c(q); q) \tag{3.2}$$

The point $q \in \mathcal{F}$ resides on a critical torus if at least one $\boldsymbol{F}_j = \pm 2$, and if,for each j for which $\boldsymbol{F}_j = \pm 2$, the associated Dirichlet eigenvalues are locked to the double point, i.e., $\mu_j = \nu_j = \zeta_j^d = \zeta_j^c$. We call the set of such j's the *critical set of j's for the critical function* q^c, and denote this set by J. Fix a point q^c on a critical torus and expand the functionals $\boldsymbol{F}_k(q)$, $k \in J$, in a neighborhood of $q = q^c$:

$$\boldsymbol{F}_j(q^c + \delta q) = \boldsymbol{F}_j(q^c) + \delta \boldsymbol{F}_j + \delta^2 \boldsymbol{F}_j + \cdots \tag{3.3}$$

Defining a critical function q^c as one for which the first variation in this expansion vanishes, one can use inverse spectral theory to show that such critical potentials exist and to represent some of them in terms of theta functions. The structure of the Hessian in expansion (3.3) then determines the nature of the critical torus. This Hessian is computed in [21] by using quadratic products of eigenfunctions. The result is that, at a real double point ζ_k^d, $\boldsymbol{F}_k$ is a max (or min) at the critical function q^c. (Note that for real $\zeta_k^c, \boldsymbol{F}_k(q)$ is interpreted as the height of Δ over the critical point ζ_k^c, and the fact that $\boldsymbol{F}_k$ is a max (or min) at the critical function q^c is simply a restatement of the fact that no gaps, only complex spines, of spectrum can bifurcate from a real double point.) Outside of the disc D, all double points are real; the functionals $\boldsymbol{F}_k$ for $|k| > N$ are maximal (or minimal) at critical tori. On the other hand, inside the disc at complex double points, the Hessian calculations of [21] show that the real part of these functionals $\boldsymbol{F}_k$ can have a saddle structure at the critical tori. This saddle structure

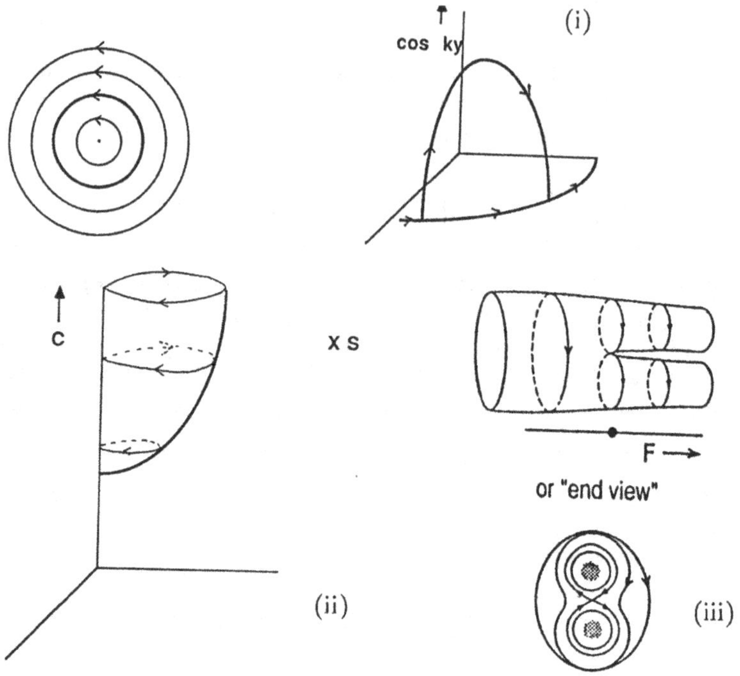

Figure 4. Three "views" of the homoclinic orbit Q_H:

(i) As it is homoclinic to a circle on the plane of constants. Note that the homoclinic orbit leaves this plane in the "$\cos ky$" direction in function space. Also note the phase shift between its "take-off" and "landing" points.

(ii) $W_u(\Pi_1)$ as coordinatized for a family of homoclinic orbits, indexed by the radius $|C|$ of the "target" circle.

(iii) The pant's leg diagram over $\mathbf{F}_1$. The homoclinic orbit is the "figure eight" of the pant.

is consistent with the fact that both crosses and gaps in the spectrum can emerge from a complex double point. Thus, critical tori with complex double points can have hyperbolic structure.

As was suggested in [11], and further developed in [21], the functionals $\mathbf{F}_k$ seem to be natural candidates Morse functions. Could they be used for a Morse theoretic foliation of the function space $\mathcal{F}$ by level sets of the completely integrable soliton equation? While unclear in the infinite dimensional setting, in low dimensional subsystems they certainly can be so used [13]; [11], producing the useful pant's leg diagrams of Figure 4.

3.2 Linearized Instabilities

We are primarily interested in temporal instabilities under the NLS time flow; that is, under the Hamiltonian flow generated by the NLS Hamiltonian. The saddle structure unveiled above by expansion (3.3) does not guarantee this NLS instability; rather, it shows which of the Hamilitonian flows generated by $\boldsymbol{F}_j$ are unstable. Comparison of the Hamiltonian vector fields J grad H and J grad $\boldsymbol{F}$ could assess the temporal instability of NLS; however, we have not done this comparison. Rather, in this section we address the NLS flow directly.

Fix a solution $q(y,\tau)$ of the NLS equation which is periodic, even in y and quasiperiodic in τ; that is, fix a q on one of the invariant tori. Linearizing NLS abour q yields

$$2i\tilde{q}_\tau \;=\; \tilde{q}_{yy} \;+\; (qq^*)\tilde{q} \;+\; 2\,q^2\tilde{q}^* \tag{3.4}$$

Quadratic products of solutions of the Zakharov-Shabat linear system,

$$\left[-\,i\sigma_3\tfrac{d}{dy} \;+\; \tfrac{1}{2}\begin{pmatrix} 0 & q \\ -q^* & 0 \end{pmatrix}\right]\vec{\psi} \;=\; \zeta\vec{\psi} \tag{3.5}$$

$$\left[-\,i\sigma_3\tfrac{d}{d\tau} \;+\; \tfrac{1}{8}\,qq^* \;-\; \tfrac{1}{2}\begin{pmatrix} 0 & -\zeta q + \frac{i}{2}q_y \\ \zeta q^* + \frac{i}{2}q^*_y & 0 \end{pmatrix}\right]\vec{\psi} \;=\; \zeta^2\vec{\psi}$$

generate a basis of solutions of the linearization (3.4) [17] [16]; [23]; [9] ; [19]. With this basis one can address the linear stability properties of the solution q. First, (in the absence of higher order multiple points), the basis splits into two parts, one labeled by simple eigenvalues and one labeled by double points. There is no exponential growth associated with that part of the basis associated to the simple eigenvalues, nor to that part associated to real double points. The only possible exponential instabilities are labeled by complex multiple points. By the counting lemma, these are at most finite in number and reside in the disc D. Typically, for each complex double point there is one exponentially growing and one exponentially decaying linearized solution. These instabilities are associated with the saddle structure described above. However, that saddle structure is associated with the topological properties of the critical level set while instabilities are associated with one particular flow, the NLS flow. It can happen that this particular flow does not "pick up" the unstable direction, and thus is accidently stable. Examples exist [12] of complex double points which are associated to instabilities and others which are associated to stable behavior for one particular flow. At present, each complex double point must be investigated individually for instability with respect to a given flow. Those solutions which do behave exponentially generate a basis for the (finite dimensional) tangent spaces of the stable and unstable manifolds of the critical torus.

3.3 Bäcklund Transformations and Global Representations of Unstable Manifolds

Using Bäcklund transformations one can exponentiate these linearized solutions to obtain global solutions of the NLS equation. Fix an even, periodic solution $q(y,\tau)$ of NLS which is quasiperiodic in τ, for which the linear operator has a complex double point ν *which is associated with an instability* and which has geometric multiplicity 2. We denote two linearly independent solutions of linear system (3.5) at $\zeta = \nu$ by $(\vec{\phi}^+, \vec{\phi}^-)$. Thus, a general solution of the linear system (3.5) at (q,ν) is given by

$$\vec{\phi}(y,\tau;\nu;c_+,c_-) \;=\; c_+\vec{\phi}^+ \;+\; c_-\vec{\phi}^- \tag{3.6}$$

We use $\vec{\phi}$ to define a transformation matrix [29] G by

$$G \;=\; G(\zeta;\nu;\vec{\phi}) \equiv N \begin{pmatrix} \zeta-\nu & 0 \\ 0 & \zeta-\nu^* \end{pmatrix} N^{-1}, \tag{3.7}$$

where

$$N \equiv \begin{bmatrix} \phi_1 & -\varphi_2^* \\ \phi_2 & \phi_1^* \end{bmatrix} \tag{3.8}$$

Then we define Q and $\vec{\Psi}$ by

$$Q(y,\tau) \equiv q(y,\tau) \;+\; 4(\nu-\nu^*)\,\frac{\phi_1\varphi_2^*}{\phi_1\varphi_1^* + \phi_2\phi_2^*} \tag{3.9}$$

and

$$\vec{\Psi}(y,\tau;\zeta) \;\equiv\; G(\zeta;\nu;\vec{\phi})\;\vec{\psi}(y,\tau;\zeta) \tag{3.10}$$

where $\vec{\psi}$ solves the linear system (3.5) at (q,ν). Formulas (3.9) and (3.10) are the Bäcklund transformations for the potential and eigenfunctions, respectively. We have the following

Theorem 1 *Let $q(y,\tau)$ denote an ℓ-periodic solution of NLS, which is linearly unstable with an exponential instability associated to a complex double point ν in $\sigma(\hat{L}(q))$. Let the complex double point ν have geometric multiplicity 2, with eigenbasis $(\vec{\phi}^+, \vec{\phi}^-)$ for linear system (3.5), and define $Q(y,\tau)$ and $\vec{\Psi}(y,\tau;\zeta)$ by (3.9) and (3.10). Then*

(i) $Q(y,\tau)$ is an ℓ-periodic solution of NLS;

(ii) $\sigma(\hat{L}(Q)) \;=\; \sigma(\hat{L}(q))$;

(iii) $Q(y,\tau)$ is homoclinic to $q(y,\tau)$, in the sense that $Q(y,\tau) \longrightarrow q_{\theta_\pm}(y,\tau)$, expontentially as $\exp(-\sigma_\nu|\tau|)$ as $\tau \longrightarrow \pm\infty$. Here $q_{\theta_\pm}$ is a "torus translate" of q, σ_ν is the nonvanishing growth rate associated to the complex double point ν, and explicit formulas exist for this growth rate and for the translation parameters $\theta_\pm$.

(iv) $\vec{\Psi}(y,\tau;\zeta)$ solves the linear system (3.5) at (Q,ζ).

This theorem is quite general, constructing homoclinic solutions from a wide class of starting solutions $q(y, \tau)$. It's proof is one of direct verification, following the sine-Gordon model [8]. Periodicity in y is achieved by choosing the transformation parameter $\zeta = \nu$ to be a double point.

In references [10] and [26], several qualitative features of these homoclinic orbits are emphasized: (i) $Q(y, \tau)$ is homoclinic to a torus which possesses rather complicated spatial and temporal structure, and is not just a fixed point; (ii) nevertheless, the homoclinic orbit typically has still more complicated spatial structure than its "target torus". (iii) When there are several complex double points, each with nonvanishing growth rate, one can iterate the Bäcklund transformations to generate more complicated homoclinic orbits. (iv) The number of complex double points with nonvanishing growth rates counts the dimension of the unstable manifold of the critical torus in that two unstable directions are coordinatized by the complex ratio c_+/c_-; under even symmetry only one real dimension satisfies the constraint of evenness.

4 An Example of a Whiskered Circle

As a concrete example of the material in Section 3, consider the spatially uniform plane wave solution of the NLS equation:

$$q = C(\tau) = c \exp\left[-i\left(\frac{c^2}{4}\tau + \gamma\right)\right] \tag{4.1}$$

Since this plane wave is independent of y and depends upon τ trivially, solutions of the linear system (3.5) can be computed explicitly:

$$\vec{\phi}^{(\pm)}(y,\tau;\zeta) = e^{\pm i\kappa(y+\zeta\tau)} \begin{pmatrix} -\frac{c}{2}e^{-i(\frac{c^2}{4}\tau+\gamma)/2} \\ (\pm\kappa - \zeta)e^{i(\frac{c^2}{4}\tau+\gamma)/2} \end{pmatrix} \tag{4.2}$$

where

$$\kappa = \kappa(\zeta) = \sqrt{\frac{1}{4}c^2 + \zeta^2}.$$

With these solutions one can construct the fundamental matrix

$$M(y;\zeta;C) = \begin{bmatrix} \cos\kappa y + i\frac{\zeta}{\kappa}\sin\kappa y & -i\frac{C}{2\kappa}\sin\kappa y \\ i\frac{C^*}{2\kappa}\sin\kappa y & \cos\kappa y - i\frac{\zeta}{\kappa}\sin\kappa y \end{bmatrix} \tag{4.3}$$

from which the Floquet discriminant can be computed:

$$\Delta(\zeta;C) = 2\cos\kappa\ell \tag{4.4}$$

From Δ, the spectrum and all other "action" quantities can be computed:

Simple Periodic Points: $\zeta^{\pm} = \pm i\,\frac{c}{2};$

Double Points: $\kappa(\zeta_j^d) = j\frac{\pi}{\ell},\ j \in Z, j \neq 0;$

Critical Points: $\zeta_j^c = \zeta_j^d, \quad j \in Z,\ j \neq 0;$

$$\zeta_0^c = 0.$$

From this spectral data, all the quantities of Section 3 can be obtained. In particular, a complete set of solutions of linearization (3.4) can be explicitly constructed from "squares" of the eigenfunctions of equation(4.2). From this complete set one see's that the purely imaginary double points are indeed associated to instabilities. There are N such purely imaginary double points,

$$(\zeta_j^d)^2 = \frac{\pi^2 j^2}{\ell^2} - \frac{c^2}{4} \quad , \quad j = 1,2,\ldots,N \tag{4.5}$$

where

$$\left[\frac{\pi^2 N^2}{\ell^2} - \frac{c^2}{4}\right] < 0 < \left[\frac{\pi^2 (N+1)^2}{\ell^2} - \frac{c^2}{4}\right]$$

with the associated temporal growth rates given by

$$\sigma_j^2 = |\zeta_j^d|^2\left(c^2 - 4\,|\zeta_j^d|^2\right), \qquad j = 1,2,\ldots,N \tag{4.6}$$

The homoclinic orbits can be explicitly computed. A single Bäcklund transformation at one purely imaginary double point yields
$Q = Q_H(y,\tau;c,\gamma;k = \pi/l, c_+/c_-)$:

$$Q_H = \left[\frac{\cos 2p \pm \sin p \,\mathrm{sech}\,(\sigma\tau+\rho)\cos 2ky - i \sin 2p \tanh(\sigma\tau+\rho)}{1 \mp \sin p \,\mathrm{sech}\,(\sigma\tau+\rho)\cos 2ky}\right] c e^{-i(c^2 4\tau+\gamma)} \tag{4.7}$$

$$\rightarrow e^{\mp 2ip} c e^{-i(\frac{c^2}{4}\tau+\gamma)} \qquad \text{as } \rho \rightarrow \pm\infty$$

where $c_+/c_- = \exp(\rho + i\beta)$ and p is defined by $k + \nu = (|c|/2)exp(ip)$.
Several points about this homoclinic orbit need to be made.

(i) The notation $Q = Q_H(y,\tau;c,\gamma;k = \pi/l, c_+/c_-)$ shows that this function of y and τ also depends upon the "target" plane wave (4.1) as parameterized by c and γ, upon the spatial period l, and upon the complex ratio c_+/c_-. The orbit depends only upon the ratio, and not upon $c_+$and c_- individually.

(ii) Q_H is homoclinic to the plane wave orbit (4.1); however, a phase shift of -4p occurs when one compares the asymptotic behavior of the orbit as $\tau \longrightarrow -\infty$ with its behavior as $\tau \longrightarrow +\infty$. This phase shift depends upon by k, $|c|$, and the double point $\nu(k,|c|)$.

(iii) For small p, the formula for Q_H becomes more transparent:

$$Q_H \simeq \left[(\cos 2p - i \sin 2p \tanh(\sigma\tau + \rho)) \pm 2 \sin p \operatorname{sech}(\sigma\tau + \rho) \cos(2ky)\right]$$

$$c e^{-i(\frac{c^2}{4}\tau + \gamma)}.$$

(iv) The complex transformation parameter $c_+/c_- = \exp(\rho + i\beta)$ can be thought of as $S \times R$; however, in formula (4.7) the evenness constraint in y has already been enforced, restricting the phase β to one of two values –

$$\beta = p \pm \frac{\pi}{2}. \qquad \text{(evenness)}$$

Evenness has reduced the formula for Q_H from $S \times R$ to two copies of R. In this manner, the even symmetry disconnects the level set and produces the "figure eight" of the "pant's leg" diagram as discussed in [11]; [26] (see Figure 4). This pant's leg is central to a description of one type of chaotic behavior which results when the NLS equation is damped and driven [26].

(v) Each lobe of the figure eight constitutes one whisker. While the target q is independent of y, each of these whiskers has y dependence through the $\cos(2ky)$. One whisker has this dependence and can be interpreted as a spatial excitation located near $y = 0$ – while the second whisker has the dependence $\cos(2k(y - \pi/2k))$, which we interpret as spatial structure located near $y = l/2$. In this example, the disconnected nature of the level set is clearly related to distinct spatial structures on the individual whiskers, which are in turn related to the chaotic behavior as illustrated in Figure 1.

Depending upon your viewpoint, the Bäcklund transformation provides an explicit of representation of (a) a pair of homoclinic orbits to a phase shift of the target q; (b) the unstable manifold of the plane wave, a whiskered torus, or (c) the figure eight of a "pant". Each of these geometric viewpoints (see Figure 4) is useful in the analysis and interpretation of chaotic responses when the NLS equation is perturbed [26].

In this example the target is always the plane wave (4.1); hence, it is always a circle of dimension one, and in this example we are really constructing only whiskered circles. On the other hand, in this example the dimension of the whiskers need not be one, but is determined by the number of purely imaginary double points which in turn is controlled by the amplitude c of the plane wave target and by the spatial period l. (The dimension of the whiskers increases linearly with l.) When there are several complex double points, Bäcklund transformations must be iterated to produce complete representations. While these iterated formulas are quite complicated, their parameterizations admit rather direct qualitative interpretations [30].

While this example is restricted to whiskered circles, in the next section we describe whiskered tori, of arbitrary (even infinite) dimension, which are nearby in phase space $\mathcal{F}$.

5 Whiskered Tori: General Setting

In integrable theory, one focuses upon an isospectral level set $\mathcal{M}$. However, when considering perturbations of integrable systems, it seems somewhat more natural to focus upon larger invariant sets. Here we will illustrate these larger invarient sets by restricting our attention to a neighborhood of the constant solution $q(y,\tau) = C(\tau) + \epsilon\tilde{q}(y,\tau)$. Fix the amplitude $|C|$ and the spatial period ℓ so that there are 2N complex double points, each of which is known to be associated to an exponential instability as discussed in the preceding section. Order the critical points as follows:

$$\textit{real critical points:} \quad ... < \zeta^c_{-N-2} < \zeta^c_{-N-1} < \zeta^c_0 < \zeta^c_{N+1} < \zeta^c_{N+2} < \cdots$$

$$\textit{purely imaginary critical points:} \quad \zeta^c_{-j} = (\zeta^c_j)^*, \; j = 1, 2, ..., N;$$

where $|\zeta^c_j| > |\zeta^c_{j+1}|, j = 1, ..., N-1$. With this ordering, we order our of functionals

$$F_j : \mathcal{F} \to C \quad \text{by} \quad F_j(q) = \Delta(\zeta^c_j(q), q), \tag{5.1}$$

in terms of which the invariant set Σ is defined:

$$\Sigma_{N,\delta} \equiv \Big\{ q \in \mathcal{F} \Big| \begin{array}{ll} (i) & F_j(q) = (-1)^j\, 2 \text{ and } \mu_j = \nu_j = \zeta^d_j, \; |j| = 1, 2, \cdots N; \\ (ii) & 0 \leq |F_0(q) - F_0(C)| < \delta_0; \\ (iii) & 0 \leq 2 - |F_j| < \delta_j, \; |j| > N \Big\}. \end{array} \tag{5.2}$$

Notice that the invariant set $\Sigma_{N,\delta}$ consists in a countable product of nested circles forming an infinite dimensional, solid torus. The "radius" of the jth circle is given by $(2 - |\mathbf{F}_j|) \in [0, \delta_j)$; thus, the jth circle can be "pinched off" by setting $|\mathbf{F}_j| = 2$. Π_N, that annulus in the plane of constants Π with N unstable modes, is a two dimensional invariant subset of $\Sigma_{N,\delta}$ which is obtained by setting $|\mathbf{F}_j| = 2$ for all $|j| > N$; schematically,

$$\Pi_N = \Sigma_{N,\vec{0}'},$$

where $\vec{\delta}' = (\delta_j, |j| > n)$. $\Sigma_{N,\delta}$ itself is infinite dimensional, requiring both action and angle information to coordinatize it. The codimension of $\Sigma_{N,\delta}$ is 2N.

The stable and unstable manifolds of $\Sigma_{N,\delta}$ coincide and are given by

$$\begin{aligned} W_u(\Sigma_{N,\delta}) &= W_s(\Sigma_{N,\delta}) \\ &= \Big\{ q \in \mathcal{F} \Big| \quad (i) \quad F_j(q) = (-1)^j\, 2,\ |j| = 1, 2, \cdots N; \\ &\qquad\qquad\quad (ii) \quad 0 \le |F_0(q) - F_0(C)| < \delta_0; \\ &\qquad\qquad\quad (iii) \quad 0 \le 2 - |F_j| < \delta_j,\ |j| > N \Big\}. \end{aligned} \tag{5.3}$$

Notice that this unstable manifold differs from $\Sigma_{N,\delta}$ itself in that the Dirichlet eigenvalues are no longer *locked* to their complex double points, but rather they are *free* to move. Because of even symmetry, the real dimension of the unstable (or stable) manifolds is given by

$$dim W_u(\Sigma_{N,\delta}) = dim(\Sigma_{N,\delta}) + N;$$

they each have codimension N.

Each choice of values for $\mathbf{F}_j, |j| > N$, fixes a torus in $\Sigma_{N,\delta}$, whose dimension is arbitrary, even infinite. The unstable manifold of these tori are the "whiskered tori" of arbitrary dimension. The reason for "fattening" these tori into the invariant set $\Sigma_{N,\delta}$ concerns likely behavior under perturbations. $\Sigma_{N,\delta}$ is normally hyperbolic; hence, in analogy to finite dimensional situations, one anticipates that the invariant set $\Sigma_{N,\delta}$ will persist under perturbations, deforming into another normally hyperbolic manifold Σ^ϵ which itself is invariant under the perturbed dynamics. Smaller invariant sets such as the tori themselves are not normally hyperbolic and expected to be very sensitive to perturbations. While unperturbed motion on $\Sigma_{N,\delta}$ will be very different from the perturbed motion on Σ^ϵ, the two invariant manifolds themselves should be very similar.

Iterated Bäcklund transformations provide explicit representations of the whiskered tori. Fix $q \in \Sigma_{N,\delta}$, and consider the N fold iteration of Bäcklund transformation (3.10):

$$Q^H = Q^H(y, \tau; c, \gamma; \mathbf{F}_j, \theta_j, |j| > N; r_1, \sigma_1, r_2, \sigma_2, ..., r_N, \sigma_N).$$

The argument list in Q^H is important as it indicates the dependence upon the target torus, with the whisker itself parametized by N real numbers $r_j = |c_j^+/c_j^-|$, $0 \le r_j \le +\infty$, together with the "parity" parameters $\sigma_j = \pm$ which determine whether the jth excitation is located in the center or at the edges. The parameters r_j and σ_j are introduced by the iterated Bäcklund transformation.

6 Conclusion

The status of periodic soliton equations, while not complete, is really quite satisfactory, particularly when the soliton equation is integrated with a 2×2 linear spectral problem. In that case, nonselfadjointness introduces instabilities and hyperbolic structure into the system, and counting lemmas ensure that these important nonselfadjoint eccentricities are finite in number. Lax pairs and Floquet discriminants provide the tools with which to identify instabilities and hyperbolic structure, and with which to understand how the level sets foliate the function space. Very natural Morse functions can be built from the Floquet discriminant, making a Morse theoretic description of the foliation of function space by isopectral level sets possible and an interesting direction for future mathematical research. In addition, Bäcklund transformations provide concrete representations of whiskered tori. Currently, these representations and their associated hyperbolic structure is being extended to discrete systems, to systems integrated with high order (than 2) linear problems, and to soliton equations in $2+1$ dimensions.

These concrete representations of whiskered tori are novel and provide fascinating models of hyperbolic structure in a rich and natural class of discrete systems and pde's. Our next challenge is to use these representations to understand behavior in near integrable equations.

References

[1] M. J. Ablowitz and B. M. Herbst. Numerically Induced Chaos in the Nonlinear Schrödinger Equation. *Phys Rev Lett*, 62:2065–2068, 1989.

[2] M. J. Ablowitz and B. M. Herbst. On Homoclinic Structure and Numerically Induced Chaos for the Nonlinear Schrödinger Equation. *SIAM Journal on Applied Mathematics*, 50:339–351, 1990.

[3] V. I. Arnol'd. Instability of Dynamical Systems with Several Degrees of Freedom. *Sov. Math. Dokl.*, 5:581–585, 1964.

[4] A. R. Bishop, M. G. Forest, D. W. McLaughlin, and E. A. Overman. A Quasiperiodic Route to Chaos in a Near-Integrable PDE. *Physica D*, 23:293–328, 1986.

[5] A. Calini, N. Ercolani, D. W. McLaughlin, and C. M. Schober. Melnikov Analysis of Conservative Perturbations of NLS. *University of Arizona*, in preparation, 1991.

[6] D. David, M. G. Forest, D. W. McLaughlin, D. Muraki, S. Sheu, and O. Wright. Several Studies of Instabilities in Third Order Systems. in preparation, 1991.

[7] N. Ercolani and M. G. Forest. The Geometry of Real Sine-Gordon Wavetrains. *Commun Math Phys*, 99:1–49, 1985.

[8] N. Ercolani, M. G. Forest, and D. W. McLaughlin. Geometry of the Modulational Instability III. *Physica D*, 43:349–384, 1990.

[9] N. Ercolani, M. G. Forest, and D. W. McLaughlin. Notes on Melnikov Integrals for Models of the Driven Pendulum Chain. *University of Arizona*, preprint, 1989.

[10] N. Ercolani, M. G. Forest, and D. W. McLaughlin. Geometry of the Modulational Instability I. II. *Memoirs of AMS*, to appear.

[11] N. Ercolani and D. W. McLaughlin. *The Geometry of Hamiltonian Systems*, chapter Toward a Topological Classification of Integrable PDE's. Springer-Verlag (New York), 1991.

[12] R. Flesch, M. G. Forest, and A. Sinha. Numerical Inverse Spectral Transform for the Periodic Sine-Gordon Equation: Theta Function Solutions and their Linearized Stability. *Physica D*, 48:169–231, 1991.

[13] A. T. Fomenko. *Topological Classification of all Integrable Hamiltonian Differential Equations of General Type with Two Degrees of Freedom*, volume 22 of *The Geometry of Hamiltonian Systems*. Springer-Verlag (New York), 1989.

[14] M. G. Forest, C. Goedde, and A. Sinha. A Numerical Study of Energy Transport in Nearly Integrable Conservative Systems. *Ohio State University*, preprint, 1991.

[15] M. G. Forest and D. W. McLaughlin. Modulations of Sine-Gordon and Sinh-Gordon Wavetrains. *Stud Appl Math*, 68:11–59, 1983.

[16] D. J. Kaup. Closure of the Squared Zakharov-Shabat Eigenstates. *J. Math. Anal. Appl.*, 54:849, 1976.

[17] D. J. Kaup. A Perturbation Expansion for the Zakharov-Shabat Inverse Scattering Transform. *SIAM J. Appl. Math.*, 31(1):121–133, July 1976.

[18] G. Kovacic and S. Wiggins. Orbits Homoclinic to Resonances, with an Application to Chaos in a Modes of the Forced and Damped Sine-Gordon Equation. *Cal Inst Tech*, preprint, 1991.

[19] I. M. Krichever. Perturbation Theory in Periodic Problems for Two-Dimensional Integrable Systems. *Sov Sci Rev C Math*, 9, 1991.

[20] Y. Li. Bäcklund Transformations and Homoclinic Structures for the Integrable Discretization of the NLS Equation. submitted to Physica D, 1991.

[21] Y. Li and D. W. McLaughlin. Instabilities, Homoclinic Orbits, Morse and Melnikov Functions for a Soliton Equation. *Princeton University*, preprint, 1991.

[22] H. P. McKean. The Sine-Gordon and Sinh-Gordon Equations on the Circle. *Comm Pure Appl Math*, 34:197–257, 1981.

[23] H. P. McKean and E. Trubowitz. Hill's Operator and Hyperelliptic Function Theory in the Presence of Infinitely Many Branch Points. *Comm Pure Appl Math*, 29(143), 1976.

[24] D. W. McLaughlin. Unpublished Notes on the Periodic NLS Equation. 1988.

[25] D. W. McLaughlin, E. Overman, S. Wiggins, and C. Xiong. Homoclinic Behavior for a 2 Mode Truncation of NLS. *Ohio State University*, preprint, 1992.

[26] D. W. McLaughlin and E. A. Overman. Whiskered Tori for Integrable Pde's: Chaotic Behavior in Near Integrable Pde's. *Surveys in Appl. Math*, 1, to appear 1992.

[27] D. W. McLaughlin and C. M. Schober. Chaotic and Homoclinic Behavior for Numerical Discretizations of the Nonlinear Schrödinger Equation. *Physica D*, to appear 1992.

[28] J. Pöschel and E. Trubowitz. *Inverse Spectral Theory.* Academic Press (New York), 1987.

[29] D. H. Sattinger and V. D. Zurkowski. Gauge Theory of Bäcklund Transformations II. *Physica D*, 26:225–250, 1987.

[30] Constance M. Schober. PhD thesis, University of Arizona, 1991.

[31] J. H. Wilkinson. *The Algebraic Eigenvalue Problem.* Clarendon Press (Oxford), 1965.

Index of Contributors